METHODOLOGICAL ADVANCES IN THE CULTURE, MANIPULATION AND UTILIZATION OF EMBRYONIC STEM CELLS FOR BASIC AND PRACTICAL APPLICATIONS

Edited by **Craig S. Atwood**

Methodological Advances in the Culture, Manipulation and Utilization of Embryonic Stem Cells for Basic and Practical Applications
Edited by Craig S. Atwood

Published by InTech
Janeza Trdine 9, 51000 Rijeka, Croatia

Publishing Process Manager Katarina Lovrecic

Technical Editor Teodora Smiljanic

Cover Designer InTech Design Team

Additional hard copies can be obtained from orders@intechopen.com

Methodological Advances in the Culture, Manipulation and Utilization of Embryonic Stem Cells for Basic and Practical Applications, Edited by Craig S. Atwood
p. cm.
ISBN 978-953-307-197-8

For EU product safety concerns:
IN TECH d.o.o., Prolaz Marije Krucifikse Kozulić 3, 51000 Rijeka, Croatia,
info@intechopen.com or visit our website at intechopen.com.

Meet the editor

Craig S. Atwood, Ph.D. is an Associate Professor of Medicine at the University of Wisconsin and a Health Science Specialist with the Geriatric Research, Education and Clinical Center at the William S. Middleton Memorial Veterans Administration Hospital in Madison. Dr. Atwood completed his Ph.D. in Biochemistry from the University of Western Australia in Perth, Australia prior to post-doctoral fellowships at the National Cancer Institute, NIH, Bethesda, and Massachusetts General Hospital, Charlestown. He held faculty positions at Harvard Medical School and Case Western Reserve University prior to his current appointment where he directs the research program of the Laboratory for Endocrinology, Aging and Disease. Dr. Atwood has broad research interests related to the endocrinology of embryogenesis, adulthood and senescence as elaborated upon in 'The Reproductive-Cell Cycle Theory of Aging'. He has published over 200 scientific articles, has served on numerous review boards and is an Editor of 20 scientific journals including the Journal of Biological Chemistry. In 2006 he received the Zenith Fellows Award from the Alzheimer's Association in recognition of his research.

Contents

Preface

The development of techniques to culture embryonic stem cells (ESCs) in a pluripotent state has been crucial for the manipulation and utilization of this special cell population for understanding basic biological questions, as well as their use in practical applications such as regenerative medicine and toxicology. The first mammalian ESC lines were derived from the mouse blastocyst 3 decades ago by two independent groups (Evans and Kaufman 1981; Martin 1981). The development of specific methods for the derivation and subsequent maintenance of ESCs in specific media conditions allowed enquiry into mechanisms of cell pluripotency, growth, differentiation, degeneration and regeneration, but also their practical application in toxicology. Some time later, the development of techniques to successfully isolate and maintain human ESCs derived from inner mass cells (ICM; (Bongso, et al. 1994; Thomson, et al. 1998)) opened up new practical applications in regenerative medicine in addition to probing basic biological mechanisms.

This book draws together reviews on the historical development of techniques, and recent advances in, the derivation, culture, maintenance and imaging of ESCs, as well as reviews on recent advances in genetic manipulation and reprogramming of ESCs for basic biological experimentation, as well as their application for the development of transgenic/chimeric mice and use in toxicological and drug discovery studies. In the first section of the book, 'Methods for Derivation and Maintenance of Embryonic Stem Cell Pluripotency', chapters by Manganelli et al. and Fagundez et al. provide an excellent historical overview of the technological advances that led to the successful maintenance of mouse and human ESCs in culture. Zhang et al. next provide detailed methodology for the isolation and dissociation of ICM for the derivation of ESCs, and their culture, passage, freezing, thawing, karyotyping and differentiation into embryoid bodies (EBs) or teratoma, as well as techniques involved in chimeric mouse production. Techniques for the preservation of ESCs are reviewed by He and includes detailed descriptions of cryopreservation at cryogenic temperatures (slow-freezing, conventional vitrification and low-cryoprotective agent vitrification) prior to discussing lyopreservation, or dry preservation at ambient temperature by evaporative drying or freeze-drying. He also discusses biophysical considerations of cell preservation. The next 5 chapters review recent advances in the development of techniques to allow more efficient and scalable culture of ESCs and induced pluripotent stem cells (iPSCs). These advances have been driven in large part by the desire to eliminate non-human cells and materials, thereby avoiding issues including immunogenicity, complex purification, heterogeneous environments, batch-to-batch (or source-to-source) variation in materials isolated from tissues, and presence of xenogenetic compounds. The development of these techniques has been within the constraints that ESC self-renewal and differentiation are known to be regulated via interactions with neighboring cells, ECM components, soluble signaling factors, and the physicochemical environment. The development of these new technologies has allowed greater control over material properties and tissue responses. The first of these chapters by D'Andrea Markert et al. describe the topographical, mechanical and chemical cues that cells receive, and the use of topographically and chemically modified surfaces for expansion or differentiation of ECSs. Haque et al. next describe recent developments in the construction of novel ECM using genetically engineered adhesion molecules and growth factors aimed at mimicking the regulatory characteristics of natural ECM and growth factors in the extracellular microenvironment. Sakuragi and Ito then describe the culture of ESCs on chemically fixed feeder cells, the use of synthetic biomaterials for ESC maintenance and EB formation and end with a description of developments in thermoresponsible polymers that allow for temperature dependent ESC detachment. Mashayekhana and Miyazakib describe how cell fate is controlled by chemical modifications to the substrate, by geometric modification and by modification of material mechanics. These authors end by describing strategies for culture surface design using glucose-displaying dendrimer substrates to control the morphology and function of ESCs. A chapter on the use of novel substrates for cell culture by Wang and Zhang emphasizes the use of three-dimensional silk fibroin scaffolds and their application to ESC culture and differentiation.

Imaging techniques are pivotal to our understanding of all aspects of ESC biochemistry and morphology as it relates to cell proliferation, differentiation and death. In the next section on 'Methods of Imaging Embryonic Stem Cells', Zheng et al. describe the use of hESCs constitutively expressing double fusion reporter genes (enhanced green fluorescent protein and firefly luciferase) for monitoring transplanted cells and their proliferation *in vivo*. In addition, these workers describe functional imaging with intravital microscopy for assessment of endothelial differentiation of hESCs *in vitro*. Not only is histological imaging important, but techniques that can non-invasively obtain information on the molecular composition and structure of intact cells can provide insights into cell function and differentiation. In this respect, Ami et al. review label-free vibrational (micro)spectroscopies such as Fourier transform infrared and Raman techniques and their use in ESC and adult stem cells for rapidly obtaining snapshots of molecular events during differentiation.

In the next section of the book, 'Methods for Reprogramming Cells', novel technologies for reprogramming cells for pluripotency and regenerative medicine are reviewed. The development of these technologies has been largely driven by either ethical considerations related to the destruction of viable embryos, and/or the incorporation of foreign genetic material (e.g. viruses or expression plasmids) into ESCs or somatic cells that can result in teratoma formation and/or immune-rejection. Although reprogramming differentiated cell types to obtain autologous human stem cell lines (iPSCs) resolves the first issue, it does not resolve the second. To circumvent these issues, Horii and Hatada in their chapter propose the use of reprogrammed parthenogenetic ESCs for cell transplantation therapy (the "third pluripotent stem cell"). The use of parthenogenetic ESCs from fertilization-failure or surplus oocytes overcomes the above issues, however as the authors point out, major challenges related to genomic imprinting must be overcome for this technology to successfully generate useable ESCs. In a similar vein, Han and Sidhu review different methodologies for reprogramming somatic cells, including nuclear transfer, somatic-stem cell fusion, cell-free extract treatment and iPSC generation, technologies that may restore regenerative capacity of terminally differentiated cells and that can be applied for transplantation and cell therapy. In particular, these authors elaborate on the use of somatic cell fusion with ESCs, but also the challenges related to stem cell nuclei in the hybrid cells, the possibility of causing immune-rejection and the genetic instability of hybrid cells, issues that must be overcome before this technology becomes a viable option. The genetic instability of reprogrammed cells and the potential for cancer is a major obstacle in these reprogramming technologies for regenerative medicine. Orbán et al. propose the use of the non-viral gene delivery techniques using transposon based methods to circumvent uncontrolled proliferation of the gene-modified stem cells, and detail the advantages and concerns of using the transposon-transposase system for stable genetic modification of ESCs.

Modulating gene function in cells is of critical importance to gain insight into the molecular underpinnings of the pluripotent state, the process of differentiation and the mechanisms underlying human disease. In this next section on 'Methods for Genetic Delivery, Manipulation and Assessment in ESC', Brafman and Willert review the pros and cons of different methods to transduce genetic elements into both ESCs and iPSCs including random transgenesis, site-specific recombination and homologous recombination. As an example of the use of transgenesis in hESCs, Kobayashi next review the use of iPSCs derived from Klinefelter syndrome (X-chromosome polysomy, typically 47, XXY) as a tool for studying the mechanism of X chromosome inactivation. In addition, the author discusses how iPSCs derived from testicular tissue might be a potential source for cell-based therapies to treat some cases of male infertility. Lee et al. discuss the use of a non-viral gene delivery system - magnet-based nanofection, and describe applications and methods of this method for use in ESC research. Tada describe chromosome engineering methods - for addition and elimination of targeted chromosomes in ESCs using human artificial chromosome vectors and chromosome elimination cassettes), respectively, for determination of chromosome function. Complimentary to these techniques, Synnergren and Sartipy review how transcriptional profiling of hESC and their derivatives offers the ability to explore global expression patterns that are activated during hESC differentiation and will allow further dissection of the molecular mechanisms that control stem cell specification. In a similar vein, Lei et al. describe novel techniques for tracking the structure of protein interaction networks via multiple genetic perturbations in mESC.

The use of ESCs for the generation of genetically modified mice has been of incredible use in mechanistic studies to determine gene function. The next section explores methods for generation of ESCs in mice, and newly, rats for use in the production of genetically modified rodents ('Methods for the Generation of Embryonic Stem Cell Lines for Production of Transgenic and Chimeric Rodents'). The first of these chapters focuses on the use of ESCs in the generation of transgenic mice and rats. Iijima et al. describe available ESCs from C57BL/6 mice and their use in production of germline chimeric mice as well as their genetic manipulation in the production of transgenic mice. In addition, these authors describe ESCs that ubiquitously expressed enhanced green fluorescent protein and the photoconvertible tandem repeat KAEDE protein for use in detecting ESCs and their daughter cells in basic research on regenerative medicine. The following 2 chapters by Kawamata and Ochiya, and Hirabayashi and Hock, respectively, describe the establishment of rat ESCs and their genetic manipulation for the creation of genetically modified rats, which as the authors point out, have several advantages over mice for biological study.

Another application of ESCs, their use in the testing of molecules for toxicity and drug discovery, dates back to 1997 (Spielmann et al.). In the final section of this book, 'Methods of Using Embryonic Stem Cells for Toxicology', 4 chapters describe the historical and current state of the art. Nakamura et al. and Kim review the development of the "embryonic stem cell test" (Spielmann et al. 1997), advantages (doesn't require animals) and disadvantages (endpoints, differences in toxicity between species, cost, guided differentiation to target cells), its utilization for the assessment of embryotoxicity and teratogenicity, and advances to this test over the last 15 years. In the next chapter, Äänismaa et al. describe the use of ESCs and their differentiation into neuronal networks for use in microelectrode array (MEA)-based platforms for studies of neuronal network development and maturation. Additionally these workers propose the use of such ESC-derived neuronal networks for toxicological assessments. Finally, research by Takezawa suggests a novel strategy for drug discovery and development by analyzing the behaviour of ESCs cultured on TOSHI (tissue/organ sections for histopathology)–substrata.

It is hoped that the research and reviews described here will help to update the ESC research community on recent methodological advances in basic and practical applications of ESCs, and where future technologies are in need of development.

References

Evans MJ & Kaufman MH 1981 Establishment in culture of pluripotential cells from mouse embryos. *Nature* 292 154-156.

Martin GR 1981 Isolation of a pluripotent cell line from early mouse embryos cultured in medium conditioned by teratocarcinoma stem cells. *Proc Natl Acad Sci U S A* 78 7634-7638.

Bongso A, Fong CY, Ng SC & Ratnam S 1994 Isolation and culture of inner cell mass cells from human blastocysts. *Hum Reprod* 9 2110-2117.

Thomson JA, Itskovitz-Eldor J, Shapiro SS, Waknitz MA, Swiergiel JJ, Marshall VS & Jones JM 1998 Embryonic stem cell lines derived from human blastocysts. *Science* 282 1145-1147.

Spielmann, H., Pohl, I., Doring, B., Liebsch, M. and Moldenhauer, F. (1997) The embryonic stem cell test (EST), an in vitro embryotoxicity test using two permanent mouse cell lines; 3t3 fibroblasts and embryonic stem cells. *Toxicology in vitro*, 10, 119-127.

Craig S. Atwood

Geriatric Research, Education and Clinical Center,
Veterans Administration Hospital,
Department of Medicine,
University of Wisconsin,
Madison, WI 53705, USA

Part 1

Methods for Derivation and Maintenance of Embryonic Stem Cell Pluripotency

1

Embryonic Stem Cells: from Blastocyst to *in vitro* Differentiation

Genesia Manganelli[1], Annalisa Fico[1,2] and Stefania Filosa[1]
[1]*Istituto di Genetica e Biofisica "A. Buzzati Traverso" CNR,*
Via Pietro Castellino 111, 80131 Napoli
[2]*Developmental Biology Institute of Marseille–Luminy (IBDML)*
Campus de Luminy, Case 907, 13288, Marseille cedex 09
[1]*Italy*
[2]*France*

1. Introduction

A stem cell is a specific kind of cell that has the unique capacity to renew itself and to give rise to specialized cell type.
In terms of potentially, stem cells can be classified in three types:

- Totipotent: is the ability to form all cell types, including the extra-embryonic tissues. In mammals, the fertilized egg, zygote and the first 2, 4, 8, 16 blastomeres from the early, are examples of totipotent cells.
- Pluripotency: is the ability to differentiate into several cell types derived from any of the three germ layers (ectoderm, mesoderm, endoderm), but they are unable to produce extra-embryonic tissues. Cells from the inner cell mass of blastocyst are pluripotent.
- Multipotent: cells can form a small number of tissues that are restricted to a particular germ layer origin: e.g. blood cells or bone cells.

In according to their source, stem cells are categorized in embryonic or adult (Fig. 1):

- Embryonic stem cells (ES cells) are derived from the inner cell mass of the blastocyst (an early stage embryo) and have a high proliferative capabilities and differ from other stem cells because they have the ability to generate derivatives of all three germ layers. Embryonic stem cells have been shown to contribute to all cell lineages, including the germ line, following microinjection studies in murine embryos which give rise to chimeras (Bradley et al., 1984; Nagy et al., 1990). *In vitro,* murine ES cells can be propagated indefinitely in an undifferentiated state, under specific culture conditions they can differentiated into specific cell types.
- Adult stem cells are undifferentiated cells found among differentiated cells of a specific tissue, including bone marrow (de Haan, 2002), skin (Watt, 2001), intestinal epithelium (Potten, 1998), liver (Theise et al., 1999), retina (Tropepe et al., 2000), central nervous system (Okano, 2002), pancreas (Ramiya et al., 2000) and skeletal muscle (Seale et al., 2001). They typically can differentiate into a relatively limited number of cell types.

There is no doubt that stem cells have the potential to treat many human afflictions, including cancer, diabetes, neurodegeneration, as well as for studying basic developmental biology, and intensive screening of drug and toxic (Watt and Driskell, 2010).

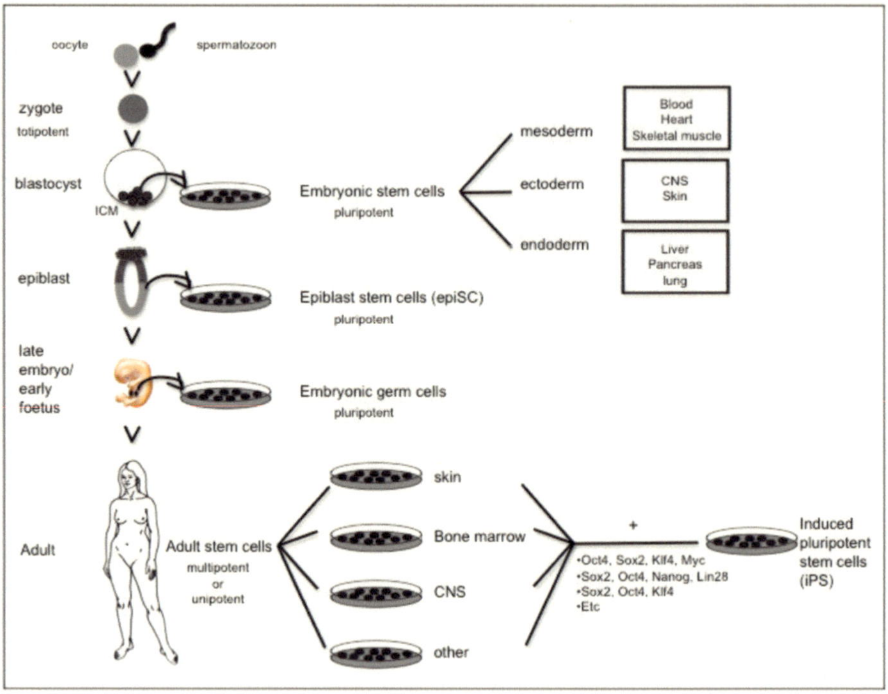

Fig. 1. Origin of embryonic and adult stem cells

2. Derivation of mouse embryonic stem cells

Embryonic stem cells are derived from the inner cell mass (ICM) of the mammalian blastocyst. The first mammalian ES cell lines were derived from mouse blastocyst in 1981 from two independent groups (Evans and Kaufman, 1981; Martin, 1981).

One distinct property of ES cells is that they remain diploid even after being cultured for many weeks. This is in contrast to other tissue culture cell lines that often do not remain diploid but spontaneously gain or lose chromosomes at high rate. A second unique property of ES cells is that they remain pluripotent and maintain the ability, like ICM cells, to form chimeras. These two properties, maintaining normal karyotype and extensive contribution in chimeras, are both necessary for ES cells to form functional germ cells in chimeras (Sedivy and Joyner, 1992) and, moreover, have made ES cells a unique tool for gene targeting and generation of genetically modified mice.

A surprising feature of mouse ES cell lines is that the majority of cell lines genetically tested are of male origin (40XY). In female (XX) ES cells, both X chromosome are active, that may result in the unsuitable propagation of ES cells (Rastan and Robertson, 1985). In either case, the XY genotype confers appreciable advantages for germ line transmission.

ES cells clonally derived from a single cell could differentiate into a variety of cell types *in vitro* and form teratocarcinomas when injected into mice (Martin, 1981). Most important, cells karyotypically normal contribute at a high frequency to a variety of tissue in chimeras, including germ cells, thus providing a practical way to introduce modifications to the mouse germline (Bradley et al., 1984).

After the first derivation of mouse ES cell lines from blastocysts, several standard protocols were developed (Robertson, 1987; Abbondanzo et al., 1993; Hogan et al., 1994; Nagy et al., 2003). The efficiency of mouse ES cell derivation is strongly influenced by genetic background. For example, ES cells can be easily derived from the inbred 129/ter-Sv strain but less efficiently from the C57BL/6 strain (Ledermann and Burki, 1991). However, mouse ES cells can be derived from some non permissive strains using modified protocols (McWhir et al., 1996; Bryja et al., 2006a; Bryja et al., 2006b). Mouse ES cells have also been derived from cleavage stage embryos and even from individual blastomeres of two- to eight-cell stage embryos (Chung et al., 2006; Wakayama et al., 2007).
ES cells or ES cell-likes have been produced in other animal models, including: medakafish from midblastulae stage (Hong et al., 1998), zebrafish from midblastulae stage (Sun et al., 1995), chickens from stage X blastoderm (Pain et al., 1996), hamsters (Doetschman et al., 1988), mink (Sukoyan et al., 1992), rabbit (Schoonjans et al., 1996), cattle (Cibelli et al., 1998; Strelchenko et al., 2004), sheep (Wells et al., 1997), and pigs (Li et al., 2003), however, only mouse and chicken ES cells are capable of colonizing the germ line.

3. Maintenance of mouse embryonic stem cells

ES cells can be stably propagated indefinitely and maintain a normal karyotype without undergoing cell senescence *in vitro* when cultured in the presence of leukemia inhibitory factor (LIF) and, depending on ES cell lines, with or without a layer of mitotically inactivated mouse embryonic fibroblasts (MEFs). LIF, a member of the IL-6 family, is known to strongly promote self-renewal in ES cells (Smith et al., 1992). LIF binds to LIF receptor (LIFR) to dimerize with interleukin 6 signal transducer (gp130), resulting in the phosphorylation of signal transducer and activator of transcription 3 (Stat3) via Janus kinase (Jak) activation (Burdon et al., 2002). Phosphorylated Stat3 dimerizes and translocates to the nucleus to activate a variety of downstream genes. Repression of Stat3 results in differentiation (Niwa et al., 2009), whereas artificial activation of Stat3 is sufficient to maintain pluripotency without LIF in the media (Matsuda et al., 1999).

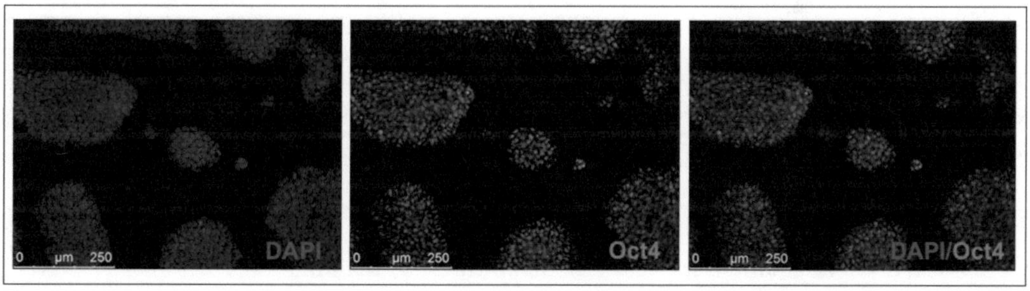

Fig. 2. Fluorescent immunostaining of undifferentiated mouse ES cells. All the undifferentiate ES cells expressed pluripotency specific marker Oct4. Immunostaining with DAPI (nuclear marker), Oct4 antibody, merge DAPI/Oct4

In combination with the LIF-Stat3 pathway, the pluripotency of ES cells is modulated by transforming growth factor β (TGFβ) superfamily members. These include Bmp and Activin, which generally play diverse roles in cellular homeostasis. In the ES cells, Bmp4 activates the MAD homolog 1 (Smad1). This upregulates the expression inhibitor of DNA-binding genes (Id), which suppress differentiation in combination with the LIF signal.

Activin/nodal signaling contributes to promote the growth of ES cells (Ying et al., 2003; Ogawa et al., 2007; Wu and Hill, 2009). Wnt signalling also contributes to the maintenance of pluripotency. In the canonical Wnt pathway, the Wnt receptor Frizzled transduces the signal to glycogen synthase kinase 3β (GSK3β) and adenomatosis polyposis coli (Apc). This enables catenin beta 1 (Ctnnb1) to traslocate into the nucleus to form the Ctnnb1/Tcf complex, which in turn activates the downstream genes (Willert and Jones, 2006). In the presence of Wnt signalling, transcription factor (Tcf3) activates the downstream genes that promote pluripotency maintenance by collaborating with the pivotal transcription factors Otc3/4 (Fig. 2), Sox2 and Nanog (Masui, 2010).

4. Derivation of human embryonic stem cells

There was a considerable delay between the derivation of mouse ES cells (1981) and the derivation of human ES cells in 1998 (Thomson et al., 1998). This delay was primarily due to species-specific ES cell differences and suboptimal human embryo culture media. In fact the first study to describe the isolation of human ICM cells was published by Bongso et al. (Bongso et al., 1994), but subsequent culture in media supplemented with LIF and serum resulted only in differentiation, not in the derivation of stable pluripotent cell lines. Human ES (hES) cells can be characterized by their immortality, expression of telomerase expression, pluripotentiality, ability to form teratomas, and maintenance of a stable karyotype and, even after prolonged undifferentiated proliferation, maintain the development potential to contribute to advanced derivatives of all three germ layers, even after clonal derivation (Amit et al., 2000). For obvious ethical reasons, experiments involving blastocyst injections and ectopic grafting in adult hosts cannot be performed in the human.

Human ES cells have been derived from morula, later blastocyst embryos (Stojkovic et al., 2004; Strelchenko et al., 2004), single blastomeres (Klimanskaya et al., 2006), and parthenogenetic embryos (Lin et al., 2007).

Previous reports suggest that the success rate in deriving hES cell lines is highly dependent on the quality of recovered blastocysts, isolation condition used and technical expertise (Pera et al., 2000; Mitalipova et al., 2003).

ES cell lines are usually derived by immunosurgery. In this process the trophoblast layer of the blastocyst is selectively removed, and the intact inner cell mass is further cultured on MEFs (Amit and Itskovitz-Eldor, 2002). Although the cloning efficiency of the hES cells was relatively poor, a several fold increase was observed when serum-free medium supplemented with basic fibroblast growth factor (βFGF) was used (Amit et al., 2000).

5. Maintenance of human embryonic stem cells

Mechanical and enzymatic transfer methods are used to maintain hES cell lines (Oh et al., 2005). The mechanical transfer method is laborious and time-consuming, although remains an efficient technique for the transfer of undifferentiated hES cells and results in similar clump sizes. The enzymatic transfer method is used when the bulk production of cells are required for various experiments and results in the more rapid growth and larger production of hES cells.

However, the cell clumps vary in size, and there is a higher probability that both differentiated and undifferentiated cell will be transferred. In the case of passaging more differentiated colonies, a combination of both methods allows mass production of hES cells by excluding differentiated colonies from passage by manual selection prior to enzyme treatment.

Another limiting factor relating to cell culture systems is that hES cells still require the presence of feeder layer. In fact, feeder-free system for hES cell culture is required if hES cell cultures are to become clinical-grade, since the use of animal feeders and/or ingredients for growth of hES cells limits the large-scale culture and medical applicability of hES cells. At present, feeder-free systems are not optimal for the derivation and growth of clinical-grade hES cell lines since the presence of animal ingredient carriers the potential risk for the cross-transfer of different infectious agents. In fact, it has been reported that hES cells embryoid body can incorporate the N-glycolylneuraminic acid (Neu5Gc) from MEFs or from conditioned medium, which resulted in an immune response (Martin et al., 2005).
The first attempt to produced feeder-free cultures of hES cells was reported by Xu et al. (Xu et al., 2001). They propaged hES cells using Matrigel, an animal based extracellular matrix (ECM) preparation, or laminin substrates in medium conditioned by MEFs. This system enabled the long term propagation of the stem cell phenotype, with strong suppression of spontaneous differentiation even at high passages (Carpenter et al., 2001).
In 2005, Prowse et al. identified 102 proteins from conditioned medium of human neonatal fibroblasts which provide invaluable information regarding the factors that may help maintain hES cells (Prowse et al., 2005).
The growth factor, ActivinA, paracrinely secreted by MEFs, is capable of supporting the growth of hES cells on laminin coated dishes for more 20 passages without the need for feeder layers (Beattie et al., 2005). Sato et al. (2004) suggest that Wnt signalling modulation can help to support the growth of hES cells cultures short-term and maintain their capacity to express some stem cell markers in the absence of a feeder cell layer (Sato et al., 2004). Another study demonstrated that noggin (BMP antagonist) combined with high βFGF concentrations in medium support the long term proliferation of undifferentiated hES cells in the absence of feeder cells and/or conditioned medium. However in this case Matrigel coated dishes were used, but this represent a problem for potential medical application of hES cells because xenogeneic pathogens can be transmitted through culture conditions (Wang et al., 2005; Xu et al., 2005).
Moreover, it has been reported that the combination of FGF2, TGFβ, LIF and a proprietary serum replacer can achieve serum-free, feeder-free maintenance of hES cells when cultured on fibronectin ECM (Amit et al., 2004).
The establishment of feeder-free system for the culture of hES cells is critical for genetic manipulation. In fact, homologous recombination could be used as a tool for the repair of specific gene defects in stem cell lines derived from patients suffering disease.

6. Comparison between human and mouse ES cells

Many of the differences between mouse and human ES cells are only beginning to be elucidated, yet it has already been demonstrated that mouse and human ES cells differ in respect to cell surface markers, with human ES cells expressing the stage specific antigens SSEA-3 and SSEA-4, the glycoproteins TRA-1-60 and TRA-1-81, and GCTM-2, none of which are detected in the mouse. In contrast, mouse ES cells express SSEA-1, which remain undetected within human ES cultures. Moreover human ES cells are insensitive to the differentiation suppressing effects of LIF pathway (Thomson et al., 1998; Reubinoff et al., 2000). However, there remain many similarities between human and murine ES cell populations. ES cells are derived from both species using very similar protocols, and same aspects of their propagation, such as the ability of MEFs to support their growth in an undifferentiated state remain almost identical (Fig. 3). Furthermore, human and mouse ES cells possess similar properties of spontaneous differentiation and expression of the pluripotent-associated transcription factor Oct-4.

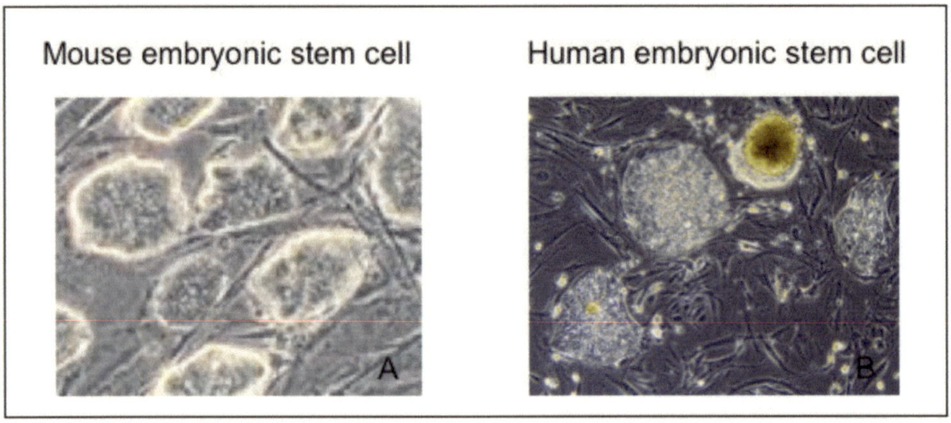

Fig. 3. Phase contrast microscopy images of mouse (A) and human (B) embryonic stem cells on mouse embryonic fibroblast

7. Differentiation of mouse embryonic stem cells *in vitro*

In the absence of feeder cells and anti-differentiating agents such as LIF, mouse ES cells spontaneously differentiate and, under appropriate conditions, generate progeny consisting of derivatives of the three embryonic germ layer: mesoderm, endoderm, and ectoderm (Keller, 1995; Smith, 2001).

Mesoderm derived lineages include the hematopoietic, vascular, and cardiac. Endoderm derivatives include pancreatic β cell and hepatocytes. Ectoderm differentiation of mouse ES cells is well established, as numerous studies have documented and characterized neuroectoderm commitment and neural differentiation.

Three general approaches are used to initiate ES cell differentiation. With the first method, the hanging drop method (Fig. 4), ES cells are allowed to aggregate and form three dimensional colonies known as embryoid bodies (EBs) (Doetschman et al., 1985; Keller, 1995). In the second method, ES cells are cultured directly on stromal cells, and differentiation takes place in contact with these cells (Nakano et al., 1994). The third protocol involves differentiating ES cells in a monolayer on extracellular matrix proteins (Nishikawa et al., 1998) or in presence of specific differentiation medium (Takahashi et al., 2003; Fico et al., 2008).

7.1 Cardiac differentiation

The development of the cardiac lineage in ES cell differentiation cultures is easily detected by the appearance of areas of contracting cells that display characteristics of cardiomyocytes. Development of the cardiomyocyte lineage progresses through distinct stages that are similar to development of the lineage *in vivo*. An ordered pattern of expression of cardiac genes is observed in the differentiation cultures, with expression of the transcription factors *gata-4* and *nkx2.5* that are required for lineage development preceding the expression of genes such as *atrial natriuretic protein* (*ANP*), *myosin light chain* (*MLC*)-*2v*, α-

myosin heavy chain (α-*MHC*), β-*myosin heavy chain* (β-*MHC*), and *connexin 43* that are indicative of distinct maturation stages within the developing organ *in vivo* (Hescheler et al., 1997; Boheler et al., 2002). Several different studies have begun to investigate the mechanisms regulating the development of the cardiac lineage in ES cell differentiation cultures. It has been demonstrated that the EGF-CFC factor Cripto, known to be essential for development *in vivo* (Ding et al., 1998; Xu et al., 1999), plays a pivotal role in differentiation of ES cells to the cardiac lineage, in fact, *Cripto-/-* ES cells display a deficiency in generating cardiomyocytes (Parisi et al., 2003). Notch signaling also plays a role in cardiac development from ES cells (Schroeder et al., 2003), in fact ES cells lacking a downstream signalling molecule of all Notch (Jk) generate more cardiac cells than wild type ES cells (Keller, 2005). However, in this case, inhibition of the pathway appears to be important for cardiac differentiation. Other factors, including BMP2 and FGF2 (Kawai et al., 2004) as well as nitric oxide (Kanno et al., 2004) and ascorbic acid (Takahashi et al., 2003), have been shown to promote or improve cardiomyocyte differentiation in ES cell cultures.

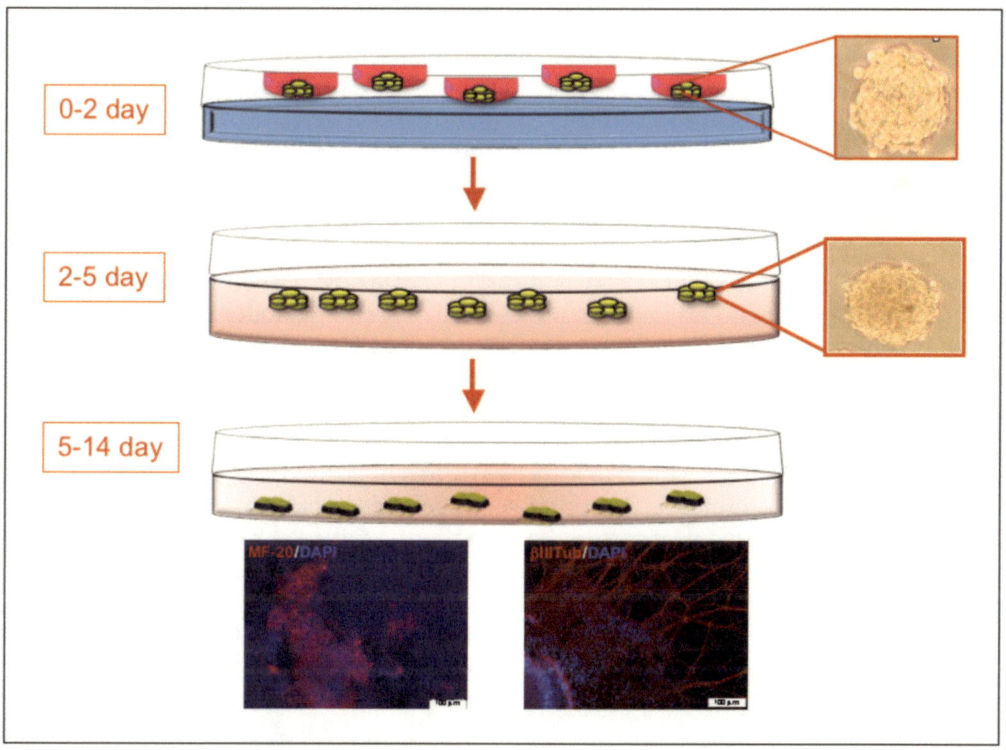

Fig. 4. Schematic rappresentation of method used to form embryoid bodies. This method is generally used to induce ES cells differentiation into cardiomyocytes or, adding retinoic acid, into neurons. MF-20 specific marker of cardiac cells, bIIITubulin (bIIITub) specific neural marker, DAPI nuclear marker

7.2 Primitive and definitive hematopoiesis

ES cells undergo hematopoietic differentiation in optimized culture conditions following serum induction (Keller, 1995). Gene expression and progenitor cell analysis revealed that

the differentiation program in these cultures closely parallels that in the early embryo, progressing through a primitive streak stage, to mesoderm, and subsequently to a yolk sac-like hematopoietic program. Detailed analysis of these early stages led to the identification of the hemangioblast, a progenitor that displays hematopoietic and vascular potential (Choi et al., 1998). After the hemangioblast appears, primitive erythroid progenitors develop in ES cells cultures, establishing the primitive erythropoiesis phase of hematopoiesis. In addition to primitive erythrocytes, other progenitors including those of the macrophage, definitive erythroid, megakaryocyte, and mast cell lineages develop in the differentiation cultures with a kinetic pattern similar to that observed in the yolk sac (Murry and Keller, 2008).
However, despite extensive efforts, to induce the formation of transplantable hematopoietic stem cells (HSCs) the development of HSCs from ESCs remains a challenge, which may reflect the complexities of embryonic hematopoietic development where different hematopoietic programs are generated at different times from different embryonic sites (Murry and Keller, 2008).

7.3 Endoderm differentiation

The generation of endoderm derivatives, in particular pancreatic β-cells and hepatocytes, has become the focus of many investigators in the field of ES cell biology. The interest in the efficient and reproducible development of these cell types derives from their clinical potential for the treatment of Type I diabetes and liver disease, respectively (Keller, 2005). Several genes used as markers of definitive endoderm (*Foxa2*, *Gata4*, and *Sox17*) (Arceci et al., 1993; Monaghan et al., 1993; Sasaki and Hogan, 1993; Laverriere et al., 1994; Kanai-Azuma et al., 2002), early liver (a-*fetoprotein* and *albumin*) (Dziadek and Adamson, 1978; Meehan et al., 1984; Sellem et al., 1984), and early pancreas (*Pdx1* and *insulin*) (McGrath and Palis, 1997) development are also expressed by visceral endoderm, a population of extraembryonic endoderm. Given the overlapping expression patterns, it can be difficult to distinguish definitive and extraembryonic endoderm in the ES cell differentiation cultures. Another problem encountered in endoderm differentiation from ES cells is the lack of specific inducers of this lineage.
It has been investigated the potential of ES cells to differentiate into endoderm derivatives and developed two different protocols that promote the generation of these cell types (Kubo et al., 2004). The first is a restricted exposure of the EBs to serum followed by a period of serum-free culture, and the second is induction with Activin A in the absence of serum. Endoderm development was quantified based on the proportion of cells that expressed Foxa2, a transcription factor found in the earliest stages of definitive endoderm development (Monaghan et al., 1993; Sasaki and Hogan, 1993). All of the Foxa2+ cells that developed in these cultures also expressed the primitive streak marker *brachyury*, a gene that is not expressed in visceral endoderm. This observation strongly suggests that the Foxa2+ cells represented definitive endoderm. Based on the number of Foxa2+ cells, the Activin A protocol was found to be the most efficient as >50% of the total population in these cultures expressed this protein, in fact, low level of Activin A promote a mesoderm fate, and high levels of Activin A induced the formation of endoderm cells (Green et al., 1992; Hudson et al., 1997).
In 2009 Borowiak et al. identified two potent small molecules, IDE1 and IDE2, that can direct mouse ES cell differentiation such that 70%–80% of cells are endoderm cells. This efficiency of induction compares favorably with published protocols employing TGF-β family members, e.g., Activin A or Nodal, which produce about 45% endoderm. The application of small molecules to differentiate mouse and human ES cells into endoderm represents a step

toward achieving a reproducible and efficient production of desired ES cell derivatives (Borowiak et al., 2009).

7.4 Neural differentiation

Several different protocols have evolved to promote neuroectoderm differentiation. The various approaches include (1) treatment of serum-stimulated EBs with retinoic acid (Bain et al., 1995), (2) sequential culture of EBs in serum followed by serum-free medium (Okabe et al., 1996), (3) differentiation of ES cells as a monolayer in serum-free medium (Tropepe et al., 2001; Ying et al., 2003; Fico et al., 2008), and (4) differentiation of ES cells directly on stromal cells in the absence of serum (Kawasaki et al., 2000; Barberi et al., 2003). As with the mesoderm and endoderm lineages, development of the ectoderm lineages in the ES differentiation cultures appears to recapitulate their development in the early embryo (Barberi et al., 2003). *In vitro* it is possible to form the three major neural cell types: neurons, astrocytes and oligodendrocytes.

The protocols for differentiation to specific types of neurons have included the sequential combination of regulators that are known to play a role in the establishment of these lineages in the early embryo. For instance, midbrain dopaminergic neurons have been generated in the EB system by overexpression in the cells of the transcription factor nuclear-receptor-related factor1 (Nurr1), and the addition to the cultures of sonic hedgehog (SHH) and FGF8 (Kim et al., 2002). Nurr1, SHH, and FGF8 are required for the development of this class of neurons in the early embryo (Ye et al., 1998; Simon et al., 2003). Other studies have demonstrated the development of cholinergic, serotonergic, and GABAergic neurons in addition to dopaminergic neurons, when differentiated on MS5 stromal cells in the presence of different combinations of cytokines (Barberi et al., 2003). Using the coculture approach together with the appropriate signaling molecules and selection steps, cells that display many of the characteristics of motor neurons has been successfully generated (Wichterle et al., 2002).

When cultured at low density in serum-free medium in the presence of LIF, ES cells generate a population that has been called primitive neural stem cells (Tropepe et al., 2001). These cells have been characterized by their ability to generate neurosphere-like colonies composed of cells that express the neural precursor cell marker, nestin (Lendahl et al., 1990). When cultured on a matrigel substrate in the presence of low amounts of serum, cells within these colonies generated neurons, astrocytes, and oligodendrocytes. In 2008, Fico et al. established a one-step protocol that allowed differenziation of mouse ES cells into a highly enriched population of neuronal cells, simply by culturing them on gelatin-coated dishes in a chemically defined serum-free medium. This differentiation method is able to generate a wide range of neural subtypes and glial cells from mouse ES cells (Fico et al., 2008).

The ability to generate different types of neurons from ES cells has dramatically raised the interest in repair of nervous system disorders by cell replacement therapy.

8. Human embryonic stem cells differentiation

Human embryonic stem cells are characterized by their ability to proliferate in the undifferentiated state in culture for a prolonged period, and by their capacity to differentiate into derivatives of all three germ layers. A variety of studies have described *in vitro* spontaneous and directed differentiation of hES cells into different lineages: cardiomyocytes (Kehat et al., 2001; Xu et al., 2002), neurons and glia (Carpenter et al., 2001; Reubinoff et al.,

2001), endothelial cells (Levenberg et al., 2002), hematopoietic precursors (Kaufman et al., 2001), trophoblast, and hepatocyte-like cells (Rambhatla et al., 2003). The most common method used for *in vitro* differentiation is to remove the hES cells from the feeder layer and culture in suspension in absence of MEFs. Following culturing in suspension, hES cells aggregate into EBs (Itskovitz-Eldor et al., 2000). The aggregation process itself triggers initial cell differentiation. It is thought that the EBs consist of derivatives of all three germ layers, which interact and cross-induce each other, resulting in complex differentiation into the various lineages. This process is considered to recapitulate early embryonic development from the blastocyst stage to the egg-cylinder stage.

8.1 Cardiac differentiation

In order to generate a cardiomyocyte-differentiating system from the hES cells, small clumps of 3–20 cells were grown in suspension for 8 days (Amit et al., 2000). The EBs were then plated on gelatin-coated culture dishes and observed microscopically for the appearance of spontaneous contraction. Rhythmically contracting areas appeared at 6 to 12 days after plating. Cells isolated from the beating areas expressed cardiac-specific structural genes, such as cardiac troponin I and brachyury (T), atrial natriuretic peptide (ANP), atrial and ventricular myosin light chains (MLCs). Immunostaining studies demonstrated the presence of the cardiac-specific sarcomeric proteins myosin heavy chain, α-actinin, desmin, and cardiac troponin I, as well as ANP (Kehat et al., 2001).

Cardiomyocyte differentiation can be enhanced in the mouse ES cell system following the addition of differentiation factors including, dimethyl sulfoxide (DMSO), retinoic acid (RA), and small molecoles. Addition of the demethylating agent 5-aza-2′-deoxycytidine to EB cultures has also been shown to be effective for mouse ES cell and human ES cell differentiation into cadiomyocytes. In contrast, RA in hES cells did not induce a higher proportion of cardiomyocytes *in vitro* (Schuldiner et al., 2000). An alternative method for deriving cardiomyocytes has been achieved following the coculture of pluripotent hES cell lines with END-2 cells (visceral-endoderm-like cell lines) (Mummery et al., 2003).

8.2 Hematopoietic differentiation

Several studies have documented hematopoietic development of hES cells using different induction schemes (Murry and Keller, 2008). As observed in the mouse system, the predominant population generated during the first 7–10 days of hES cell differentiation is primitive erythroid progenitors, indicating that the equivalent of yolk-sac hematopoiesis develops first in these cultures (Zambidis et al., 2005; Kennedy et al., 2007). As observed with mouse ES cell and the mouse embryo, the onset of hematopoiesis in hES cell cultures is marked by development of the hemangioblast between days 2 and 4 of differentiation, prior to establishment of the primitive erythroid lineage (Kennedy et al., 2007; Lu et al., 2007; Davis et al., 2008)

8.3 Neural differentiation

In 2001, Reubinoff and Zhang highlighted the potential of hES cells to generated neural cells (Reubinoff et al., 2001; Zhang et al., 2001). Zhang et al. have combined the techniques which were initially developed for the neural differentiation of mouse ES cells and adapted these to produce human neural stem cells. This occurs via a successive stepwise approach, which consists of inducing the formation of EBs and from these generating neural rosettes, which

are proliferating structures that mimic neural tube formation. Rosettes are subsequently harvested by selective dissociation and are cultured as free-floating aggregates of neural precursors, capable of generating neurons and glia (Zhang et al., 2001).
Reubinoff demonstrated that neural differentiation was induced by overgrowth of undifferentiated ES cells. Maintaining hES cells in culture without passage or replenishing feeder cells led to spontaneous neural differentiation within a heterogeneous population of hES cell progeny. Individual clusters of presumptive neural progenitors were identified by phase contrast microscopy and manually transferred onto uncoated dishes. Following culture in defined medium supplemented with βFGF and epidermal growth factor (EGF), these cells formed aggregates highly enriched with neural precursor cells. After withdrawal of βFGF and EGF, downregulation of nestin and mash-1 is followed by upregulated expression of neuron-specific NFM, synaptophysin, Nurr1, and tyrosine hydroxylase (TH) genes. A decreased formation of nestin-positive cells is assimilated with an increased number of neuronal cells expressing neuron-specific protein. Mature neuronal cells are evidenced by the production of neurotransmitters such as dopamine, serotonin, GABA, and glutamate. These results suggest that in presence of neuronal differentiation factors, such as retinoic acid, FGF4, FGF8, or βFGF, hES-derived cells, led to the enrichment of cholinergic, serotinergic, dopaminergic and GABAergic neurons, respectively (Okabe et al., 1996; Lee et al., 2000; Rolletschek et al., 2001; Barberi et al., 2003).
Li et al. (2005) differentiated hES cells into spinal motoneurons using retinoic acid and in the presence of SHH (Li et al., 2005).

8.4 Pancreatic β -islet cells

1–3% of cells within 60–70% of human EBs produced from hES cells have been observed to stain positively for insulin (Assady et al., 2001).
A modification of Lumelskey and colleagues (2001) method resulted in the production of insulin-secreting cells derived from hES cells (Lumelsky et al., 2001). This was achieved following an additional step of culture including, a lowering of the glucose concentration in the medium, removal of βFGF and addition of nicotinamide. Dissociating the cells and growing them in suspension resulted in the formation of clusters, which secreted higher levels of insulin than their *in vivo* counterparts and could be maintained *in vitro*. These cells expressed pancreatic genes and following immunofluorescence and in situ hybridization studies, it was confirmed that a high percentage of insulin-expressing cells were located within these cell clusters (Segev et al., 2004).

9. A new age for ES cells: induced pluripotent stem cells

Takahashi and Yamanaka recently achieved a significant breakthrough in reprogramming somatic cells back to an ES like state (Takahashi and Yamanaka, 2006). They successfully reprogrammed mouse embryonic fibroblasts and adult fibroblasts to pluripotent ES-like cells after viral-mediated transduction of the four transcription factors Oct4, Sox2, c-myc and Klf4 followed by selection for activation of the Oct4 target gene Fbx15. Cells that had activated Fbx15 were designated with a coined expression "induced pluripotent stem" (iPS) cells. These cells were shown to be pluripotent by their ability to form teratomas although they were unable to generate live chimeras. In subsequent experiments when activation of the endogenous Oct4 or Nanog genes was used as a more stringent selection criterion for pluripotency, the resulting Oct4-iPS or Nanog-iPS cells, in contrast to Fbx15-iPS cells, were fully reprogrammed to a pluripotent ES cell state by molecular and biological criteria (Maherali et al., 2007; Wernig et al., 2007). Shortly after the reprogramming of mouse cells had been achieved the generation of iPS cells from human fibroblasts was reported (Takahashi et al., 2007; Yu et al., 2007).

While genetic experiments have established that Oct4 and Sox2 are essential for pluripotency (Chambers and Smith, 2004), the role of the two oncogenes, c-myc and Klf4, in reprogramming is less clear. Some of these oncogenes may, in fact, be dispensable for reprogramming as both mouse and human iPS cells have been obtained in the absence of c-myc transduction, although with low efficiency (Nakagawa et al., 2008; Wernig et al., 2008).

One of the promises of patient-specific ES cells is the potential for customized therapy of diseases. Previous studies have shown that disease-specific ES cells produced by nuclear cloning in combination with gene correction can be used to correct an immunologic disorder in a proof-of-principle experiment in mice (Rideout et al., 2002). In a similar approach, by using a humanized sickle cell anemia mouse model, it has been shown that mice can be rescued after transplantation with hematopoietic progenitors obtained *in vitro* from autologous iPS cells (Hanna et al., 2007). Finally, it has been shown that iPS cells can be efficiently differentiated into neural precursor cells giving rise to neuronal and glial cell types in culture. Neural precursors derived from iPS cell were able to improve behaviour in a rat model of Parkinson's disease upon transplantation into the adult brain demonstrating the therapeutic potential of directly reprogrammed fibroblasts for neuronal cell replacement in an animal model (Wernig et al., 2008; Jaenisch, 2009).

10. Conclusion

Embryonic stem cells represent a powerful tool for future regenerative medicine due to their capacity of self-renewal and pluripotency. Studies in animal models have shown that transplantation of fetal stem cell, ES cells, or pluripotent stem cell derivatives can successfully treat many chronic diseases, such as Parkinson's disease, diabetes, traumatic spinal cord injury, Purkinje cell degeneration, Duchenne's muscular dystrophy, liver or heart failure, and osteogenesis imperfecta (Zhang et al., 1996; Horwitz et al., 1999; McDonald et al., 1999; Kobayashi et al., 2000; Li et al., 2000; Soria et al., 2000; Kim et al., 2002).

Almost every day there are reports in the media of new stem cell therapies. There is no doubt that stem cells have the potential to treat many human afflictions, including ageing, cancer, diabetes, blindness and neurodegeneration. In January 2009, the US Food and Drug Administration approved the first clinical trial involving human ES cells, just over 10 years after they were first isolated. In this trial, the safety of ES cell-derived oligodendrocytes in repair of spinal cord injury will be evaluated. Nevertheless, one of the attractions of transplanting iPS cells is that the patient's own cells can be used, obviating the need for immunosuppression (Watt and Driskell, 2010).

Adult tissue stem cells, ES cells and iPS cells can all be used to screen for compounds that stimulate selfrenewal or promote specific differentiation programmes. Finding drugs that selectively target cancer stem cells offers the potential to develop cancer treatments that are not only more effective, but also cause less collateral damage to the patient's normal tissues than drugs currently in use (Watt and Driskell, 2010).

11. References

Abbondanzo, S.J.; Gadi, I. & Stewart, C.L. (1993) Derivation of embryonic stem cell lines. *Methods Enzymol,* 225, 1993) 803-23, ISSN: 0076-6879

Amit, M.; Carpenter, M.K.; Inokuma, M.S.; Chiu, C.P.; Harris, C.P.; Waknitz, M.A.; Itskovitz-Eldor, J. & Thomson, J.A. (2000) Clonally derived human embryonic stem cell lines maintain pluripotency and proliferative potential for prolonged periods of culture. *Dev Biol,* 227, 2, (Nov 2000) 271-8, ISSN: 0012-1606

Amit, M. & Itskovitz-Eldor, J. (2002) Derivation and spontaneous differentiation of human embryonic stem cells. *J Anat,* 200, Pt 3, (Mar 2002) 225-32, ISSN: 0021-8782

Amit, M.; Shariki, C.; Margulets, V. & Itskovitz-Eldor, J. (2004) Feeder layer- and serum-free culture of human embryonic stem cells. *Biol Reprod,* 70, 3, (Mar 2004) 837-45, ISSN: 0006-3363

Arceci, R.J.; King, A.A.; Simon, M.C.; Orkin, S.H. & Wilson, D.B. (1993) Mouse GATA-4: a retinoic acid-inducible GATA-binding transcription factor expressed in endodermally derived tissues and heart. *Mol Cell Biol,* 13, 4, (Apr 1993) 2235-46, ISSN: 0270-7306

Assady, S.; Maor, G.; Amit, M.; Itskovitz-Eldor, J.; Skorecki, K.L. & Tzukerman, M. (2001) Insulin production by human embryonic stem cells. *Diabetes,* 50, 8, (Aug 2001) 1691-7, ISSN: 0012-1797

Bain, G.; Kitchens, D.; Yao, M.; Huettner, J.E. & Gottlieb, D.I. (1995) Embryonic stem cells express neuronal properties *in vitro. Dev Biol,* 168, 2, (Apr 1995) 342-57, ISSN: 0012-1606

Barberi, T.; Klivenyi, P.; Calingasan, N.Y.; Lee, H.; Kawamata, H.; Loonam, K.; Perrier, A.L.; Bruses, J.; Rubio, M.E.; Topf, N.; Tabar, V.; Harrison, N.L.; Beal, M.F.; Moore, M.A. & Studer, L. (2003) Neural subtype specification of fertilization and nuclear transfer embryonic stem cells and application in parkinsonian mice. *Nat Biotechnol,* 21, 10, (Oct 2003) 1200-7, ISSN: 1087-0156

Beattie, G.M.; Lopez, A.D.; Bucay, N.; Hinton, A.; Firpo, M.T.; King, C.C. & Hayek, A. (2005) Activin A maintains pluripotency of human embryonic stem cells in the absence of feeder layers. *Stem Cells,* 23, 4, (Apr 2005) 489-95, ISSN: 1066-5099

Boheler, K.R.; Czyz, J.; Tweedie, D.; Yang, H.T.; Anisimov, S.V. & Wobus, A.M. (2002) Differentiation of pluripotent embryonic stem cells into cardiomyocytes. *Circ Res,* 91, 3, (Aug 2002) 189-201, ISSN: 0009-7330

Bongso, A.; Fong, C.Y.; Ng, S.C. & Ratnam, S. (1994) Isolation and culture of inner cell mass cells from human blastocysts. *Hum Reprod,* 9, 11, (Nov 1994) 2110-7, ISSN: 0268-1161

Borowiak, M.; Maehr, R.; Chen, S.; Chen, A.E.; Tang, W.; Fox, J.L.; Schreiber, S.L. & Melton, D.A. (2009) Small molecules efficiently direct endodermal differentiation of mouse and human embryonic stem cells. *Cell Stem Cell,* 4, 4, (Apr 3 2009) 348-58, ISSN: 1934-5909

Bradley, A.; Evans, M.; Kaufman, M.H. & Robertson, E. (1984) Formation of germ-line chimaeras from embryo-derived teratocarcinoma cell lines. *Nature,* 309, 5965, (May 1984) 255-6, ISSN: 0028-0836

Bryja, V.; Bonilla, S. & Arenas, E. (2006a) Derivation of mouse embryonic stem cells. *Nat Protoc,* 1, 4, 2006a) 2082-7, ISSN: 1754-2189

Bryja, V.; Bonilla, S.; Cajanek, L.; Parish, C.L.; Schwartz, C.M.; Luo, Y.; Rao, M.S. & Arenas, E. (2006b) An efficient method for the derivation of mouse embryonic stem cells. *Stem Cells,* 24, 4, (Apr 2006b) 844-9, ISSN: 1066-5099

Burdon, T.; Smith, A. & Savatier, P. (2002) Signalling, cell cycle and pluripotency in embryonic stem cells. *Trends Cell Biol,* 12, 9, (Sep 2002) 432-8, ISSN: 0962-8924

Carpenter, M.K.; Inokuma, M.S.; Denham, J.; Mujtaba, T.; Chiu, C.P. & Rao, M.S. (2001) Enrichment of neurons and neural precursors from human embryonic stem cells. *Exp Neurol,* 172, 2, (Dec 2001) 383-97, ISSN: 0014-4886

Chambers, I. & Smith, A. (2004) Self-renewal of teratocarcinoma and embryonic stem cells. *Oncogene,* 23, 43, (Sep 2004) 7150-60, ISSN: 0950-9232

Choi, K.; Kennedy, M.; Kazarov, A.; Papadimitriou, J.C. & Keller, G. (1998) A common precursor for hematopoietic and endothelial cells. *Development,* 125, 4, (Feb 1998) 725-32, ISSN: 0950-1991

Chung, Y.; Klimanskaya, I.; Becker, S.; Marh, J.; Lu, S.J.; Johnson, J.; Meisner, L. & Lanza, R. (2006) Embryonic and extraembryonic stem cell lines derived from single mouse blastomeres. *Nature,* 439, 7073, (Jan 2006) 216-9, ISSN: 0028-0836

Cibelli, J.B.; Stice, S.L.; Golueke, P.J.; Kane, J.J.; Jerry, J.; Blackwell, C.; Ponce de Leon, F.A. & Robl, J.M. (1998) Transgenic bovine chimeric offspring produced from somatic cell-derived stem-like cells. *Nat Biotechnol,* 16, 7, (Jul 1998) 642-6, ISSN: 1087-0156

Davis, R.P.; Ng, E.S.; Costa, M.; Mossman, A.K.; Sourris, K.; Elefanty, A.G. & Stanley, E.G. (2008) Targeting a GFP reporter gene to the MIXL1 locus of human embryonic stem cells identifies human primitive streak-like cells and enables isolation of primitive hematopoietic precursors. *Blood,* 111, 4, (Feb 15 2008) 1876-84, ISSN: 0006-4971

de Haan, G. (2002) Hematopoietic stem cells: self-renewing or aging? *Cells Tissues Organs,* 171, 1, 2002) 27-37, ISSN: 1422-6405

Ding, J.; Yang, L.; Yan, Y.T.; Chen, A.; Desai, N.; Wynshaw-Boris, A. & Shen, M.M. (1998) Cripto is required for correct orientation of the anterior-posterior axis in the mouse embryo. *Nature,* 395, 6703, (Oct 1998) 702-7, ISSN: 0028-0836

Doetschman, T.; Williams, P. & Maeda, N. (1988) Establishment of hamster blastocyst-derived embryonic stem (ES) cells. *Dev Biol,* 127, 1, (May 1988) 224-7, ISSN: 0012-1606

Doetschman, T.C.; Eistetter, H.; Katz, M.; Schmidt, W. & Kemler, R. (1985) The *in vitro* development of blastocyst-derived embryonic stem cell lines: formation of visceral yolk sac, blood islands and myocardium. *J Embryol Exp Morphol,* 87, (Jun 1985) 27-45, ISSN: 0022-0752

Dziadek, M. & Adamson, E. (1978) Localization and synthesis of alphafoetoprotein in post-implantation mouse embryos. *J Embryol Exp Morphol,* 43, (Feb 1978) 289-313, ISSN: 0022-0752

Evans, M.J. & Kaufman, M.H. (1981) Establishment in culture of pluripotential cells from mouse embryos. *Nature,* 292, 5819, (Jul 1981) 154-6, ISSN: 0028-0836

Fico, A.; Manganelli, G.; Simeone, M.; Guido, S.; Minchiotti, G. & Filosa, S. (2008) High-throughput screening-compatible single-step protocol to differentiate embryonic stem cells in neurons. *Stem Cells Dev,* 17, 3, (Jun 2008) 573-84, ISSN: 1547-3287

Green, J.B.; New, H.V. & Smith, J.C. (1992) Responses of embryonic Xenopus cells to activin and FGF are separated by multiple dose thresholds and correspond to distinct axes of the mesoderm. *Cell,* 71, 5, (Nov 1992) 731-9, ISSN: 0092-8674

Hanna, J.; Wernig, M.; Markoulaki, S.; Sun, C.W.; Meissner, A.; Cassady, J.P.; Beard, C.; Brambrink, T.; Wu, L.C.; Townes, T.M. & Jaenisch, R. (2007) Treatment of sickle cell anemia mouse model with iPS cells generated from autologous skin. *Science,* 318, 5858, (Dec 2007) 1920-3, ISSN: 0036-8075

Hescheler, J.; Fleischmann, B.K.; Lentini, S.; Maltsev, V.A.; Rohwedel, J.; Wobus, A.M. & Addicks, K. (1997) Embryonic stem cells: a model to study structural and functional properties in cardiomyogenesis. *Cardiovasc Res,* 36, 2, (Nov 1997) 149-62, ISSN: 0008-6363

Hogan, B.; Beddington, R.; Costantini, F. & Lacy, E. (1994) *Manipulating the Mouse Embryo: A Laboratory Manual (Second Edition).* Cold Spring Harbor Laboratory Press, ISBN: 0-87969-384-3, New York, USA

Hong, Y.; Winkler, C. & Schartl, M. (1998) Production of medakafish chimeras from a stable embryonic stem cell line. *Proc Natl Acad Sci U S A,* 95, 7, (Mar 1998) 3679-84, ISSN: 0027-8424

Horwitz, E.M.; Prockop, D.J.; Fitzpatrick, L.A.; Koo, W.W.; Gordon, P.L.; Neel, M.; Sussman, M.; Orchard, P.; Marx, J.C.; Pyeritz, R.E. & Brenner, M.K. (1999) Transplantability and therapeutic effects of bone marrow-derived mesenchymal cells in children with osteogenesis imperfecta. *Nat Med,* 5, 3, (Mar 1999) 309-13, ISSN: 1078-8956

Hudson, C.; Clements, D.; Friday, R.V.; Stott, D. & Woodland, H.R. (1997) Xsox17alpha and -beta mediate endoderm formation in Xenopus. *Cell,* 91, 3, (Oct 1997) 397-405, ISSN: 0092-8674

Itskovitz-Eldor, J.; Schuldiner, M.; Karsenti, D.; Eden, A.; Yanuka, O.; Amit, M.; Soreq, H. & Benvenisty, N. (2000) Differentiation of human embryonic stem cells into embryoid bodies compromising the three embryonic germ layers. *Mol Med,* 6, 2, (Feb 2000) 88-95, ISSN: 1076-1551

Jaenisch, R. (2009) Stem cells, pluripotency and nuclear reprogramming. *J Thromb Haemost,* 7 Suppl 1, (Jul 2009) 21-3, ISSN: 1538-7933

Kanai-Azuma, M.; Kanai, Y.; Gad, J.M.; Tajima, Y.; Taya, C.; Kurohmaru, M.; Sanai, Y.; Yonekawa, H.; Yazaki, K.; Tam, P.P. & Hayashi, Y. (2002) Depletion of definitive gut endoderm in Sox17-null mutant mice. *Development,* 129, 10, (May 2002) 2367-79, ISSN: 0950-1991

Kanno, S.; Kim, P.K.; Sallam, K.; Lei, J.; Billiar, T.R. & Shears, L.L., 2nd (2004) Nitric oxide facilitates cardiomyogenesis in mouse embryonic stem cells. *Proc Natl Acad Sci U S A,* 101, 33, (Aug 2004) 12277-81, ISSN: 0027-8424

Kaufman, D.S.; Hanson, E.T.; Lewis, R.L.; Auerbach, R. & Thomson, J.A. (2001) Hematopoietic colony-forming cells derived from human embryonic stem cells. *Proc Natl Acad Sci U S A,* 98, 19, (Sep 2001) 10716-21, ISSN: 0027-8424

Kawai, T.; Takahashi, T.; Esaki, M.; Ushikoshi, H.; Nagano, S.; Fujiwara, H. & Kosai, K. (2004) Efficient cardiomyogenic differentiation of embryonic stem cell by fibroblast growth factor 2 and bone morphogenetic protein 2. *Circ J,* 68, 7, (Jul 2004) 691-702, ISSN: 1346-9843

Kawasaki, H.; Mizuseki, K.; Nishikawa, S.; Kaneko, S.; Kuwana, Y.; Nakanishi, S.; Nishikawa, S.I. & Sasai, Y. (2000) Induction of midbrain dopaminergic neurons from ES cells by stromal cell-derived inducing activity. *Neuron,* 28, 1, (Oct 2000) 31-40, ISSN: 0896-6273

Kehat, I.; Kenyagin-Karsenti, D.; Snir, M.; Segev, H.; Amit, M.; Gepstein, A.; Livne, E.; Binah, O.; Itskovitz-Eldor, J. & Gepstein, L. (2001) Human embryonic stem cells can differentiate into myocytes with structural and functional properties of cardiomyocytes. *J Clin Invest,* 108, 3, (Aug 2001) 407-14, ISSN: 0021-9738

Keller, G. (2005) Embryonic stem cell differentiation: emergence of a new era in biology and medicine. *Genes Dev,* 19, 10, (May 2005) 1129-55, ISSN: 0890-9369

Keller, G.M. (1995) *In vitro* differentiation of embryonic stem cells. *Curr Opin Cell Biol,* 7, 6, (Dec 1995) 862-9, ISSN: 0955-0674

Kennedy, M.; D'Souza, S.L.; Lynch-Kattman, M.; Schwantz, S. & Keller, G. (2007) Development of the hemangioblast defines the onset of hematopoiesis in human ES cell differentiation cultures. *Blood,* 109, 7, (Apr 1 2007) 2679-87, ISSN: 0006-4971

Kim, J.H.; Auerbach, J.M.; Rodriguez-Gomez, J.A.; Velasco, I.; Gavin, D.; Lumelsky, N.; Lee, S.H.; Nguyen, J.; Sanchez-Pernaute, R.; Bankiewicz, K. & McKay, R. (2002) Dopamine neurons derived from embryonic stem cells function in an animal model of Parkinson's disease. *Nature,* 418, 6893, (Jul 2002) 50-6, ISSN: 0028-0836

Klimanskaya, I.; Chung, Y.; Becker, S.; Lu, S.J. & Lanza, R. (2006) Human embryonic stem cell lines derived from single blastomeres. *Nature,* 444, 7118, (Nov 2006) 481-5, ISSN: 0028-0836

Kobayashi, N.; Miyazaki, M.; Fukaya, K.; Inoue, Y.; Sakaguchi, M.; Uemura, T.; Noguchi, H.; Kondo, A.; Tanaka, N. & Namba, M. (2000) Transplantation of highly differentiated immortalized human hepatocytes to treat acute liver failure. *Transplantation,* 69, 2, (Jan 2000) 202-7, ISSN: 0041-1337

Kubo, A.; Shinozaki, K.; Shannon, J.M.; Kouskoff, V.; Kennedy, M.; Woo, S.; Fehling, H.J. & Keller, G. (2004) Development of definitive endoderm from embryonic stem cells in culture. *Development,* 131, 7, (Apr 2004) 1651-62, ISSN: 0950-1991

Laverriere, A.C.; MacNeill, C.; Mueller, C.; Poelmann, R.E.; Burch, J.B. & Evans, T. (1994) GATA-4/5/6, a subfamily of three transcription factors transcribed in developing heart and gut. *J Biol Chem,* 269, 37, (Sep 1994) 23177-84, ISSN: 0021-9258

Ledermann, B. & Burki, K. (1991) Establishment of a germ-line competent C57BL/6 embryonic stem cell line. *Exp Cell Res,* 197, 2, (Dec 1991) 254-8, ISSN: 0014-4827

Lee, S.H.; Lumelsky, N.; Studer, L.; Auerbach, J.M. & McKay, R.D. (2000) Efficient generation of midbrain and hindbrain neurons from mouse embryonic stem cells. *Nat Biotechnol,* 18, 6, (Jun 2000) 675-9, ISSN: 1087-0156

Lendahl, U.; Zimmerman, L.B. & McKay, R.D. (1990) CNS stem cells express a new class of intermediate filament protein. *Cell,* 60, 4, (Feb 1990) 585-95, ISSN: 0092-8674

Levenberg, S.; Golub, J.S.; Amit, M.; Itskovitz-Eldor, J. & Langer, R. (2002) Endothelial cells derived from human embryonic stem cells. *Proc Natl Acad Sci U S A,* 99, 7, (Apr 2002) 4391-6, ISSN: 0027-8424

Li, M.; Zhang, D.; Hou, Y.; Jiao, L.; Zheng, X. & Wang, W.H. (2003) Isolation and culture of embryonic stem cells from porcine blastocysts. *Mol Reprod Dev,* 65, 4, (Aug 2003) 429-34, ISSN: 1040-452X

Li, R.K.; Weisel, R.D.; Mickle, D.A.; Jia, Z.Q.; Kim, E.J.; Sakai, T.; Tomita, S.; Schwartz, L.; Iwanochko, M.; Husain, M.; Cusimano, R.J.; Burns, R.J. & Yau, T.M. (2000) Autologous porcine heart cell transplantation improved heart function after a myocardial infarction. *J Thorac Cardiovasc Surg,* 119, 1, (Jan 2000) 62-8, ISSN: 0022-5223

Li, X.J.; Du, Z.W.; Zarnowska, E.D.; Pankratz, M.; Hansen, L.O.; Pearce, R.A. & Zhang, S.C. (2005) Specification of motoneurons from human embryonic stem cells. *Nat Biotechnol,* 23, 2, (Feb 2005) 215-21, ISSN: 1087-0156

Lin, G.; OuYang, Q.; Zhou, X.; Gu, Y.; Yuan, D.; Li, W.; Liu, G.; Liu, T. & Lu, G. (2007) A highly homozygous and parthenogenetic human embryonic stem cell line derived from a one-pronuclear oocyte following *in vitro* fertilization procedure. *Cell Res,* 17, 12, (Dec 2007) 999-1007, ISSN: 1001-0602

Lu, S.J.; Feng, Q.; Caballero, S.; Chen, Y.; Moore, M.A.; Grant, M.B. & Lanza, R. (2007) Generation of functional hemangioblasts from human embryonic stem cells. *Nat Methods,* 4, 6, (Jun 2007) 501-9, ISSN: 1548-7091

Lumelsky, N.; Blondel, O.; Laeng, P.; Velasco, I.; Ravin, R. & McKay, R. (2001) Differentiation of embryonic stem cells to insulin-secreting structures similar to pancreatic islets. *Science,* 292, 5520, (May 2001) 1389-94, ISSN: 0036-8075

Maherali, N.; Sridharan, R.; Xie, W.; Utikal, J.; Eminli, S.; Arnold, K.; Stadtfeld, M.; Yachechko, R.; Tchieu, J.; Jaenisch, R.; Plath, K. & Hochedlinger, K. (2007) Directly reprogrammed fibroblasts show global epigenetic remodeling and widespread tissue contribution. *Cell Stem Cell,* 1, 1, (Jun 2007) 55-70, ISSN: 1934-5909

Martin, G.R. (1981) Isolation of a pluripotent cell line from early mouse embryos cultured in medium conditioned by teratocarcinoma stem cells. *Proc Natl Acad Sci U S A,* 78, 12, (Dec 1981) 7634-8, ISSN: 0027-8424

Martin, M.J.; Muotri, A.; Gage, F. & Varki, A. (2005) Human embryonic stem cells express an immunogenic nonhuman sialic acid. *Nat Med,* 11, 2, (Feb 2005) 228-32, ISSN: 1078-8956

Masui, S. (2010) Pluripotency maintenance mechanism of embryonic stem cells and reprogramming. *Int J Hematol,* 91, 3, (Apr 2010) 360-72, ISSN: 0925-5710

Matsuda, T.; Nakamura, T.; Nakao, K.; Arai, T.; Katsuki, M.; Heike, T. & Yokota, T. (1999) STAT3 activation is sufficient to maintain an undifferentiated state of mouse embryonic stem cells. *Embo J,* 18, 15, (Aug 1999) 4261-9, ISSN: 0261-4189

McDonald, J.W.; Liu, X.Z.; Qu, Y.; Liu, S.; Mickey, S.K.; Turetsky, D.; Gottlieb, D.I. & Choi, D.W. (1999) Transplanted embryonic stem cells survive, differentiate and promote recovery in injured rat spinal cord. *Nat Med,* 5, 12, (Dec 1999) 1410-2, ISSN: 1078-8956

McGrath, K.E. & Palis, J. (1997) Expression of homeobox genes, including an insulin promoting factor, in the murine yolk sac at the time of hematopoietic initiation. *Mol Reprod Dev,* 48, 2, (Oct 1997) 145-53, ISSN: 1040-452X

McWhir, J.; Schnieke, A.E.; Ansell, R.; Wallace, H.; Colman, A.; Scott, A.R. & Kind, A.J. (1996) Selective ablation of differentiated cells permits isolation of embryonic stem cell lines from murine embryos with a non-permissive genetic background. *Nat Genet,* 14, 2, (Oct 1996) 223-6, ISSN: 1061-4036

Meehan, R.R.; Barlow, D.P.; Hill, R.E.; Hogan, B.L. & Hastie, N.D. (1984) Pattern of serum protein gene expression in mouse visceral yolk sac and foetal liver. *Embo J,* 3, 8, (Aug 1984) 1881-5, ISSN: 0261-4189

Mitalipova, M.; Calhoun, J.; Shin, S.; Wininger, D.; Schulz, T.; Noggle, S.; Venable, A.; Lyons, I.; Robins, A. & Stice, S. (2003) Human embryonic stem cell lines derived from discarded embryos. *Stem Cells,* 21, 5, 2003) 521-6, ISSN: 1066-5099

Monaghan, A.P.; Kaestner, K.H.; Grau, E. & Schutz, G. (1993) Postimplantation expression patterns indicate a role for the mouse forkhead/HNF-3 alpha, beta and gamma genes in determination of the definitive endoderm, chordamesoderm and neuroectoderm. *Development*, 119, 3, (Nov 1993) 567-78, ISSN: 0950-1991

Mummery, C.; Ward-van Oostwaard, D.; Doevendans, P.; Spijker, R.; van den Brink, S.; Hassink, R.; van der Heyden, M.; Opthof, T.; Pera, M.; de la Riviere, A.B.; Passier, R. & Tertoolen, L. (2003) Differentiation of human embryonic stem cells to cardiomyocytes: role of coculture with visceral endoderm-like cells. *Circulation*, 107, 21, (Jun 2003) 2733-40, ISSN: 0009-7322

Murry, C.E. & Keller, G. (2008) Differentiation of embryonic stem cells to clinically relevant populations: lessons from embryonic development. *Cell*, 132, 4, (Feb 22 2008) 661-80, ISSN: 0092-8674

Nagy, A.; Gertsenstein, M.; Vintersten, K. & Behringer, R. (2003) *Manipulating the Mouse Embryo: A Laboratory Manual (Third Edition)*. Cold Spring Harbor Laboratory Press, ISBN: 978-087969591-0, New York, USA

Nagy, A.; Gocza, E.; Diaz, E.M.; Prideaux, V.R.; Ivanyi, E.; Markkula, M. & Rossant, J. (1990) Embryonic stem cells alone are able to support fetal development in the mouse. *Development*, 110, 3, (Nov 1990) 815-21, ISSN: 0950-1991

Nakagawa, M.; Koyanagi, M.; Tanabe, K.; Takahashi, K.; Ichisaka, T.; Aoi, T.; Okita, K.; Mochiduki, Y.; Takizawa, N. & Yamanaka, S. (2008) Generation of induced pluripotent stem cells without Myc from mouse and human fibroblasts. *Nat Biotechnol*, 26, 1, (Jan 2008) 101-6, ISSN: 1087-0156

Nakano, T.; Kodama, H. & Honjo, T. (1994) Generation of lymphohematopoietic cells from embryonic stem cells in culture. *Science*, 265, 5175, (Aug 1994) 1098-101, ISSN: 0036-8075

Nishikawa, S.I.; Nishikawa, S.; Hirashima, M.; Matsuyoshi, N. & Kodama, H. (1998) Progressive lineage analysis by cell sorting and culture identifies FLK1+VE-cadherin+ cells at a diverging point of endothelial and hemopoietic lineages. *Development*, 125, 9, (May 1998) 1747-57, ISSN: 0950-1991

Niwa, H.; Ogawa, K.; Shimosato, D. & Adachi, K. (2009) A parallel circuit of LIF signalling pathways maintains pluripotency of mouse ES cells. *Nature*, 460, 7251, (Jul 2009) 118-22, ISSN: 0028-0836

Ogawa, K.; Saito, A.; Matsui, H.; Suzuki, H.; Ohtsuka, S.; Shimosato, D.; Morishita, Y.; Watabe, T.; Niwa, H. & Miyazono, K. (2007) Activin-Nodal signaling is involved in propagation of mouse embryonic stem cells. *J Cell Sci*, 120, Pt 1, (Jan 2007) 55-65, ISSN: 0021-9533

Oh, S.K.; Kim, H.S.; Park, Y.B.; Seol, H.W.; Kim, Y.Y.; Cho, M.S.; Ku, S.Y.; Choi, Y.M.; Kim, D.W. & Moon, S.Y. (2005) Methods for expansion of human embryonic stem cells. *Stem Cells*, 23, 5, (May 2005) 605-9, ISSN: 1066-5099

Okabe, S.; Forsberg-Nilsson, K.; Spiro, A.C.; Segal, M. & McKay, R.D. (1996) Development of neuronal precursor cells and functional postmitotic neurons from embryonic stem cells *in vitro*. *Mech Dev*, 59, 1, (Sep 1996) 89-102, ISSN: 0925-4773

Okano, H. (2002) Neural stem cells: progression of basic research and perspective for clinical application. *Keio J Med*, 51, 3, (Sep 2002) 115-28, ISSN: 1880-1293

Pain, B.; Clark, M.E.; Shen, M.; Nakazawa, H.; Sakurai, M.; Samarut, J. & Etches, R.J. (1996) Long-term *in vitro* culture and characterisation of avian embryonic stem cells with multiple morphogenetic potentialities. *Development,* 122, 8, (Aug 1996) 2339-48, ISSN: 0950-1991

Parisi, S.; D'Andrea, D.; Lago, C.T.; Adamson, E.D.; Persico, M.G. & Minchiotti, G. (2003) Nodal-dependent Cripto signaling promotes cardiomyogenesis and redirects the neural fate of embryonic stem cells. *J Cell Biol,* 163, 2, (Oct 2003) 303-14, ISSN: 0021-9525

Pera, M.F.; Reubinoff, B. & Trounson, A. (2000) Human embryonic stem cells. *J Cell Sci,* 113 (Pt 1), (Jan 2000) 5-10, ISSN: 0021-9533

Potten, C.S. (1998) Stem cells in gastrointestinal epithelium: numbers, characteristics and death. *Philos Trans R Soc Lond B Biol Sci,* 353, 1370, (Jun 1998) 821-30, ISSN: 0962-8436

Prowse, A.B.; McQuade, L.R.; Bryant, K.J.; Van Dyk, D.D.; Tuch, B.E. & Gray, P.P. (2005) A proteome analysis of conditioned media from human neonatal fibroblasts used in the maintenance of human embryonic stem cells. *Proteomics,* 5, 4, (Mar 2005) 978-89, ISSN: 1615-9853

Rambhatla, L.; Chiu, C.P.; Kundu, P.; Peng, Y. & Carpenter, M.K. (2003) Generation of hepatocyte-like cells from human embryonic stem cells. *Cell Transplant,* 12, 1, 2003) 1-11, ISSN: 0963-6897

Ramiya, V.K.; Maraist, M.; Arfors, K.E.; Schatz, D.A.; Peck, A.B. & Cornelius, J.G. (2000) Reversal of insulin-dependent diabetes using islets generated *in vitro* from pancreatic stem cells. *Nat Med,* 6, 3, (Mar 2000) 278-82, ISSN: 1078-8956

Rastan, S. & Robertson, E.J. (1985) X-chromosome deletions in embryo-derived (EK) cell lines associated with lack of X-chromosome inactivation. *J Embryol Exp Morphol,* 90, (Dec 1985) 379-88, ISSN: 0022-0752

Reubinoff, B.E.; Itsykson, P.; Turetsky, T.; Pera, M.F.; Reinhartz, E.; Itzik, A. & Ben-Hur, T. (2001) Neural progenitors from human embryonic stem cells. *Nat Biotechnol,* 19, 12, (Dec 2001) 1134-40, ISSN: 1087-0156

Reubinoff, B.E.; Pera, M.F.; Fong, C.Y.; Trounson, A. & Bongso, A. (2000) Embryonic stem cell lines from human blastocysts: somatic differentiation *in vitro. Nat Biotechnol,* 18, 4, (Apr 2000) 399-404, ISSN: 1087-0156

Rideout, W.M., 3rd; Hochedlinger, K.; Kyba, M.; Daley, G.Q. & Jaenisch, R. (2002) Correction of a genetic defect by nuclear transplantation and combined cell and gene therapy. *Cell,* 109, 1, (Apr 2002) 17-27, ISSN: 0092-8674

Robertson, E.J. (1987) Embryo derived stam cell lines. In: *Teratocarcinomas and Embryonic Stem Cells: A Practical Approach* Robertson, E.J., 71-112, IRL Press, ISBN: 978-185221005, Oxford.

Rolletschek, A.; Chang, H.; Guan, K.; Czyz, J.; Meyer, M. & Wobus, A.M. (2001) Differentiation of embryonic stem cell-derived dopaminergic neurons is enhanced by survival-promoting factors. *Mech Dev,* 105, 1-2, (Jul 2001) 93-104, ISSN: 0925-4773

Sasaki, H. & Hogan, B.L. (1993) Differential expression of multiple fork head related genes during gastrulation and axial pattern formation in the mouse embryo. *Development,* 118, 1, (May 1993) 47-59, ISSN: 0950-1991

Sato, N.; Meijer, L.; Skaltsounis, L.; Greengard, P. & Brivanlou, A.H. (2004) Maintenance of pluripotency in human and mouse embryonic stem cells through activation of Wnt signaling by a pharmacological GSK-3-specific inhibitor. *Nat Med,* 10, 1, (Jan 2004) 55-63, ISSN: 1078-8956

Schoonjans, L.; Albright, G.M.; Li, J.L.; Collen, D. & Moreadith, R.W. (1996) Pluripotential rabbit embryonic stem (ES) cells are capable of forming overt coat color chimeras following injection into blastocysts. *Mol Reprod Dev,* 45, 4, (Dec 1996) 439-43, ISSN: 1040-452X

Schroeder, T.; Fraser, S.T.; Ogawa, M.; Nishikawa, S.; Oka, C.; Bornkamm, G.W.; Honjo, T. & Just, U. (2003) Recombination signal sequence-binding protein Jkappa alters mesodermal cell fate decisions by suppressing cardiomyogenesis. *Proc Natl Acad Sci U S A,* 100, 7, (Apr 2003) 4018-23, ISSN: 0027-8424

Schuldiner, M.; Yanuka, O.; Itskovitz-Eldor, J.; Melton, D.A. & Benvenisty, N. (2000) Effects of eight growth factors on the differentiation of cells derived from human embryonic stem cells. *Proc Natl Acad Sci U S A,* 97, 21, (Oct 2000) 11307-12, ISSN: 0027-8424

Seale, P.; Asakura, A. & Rudnicki, M.A. (2001) The potential of muscle stem cells. *Dev Cell,* 1, 3, (Sep 2001) 333-42, ISSN: 1534-5807

Sedivy, J.M. & Joyner, A.L. (1992) *Gene Targeting.* W.H. Freeman and company, ISBN: 0-7167-7013-X, New York

Segev, H.; Fishman, B.; Ziskind, A.; Shulman, M. & Itskovitz-Eldor, J. (2004) Differentiation of human embryonic stem cells into insulin-producing clusters. *Stem Cells,* 22, 3, 2004) 265-74, ISSN: 1066-5099

Sellem, C.H.; Frain, M.; Erdos, T. & Sala-Trepat, J.M. (1984) Differential expression of albumin and alpha-fetoprotein genes in fetal tissues of mouse and rat. *Dev Biol,* 102, 1, (Mar 1984) 51-60, ISSN: 0012-1606

Simon, H.H.; Bhatt, L.; Gherbassi, D.; Sgado, P. & Alberi, L. (2003) Midbrain dopaminergic neurons: determination of their developmental fate by transcription factors. *Ann N Y Acad Sci,* 991, (Jun 2003) 36-47, ISSN: 0077-8923

Smith, A.G. (2001) Embryo-derived stem cells: of mice and men. *Annu Rev Cell Dev Biol,* 17, 2001) 435-62, ISSN: 1081-0706

Smith, A.G.; Nichols, J.; Robertson, M. & Rathjen, P.D. (1992) Differentiation inhibiting activity (DIA/LIF) and mouse development. *Dev Biol,* 151, 2, (Jun 1992) 339-51, ISSN: 0012-1606

Soria, B.; Roche, E.; Berna, G.; Leon-Quinto, T.; Reig, J.A. & Martin, F. (2000) Insulin-secreting cells derived from embryonic stem cells normalize glycemia in streptozotocin-induced diabetic mice. *Diabetes,* 49, 2, (Feb 2000) 157-62, ISSN: 0012-1797

Stojkovic, M.; Lako, M.; Stojkovic, P.; Stewart, R.; Przyborski, S.; Armstrong, L.; Evans, J.; Herbert, M.; Hyslop, L.; Ahmad, S.; Murdoch, A. & Strachan, T. (2004) Derivation of human embryonic stem cells from day-8 blastocysts recovered after three-step *in vitro* culture. *Stem Cells,* 22, 5, 2004) 790-7, ISSN: 1066-5099

Strelchenko, N.; Verlinsky, O.; Kukharenko, V. & Verlinsky, Y. (2004) Morula-derived human embryonic stem cells. *Reprod Biomed Online,* 9, 6, (Dec 2004) 623-9, ISSN: 1472-6483

Sukoyan, M.A.; Golubitsa, A.N.; Zhelezova, A.I.; Shilov, A.G.; Vatolin, S.Y.; Maximovsky, L.P.; Andreeva, L.E.; McWhir, J.; Pack, S.D.; Bayborodin, S.I. & et al. (1992) Isolation and cultivation of blastocyst-derived stem cell lines from American mink (Mustela vison). *Mol Reprod Dev,* 33, 4, (Dec 1992) 418-31, ISSN: 1040-452X

Sun, L.; Bradford, C.S.; Ghosh, C.; Collodi, P. & Barnes, D.W. (1995) ES-like cell cultures derived from early zebrafish embryos. *Mol Mar Biol Biotechnol,* 4, 3, (Sep 1995) 193-9, ISSN: 1053-6426

Takahashi, K.; Tanabe, K.; Ohnuki, M.; Narita, M.; Ichisaka, T.; Tomoda, K. & Yamanaka, S. (2007) Induction of pluripotent stem cells from adult human fibroblasts by defined factors. *Cell,* 131, 5, (Nov 2007) 861-72, ISSN: 0092-8674

Takahashi, K. & Yamanaka, S. (2006) Induction of pluripotent stem cells from mouse embryonic and adult fibroblast cultures by defined factors. *Cell,* 126, 4, (Aug 2006) 663-76, ISSN: 0092-8674

Takahashi, T.; Lord, B.; Schulze, P.C.; Fryer, R.M.; Sarang, S.S.; Gullans, S.R. & Lee, R.T. (2003) Ascorbic acid enhances differentiation of embryonic stem cells into cardiac myocytes. *Circulation,* 107, 14, (Apr 2003) 1912-6, ISSN: 0009-7322

Theise, N.D.; Saxena, R.; Portmann, B.C.; Thung, S.N.; Yee, H.; Chiriboga, L.; Kumar, A. & Crawford, J.M. (1999) The canals of Hering and hepatic stem cells in humans. *Hepatology,* 30, 6, (Dec 1999) 1425-33, ISSN: 0270-9139

Thomson, J.A.; Itskovitz-Eldor, J.; Shapiro, S.S.; Waknitz, M.A.; Swiergiel, J.J.; Marshall, V.S. & Jones, J.M. (1998) Embryonic stem cell lines derived from human blastocysts. *Science,* 282, 5391, (Nov 1998) 1145-7, ISSN: 0036-8075

Tropepe, V.; Coles, B.L.; Chiasson, B.J.; Horsford, D.J.; Elia, A.J.; McInnes, R.R. & van der Kooy, D. (2000) Retinal stem cells in the adult mammalian eye. *Science,* 287, 5460, (Mar 2000) 2032-6, ISSN: 0036-8075

Tropepe, V.; Hitoshi, S.; Sirard, C.; Mak, T.W.; Rossant, J. & van der Kooy, D. (2001) Direct neural fate specification from embryonic stem cells: a primitive mammalian neural stem cell stage acquired through a default mechanism. *Neuron,* 30, 1, (Apr 2001) 65-78, ISSN: 0896-6273

Wakayama, S.; Hikichi, T.; Suetsugu, R.; Sakaide, Y.; Bui, H.T.; Mizutani, E. & Wakayama, T. (2007) Efficient establishment of mouse embryonic stem cell lines from single blastomeres and polar bodies. *Stem Cells,* 25, 4, (Apr 2007) 986-93, ISSN: 1066-5099

Wang, G.; Zhang, H.; Zhao, Y.; Li, J.; Cai, J.; Wang, P.; Meng, S.; Feng, J.; Miao, C.; Ding, M.; Li, D. & Deng, H. (2005) Noggin and bFGF cooperate to maintain the pluripotency of human embryonic stem cells in the absence of feeder layers. *Biochem Biophys Res Commun,* 330, 3, (May 2005) 934-42, ISSN: 0006-291X

Watt, F.M. (2001) Stem cell fate and patterning in mammalian epidermis. *Curr Opin Genet Dev,* 11, 4, (Aug 2001) 410-7, ISSN: 0959-437X

Watt, F.M. & Driskell, R.R. (2010) The therapeutic potential of stem cells. *Philos Trans R Soc Lond B Biol Sci,* 365, 1537, (Jan 2010) 155-63, ISSN: 0962-8436

Wells, D.N.; Misica, P.M.; Day, T.A. & Tervit, H.R. (1997) Production of cloned lambs from an established embryonic cell line: a comparison between *in vivo*- and *in vitro*-matured cytoplasts. *Biol Reprod,* 57, 2, (Aug 1997) 385-93, ISSN: 0006-3363

Wernig, M.; Meissner, A.; Cassady, J.P. & Jaenisch, R. (2008) c-Myc is dispensable for direct reprogramming of mouse fibroblasts. *Cell Stem Cell,* 2, 1, (Jan 2008) 10-2, ISSN: 1934-5909

Wernig, M.; Meissner, A.; Foreman, R.; Brambrink, T.; Ku, M.; Hochedlinger, K.; Bernstein, B.E. & Jaenisch, R. (2007) *In vitro* reprogramming of fibroblasts into a pluripotent ES-cell-like state. *Nature,* 448, 7151, (Jul 2007) 318-24, ISSN: 0028-0836

Wichterle, H.; Lieberam, I.; Porter, J.A. & Jessell, T.M. (2002) Directed differentiation of embryonic stem cells into motor neurons. *Cell,* 110, 3, (Aug 2002) 385-97, ISSN: 0092-8674

Willert, K. & Jones, K.A. (2006) Wnt signaling: is the party in the nucleus? *Genes Dev,* 20, 11, (Jun 2006) 1394-404, ISSN: 0890-9369

Wu, M.Y. & Hill, C.S. (2009) Tgf-beta superfamily signaling in embryonic development and homeostasis. *Dev Cell,* 16, 3, (Mar 2009) 329-43, ISSN: 1534-5807

Xu, C.; Inokuma, M.S.; Denham, J.; Golds, K.; Kundu, P.; Gold, J.D. & Carpenter, M.K. (2001) Feeder-free growth of undifferentiated human embryonic stem cells. *Nat Biotechnol,* 19, 10, (Oct 2001) 971-4, ISSN: 1087-0156

Xu, C.; Liguori, G.; Persico, M.G. & Adamson, E.D. (1999) Abrogation of the Cripto gene in mouse leads to failure of postgastrulation morphogenesis and lack of differentiation of cardiomyocytes. *Development,* 126, 3, (Feb 1999) 483-94, ISSN: 0950-1991

Xu, C.; Police, S.; Rao, N. & Carpenter, M.K. (2002) Characterization and enrichment of cardiomyocytes derived from human embryonic stem cells. *Circ Res,* 91, 6, (Sep 2002) 501-8, ISSN: 0009-7330

Xu, R.H.; Peck, R.M.; Li, D.S.; Feng, X.; Ludwig, T. & Thomson, J.A. (2005) Basic FGF and suppression of BMP signaling sustain undifferentiated proliferation of human ES cells. *Nat Methods,* 2, 3, (Mar 2005) 185-90, ISSN: 1548-7091

Ye, W.; Shimamura, K.; Rubenstein, J.L.; Hynes, M.A. & Rosenthal, A. (1998) FGF and Shh signals control dopaminergic and serotonergic cell fate in the anterior neural plate. *Cell,* 93, 5, (May 1998) 755-66, ISSN: 0092-8674

Ying, Q.L.; Stavridis, M.; Griffiths, D.; Li, M. & Smith, A. (2003) Conversion of embryonic stem cells into neuroectodermal precursors in adherent monoculture. *Nat Biotechnol,* 21, 2, (Feb 2003) 183-6, ISSN: 1087-0156

Yu, J.; Vodyanik, M.A.; Smuga-Otto, K.; Antosiewicz-Bourget, J.; Frane, J.L.; Tian, S.; Nie, J.; Jonsdottir, G.A.; Ruotti, V.; Stewart, R.; Slukvin, II & Thomson, J.A. (2007) Induced pluripotent stem cell lines derived from human somatic cells. *Science,* 318, 5858, (Dec 2007) 1917-20, ISSN: 0036-8075

Zambidis, E.T.; Peault, B.; Park, T.S.; Bunz, F. & Civin, C.I. (2005) Hematopoietic differentiation of human embryonic stem cells progresses through sequential hematoendothelial, primitive, and definitive stages resembling human yolk sac development. *Blood,* 106, 3, (Aug 1 2005) 860-70, ISSN: 0006-4971

Zhang, S.C.; Wernig, M.; Duncan, I.D.; Brustle, O. & Thomson, J.A. (2001) *In vitro* differentiation of transplantable neural precursors from human embryonic stem cells. *Nat Biotechnol,* 19, 12, (Dec 2001) 1129-33, ISSN: 1087-0156

Zhang, W.; Lee, W.H. & Triarhou, L.C. (1996) Grafted cerebellar cells in a mouse model of hereditary ataxia express IGF-I system genes and partially restore behavioral function. *Nat Med,* 2, 1, (Jan 1996) 65-71, ISSN: 1078-8956

2

Mouse Embryonic Stem Cells Basics from a Fertilized Zygote to These Promising Pluripotent Stem Cells

Carol Fagundez, Mónica Loresi, Stella Delcourt and Pablo Argibay
Instituto de Ciencias Básicas y Medicina Experimental – Hospital Italiano de Buenos Aires
Argentina

1. Introduction

The intention of this chapter is to provide an overview of mouse embryonic stem cells, presenting useful and practical advice to those working in this field, including some personal experience of the authors in this topic.

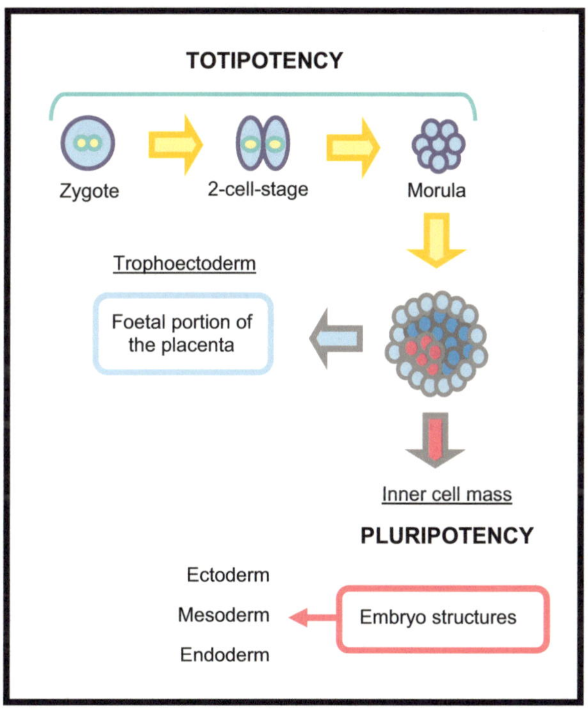

Fig. 1. Origin of mouse embryonic stem cells

In mammals, the fertilized oocyte generates a complex organism containing about 200 types of specialized somatic cells (Wobus, 2001). The zygote is considered the first entity of life and has the ability to produce an entire organism. This property, known as totipotency, is maintained until morula stage (Wobus and Boheler, 2005). By the time the embryo reaches blastocysts stage, at 3.5 days post coitum (dpc), it gets through its first differentiation giving rise to two well-defined cell populations, the trophoectoderm (TE) and the inner cell mass (ICM). The TE is an epithelial monolayer of cells externally located responsible for the foetal portion of the placenta and enclosures a hollowed fluid-filled cavity called blastocoele which contains the ICM, a compacted group of cells known to be the source of stem cells. The ICM will in turn originate every part of the embryo along with extraembryonic membranes like the allantois and the amnion. Before implantation occurs, the primitive endoderm differentiates on the blastocoelic surface of the ICM, producing the endoderm layer of the yolk sac (Fig.1) (Rossant, 2001; Sherman, 1975a).

2. First attempts to isolate pluripotent cells

Some decades ago, researchers used to think that the uterus emitted signals which triggered various events responsible for the differentiation of the embryo. Early efforts to investigate this matter involved removing embryos from the mice female tract prior to implantation and reimplanting them in a variety of foreign sites. Some of these experiments were successful and suggested that differentiation of a number of embryonic cell types could be carried out outside the uterus demonstrating that these events do not necessarily depend on the maternal environment. Posterior efforts were made to achieve proper conditions for the establishment of blastocysts culture. Short-term blastocysts culture was achieved by some researchers from 1965 to 1966 (Gwatkin, 1966; Cole and Paul, 1965). These authors reported that the trophoectoderm layer developed into giant cells resembling the *in vivo* trophoblast, but the ICM failed to proliferate or even survive beyond a few days, but in some isolated cases a limited proliferation of ICM cells in 5-20% of the blastocysts cultured was obtained (Sherman, 1975a).

The first success at long-term culture of mouse blastocysts was reported in 1975 by Sherman (Sherman, 1975b), who was able to grow trophoblast cells and even ICM cells in culture for several months. A cell line from these cells was obtained, but it was poorly tumorigenic, and when tumours were produced, they showed a low degree of differentiation. Moreover, blastocyst-derived cell lines did not possess stable diploid chromosome number and they eventually became hypotetraploid (Sherman, 1975a).

The first pluripotent cell lines to be established were embryonal carcinoma cell (ECC) lines, derived from the undifferentiated compartment of murine and human germ cell tumours. These cells shared morphological, biochemical and immunological properties with pluripotent embryonic stem cells. In some cases, these properties included the ability to differentiate *in vitro* into derivatives of the three germ layers: endoderm, mesoderm and ectoderm, similarly to the isolated embryonic inner cell mass (ICM). The stem cells taken from these lines could participate in the development of completely normal adult mice when injected into blastocyst-stage embryos (Martin, 1981). However, ECCs showed chromosomal aberrations, they also lost their ability to differentiate and chemical inducers were necessary to trigger differentiation under *in vitro* conditions. Another disadvantage was the fact that ECCs sporadically colonized the germ cell line when performing the blastocyst injection technique. These data suggested that ECCs did not retain the pluripotency shown by early embryonic cells. It was necessary then, to find a proper source of stable pluripotent cells destined to a diversity of applications (Rippon and Bishop, 2004; Wobus and Boheler, 2005).

In 1981, Evans and Kaufman successfully isolated pluripotent cells of pre-implantation embryos from 129 SvE strain mice, obtaining a line with normal karyotype, capable of producing teratocarcinomas and forming embryoid bodies (EBs) which eventually, differentiated into a complex of tissues. The authors termed these cells "EK", to distinguish them from ECCs, of similar properties (Evans and Kaufman, 1981). A few months later, Martin published his results about the establishment of a pluripotent cell line derived from the isolated ICM of mouse embryos cultured in presence of ECC conditioned medium. These cells showed all the essential features of teratocarcinoma stem cells. These results lead to researchers to think that the medium conditioned by teratocarcinoma stem cells owned a factor, perhaps identical to a normal endogenous embryonic growth factor, capable of stimulating the proliferation of a small population of pluripotent cells in the normal embryo. The term "embryonic stem cell" (ESC) was introduced by Martin to denote the origin of these cells from embryos and to make a distinction from the ECC derived from teratocarcinomas (Martin, 1981).

In 1984, Wobus *et al.* reproduced some experiments with mouse embryonic stem cells (mESCs) demonstrating its requirements of a fresh and good quality feeder layer of murine embryonic fibroblasts (MEFs), since long-term culture without its presence was not possible. In this work, the authors also proved that these cells express high alkaline phosphatase activity (Wobus et al., 1984).

3. Considerations at the moment of working with mESCs

There are several factors to consider for successful isolation and maintenance of good quality ESCs, such as the culture media employed, the reagents used for supplementation and the feeder cells, among others.

Supplements for the culture medium are very important for the establishment of mESCs, and the leukaemia inhibitory factor (LIF) is known to be essential (Deacon et al., 1998). Maintenance of the stem cell phenotype *in vitro* requires the presence of a feeder layer of fibroblasts or a soluble factor with a differentiation inhibitory activity (DIA). In the absence of DIA the embryonic stem cells differentiate into a wide range of cell types. In 1988, Williams *et al.* purified the myeloid leukaemia inhibitory factor, a haematopoietic regulator, which shows several similarities with DIA (Smith et al., 1988; Williams et al., 1988), and could substitute it in the conservation of pluripotent ESCs lines. These authors assayed different ESCs lines for their performance under the presence of LIF and noted that the proportion of mESCs colonies having a stem cell phenotype was related to the concentration of LIF in the culture (Williams et al., 1988). The examination of ESCs over a range of LIF concentrations has demonstrated that supplementation of ESC medium with LIF had little effect on growth rates, but it significantly diminished the probability of cells to undergo differentiation (Zandstra et al., 2000). Apparently, the volume of LIF necessary for the efficient derivation of mESCs is influenced by genetic diversity (Baharvand and Matthaei, 2004; Tanimoto et al., 2008).

The leukaemia inhibitory factor is a secreted polypeptide cytokine that inhibits the spontaneous differentiation of ESCs and is produced by both, MEFs and STO cells. Mouse ESCs can be cultured without feeder cells in gelatine-coated plates when medium is supplemented with purified LIF. Under such conditions, mESCs may go through a "crisis" in which colonies flatten and differentiation begins at the edges of the colony. After a few passages, a good morphology may appear and sublines from these cells can be derived with normal morphology under these circumstances. However, since MEF and STO cells are probably secreting other factors, besides LIF, that enhance the survival or growth of ESCs, most laboratories continue to use feeder cells and medium supplemented with LIF (Nagy et al., 2003).

The culture media used for the establishment of ESCs usually contain serum, which includes various unknown substrates that affect cell viability and properties (Lee et al., 2006). Foetal bovine serum (FBS) is a vital supplement, but contains a variety of pro-differentiation factors which can affect the propagation of undifferentiated ESCs (Bryja et al., 2006; Tanimoto et al., 2008). These factors potentially generate heterogeneity in ESCs cultures, which results in differential responses of individual stem cells batches to the same culture environment and differentiation signals. For this reason many researchers have decided to introduce a synthetic serum (KoSR, knockout serum replacement) in their protocols to substitute FBS. It is known that KoSR does not contain any undefined growth factors or differentiation promoting factors. Therefore, the use of this serum instead of FBS eliminates some of the variability of the ESCs establishment system (Lee et al., 2006).

According to Lee *et al.* partial or whole replacement of FBS with KoSR not only suppresses ICM cell differentiation but also enhances colony formation. These results imply that FBS replacement reduces or eliminates the differentiation factors from the culture environment, which might be included in the FBS. Moreover, KoSR may contain a factor that promotes ICM cell proliferation and results in efficiency colony formation as shown by Tanimoto *et al.* who obtained a higher rate of ES-like colony generation when ES medium was supplemented with KoSR (Lee et al., 2006; Tanimoto et al., 2008).

Bryja *et al.* observed that ES medium containing KoSR did not support growth of trophoblast cells. It seems that the slower growth and increased degeneration of trophoblastic cells can be an important factor favouring growth of ESCs within the blastocyst, and this synthetic serum appears to selectively favour the growth of embryonic stem cells at the expense of other cell types within the blastocyst (Bryja et al., 2006).

Apparently, mESCs cultured under KoSR show low Erk kinase activation. Some experiments have shown that loss of Erk2 suppresses differentiation in embryonic stem cells. Erk inhibition was reported to allow the derivation of mESCs even from non-permissive mouse strains, and one possibility is that KoSR maintains the activity of this enzyme at low levels, facilitating mES cells isolation (Bryja et al., 2006; Buehr and Smith, 2003; Masui, 2010).

Some authors have attempted supplementation of ES medium with a FBS-KoSR mixture in different ratios to improve the isolation and the subsequent establishment of ESCs (Lee et al., 2006; Tanimoto et al., 2008). However, culture in ES medium without complete or partial FBS replacement do not always support ESCs derivation, but could be beneficial to increase cell growth rate of the colonies already established, while preserving the stemness state (Fagundez et al., 2009; Lee et al., 2006).

The use of KoSR could be useful to easily isolate mESCs from cultures of whole blastocyst embryos, since it prevents development of trophoblastic cells which tend to colonize the culture, avoiding also the application of immunosurgery, a complex technique involving the use of cytotoxic antibodies, that is not always efficient at the moment of eliminating all the trophoblastic cell population of the embryo (Fagundez et al., 2009; Gardner, 1985).

The original protocols of mESCs isolation involve the employment of high-glucose DMEM as culture medium. In the recent years many researchers have replaced this medium by a knockout version (Ko-DMEM), of reduce osmolarity. The difference between DMEM and Ko-

DMEM in supporting the subculturing of ICM cell-derived colonies probably implies that the inorganic salt composition of the basal medium affects ES cell maintenance (Lee et al., 2006).
Tanaka *et al.* isolated the ICM of blastocysts by lasser dissection and cultured them under the standard culture medium (DMEM + FBS) and a knockout medium (Ko-DMEM + KoSR) obtaining a higher efficiency employing the second alternative (17.5% over 1.6%). This alternative culture media, also employed by Bryja *et al.*, demonstrated to be suitable for the establishment of ESCs lines in a feeder cell-free system. The authors suggested that the beneficial effects of this medium may be due to the relatively low osmolarity and the absence of differentiating factors present in bovine sera (Tanaka et al., 2006).
The adrenocorticotropic hormone (ACTH) has been employed by some to improve mESCs culture conditions. Ogawa *et al.* found that small concentrations of ACTH favoured clonal propagation of mESCs when added to the medium containing KoSR, and this compound was also able to maintain pluripotency of embryonic stem cells during several passages (Ogawa et al., 2004). In 2006, Wakayama *et al.* used the beneficial properties shown by ACTH and demonstrated that this factor facilitated the establishment of ESCs lines from isolated blastomeres with high success rates (Wakayama et al., 2007).
One important point to consider when culturing mESCs is that they should be seeded in high density in order to favour communication between them; but care should be taken to avoid overgrowth, which leads to differentiation. Medium needs to be change daily or when it turned yellowish, which indicates acidification. The culture medium should be stored at 4°C, protected from the light and employed within the month of preparation.
When a cell line is obtained from another laboratory, it is advisable to culture them under the same conditions that they were derived because medium requirements can vary between different ESC lines. Variations in the original protocol could lead to eventual alterations of ESCs features. When conditions are not optimal, ESCs will in turn, acquire genetic lesions that will compromise their germ-line potential. Suboptimal conditions may include insufficient nutrients and growth factors, as well as leaving the cells for too long at high density without splitting them, which favours the differentiation into endoderm-like cells. To avoid loss of stemness state ESCs should be split every 2-3 days, and it is convenient to dissociate the stem cells colonies into single cells, since clumps of cells tend to differentiate (Nagy et al., 2003).
To maintain the undifferentiated state of ESCs, two types of feeder cells are most commonly used: primary culture of murine embryonic fibroblasts (MEFs) and STO mouse fibroblasts cell line. MEFs are obtained from 14.5-15.5 dpc mouse embryos. As primary cells, they have the advantage of being a very potent and reliable source of feeder cells. However, these cells show a limited life span in culture and must be replenish continuously from frozen stocks. What is more, they show a batch to batch variation, as not all embryos will produce good-quality fibroblasts (Nagy et al., 2003). Bryja *et al.* used MEFs from several strains of mice as feeder cells for mESCs isolation, but found no difference in their ability to support ESCs derivation; but what they did find was that earlier passages of MEFs performed better during the ESCs derivation protocols, since later passages of MEFs (P 3-5) showed lower viability and as a result, did not support the ESCs derivation as earlier passages (Bryja et al., 2006).
The STO cells were derived from SIM mice and are a *thioguanine-* and *ouabain-* resistant subline of fibroblasts. They have been used as a feeder layer for establishing ESCs from human, mouse and bovine origin as well as for embryonic germ cells. The advantage of using a cell line instead of MEFs for preparing the feeder layer is that these cells are easily grown and do not require a lengthy and tedious method to replenish frozen stocks.

Feeder cells must be mitotically inactivated before use. This can be achieved in two ways: by treatment with Mitomycin C, a substance which crosslinks DNA and blocks cell proliferation; or by gamma-irradiation, that avoids the risk of carrying over mitomycin C from the feeder layer to the ESCs. The proper radiation dose must be determined in order to just arrest feeder cells growth without affecting seriously their viability, as ESCs need the growth factors released by fibroblasts (Zhou et al., 2005; Nagy et al., 2003).

The cells employed as feeder layer should be seeded preferentially 24 hours before use and its quality should be checked. Feeder cells are very important to achieve optimal growth conditions of ESCs, they should be plated in a good number but not in complete confluence (around 80%), so they can spread across the culture plate.

4. The incidence of genetics

It is generally accepted that establishment of ESCs is influenced by species, strain within the same species and even retrieval methods. Even during the very first attempts to obtain mESCs, the strain of mice chosen was considered an important factor involved in the successful isolation of these cells (Sherman, 1975a).

Almost all the mESCs lines in current use have been derived from the 129 strain, known by bibliography as the most permissive strain, with a 30% frequency of establishment. The low frequency of ES cell production employing other mouse strains compared with the 129 suggests an undefined genetic difference in this strain, possibly related to the high incidence of spontaneous testicular teratomas and teratocarcinomas that presents (Baharvand and Matthaei, 2004; Tanimoto et al., 2008). Derivation of ESCs is strongly mouse strain-dependent, and usually, the efficiency of derivation in strains other than 129 strain does not exceed 10% (McWhir et al., 1996).

Genetic background is very important in several studies using experimental animals and so, establishment of mESCs lines from a variety of inbred and mutant mouse strains is desirable for wider utilization of ESCs (Kawase et al., 1994). Since the use of aged ESCs lines can lead to poor germ-line contribution in chimeras, the isolation of novel mESCs lines is also important (Baharvand and Matthaei, 2004).

Although the genetic background of the strain chosen for the obtaining of mESCs plays an important role, the establishment of a line from a particular strain also depends on the culture conditions employed. Kawase *et al.* could establish many ESCs lines from the mouse strain C57BL/6, but they had difficulties when it came to obtain lines from Balb-c employing the same culture medium. Similarly, Baharvand and Matthaei produced two lines from C57BL/6 but needed to use five times the concentration of LIF in the ES medium to obtain embryonic stem cells from Balb-c mice (Baharvand and Matthaei, 2004; Kawase et al., 1994). In the case of F1 hybrids or mixed hybrid genetic background, the success rate of establishing ESCs lines is very high (Nagy et al., 2003). Improving culture conditions, Lee *et al.* obtained mESCs lines from C57BL/6xDBA2 hybrid mice (B6D2F1), outbreed ICR strain and parthenogenetically activated embryos employing Ko-DMEM supplemented with a mixture of FBS and KoSR (Lee et al., 2006).

5. Features of mouse embryonic stem cells

The pluripotency of mESCs is evident from three main characteristics: 1) the cells can be injected into blastocysts to produce chimeric animals which carry these cells even at the germ-line level; 2) *in vitro* aggregation of ESCs leads to EBs formation producing derivatives of the three germ layers and; 3) inoculation of these pluripotent cells into immuno-deficient mice induces formation of teratomas.

Mouse ESCs self-renewal can be corroborated by staining the colonies for specific markers, such as SSEA-1, Oct4, Nanog or Sox2 (Fig. 2D-F). Presence of alkaline phosphatase activity and telomerase can also be measured as an indicator of the stemness state (Fig. 2B). Alkaline phosphatase (ALP) is an enzyme of restricted tissue distribution. In the adult mammal it is present in most tissues, but high activities are found only in: small intestine, kidney, bone and placenta. In the embryo, high activities are more widely distributed and all tissues are rich in ALP at early stages of development. It has been seen that high levels of ALP are characteristics of ECC, whereas the somatic cells that differentiate from the ALP-rich stem cells exhibit low activity (Berstine et al., 1973). In a similar manner, telomerase activity is diminished in adult cells but it is upregulated in embryonic stem cells. Telomerase activity and telomere maintenance are associated with immortality of cells. As stem cells have elongated proliferative capacity, they need a mechanism to maintain telomere length through many cell divisions (Hiyama and Hiyama, 2007).

When mESCs are observed under the light microscope, they are seen as part of spindle to round-shape colonies with smooth edges. The colonies are compacted; a feature apparently mediated by desmosome-like junctions, being thickened at the centre and thinned out towards the periphery (Fig. 2A). These cells grow in a multilayer manner, lack contact inhibition and anchorage dependence, and possess big nuclei surrounded by scarce cytoplasm, with prominent dark nucleoli (Fig. 2C) (Baharvand and Matthaei, 2003; Wobus and Boheler, 2005).

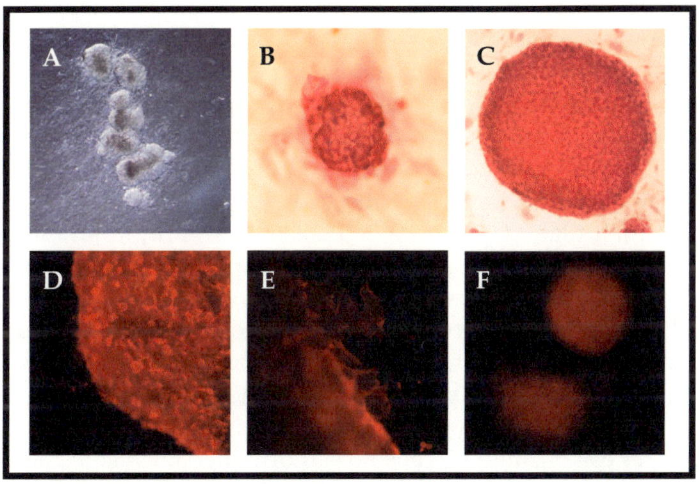

Fig. 2. Features of murine embryonic stem cells.
A) Mouse ESCs colonies; B) ALP staining; C) Mouse ESCs colony stained with hematoxylin-eosin showing the high nucleus/cytoplasm ratio; D) Expression of the cell surface marker SSEA-1; E) Extensions emitted by cells located at the edge of the mESCs colony; F) Nuclear staining pattern of Oct4 in mESCs colonies

Baharvand and Matthaei have brought light into mESCs structure by means of electron microscopy analysis. They have seen that the ultrastructural characteristics of mESCs are similar to the ICM of mouse blastocysts.

The mESCs cultured on a feeder layer of MEFs were seen well polarized, with its organelles in close proximity to the MEFs and some extensions of different length and microvilli projected to the feeder cells. Coated vesicles were abundant and were probably related to protein absorption. When attention was directed to MEFs, it was seen that they showed prominent Golgi complexes, and rough endoplasmic reticulum, which may be indicative of increased protein synthesis and secretion. Since MEFs prevent mESCs from differentiating, it is assumed that they secrete proteins that are important to maintain an undifferentiated state (Baharvand and Matthaei, 2003). When staining of mESCs with the cell surface marker SSEA-1 was carried out in our laboratory, the extensions projected to the MEFs mentioned above were observed, supporting this information (Fig. 2E) (Fagundez et al., 2009).

6. Embryo stages capable of giving rise to mESCs

The obtainment of mESCs is not only restricted to the blastocysts stage, since morulae have demonstrated being capable of forming mESCs colonies when co-culture with feeder cells. When Tesar cultured morulae, the ones that attached to the feeder layer produced outgrowths within a few days; however, the ones that did not continued with its development to blastocyst-stage. When morula-stage embryos were seeded without feeder layers onto gelatine-coated culture dishes, they failed to produce stem cells colonies. The embryos flattened out showing differentiation with presence of trophoblast giant cells. These findings show that employment of certain feeder layers may impose some restraints on further differentiation (Tesar, 2005).

In 1996, Delhaise *et al.* first reported the establishment of one ESCs line from isolated blastomeres of pre-compacted 8-cell stage embryos separated by disaggregation, but it was not able to germ-line transmission in chimeric mice. Ten years later, Chung *et al.* could isolate blastomeres from 8-cell embryos and culture them over a feeder layer of mESCs, obtaining 5 lines. The authors assured that co-culture with feeder cells was critical for the success of the system employed. Despite being successful in the isolation, the establishment rate was quite low, of about 4% (Chung et al., 2006). However, Wakayama *et al.* derived mESCs lines from single blastomeres obtained by biopsy from 2-, 4- and 8-cell stage mouse hybrids embryos with a high success rate of 14-69%. Contrarily to the results obtained by Tesar with morulae seeding, all the blastomeres developed to blastocysts when plated onto MEFs before forming mESCs colonies (Wakayama et al., 2007). The efficiency in establishing ESCs lines from blastomeres of 2-, 4- and 8-cell embryos decreased according to the stage, being higher when they came from 2-cell stage embryos. These results demonstrate a pattern indicating that only 1 or 2, but not all blastomeres from each embryonic stage are capable of deriving ESCs, which might suggest that sister blastomeres are in general not equally competent for establishing ES cells lines (Lorthongpanich et al., 2008; Wakayama et al., 2007). Lorthongpanich *et al.* could establish ESCs lines from isolated blastomeres of 2-cell stage embryos from CD-1 outbred mice strain but were unable to obtain lines from blastomeres coming from 4-cell stage embryos (Lorthongpanich et al., 2008).

The pre-blastocyst period in the mouse development constitutes a stage in which blastomeres are not yet committed to either the ICM or the trophoectodermal lineage. It is therefore conceivable that ESCs lines derived from morulae or earlier cleavage stage embryos might originate before cells were committed to one or other of these two lineages and thus exhibit a wider developmental potential than those of ICM or epiblast origin. Alternatively, permissiveness for ESCs derivation could be restricted to a specific stage of

early development to which embryos must progress irrespective of the time they are explanted (Tesar, 2005).
In many of the cases when karyotype was perfomed, the established ESCs lines showed that they were male. One possible reason is that XY lines are relatively more stable, while XX lines are unstable and have a tendency to become XO (Baharvand and Matthaei, 2004; Kawase et al., 1994). However, the majority of the lines obtained by Tesar were found to be XX (Tesar, 2005).

7. Differentiation of mESCs

When factors that maintain the stemness of ESCs are removed, they are capable of differentiate into a wide range of cell types derived from the germ layers: ectoderm, mesoderm and endoderm. This differentiation can be achieved by formation of embryoid bodies (EBs), which were defined by Dang *et al.* as aggregates of differentiating cells derived from one or more pluripotent stem cells that form or has the potential to form cells from each of the three embryonic germ layers (Dang et al., 2002).
The differentiation state acquired by EBs is often affected by the culture conditions at which they are exposed to. In some occasions, growth factors are added to the medium to promote differentiation to the desired cell type, such as: basic fibroblast growth factor (bFGF), transforming growth factor β (TGFβ), activin-A, bone morphogenetic protein 4 (BMP4), hepatocyte growth factor (HGF), epidermal growth factor (EGF), nerve growth factor (NGF), and retinoc acid (RA) (Kurosawa, 2007).
The three-dimensional cell aggregate formed by the EB facilitates multicellular interactions where cell to cell contact and gap junctions are established. An EB can model the processes of cell differentiation produced during early mammalian development (Kurosawa, 2007). Synchronous differentiation of EBs recapitulates organ-specific development and it is achieved in culture by formation of uniformly sized EBs (Ezekiel et al., 2007). There are some results which indicate that prior to day 3, EBs in suspension culture are, in terms of development, equivalent to pregastrulation-stage embryos. Between days 3 and 5, the EBs contain cell types present in embryos during gastrulation. Since day 6, these cell aggregates show similarities to embryos in the stage of early organogenesis (Ezekiel et al., 2007). However, EB differentiation does not reconstitute the complete array of embryonic development showing no "body plan", and formation of the primitive streak and cell movements associated to gastrulation do not occur (Jackson et al., 2010; Rippon and Bishop, 2004).
There are a variety of methods proposed to induce EB formation. All of these protocols were created to promote the efficient and reproducible obtainment of the cell types desired, as the pattern and efficiency of differentiation are affected by parameters like ESCs density, media components (aminoacids, growth factors and extracellular matrix proteins), pH and osmolarity, as well as the quality and type of serum employ (Wobus and Boheler, 2005).
A common technique is to simply deprive ESCs from the feeder cells or the presence of LIF, and culture them under conditions that initially prevent their adherence to the culture dishes. However, this is not always sufficient to obtain the expected results as will be discussed later.
The embryoid bodies are classified as simple or cystic according to the stage of differentiation achieved. In general, after 2 to 4 days of culture in suspension, EBs resemble a morula-like structure and are called simple EBs. By days 4-5, a cavity is formed inside the cell aggregates giving rise to cystic EBs, which look like an embryo in the blastula or egg

cylinder stage. This kind of EBs consists of a double-layered structure, with an inner ectodermal layer and an outer layer of endoderm, enclosing the cavity (Fig. 3A). Commonly, after 8-10 days in suspension culture, structures homologous to the visceral yolk sac of postimplantation embryos appear in the cystics EBs (Fig. 3B,C) (Koike et al., 2007; Kurosawa, 2007).

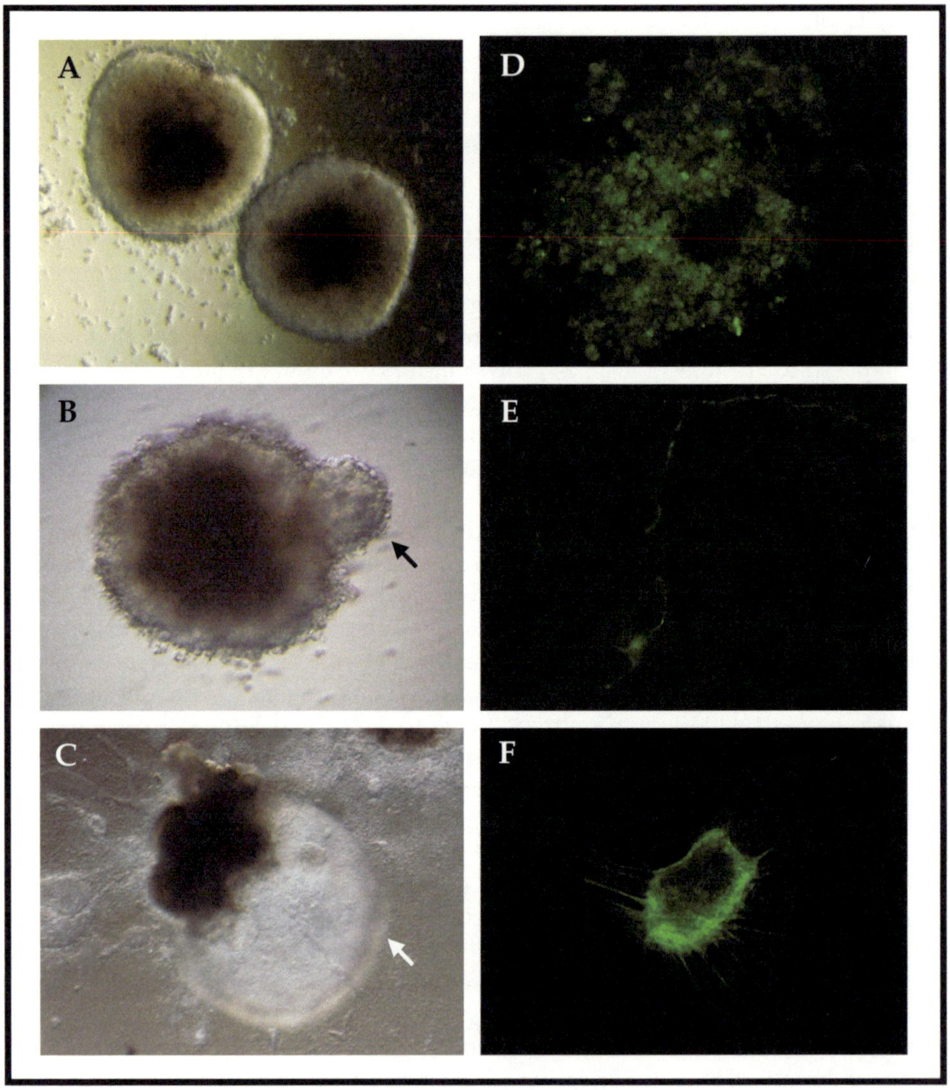

Fig. 3. Differentiation of mESC by means of embryoid bodies.
A) EBs of well-defined borders; B) and C) EBs showing yolk sac-like structures; D) – F) Immunocytochemistry of EB-derived cells stained for presence of α-fetoprotein, β-III-tubulin and muscle actin, respectively

The majority of the methods utilized for EB differentiation studies involve: 1) production of EBs in suspension culture in absence of anti-differentiation factors; 2) growth of EBs on gelatine-coated culture plates; and 3) exposure of EBs to inducer agents to obtain a specific cell type.

The differentiation status acquired by EBs is markedly affected by the culture conditions employed and this is the reason why researches have been working in the development of new methods to improve EBs obtainment. A short description of the most common procedures is provided next.

Hanging drop culture. It is the most popular technique as it allows the formation of homogenous-sized EBs from a known number of ESCs. Generally, small drops containing 400-1000 ESCs are placed on the lid of a 100mm Petri dish. The lid is then inverted to permit the aggregation of the cells at the bottom of the drop, and placed over a culture dish filled with PBS to avoid drops from drying out. After a few days, EBs are collected and cultured in suspension in a non-adherent surface dish. The EBs can be left there to further differentiation or transferred onto gelatine-coated plates (Kurosawa, 2007).

Suspension culture in bacterial-grade dishes. A suspension with a minimum of 10^3 ESCs is seeded into a bacterial grade-dish. The cells do not attach to the plastic surface of the dish; instead they stick to each other forming aggregates that will give rise to EBs (Kurosawa, 2007). Another way to form EBs employing these plates is to culture the whole colonies of mESCs during varying periods of time (Fagundez et al., 2009). Unfortunately, both of these methods produce heterogeneous EBs with distinct degrees of differentiation. However, when the objective of EB formation is just to corroborate *in vitro* pluripotency of mESCs and the differentiation status acquired is not of great importance, this seems to be a good approach since the hanging drop (HD) method is not always easy to perform, and the results could also be variable. In our laboratory we were not able to form EBs with HD technique, as cells were unable to group and failed to produce three-dimensional aggregates.

Although generation of EBs in suspension culture employing whole ESCs colonies has some disadvantages, it seems possible that the proximity established among cells when growing together could be an important factor favouring the differentiation process, since cell to cell contact plays a major role in the early stages of embryonic development (Choi et al., 2005; Fagundez et al., 2009).

Semisolid suspension culture. It was originally employed to obtain cell aggregates of clonal origin. A medium with methylcellulose is used to obtain a semisolid consistence which permits ESCs to remain in isolation. In this way, single cells are not able to coalesce and every one of them develops into an EB. When the culture medium is changed for a standard one, without methylcellulose, the resulting EBs fuse into larger structures. This method has been used for the study of haematopoietic differentiation, endothelial cells and in haematopoietic colony-forming assays (Kurosawa, 2007).

Suspension culture in low-adherence vessels. This technique involves the use of low-adherence containers to produce EBs. A defined number of ESCs are placed in a 96-well plate of round bottom or in a polypropylene 1.5 ml-conical tube. The vessels concentrate the cells at the bottom, promoting cell aggregation and favouring contact with each other. After a few days of culture, a large EB is formed. Similarly to the hanging drop method, these two techniques allow the formation of uniform EBs from a determined number of cells (Kurosawa, 2007; Kurosawa et al., 2003; Ezekiel et al., 2007).

Generation of EBs on a STO feeder layer. Zhou *et al.* demonstrated that EBs can be formed from mESCs cultured on a low density of STO feeder cells, indicating that they can not only maintain undifferentiated mESCs, but also allow the formation of typical EBs. The generation of EBs by this method involves three steps: formation of compact cell aggregates, production of simple EBs, and generation of cystic EBs in suspension culture on bacterial-grade dishes. The embryoid bodies-derived cells obtained differentiated into cardiac, hepatic and neural cells with specific inducers (Zhou et al., 2005).

Embryoid bodies recapitulate many aspects of lineage-specific differentiation programmes as well as temporal and spatial gene expression patterns of early embryogenesis. Therefore, the EBs provide a valuable tool for investigating embryonic development at the cellular and molecular level (Zhou et al., 2005). During their *in vitro* maturation, mESCs undergo morphological changes and acquire various molecular markers of differentiated cells (Fig. 3D-F) (Itskovitz-Eldor et al., 2000). Leahy *et al.* employed developmentally regulated marker genes to compare the temporal and spatial relationship between differentiation of EBs and mouse postimplantation embryos. Their experiments demonstrated that the appearance of developmental markers and their overall distribution in EBs is strictly correlated with their expression patterns observed during specific stages of embryogenesis (Leahy et al., 1999).

Koike *et al.* studied the expression shifts of two genes that are highly expressed in undifferentiated ESCs, *Oct4* and *Rex-1*, during formation of EBs. The expression levels of these two genes almost disappeared by day 5 of EB culture. These authors induced EB formation under different conditions starting from the same cell number of ESCs and found variations in the morphological appearance of the cell aggregates. They also found that the expression of *GATA-4*, a differentiation marker, increased during culture, but its expression profile varied more widely than *Oct4* and *Rex-1* according to the types of EBs. This suggests that the progress of differentiation in the distinct EBs is not homogenous. The culture conditions for EB formation, such as cell number and density, and culture configurations seem to be involved in the differentiation status of the resulting EBs (Koike et al., 2007).

As mentioned above, differentiation of ESCs by means of EB formation recapitulates changes in the embryonic development, and factors with main roles during early embryogenesis might be involved in the production of EBs. Mutual cell to cell interactions are important for the differentiation of stem cells into the cell types derived from the three germ layers. Cell-cell contact and apoptosis are very important during the early developmental stages and are expected to be found during the formation of embryoid bodies (Choi et al., 2005; Doetschman et al., 1985). Apoptosis is a natural process of development and several studies have revealed that apoptosis-related genes are involved in the differentiation of ESCs to EBs (Choi et al., 2005). During EB formation, multiple foci of cell death appear to finally merge into a single central cavity surrounded by a monolayer of primitive ectoderm cells (Fig. 4C,D). After aggregation of ESCs, the more external cells differentiate to primitive endoderm followed by visceral and parietal endoderm. Unlike normal embryos, endodermal cells are located in the outer surface of the EBs and are connected via junctional complexes showing numerous microvilli (Karbanova and Mokry, 2002; Rodda et al., 2002). Along with the establishment of the extraembryonic endoderm lineage a basement membrane is deposited between the pluripotent cells and the endoderm. It is proposed that this membrane induces an alteration in the pluripotent state of cells, such that the ones in contact with the membrane survive and reorganize to a columnar epithelium of primitive ectoderm, whereas the cells located at the centre and apart from the extracellular matrix undergo cell death and form a cavity (Rodda et al., 2002). In mice, this basal membrane is called Reichert´s membrane and is composed primarily by collagen IV and glycoproteins, and it is located between parietal endoderm cells and trophoblast cells in rodent embryos. Reichert-like basement membrane deposits are often seen during the first days of EB culture and its presence can be identified by Periodic Acid Shiff (PAS) reaction (Fig. 4A,B) (Fagundez et al., 2009; Zhou et al., 2005).

There are different types of culture media that can be employed to induce formation of EBs and results vary according to the medium chosen and whether they are supplemented either with FBS or Ko-SR. As mentioned earlier, FBS contains pro-differentiation factors which make it a useful tool in differentiation protocols (Fagundez et al., 2009). Although some researchers were able to obtain EBs in a serum-free media (Bettiol et al., 2007; Taha and Valojerdi, 2008) it is generally accepted that at least small amounts of FBS are needed to favour differentiation, and no EBs are obtained in complete absence of serum (Mansergh et al., 2009).

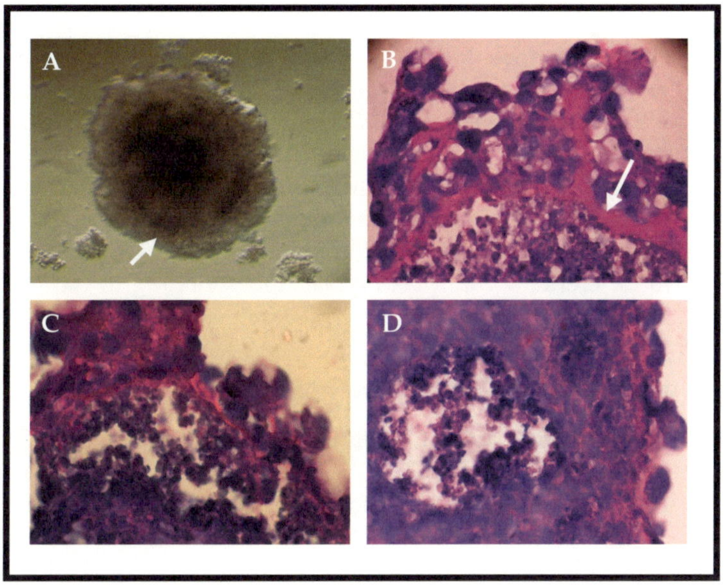

Fig. 4. Histology of embryoid bodies.
A) Deposits of Reichert-like basement membrane (arrow); B) PAS staining showing these deposits (arrow); C) and D) Peripheral and central cavities originated during EB formation

8. Teratoma formation

Teratomas are naturally occurring tumours that contain chaotic associations of structures resembling normal tissues of the three germ layers including glandular epithelium, cartilage, bone, smooth and striated muscle, as well as neural and stratified squamous epithelium; nevertheless they do not show the ordered and appropriate location of tissues as would be found in normal embryogenesis (Aleckovic and Simon, 2008; Berstine et al., 1973).

Teratomas are classified by their content in solid, containing only tissues; cystic, with pockets of fluid; or mixed, containing a combination of both solid and cystic components (Aleckovic and Simon, 2008).

Teratomas can also be experimentally induced in mice injecting pluripotent stem cells into growth permissive ectopic sites. Embryonic stem cells are generally transplanted into immuno-deficient mice leading to the formation of palpable tumours at the graft site, which are normally allowed to develop for 6-10 weeks before its remotion for analysis. Different strains of mice have been used to study the *in vivo* potential of ESCs, but little is known about the influence of the strain of the immuno-deficient mice on teratoma formation. There is also no evidence that the specific number of ESCs implanted has any effect on developmental processes in teratomas; however, no teratoma was observed when an inferior number from 50000 mESCs were transplanted into the hearts of immuno-deficient mice (Nussbaum et al., 2007). Generally, a number of 10^3-10^6 embryonic stem cells are injected in order to ensure teratoma formation (Aleckovic and Simon, 2008). Experiments of ESCs injections have demonstrated that these cells behave differently according to the site of transplantation, showing variation on growth rate and cell differentiation of the teratomas obtained. For example, cells grafted into the intrahepatic location produced larger tumours in a short period of time consisting of a mixture of differentiated and immature tissues, whereas cells transplanted subcutaneously formed smaller teratomas composed basically of well differentiated cells, but took longer to develop (Cooke et al., 2006). Some studies indicate that local environment cues from the tissue in which ESCs are transplanted may influence in the ability of these cells to differentiate. It has been reported that mESCs produce a greater ratio of cartilaginous tissues in teratomas formed in the knee comparing to cells grafted into the subcutaneous space. Therefore, it seems possible that implanted cells are more likely to differentiate into tissues resembling their surroundings, probably due to local environmental signals (Przyborski, 2005; Wakitani et al., 2003). However, transplanted ESCs do not always differentiate into phenotypes corresponding to those of implantation site (Deacon et al., 1998).

9. Molecular aspects of mESCs pluripotency

Pluripotency is the ability to give rise to all cells of the organism (Keller, 2005). In embryonic stem cells it is mainly orchestrated by *Oct4*, *Nanog* and *Sox2*, the earliest-expressed set of genes known to maintain pluripotency. They work in conjunction with other transcription factors to specify the pluripotent state and thus constitute the basis of a transcription hierarchy (Boyer et al., 2006).

The property of self-renewal shown by ESCs depends on a stoichiometric balance among various signalling molecules and if an imbalance in any of them occurs, ESCs identity can be lost (Wobus and Boheler, 2005). Deprivation of *Oct4* or a reduction to less than 50% of its normal expression level causes inappropriate differentiation of ESCs into trophoectoderm through upregulation of *Cdx2*, one of its target genes; whereas a less than two-fold increase results in formation of primitive endoderm and mesoderm (Johnson et al., 2006; Masui, 2010; Nichols et al., 1998; Niwa, 2001; Niwa et al., 2000). *Oct4* deficient embryos (*Oct4* $^{-/-}$) develop to a stage that resembles a blastocyst with cells allocated inside them but which not constitute a real ICM since these cells lack pluripotency and are composed of trophoectoderm cells, resulting in peri-implantation developmental arrest. As expected, these structures cannot be employed for production of ESCs lines (Boiani and Scholer, 2005; Johnson et al., 2006). Different experiments have shown that *Oct4* is not sufficient for the maintenance of pluripontency, because forced expression fails to give rise to mESCs (Johnson et al., 2006). However, this is not the case of *Nanog*, as forced expression of this gene seems to make mESCs more resistant to spontaneous differentiation in the absence of LIF. The persistence of *Nanog* apparently delays the differentiation of ESCs increasing the threshold of differentiation, but do not completely abolish it. *Nanog*-null ESCs can be established and still preserve pluripotency but with an increasing tendency to differentiate (Chambers et al., 2007). *Nanog* is expressed throughout the pluripotent cells of the ICM but its expression is decreased in extraembryonic lineages and in pluripotent cells of the peri-implantation embryo. This leads to the hypothesis that one of the major roles of *Nanog* might be repression of *GATA-6*, responsible for the differentiation into extraembryonic endoderm (Chazaud et al., 2006). *Nanog* -/- embryos cannot preserve the pluripotent lineage, and as a consequence, arrest at peri-implantation. The ICM of these embryos, as well as *Nanog*-deficient ESCs, turn into parietal endoderm cells (Boiani and Scholer, 2005; Johnson et al., 2006). *Sox2* -/- embryos arrest at similar time that *Oct4* -/- and *Nanog* -/- embryos do. Blastocyst-like structures are formed, but lack primitive ectoderm. Therefore, *Sox2* -/- blastocysts are incapable of ESCs derivation. As *Oct4* and *Nanog*, *Sox2* plays an important role in the conservation of pluripotency, but its expression is not restricted to pluripotent cells; early primitive ectoderm, anterior primitive endoderm, germ cells and multipotent extraembryonic ectoderm cells also count with the presence of this gene (Boyer et al., 2006; Johnson et al., 2006).

In mouse embryonic stem cells, the maintenance of pluripotency is also regulated by the signal transducer and activator of transcription 3 (STAT3), whose function is triggered by LIF, which binds to the LIF receptor to dimerize with gp130, resulting in the phosphorylation of STAT3 via Janus kinase (JAK) activation. The phosphorylated STAT3 dimerizes and translocates to the nucleus, acting as a transcription factor to activate target genes which include essential genes to maintain the pluripotent cell phenotype (Boiani and Scholer, 2005; Johnson et al., 2006; Masui, 2010; Niwa, 2001). Repression of STAT3 results in differentiation, whereas its artificial activation is sufficient to maintain pluripotency of the cells in absence of LIF (Masui, 2010). Apart from the LIF-STAT3 pathway, the pluripotency of mESCs is also regulated by members of the superfamily of the TGFβ, which include activin and BMP. The BMP4, for example, activates Smad1, which upregulates the expression of the inhibitor of DNA-binding genes (Id) which, in turn, suppresses differentiation along with the LIF signal (Masui, 2010).

The *Wnt* signalling also has an important function, since administration of Wnt proteins can contribute to preserving the pluripotency of ESCs (Masui, 2010).

The chromatin of ESCs displays characteristics of transcriptionally permissive euchromatin, like acetylated histone modifications and increased accessibility to nucleases. On the contrary, lineage specification is distinguished by a decreased in acetylation and an increased in heterochromatin formation, indicating that a restriction in developmental potential is associated with a marked decrease in genomic plasticity. It is possible then that the balance between pluripotency and lineage commitment is regulated by genetic and epigenetic factors as well (Boyer et al., 2006).

10. The use of mESCs for research purposes

Mouse ESCs have had a great impact on many fields of research. One of the advantages of mESCs is that they can be genetically modified. When ESCs are placed into a blastocyst, they can contribute to all the lineages of the embryo including the germ line. This property makes mESCs a useful tool to produce precise alterations in the genome to study the phenotypic effects of these alterations *in vivo*, generating "knock out" and "knock in" transgenic animals (Jackson et al., 2010; Rippon and Bishop, 2004; Williams et al., 1988). By means of transgenic techniques, they have also proved their potential for modelling human diseases, providing an opportunity to unravel biochemical, genetic and/or physiological causes of a plethora of pathologies.

Murine embryonic stem cells have the potential to produce unlimited numbers of cells in culture which can be used to elucidate the underlying mechanisms of pluripotency and cell lineage specification, allowing comprehension of processes in developmental biology and regulation of genes during mammalian embryogenesis.

The plasticity of these cells makes them useful to study spatial and temporal relationships that determine cell, tissue and organ development.

Mouse embryonic stem cells have become very useful for biomedical research and have played an important part in clarifying many of the signal transduction pathways that regulate progression through the cell cycle.

Since its isolation in 1981, mouse embryonic stem cells have served as an invaluable instrument to explain many processes of the organism extrapolating the knowledge acquired to the human field, and they will continue to do so in the times to come.

11. References

Aleckovic, M. and Simon, C. (2008). Is teratoma formation in stem cell research a characterization tool or a window to developmental biology? *Reproductive Biomedicine Online*, 17, 2, (Aug) (270-280), ISSN 1472-6483

Baharvand, H. and Matthaei, K. I. (2003). The ultrastructure of mouse embryonic stem cells. *Reproductive Biomedicine Online*, 7, 3, (Oct) (330-335), ISSN 1472-6483

Baharvand, H. and Matthaei, K. I. (2004). Culture condition difference for establishment of new embryonic stem cell lines from the C57BL/6 and BALB/c mouse strains. *In Vitro Cell Dev Biol Anim*, 40, 3-4, (Mar-Apr) (76-81), ISSN 1071-2690

Berstine, E. G.; Hooper, M. L.; Grandchamp, S. and Ephrussi, B. (1973). Alkaline phosphatase activity in mouse teratoma. *Proc Natl Acad Sci U S A*, 70, 12, (Dec) (3899-3903), ISSN 1091-6490

Bettiol, E.; Sartiani, L.; Chicha, L.; Krause, K. H.; Cerbai, E. and Jaconi, M. E. (2007). Fetal bovine serum enables cardiac differentiation of human embryonic stem cells. *Differentiation*, 75, 8, (Oct) (669-681), ISSN 0301-4681

Boiani, M. and Scholer, H. R. (2005). Regulatory networks in embryo-derived pluripotent stem cells. *Nat Rev Mol Cell Biol*, 6, 11, (Nov) (872-884), ISSN 1471-0072

Boyer, L. A.; Mathur, D. and Jaenisch, R. (2006). Molecular control of pluripotency. *Curr Opin Genet Dev*, 16, 5, (Oct) (455-462), ISSN 0959-437X

Bryja, V.; Bonilla, S.; Cajanek, L.; Parish, C. L.; Schwartz, C. M.; Luo, Y.; Rao, M. S. and Arenas, E. (2006). An efficient method for the derivation of mouse embryonic stem cells. *Stem Cells*, 24, 4, (Apr) (844-849), ISSN 1549-4918

Buehr, M. and Smith, A. (2003). Genesis of embryonic stem cells. *Philos Trans R Soc Lond B Biol Sci*, 358, 1436, (Aug) (1397-1402; discussion 1402), ISSN 0962-8436

Chambers, I.; Silva, J.; Colby, D.; Nichols, J.; Nijmeijer, B.; Robertson, M.; Vrana, J.; Jones, K.; Grotewold, L. and Smith, A. (2007). Nanog safeguards pluripotency and mediates germline development. *Nature*, 450, 7173, (Dec) (1230-1234), ISSN 0028-0836

Chazaud, C.; Yamanaka, Y.; Pawson, T. and Rossant, J. (2006). Early lineage segregation between epiblast and primitive endoderm in mouse blastocysts through the Grb2-MAPK pathway. *Dev Cell*, 10, 5, (May) (615-624), ISSN 1534-5807

Choi, D.; Lee, H. J.; Jee, S.; Jin, S.; Koo, S. K.; Paik, S. S.; Jung, S. C.; Hwang, S. Y.; Lee, K. S. and Oh, B. (2005). In vitro differentiation of mouse embryonic stem cells: enrichment of endodermal cells in the embryoid body. *Stem Cells*, 23, 6, (Jun-Jul) (817-827), ISSN 1549-4918

Chung, Y.; Klimanskaya, I.; Becker, S.; Marh, J.; Lu, S. J.; Johnson, J.; Meisner, L. and Lanza, R. (2006). Embryonic and extraembryonic stem cell lines derived from single mouse blastomeres. *Nature*, 439, 7073, (Jan) (216-219), ISSN 0028-0836

Cole, R. and Paul J. (1965). Properties of cultured preimplantation mouse and rabbit embryos and cell strains derived from them, In: *Preimplantation Stages of Pregnancy: a Ciba Foundation Symposium*, Wolstenholme G. and O´Connor M. (Eds), (82-112), Little Brown and Company, London.

Cooke, M. J.; Stojkovic, M. and Przyborski, S. A. (2006). Growth of teratomas derived from human pluripotent stem cells is influenced by the graft site. *Stem Cells Dev*, 15, 2, (Apr) (254-259), ISSN 1547-3287

Dang, S. M.; Kyba, M.; Perlingeiro, R.; Daley, G. Q. and Zandstra, P. W. (2002). Efficiency of embryoid body formation and hematopoietic development from embryonic stem cells in different culture systems. *Biotechnol Bioeng*, 78, 4, (May) (442-453), ISSN 0006-3592

Deacon, T.; Dinsmore, J.; Costantini, L. C.; Ratliff, J. and Isacson, O. (1998). Blastula-stage stem cells can differentiate into dopaminergic and serotonergic neurons after transplantation. *Exp Neurol*, 149, 1, (Jan) (28-41), ISSN 0014-4886

Doetschman, T. C.; Eistetter, H.; Katz, M.; Schmidt, W. and Kemler, R. (1985). The in vitro development of blastocyst-derived embryonic stem cell lines: formation of visceral yolk sac, blood islands and myocardium. *J Embryol Exp Morphol*, 87, (Jun) (27-45), ISSN 0022-0752

Evans, M. J. and Kaufman, M. H. (1981). Establishment in culture of pluripotential cells from mouse embryos. *Nature*, 292, 5819, (Jul) (154-156), ISSN 0028-0836

Ezekiel, U.; Muthuchamy, M.; Ryerse, J. and Heuertz, R. (2007). Single embryoid body formation in a multi-well plate. *Electron J Biotechnol*, 10, 2, (Apr) (328-335), ISSN 0717-3458

Fagundez, C. B.; Loresi, M. A.; Ojea Quintana, M. E.; Delcourt, S. M.; Testa, R.; Gogorza, S. J. and Argibay, P. F. (2009). A simple approach for mouse embryonic stem cells isolation and differentiation inducing embryoid body formation. *Cell Biol Int*, 33, 11, (Nov) (1196-1200), ISSN 1065-6995

Gardner, R. L. (1985). Regeneration of endoderm from primitive ectoderm in the mouse embryo: fact or artifact? *J Embryol Exp Morphol*, 88, (Aug) (303-326), ISSN 0022-0752

Gwatkin, R. B. (1966). Defined media and development of mammalian eggs in vitro. *Ann N Y Acad Sci*, 139, 1, (Oct) (79-90), ISSN 0077-8923

Hiyama, E. and Hiyama, K. (2007). Telomere and telomerase in stem cells. *Br J Cancer*, 96, 7, (Apr) (1020-1024), ISSN 0007-0920

Itskovitz-Eldor, J.; Schuldiner, M.; Karsenti, D.; Eden, A.; Yanuka, O.; Amit, M.; Soreq, H. and Benvenisty, N. (2000). Differentiation of human embryonic stem cells into embryoid bodies compromising the three embryonic germ layers. *Mol Med*, 6, 2, (Feb) (88-95), ISSN 1076-1551

Jackson, M.; Taylor, A. H.; Jones, E. A. and Forrester, L. M. (2010). The culture of mouse embryonic stem cells and formation of embryoid bodies, In: *Mouse Cell Culture:*

Methods in Molecular Biology 633, Ward A and Tosh D (Eds), (1-18), ISBN 1588297721, United Kingdom

Johnson, B. V.; Rathjen, J. and Rathjen, P. D. (2006). Transcriptional control of pluripotency: decisions in early development. *Curr Opin Genet Dev*, 16, 5, (Oct) (447-454), ISSN 0959-437X

Karbanova, J. and Mokry, J. (2002). Histological and histochemical analysis of embryoid bodies. *Acta Histochem*, 104, 4, (Feb) (361-365), ISSN 0065-1281

Kawase, E.; Suemori, H.; Takahashi, N.; Okazaki, K.; Hashimoto, K. and Nakatsuji, N. (1994). Strain difference in establishment of mouse embryonic stem (ES) cell lines. *Int J Dev Biol*, 38, 2, (Jun) (385-390), ISSN 0214-6282

Keller, G. (2005). Embryonic stem cell differentiation: emergence of a new era in biology and medicine. *Genes Dev*, 19, 10, (May) (1129-1155), ISSN 0890-9369

Koike, M.; Sakaki, S.; Amano, Y. and Kurosawa, H. (2007). Characterization of embryoid bodies of mouse embryonic stem cells formed under various culture conditions and estimation of differentiation status of such bodies. *J Biosci Bioeng*, 104, 4, (Oct) (294-299), ISSN 1389-1723

Kurosawa, H. (2007). Methods for inducing embryoid body formation: in vitro differentiation system of embryonic stem cells. *J Biosci Bioeng*, 103, 5, (May) (389-398), ISSN 1389-1723

Kurosawa, H.; Imamura, T.; Koike, M.; Sasaki, K. and Amano, Y. (2003). A simple method for forming embryoid body from mouse embryonic stem cells. *J Biosci Bioeng*, 96, 4, (Dec) (409-411), ISSN 1389-1723

Leahy, A.; Xiong, J. W.; Kuhnert, F. and Stuhlmann, H. (1999). Use of developmental marker genes to define temporal and spatial patterns of differentiation during embryoid body formation. *J Exp Zool*, 284, 1, (Jun) (67-81), ISSN 1932-5231

Lee, S. T.; Oh, S. W.; Kim, D. Y.; Han, J. Y.; Moon, S. Y. and Lim, J. M. (2006). Serum replacement with a growth factor-free synthetic substance in culture medium contributes to effective establishment of mouse embryonic stem cells of various origins. *Fertil Steril*, 86, 4 Suppl, (Oct) (1137-1145), ISSN 0015-0282

Lorthongpanich, C.; Yang, S. H.; Piotrowska-Nitsche, K.; Parnpai, R. and Chan, A. W. (2008). Development of single mouse blastomeres into blastocysts, outgrowths and the establishment of embryonic stem cells. *Reproduction*, 135, 6, (Jun) (805-813), ISSN 1470-1626

Mansergh, F. C.; Daly, C. S.; Hurley, A. L.; Wride, M. A.; Hunter, S. M. and Evans, M. J. (2009). Gene expression profiles during early differentiation of mouse embryonic stem cells. *BMC Dev Biol*, 9, (Jan), (5), ISSN 1471-213X

Martin, G. R. (1981). Isolation of a pluripotent cell line from early mouse embryos cultured in medium conditioned by teratocarcinoma stem cells. *Proc Natl Acad Sci U S A*, 78, 12, (Dec) (7634-7638), ISSN 1091-6490

Masui, S. (2010). Pluripotency maintenance mechanism of embryonic stem cells and reprogramming. *Int J Hematol*, 91, 3, (Apr) (360-372), ISSN 0925-5710

McWhir, J.; Schnieke, A. E.; Ansell, R.; Wallace, H.; Colman, A.; Scott, A. R. and Kind, A. J. (1996). Selective ablation of differentiated cells permits isolation of embryonic stem cell lines from murine embryos with a non-permissive genetic background. *Nat Genet*, 14, 2, (Oct) (223-226), ISSN 1061-4036

Nagy, A.; Gertsenstein, M.; Vintersten, K. and Behringer, R. (2003). *Manipulating the Mouse Embryo. A Laboratory Manual*, Cold Spring Harbor Laboratory Press, ISBN 978-087969591-0, United States

Nichols, J.; Zevnik, B.; Anastassiadis, K.; Niwa, H.; Klewe-Nebenius, D.; Chambers, I.; Scholer, H. and Smith, A. (1998). Formation of pluripotent stem cells in the mammalian embryo depends on the POU transcription factor Oct4. *Cell*, 95, 3, (Oct) (379-391), ISSN 0092-8674

Niwa, H. (2001). Molecular mechanism to maintain stem cell renewal of ES cells. *Cell Struct Funct*, 26, 3, (Jun) (137-148), ISSN 0386-7196

Niwa, H.; Miyazaki, J. and Smith, A. G. (2000). Quantitative expression of Oct-3/4 defines differentiation, dedifferentiation or self-renewal of ES cells. *Nat Genet*, 24, 4, (Apr) (372-376), ISSN 1061-4036

Nussbaum, J.; Minami, E.; Laflamme, M. A.; Virag, J. A.; Ware, C. B.; Masino, A.; Muskheli, V.; Pabon, L.; Reinecke, H. and Murry, C. E. (2007). Transplantation of undifferentiated murine embryonic stem cells in the heart: teratoma formation and immune response. *Faseb J*, 21, 7, (May) (1345-1357), ISSN 0892-6638

Ogawa, K.; Matsui, H.; Ohtsuka, S. and Niwa, H. (2004). A novel mechanism for regulating clonal propagation of mouse ES cells. *Genes Cells*, 9, 5, (May) (471-477), ISSN 1365-2443

Przyborski, S. A. (2005). Differentiation of human embryonic stem cells after transplantation in immune-deficient mice. *Stem Cells*, 23, 9, (Oct) (1242-1250), ISSN 1549-4918

Rippon, H. J. and Bishop, A. E. (2004). Embryonic stem cells. *Cell Prolif*, 37, 1, (Feb) (23-34), ISSN 0960-7722

Rodda, S. J.; Kavanagh, S. J.; Rathjen, J. and Rathjen, P. D. (2002). Embryonic stem cell differentiation and the analysis of mammalian development. *Int J Dev Biol*, 46, 4, (449-458), ISSN 0214-6282

Rossant, J. (2001). Stem cells from the Mammalian blastocyst. *Stem Cells*, 19, 6, (Nov) (477-482),

Sherman, M. I. (1975a). The culture of cells derived from mouse blastocysts. *Cell*, 5, 4, (Aug) (343-349), ISSN 1549-4918

Sherman, M. I. (1975b). Long term culture of cells derived from mouse blastocysts. *Differentiation*, 3, 1-3, (Aug) (51-67), ISSN 0301-4681

Smith, A. G.; Heath, J. K.; Donaldson, D. D.; Wong, G. G.; Moreau, J.; Stahl, M. and Rogers, D. (1988). Inhibition of pluripotential embryonic stem cell differentiation by purified polypeptides. *Nature*, 336, 6200, (Dec) (688-690), ISSN 0028-0836

Taha, M. F. and Valojerdi, M. R. (2008). Effect of bone morphogenetic protein-4 on cardiac differentiation from mouse embryonic stem cells in serum-free and low-serum media. *Int J Cardiol*, 127, 1, (Jun) (78-87), ISSN 0167-5273

Tanaka, N.; Takeuchi, T.; Neri, Q. V.; Sills, E. S. and Palermo, G. D. (2006). Laser-assisted blastocyst dissection and subsequent cultivation of embryonic stem cells in a serum/cell free culture system: applications and preliminary results in a murine model. *J Transl Med*, 4, (May) (20), ISSN 1479-5876

Tanimoto, Y.; Iijima, S.; Hasegawa, Y.; Suzuki, Y.; Daitoku, Y.; Mizuno, S.; Ishige, T.; Kudo, T.; Takahashi, S.; Kunita, S.; Sugiyama, F. and Yagami, K. (2008). Embryonic stem cells derived from C57BL/6J and C57BL/6N mice. *Comp Med*, 58, 4, (Aug) (347-352), ISSN 1532-0820

Tesar, P. J. (2005). Derivation of germ-line-competent embryonic stem cell lines from preblastocyst mouse embryos. *Proc Natl Acad Sci U S A*, 102, 23, (Jun) (8239-8244), ISSN 1091-6490

Wakayama, S.; Hikichi, T.; Suetsugu, R.; Sakaide, Y.; Bui, H. T.; Mizutani, E. and Wakayama, T. (2007). Efficient establishment of mouse embryonic stem cell lines from single blastomeres and polar bodies. *Stem Cells*, 25, 4, (Apr) (986-993), ISSN 1549-4918

Wakitani, S.; Takaoka, K.; Hattori, T.; Miyazawa, N.; Iwanaga, T.; Takeda, S.; Watanabe, T. K. and Tanigami, A. (2003). Embryonic stem cells injected into the mouse knee joint form teratomas and subsequently destroy the joint. *Rheumatology (Oxford)*, 42, 1, (Jan) (162-165), ISSN 1462-0324

Williams, R. L.; Hilton, D. J.; Pease, S.; Willson, T. A.; Stewart, C. L.; Gearing, D. P.; Wagner, E. F.; Metcalf, D.; Nicola, N. A. and Gough, N. M. (1988). Myeloid leukaemia inhibitory factor maintains the developmental potential of embryonic stem cells. *Nature*, 336, 6200, (Dec) (684-687), ISSN 0028-0836

Wobus, A. M. (2001). Potential of embryonic stem cells. *Mol Aspects Med*, 22, 3, (Jun) (149-164), ISSN 0098-2997

Wobus, A. M. and Boheler, K. R. (2005). Embryonic stem cells: prospects for developmental biology and cell therapy. *Physiol Rev*, 85, 2, (Apr) (635-678), ISSN 0031-9333

Wobus, A. M.; Holzhausen, H.; Jakel, P. and Schoneich, J. (1984). Characterization of a pluripotent stem cell line derived from a mouse embryo. *Exp Cell Res*, 152, 1, (May) (212-219), ISSN 0014-4827

Zandstra, P. W.; Le, H. V.; Daley, G. Q.; Griffith, L. G. and Lauffenburger, D. A. (2000). Leukemia inhibitory factor (LIF) concentration modulates embryonic stem cell self-renewal and differentiation independently of proliferation. *Biotechnol Bioeng*, 69, 6, (Sep) (607-617), ISSN 0006-3592

Zhou, Q. J.; Shao, J. Z.; Xiang, L. X.; Hu, R. Z.; Lu, Y. L.; Yao, H. and Dai, L. C. (2005). Generation of embryoid bodies from mouse embryonic stem cells cultured on STO feeder cells. *Cell Biol Int*, 29, 9, (Sep) (817-825), ISSN 1065-6995

3

Effective Derivation and Manipulation of Mouse Embryonic Stem Cells

Peng Zhang, Xinglong Wu, Pengbo Wang and Xiangyun Li
Faculty of Animal Science and Technology
Agricultural University of Hebei
P. R. China

1. Introduction

Mouse embryonic stem (ES) cells were first isolated in 1981 from the inner cell mass of the blastocyst before implantation within the uterine wall. ES cells have the unique property of pluripotency and also of capable of infinite self-renewal. They can be maintained in undifferentiated state in culture or be differentiated to multilineage cell types from all three embryonic layers, both in vivo and in vitro (Evans & Kaufman, 1981; Martin, 1981; Brook & Gardner 1997). These properties of ES cells have made them an extremely interesting and important tool for basic and applied research, especially for the studies of embryogenesis, gene function, and development. However, current protocols for mouse ES cell derivation are often very inefficient and require a great deal of specialized training and expertise, the potential of mouse ES cells has not yet been fully and systematically exploited.
There are numerous protocols available for mouse ES cell line derivation from blastocysts, and interestingly the effectiveness of derivation of ES cells is largely based on the mouse strain. In practice the efficiency of derivation in strains other than 129 does not usually exceed 10% (Bryja et al., 2006; Batlle-Morera et al., 2008). Moreover, some protocols require the use of sophisticated techniques, such as isolation of inner cell mass via immunosurgery of intact blastocysts, isolation of epiblast cells from implanted the egg cylinderstage embryos, or selective ablation of differentiated cells. Other variations could include derivation on feeder layers, the absence of supporting feeders all together, or the use of conditioned media or the use of serum replacement (McWhir et al., 1996; Schoonjans et al., 2003; Cheng et al., 2004; Tesar, 2005; Bryja et al., 2006; Doungpunta, et al., 2009). Rho kinase inhibitor Y-27632 and the dissociation reagent Accutase were reported to significantly inhibit apoptosis of human ES cells during passaging (Watanabe, et al., 2007; Ruchi, et al., 2008), and since then we have adapted these methods in our mouse ES cell derivation protocol. Our data demonstrates that Y-27632 and Accutase increase the efficiency of mouse ES cell derivation, and the resultant ES cells retain developmental pluripotency (including stable karyotype, surface markers, teratoma formation, and the ability to undergo germline transmission). In this chapter, we describe a simple and efficient protocol for derivation of mouse ES cells and provide details on how to culture and manipulate the resultant cells. As compared to other available protocols, this method does not require special equipment, genetic modification, or advanced training other than regular tissue culture and animal handling skills. It is our hope that this protocol will allow investigators new to the ES field

to efficiently derive mouse ES cell lines even if they do not have previous experience in this area.

2. Materials

2.1 Mice

Mice were purchased from Beijing Vitalriver Laboratory Animal Technology (Beijing, China). The mice were housed at 25°C under 50%~60% relative humidity with a 12h light:12h dark photoperiod (lights on at 06:00) until they were required. Mice were fed with commercial pelleted food and water *ad libitum*. All experimental protocols and animal handling procedures were reviewed and approved by the Laboratory Animal Care and Use Committee of Hebei Province.

2.2 Equipment

Special care should be taken to assure that the following equipment is decontaminated before use and periodically check that the equipment is functioning properly.

1. Autoclaves (Boxun Apparatus YXQ-LS-30II, Shanghai)
2. Dissecting microscope (Nikon SMZ645)
3. Freezers and refrigerators (Xinfei BCD-213KA, Henan)
4. Inverted microscope with 4×, 10× and 20× objectives (Olympus IX71)
5. Liquid nitrogen storage tanks (Dongya YDS-50B-125, Sichuan)
6. Micro fusion chamber (gap between electrodes: 0.2mm) (Eppendorf, 4308 030.003)
7. Microforge (Narishige MF-900)
8. Micromanipulator set (Narishige MM-89)
9. Multiporator (Eppendorf AG 22331)
10. Piezo impact drive system (Prime Tech, PMM-150FU)
11. Pipette puller (PN-30) and pipette (B100-75-10) (Sutter Instrument)
12. Sterile horizontal flow hood with UV light (Boxun Apparatus VS-840-1, Shanghai)
13. Table-top centrifuge (Guohua TGL-16, Changzhou)
14. Tissue culture incubator: settings: 37°C, 5% CO_2, normal atmospheric concentration, and a saturated aqueous atmosphere (Sanyo MCO-15AC)
15. Water bath: Set at 37°C for regular use but can also be used at 56°C to heat inactivate serum (BHW2, GB11241-89, Beijing)
16. Water purification equipment and medium filtration devices (Pall PL 5123, America)

2.3 Plasticware and other materials

It is recommended that disposable plastic material should be used for all tissue culture work.

1. Cell counter
2. Centrifuge tubes (Nunc, 15ml and 50ml)
3. Cryovials (Nalgene, 1.8ml)
4. Embryo handling pipette
5. Eppendorf tubes (1ml)
6. Equipment for dissection (razor blades, scissors, micro dissecting scissors, straight and curved forceps, tweezers)
7. Four-well plates (Nunc)
8. Gas burner

9. High-quality CO_2 (Beijing Oxygen Plant Specialty Gases Institute Company, Beijing)
10. Microcapillaries harvard GC100T-10 (Harvard Apparatus LTD)
11. Pasteur pipettes
12. Petri dishes (Nunc, 35, 60 and 90mm diameter)
13. Repeat pipettman (Eppendorf)
14. Sterile filter with 0.22μm membrane with low protein binding (Millipore)
15. Syringes (Weigao, Shandong, 1ml, 10ml, and 20ml)

2.4 Reagents and solutions

All chemicals, reagents and solutions should be cell culture tested or of analytical grade.

1. 2, 2, 2-tribromoethyl alcohol (Sigma, T48402)
2. 2-Mercaptoethanol (Invitrogen 21985-023)
3. 70% Ethanol (Hengxing Chemical Preparation Company, Tianjin)
4. Acetic acid (Zhongliante Chemical Preparation Company, Beijing)
5. Agar (Sigma, A1296)
6. Albumin from bovine serum (BSA) (Sigma, A3311)
7. D-Glucose (Sigma, G8769)
8. Dimethyl sulfoxide (DMSO) (Sigma, D2650)
9. Distilled water, embryo tested (Invitrogen, 15230-162)
10. Dulbecco's Modified Eagle's Medium (DMEM), (1×), liquid, with L-glutamine, 4500 mg/L D-glucose, without sodium pyruvate (Invitrogen, 12430-054)
11. Dulbecco's phosphate buffered saline without Ca^{2+} and Mg^{2+} (DPBS) 1×, liquid (Invitrogen 14190-144)
12. ES cell medium: DMEM supplemented with 15% fetal bovine serum, 1% non-essential amino acids, 1% penicillin/streptomycin, 1% L-glutamine, 0.1mM 2-mercaptoethanol, 1000 IU/ml LIF, Filter sterilize and store at 4°C up to 3-4 weeks.
13. ESGRO Complete Accutase (Millipore, SF006)
14. Ethylenediaminetetraacetic acid disodium (EDTA) (Sigma, ED2SS)
15. Fetal bovine serum (FBS) (Invitrogen, 12483-020)
16. Fluorinert FC-77 (Sigma, F4758)
17. Freezing medium: 90% serum plus 10% DMSO
18. Fusion buffer: 0.3M mannitol (Sigma, M-9546) containing 0.1mM $MgCl_2$, 0.05mM $CaCl_2$, 0.5mM HEPES and 0.1% BSA
19. Giemsa stain (SSS Reagent Company, Shanghai)
20. GlutaMAX-I Supplement (Invitrogen, 35050-061)
21. HEPES sodium salt (Sigma, H3784)
22. Human chorionic gonadotrophin (HCG) (Sansheng, Ningbo, 10000 IU): add 200ml 0.9% NaCl for 50 IU/ml aliquots and store at -20°C
23. KaryoMAX® colcemid® solution, liquid (10μg/ml) in DPBS (Invitrogen, 15212-012)
24. KSOM with / 1/2 amino acids, glucose and phenol red (Millipore, MR-121-D)
25. Leukaemia inhibitory factor (LIF) (Millipore, LIF2010)
26. M2 media (Sigma, M7167)
27. Methanol (Hengxing Chemical Preparation Company, Tianjin)
28. Mineral oil (Sigma, M8410)
29. Mitomycin C (Sigma, M-4287)
30. Mouse embryonic fibroblast (MEF) medium: DMEM supplemented with 10% fetal bovine serum, 1% non-essential amino acids, 1% penicillin/streptomycin, 1%

L-glutamine, 0.1mM 2-mercaptoethanol. Filter sterilize and store at 4°C for up to 3-4 weeks.
31. Paraformaldehyde (Kermel, Tianjin)
32. Penicillin/Streptomycin (Invitrogen, 15140-122)
33. Pregnant mare serum gonadotrophi (PMSG) (Sansheng, Ningbo, and 5000 IU): add 100 ml 0.9% NaCl for 50 IU/ml aliquots and store at -20°C.
34. Protease (Sigma, P8811)
35. Tert-amyl alcohol (Sigma, 240486)
36. Triton X-100 (Sigma, T9284)
37. Trypsin-EDTA (0.05% with EDTA· 4Na) (Invitrogen, 25300-054)
38. Y-27632 dihydrochloride monohydrate (Sigma, Y0503)

3. Methods

3.1 Isolation of mouse embryonic fibroblasts

1. Quick and humane sacrifice of pregnant female CD1 mice (E13.5) by cervical dislocation.
2. Dissect out uterine horns and transfer into 90mm dish.
3. Carefully remove fetuses from the uterus using forceps and scissors and rinse in DPBS (see Fig.1A).

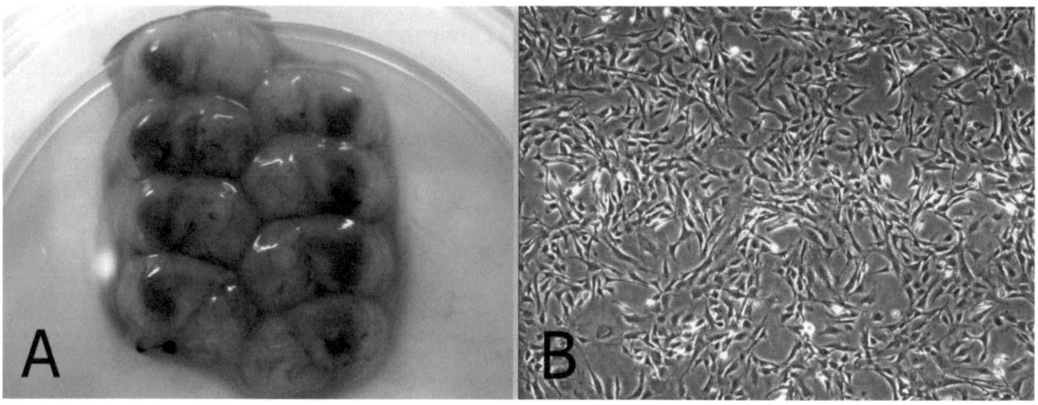

Fig. 1. E13.5 mouse fetuses (A) and primary mouse embryonic fibroblasts (B)

4. Cut away brain and dark red organs, wash with fresh DPBS, and remove as much blood clots as possible.
5. Place tissue into a 60mm dish. Using an elbow iris scissors finely mince the fetuses into 1mm^3 pieces.
6. Wash with fresh DPBS three times.
7. Add 2ml of 0.05% trypsin/EDTA and incubate at 37°C for 5-10min.
8. Inactivate trypsin/EDTA by adding equal volume of MEF medium. Using a syringe pipette up and down several times to dissociate the tissue clumps into single cells.
9. Collect the cell suspensions in a fresh 15ml centrifuge tube and pellet cells by centrifugation at 200g for 5-10min.
10. Carefully take off the supernatant and resuspend cells in 1ml MEF medium.
11. Seed cells into approximately two 60mm dishes containing 6ml MEF medium. Incubate dishes in a 37°C incubator with 5% CO_2 for approximately 3h.

12. Change medium after cells attach.
13. Allow cells to grow until confluence (repeat step 6-10). Split the cells 1:3-4 (label these cells as P1) (see Fig.1B).
14. When cells reach confluence, repeat step 6-10. Resuspend cells in freezing medium and freeze each plate in three cryovials.
15. Transfer the cryovials to -80°C freezer, the next day transfer cells to a liquid nitrogen storage tank.

3.2 Mouse embryonic fibroblast feeder layer preparation

1. Remove cells from liquid nitrogen and thaw quickly in 37°C H_2O bath.
2. Transfer cells to a fresh 15ml centrifuge tube and pellet cells by centrifugation at 200g for 5-10min.
3. Carefully take off the supernatant and resuspend cells in 1ml MEF medium.
4. Seed cells into 60mm dish containing 6ml MEF medium.
5. Allow cells to grow until confluence, discard the medium and add 2ml MEF medium containing 10μg/ml mitomycin C.
6. Incubate dishes in a 37°C incubator with 5% CO_2 for 3h.
7. Carefully take off the medium and wash four times with DPBS (see Note 1).
8. Add 0.05% Trypsin/EDTA for 1min and then add an equal volume of MEF medium for inactivation.
9. Using mechanical force through pipetting, recover the cells from the dish, centrifuge and resuspend the cell pellet in MEF medium.
10. Seed 5×10^5 cells into a 35mm dish containing 2ml MEF medium.
11. Put dishes in a 37°C incubator with 5% CO_2 until use.

3.3 Isolation of blastocysts

1. Mouse embryonic fibroblast feeder layers should be changed into ES medium 2h prior to blastocyst isolation.
2. Quick and humane sacrifice of pregnant female C57BL/6 mice (E3.5) mated to 129/Sv by cervical dislocation.
3. Open the abdominal cavity, the uteri are removed by cutting across the cervix and are cut below the junction with the oviducts.
4. Place the uteri in a small volume of M2 medium in a 35mm dish and flush each horn with M2 medium.
5. Collect and wash blastocysts three times in M2 medium.
6. Transfer the blastocysts onto the prepared MEF feeder layer and culture at 37°C within a CO_2 incubator.

3.4 Isolation and dissociation of ICM outgrowth

1. After 4 to 5 days, gently circle the ICM outgrowths with a finely drawn glass probe, removing the ICM from the surrounding trophoblast cells.
2. Take a sterile non-coated Petri dish and add several small drops (30μl) of DPBS and Accutase (see Note 2).
3. Transfer the ICM outgrowths to the drops of DPBS, and then repeat this procedure in the Accutase drops and incubate at 37°C for 15-20 min.

4. Using a P200 pipette and yellow tip, transfer the ICM outgrowths to a small drop of ES cell medium and pipet outgrowths into small cell clumps of 5-10 cells using mechanical force (see Note 3).
5. Transfer the cell clumps into 4-well plates with fresh MEF feeders and ES cell medium supplemented with 10μM Y-27632 (see Note 4).
6. After 24h of culture, change replace the medium with normal ES cell medium without Y-27632.
7. Change ES medium every other day, and observe the cultures. After 4 to 6 days ES cell colonies will appear (see Fig. 2).
8. Definitive ES cell colonies are collected and dissociated into individual cells using the method above for ICM outgrowth dissociation. The cells are seeded again in 35mm dishes with fresh feeder layers. This is considered passage 1.

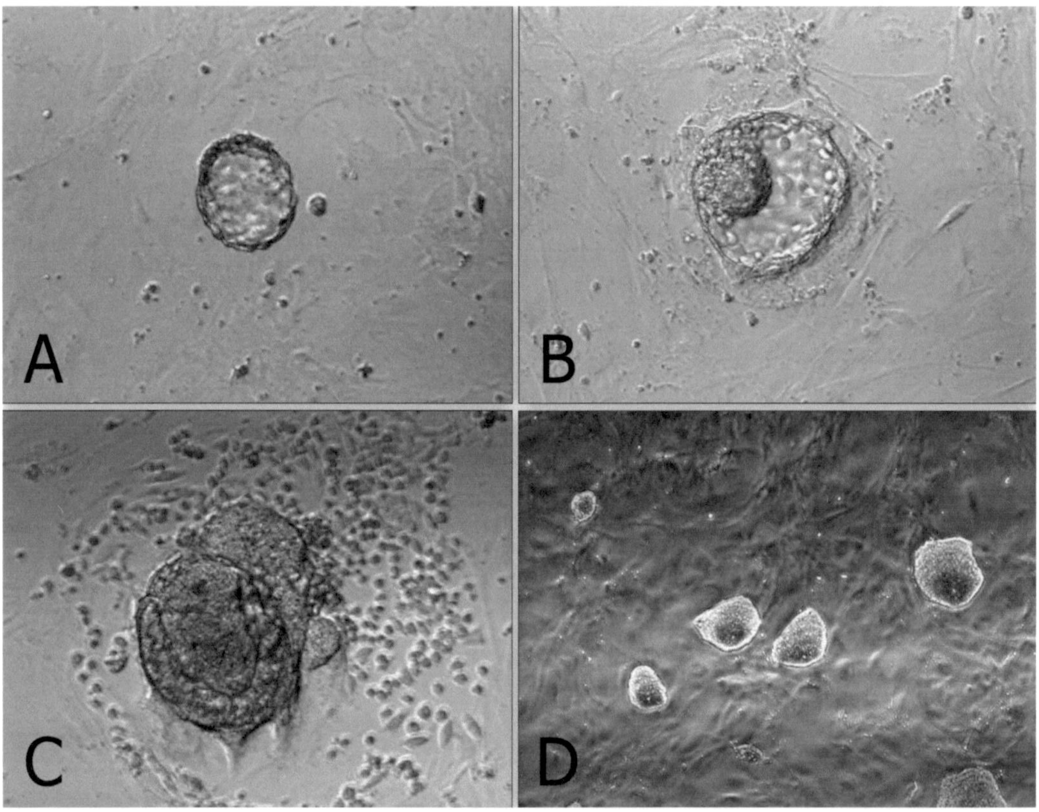

Fig. 2. Derivation of ES cells from mouse blastocysts. A. Blastocyst (E3.5). B. Blastocyst attach and spread on MEF feeder layer. C. Expanded ICM outgrowth, ready for the first dissociation. D. ES cell colonies.

3.5 Culture and passage of ES Cells

1. When ES cell colonies become large and within proximity of other colonies it is necessary to passage
2. Change fresh ES cell medium 2h prior to passaging.

3. Aspirate the old medium and wash the cells twice with DPBS.
4. Add appropriate amount of 0.05% trypsin/EDTA solution, and incubate for 1-2min at room temperature.
5. Once the cell colonies begin to detach, carefully remove trypsin/EDTA solution, and add 1ml ES cell medium, and further dissociate the detached ES colonies into single cell suspension by pipetting several times.
6. Adjust the concentration of ES cells, and transfer the suspensions into 35mm dishes at a rate of 1:5.

3.6 Freezing of ES Cells

1. Change fresh ES cell medium 2h prior to freezing.
2. Trypsinize cells and harvest as described earlier.
3. Collect the cells by centrifugation at 200g for 5min.
4. Remove the supernatant and resuspend the pellet in freezing medium.
5. Aseptically aliquot the suspension into sterile freezing vials, label each vial with the date and cell type/clone number, and place the vials into a thermos cup.
6. Freeze the cells overnight at -80°C, then transfer to the liquid nitrogen.

3.7 Thawing of ES cells

1. Remove a vial of frozen cells from the liquid nitrogen, and transfer to 37°C water bath.
2. Transfer cell suspension to a 15ml centrifuge tube, and add ES cell medium to 5ml.
3. Pellet cells by centrifugation at 200g for 5min.
4. Carefully take off the supernatant and resuspend cells in 1ml ES cell medium.
5. Plate ES cell suspension onto a prepared feeder layer in a 35mm dish at a suitable density.

3.8 Karyotyping of ES Cells

1. Add fresh ES cell media containing colcemid at a final concentration of 0.1μg/ml to an exponentially growing ES cell cultures. Return to the incubator for 40min.
2. Wash slides in fresh fixation solution (3:1 methanol : acetic acid) and then soak them in ice cold water until ready to use (distilled water plus some ice). It is important for the slides to be both cold and wet when ready for use.
3. Wash cells twice with DPBS. Completely dissociate colonies into single cell with trypsin. Add MEF media and resuspend the cells in a 15ml tube. Spin down at 200g for 10min. Remove the supernatant.
4. Add 1ml 0.56% KCl dropwise. Flick the tube to loose the pellet again to a single cell suspension (no big chucks). Add 9ml of 0.56% KCl and incubate for 15min.
5. Add 2-3 drops of fresh fixation solution to the tube, and invert it several times, Spin down the cells at 200g for 5min.
6. Remove the supernatant. Add 1ml fixation solution. Flick tube to resuspend pellet, add 9ml fixation solution. Spin down the cells at 200g for 5min.
7. Repeat step 6 three times and resuspend cells in appropriate fixation solution. Adjust cell density to 1×10^6/ml.
8. Remove slide from the water, blot edges to remove excess liquid and drop the cell suspension (dropwise) from at least one foot above the surface of the slide (2-3 drops/slide). Allow the cells spread and then place slides on heat stage (60°C) to speed

up drying). Prepare 3 or 4 slides for each sample. The remaining cells can be stored at -20°C for several years.

9. Stain slides with Giemsa solution for 15min. Rinse slides with running water.
10. Photograph chromosome spreads and count (see Fig. 3A)

3.9 Alkaline phosphatase stain

Staining for alkaline phosphatase should be performed at room temperature using an alkaline phosphatase detection kit (Boster Bio-Tech AR1023, Wuhan).

1. Aspirate the medium and wash the ES cell cultures twice with DPBS for 1min each.
2. Fix ES cells with 4% paraformaldehyde in DPBS at 4°C for 15min.
3. Remove paraformaldehyde and add freshly prepared alkaline phosphatase staining solution.
4. Keep reactions in the dark, and incubate at room temperature until desired degree of color development has occurred (This process usually takes 20-40min).
5. Terminate the color reaction by washing with running water.
6. Observe the images under an inverted microscope (see Fig. 3B).

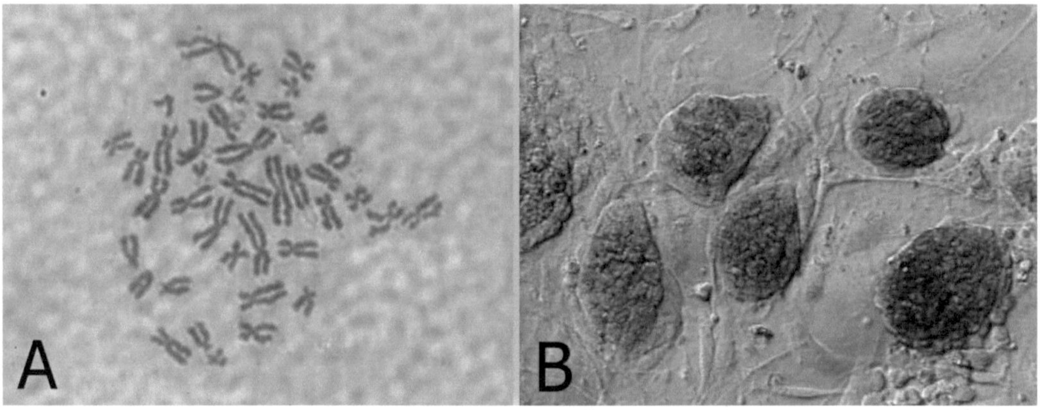

Fig. 3. A. Karyotype of ES cells. B. ES cell colonies with positive alkaline phosphatase activity.

3.10 Immunostaining of ES Cells

1. Seed ES cells on feeder layers in 4- well plate until near confluence.
2. Remove media and fix cells with 4% paraformaldehyde for 10min at room temperature.
3. Wash with DPBS for three times.
4. Add 0.2% Triton X-100 in DPBS for 5min, and then wash with DPBS for three times.
5. Add DPBS with 10% FBS for 30min.
6. Dilute primary antibodies (rabbit anti-oct-4 polyclonal IgG, rabbit anti-nanog polyclonal IgG, rabbit anti-sox-2 polyclonal IgG, mouse anti-ssea-1 monoclonal IgG) (Santa Cruz, sc-9081, sc-33760, sc-20088, sc-21702) at 1:200 dilution rate in DPBS with 10% FBS.
7. Add primary antibodies and incubate cells overnight at 4°C refrigerator.
8. Wash with DPBS three times.

9. Dilute second antibodies (donkey anti rabbit IgG-FITC, donkey anti mouse IgG-FITC) (Santa Cruz, sc-2090, sc-2099) at 1:500 dilution rate in DPBS with 10% FBS.
10. Add appropriate second antibodies into cells and incubate 2h at room temperate.
11. Wash with DPBS three times and observe under an inverted fluorescent microscope (see Fig. 4).

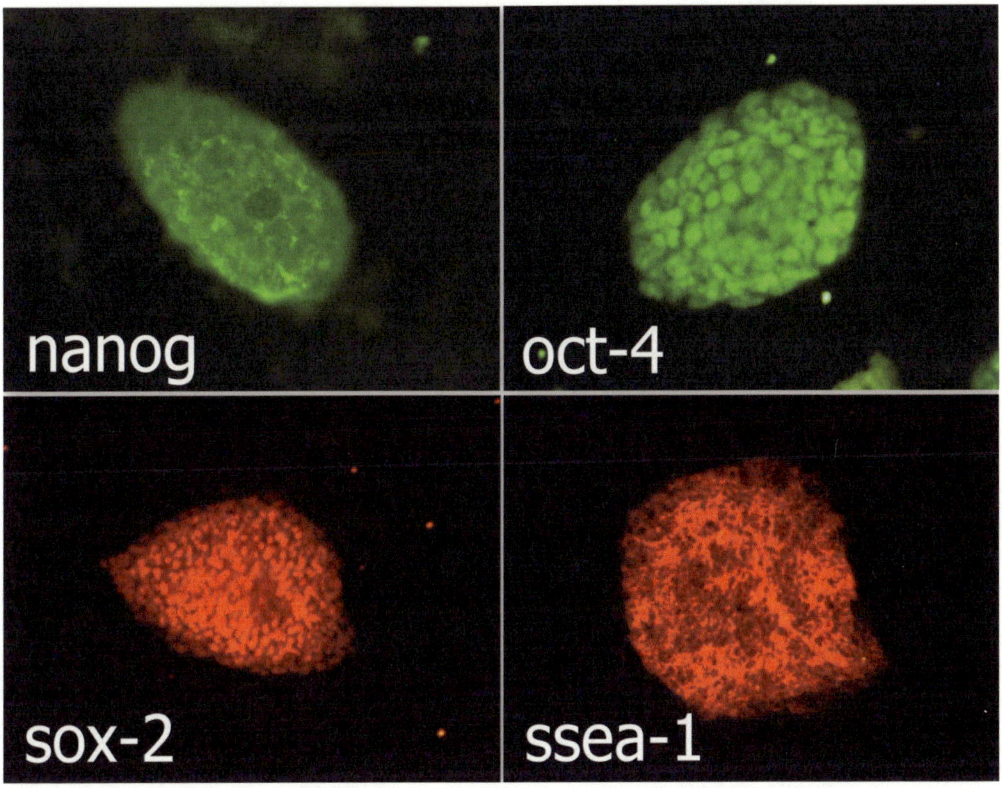

Fig. 4. Immunostaining of ES cell pluripotent markers (nanog, oct-4, sox-2, ssea-1).

3.11 Formation of embryoid bodies

1. Trypsinize and dissociate ES cell colonies into single cells.
2. Collect by centrifugation and resuspend the cells at a density of 5×10^5 cells/ml in MEF medium supplemented with 10μM Y-27632.
3. Transfer 2ml cell suspension into a 35mm dish coated with 1% agar (see Note 5).
4. Culture overnight at incubator, carefully collect cell aggregates and transfer them into a new 35mm dish coated agar. Add 2ml MEF medium without Y-27632 for further culture.
5. After 3-5 days, many cystic embryoid bodies appear (see Fig.5).

3.12 Formation of teratoma

1. Pretreat ES colonies with 10μM Y-27632 for 1h prior to dissociation.
2. Dissociate with Accutase, harvest the cells by centrifuge at 200g for 10min and remove the supernatant.

3. Add DPBS and resuspend cells to a final density of 2×10^6 cells/ml.
4. Subcutaneously inject 0.2 ml cell suspension into the CD1 mouse groin (see Note 6).
5. Teratoma will be observed approximately 6 weeks later (see Fig. 6).

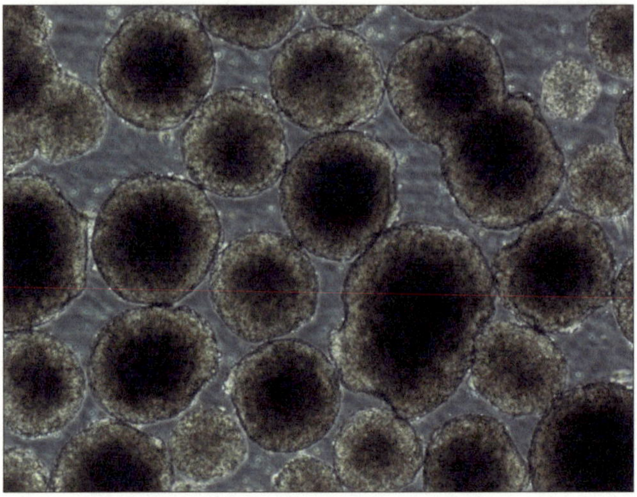

Fig. 5. Cystic embryoid bodies by ES cell suspension culture

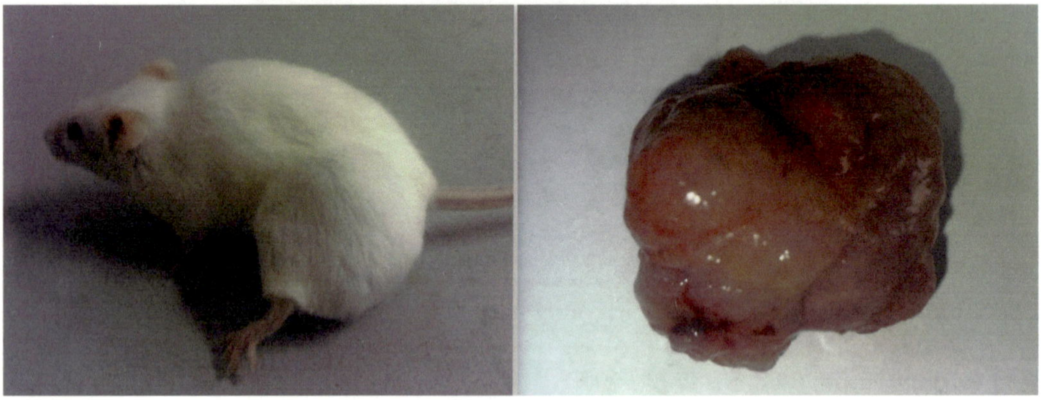

Fig. 6. Formation of teratoma by injecting ES cells into CD1 mouse groin

3.13 Production of chimeric mice by aggregation with eight-cell stage embryos

1. Prepare several drops of KSOM culture medium containing 10μM Y-27632 in a 35mm Petri dish and cover with mineral oil.
2. Sterilize a darning needle by dipping in ethanol. Make a series of 10 small depressions in each microdrop by pressing a darning needle into the Petri dish surface (see Note 7). Keep the plate in a 37°C incubator at 5% CO_2.
3. Oviducts from pregnant female mice (E2.5) are flushed with M2 to collect eight-cell embryos.
4. Place several drops of M2, 0.5% protease and KSOM in 35mm Petri dish (do not mix).
5. Wash the embryos with three drops of protease medium and transfer them into drops of protease. Incubate embryos at 37°C for 5-7min.

6. Wash embryos three times with KSOM and when zona pellucida begins to disappear transfer them into drops of KSOM.
7. Transfer embryos into an aggregation plate (one embryo in each depression). Return plate to the incubator.
8. Pull a glass capillary tube using a pipette puller, this will be used for ES cell clump preparation.
9. Pick ES cell colonies from Petri dish and wash them three times with fresh DPBS.
10. Transfer colonies to trypsin/EDTA and incubate at 37°C for 3-5min.
11. Immediately transfer the colonies into a drop of KSOM and wash them three times.
12. Colonies should be dissociated into 15-20 cell clumps using a glass capillary tube and mechanical force.
13. Using a pipette, place one clump of ES cells into each depression.
14. Return the plate to incubator and culture overnight (see Fig. 7).
15. Transfer 8-10 aggregation blastocysts to each uterine horn of pseudopregnant CD1 females (2.5dpc). The female CD1 mouse will deliver approximately 17 days later.

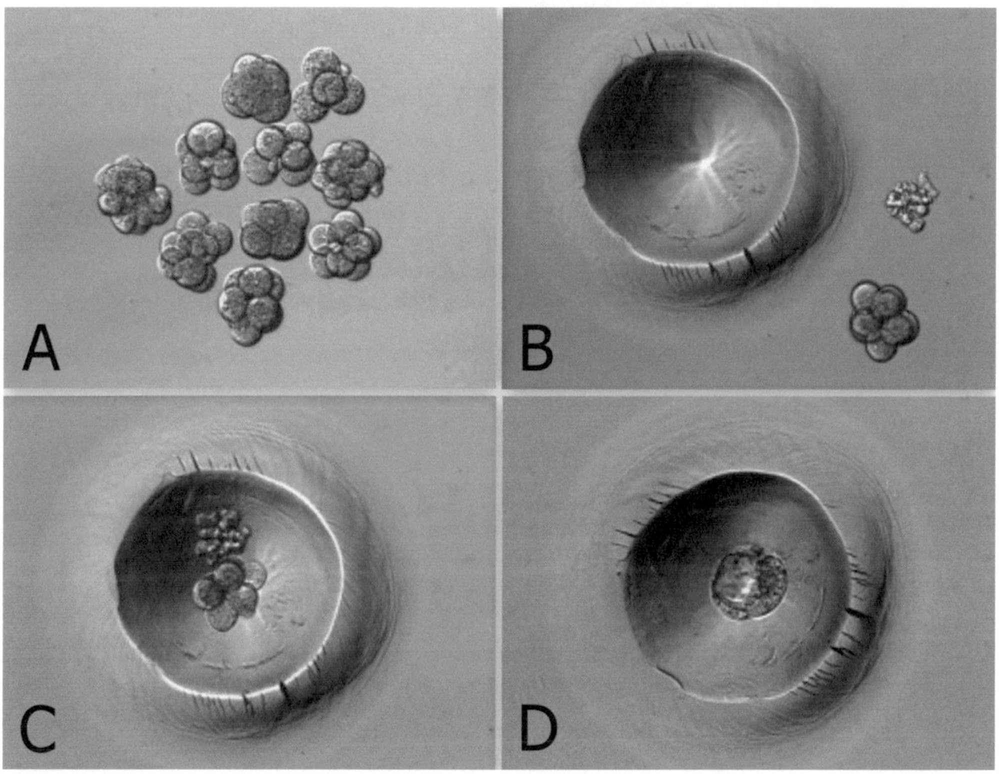

Fig. 7. Production of chimeras by aggregation. A. 8-cell embryos without zona pellucida. B, C. Aggregation depression, ES cells and embryos. D. Aggregate of ES cells and embryo

3.14 Production of chimeric mice by tetraploid blastocyst injection

Production of tetraploid mouse embryos

1. Collect 2-cell embryos (E1.5) of the CD1 outbred strain by flushing the oviducts and wash them using several drops of M2 medium.

2. Transfer them into KSOM drops and keep at incubator until electrofusion.
3. Wash 2-cell embryos through several drops of fusion buffer and place 20-30 embryos into the fusion buffer between the electrodes of the 0.2mm micro fusion chamber (see Fig. 8A).
4. Set Eppendorf Multiporator mode to electrofusion: 2V voltage and 15s duration for embryo prealignment; 20V voltage, 50µs duration and two pulses for fusion.
5. Transfer the embryos to the M2 drop and wash them through several drops of M2 medium.
6. Transfer the embryos into KSOM drops under mineral oil and culture at 37°C incubator.
7. In the next 30-60min, choose all properly fused embryos and incubate them in KSOM medium under mineral oil at 37°C, 5 % CO_2 (see Fig. 8B).
8. After 24h culture, most of embryos should develop to blastocyst stage and are ready to be injected with ES cells.

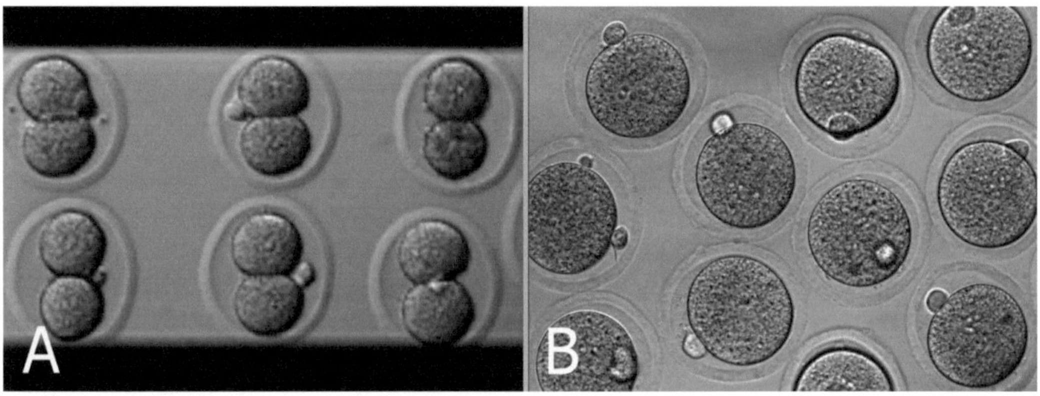

Fig. 8. A. 2-cell stage embryos aligned in fusion chamber. B. Fused embryos.

Preparation of ES cells for blastocyst injection

1. Replace fresh ES medium 3h prior to injection.
2. Pick ES cell colonies and rinse them with DPBS.
3. Transfer the colonies into Accutase drops and incubate at 37°C for 5 to 7min.
4. Using a P200 pipette, transfer the colonies to a small drop of ES medium with 10µM Y-27632, and break up the colonies into single cell suspension.
5. Incubate the single cell suspension with 10µM Y-27632 for 1h at room temperature prior to injection to remove cell membrane blebs (see Fig. 9) (see Note 8).

Microinjection needles preparation using a PN-30 micropipette puller and a MF-900 microforge

1. Turn on the machine and choose appropriate settings. For the PN-30 puller, use magnet (main) 90, magnet (sub) 30, and heater 80, as the starting values.
2. Fix one glass capillary (100mm) on the needle puller.
3. Start the pulling process of the needle.
4. Remove the needle carefully and check the tip under a stereomicroscope.
5. Mount a capillary in the microforge and break this at the required diameter by fusing the glass onto the glass bead on the microforge and turning off the heat while drawing it away.

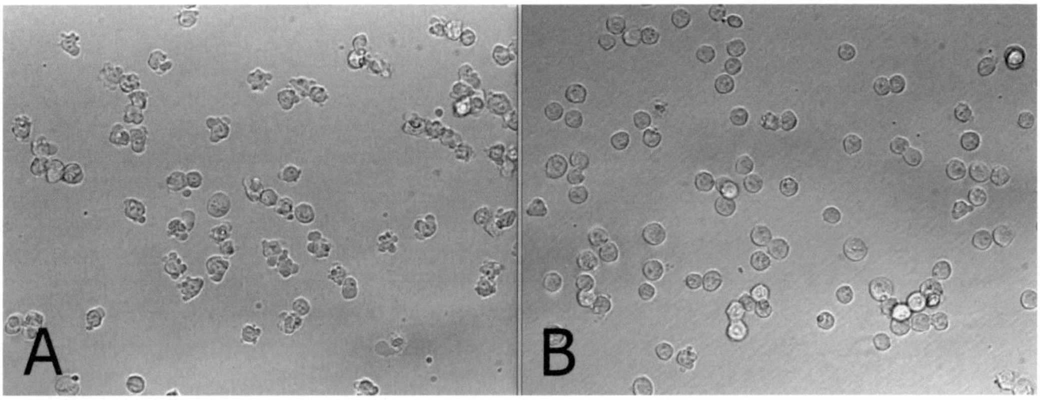

Fig. 9. Compare to the control cells without Y-27632 treatment (A), the treatment of Y-27632 make ES cells smooth and soft (B) so that they can easily be picked up by injection pipette.

6. For making holding pipets: break the holding pipet at a point where the outside diameter is approx 80µm and the inside diameter approx 60µm, and then move the holding pipet toward the glowing glass bead, and the cut will start to melt inward creating a smoother, narrower tip.
7. For making injection pipets: break the injection pipet at a point where the outside diameter is approx 15µm and the inside diameter approx 12µm.
8. Hold the pipet above the glass bead and heat a point about 5mm from the pipet tip. The pipet will bend under gravity as the glass melts. Allow the pipet to bend approx 60°.

Piezo-supported microinjection

1. Prepare dishes for injection using the lids of standard 35mm plastic disposable Petri dishes.
2. Place a 400µl drop of M2 medium on the lid and cover it with mineral oil.
3. Fluorinert FC-77 in a 15mm length was back-loaded into an injection pipette. Push Fluorinert FC-77 through the shoulder to near the tip to empty the air in the pipette
4. Attach the injection pipette to the pipette holder of the piezo unit, and hang the piezo unit on the micromanipulator, and aspirate 1mm M2 into the pipette to keep ES cells away from Fluorinert FC-77.
5. Attach the holding pipette on the other side of the micromanipulator.
6. Transfer embryos and ES cells into the drop of M2 under mineral oil.
7. Aspirate several ES cells, and use several piezo pulses (e.g., intensity=3, frequency=3) to penetrate the zona and trophectoderm layer (see Note 9).
8. Advance the tip of the pipette near the opposite side of the blastocysts from where it is held by the holding pipette.
9. Inject 15-20 ES cells into the blastocoel cavity (see Fig. 10A, B).
10. Culture injected blastocysts in KSOM medium at 37°C, 5 % CO_2 for 2h, and allow them recovery.
11. Transfer 8-10 injected blastocysts to each uterine horn of 2.5dpc pseudopregnant CD1 femals.
12. Recipient mice will give birth after 17 days (see Fig. 10C, D).

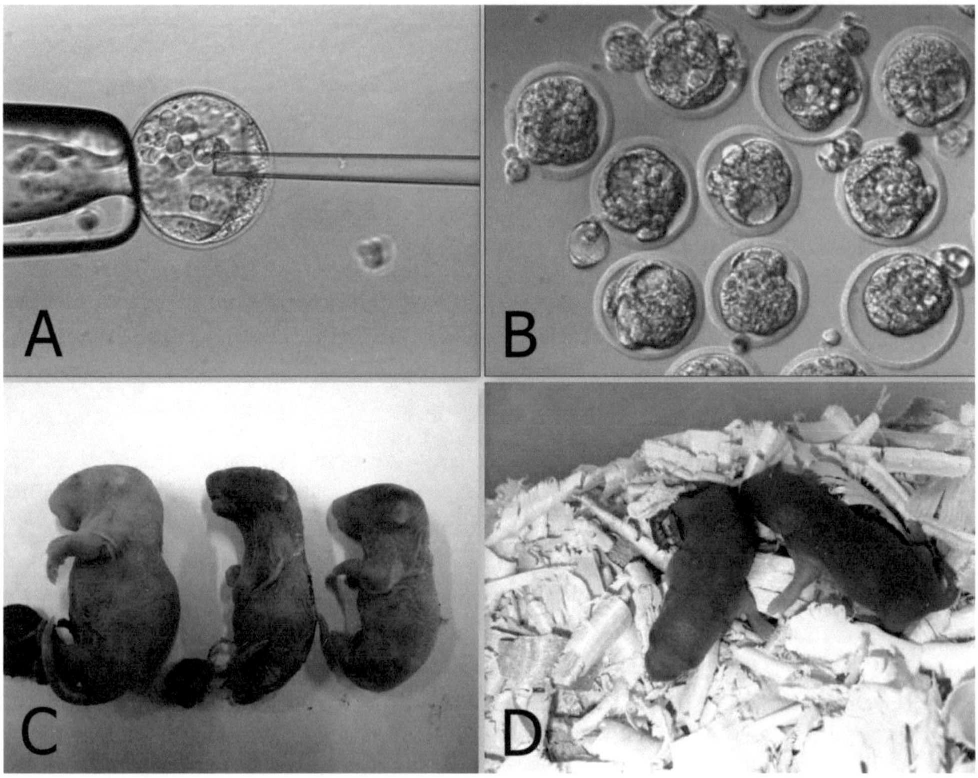

Fig. 10. A. Piezo-actuated tetraploid blastocyst injection. B. Injected tetraploid blastocysts. C. Dead ES mouse pups. D. Survival ES mouse pups.

3.15 Embryo transfer and Caesarean sections

1. Select a pseudopregnant CD1 female (2.5dpc) as an embryo recipient.
2. Prepare the transfer pipette, forceps and scissors etc.
3. Anesthetized mice by intraperitoneal injection of Avertin, after drug delivery quickly put the mouse back into the cage (see Note 10).
4. After the mouse is fully anesthetized place the mouse in the prone position on a lid of a Petri dish. Using an absorbent ball with 75% ethanol sterilize the area and shave the animal.
5. Make a 0.5-1cm transverse dorsal incision at the below of the last ribs.
6. Uncover incision to find ovarian fat pad, pull it out using forceps and hold it using clamps.
7. Collect the embryos carefully one by one lined up in the transfer pipette with as little medium in between as possible.
8. Make a hole through the uterine wall into the lumen with a 25G needle. The site of this hole should be near the last loop of the oviduct where it descends into the top of the uterine horn at the UT junction.
9. Insert the transfer pipette into a hole and carefully expel the embryos. Check the transfer pipette to ensure that all embryos have been transferred. 8-10 embryos are usually transferred into each horn.
10. When embryos are to be transferred into uteri, put the ovary back into the abdominal cavity and suture the incision.

11. Place the mouse in clean cage. Lighting to heat until it regaines consciousness.
12. After 17 days, the pups will be born.

4. Notes

1. Mitomycin C (Sigma, M4287): Dissolve the powder into sterile distilled DMEM and adjust the concentration to 10µg/ml just before use. Wash layers several times with DPBS to completely remove any mitomycin C residue from the feeder layers.
2. Accutase is a ready to use cell detachment solution of proteolytic and collagenolytic enzymes, and it does not contain mammalian or bacterial derived products. Cell lines tested for Accutase application includes fibroblasts, keratinocytes, vascular smooth muscle cells, primary chick embryo neuronal cells, bone marrow stem cells, 293 cells, 3T3 cells, HeLa cells, insect cells, human embryonic stem cells and human neural stem cells. Moreover, Accutase is quite gentle on cells and will not induce cell death if cells are treated a little longer than the optimal time period. Thus, we use Acutase instead of trypsin for enzymatic dissociation of ICM.
3. This is a key step for ES cell isolation. Watch the whole dissociation process under a microscope. Careful not to pipette with too much force, as single cells will not survive. Compared to trypsin, Accutase treated ICM outgrowths are easier to dissociate into small cell clumps.
4. The ROCK inhibitor Y-27632 permits survival of dissociated human embryonic stem cells. Similar to human ES cells, we find that mouse ICM cells show sensitivity to trypsin. They undergo cell death after dissociation, and the cloning efficiency of dissociated ICM cells is generally very low. Our data indicates that increased cellular adhesion induced by Y-27632 enhances the survival of dissociated ICM cells. With more ES-like colonies appearing when the dissociated ICM cells are seeded in medium supplemented with 10 µM Y-27632 for 24h. Moreover, we have not observed adverse effects of Y-27632 treatment on pluripotency in maintenance culture even after a number of passages.
5. Dishes are coated with agar to completely prevent cell attachment. This can increase cell aggregation and embryonic body formation. In brief, add 1g agar into 100ml DPBS and autoclave for 30min. Cool to 50°C at room temperature. Rapidly cover the internal wall and bottom of 35mm dishes with agar solution. Remove the extra solution and place dishes in hood until the remaining agar solidification. In addition, adding Y-27632 in the embryonic body medium can also enhance suspending cells to aggregate and embryonic body formation.
6. Typically, severe combined immunodeficient mice are used for making teratoma. This strain of mouse is expensive and not easily raised. Our data indicates that most mouse strains could be recipient for teratoma production by inguinal subcutaneous injection, such as CD1, C57BL/6×129/Sv.
7. Carefully make small depressions using needle and appropriate force in the Petri dish surface. Too much force may break the dish and insufficient force will not produce the required depressions, which can easily lead to the loss of embryos and ES cells.
8. Prior to blastocyst injection, the pretreatment of 10µM Y-27632 for 1h at room temperature can alter the membrane architecture of ES cells so that they can be easily collected by injection pipet. Otherwise, the ES cell surface is covered with blebs (see Fig. 10).
9. Within the microinjection pipette, ES cells should be kept approximately 100µm away from the pipette tip. Damage to ES cells may occur if the cells are very close to the tip when the piezo pulses are applied.

10. Avertin anesthetic: 1.25% stocks avertin is prepared by mixing 2.5g of 2, 2, 2-tribromoethyl alcohol (Sigma), 5ml of tert-amyl alcohol (Sigma) with 200ml water. Avertin is intraperitoneally injected at a dose of 0.02ml per gram of body weight for anesthetization.

5. Acknowledgements

We would like to thank Roman Krawetz (The University of Calgary, Canada) for providing his expert and critical reading on the manuscript. This work was supported by grants from National Natural Science Foundation of China (No. 30871790) and Natural Science Foundation of Hebei (No. C2009000600).

6. References

Batlle-Morera, L.; Smith, A. & Nichols, J. (2008). Parameters influencing derivation of embryonic stem cells from murine embryos. *Genesis* 46(12):758-767, ISSN 1526-954X.

Brook, F.A. & Gardner, R.L. (1997). The origin and efficient derivation of embryonic stem cells in the mouse. *Proc Natl Acad Sci USA* 94:5709-5712, ISSN 0027-8424.

Bryja, V.; Bonilla, S.; Cajánek, L.; Parish, C.L.; Schwartz, C.M.; Luo, Y.; Rao, M.S. & Arenas, E. (2006). An efficient method for the derivation of mouse embryonic stem cells. *Stem Cells* 24(4):844-849, ISSN 1066-5099.

Cheng, J.; Dutra, A.; Takesono, A.; Garrett-Beal, L. & Schwartzberg, P.L. (2004). Improved generation of C57BL/6J mouse embryonic stem cells in a defined serum-free media. *Genesis* 39:100-104, ISSN 1526-954X.

Doungpunta, J.; Santhi, A.; Sathanawongs, A.; Jarujinda, Y. & Oranratnachai, A. (2009). Fivefold increase in derivation rates of mouse embryonic stem cellsafter supplementation of the media with multiple factors. *Theriogenology* 72(2):232-242, ISSN 0093-691X.

Evans, M. J. & Kaufman, M. H. (1981). Establishment in culture of pluripotential cells from mouse embryos. *Nature* 292(5819):154-156, ISSN 0028-0836.

Martin, G. R. (1981). Isolation of a pluripotent cell line from early mouse embryos cultured in medium conditioned by teratocarcinoma stem cells. *Proc Natl Acad Sci USA* 78:7634-7638, ISSN 0027-8424.

McWhir, J.; Schnieke, A.E.; Ansell, R.; Wallace, H.; Colman, A.; Scott, A.R.& Kind, A.J. (1996). Selective ablation of differentiated cells permits isolation of embryonic stem cell lines from murine embryos with a non-permissive genetic background. *Nat Genet* 14:223-226, ISSN 1061-4036.

Ruchi, B.; Jacqueline, L.; Min, K. & Alexey, V.T. (2008). Efficient propagation of single cells accutase-dissociated human embryonic stem cells. *Mol Reprod Dev* 75(5):818-827, ISSN 1040-452X.

Schoonjans, L.; Kreemers, V.; Danloy, S.; Moreadith, R.W.; Laroche, Y. & Collen, D. (2003). Improved generation of germline competent embryonic stem cell lines from inbred mouse strains. *Stem Cells* 21(1):90-97, ISSN 1066-5099.

Tesar, P.J. (2005). Derivation of germ-line-competent embryonic stemcell lines from preblastocyst mouse embryos. *Proc Natl Acad Sci USA* 102(23):8239-8244, ISSN 0027-8424.

Watanabe, K.; Ueno, M.; Kamiya, D.; Nishiyama, A.; Matsumura, M.; Wataya, T.; Takahashi, J.B.; Nishikawa, S.; Nishikawa, S.; Muguruma, K. & Sasai, Y. (2007). A ROCK inhibitor permits survival of dissociated human embryonic stem cells. *Nat Biotechnol* 25:681-686, ISSN 1087-0156.

4

Artificial Acellular Feeder Layer: An Advanced Engineered Extracellular Matrix for Stem Cell Culture

Amranul Haque[1], Masato Nagaoka[1,2], Xiao-Shan Yue[1], Stephen A. Duncan[2] and Toshihiro Akaike[1]
[1]*Tokyo Institute of Technology, Yokohama*
[2]*Medical College of Wisconsin, Milwaukee, Wisconsin*
[1]*Japan*
[2]*USA*

1. Introduction

Stem cell behavior and function is influenced by a complicated 3-D microenvironment consisting of extracellular matrix (ECM), neighboring cells, growth factors, hormones, and nutrients (Discher et al., 2009). Among them cell-ECM interaction regulates many aspects of cell behavior, including cell survival, growth, proliferation, differentiation, migration, and morphogenesis (Fig. 1) (Hynes, 1992). Tissue engineering strives to replace damaged tissues with stem cells seeded onto biologically derived or synthetic materials to mimic the regulatory characteristics of ECM and thus restore the normal control of cell function. In general, materials from natural sources (e.g., collagen, laminin or fibronectin) are advantageous for cell culture because of the presence of cell recognizable receptors (e.g., ECM molecule, galactose, can specifically recognize asialoglycoprotein receptor (ASGPR) on the hepatocytes) (Cho et al., 2006; Lutolf & Hubbell 2005). However, critical problems in biocompatibility, mechanical properties, degradation, pathogen transmission and numerous other areas remain. For tissue-engineering strategies to be successful, the complicated relationship between cells and the ECM must be simplified in a way to understand appropriate cell behavior. Through the design and expression of artificial genes using recombinant DNA technology, it is now possible to prepare artificial ECM proteins with controlled mechanical properties and with domains chosen to modulate cellular behaviour (Nagaoka et al., 2002; Ogwara et al., 2005; Azuma et al., 2010; Yue et al., 2010). This approach avoids several important limitations encountered in the use of natural ECM proteins, including complex purification, immunogenicity, heterogeneous environment, batch-to-batch (or source-to-source) variation in materials isolated from tissues, and presence of xenogenetic compounds. Moreover, the designing of artificial extracellular matrix should enable more efficient and scalable culture of embryonic stem (ES) or induced pluripotent stem (iPS) cells, as well as greater control over material properties and tissue responses (Haque et al., 2010; Nagaoka et al., 2010a).

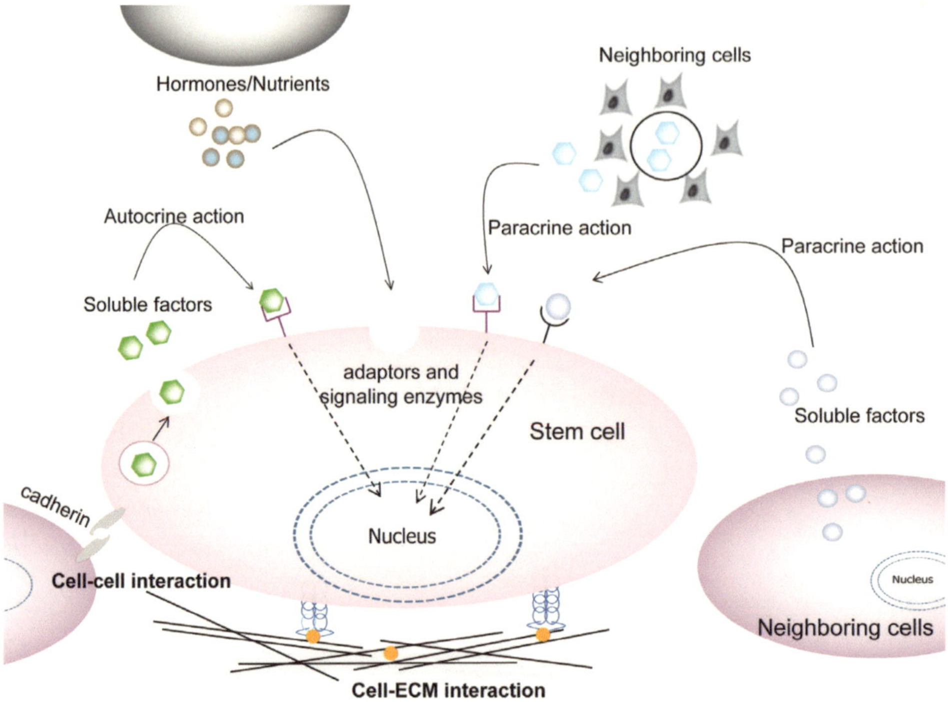

Fig. 1. Components of stem cell microenvironment that determine cell fate. Stem cell behavior that is proliferation, differentiation, migration and apoptosis is influenced by coordinated interaction of soluble factors, extracellular matrix and signals from neighboring cells. Specific binding of signal molecules with cell-surface receptors induces complex intracellular signaling pathways with subsequent effect on gene expression, self-renewal, morphogenesis and differentiation (Lutolf & Hubbell 2005)

2. Classification of extracellular matrix for ES cell

The ECM is a complex composite of fibrous structural proteins (e.g. collagen, laminin, elastin, or fibronectin) and specialized proteins (e.g. growth factors, glycoproteins and proteoglycans. It provides an important model for biomaterial design (Langer & Tirrell, 2004). ECM can be broadly categorized into natural and recombinant considering ES cell culture, self-renewal and differentiation.

2.1 Natural extracellular matrix

Various amounts and types of collagens, laminin, elastin, matrigel, fibronectin, vitronectin, or proteoglycans have been used as natural extracellular matrices (Nagaoka et al., 2010a). Generally, gelatin, that has a history of extensive use in myocardial and neural tissue engineering, can provoke an unspecific inflammatory response (Akhyari et al., 2008). Recently, it has been shown that natural matrix molecule (fibronectin-coated collagen gel) can induce the directed differentiation of endoderm from mouse embryonic stem (mES) cells without any requirement of cell sorting (Parashurama et al., 2008). But due to less control on structural composition of ECM, cell-matrix interaction and use of complex serum-supplementation, it was difficult to find the exact mechanism of collagen gel on directed differentiation of endoderm cells. Moreover, Eschenhagen et al. (1997) have established one of the convincing models of 3-D cardiac cell cultures using matrigel, where differentiation status and functional parameters reach a quality close to that of native myocardium. However, still there has been some controversy about matrigel and the question of the growth factor content within matrigel has been partially regarded as a potential contaminant. Furthermore, some of these components have no regulatory approval for use in human patients (Akhyari et al., 2008).

Categories	Targets	Fused domain	Advantages	References
Cell-cell interacton molecules	E- or N-cadherin	IgG Fc	Increased cell adhesion, growth and transfection, efficient differantiation	Nagaoka et al., 2006; Yue et al., 2010
Growth factor	LIF	IgG Fc	Strong and stable signaling, controlled presentation	Nagaoka et al., 2008b
Co-immobilized	E-cadherin and LIF	IgG Fc	Less dependency on LIF, single cell morphology, controlled pluripotency	Nagaoka et al., 2008b

Table 1. Characteristics of recombinant fusion proteins for stem cell culture and differentiation (Nagaoka et al., 2010a)

2.2 Recombinant extracellular matrix

In recent years, artificial ECMs gained special interest due to their excellent mechanical properties, process ability and low cost (Gupta et al., 2002). It is expected that synthetic ECMs can mimic many functions of the natural ECM including cellular 3-D architecture, mechanical integrity to the new tissue and the space for the diffusion of nutrients and metabolites (Putnam and Mooney, 1996). Among synthetic ECM molecules, the use of functional fusion proteins shows great promise in biology and biomedical technology due to simple and cheap production and purification strategy. Mostly three categories of synthetic ECM using recombinant fusion proteins have been explored by Akaike group in stem cell culture research (Table 1) (Nagaoka et al., 2010a; Haque et al., 2010). First, fusion proteins targeting cell-cell interacton molecules, such as cadherins (Nagaoka et al., 2006; Yue et al., 2010). Nagaoka and coworkers has also been reported second categories of fusion proteins targeting cell-bound growth factor. A co-immobilized matrix consisting both types of fusion proteins can also used for stem cell culture and pluripotency (Nagaoka et al., 2008b).

2.2.1 Cell-cell adhesion molecule based extracellular matrix

Cadherins are most important family of Ca2+-dependent intercellular adhesion molecules that may play a role in selective cell adhesion (Miyatani et al., 1989). Akaike and coworkers used two members of this family, epithelial cadherin (E-cadherin) and neural cadherin (N-cadherin) as ECM molecule to culture and differentiate ES cells (Nagaoka et al., 2006; Yue et al., 2010). The different cadherin types exhibit distinct tissue distribution pattems (Takeichi et al., 1981).

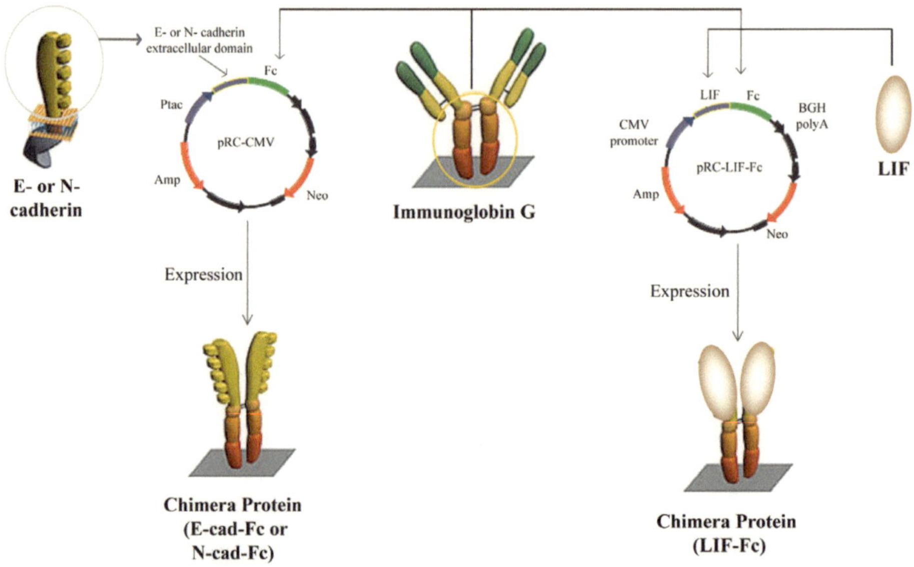

Fig. 2. Construction of fusion proteins for artificial extracellular matrix. Generated segments of an extracellular domain of mouse E-cadherin, N-cadherin or LIF and IgG Fc region were subcloned into a eukaryotic expression vector pRC/CMV. CMV promoter, human cytomegalovirus immediate-early promoter/enhancer; BGH poly A, bovine growth hormone polyadenylation signal; Amp, ampicillin resistance gene; Neo, neomycin resistance gene (Nagaoka et al., 2002; 2006; 2008b; Yue et al., 2010)

It has been known that E-cadherin mediated cell-cell interaction is essential for tissue morphogenesis and is also required for maintenance of organized solid tissues (Ozawa et al., 1990). The C-terminal domain of E-cadherin is associated with cytoplasmic proteins (α-, β-, γ -catenin and p120^{ctn}) and this molecular complex regulates homophilic interaction and dimerization of E-cadherin in presence of Ca^{2+} (Ozawa et al., 1990, Takeichi, 1995; Brieher et al., 1996). The temporal and spatial expression of E-cadherin on the surfaces of different types of cells is essential for different cellular regulatory functions. For example, E-cadherin plays a central role during preimplantation embryonic development. In its absence, proper cell polarization is absent and trophectoderm formation is impaired (Larue et al., 1994; Reithmacher et al., 1995). Moreover, E-cadherin may fine-tune ES cell pluripotency by modulating intracellular β-catenin levels or Nanog expression (Chuo et al., 2008). E-cadherin can also regulate differentiation of ES cells by modulating its expression or expression of transcription factors (Nishikawa et al., 2005). Nagaoka et al. showed that the use of artificial ECM in absence of these natural adhesion molecules can also support both undifferentiated and differentiated cells to function according to their native counterparts. Designing of artificial ECM with immobilized extracellular domain of E-cadherin as insoluble ligands not only facilitated efficient proliferation of mES cells with scattering behavior (Nagaoka et al., 2008a) but also showed enhance cell attachment and differentiation of primary hepatocytes (Nagaoka et al., 2006; 2008a). In chimeric protein of E-cadherin extracellular domain and IgG-Fc region (abbreviated as E-cad-Fc) (Fig. 2), the E-cadherin primarily attach to ECM through Fc region that has the potentiality to stably adsorb to a plastic surface such as polystyrene and dimerize via the hinge region. On the other hand, the extracellular domain of E-cadherin holds cells through homophilic interaction and thereby, activates specific signaling pathways. The creation of such type of highly defined artificial matrix molecules can improve mechanical performance and help in finding the complexity of signaling in cell-ECM interactions (Fig. 4) (Nagaoka et al., 2010a).

The scond type of cadherin molecule, N-cadherin, is expressed in various neural tissues and has been implicated in the attachment of axons to other cells, raising the possibility that this molecule is involved in neuronal recognition mechanism (Miyatani et al., 1989). Especially in central neural system, N-cadherin mediated the interaction between neurons and astrocytes. Also, the interaction between fibroblast growth factor receptor (FGFR) and the extracellular domain 4 of N-cadherin promoted the extension of neurites, indicating the very close relationship between N-cadherin and neural system formation and function (Williams et al., 2001). Yue and coworkers (2010) also constructed another fusion protein, N-cadherin fused to IgG-Fc (abbreviated as N-cad-Fc), by fusing the extracellular domain of N-cadherin and the Fc fragment of mouse IgG protein (Fig. 2). The development of this type of defined extracellular matrix favors the development of homogeneous culture conditions for ES cell differentiation under single cell level.

2.2.2 ECMs with immobilized fusion protein of soluble factors

Growth factors play an important role in regulating cellular behavior and function, such as proliferation, migration and differentiation. The biological signals of growth factors and cytokines are mediated by two different forms, the secreted form and the cell membrane- or matrix-anchored form, which release different signal transduction cascades (Tanaka et al., 1998). Controlled release of growth factors from engineered ECM can facilitate analysis of cellular morphogenesis, cell-cell interaction and monitoring of signaling pathways (Nagaoka et al., 2002). Ogiwara et al. (2005) proposed a novel artificial extracellular matrix with immobilized recombinant epidermal growth factor (EGF) fused to an IgG-Fc region (abbreviated as EGF-Fc) to investigate the effect of this matrix on phosphorylation of EGF receptor, the activation of mitogen-activated protein kinase (MAPK) and morphological changes in the cytoskeleton. Compared to soluble EGF, immobilized EGF-Fc showed more strong and stable activation of MAPK in A431 cells. Moreover, activation of Ral and cdc42 was found in A431 cells on immobilized EGF-Fc coated surface, but absent on collagen gels (Ogiwara et al., 2005). Besides, as growth factors are required in only very tiny quantities to elicit biological response, designing artificial matrices for controlled growth factor presentation is necessary. Recently, many studies showed the difficulties in using soluble leukemia inhibitory factor (LIF) to control self-renewal of mES cells without differentiation (Ying et al., 2003a). Considering these limitations together with analysis of cell behavior and function, Nagaoka and coworkers proposed matrix anchored form of LIF to design artificial acellular feeder layer for mES cells. The undifferentiated state of mES cells was maintained on the surface coated with chimeric protein, LIF-Fc (Fig. 2). Furthermore, when cultured on the co-immobilized surface with LIF-Fc and E-cad-Fc, mES cells showed undifferentiated state and pluripotency without additional LIF supplementation. This study showed that immobilized LIF and E-cadherin can maintain mES cells efficiently and that the immobilzable LIF-Fc fusion protein is useful for the investigation of signaling pathways of an immobilized form of LIF in the maintenance ES cell pluripotency (Nagaoka et al., 2008b).

2.2.3 Extracellular matrix with co-immobilized fusion proteins

The combination of cell-adhesive matrix and a controlled-signaling scheme for a growth factor could allow stem cells to control cultivation and self-renewal in an efficient way (Fig. 3). Moreover, the spatial and temporal expression of genes at different developmental stages further necessities the use of co-immobilized matrix to induce differentiation of embryonic stem cells towards specific lineages of populations. Nagaoka and co-workers applied two immobilizable model proteins, E-cad-Fc and LIF-Fc, to establish a new culture system of mouse ES cells, and demonstrated that mouse ES cells can be effectively maintained on a co-immobilized surface with the model proteins (Nagaoka et al 2008b).

3. Culture of ES cells on recombinant extracellular matrix

The purpose of using extracellular matrices to culture ES cells is many folds: (i) proliferation of cells under fully defined and homogeneous culture conditions, (ii) to mimic the process of cell differentiation, and (iii) to generate a specific type of mature cells at a high purity (Nagaoka et al., 2006; 2008b; Nishikawa et al., 2007). Although some progress has been made, this is still difficult to achieve homogeneous and fully defined culture environment. Despite this intractable problem, continuous attempts have been made to proliferate and differentiate ES-cell under homogeneous conditions in the absence of EB formation. For proliferation of ES cells in monolayer cultures with homogeneous environment, only few successes have been reported.

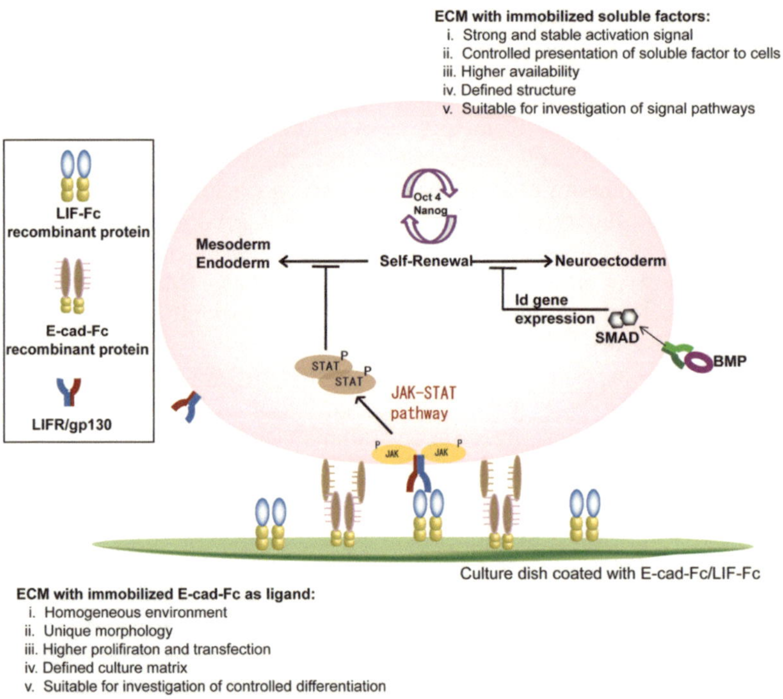

Fig. 3. A proposed model of ES cell pluripotency on artificial extracellular matrix using co-immobilized recombinant fusion protein of E-cadherin and leukemia inhibitory factor (LIF). LIF-dependent activation of STAT3 blocks ES cell differentiation and promotes self-renewal. Oct3/4 and Nanog are the most important transcription factor in this pathway. BMP signal is important to block neurogenesis. Cells on recombinant matrices with single cell morphology with uniform culture conditions and favors homogeneous distribution of signals inside the cell. Specific merits are enumerated in the figure for this culture condition (Friel et al., 2005; Nagaoka et al., 2006; 2008b)

3.1 Pluripotency

Pluripotency is characterized by the ability of a stem cell to self-renew indefinitely while maintaining the capacity to differentiate into derivatives of all three germ layers (ectoderm, endoderm and mesoderm). Pluripotency is the key for ES cells as they can be used for appropriate cell proliferation, cell cycle regulation, gene expression and differentiation. Oct3/4, Nanog and Sox2 are three important transcription factors that can control the maintenance of pluripotency of mouse ES cells by inhibiting differentiation to three germ layers (Nichols et al., 1998; Mitsui et al., 2003; Avilion et al., 2003). Despite clear morphological differences and different growth factor requirements, human ES cells are

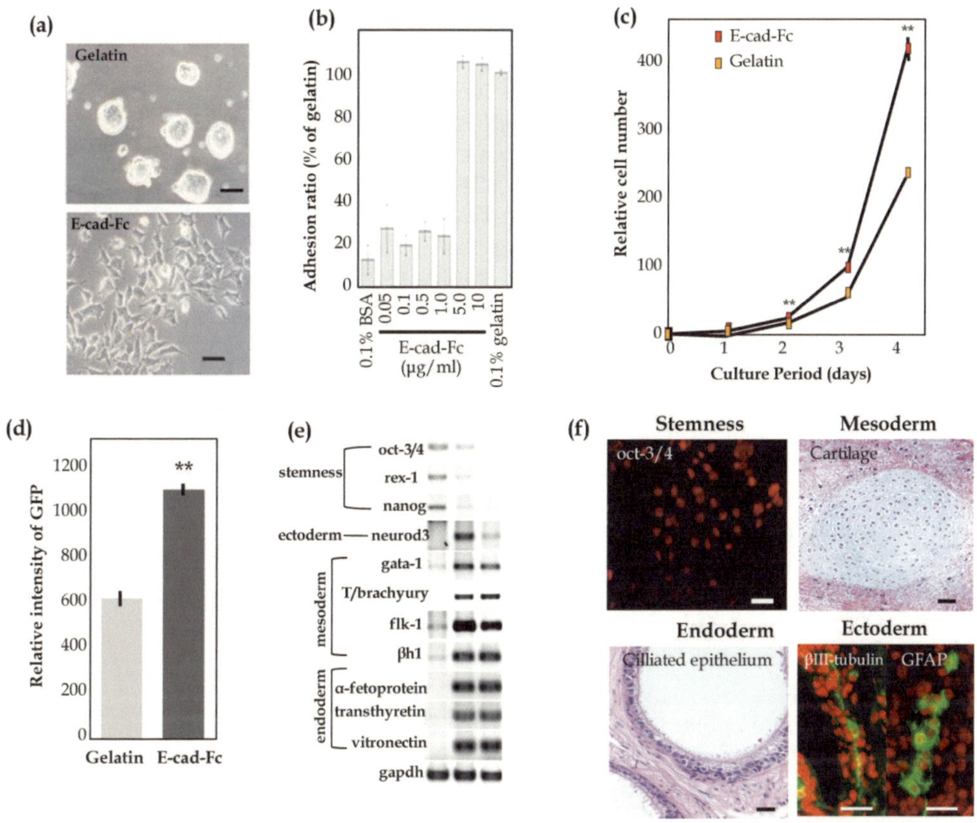

Fig. 4. Advantageous features of ES cells on the E-cad-Fc fusion protein-immobilized surface. (a) Single cell morphology of ES cells (EB3) on E-cad-Fc-coated plate. (b) ES cells adhered to E-cad-Fc-coated dishes with equivalent efficiency as to 0.1% gelatin-coated dishes after 3 hours of incubation. (c) ES cells show higher proliferation efficiency on the E-cad-Fc-coated surface. (d) Transfection efficiency of ES3 cells cultured on gelatin- or E-cad-Fc-coated surface. Relative expression of GFP was evaluated. (e, f) Pluripotency of ES cells on E-cad-Fc-coated surface. (e) R1 cells were maintained on gelatin or E-cad-Fc for 26 days, and then were cultured to form embryoid bodies. After 14 days culture of embryoid bodies, expression of marker genes was analyzed by RT-PCR. Lane 1: undifferentiated cells; lane 2: on gelatin; lane 3: on E-cad-Fc. (f) Characterization of teratomas from ES cells (EB3) cultured on an E-cad-Fc-coated surface. The data indicate means ± SEM. **:$P<0.001$ versus gelatinized plates. Scale bar: 50 μm. Courtesy: Nagaoka et al., 2006

thought to be equivalent to mouse ES cells at molecular level (Chou et al., 2008). To understand the molecular basis of pluripotency of mouse and human stem cells and to find the common regulatory mechanism, it is necessary to analyze the cells under homogeneous environmental condition at single cell level.

Nagaoka and coworkers (2006, 2008b and 2010b) also showed that both human and mouse ES cells cultured on E-cad-Fc coated surface could be maintained with unique morphological character and complete ES cell features, and that they showed higher proliferative ability and transfection efficiency than those grown under conventional conditions (Fig. 4). Furthermore, they require less LIF, probably due to the homogeneous exposure of cells to LIF that was achieved in this culture system (Nagaoka et al., 2006). Moreover, Cui et al. (2004) reported the heterogeneous distribution of cell-cell adhesion molecules in undifferentiated ES cell colonies. These observations indicate that an aggregated colony formation with close cell-cell communications may generate a heterogeneous environment within the colonies, which potentially inhibit the proliferation of ES cells and the distribution of soluble factors.

A number of new culture methods for ES cells have recently been developed that do not require feeder cells or serum (Xu et al., 2001; Ying et al., 2003b; Sato et al., 2004; James et al., 2005), but a drawback of these methods is that the undifferentiated ES cells form aggregates, which have a potency to inhibit diffusion of soluble factors to cells, and thereby may affect the pluripotency of the cultured cells. Close contact between cells in these aggregates may lead to paracrine interactions with neighboring cells that could generate heterogeneity and initiate differentiation. Furthermore, aggregated cells within the colony cannot be isolated without trypsin treatment which imparts a strong stress on human ES cells and cause massive cell death. It has been reported that culture of human ES cells on E-cad-Fc stabilizes E-cadhein on cell surface by inhibiting endocytosis mediated degradation and thereby, improves ES cell survival after single cell dissociation. Moreover, Xu et al. (2010) also reported that both mouse and human ES cells are much more dependent on E-cadherin than integrin for survival after single cell dissociation in presence of LIF. Therefore, the application of homogeneous single cell culture system using E-cad-Fc and LIF-Fc could be expected to improve the difficulty in culture of ES cells (Fig. 5).

3.2 Differentiation

ES cells have enormous potential for the study of embryogenesis at the cell level. However, compared with the generation of pluripotent ES cells, little is known about molecular mechanisms that govern ES cell differentiation. This might be due to lack of homogeneous and defined culture condition to monitor the developmental pathway under single cell level. Until recently, differentiation of ES cells into the three germ layers is attained through the use of specific culture conditions: embryoid-body (EB) formation, cultures on feeder cells, and cultures on matrices. Most cell lineages, including the neuronal, mesoderm and hepatocyte lineages have been shown to be generated within the EB (Nishikawa et al., 2007). However, monitoring cell behavior and guided differentiation to specific cell lineages was not possible without the use of laborious cell sorting (Haque et al., 2010). In contrast, monolayers of freshly isolated cells or cell lines are often used as feeder cells to support ES-cell activities. The maintenance of human ES cells still requires feeder cells because these cells do not respond to LIF (Thomson et al., 1996). However, the difficulties for using feeder cell lines include maintenance of stromal cell, undefined cultural condition and possibility of pathogen transmission. Considering these limitations, scientists are now focused on the development feeder free culture condition for ES cell culture.

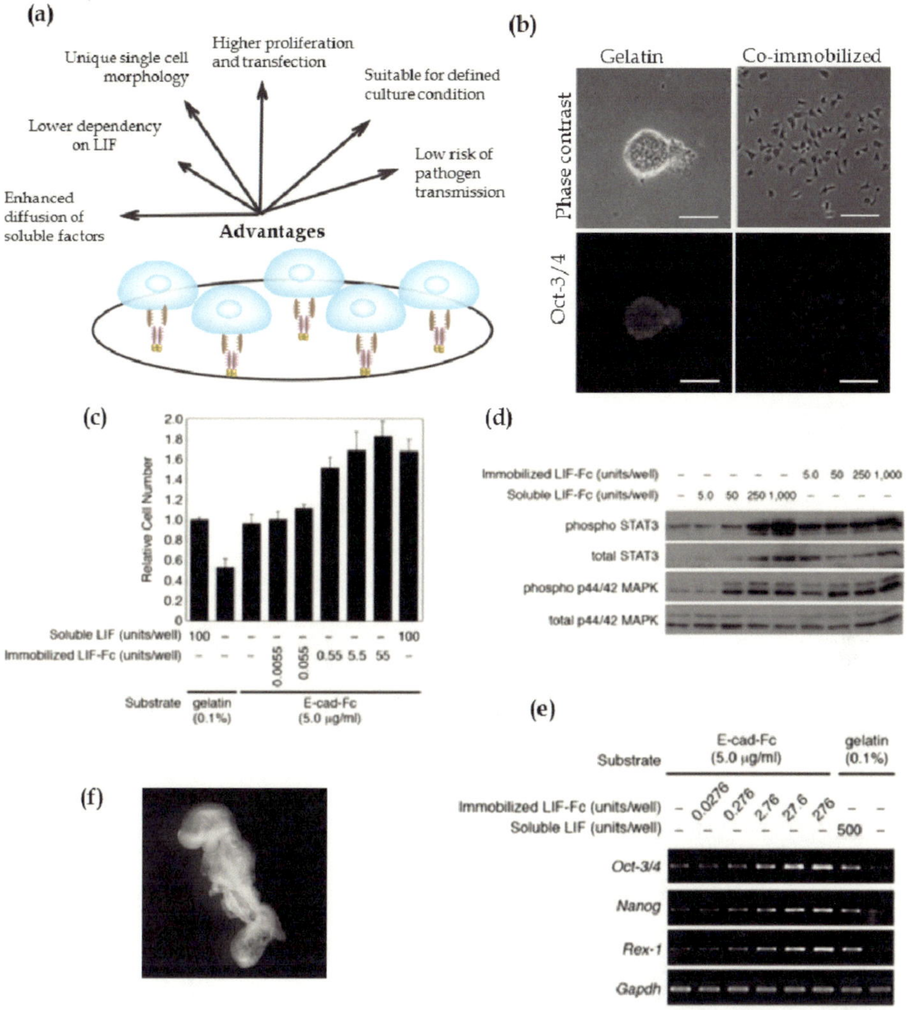

Fig. 5. Cell morphology, adhesion and maintenance of ES cell features on the co-immobilized surface of LIF-Fc and E-cad-Fc. (a) Cells on acellular feeder layer favor single cell culture with homogeneous culture conditions. Specific merits are enumerated in the figure. (b) Morphological observation and expression of pluripotency marker, Oct-3/4 by mouse ES cells (EB3) on a gelatin- or E-cad-Fc-coated surface. (c) The higher proliferative activity of EB3 cells on 5μg ml^{-1} of E-cad-Fc coated matrix compared to 0.1% gelatin coated surfaces in the presence of LIF for 2 days. Even at lower concentration of immobilized LIF (LIF-Fc) at 55 units well^{-1}, the ES cells showed higher proliferation ability on a co-immobilized surface than on a gelatinized surface. (d) After culturing for 3 days, the activation of STAT3 and MAPK by immobilized LIF-Fc was analyzed by Western blotting. (e) Expression of three genes that are markers of the undifferentiated state was analyzed by RT-PCR. (f) Feeder-dependent R1 cells were maintained on the co-immobilized surface of E-cad-Fc and LIF-Fc for 10 passages, and then cells were aggregated with four-cell stage tetraploid CD-1 embryos. The embryos were observed at the stage E9.5. Scale bar: 50 μm. Courtesy: Nagaoka et al., 2008b

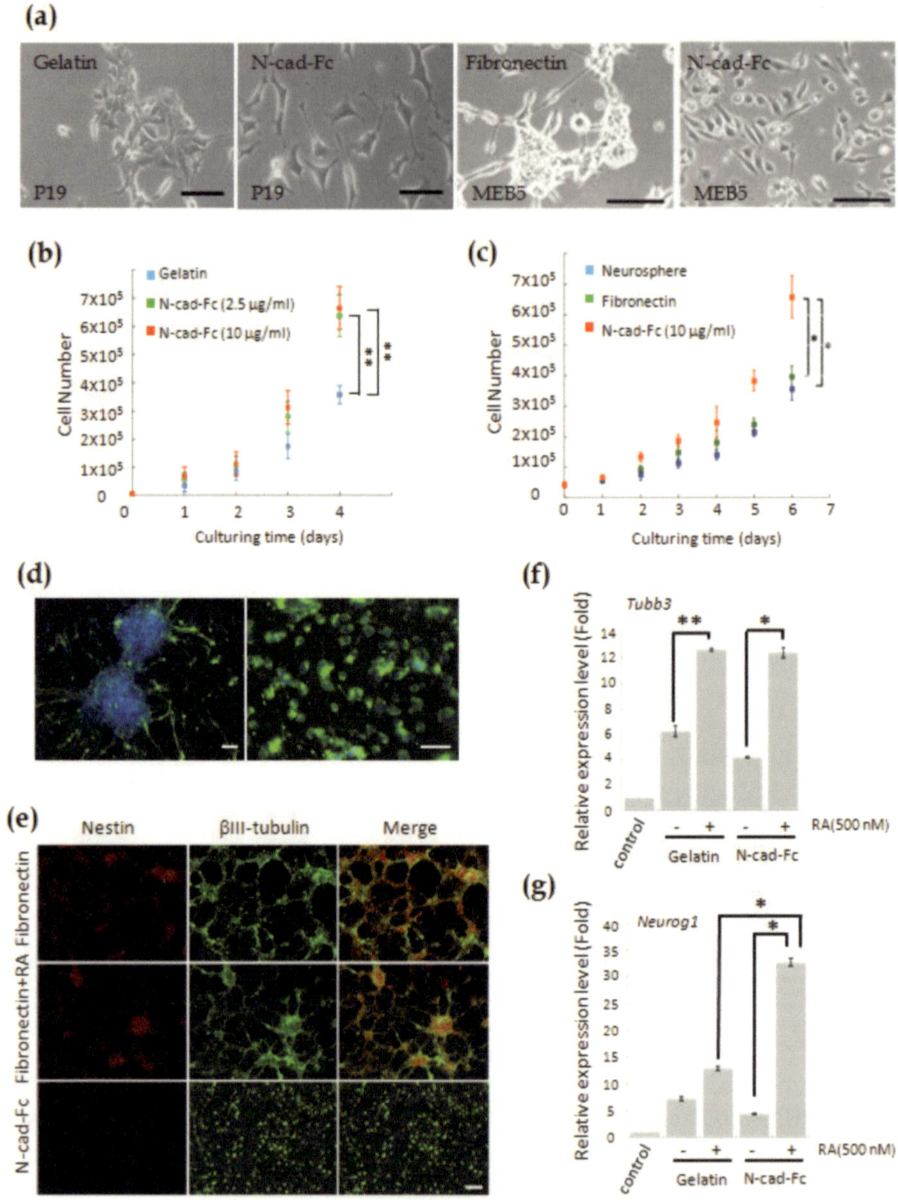

Fig. 6. Morphological observation, cell growth and differentiation P19 embryonal carcinoma and MEB5 neural stem cells. (a) Morphology of P19 and MEB5 cells at day 2 on three different culture matrix. Scale bar: 50 μm (left, P19) and 100 μm (right, MEB5). The proliferative activity of P19 (b) and MEB5 (c) cells were evaluated. Immunostaining images showing expression of specific neural markers, βIII-tubulin (green) and Nestin (red) by P19 (d) and MEB5 (e) cells at 6 days of differentiation in presence or absence of retinoic acid (RA). Scale bar: 50 μm. The quantitative expression level of neural markers, Tubb3 (f) and Neurog1 (g), showing the efficient differentiation potential of P19 cells on N-cad-Fc-coated matrix. The data indicates means ± SD (n= 3). *p < 0.01, **p < 0.001. Courtesy: Yue et al., 2010

For ES-cell differentiation in monolayer cultures under chemically defined conditions, only a few successes have been reported. Ying and coworkers showed that differentiation of ES cells to neuronal cells can be induced on gelatin-coated dishes with a conventional serum-free medium. Researchers also reported two serum-free conditions for definitive endoderm and visceral endoderm that are based on the serum-free medium in the presence of insulin, transferring and bovine serum albumin (Yasunaga et al., 2005). Under these conditions, the purity of the cells was achieved after laborious cell sorting using flow cytometry, which is practically impossible to billions of differentiated cells for therapeutic applications. Therefore, single cell culture on artificial ECM using the chimeric proteins of E-cadherin or growth factors could be the ultimate target to obtain homogeneous population of non-stressed differentiated cells at high efficiencies. Yue et al. (2010) showed that differentiation of ES cells to neuronal cells can be achieved on N-cadherin dependent artificial ECM (N-cad-Fc). Both P19 embryonal carcinoma cells and MEB5 neural stem cells cultured on N-cad-Fc-coated surface showed scattering morphologies without colony formation and higher proliferating potency than conventional culture systems with maintenance of undifferentiated state (Fig. 6). Both of two cell lines cultured on N-cad-Fc-coated surface differentiated into neural cells under single cell level. Furthermore, the expression of neuron-related markers, Tubb3, Nestin and Neurog1, in two cell lines cultured on N-cad-Fc-coated surface was much higher than colony forming conditions on gelatin-coated plates. Therefore, it will be expected that the constructed N-cad-Fc can be used as an artificall ECM for neural differentiation of stem cells to overcome the existing problems of convensional culture systems.

4. Conclusion and future perspectives

Biomaterials have already had an enormous impact on health care, and are already widely used in biomedical applications. However, relatively less effort has been made to use synthetic nanobiomaterials in embryonic stem cell research. Until recently, cell and matrix biologists have almost exclusively used natural ECM-derived materials as 3-D model systems. Precise control over extracellular microenvironment using engineered ECM would be useful to fulfill the huge demand of cells for transplantation in tissue engineering and regenerative medicine for development of bioartificial organs. Also, there will be increasing demands that 3-D recombinant fusion proteins provide better model systems for physiologic situations because 3-D fusion proteins induce more effctively cellular functions than 2-D ones although 2-D *in vitro* assay on 2-D recombinant fusion proteins are still applied in many cell culture studies. Moreover, the construction of lineage specific ECM to enrich specific population of cells for transplantation has not yet been developed. Matrix dependent selective induction of ES cells to uniform population of cells can significantly enrich specific population of differentiated cells for biomedical applications without the need for any laborious cell sorting and purification to eliminate contaminating cells. Furthermore, future research will be aimed to the design of hybrid materials consisting of intelligent recombinant proteins and other biomaterials with high mechanical characteristics for understanding developmental pathways of human ES or iPS cell-derived differentiated cells.

5. References

Akhyari, P.; Kamiya, H.; Haverich, A.; Karck, M. & Lichtenberg, A. (2008). Myocardial tissue engineering: the extracellular matrix. *European Journal of Cardio-thoracic Surgery*, 34, 2, (229-241).

Avilion, AA.; Nicolis, SK.; Pevny, LH.; Perez, L.; Vivian, N. & Lovell-Badge, R. (2003). Multipotent cell lineages in early mouse development depend on SOX2 function. *Genes and Development,* 17, 1, (126-140).

Azuma, K.; Nagaoka, M.; Cho, CS. & Toshihiro Akaike. (2010). An artificial extracellular matrix created by hepatocyte growth factor fused to IgG-Fc. *Biomaterials,* 31, 5, (802–809).

Brieher, WM.; Yap, AS. & Gumbiner, BM. (1996). Lateral dimerization is required for the homophilic binding activity of C-cadherin. *Journal of Cell Biology,* 135, 2, (487-496).

Cho, CS.; Seo, SJ.; Park, IK.; Kim, SH.; Kim, TH.; Hoshiba, T.; Harada, I. & Akaike, T. (2006). Galactose-carrying polymers as extracellular matrices for liver tissue engineering. *Biomaterials,* 27, 4, (576-585).

Chou, YF.; Chen, HH.; Eijpe, M.; Yabuuchi, A.; Chenoweth, JG.; Tesar, P.; Lu, J.; McKay, RDG. & Geijsen, N. (2008). The growth factor environment defines distinct pluripotent ground states in novel blastocyst-derived stem cells. *Cell,* 135, 3, (449-461).

Cui, L.; Johkura, K.; Yue, F.; Ogiwara, N.; Okouchi, Y.; Asanuma, K. & Sasaki, K. (2004). Spatial distribution and initial changes of SSEA-1 and other cell adhesion-related molecules on mouse embryonic stem cells before and during differentiation. *The Journal of Histochemistry and Cytochemistry,* 52, 11, (1447-1457).

Discher, DE.; Mooney DJ. & Zandstra PW. (2009). Growth factors, matrices, and forces combine and control stem cells. *Science,* 324, 5935, (1673-1677).

Eschenhagen, T.; Fink, C.; Remmers, U.; Scholz, H.; Wattchow, J.; Weil, J.; Zimmermann, W.; Dohmen, HH.; Schafer, H.; Bishopric, N.; Wakatsuki, T. & Elson, EL. (1997). Three-dimensional reconstitution of embryonic cardiomyocytes in a collagen matrix: a new heart muscle model system. *FASEB Journal,* 11, 8, 683-694.

Friel, R.; Sar, S. & Mee, PJ. (2005). Embryonic stem cells: Understanding their history, cell biology and signalling. *Advanced Drug Delivery Reviews,* 57, 13, (1894– 1903).

Gupta, B.; Plummer, C.; Bisson, I.; Frey, P. & Hilborn, J. (2002). Plasma induced graft polymerization of acrylic acid onto poly(ethyleneterephthalate) films: characterization and human smooth muscle cell growth on grafted films. *Biomaterials,* 23, 3, (863-871).

Haque, MA.; Nagaoka, M.; Hexig, B. & Akaike, T. (2010). Artificial extracellular matrix for embryonic stem cell cultures: a new frontier of nanobiomaterials. *Science and Technology of Advanced Materials,* 11, 014106 (1-9).

Hynes RO. (1999). Cell adhesion: old and new questions. *Trends in Cell Biology,* 9, 12, (M33-37).

James, D.; Levine, AJ.; Besser, D. & Hemmati-Brivanlou, A. (2005). TGFβ/activin/ nodal signaling is necessary for the maintenance of pluripotency in human embryonic stem cells. *Development,* 132, 6, (1273-1282).

Langer, R. & Tirrell, DA. Designing materials for biology and medicine. (2004). Designing materials for biology and medicine. *Nature,* 428, 6982 (487-492).

Larue, L.; Ohsugi, M.; Hirchenhain, J. & Kemler, R. (1994). E-cadherin null mutant embryos fail to form a trophectoderm epithelium. *Proceedings of the National Academy of Sciences of the USA,* 91, 17, (8263-8267).

Lutolf, MP. & Hubbell. JA. (2005). Synthetic biomaterials as instructive extracellular microenvironment for morphogenesis in tissue engineering. *Nature Biotechnology,* 23, 1, (47-55).

Mitsui, K.; Tokuzawa, Y.; Itoh, H.; Segawa, K.; Murakami, M.; Takahashi, K.; Maruyama, M.; Mayeda, M. & Tamanaka, S. *Cell.* 113, 5, (631-642).

Miyatani, S.; Shimamura, K.; Hatra, M; Nagafuchi, A.; Nose, A.; Matsunaga, M.; Hatra, K. & Takeichi, M. (1989). Neural cadherin: role in selective cell-cell adhesion. *Science,* 245, 4918, (631-635).

Nagaoka, M.; Ise, H. & Akaike T. (2002). Immobilized E-cadherin model can enhance cell attachment and differentiation of primary hepatocytes but not proliferation. *Biotechnology Letters,* 24, (1857-1862).

Nagaoka, M.; Koshimizu, U.; Yuasa, S.; Hattori, F.; Chen, H.; Tanaka, T.; Okabe, M.; Fukuda, K. & Akaike, T. (2006). E-cadherin-coated plates maintain pluripotent ES cells without colony formation. *PLoS One,* 1, e15.

Nagaoka, M.; Ise, H.; Harada, I.; Koshimizu, U.; Maruyama, A. & Akaike, T. (2008a). Embryonic undifferentiated cells show scattering activity on a surface coated with immobilized E-cadherin. *Journal of Cellular Biochemistry,* 103, 1, (296-310).

Nagaoka, M.; Hagiwara, Y.; Takemura, K.; Murakami, Y.; Li, J.; Duncan, SA. & Akaike, T. (2008b). Design of the artificial acellular feeder layer for the efficient propagation of mouse embryonic stem cells *J Biol Chem,* 283, 39, (26468-26476)..

Nagaoka, M.; Jiang, HL.; Hoshiba, T.; Akaike, T. & Cho CS. (2010a). Application of recombinant fusion proteins for tissue engineering. *Annals of Biomedical Engineering,* 38, 3, (683–693).

Nagaoka, M.; Si-Tayeb, K.; Akaike, T. & Duncan, SA. (2010b). Methodology article Culture of human pluripotent stem cells using completely defined conditions on a recombinant E-cadherin substratum. *BMC Developmental Biology,* 10, 60, (1-12).

Nichols, J.; Zevnik, B.; Anastassiadis, K.; Niwa, H.; Klewe-Nebenius, D.; Chambers, I.; Scholer, H. & Smith, A. Formation of pluripotent stem cells in the mammalian embryo depends of the POU transcription factor Oct4. *Cell,* 95, 3, (379-391).

Nishikawa, SI.; Jakt, LM. & Era, T. (2007). Embryonic stem-cell culture as a tool for developmental cell biology. *Nature Reviews: Molecular Cell Biology,* 8, 6, (502-507).

Ogiwara, K.; Nagaoka, M.; Cho, CS. & Akaike, T. (2005). Construction of a novel extracellular matrix using a new genetically engineered epidermal growth factor fused to IgG-Fc. *Biotechnology Letters,* 27, (1633-1637).

Ozawa, M.; Ringwald, M. & Kemler, R. (1990). Uvomorulin-catenin complex formation is regulated by a specific domain in the cytoplasmic region of the cell adhesion molecule. *Proceedings of the National Academy of Sciences of the USA,* 87, 11, (4246-4250).

Parashurama, N.; Nahmias, Y.; Cho, CH.; Poll, DV.; Tilles, AW.; Berthiaume, FO. & Yarmush, ML. Activin alters the kinetics of endoderm induction in embryonic stem cells cultured on collagen gels. *Stem Cells,* 26, 2, (474-484).

Putnam, AJ. & Mooney, DJ. (1996). Tissue engineering using synthetic extracellular matrices. *Nature Medicine,* 2, 7, (824-6).

Riethmacher, D.; Brinkmann, V. & Birchmeier, C. (1995). A targeted mutation in the mouse E-cadherin gene results in defective preimplantation development. *Proceedings of the National Academy of Sciences of the USA,* 92, 3, (855-859).

Sato, N.; Meijer, L.; Skaltsounis, L.; Greengard, P. & Brivanlou, AH. (2004). Maintenance of pluripotency in human and mouse embryonic stem cells through activation of Wnt signaling by a pharmacological GSK-3-specific inhibitor. *Nature Medicine,* 10, 1, (55-63).

Takeichi, M.; Atsumi, T.; Yoshida, C.; Uno, K. & Okada, TS. (1981). Selective adhesion of embryonal carcinoma cells and differentiated cells by Ca^{2+}-dependent sites. *Developmental Biology,* 87, 2, (340-350).

Takeichi, M. Morphogenetic roles of classic cadherins. (1995). *Current Opinion in Cell Biology,* 7, 5, (619-627).

Tanaka, Y.; Kimata, K.; Adams, DH. & Eto, S. (1998). Modulation of cytokine function by heparan sulfate proteoglycans: sophisticated models for the regulation of cellular responses to cytokines. *Proceedings of the Association of American Physicians,* 110, 2, (118-125).

Thomson, JA.; Kalishman, J.; Golos, TG.; Durning, M.; Harris, CP. & Hearn, JP. (1996). Pluripotent cell lines derived from common marmoset (*Callithrix jacchus*) blastocysts. *Biology of Reproduction,* 55, 2, (254-259).

Williams, EJ.; Williams, G.; Howell, FV.; Skaper, SD.; Walsh, FS. & Doherty, P. (2001). Identification of an N-cadherin motif that can interact with the fi broblast growth factor receptor and is required for axonal growth. *Journal of Biological Chemistry,* 276, 47, 43879-86.

Xu, C.; Inokuma, MS.; Denham, J.; Golds, K.; Kundu, P.; Gold, JD. & Carpenter, MK. (2001). Feeder-free growth of undifferentiated human embryonic stem cells. *Nature Biotechnology,* 19, 10, (971-974).

Xu, Y.; Zhu, X.; Hahm, SK.; Wei, W.; Hao, E.; Hayek, A. & Ding, S. (2010). Revealing a core signaling regulatory mechanism for pluripotent stem cell survival and self-renewal by small molecules. *Proceedings of the National Academy of Sciences of the USA,* 107, 18, (8129–8134).

Yasunaga, M.; Tada, S.; Nishikawa, ST.; Nakano, Y.; Okada, M.; Jakt, LM.; Nishikawa, S.; Chiba, T.; Era, T. & Nishikawa, SI. (2005). Induction and monitoring of definitive and visceral endoderm differentiation of mouse ES cells. *Nature Biotechnology,* 23, 12, (1542-1550).

Ying, QL.; Nichols, J.; Chambers, I. & Smith, A. (2003a). BMP induction of Id proteins suppresses differentiation and sustains embryonic stem cell self-renewal in collaboration with STAT3. *Cell,* 115, (281-292).

Ying, QL.; Stavridis, M.; Griffiths, D.; Li, M. & Smith, A. (2003b). Conversion of embryonic stem cells into neuroectodermal precursors in adherent monoculture. *Nature Biotechnology,* 21, 2, (183–186).

Yue, XS.; Murakami, Y.; Tamai, T.; Nagaoka, M.; Cho CS..; Ito, Y. & Akaike T. (2010). A fusion protein N-cadherin-Fc as an artificial extracellular matrix surface for maintenance of stem cell features. *Biomaterials,* 31, 20, (5287-5296).

5

Biomaterials for *In Vitro* Expansion of Embryonic Stem Cells

Makoto Sakuragi[1] and Yoshihiro Ito[1]
[1]*Nano Medical Engineering Laboratory, RIKEN Advanced Science Institute, Saitama Japan*

1. Introduction

In vitro expansion of stem cells is very important for achievement of regenerative medicine. For stem cell expansion methods, several demands are needed to be fulfilled, such as, efficient support of cell proliferation, uniform maintenance of pluripotency, appropriate purity of the stem cells which means less contamination of other cells or potent immunogen, efficient viability after the procedures and so on. However, it is very difficult to efficiently and safely culture some stem cells, such as embryonic stem cells or iPS cells. For example, human embryonic stem (hES) cells cultured on mouse feeder cells expressed an immunogenic nonhuman sialic acid [Martin, 2005]. Human iPS cell developed by Yamanaka's group is cultured on Matrigel™ which is a solubilized basement membrane extracted from the Engelbreth-Holm-Swarm mouse sarcoma, a tumor rich in extracellular matrix proteins [Takahashi, 2007]. To eliminate non-human cells and materials, some substitutional methods are devised. Human derived cells [Miyamoto, 2004], human derived materials [Ueno, 2006; Nagase, 2009; Furue, 2008], recombinant proteins [Nagaoka, 2010; Rodin, 2010] and culture medium which enable feeder free culture are investigated by many researchers. Here our strategies for development of biomaterials for expansion of ES cells are discussed.

2. Culture of ES cells on chemically fixed feeder cells

ES cells were first derived from the inner cell mass of mouse blastocysts in the early 1980s [Evans, 1981]. ES cells have the unique ability to give rise to any type of somatic cell lineage. Even after the establishment of iPS cells, ES cell features are the standards of pluripotent stem cells, such as, molecular marker expressions and potency to respond to differentiation signals. Therefore, ES cells are still important for analyzing the nature of purlipotent stem cells such as, molecular mechanisms of self-renewal and in vitro model of embryogenesis. In addition, the implementation of cell-based or regenerative therapies requires pluripotent stem cells as a renewable source of cells. Efficient *in vitro* expansion of stem cells is therefore important for the development of ES cell technologies and other pluripotent stem cell technologies.

When ES cells are cultured, the environment in which they are propagated has the potential to influence their capacity to act as therapeutic agents in tissue engineering. Animal derived materials may change the potent immunogenic features. Artificial materials may alter the purlipotency or differentiation state. It is therefore important to understand the response of ES cells to synthetic or biological matrices to ascertain their usefulness as implant materials. Previous studies demonstrate that mouse embryonic fibroblasts (MEFs) support the continued propagation of ES cells in the primitive undifferentiated state while retaining their pluripotency. In the absence of MEFs, gelatin can similarly support the growth and propagation of mouse ES cells in the presence of cytokine leukemia inhibitory factor (LIF), as described by Hamazaki et al. [Hamazaki, 2004]

In the case of human ES cells, bFGF is reported to be useful for growth of human ES cells with keeping the undifferentiated state [Xu, 2005]. However, these replacements cannot be generalized for all lines of ES cells. For culture of primate ES cells including human ES cells, some types of human feeder cells were reported [Cheng, 2003; Richard, 2002]. Among them, human amniotic endothelial (HAE) cell, which was reported by Miyamoto et al. [Miyamoto, 2004], are useful by taking into consideration the easiness of acquisition. However, there are two disadvantages of usage of these feeder cells. One is troublesome for preparation before each ES cell culture. In addition, the feeder cells cannot be completely removed for the next culture after trypsinization.

As we reported, chemically fixed feeder cells supported the growth of hematopoietic stem cells [Ito, 2006]. Therefore, the feeder cells were chemically fixed for supporting the growth of ES cells with keeping the undifferentiated state (Figure 1) [Ito, 2007]. Glutaraldehyde (GA)- or paraformaldehyde (PFA)-fixed MEF cells and HAE cells were prepared for culture of mouse and primate (monkey) ES cells, respectively. HAE was immortalized by infection with hTert cDNA.

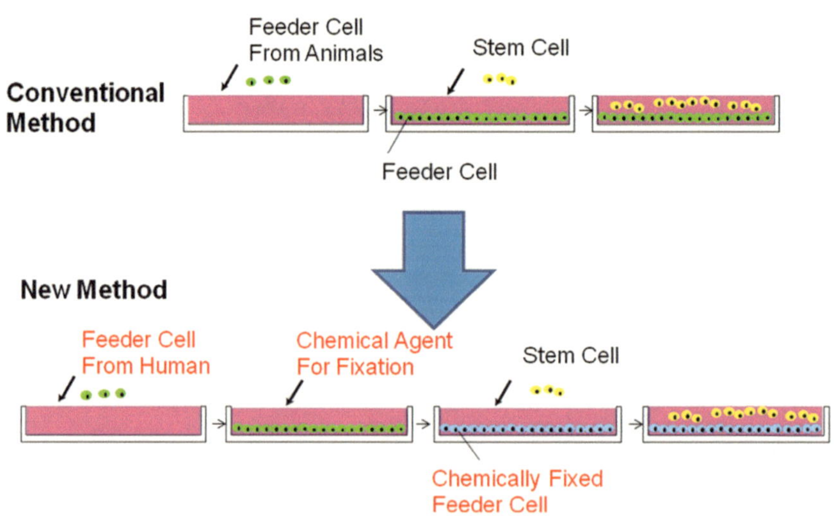

Fig. 1. Chemically fixed feeder cells were prepared as a new technology for culture of stem cells

Figure 2 shows the micrographs of mouse ES cells cultured on several materials. Although the colony size of mouse ES cells cultured on chemically fixed MEF cells was a little smaller than that on MEF which was treated with mitomycine C, they have clear outlines. This result indicated that the mouse ES cells grew with keeping the undifferentiated state. Freeze-dried chemically fixed MEF cells also supported colony formation of mouse ES cells.

On the other hand, on gelatin-coated surface, colony formation was not sufficient and ES cells spread out from the colonies. This indicates that the differentiation began on the gelatin-coated surface. On freeze-dried non-fixed MEF cells ES cells formed colonies but spread out a little.

To confirm the undifferentiation state, the activity of alkaline phosphatase of mouse ES cells was measured by staining (Figure 3). The cells on chemically-fixed cells were stained stronger, even with and without freeze-drying, than that cultured on MEF treated with mitomycine C. On the other hand, the cells cultured only on the gelatin-coated dish were not stained so much.

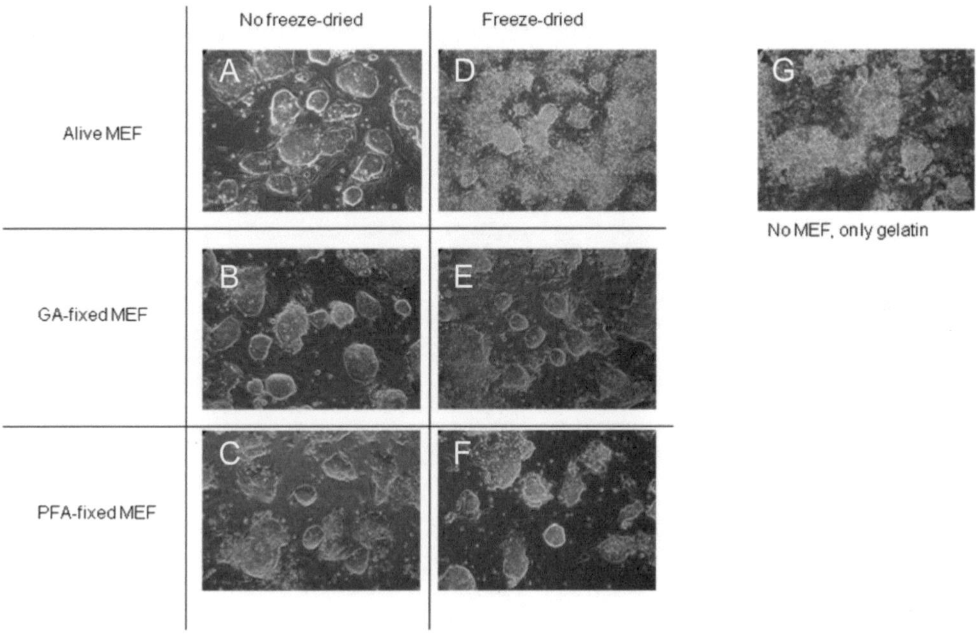

Fig. 2. Phase contrast micrographs of mouse ES cells cultured on different materials for 4 days. A: mitomycine-C-treated MEF, B: GA-fixed MEF, C: PFA-fixed MEF, D: freeze-dried MEF, E: freeze-dried GA-fixed MEF, F: freeze-dried PFA-fixed MEF, G: Gelatin. Reproduced with permission [Ito, 2007]

Immunostaining of SSEA-1 and SSEA-4, which are known to be the markers of undifferentiation and differentiation of mouse ES cells, respectively, were performed and the unidifferentiation of mouse ES cells on chemically fixed cells was checked. Expression of transcription factor Oct-3/4 of mouse ES cells was investigated by RT-PCR. On all materials, mouse ES cells expressed SSEA-1 and Oct-3/4, but not SSEA-4. Considering that LIF was added into all culture media, the expression was considered to be natural.

Growth rates of mouse ES cells on different materials were investigated and no significant difference among mouse ES cells on the surfaces was observed. These results indicate that the chemically-fixed MEF cells supported the growth of mouse ES cells with keeping undifferentiated state as well as native MEF cells even with freeze-drying.

Monkey ES cells were also investigated if they can retain the undifferentiation state on fixed feeder cells. Although the monkey ES cells spread out on gelatin-coated dish, the cells on MEF treated with mitomycine C or on chemically fixed HAE cells grew with forming colonies (Figure 4). On the MEF cells, the shape of colonies was round, but on the fixed and immortalized HAE, the shape of colonies were spindle. It seemed that the monkey ES cells interacted with HAE more than MEF. In this investigation, because the immortalized HAE did not survive in the presence of mitomycine C, the cells were not prepared as a control. Therefore, MEF treated with mitomycine C was employed as the control in the present study.

When the cells were stained by using the activity of alkaline phosphatase, all of the colonies were stained except for the cells on gelatin-coated dish as shown in Figure 5. Staining of SSEA-1 and SSEA-4 was performed and the result indicates the cells kept the undifferentiated state on these materials. This was also supported from the data on Oct-4 expression. Freeze-drying did not reduce the supporting activity. These results confirm the difference between mouse ES cells and primate ES cells in an aspect of self undiferentioation state retaining mechanism.

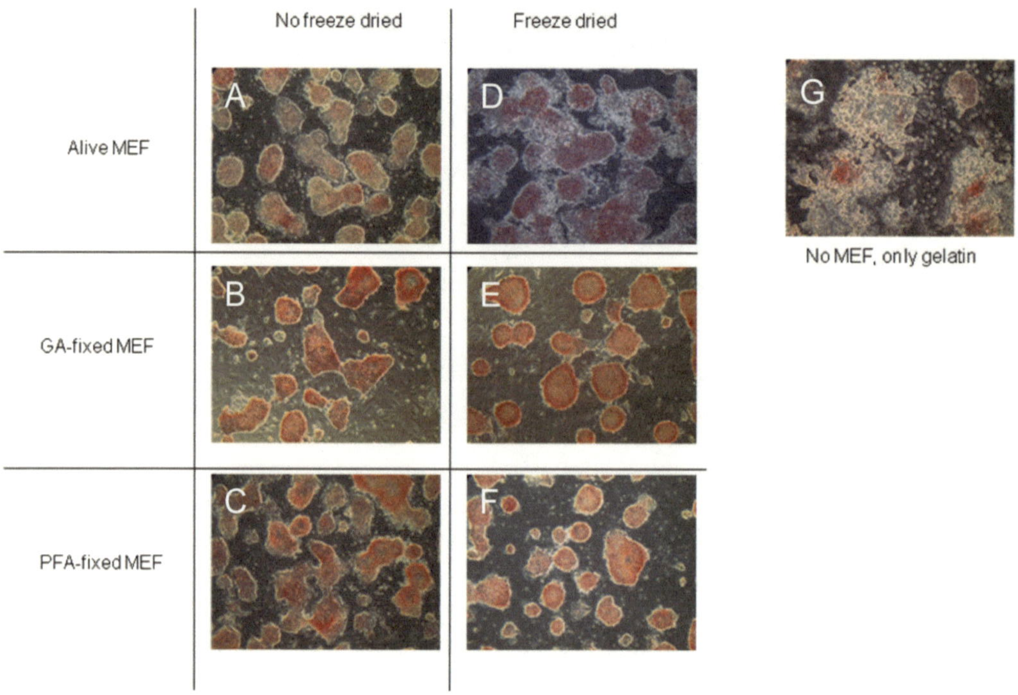

Fig. 3. Staining for Alkaline phosphatase of mouse ES cells cultured on different materials for 6 days. A: mitomycine-C-treated MEF, B: GA-fixed MEF, C: PFA-fixed MEF, D: freeze-dried MEF, E: freeze-dried GA-fixed MEF, F: freeze-dried PFA-fixed. Reproduced with permission [Ito, 2007]

Although fixed HAE cells have incompleteness supporting ability of monkey ES cells, they support the growth and colony formation. Neither chemically fixation nor freeze drying significantly affected the supporting activity, although the growth rates were different from each other. Considering fixation reagent, growth supporting ability of PFA-fixed HAE was a little higher than that of GA-fixed one even before and after freeze-drying. On gelatin-coated surface monkey ES cells grew with differentiation.

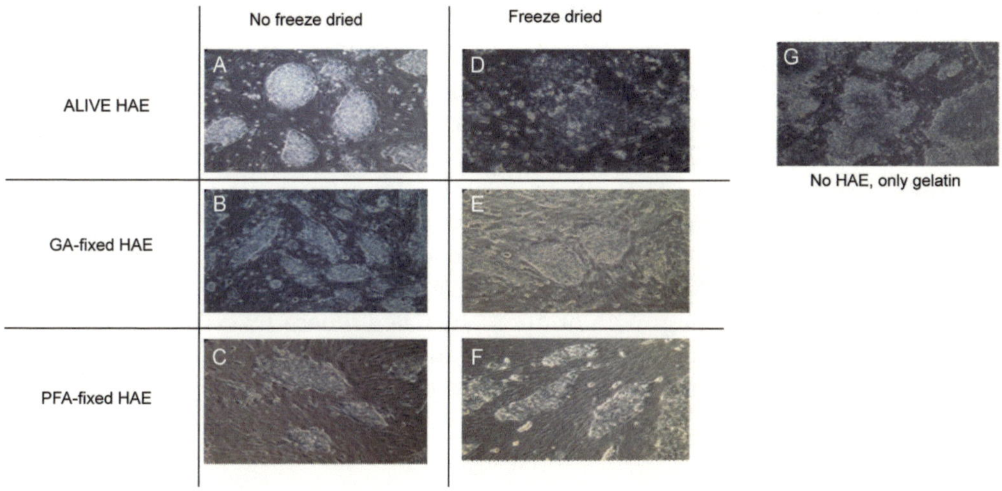

Fig. 4. Phase contrast micrographs of monkey ES cells cultured on mitomycine-C-treated MEF (A), GA-treated HAE (B), PFA-treated HAE (C), freeze-dried HAE (D), freeze-dried GA-fixed HAE (E), freeze-dried PFA-fixed HAE (F), and gelatin (G) for 4 days. Reproduced with permission [Ito, 2007]

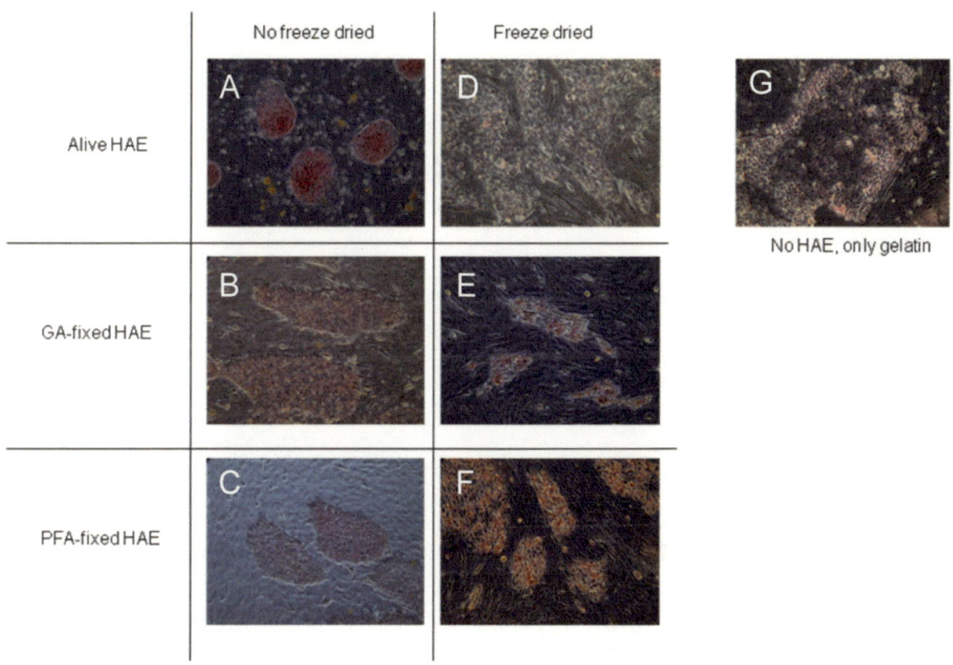

Fig. 5. Staining for Alkaline phosphatase of monkey ES cells cultured on mitomycine-C-treated MEF (A), GA-treated HAE (B), PFA-treated HAE (C), and freeze-dried HAE (D) for 3 days, and freeze-dried GA-fixed HAE (E), freeze-dried PFA-fixed HAE (F), and gelatin (G) for 8 days. Reproduced with permission [Ito, 2007]

3. Synthetic biomateirals for ES cells maintenance and EB formation

Synthetic culture substrates for ES cells were also reported. Harrison et al. [Harison, 2004] reported that fluoride-containing hydroxyapatite supported mouse ES cell growth and the biodegradable substrate affected the pluripotency. Horak et al. [Horak, 2004] cultured mouse ES cells on poly(2-hydroxyethyl methacrylate)-based slabs. Langer's group [Levenberg, 2003; Anderson, 2004] cultured human ES cells in the presence of various materials. A Germany group [Neuss, 2008] also investigated the combination of biomaterials with stem cells. Considering the importance of immobilized biosignal molecules [Ito, 2008], Makino et al. [Makino, 2004] and Nagaoka et al. [Nagaoka, 2008] reported that immobilized LIF and cadherin support the growth of undifferentiated mouse ES cells, respectively. On the other hand, Kurosawa et al. [Kurosawa, 2003] and Konno et al. [Konno, 2005] employed polypropylene tubes and phospholipid polymers, respectively, for the formation of embryoid bodies.

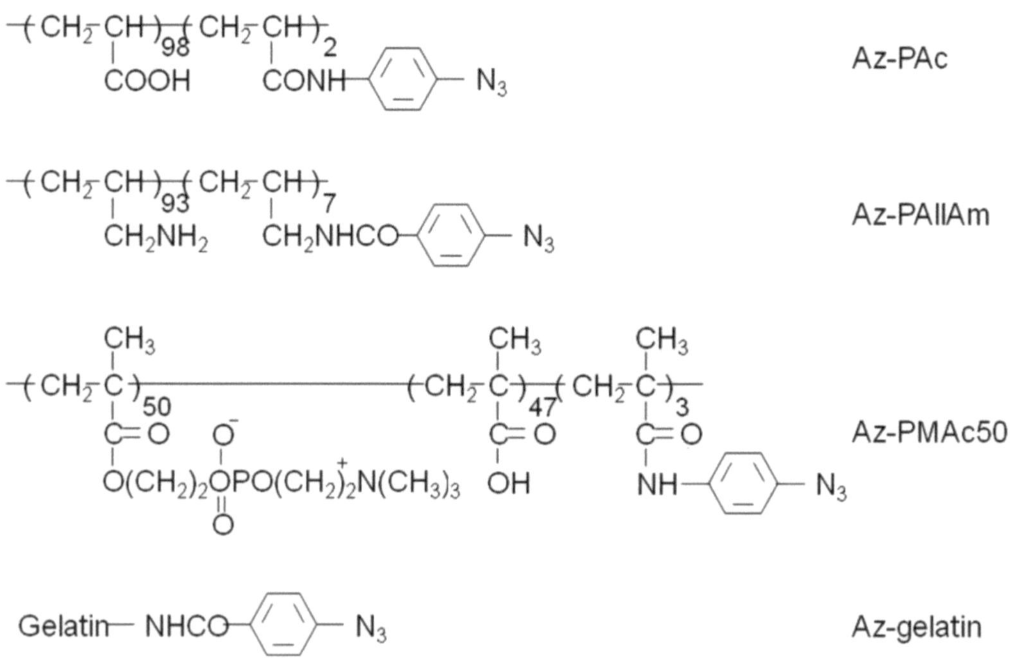

Fig. 6. Chemical structures of biomaterials used for ES cells in this investigation.
Reproduced with permission

We prepared different types of polymers and covalently immobilized them on conventional polystyrene tissue culture dishes to investigate the effects of polymer surface properties on the culture of mouse ES cells (Figure 6) [Konno, 2006]. Four types of photoreactive polymer were prepared and photoimmobilized: azidophenyl-derivatized poly(acrylic acid) (Az-PAc) as an anionic polymer, azidophenyl-derivatized polyallylamine (Az-PAllAm) as a cationic polymer, azidophenyl-derivatized poly(2-methacryloyloxyethyl phosphorylcholine-co-methacrylic acid) (Az-PMAc50) as a zwitterionic polymer, and azidophenyl-derivatized gelatin (Az-gelatin) as a biological polymer.

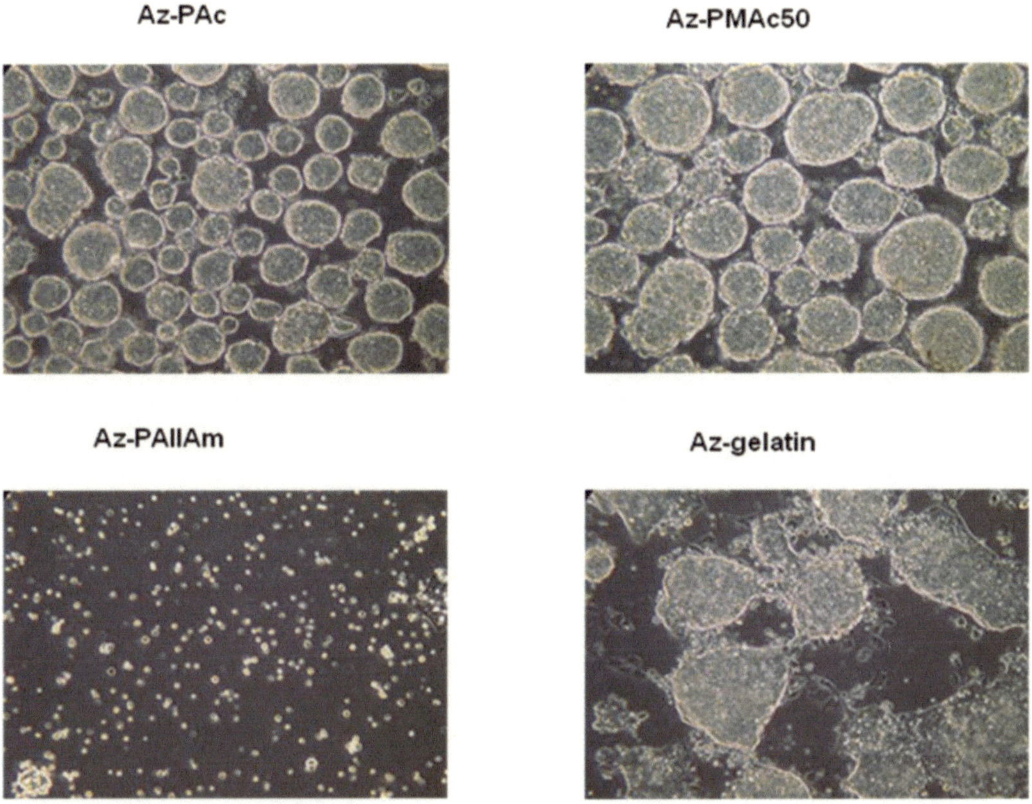

Fig. 7. Mouse ES cells cultured for 3 days on synthetic biomaterials. Reproduced with permission [Konno, 2006]

Phase contrast microscopic images of ES cells after three days of culture on the photoimmobilized polymers are shown in Figure 7. The ES cells immediately and spontaneously aggregated on the Az-PMAc50 surface, which comprised concentrated phospholipid polar groups. The Az-PAc surface, which was anionically charged, also induced ES cell aggregation. The ES cells appeared to form embryoid body (EB)-like cell aggregations that did not adhere to the Az-PMAc50 surface or the Az-PAc surface. The cell aggregates moved randomly when the Az-PMAc50 and the Az-PAc culture plates were shaken. The ES cell aggregates that formed on the Az-PMAc50 surface were larger than the cell aggregates formed on the Az-PAc surface. In contrast, the ES cells strongly adhered to the Az-PAllAm surface, which was positively charged, but did not form aggregates. On the Az-gelatin surface, the ES cells adhered and formed colonies. The cells did not move when the Az-PAllAm and Az-gelatin culture plates were shaken.

Figure 8 shows the results of staining for ALP, a marker of the undifferentiated state of the cells, after four days of culture. On the Az-PAc and Az-PMAc50 surfaces, the cell aggregates that formed readily stained for ALP, although on the Az-PAllAm it was difficult to detect the staining of the colonies of cells, because the cells did not form colonies. On the Az-gelatin surface, ES cell colonies were not densely stained for ALP, which indicates partial differentiation. It is considered that this result related to the cell spreading from colonies or the deformation of colonies, as shown in Figure 7.

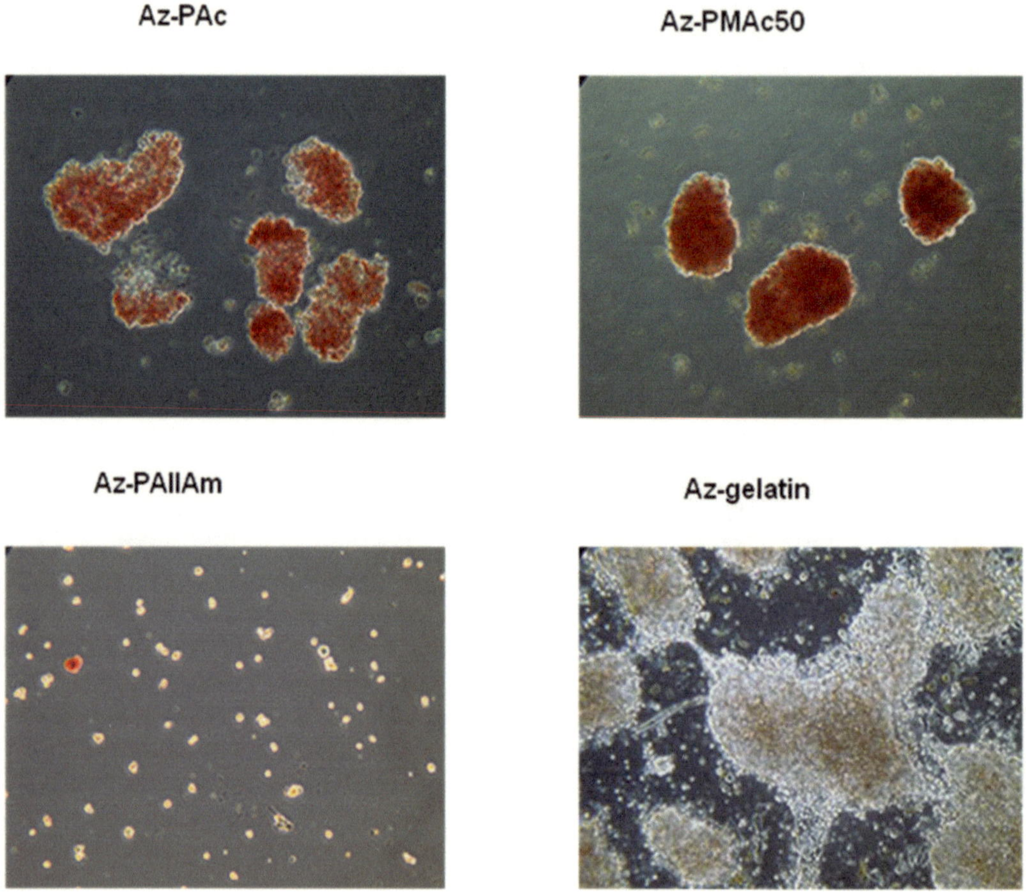

Fig. 8. ALP staining of mouse ES cells cultured on polymers for 4 days. Reproduced with permission [Konno, 2006]

Figure 9 shows the increase in cell numbers on the photoimmobilized polymer surfaces after five days. The highest number of cells occurred on the Az-gelatin surface. The growth of cells on the Az-PMAc50 and Az-PAc surfaces, to which the ES cells did not adhere, was lower than the growth of cells on the Az-gelatin surface. The lowest growth rate was observed on the Az-PAllAm surface, although the cells adhered to this surface. These results indicate that the biological interaction of gelatin is important for cell growth, rather than the simple adhesive interaction.

The LIF/stat3 signaling cascade is involved in maintenance mechanism of mouse ES cell undifferentiation state [Matsuda, 1999; Sekkai, 2005]. The cytoplasmic protein stat3 is activated by the binding of LIF to the LIF and gp130 receptors, which both reside on the mouse ES cell membrane. Activated stat3 (phosphorylated stat3) induces the expression of the transcription factor Oct3/4. In this study, phosphorylated stat3 (p-stat3) and total stat3 (t-stat3) in cells on the immobilized photoreactive polymer surfaces were detected using western blotting. The relative intensity of p-stat3 to t-stat3 (p-stat3/t-stat3), which indicates the activation of signal transduction, was investigated. As a result, phosphorylation of stat3 was observed on all polymer surfaces, and no significant difference between the surfaces was found. Although ALP staining was not observed in the cells on the Az-PAllAm surface by microscopy, at the molecular level of detection, the undifferentiated state of the cells was revealed.

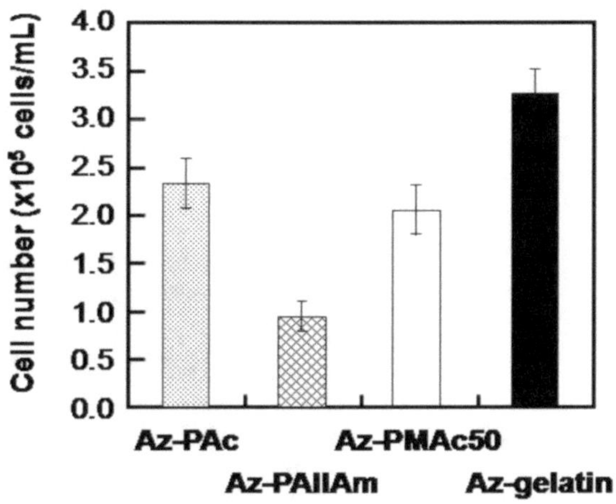

Fig. 9. Growth of mouse ES cells on polymers after 5 days culture (n = 5). Reproduced with permission [Konno, 2006]

Expression of the transcription factor Oct3/4 is necessary for ES cells to maintain an undifferentiated state [Niwa, 2005]. Phosphorylation of stat3 induces the expression of Oct3/4. We confirmed the expression of Oct3/4 by RT-PCR. When the relative expression of Oct3/4 to G3PDH (HKG) was calculated using image analysis, the same level of expression was observed in cells on all polymer surfaces (Figure 10).

The expression of GATA4, which is a molecular marker for the early endodermal state or differentiation to the EB [Kelly, 1993; Grepin, 1997], was also measured. Using the data in Figure 10, the relative intensity of GATA4 to G3PDH was calculated. The values were 0.4, 0.2, 0.18, and 0.02 on Az-PMAc, Az-PAc, Az-gelatin, and Az-PAllAm, respectively. High expression of GATA4 occurred in cells on the Az-PMAc50 surface. This indicates that the Az-PMAc50 surface effectively induced EB formation and differentiation of ES cells. Strong expression of GATA4 was not observed in the cells on the Az-PAc surface, although the cells aggregated on this surface. Similar expression was observed in the cells on the Az-gelatin, but, in this case, the colonies were spread out. In contrast, cells grown on the Az-PAllAm surface did not express GATA4, which we considered was due to the lack of cell aggregation.

Many researchers have investigated the behavior of ES cells cultured on various biomaterials. Properties such as hydrophilicity / hydrophobicity, electrostatic charge, and topographical roughness affect cell adhesion, growth, and differentiation. In this study, the effect of electrostatic charge on mouse ES cells was investigated in the presence of LIF. Considering that it is known that LIF activates stat 3 [Matsuda, 1999; Sekkai, 2005], and expression of Oct3/4 [Niwa, 2005], and ALP [Singla, 2006], it is reasonable that the undifferentiated state of cells was maintained on all surfaces in the presence of LIF.

However, the morphology, growth, and differentiation of ES cells to EBs depended on polymer surface properties. Neither Az-PAc nor Az-PMc50 surfaces induced adhesion of ES cells, which led to EB formation. When an EB did form, the growth of cells was reduced. Negative charges or zwitterions reduced the interaction of ES cells with the polymer surface and the cells aggregated to form an EB. EB size and expression of GATA4 (an indicator of EB formation) in cells grown on the Az-PMAc50 surface were greater compared with cells grown on the Az-PAc surface. Recently, we reported that EB formation was efficiently induced on a phosphorylcholine-derived polymer-coated surface [Konno, 2005]. Therefore, the reduction of cell-surface interactions with this polymer would enhance the cell–cell interactions. The lower interaction leads formation of EBs, and thus that explain the larger size of the EBs and stronger expression of GATA4 on the Az-PMAc50 surface.

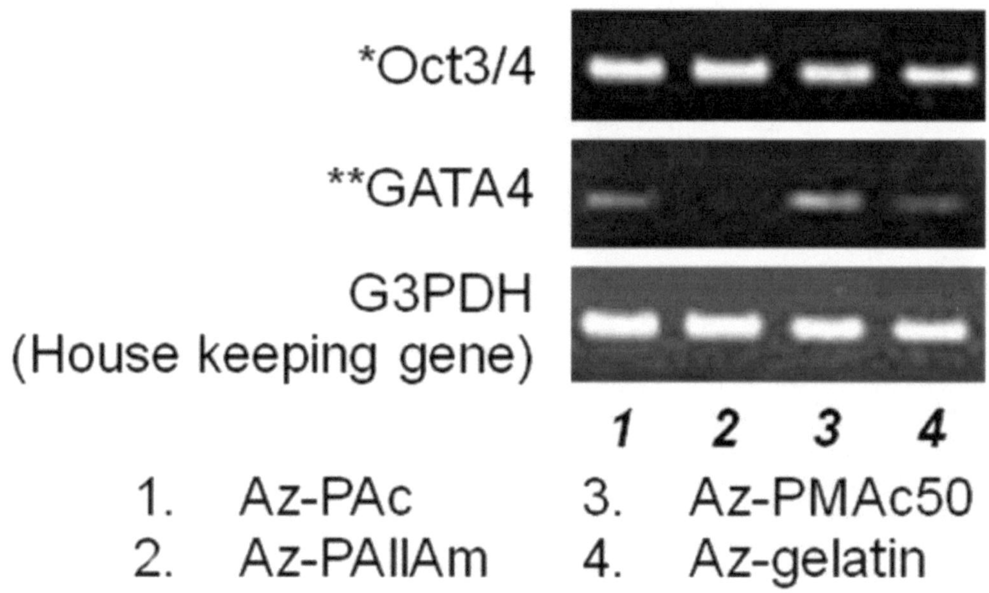

Fig. 10. RT-PCR detection of G3PDH, Oct3/4, and GATA4 from mouse ES cells cultured on polymers for 6 days. Reproduced with permission [Konno, 2006]

ES cells adhere to cationic or gelatin surfaces. However, mouse ES cells formed colonies on the gelatin surface, but not on the Az-PAllAm surface. Some biological interactions are required to induce ES cells to aggregate and adhere to a surface. The growth of ES cells was significantly enhanced on the Az-gelatin surface, and GATA4 expression was slightly induced. These phenomena are the same as those for normal gelatin-coated surfaces.

4. Thermoresponsive polymer for ES cells detachment

Among the procedures of in vitro expansion of ES cell, enzymatic detachment of the cells from the culture plate is one of the most deleterious procedures to the cells. It is known that enzymatic separation of the ES cell colonies into single cells causes massive cell death for human ES cells, and chemical inhibitor of the massive cell death was investigated [Watanabe, 2007; Ohgushi, 2010].

An approach known as cell sheet engineering has been proposed for culturing and passaging cells without the use of trypsin [Okano, 1993]. Using a coating of a thermally responsive polymer, the surface properties of the substrate can be changed by changing the temperature of the environment. Poly(N-isopropylacrylamide) (PNIPAAm) is a popular polymer of this type. It exhibits a lower critical solution temperature (LCST) of 32–33 °C, being hydrophilic at low temperatures and precipitating above the critical phase transition temperature [Xia, 2005]. This unique physical property has been exploited in the fabrication of thermally responsive surfaces for cell sheet engineering [Okano, 1993; Yamato, 2001; Liu, 2002]. At cell culture temperatures above the LCST, the surface is hydrophobic, and cells or tissues attach to the substrate. When the temperature is lowered below the LCST, the surface becomes hydrophilic, and they detach. This mild technique of cell detachment preserves cell–cell and cell–extracellular matrix (ECM) interactions, unlike trypsinization [Kushida, 1999; Shimizu, 2003].

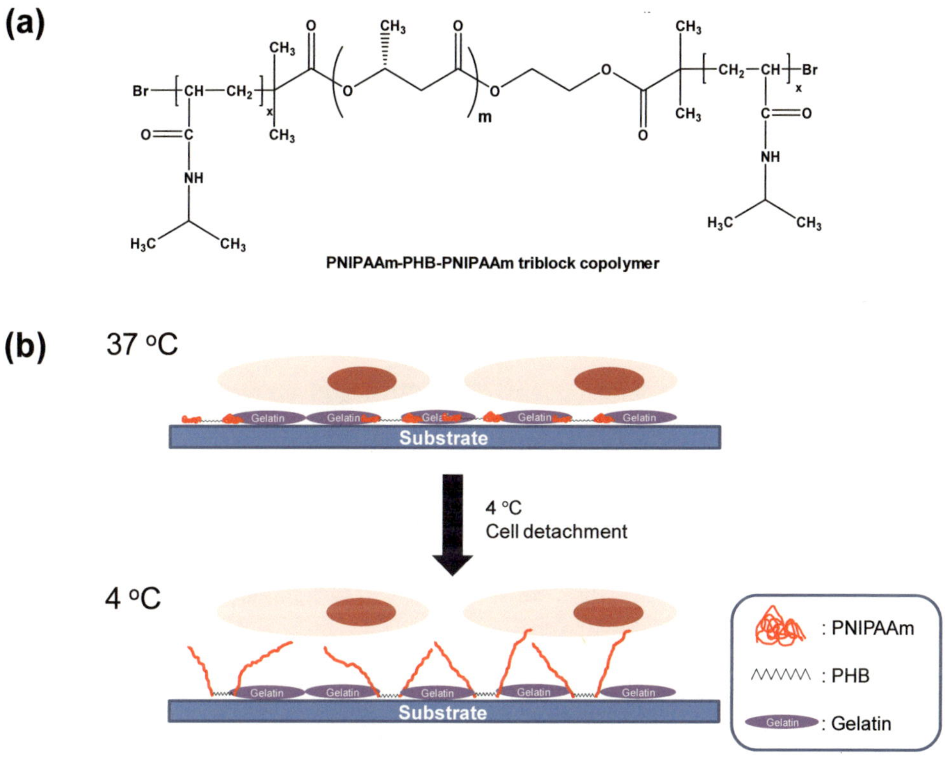

Fig. 11. Chemically fixed feeder cells were prepared as a new technology for culture of stem cells. Reproduced with permission [Loh, 2009]

Several papers have reported that the formation of ES cell aggregates is essential prior to the differentiation of these cells [Lee, 2000; Dang, 2004; Konno, 2005]. The formation of ES cell aggregates takes a minimum of 3 days, thus cells which are freshly trypsinized cannot be used for immediate differentiation. So, there are some advantages to detach the cells without trypsinization, and not to separate the colony into single cells prior to the differentiation.

We synthesized thermoresponsive polymer to recovering ES cells without enzymatic detachment or separation into single cells. We used a triblock copolymer of poly(N-isopropylacrylamide)-poly[(R)-3-hydroxybutyrate]-poly(N-isopropylacrylamide) (PNIPAAm-PHB-PNIPAAm) [Loh, 2009] and prepared a mixture of the thermoresponsive polymer [Loh, 2009] and gelatin with simple drop-casting technique (Figure 11). As the thermoresponsive polymer PHB belongs to a class of naturally derived biologically synthesized polyesters known as poly[(R)-3-hydroxyalkanoate]s [Loh, 2007]. PHB has also been extracted from genetically modified plants [Petrasovits, 2007; Purnell, 2007]. Based on its advantageous properties, PHB might be suitable for a variety of biomedical applications such as tissue engineering scaffolds, and these have been reported to be suitable for enabling cell adhesion [Chen, 2003; Wang, 2005].

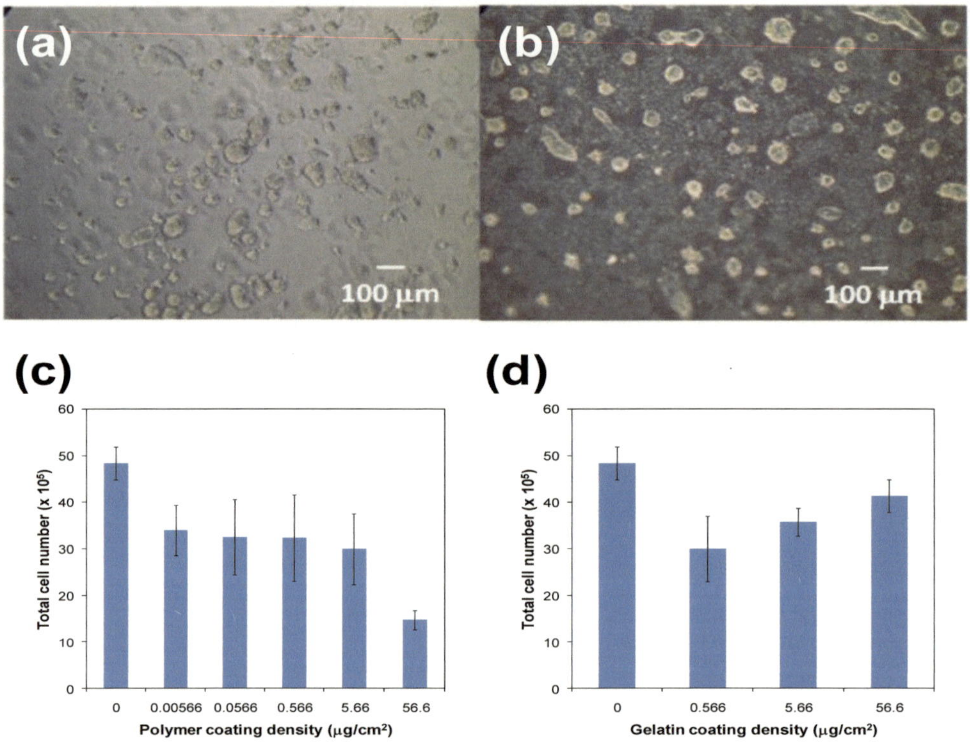

Fig. 12. Morphology of mouse ES cells cultured on gelatin/PNIPAAm-PHB-PNIPAAm surfaces of different thicknesses: (a) 5.66 μg/cm^2 and (b) 56.6 μg/cm^2. (c) Cell growth on different copolymer coating densities of gelatin/PNIPAAm-PHB-PNIPAAm after 3 days. (Gelatin coating density = 0.566 μg/cm^2). (d) Cell growth on different gelatin coating densities of gelatin/PNIPAAm-PHB-PNIPAAm after 3 days (Copolymer coating density = 5.66 μg/cm^2). Reproduced with permission [Loh, 2009]

As coating method, we used a simple drop-casting technique for the preparation of a homogeneous thermoresponsive surface, instead of using spin coating.

Figure 12 shows the micrographs of mouse ES cells cultured on thermoresponsive polymer at different densities. On the surfaces at 56.6 μg/cm^2, colonies had clear outlines. While, the colonies on the surface at 5.66 μg/cm^2 had less (compare Figure 12a and Figure 12b), these result suggested that the mouse ES cells grew with keeping the undifferentiated state on the surfaces at 56.6 μg/cm^2.

Cell numbers were counted for the layers with different surface densities and are shown in Figure 12c. The cell numbers were generally similar up to 5.66 μg/cm^2. Cells grew on the layers at 56.6 μg/cm^2, but the cell proliferation rate was much lower than on layers of lower coating density (Figure 12c). The amount of gelatin incorporated into the coating also affected cell proliferation (Figure 12d).

Considering that our aim was to achieve a high growth rate of ES cells as well as a maximal thermal response, we decided to use substrates with gelatin and polymer coating densities of 0.566 and 5.66 μg/cm^2, respectively. Mouse ES cells were cultured on five different coated surfaces. The cell growth was monitored over 4 days. The growth rate was found to be in the following order: gelatin > gelatin/PNIPAAm-PHB-PNIPAAm > gelatin/PNIPAAm homopolymer > PNIPAAm-PHB-PNIPAAm > PNIPAAm homopolymer. It is known that gelatin is essential for culture of ES cells. Gelatin immobilized on acrylic acid grafted poly(L-lactide-co-ε-caprolactone) aided in the growth and adhesion of human mesenchymal stem cells, compared with the nonimmobilized polymer [Shin, 2008]. Our results showed that although cells could be cultured on the plain polymer surfaces, incorporation of the gelatin coating significantly improved both cell growth and adhesion. On the other hand, it appeared that incorporation of the polymers stunted the growth of the cells compared with the substrate coated with gelatin.

The highest growth rate of ES cell was obtained with the gelatin/PNIPAAm-PHB-PNIPAAm coating. This was more effective than the PNIPAAm coating, probably because of the presence of the PHB segment. As PHB-based scaffolds are suitable for cell adhesion [Chen, 2003; Wang; 2005; Wu, 2008], the PHB segment might contribute to the growth enhancement of anchorage-dependent cells although the effect was less than that of gelatin.

Cell detachment was tested by incubating the culture dish at 4 °C for 20 min. As a control, cells growing on the gelatin-coated dish were also incubated at 4 °C for the same period. Figure 13a shows the thermoresponsive cell detachment from the cooled gelatin/PNIPAAm-PHB-PNIPAAm surface. The cell colonies eventually detached after 20 min. On the other hand, as shown in Figure 13b, cells grown on the control gelatin surface did not detach.

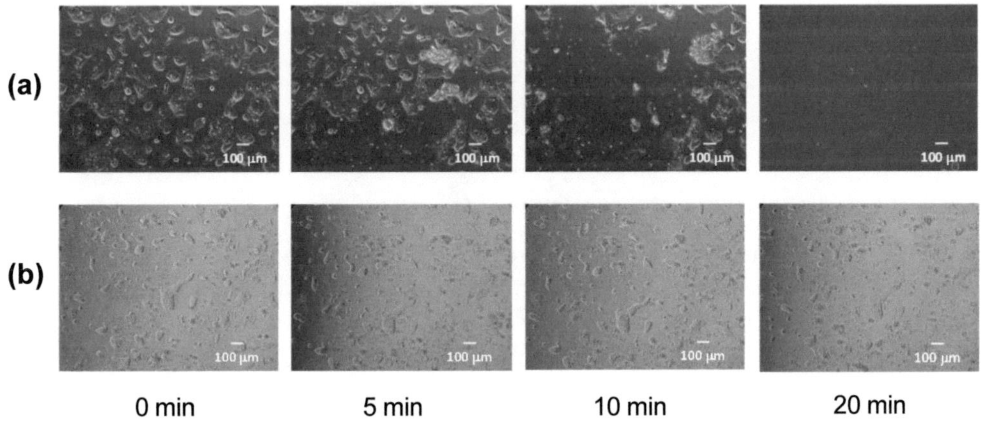

Fig. 13. Mouse ES cell detachment demonstrated on (a) a gelatin/PNIPAAm-PHB-PNIPAAm surface compared with (b) a gelatin-coated substrate. Reproduced with permission [Loh, 2009]

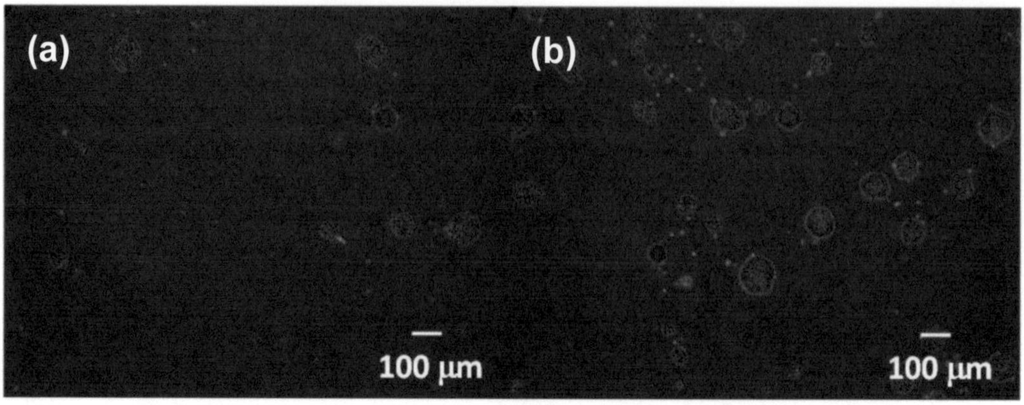

Fig. 14. Phase contrast microscope images mouse ES cells stained for alkaline phosphatase after being cultured for 3 days on the different surfaces: (a) Gelatin/PNIPAAm and (b) Gelatin/PNIPAAm-PHB-PNIPAAm. Reproduced with permission [Loh, 2009]

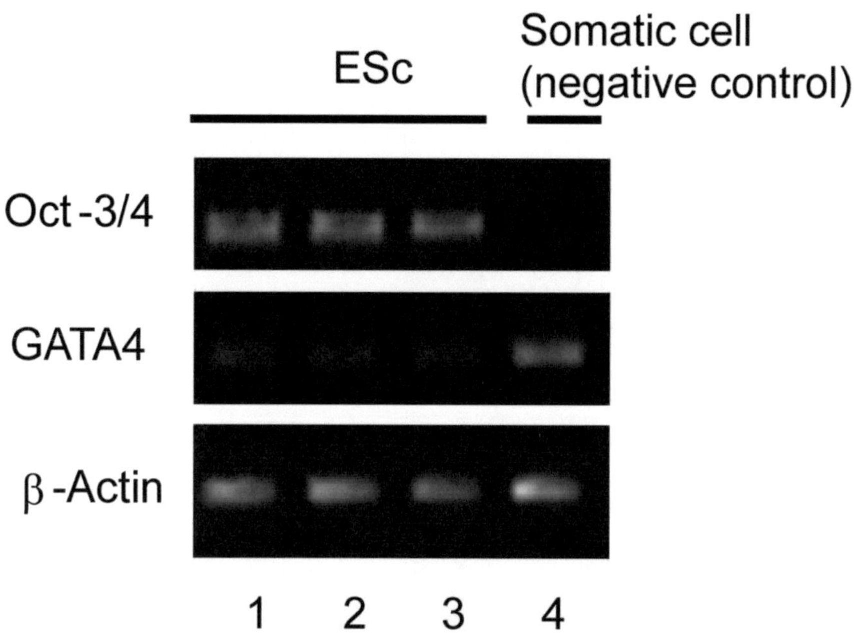

Fig. 15. Expression levels of β-actin, Oct3/4 and GATA4 in mouse ES cells cultured on the different substrates, measured by reverse transcription polymerase chain reaction. Lane 1, gelatin control; lane 2, gelatin/PNIPAAm-PHB-PNIPAAm; lane 3, gelatin/PNIPAAm and lane 4, STO mouse ES cells (negative control). Reproduced with permission [Loh, 2009]

In Figure 11b, we show a schematic diagram of the likely cell detachment process. The cells adhere on the substrate coated with copolymer micelles, which have a hydrophobic core and a collapsed PNIPAAm corona at 37 °C. When the substrate is cooled, the PNIPAAm segment relaxes and becomes more hydrophilic. This hydrophilic surface is not suitable for the attachment of the cells, and so they detach.

To confirm the undifferentiation state, first, the activity of alkaline phosphatase of mouse ES cells was measured by staining (Figure 14). Cells cultured on the different surfaces were tested for ALP activity after three days of culture. Staining for ALP was positive, indicating the undifferentiated state of the ES cells when cultured on the coated substrates.
Second, RT-PCR was performed to check the expression level of Oct3/4 and thus to confirm the undifferentiated state of the ES cells. The Oct3/4 bands of the cells cultured on the gelatin, gelatin/PNIPAAm-PHB-PNIPAAm and the gelatin/PNIPAAm surfaces were of almost equal intensity, as shown in Figure 15. The negative control used in this experiment was mRNA extracted from STO, mouse embryonic fibroblast cells. Next, we tested for the expression of GATA4. Cells express GATA4 during the early endodermal state or while differentiating to embryoid bodies [Kelly, 1993; Grepin,1997]. We detected a weak band of GATA4 from the ES cells cultured on the gelatin, gelatin/PNIPAAm-PHB-PNIPAAm and gelatin/PNIPAAm surfaces. Overall, the ES cells showed a lower expression of GATA4 than the STO cells. Incorporation of the polymers as a surface coating reduced the intensity of the band compared with the pure gelatin surface. Taken in total, the ALP, western blotting and the RT-PCR results show that the mouse ES cells remained in an undifferentiated state after culture on these thermosensitive coatings.

5. Conclusion

We devised three kinds of biomaterials for *in vitro* expansion of ES cells. The first is chemically fixed feeder cells. The fixed feeder cells are easy for handling. With fixed feeder cells, ES cells could be cultured without contamination of other living cell, metabolism and alternation of feeder cells during the culture. The second is photoreactive polymers. By using the polymers Embryoid body like structure could be prepared. The third is thermoresponsive polymer. By using the polymer combined with gelatin, ES cell colonies could be detached from culture dish without deleterious enzymatic separation or separation into single cells or loss of extra cellar matirics. Each of the materials retained the ability to support cell proliferation and puluripotency based on molecular markers expressions.
There is no golden standard of feeder cells or coating material of culture dish yet, for regenerative medicine or stem cell research. Each of feeder cells or materials has advantages and disadvantages. Ideally a defined artificial substrate without any biologically derived components is desired for stem cell culture. Therefore, as a fundamental research, some different surfaces were prepared and biological interactions between ES cells were investigated. Not only biological properties but also physico-chemical properties significantly affected the behavior of stem cells. In future, by consideration on these results, new biomaterials should be designed and developed for stem cell culture.

6. References

Anderson D. G., Levenberg S., Langer R. (2004). Nanoliter-scale synthesis of arrayed biomaterials and application to human embryonic stem cells. *Nat Biotechnol* 22. 863-866.

Cheng L., Hammond H., Ye Z., Zhan X., Dravid G. (2003). Human adult marrow cells support prolonged expansion of human embryonic stem cells in culture. *Stem Cells* 21. 131-142.

Chen X. Y., Zhang X. F., Zhu Y., Zhang J. Z., Hu P. (2003). Surface modification of polyhydroxyalkanoates by ion implantation. Characterization and cytocompatibility improvement. *Polym J* 35. 148-154.

Dang S. M., Gerecht-Nir S., Chen J., Itskovitz-Eldor J., Zandstra P. W.(2004). Controlled, scalable embryonic stem cell differentiation culture. *Stem Cells* 22. 275-282.

Evans M. J., Kaufman M. H. (1981). Establishment in culture of pluripotential cells from mouse embryos. *Nature* 292. 154-156.

Furue M. K., Na J., Jackson J. P., Okamoto T., Jones M., Baker D., Hata R., Moore H. D., Sato J. D., Andrews P. W. (2008). Heparin promotes the growth of human embryonic stem cells in a defined serum-free medium. *Proc Natl Acad Sci U S A* 105. 13409-13414.

Grepin C., Nemer G., Nemer M. (1997). Enhanced cardiogenesis in embryonic stem cells overexpressing the GATA–4 transcription factor. *Development* 124. 2387-2395.

Hamazaki T., Oka M., Yamanaka S., Terada N. (2004). Aggregation of embryonic stem cells induces Nanog repression and primitive endoderm differentiation. *J Cell Sci* 117. 5681-5686.

Harrison J., Melville A. J., Forsythe J. S., Muddle B. C., Trounson A. O., Gross K. A., Mollard R. (2004). Sintered hydroxyfluoroapatites-IV: the effect of fluoride substitutions upon colonization of hydroxyapatites by mouse embryonic stem cells. *Biomaterials* 25. 4977-4986.

Horak D., Kroupova J., Slouf M., Dvorak P. (2004). Poly(2-hydroxyethyl methacrylate)-based slabs as a mouse embryonic stem cell support. *Biomaterials* 25. 5249-5260.

Ito Y. (2008). Covalently immobilized biosignal molecule materials for tissue engineering. *Soft Matter* 4. 46-56.

Ito Y., Hasuda H., Kitajima T., Kiyono T. (2006). Ex vivo expansion of human cord blood hematopoietic progenitor cells using glutaraldehyde-fixed human bone marrow stromal cells. *J Biosci Bioeng* 102. 467-469.

Ito Y., Kawamorita M., Yamabe T., Kiyono T., Miyamoto K. (2007). Chemically fixed nurse cells for culturing mouse or primate embryonic stem cells. *J Biosci Bioeng* 103. 113-121.

Kushida A., Yamato M., Konno C., Kikuchi A., Sakurai Y., Okano T. (1999). Decrease in culture temperature releases monolayer endothelial cell sheets together with deposited fibronectin matrix from temperature-responsive culture surfaces. *J Biomed Mater Res* 45. 355-362.

Kelly C., Blumberg H., Zon L. I., Evans T. (1993). GATA-4 is a novel transcription factor expressed in endocardium of the developing heart. *Development* 118. 817-827.

Konno T., Akita K., Kurita K., Ito Y. (2005). Formation of embryoid bodies by mouse embryonic stem cells on plastic surfaces. *J Biosci Bioeng* 100. 88-93.

Konno T., Kawazoe N., Chen G., Ito Y. (2006). Culture of mouse embryonic stem cells on photoimmobilized polymers. *J Biosci Bioeng* 102. 304-310.

Kurosawa H., Imamura T., Koike M., Sasaki K., Amano Y. (2003).A simple method for forming embryoid body from mouse embryonic stem cells. *J Biosci Bioeng* 96. 409-411.

Lee S. H., Lumelsky N., Studer L., Auerbach J. M., McKay R. D. (2000). Efficient generation of midbrain and hindbrain neurons from mouse embryonic stem cells. *Nat Biotechnol 18*. 675-659

Levenberg S., Huang N. F., Lavik E., Rogers A. B., Itskovitz-Eldor J., Langer R. (2003). Differentiation of human embryonic stem cells on three-dimensional polymer scaffolds. *Proc Natl Acad Sci USA* 100. 12741-12746.

Liu H. C., Ito Y. (2002). Cell attachment and detachment on micropattern-immobilized poly(N-isopropylacrylamide) with gelatin. *Lab Chip* 2. 175-178.

Loh X. J., Goh S. H., Li J. (2007). Hydrolytic degradation and protein release studies of thermogelling polyurethane copolymers consisting of poly[(R)-3-hydroxybutyrate], poly(ethylene glycol), and poly(propylene glycol). *Biomaterials* 28. 4113-4123.

Loh X. J., Gong J, Sakuragi M., Kitajima T., Liu M., Li J., Ito Y. (2009). Surface coating with a thermoresponsive copolymer for the culture and non-enzymatic recovery of mouse embryonic stem cells. *Macromol Biosci* 9. 1069-79.

Makino H., Hasuda H., Ito Y. (2004). Immobilization of leukemia inhibitory factor (LIF) to culture mouse embryonic stem cells. *J Biosci Bioeng* 98. 374-379.

Martin M. J., Muotri A., Gage F., Varki A. (2005). Human embryonic stem cells express an immunogenic nonhuman sialic acid. *Nat Med* 11. 228-232.

Matsuda T., Nakamura T., Nakao K., Arai T., Katsuki M., Heike T., Yokota T. (1999). STAT3 activation is sufficient to maintain an undifferentiated state of mouse embryonic stem cells. *EMBO J* 18. 4261-4269.

Miyamoto K., Hayashi K., Suzuki T., Ichihara S., Yamada T., Kano Y., Yamabe T., Ito Y. (2004).Useful human amniotic epithelial feeder cells for undifferentiated growth of primate embryonic stem cells. *Stem Cells* 22. 433-440.

Nagaoka M., Ise H., Harada I., Koshimizu U., Maruyama A., Akaike T. (2008). Embryonic undifferentiated cells show scattering activity on a surface coated with immobilized E-cadherin. *J Cell Biochem* 103. 296-310.

Nagaoka M., Koshimizu U., Yuasa S., Hattori F., Chen H., Tanaka T., Okabe M., Fukuda K., Akaike T. (2006). E-cadherin-coated plates maintain pluripotent ES cells without colony formation. *PLoS ONE* 1 e15.

Nagaoka M., Si-Tayeb K., Akaike T., Duncan S., A. (2010). Culture of human pluripotent stem cells using completely defined conditions on a recombinant E-cadherin substratum. *BMC Dev Biol.* 2. 10:60.

Nagase T., Ueno M., Matsumura M., Muguruma K., Ohgushi M., Kondo N., Kanematsu D., Kanemura Y., Sasai Y., (2009). Pericellular matrix of decidua-derived mesenchymal cells: a potent human-derived substrate for the maintenance culture of human ES cells. *Dev Dyn* 23. 1118-11130.

Neuss S., Apel C., Buttler P., Denecke B., Dhanasingh A., Ding X., Grafahrend D., Groger A., Hemmrich K., Herr A., Jahnen-Dechent W., Mastitskaya S., Perez-Bouza A., Rosewick S., Salber J., Woeltje M., Zenke M. M. (2008). Assessment of stem cell/biomaterials combinations for stem cell-base tissue engineering. *Biomaterials* 29. 302-313.

Niwa H., Toyooka T., Shimosato D., Strumpf D., Takahashi K., Yagi R., Rossant J. (2005). Interaction between Oct3/4 and Cdx2 determines trophectoderm differentiation. *Cell* 123. 917-929.

Ohgushi M., Matsumura M., Eiraku M., Murakami K., Aramaki T., Nishiyama A., Muguruma K., Nakano T., Suga H., Ueno M., Ishizaki T., Suemori H., Narumiya S., Niwa H., Sasai Y., (2010). Molecular pathway and cell state responsible for dissociation-induced apoptosis in human pluripotent stem cells. *Cell Stem Cell* 6. 225-239.

Okano T. (1993). Molecular design of temperature-responsive polymers as intelligent materials. *Adv Polym Sci* 110. 179-197.

Petrasovits L. A., Purnell M. P., Nielsen L. K., Brumbley S. M. (2007). Production of polyhydroxybutyrate in sugarcane.*Plant Biotechnol J* 5. 162-172.

Purnell M. P., Petrasovits L. A., Nielsen L. K., Brumbley S. M. (2007). Spatio-temporal characterization of polyhydroxybutyrate accumulation in sugarcane. *Plant Biotechnol J* 5. 173-184.

Richards M., Fong C. Y., Chan W. K., Wong P. C., Bongso A. (2002). Human feeders support prolonged undifferentiated growth of human inner cell masses and embryonic stem cells. *Nat Biotechnol* 20. 933–936.

Rodin S., Domogatskaya A., Ström S., Hansson E., M., Chien K., R., Inzunza J., Hovatta O., Tryggvason K., (2010). Long-term self-renewal of human pluripotent stem cells on human recombinant laminin-511. *Nat Biotechnol.* 6. 611-615.

Sekkai D., Gruel G., Herry M., Moucadel V., Constaninescu S. N., Albagli O., Tronik-Le Roux D., Vainchenker W., Bennaceur-Griscelli A. (2005). Microarray analysis of LIF/Stat3 transcriptional targets in embryonic stem cells. *Stem Cells* 23. 1634-1642.

Shimizu T., Yamato M., Kikuchi A., Okano T. (2003). Cell sheet engineering for myocardial tissue reconstruction. *Biomaterials* 24. 2309-2316.

Shin Y. M., Kim K. S., Lim Y. M, Nho Y. C., Shin H (2008) Modulation of spreading, proliferation, and differentiation of human mesenchymal stem cells on gelatin-immobilized poly(l-lactide-co-ε-caprolactone) substrates. *Biomacromolecules* 9. 1772–1781.

Singla S. D., Schneider D. J., Lewinter M. M., Sobel B. E. (2006). Wnt3a but not wnt11 supports self-renewal of embryonic stem cells. *Biochem Biophys Res Commun* 345. 789-795.

Takahashi K., Tanabe K., Ohnuki M., Narita M., Ichisaka T., Tomoda K., Yamanaka S. (2007). Induction of pluripotent stem cells from adult human fibroblasts by defined factors. *Cell* 131. 861-872.

Ueno M., Matsumura M., Watanabe K., Nakamura T., Osakada F., Takahashi M., Kawasaki H., Kinoshita S., Sasai Y., (2006). Neural conversion of ES cells by an inductive activity on human amniotic membrane matrix. *Proc Natl Acad Sci U S A* 103. 9554-9559.

Wang Y. W., Yang F., Wu Q., Cheng Y. C., Yu P. H. F., Chen J., Chen G. Q. (2005). Effect of composition of poly(3-hydroxybutyrate-co-3-hydroxyhexanoate) on growth of fibroblast and osteoblast. *Biomaterials* 26. 755-761.

Watanabe K.,Ueno M., Kamiya D., Nishiyama A., Matsumura M., Wataya T., Takahashi J. B., Nishikawa S., Nishikawa S., Muguruma K., Sasai Y. (2007). ROCK inhibitor permits survival of dissociated human embryonic stem cells. *Nat Biotechnol* 25. 681-686.

Wu L. P., Chen S. T., Li Z. B., Xu K. T., Chen G. Q. (2008). Synthesis, characterization and biocompatibility of novel biodegradable poly[((R)-3-hydroxybutyrate)-block-(D,L-lactide)-block-(ε-caprolactone)] triblock copolymers. *Polym Intern* 57. 939-949.

Xia Y., Yin X. C., Burke N. A. D., Stover H. D. H. (2005). Thermal Response of Narrow-Disperse Poly(N-isopropylacrylamide) Prepared by Atom Transfer Radical Polymerization. *Macromolecules* 38. 5937-5943.

Xu C., Rosler E., Jiang J., Lebkowski J. S., Gold J. D., O'Sullivan C., Delavan-Boorsma K., Mok M., Bronstein A., Carpenter M. K. (2005). Basic fibroblast growth factor supports undifferentiated human embryonic stem cell growth without conditioned medium. *Stem Cells* 23. 315-323.

Yamato M., Utsumi M., Kushida A., Konno C., Kikuchi A., Okano T. (2001). Thermo-responsive culture dishes allow the intact harvest of multilayered keratinocyte sheets without dispase by reducing temperature. *Tissue Eng* 7. 473-480.

6

Surface Engineering to Control Embryonic Stem Cell Fate

Shohreh Mashayekhan[1] and Jun-ichi Miyazaki[2]
[1]*Department of Chemical & Petroleum Engineering, Sharif University of Technology, Azadi Ave, Tehran, 11365-8639,*
[2]*Division of Stem Cell Regulation Research (G6), Graduate School of Medicine, Osaka University, 2-2 Yamadaoka, Suita, Osaka, 565-0871*
[1]*Iran*
[2]*Japan*

1. Introduction

The niche established by supportive cells and the extracellular polymeric matrix (ECM) probably regulates stem cell fate through multiple, complimentary mechanisms, including the spatiotemporally defined presentation of immobilized signaling molecules, the modulation of matrix stiffness, the physicochemical characteristics of the environment, and the creation of cytokine gradients. In contrast to tissue-specific stem cells, embryonic stem (ES) cells are present only transiently in the developing embryo, and therefore, do not have a stable niche in vivo. ES cells also differ from tissue-specific stem cells in their ability to be readily expanded in culture over long time periods. However, the culture systems that have been used successfully for ES cell expansion suggest that ES cell self-renewal versus differentiation is regulated in a similar manner to tissue-specific stem cells, via interactions with other cells, ECM components, soluble factors, and the physicochemical environment (McDevitt & Palecek, 2008). ES cells commute between metastable states from the inner cell mass (ICM) to the epiblast stage, and these reversible states are associated with distinct differentiation potentials (Toyooka et al., 2008; Hayashi et al., 2008; Pelton et al., 2002). Thus, ES cells represent a highly dynamic, self-renewing population that responds to environmental cues to maintain its pluripotency or to differentiate. In ES cell cultures, these cues include growth factors in the culture medium surrounding the ES cell colonies or secreted by the colonies themselves, and signals arising from the ES cells' adhesion to the substrate and the stiffness of the substrate (Discher et al., 2009).
ES cells are anticipated to serve as an unlimited cell source for cell transplantation therapy. However, the most common techniques for controlling ES cell fate using soluble biochemical and biological factors (cytokines and growth/differentiation factors) in the growth medium are often inefficient, and the resulting cell population (either undifferentiated or differentiated) is not homogenous. The idea that ES cell populations are homogenous was first challenged by Cui et al., who observed differential spatial distributions of adhesion molecules within ES cell colonies (Cui et al., 2004), and more recently by the derivation of epiblast stem cells from ES cell (Brons et al., 2007) and the identification of ES cell subpopulations in mouse ES cell cultures (Toyooka et al., 2008).

To improve the efficiency for controlling the ES cell fate, researchers have recently focused on the stimulation of receptors on the ES cell membrane through interactions with solid surfaces. In particular, the interactions of biologically active components with cells can be strengthened by fixing the signals on a surface in close contact with their targets on the cell membrane, because when the signaling components are dispersed in a bulk liquid (medium), they are less likely to encounter their targets.

This chapter will present various surface design strategies for regulating ES cell morphology and function that use micro/nanoscale technologies and a wide range of natural and synthetic materials. First, we will introduce the principles for modifying the culture surface with reference to recent studies that have used various surface design strategies (reviewed in Dellatore et al., 2008; Keung et al., 2010; Saha et al., 2007) and their corresponding effects on ES cell behavior. The latter part of the chapter will describe dendrimer-immobilized surfaces designed in the authors' studies and their effects on the in vitro culture of mouse ES cells.

2. Surface-based control of the morphology and function of cultured ES cells

In this section, we provide an overview for designing the culture surface, as categorized into four general approaches for controlling ES cell fate (Table 1).

Modification	Examples	Observations	Reference
Chemical modification	Plasma etched polystyrene	Maintenance of hESC pluripotency	Mahlstedt et al., 2010
	Plasma-deposited gradients of octadiene to acrylic acid	Effect on mESC adhesion and differentiation	Wells et al., 2009
	PDMS or SAM surfaces presenting terminal hydrophobic moieties	Enhancement of the differentiation yields of hESCs	Valamehr et al., 2008
	Combinatorial library of biomaterials formed from acrylate and methacrylate monomers	Uniform hESC differentiation into epithelial cells	Anderson et al., 2004
Biofunctionalization	ECM, such as matrigel, laminin, fibronectin	Expansion and maintenance of hESCs and mESCs	Stewart et al., 2008; Meng et al., 2010; Flaim et al., 2008
	Laminin, fibronectin, and gelatin	Promotion of mESC differentiation toward neural lineages	Goetz et al., 2006
	Decellularized bone-specific ECM	Promotion of mESC differentiation toward the osteogenic lineage	Evans et al., 2010

Table 1. Various strategies for surface engineering to control ES cell fate

Modification	Examples	Observations	Reference
Biofunctionalization (Continued)	ECM molecules on a Layer-by-layer self-assembled surface of HA and chitosan	Efficient attachment of hESCs	Doran et al., 2010
	E-cadherin-coated surface	Increased proliferative ability and transfection efficiency for mESCs	Nagaoka et al., 2006
	Laminin peptides presented in SAMs on gold	Support of hESC expansion by different peptides from the laminin γ and β chain	Derda et al., 2007
	RGD-modified materials	Promotion of hESC differentiation toward the chondrogenic lineage	Hwang et al., 2006
	RGD and CRGDC-modified materials	Support of hESC culture	Kolhar et al., 2010
	Random peptide libraries using phage display	Expansion and maintenance of hESCs on SAMs presenting specific peptide sequences	Derda et al., 2010
	Immobilized LIF	Expansion and maintenance of ESCs	Nagaoka et al., 2008; Makino et al., 2004; Alberti et al., 2008
	Covalent binding of FGF-2 to polyamide nanofibrillar surfaces	Support of hESC expansion and colony formation	Nur-E-Kamal et al., 2008
	Immobilized VEGF	Promotion of mESC differentiation toward endothelial cells	Chiang et al., 2010
Geometric modification	Topographically microstructured surface libraries	Effect on proliferation and differentiation of mESCs	Markert et al., 2009
	Electrospun polyamide nanofibers	Expansion and maintenance of mESCs	Nur-E-Kamal et al., 2006
	Nanoscale ridge/groove pattern arrays	Promotion of hESC differentiation toward the neuronal lineage	Lee et al., 2010
	Electrospun fibrous scaffolds	Promotion of mESC differentiation toward the neuronal lineage	Xie et al., 2009

Table. 1. (Continued)

Modification	Examples	Observations	Reference
Geometric modification (Continued)	Nanofibrous architecture	Promotion of hESC differentiation toward the osteogenic lineage	Smith et al., 2010
Mechanical modification	Nanofilms made of PLL and HA	Promotion of mESC differentiation toward the epiblast lineage by surface stiffness	Blin et al., 2010
	PDMS substrates	Promotion of mESC differentiation toward the osteoblast lineage by surface stiffness	Evans et al., 2009

Acronyms:
mESC: mouse embryonic stem cell; hESC: human embryonic stem cell; PDMS: Polydimethylsiloxane; SAM: Self-assembled monolayer; ECM: Extracellular polymeric matrix; HA: Hyaluronic acid; RGD: Integrin-binding Arg-Gly-Asp; CRGDC: Cyclic RGD; LIF: leukemia inhibitory factor; FGF-2: Fibroblast growth factor; VEGF: Vascular endothelial growth factor; PLL: poly (L-lysine).

Table. 1. (Continued)

2.1 Control of cells by chemical modification of the substrate

The chemical properties of substrates (e.g., hydrophobicity) play an important role in the kinetics of protein adsorption and folding, which in turn influence cellular activities. Mahlstedt et al. demonstrated that the physicochemical modification of polystyrene by plasma etching can improve the culture surface's ability to maintain human ES cell pluripotency (Mahlstedt et al., 2010). Elsewhere, plasma-deposited gradients of octadiene to acrylic acid were fabricated to investigate the effect of carboxylic group (COOH) concentration on mouse ES cell adhesivity and differentiation status (Wells et al., 2009). In addition, by altering the hydrophobicity of a surface, the formation and differentiation potential of ES cells within embryoid bodies (EBs) can be tuned to promote a desirable EB size and composition (Valamehr et al., 2008).

Because it is often difficult to predict how a stem cell will respond to environmental cues, methods have been developed for the rapid screening of interactions between biomaterials and stem cells. A combinatorial library of biomaterials formed from different acrylate and methacrylate monomers has proved to be useful for identifying environments suitable for the uniform differentiation of ES cells into epithelial cells (Anderson et al., 2004).

2.2 Control of cells by biofunctionalization

Artificial materials can be endowed with precise biological functionalities by immobilizing bioactive molecules such as cytokines, growth factors, ECM proteins, and adhesive peptides on their surface. These biomolecules can be simply adsorbed onto the material's surface or covalently linked via chemical groups previously created on the surface. The biological response following the surface biomodification of a material depends on structural parameters, such as the density of the ligands, their spatial distribution, their colocalization with synergistic ligands, etc.

2.2.1 Cell-adhesive peptides and proteins

Specific ECM-cell and cell-cell interactions are important for providing spatial anchors as well as signals that regulate stem cell maintenance, survival, and differentiation. Cell adhesion is also required for a cell to sense other contextual information, such as the mechanical properties of the microenvironment. Here we review the ways that engineered systems have been used to identify functional adhesive peptide sequences or proteins and to investigate their interactions with ES cells.

ECMs can be used either for feeder-free culturing or for stimulating ES cell differentiation toward a desired cell lineage by incorporating tissue-specific ECM signals. Stewart et al. reviewed the feeder-free conditions that have been successfully applied to culture human ES cells using various types of ECM, such as matrigel, laminin, and fibronectin (Stewart et al., 2008). For example, matrigel, a complex mixture of hundreds of ECM and other proteins, has emerged as a common substrate for human ES cell and human induced-pluripotent stem (iPS) cell culture. Meng and colleagues (Meng et al., 2010) investigated the adhesive interactions in matrigel involved in the maintenance of human ES cell pluripotency. They found that whereas three peptides were able, individually, to support human ES cell growth and pluripotency for short periods of time, their combination enhanced the quality of the culture and the duration of the cells' pluripotency. This finding illustrates how engineered systems can be used to parse out the synergistic contribution of individual motifs within full-length natural proteins, which may inspire future mechanistic studies.

Flaim and co-workers (Flaim et al., 2008) analyzed combinatorial mixtures of ECM molecules to understand their cooperative control of murine ES cell differentiation, and rapidly identified key mixtures with synergistic properties. Other groups have directed stem cell differentiation toward neural lineages by using laminin, fibronectin, and gelatin (Goetz et al., 2006). In another report, decellularized bone-specific ECM promoted the osteogenic differentiation of ES cells (Evans et al., 2010). Recently, Doran et al. used a simple, effective, and efficient method to design a defined high-protein-content surface for stem cell culture (Doran et al., 2010). They demonstrated the highly efficient attachment of human ES cells to various extracted and recombinant ECM molecules presented on a layer-by-layer self-assembled surface of hyaluronic acid and chitosan.

In another study, Nagaoka et al. demonstrated that mouse ES cells cultured on an E-cadherin-coated surface maintained unique morphological characteristics, retained the full complement of ES cell features, and showed a higher proliferative ability and transfection efficiency than those grown under conventional conditions. Furthermore, when grown on the E-cadherin-coated surface, the ES cells also required less leukemia inhibitory factor (LIF) than those grown under conventional conditions, probably due to the homogenous exposure to LIF achieved in this culture system (Nagaoka et al., 2006).

Cell-adhesive ligands can, when incorporated into biomaterials, be used to mediate specific receptor–ligand interactions, and thereby to activate selected receptor-mediated signaling pathways to control cell behavior and differentiation. Several cell-adhesive peptides, such as the integrin-binding Arg-Gly-Asp (RGD) motif, have been incorporated into materials to enhance the cell–matrix interaction. For instance, RGD promotes the chondrogenic differentiation of human ES cells (Hwang et al., 2006). In another study, Kolhar et al. demonstrated that both RGD and cyclic RGD (CRGDC) can support the culture of human ES cells, with CRGDC increasing their adhesion 4-fold over the linear RGD peptide (Kolhar et al., 2010). The identification of peptide sequences such as RGD has been pivotal in advancing biomaterial research, because of the ease of synthesizing, manipulating, and tuning the properties of such materials (Hersel et al., 2003). Nevertheless, only a few adhesive peptide sequences have been found in natural proteins. It is likely that the identification of cell growth substrates would be accelerated by the discovery of new peptide ligands for cell-surface receptors.

In addition, several peptide mimics of the laminin cell-binding domain have been evaluated in stem cell cultures. Derda et al. evaluated a wide variety of laminin peptides presented in self-assembled monolayers (SAMs) on gold for their ability to support human ES cell adhesion and proliferation (Derda et al., 2007). Four different peptides from the laminin γ chain and one peptide from the β chain supported ES cell expansion and the expression of the primitive markers Oct4, alkaline phosphatase, and SSEA4, to a similar extent as matrigel in six-day cultures (Derda et al., 2007). In another recent study, Derda et al. screened random peptide libraries using phage display to identify novel ligands to support the proliferation of pluripotent cells. When human ES cells were cultured on SAMs presenting the sequence TVKHRPDALHPQ or LTTAPKLPKVTR in a chemically defined medium, they expressed pluripotency markers at levels similar to those of cells cultured on matrigel (Derda et al., 2010). These results indicate that this screening strategy is a productive avenue for generating new materials that control the growth and differentiation of cells.

The combined use of rational and library-based screening methods should provide an increasing number of ligands for the functionalization of synthetic systems, and may aid the mechanistic investigation of specific receptors and signaling events that regulate the responses of stem cells to their microenvironment.

2.2.2 Cytokines and growth factors

The ECM not only offers sites for cell adhesion, but it can also serve as a platform for the presentation of other biochemical factors that orchestrate cell-cell interactions. Whereas stem cell researchers have often investigated growth factors and cytokines as soluble factors, many of these proteins have matrix-binding domains that may enable them to be presented within the niche as "solid phase" ligands.

For example, several studies have immobilized LIF by various strategies to maintain ES cells in an undifferentiated state. An immobilizable fusion protein consisting of LIF and the IgG-Fc region, named LIF-FC, can maintain the ES cells in the undifferentiated state (Nagaoka et al., 2008). Similarly, a photo-immobilized LIF stimulates the activation of STAT3 for a longer time than does soluble LIF, and as a result, maintains ES cells in an undifferentiated state (Makino et al., 2004). In another study, both LIF and stem cell factor (SCF) were immobilized, and the threshold effects of these factors on stem cell maintenance were analyzed (Alberti et al., 2008). These studies demonstrated that using immobilized LIF reduces the need to add soluble LIF frequently to the medium.

Finally, the covalent binding of growth factors has proved to be helpful in controlling human ES cell growth and differentiation. Fibroblast growth factor (FGF)-2 immobilized on polyamide nanofibrillar surfaces inhibits the rapid degradation of FGF-2 in solution and supports the expansion and colony formation of human ES cells (Nur-E-Kamal et al., 2008). Another study demonstrated that the cultivation of mouse ES cells on surfaces with immobilized vascular endothelial growth factor-A (VEGF) yields primarily endothelial cells, whereas their cultivation on such surfaces without VEGF yields primarily vascular smooth muscle-like cells (Chiang et al., 2010).

2.3 Control of cells by geometric modification

Topographical structures such as grooves, ridges, and pits are present in many natural structures at the nanoscale level, as in the fibrous structure of collagen and other ECM proteins, and at the microscale level, as in the pores in bone marrow and the undulating basement membranes in the epidermis. The presence of topographical information in natural systems has motivated the use of technologies such as soft lithography, microfluidics, electrospinning, and the deposition of nanostructures (Khademhosseini et al., 2006; Pirone & Chen, 2004; Yang et al., 2005) to engineer substrate materials' topography to affect stem cell responses at both the nano and micro levels.

How cells sense topographical cues from the environment has been debated, but the cellular response to surface topographies is known to involve cytoskeletal changes and the modulation of focal adhesion formation (Lim & Donahue, 2007; Biggs et al., 2008). A recent study indicated that integrins may be involved in these cellular responses (Wood et al., 2008), suggesting that established adhesion signaling pathways are involved.

Little is known about the effect of artificial micro- and nanoscale topographical surfaces on the ES cell differentiation state. Recently, Markert et al. investigated the influence of topographical microstructures on the proliferation and differentiation of mouse ES cells. Their findings indicated that one class of microstructures sustains the feeder-free proliferation of undifferentiated ES cells and another class enforces differentiation, as indicated by the spreading of the cells (Markert et al., 2009). Murine ES cells cultured on electrospun polyamide nanofibers that mimic the basement membrane texture showed twice the cell expansion of those cultured on coverslips, while retaining their Nanog expression and differentiation potential (Nur-E-Kamal et al., 2006). Lee et al demonstrated that nanoscale ridge/groove pattern arrays alone can effectively and rapidly induce the differentiation of human ES cells into a neuronal lineage, without the use of any differentiation-inducing agents. They proposed that elongation of the cytoskeleton during the morphological changes in cells guided by ridge/groove patterns results in a transfer of tensional force to the nucleus, which influences gene expression and signal transduction (Lee et al., 2010). Similarly, another study demonstrated that mouse ES cells can be induced to differentiate into specific neural lineages, that is, neurons, oligodendrocytes, and astrocytes, when seeded onto electrospun fibrous scaffolds (Xie et al., 2009). In another study, the nanofibrous architecture of the substrate enhanced the osteogenic differentiation of human ES cells compared to a more traditional scaffolding architecture (Smith et al., 2010).

Thus, surface engineering approaches that alter the topographical structure of the substrate surface can be used to modulate ES cell behavior and fate.

2.4 Control of cells by modification of material mechanics

Of the many mechanical properties of biological systems, stiffness or rigidity is perhaps the most apparent and widely studied. Mechanical stiffness reflects a material's ability to store and frictionally dissipate applied mechanical energy, as reflected by storage (elastic) modulus and loss (viscous) modulus, respectively. The elastic modulus is the measure of the stress required to achieve a specific strain in a material without permanent deformation, and has emerged as an important regulator of stem cell function. Upon mechanical stimulation, cells convert mechanical signals into biochemical responses through a mechanism called, "mechano-transduction" (Orr et al., 2006). Cells interact with their surroundings via ECM receptors such as integrins and laminin receptors. Specifically, the ECM dynamics and matrix stiffness are translated into cytoskeletal tension mediated by integrin–ECM interactions (Katsumi et al., 2004). Integrin signaling is principally mediated by focal adhesion kinases, and the cell's responses to these signals can modulate a number of intracellular pathways that may cooperatively affect the activation SMADs, Rho GTPases, ERK, and other downstream signaling pathways that lead to transcriptional and epigenetic changes (Miyamoto et al., 1995). For example, integrin-mediated adhesion signaling cooperates with soluble-factor signaling to regulate Rho GTPases and generate actin cytoskeletal tension (Clark et al., 1998).

Recently, Blin et al used nanofilms made of poly(L-lysine) and hyaluronan (HA), named PLL/HA, which were cross-linked to various extents, to modulate the nanoenvironment of ES cells. The adhesion of ES cells to the nanofilms increased from the native film to the highly cross-linked films. The adhesion process was associated with cell proliferation. The dynamic balance of the ES cells between the ICM and the epiblast states was also dependent on the cross-linking of the nanofilms. The more cross-linked and thus stiffer the film was, the more cells were driven toward the epiblast fate. This finding suggests that the stiffness of the nanofilm can play a key role in modulating the ES cell niche to govern the ES cell self-renewal and fate (Blin et al., 2010).

Similarly, in another study, the behavior of ES cells grown on a flexible polydimethylsiloxane substrate of varying stiffness was examined. While cell attachment was unaffected by the stiffness of the growth substrate, both cell spreading and cell growth increased with increasing substrate stiffness. Moreover, several genes expressed in the primitive streak during gastrulation and implicated in early mesendodermal differentiation were upregulated in cell cultures on the stiffer substrates than on the softer ones. Finally, the osteogenic differentiation of ES cells was enhanced on stiff substrates compared to soft ones, demonstrating that the mechanical environment can play a role in both early and terminal ES cell differentiation (Evans et al., 2009).

3. Strategies for culture surface design using glucose-displaying dendrimer substrates

3.1 Surface design and characterization

A schematic illustration showing preparation of culture surfaces based on dendrimer substrates is shown in Fig. 1.

Starburst polyamidoamine (PAMAM) dendrimers are highly branched spherical polymers with well-defined structures and primary amino groups at their terminals. It is quite easy to modify the chemical properties of dendrimers by adjusting their terminal groups (Kawase et al., 2000; Tomalia et al., 2003). When an additional layer or generation is polymerized on the dendrimer molecules, the number of terminal amino groups is doubled. The defined dendrimer structure and large number of terminal amino groups allow great flexibility in the design variables, including the ligand species presented on the terminal groups, dendrimer size, and ligand density, making these polymers suitable for use as biocompatible nanometer-sized capsules in gene- or drug-delivery systems, as well as in scaffolds for cell culturing (Tomalia et al., 2003).

Dendrimers deposited on a solid surface have unique properties that yield physical and chemical variations in the surface; these properties are also affected by the ligand species and amounts displayed on the dendrimers, and the locations of the displayed ligands.

Surfaces with different topographies can be obtained by changing the dendrimer density and generation number, as illustrated in Fig. 1. In addition, dendrimers can offer extended design parameters, such as an altered ligand ratio of D- to L-glucose isomers (termed one-ligand display for cell anchoring) or the co-display of an adhesive ligand (D-glucose) and a functional ligand (e.g., growth factor) on the surface (termed multi-ligand display for cell anchoring and stimulation) (Fig. 1).

D-Glucose molecules on the culture surface and glucose transporters (GLUTs) on the cytoplasmic membrane are assumed to function as binding and receptor sites, respectively. GLUTs show sharp specificity in their binding affinity for glucose isomers: they exhibit high affinities for D-glucose but extremely low affinities for L-glucose. D-Glucose itself does not induce cell signaling. However, it is likely that such high-affinity GLUTs can act as a cell-anchoring mechanism by binding D-glucose molecules displayed on the surface.

Evidence suggests that the nanoscale geometry of dendrimer substrates plays crucial roles in determining cellular responses to the substrate. The generation number of dendrimers and their density yield varying, cell-specific responses. Kim et al characterized various dendrimer-immobilized surfaces with different architectures in terms of their surface roughness using an atomic force microscope, and found their mean roughness to range from 1.8-11.0 nm. The combination of displayed D-glucose and roughness promoted cytoskeletal formation, accompanied by the elongation of cells on the culture surface. The authors concluded that a dendrimer substrate with a D-glucose display offers a solid environment that permits the partial anchoring of the cells via the temporarily grasping of the GLUTs by D-glucose (Kim et al., 2007a).

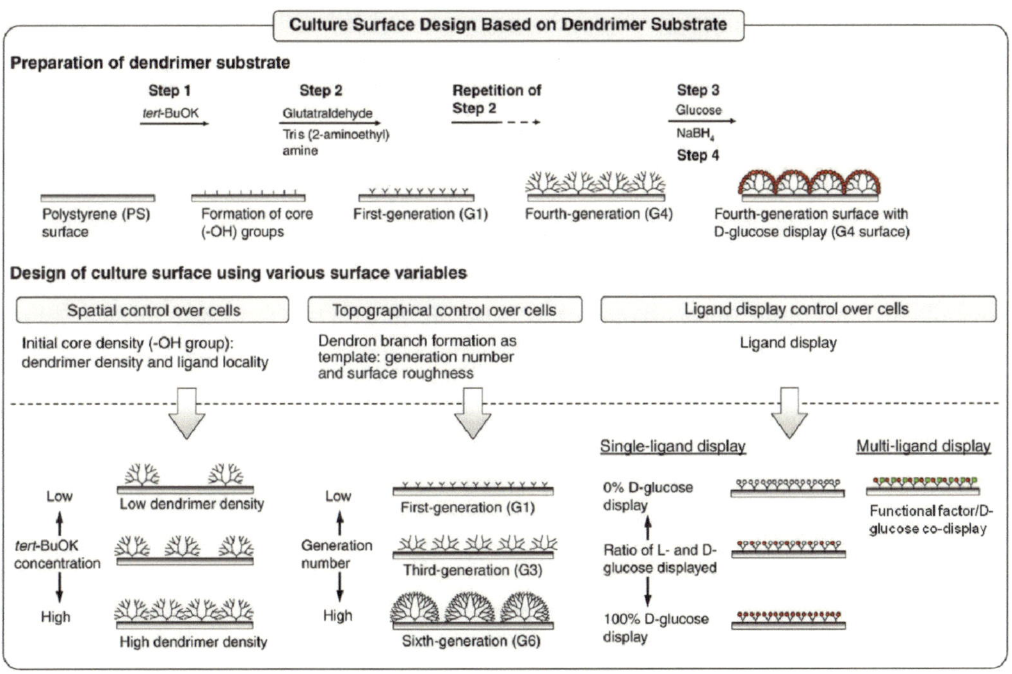

Fig. 1. Schematic illustrations showing the preparation of culture surfaces based on dendrimer substrates (reproduced with permission from Kim et al., 2010a)

In another study, an extended substrate design with improved cell anchoring and migration using the concurrent display of D-glucose and EGF was reported (Kim et al., 2007b). The displayed D-glucose molecules permit the cells to be in close contact with the surface via the grasping of GLUTs on the cytoplasmic membrane, thereby leading to increased focal contacts that can induce the up-regulation of EGF receptor signaling. This study used an advanced design to target cells by plating them on dendrimer-immobilized substrates that strongly stimulated cell behaviors.

These studies demonstrated the potential for dendrimer-immobilized surfaces to regulate cell morphology and subsequently cell functions, via morphologic priming. Recent strategies and concepts for culture surface designs based on cell anchoring mechanisms, and using glucose-displaying dendrimer substrates to regulate cell morphology and function, are reviewed elsewhere (Kim et al., 2010a). In the next section, we will describe the morphological and functional responses of mouse ES cells cultured on a D-glucose-displaying dendrimer (GLU/D) surface.

3.2 Enrichment of undifferentiated mouse ES cells on dendrimer-immobilized surface

ES cells are pluripotent cells that are characterized by their ability to propagate indefinitely in culture as undifferentiated cells with a normal karyotype, and to differentiate into derivatives of the three primary germ layers. Although ES cells are expected to serve as an unlimited cell source for cell-transplantation therapy, great care is required to maintain undifferentiated ES cell cultures, since the cells can spontaneously differentiate via seemingly random pathways under normal ES cell culture conditions, especially in the course of expanding the colony density and size (Watt & Hogan, 2000). Therefore, cultured ES cells may develop into colonies of heterogeneous cell types that include cells with less pluripotency. Our group has been investigating the possibility of using the dendrimer surface as a tool for obtaining cell preparations enriched in undifferentiated ES cells (Mashayekhan et al., 2008).

Here we present our results showing the enrichment of undifferentiated ES cells by serial passaging on a fourth-generation GLU/D surface. The morphologies of the ES single cells as well as the ES cell colonies on different culture surfaces were compared as indicated in Fig. 2. The single-cell observation on day 1 showed that most of the cells on the GLU/D surface were round, while those on the gelatinized surface exclusively showed a stretched morphology (Fig. 2 A, B). Moreover, the cells on the GLU/D surface formed loosely attached spherical colonies, while those on the conventional surface formed flatter colonies that were firmly attached to the surface (Fig. 2 C, D).

Time-lapse observations showed that on the gelatinized surface, the cells started to divide while spreading, and they experienced contact inhibition upon becoming confluent on the surface, resulting in the formation of dome-shaped colonies. In contrast, the cells on the GLU/D surface made spherical colonies as they divided, probably because of the increased frequency of cell-cell contacts. As shown in Fig. 2E, the outermost layer of the spherical colonies near the GLU/D surface consisted of much fewer cells than in the colonies (either flat or compact) on the gelatinized surface, which can explain the difference in the colonies' attachment strength to the two surfaces.

Cell morphology is one of the most important parameters in the regulation of stem cell growth and differentiation, and is determined through signaling that reorganizes the actin cytoskeleton. The cytoskeleton is implicated in mechanotransduction, since it links the stimulation from an extracellular environment (e.g., solid surface) with an intracellular signaling mechanism that regulates cell functions. Cellular mechanotransduction requires the rearrangement of membrane constituents, focal contact formation, and an association with a dynamic actin cytoskeleton and Rho family GTPase-mediated signal pathways, which have emerged as key regulators of cadherin-mediated cell-cell adhesion (Fukata & Kaibuchi, 2001).

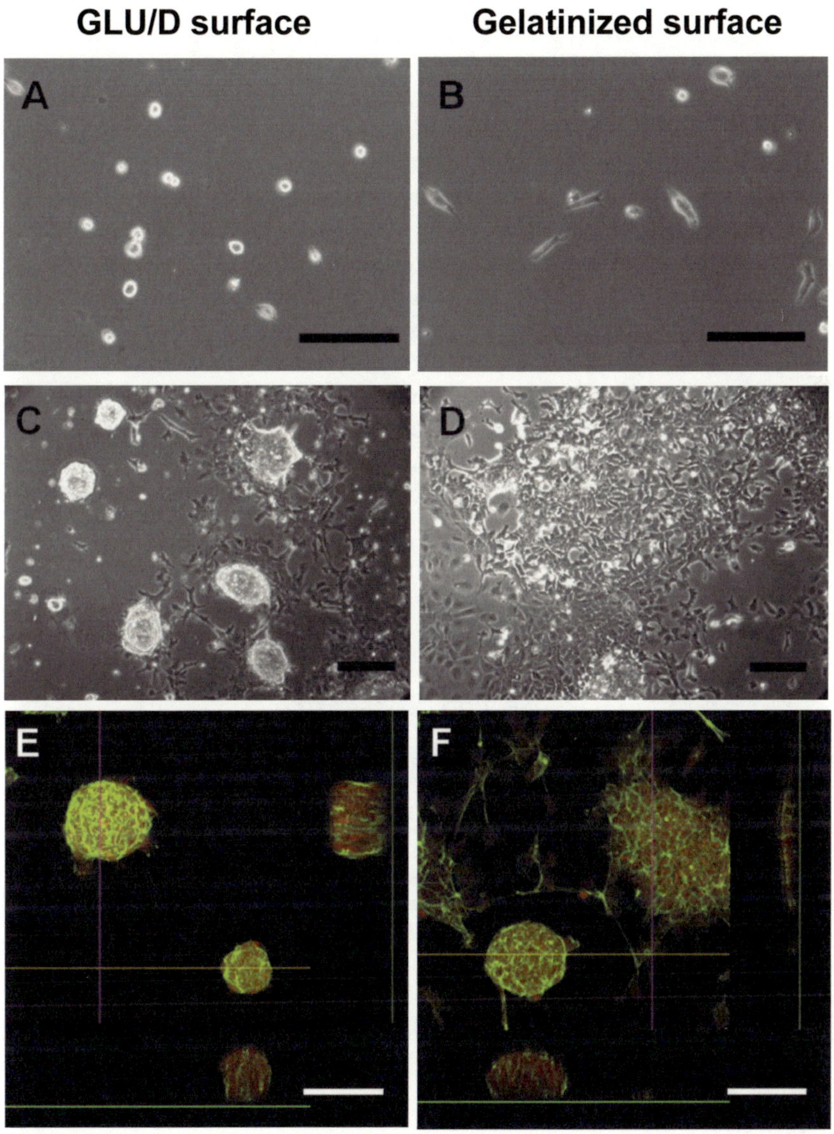

Fig. 2. Morphology of ES cells on different surfaces. ES cell colonies in A and B are shown 1 day after seeding, and those in C, D, E, and F are shown 4 days after seeding. The images at the bottom and right sides in E and F show the tomograms sectioned at the x-z (yellow line) and y-z (pink line) planes, respectively. The scale bars represent 100 μm

Kim et al. suggested that dendrimer-immobilized surfaces with a D-glucose display can induce a moderate activation of Rho family GTPases during the induced migration of rabbit chondrocytes. The activated Rho family GTPases can consequently promote cell-cell interactions via N-cadherin-mediated adhesion during cell aggregation to facilitate the development of chondrogenic phenotypes (Kim et al., 2009). Moreover, Kim et al. observed the spatiotemporal activation of N-cadherin expression when they altered the Rho family GTPase activity in human mesenchymal stem (hMS) cells by plating them on a GLU/D surface; this change promoted the formation of cell aggregates, which in turn directed hMS cell differentiation toward a cardiomyocyte phenotype (Kim et al., 2010b).
Recent studies showed that the morphology of single cells and of loosely attached spherical colonies of ES cells on a fourth-generation GLU/D surface were similar to those observed in hMS cells cultured on a fifth-generation GLU/D surface. Moreover, examination of the cytoskeletal and focal adhesion formation revealed that the development of stress fibers and vinculin plaques was suppressed for both ES and hMS cells cultured on GLU/D surfaces (Mashayekhan et al, 2008; Kim et al., 2010b). Although the detailed mechanism for the formation of ES cell aggregates on GLU/D is still unclear, we suggest that the mounded shape of the cell clusters that forms on dendrimer-immobilized surfaces promotes the expression of E-cadherin, a crucial cell-cell adhesion element in ES cells (Larue et al., 1996), which leads to the formation of spherical colonies.
Since the majority of colonies that formed on the GLU/D surface showed a morphology typical of undifferentiated cells (round and compact colonies with poorly delineated cell-cell borders), and were loosely attached to the surface, we tested whether preparations enriched in undifferentiated ES cells could be obtained by performing several passages of the cells on the GLU/D surface.

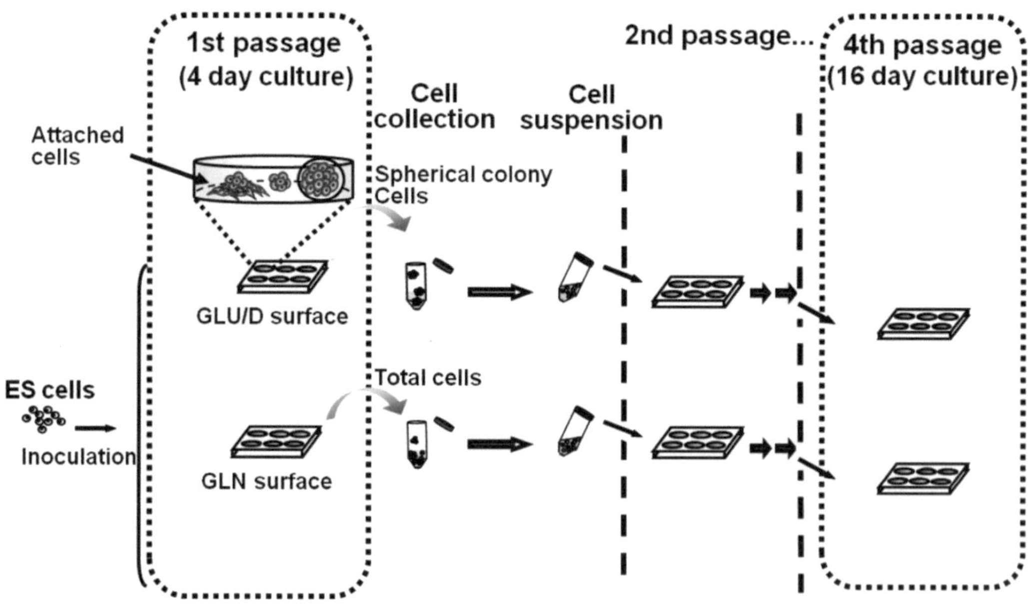

Fig. 3. Passaging protocol for the enrichment of ES cells in the undifferentiated state

The passaging protocol is illustrated in Fig. 3. The spherical colonies that were loosely attached to the GLU/D surface were harvested by tapping on day 4, dissociated into single cells by trypsin/EDTA treatment, and replated. These procedures were repeated every 4 days. For comparison, ES cells were also cultured on a gelatinized surface; in this case, the entire cell population was collected on day 4, and subjected to the enzymatic treatment for replating. The differentiation states of the cells cultured on the different surfaces were then compared by alkaline phosphatase (ALP) staining and gene expression analysis. For the ALP analysis, the spherical colony cells grown on the GLU/D surface and the cells grown on the gelatinized surface at passages 1 (4-day culture) and 4 (16-day total culture) were harvested, trypsinized, and replated onto gelatinized plates.

During the long-term passaging, the frequency of colonies with a spherical shape and the ALP activity of the spherical colony cells grown on the GLU/D surface increased gradually with the number of passages. Moreover, at passage 4, the percentage of ALP-positive colonies was significantly greater on the GLU/D surface than that on the gelatinized surface (Mashayekhan et al., 2008).

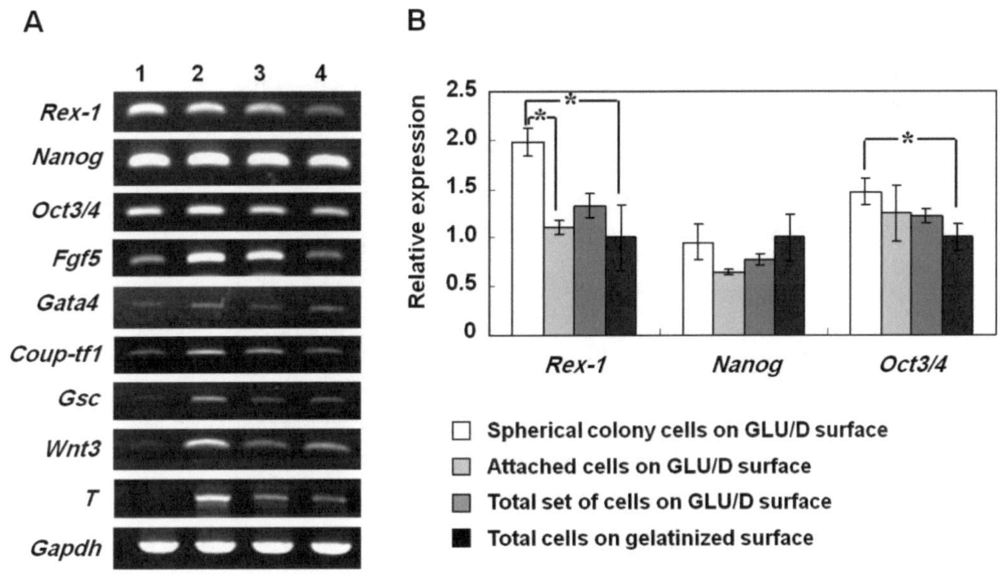

Fig. 4. RT-PCR analysis of ES cells cultured on different surfaces after four passages. (A) Conventional RT-PCR analysis for three markers of undifferentiated stem cells (Rex-1, Nanog, and Oct3/4) and six markers of early differentiation (Fgf5, Gata4, Coup-tf1, Gsc, Wnt3, and T). Lanes 1, 2, 3, and 4 correspond to the spherical colony cells, attached cells, and total cells cultured on the GLU/D surface, and the total cells collected from the gelatinized surface, respectively. (B) Quantitative RT-PCR analyses for the three stem cell markers (Rex-1, Nanog, and Oct3/4) in ES cells cultured on the GLU/D or gelatinized surface. The data were obtained from three independent experiments. The vertical bars show the standard deviation (*$p < 0.05$)

As shown in Fig. 4, we performed RT-PCR on cells from the GLU/D cultures, separating them into three groups: spherical colony cells, cells that remained attached to the surface after tapping, and cells belonging to both types of colonies. We used cells grown on a gelatinized surface for comparisons. First we found that the markers for undifferentiated cells, Rex-1 and Oct3/4, were more highly expressed in the spherical colony cells grown on the GLU/D surface than in the other sets of cells or those grown on the gelatinized surface. Quantitative RT-PCR analysis confirmed that the cells from the spherical colonies on the GLU/D surface had higher expression levels of Rex-1 and Oct3/4 than the other cells. We also tested the different cell groups for the expression of early differentiation markers, by conventional RT-PCR. We found that early endodermal (Gata4), mesendodermal (Gsc), and mesodermal (T and Wnt3) differentiation markers were expressed at lower levels in the spherical colony cells from the GLU/D surface than in the attached cells or those grown on the gelatinized surface. Among all the cells of the different states tested, the expression levels of all the early differentiation markers were highest in the cells that were attached to the GLU/D surface, which appeared as flattened colonies. In contrast, the markers of undifferentiated cells, Rex-1 and Nanog, were expressed at significantly lower levels in the attached cells than in the spherical colony cells cultured on the GLU/D surface (Fig. 4). Thus, ES cells exhibiting various degrees of differentiation existed on the GLU/D surface, in a localized or enriched manner.

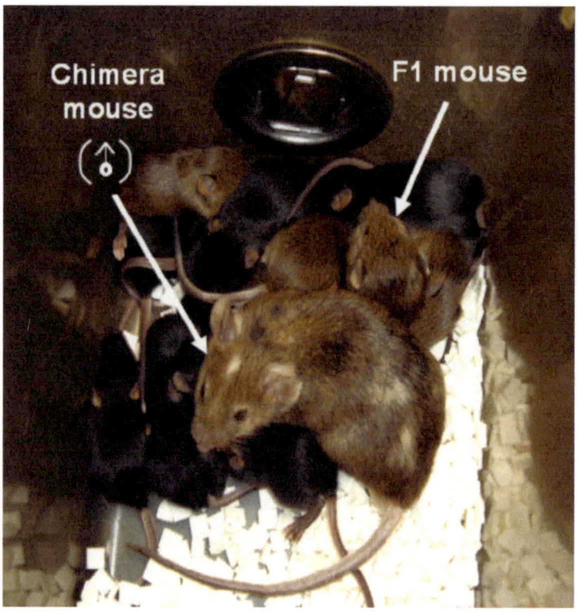

Fig. 5. Chimeric mice generated by blastocyte injection of ES cells cultured on the GLU/D surface. Dissociated ES cells from the spherical colonies on the GLU/D surface were injected into the blastocysts of C57BL/6J mice. The blastocysts were then transferred into the uteri of pseudopregnant MCH/ICR female mice. The resultant chimeric males with a white/agouti coat color ratio greater than 50% were bred with C57B/6J females to test for germ-line transmission

Overall, our RT-PCR analysis revealed that the markers for the undifferentiated state and for early differentiation were expressed at higher and lower levels, respectively, in the spherical colony cells passaged on the GLU/D surface, than in the cells grown on gelatinized surface. These results support the view that the GLU/D surface is more effective than the gelatinized surface for maintaining ES cells in an undifferentiated state. In the flattened colonies on the GLU/D surface, all the markers of early differentiation were detected at much higher levels than in the cells from the spherical colonies. Thus, by using the proposed protocol of serial passaging on the GLU/D surface, which excluded the cells with a relatively stretched shape and flattened colonies and selectively transferred the loosely attached spherical colony cells to the next passage, the ES cells could be maintained in the undifferentiated state.

Finally, to confirm that the pluripotency of the ES cells grown on the GLU/D surface was maintained, we generated chimeric mice and checked the germ-line transmissibility of these cells. The ES cells were passaged four times on the GLU/D surface, and the spherical colonies were then dissociated into single cells prior to blastocyst injection. Among 43 progeny mice, 16 had the agouti coat color, indicating successful germ-line transmission, as typically shown in Fig. 5.

Differentiated cells are known to appear spontaneously on a gelatinized surface even in a complete ES medium containing serum, and the expression of mesodermal and extra-embryonic marker genes is slightly up-regulated under these conditions, due to the activation of integrin signaling, which is known to inhibit mouse ES cell self-renewal by increasing the LIF-induced activation of ERK1/2 (Hayashi et al., 2007). Considering the difficulty in culturing undifferentiated mouse ES cells without feeder cells in serum-containing medium, the GLU/D surface used in this study may be a useful biomaterial for culturing mouse ES cells. In the case of human ES cells, it is especially desirable to exclude foreign culture components like feeder cells and nonhuman-derived serum, to minimize the risk of pathogens such as retroviruses in therapeutic applications (Beattie et al., 2005; Amit & Itskovitz-Eldor, 2006; Chin et al., 2007). In this context, the application of the dendrimer-immobilized surface is a promising novel strategy for overcoming the difficulties in propagating human ES cells.

4. Conclusions and outlook for the future

The current chapter described general strategies for designing culture surfaces to control the morphology and function of ES cells. In addition, we introduced our approach to designing a culture surface using dendrimer substrates displaying D-glucose as a ligand to enrich the undifferentiated state of ES cells. The results suggest that the GLU/D surface is a potential tool for changing both the topography and the biochemistry of the surface, which play key roles in modulating the niche of ES cells and in turn govern their morphology and fate.

Although ES cells are potentially powerful tools in therapeutic applications for tissue regeneration, we still have little understanding of the microenvironment-specified molecular mechanisms and signaling pathways that lead to their efficient differentiation and to tissue formation. Identifying specific cues in the microenvironment and understanding how neighboring cells and the ECM control developmental fates will be required to promote the differentiation of ES cells into targeted cell lineages. As bioengineers learn more about how the microenvironment directs stem cell fate decisions, these factors can be incorporated into the culture conditions to better control ES cell growth and differentiation.

In general, the knowledge garnered using engineered systems will advance stem cell biology and provide prototypes for tissue engineering and strategies for therapeutics.

5. References

Alberti K.; Davey RE.; Onishi K.; George S.; Salchert K.; Seib FP.; Bornhäuser M.; Pompe T.; Nagy A.; Werner C. & Zandstra PW. (2008) Functional immobilization of signaling proteins enables control of stem cell fate. *Nat. Methods*, 5(7), 645-50.

Amit M. & Itskovitz-Eldor J. (2006) Feeder-free culture of human embryonic stem cells. *Methods Enzymol*, 420:37–49.

Anderson DG.; Levenberg S. & Langer R. (2004) Nanoliter-scale synthesis of arrayed biomaterials and application to human embryonic stem cells. *Nat. Biotechnol.* 22(7), 863-66.

Beattie GM.; Lopez AD.; Bucay N.; Hinton A.; Firpo MT.; King CC. & Hayek A. (2005) Activin A maintains pluripotency of human embryonic stem cells in the absence of feeder layers. *Stem Cells*, 23(4):489–95.

Biggs MJ.; Richards RG.; McFarlane S.; Wilkinson CD.; Oreffo RO. & Dalby MJ. (2008). Adhesion formation of primary human osteoblasts and the functional response of mesenchymal stem cells to 330 nm deep microgrooves. *J R Soc Interface*, 5(27):1231–42.

Blin G.; Lablack N.; Louis-Tisserand M.; Nicolas C.; Picart C. & Pucéat M. (2010) Nano-scale control of cellular environment to drive embryonic stem cells selfrenewal and fate. *Biomaterials*, 31(7):1742-50.

Brons IG.; Smithers LE.; Trotter MW.; Rugg-Gunn P.; Sun B.; Chuva de Sousa Lopes SM.; Howlett SK.; Clarkson A.; Ahrlund-Richter L.; Pedersen RA. & Vallier L. (2007) Derivation of pluripotent epiblast stem cells from mammalian embryos. *Nature*, 448 (7150):191–5.

Chiang CK.; Chowdhury MF.; Iyer RK.; Stanford WL. & Radisic M. (2010) Engineering surfaces for site-specific vascular differentiation of mouse embryonic stem cells. *Acta Biomater.* 6(6):1904-16.

Chin AC.; Fong WJ.; Goh LT.; Philp R.; Oh SK. & Choo AB. (2007) Identification of proteins from feeder conditioned medium that support human embryonic stem cells. *J Biotechnol.*, 130(3):320–8.

Clark EA.; King WG.; Brugge JS.; Symons M. & Hynes RO. (1998) Integrin-mediated signals regulated by members of the rho family of GTPases. *J Cell Biol.*, 142(2):573–586.

Cui L.; Johkura K.; Yue F.; Ogiwara N.; Okouchi Y.; Asanuma K. & Sasaki K. (2004) Spatial distribution and initial changes of SSEA-1 and other cell adhesion-related molecules on mouse embryonic stem cells before and during differentiation. *J Histochem Cytochem*, 52 (11):1447–57.

Dellatore SM.; Garcia AS. & Miller WM. (2008) Mimicking stem cell niches to increase stem cell expansion. *Curr Opin Biotechnol.*, 19(5):534-40.

Derda R.; Li LY.; Orner BP.; Lewis RL.; Thomson JA. & Kiessling LL. (2007) Defined substrates for human embryonic stem cell growth identified from surface arrays. *ACS Chem Biol.*, 2(5):347-55.

Derda R.; Musah S.; Orner BP.; Klim JR.; Li L. & Kiessling LL. (2010) High-Throughput Discovery of Synthetic Surfaces That Support Proliferation of Pluripotent Cells. *J Am Chem Soc.*, 132(4):1289-95.

Discher DE.; Mooney DJ. & Zandstra PW. (2009) Growth factors, matrices, and forces combine and control stem cells. *Science*, 324 (5935):1673–7.

Doran MR.; Frith JE.; Prowse AB.; Fitzpatrick J.; Wolvetang EJ.; Munro TP.; Gray PP. & Cooper-White JJ. (2010) Defined high protein content surfaces for stem cell culture. *Biomaterials*, 31(19):5137-42.

Evans ND.; Minelli C.; Gentleman E.; LaPointe V.; Patankar SN.; Kallivretaki M.; Chen X.; Roberts CJ. & Stevens MM. (2009) Substrate stiffness affects early differentiation events in embryonic stem cells. *Eur Cell Mater.*, 18:1-13.

Evans ND.; Gentleman E.; Chen X.; Roberts CJ.; Polak JM. & Stevens MM. (2010) Extracellular matrix-mediated osteogenic differentiation of murine embryonic stem cells. *Biomaterials*, 31(12):3244-52.

Flaim C J.; Teng D.; Chien S. & Bhatia S. N. (2008) Combinatorial signaling microenvironments for studying stem cell fate. *Stem Cells Dev.*, 17(1), 29–39.

Fukata M. & Kaibuchi K. (2001) Rho-family GTPases in cadherin-mediated cell-cell adhesion, *Nat Rev Mol Cell Biol.*, 2(12), 887-97.

Goetz AK.; Scheffler B.; Chen HX.; Wang S.; Suslov O.; Xiang H.; Brüstle O.; Roper SN. & Steindler DA. (2006) Temporally restricted substrate interactions direct fate and specification of neural precursors derived from embryonic stem cells. *Proc Natl Acad Sci U S A.*, 103(29):11063-8.

Hayashi K.; Lopes SM.; Tang F. & Surani MA. (2008) Dynamic equilibrium and heterogeneity of mouse pluripotent stem cells with distinct functional and epigenetic states. *Cell Stem Cell*, 3(4):391–401.

Hayashi Y.; Furue MK.; Okamoto T.; Ohnuma K.; Myoishi Y. & Fukuhara Y. (2007) Integrins regulate mouse embryonic stem cell self renewal. *Stem Cells*, 25(12):3005–15.

Hersel U.; Dahmen C. & Kessler H. (2003) RGD modified polymers: biomaterials for stimulated cell adhesion and beyond. *Biomaterials*, 24 (24), 4385-415.

Hwang NS.; Varghese S.; Zhang Z. & Elisseeff J. (2006) Chondrogenic differentiation of human embryonic stem cell-derived cells in arginine-glycine-aspartate-modified hydrogels. *Tissue Eng.*, 12(9):2695–706.

Katsumi A.; Orr AW.; Tzima E. & Schwartz MA. (2004) Integrins in mechanotransduction. *J Biol Chem.*, 279(13):12001–4.

Kawase M.; Kurikawa N.; Miura N.; Shiomi T.; Ozawa C.; Higashiyama S.; Mizoguchi T. & Yagi K. (2000) Immobilization of ligand-modified polyamidoamine dendrimer for cultivation of hepatoma cells. *Artif Org.*; 24(1):18–22.

Keung AJ.; Kumar S. & Schaffer DV. (2010) Presentation counts: microenvironmental regulation of stem cells by biophysical and material cues. *Annu Rev Cell Dev Biol.*, 26:533-56.

Khademhosseini A.; Langer R.; Borenstein J. & Vacanti JP. (2006) Microscale technologies for tissue engineering and biology. *Proc. Natl. Acad. Sci. USA*, 103(8):2480–87.

Kim MH.; Kino-oka M.; Kawase M.; Yagi K. & Taya M. (2007a) Response of human epithelial cells to culture surfaces with varied roughness prepared by immobilizing dendrimer with/without D-glucose display. *J Biosci Bioeng.*, 103(2):192–9.

Kim MH.; Kino-oka M.; Kawase M.; Yagi K. & Taya M. (2007b) Synergistic effect of D-glucose and epidermal growth factor display on dynamic behaviors of human epithelial cells. *J. Biosci. Bioeng.*, 104(5):428–31.

Kim MH.; Kino-oka M.; Morinaga Y.; Sawada Y.; Kawase M.; Yagi K. & Taya M. (2009) Morphological regulation and aggregate formation of rabbit chondrocytes on dendrimer-immobilized surfaces with D-glucose display. *J. Biosci. Bioeng.* 107(2), 196–205.

Kim MH.; Kino-oka M. & Taya M. (2010a) Designing culture surfaces based on cell anchoring mechanisms to regulate cell morphologies and functions. *Biotechnol Adv.*, 28(1):7-16.

Kim MH.; Kino-oka M.; Maruyama N.; Saito A.; Sawa Y.; & Taya M. (2010b) Cardiomyogenic Induction Of Human Mesenchymal Stem Cells By Altered Rho Family GTPase Expression On Dendrimer-Immobilized Surface With D-Glucose Display. *Biomaterials*, 31(30):7666-77.

Kolhar P.; Kotamraju VR.; Hikita ST.; Clegg DO. & Ruoslahti E. (2010) Synthetic surfaces for human embryonic stem cell culture. *J Biotechnol.*, 146(3):143-6.

Larue L.; Antos C.; Butz S.; Huber O.; Delmas V.; Dominis M, & Kemler R. (1996) A role for cadherin in tissue formation. *Development.*, 122(10):3185-94.

Lee MR,; Kwon KW.; Jung H.; Kim HN.; Suh KY.; Kim K. & Kim KS. (2010) Direct differentiation of human embryonic stem cells into selective neurons on nanoscale ridge/groove pattern arrays. *Biomaterials*, 31(15):4360-6.

Lim JY. & Donahue HJ. (2007). Cell sensing and response to micro- and nanostructured surfaces produced by chemical and topographic patterning. *Tissue Eng.*, 13(8):1879–91.

Mahlstedt MM.; Anderson D.; Sharp JS.; McGilvray R.; Muñoz MD.; Buttery LD.; Alexander MR.; Rose FR. & Denning C. (2010) Maintenance of pluripotency in human embryonic stem cells cultured on a synthetic substrate in conditioned medium. *Biotechnol Bioeng.* 105(1):130-40.

Makino H.; Hasuda H. & Ito Y. (2004) Immobilization of Leukemia Inhibitory Factor (LIF) to culture murine embryonic stem cells. *J Biosci Bioeng.*, 98(5):374-9.

Markert LD.; Lovmand J.; Foss M.; Lauridsen RH.; Lovmand M.; Füchtbauer EM.; Füchtbauer A.; Wertz K.; Besenbacher F.; Pedersen FS. & Duch M. (2009) Identification of Distinct Topographical Surface Microstructures Favoring Either Undifferentiated Expansion or Differentiation of Murine Embryonic Stem Cells. *Stem Cells Dev.*,18(9):1331-42.

Mashayekhan S.; Kim MH.; Miyazaki S.; Tashiro F.; Kino-oka M.; Taya M. & Miyazaki J. (2008) Enrichment of undifferentiated mouse embryonic stem cells on a culture surface with a glucose displaying dendrimer. *Biomaterials*,29(31):4236-43.

McDevitt T. & Palecek S. (2008). Innovation in the culture and derivation of pluripotent human stem cells. *Curr Opin Biotechnol.*, 19(5):527-33.

Meng Y.; Eshghi S.; Li YJ.; Schmidt R.; Schaffer DV. & Healy KE. (2010) Characterization of integrin engagement during defined human embryonic stem cell culture. *FASEB J.* 24(4), 1056–65.

Miyamoto S.; Teramoto H.; Coso OA.; Gutkind JS.; Burbelo PD.; Akiyama SK. & Yamada KM. (1995) Integrin function: molecular hierarchies of cytoskeletal and signaling molecules. *J Cell Biol.*,131(3):791-805.

Nagaoka M.; Koshimizu U.; Yuasa S.; Hattori F.; Chen H.; Tanaka T.; Okabe M.; Fukuda K. & Akaike T. (2006) E-Cadherin-Coated Plates Maintain Pluripotent ES Cells without Colony Formation. *PLoS One.*, 20;1:e15.

Nagaoka M.; Hagiwara Y.; Takemura K.; Murakami Y.; Li J.; Duncan SA. & Akaike T. (2008) Design of the Artificial Acellular Feeder Layer for the Efficient Propagation of Mouse Embryonic Stem Cells. *J Biol Chem.*, 283(39):26468-76.

Nur-E-Kamal A.; Ahmed I.; Kamal J.; Schindler M. & Meiners S. (2006) Three dimensional nanofibrillar surfaces promote self renewal in mouse embryonic stem cells. *Stem Cells*, 24(2):426-33.

Nur-E-Kamal A.; Ahmed I.; Kamal J.; Babu AN.; Schindler M. & Meiners S. (2008) Covalently attached FGF-2 to three-dimensional polyamide nanofibrillar surfaces demonstrates enhanced biological stability and activity. *Mol Cell Biochem.* 309(1-2):157-66.

Orr AW.; Helmke BP.; Blackman BR. & Schwartz MA. (2006) Mechanisms of mechanotransduction. *Dev Cell*, 10(1):11–20.

Pelton TA.; Sharma S.; Schulz TC.; Rathjen J. & Rathjen PD. (2002) Transient pluripotent cell populations during primitive ectoderm formation: correlation of in vivo and in vitro pluripotent cell development. *Journal of Cell Science*, 115 (Pt 2):329–39.

Pirone DM. & Chen CS. (2004) Strategies for engineering the adhesive microenvironment. *J. Mammary Gland Biol. Neoplasia* 9(4):405–17.

Saha K.; Pollock JF.; Schaffer DV. & Healy KE. (2007) Designing synthetic materials to control stem cell phenotype. *Curr Opin Chem Biol.*, 11(4):381-7.

Smith LA.; Liu X.; Hu J. & Ma PX. (2010) The Enhancement of human embryonic stem cell osteogenic differentiation with nano-fibrous scaffolding. *Biomaterials*, 31(21):5526-35.

Stewart MH.; Bendall SC. & Bhatia M. (2008) Deconstructing human embryonic stem cell cultures: niche regulation of self-renewal and pluripotency. *J Mol Med.* 86(8), 875-86.

Tomalia DA.; Huang B.; Swanson DR.; Brothers HM. & Klimash JW. (2003) Structure control within poly (amidoamine) dendrimers: size, shape and region chemical mimicry of globular proteins. *Tetrahedron*, 59(22):3799–813.

Toyooka Y.; Shimosato D.; Murakami K.; Takahashi K. & Niwa H. (2008) Identification and characterization of subpopulations in undifferentiated ES cell culture. *Development*, 135(5):909–18.

Valamehr B.; Jonas SJ.; Polleux J.; Qiao R.; Guo S.; Gschweng EH.; Stiles B.; Kam K.; Luo TJ.; Witte ON.; Liu X.; Dunn B. & Wu H. (2008) Hydrophobic surfaces for enhanced differentiation of embryonic stem cell-derived embryoid bodies. *Acad. Sci. USA*, 105(38), 14459-64.

Watt FM. & Hogan BL. (2000) Out of Eden: stem cells and their niches. *Science*, 287(5457):1427–30.

Wells N.; Baxter MA.; Turnbull JE.; Murray P.; Edgar D.; Parry KL.; Steele DA. & Short RD. (2009) The geometric control of E14 and R1 mouse embryonic stem cell pluripotency by plasma polymer surface chemical gradients. *Biomaterials*, 30(6), 1066-70.

Wood MA.; Bagnaninchi P. & Dalby MJ. (2008) The [beta] integrins and cytoskeletal nanoimprinting. *Exp Cell Res.*, 314(4):927-35.

Xie J.; Willerth SM.; Li X.; Macewan MR.; Rader A.; Sakiyama-Elbert SE. & Xia Y. (2009) The differentiation of embryonic stem cells seeded on electrospun nanofibers into neural lineages. *Biomaterials*, 30(3):354-62.

Yang F.; Murugan R.; Wang S. & Ramakrishna S. (2005) Electrospinning of nano/micro scale poly(L-lactic acid) aligned fibers and their potential in neural tissue engineering. *Biomaterials*, 26(15):2603-10.

7

Preservation of Embryonic Stem Cells

Xiaoming He, Ph.D.
Department of Mechanical Engineering and
Biomedical Engineering program
University of South Carolina, Columbia, SC
United States of America

1. Introduction

With recent advances in tissue engineering, regenerative medicine, cell transplantation, stem cell therapy, and assisted reproduction, the living cell is becoming increasingly important as a tool for drug screening and therapy in modern medicine (Gearhart 1998; Langer and Vacanti 1993). As a result of their capability of differentiating into any type of cells, the pluripotent embryonic stem (ES) cells are of particular importance to the modern cell-based medicine (Gearhart 1998). However, ES cells may differentiate gradually during passaging when cultured at 37 °C. Therefore, for the eventual success of using ES cells in the emerging cell-based medicine, it is of great importance to maintain their pluripotency in the long term without passaging and in a cost effective way so that the cells can be widely distributed and readily available to end users in both research and clinical settings. This can be done by cell preservation to put the cells in a state of suspended animation, which can be achieved by either cooling the cells to preserve (cryopreservation) at a cryogenic temperature and/or drying the cells to preserve (lyopreservation) at ambient temperature (Acker 2004; Blow 2009; Coger and Toner 2000; Toner and Kocsis 2002). In either case, the cells must enter (before being damaged) an amorphous (or glassy) phase, a thermodynamically metastable state with an extremely high viscosity and low molecular mobility and activity to arrest any biophysical and biochemical activities within the cells. Although contemporary methods for cell prservation still rely on the use of cryogenic tempertaure (cryopreservation), cell lyopreservation at ambient temperature is gaining more and more attention, due to the relatively high cost of maintaining and difficulty of transporting cryopreserved cells in cryogenic fluids such as liquid nitrogen (Acker 2004; Blow 2009; Deb 2009; Kanias and Acker 2006; Meyers 2006). In this chapter, the fundamentals and recent advancement of both cryo and lyopreservation are first summarized, followed by a critical review of the progress and challenges in applying the various cell preservation strategies to maintain the pluripotent properties of embryonic stem cells in the long term. This chapter is concluded with an outlook of the future directions of embryonic stem cell preservation.

2. Cryopreservation at cryogenic temperatures

Cell cryopreservation can be achieved by either slow-freezing or vitrification (Coger and Toner 2000; Fahy et al. 1984; Mazur 1984; Rall and Fahy 1985). The former relies on the formation of extracellular ice (the crystalized state of water) to freeze concentrate the extracellular solution in the presence of a low concentration (< ~ 2 M) of cryoprotectants (or cryoprotective agents, or CPAs for short) such as glycerol, ethylene glycol, PROH (1,2-propanediol or propylene glycol), and DMSO (dimethylsulfoxide). As a result, the cells gradually dehydrate due to osmosis during freeezing and enter a glassy phase suitable for long-term storage. While slow-freezing is to vitrify what is within the cells in the presence of extracellular ice, the goal of vitirfication is to have both intra and extracellular water enter the amorphous, glassy phase to eliminate any potential damaging (either mechanical or physicochemical) effect of ice foramtion. Conventionally, vitrification is achieved by using an unusually high concentration (up to 7 M) of CPAs, which can cause significant metabolic and osmotic injury to living cells. As a result, various approaches have been investigated to achieve vitrification of living cells at a low concentration of CPAs such as that used for slow-freezing, which is called low-CPA vitrification. The three approaches mentioned above for cell cryopreservation can be best demonstrated in an extended phase diagram of temperature versus solute concentration (Fig. 1), which consists of four regimens representing four probable phases (liquid, subcooled liquid, supersaturated liquid, and the solid-like amorphous or glassy phase) of an aqueous solution. The four lines that separate the four regimens are called liquidus (between liquid and subcooled liquid), extended

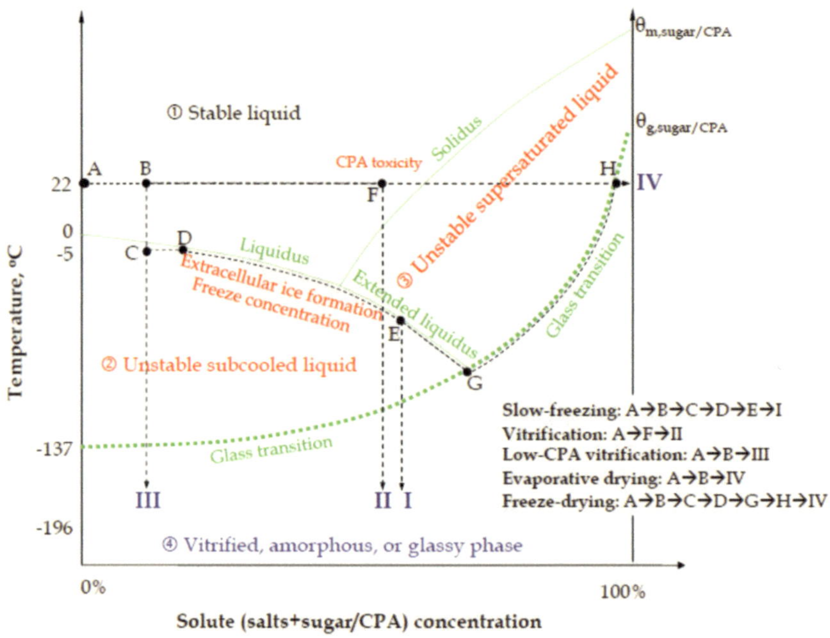

Fig. 1. An illustration on the extended phase diagram of the various methods for cell preservation at either cryogenic or ambient temperature from an initial (stable) liquid state (A) to a final glassy phase (I, II, III, or IV): The phase diagram is divided into four thermodynamic regimens by the liquidus, extended liquidus, solidus, and the line of glass transition; the four regimens are the liquid, subcooled liquid, supersaturated liquid, and the glassy phase; cells must enter the glassy phase for long-term storage; CPA represents cryoprotectant, and θ_g and θ_m represent the glass transition and melting temperature of pure sugar/CPA used as lyoprotectant or CPA, respectively. Of note, the diagram is not to scale (for example, the melting temperature of pure CPAs is usually below 20 °C)

liquidus (between subcooled liquid and supersaturated liquid), solidus (between liquid and supersaturated liquid), and the glass transition line (between the glassy phase and the unstable liquid that is either subcooled or supersaturated). A detailed description of the three processes of cell cryopreservation is given below.

2.1 Slow-freezing

For the conventional slow-freezing approach, the following steps are typically used (A→B→C→D→E→I in Fig. 1): (1), cells in an aqueous solution (state A) are first loaded with CPAs at a concentration usually up to 15 wt% (or up to ~ 2 M, state B); (2), the cells are then subcooled usually down to between -2 and -7 ^{o}C (C) to seed ice in the extracellular space by touching the sample contained usually in a cryovial with a deeply cooled (e.g., in liquid nitrogen) object (C→D); (3), the cells are further cooled slowly (typically, < 10 ^{o}C/min) along the liquidus and (if necessary) extended liquidus to between -40 and -100 ^{o}C, a process called freeze concentration (D→E); and (4), the cells in the frozen sample are transferred into liquid nitrogen for long-term storage (E→I).

During the slow-freezing process, the formation of extracellular ice after ice-seeding leads to freeze concentration of the unfrozen solutions by ejecting solutes and cells from the frozen to unfrozen phase (note: unlike pure water that becomes frozen at a specific tempertaure, solutions become frozen over a temperature range that is dependent on the types of solutes in the solution). As a result, dehydration of cells in the unfrozen phase driven by osmosis ensues, which minimizes intracellular water available for ice formation inside the cells so that the cells can enter the glassy phase (I) when transferred into liquid nitrogen. This approach typically requires a specialized machine usually called controlled rate freezer (CRF) to achieve freezing in a controllable manner. The time required for the slow-freezing process is typically in hours.

2.2 Conventional vitrification

Vitrification by definition is ice free. In other words, no (or negligible) ice formation or freezing will occur in the sample during cooling (Fahy et al. 1984; Fahy et al. 2004b; Rall and Fahy 1985). Conventional vitrification (A→F→II in Fig. 1) has also been studied for cryopreservation of both cells and tissue. In this approach, biological samples (state A) are first loaded with a very high concentration of CPAs (up to ~ 7 M, state F) (Fahy et al. 1984; Fahy et al. 2004b; Rall and Fahy 1985). The samples are then cooled directly from ambient temperature to a cryogenic temperature usually in liquid nitrogen (state II) and stored there for future use.

Although the conventional vitrification approach can be used to eliminate the detrimental effect of ice formation altogether, the unusually high CPA concentration required by the approach is toxic to most mammalian cells even in a short period of exposure (ranging from seconds to minutes dependent on the specific cells and tissues) (Chen et al. 2000; Chen et al. 2001a; Fahy et al. 2004a; Fowler and Toner 2005; Heng et al. 2005; Hunt et al. 2006). Therefore, the samples should be cooled as soon as possible after loading with CPAs. Oftentimes, a mixture of multiple CPAs is used to reduce the cytotoxicity of the high CPA concentration required (Fuller 2004). In addition, large, membrane impermeable molecules such as sugars (typically sucrose and trehalose) have been used to dehydrate the cells somewhat before cooling and protect cell membrane from injury during cooling (Beattie et al. 1997; He et al. 2008b). Vitrification can be done without a specialized machine and the time required is much shorter than that for slow-freezing.

2.3 Low-CPA vitrification

Low-CPA vitrification (A→B→III in Fig. 1) is a further advancement of the conventional vitrification with the goal to reduce the CPA concentration (e.g., at state F versus B) required for vitrification to a low, nontoxic level (similar to that used in slow-freezing). This can be done by creating an ultrafast cooling rate to cool the cells for cryopreservation. This is because the higher the cooling rate, the less the amount of cryoprotectants is required for achieving vitrification (Berejnov et al. 2006; Boutron 1986; He et al. 2008b; Karlsson et al. 1994; Toner et al. 1990; Yang et al. 2009). For example, even pure water can be vitrified without any ice formation when the cooling rate is approximately one million degree Celsius per second (Bhat et al. 2005; Bruggeller and Mayer 1980; Yang et al. 2009).

Various devices have been utilized to achieve fast cooling rates (up to ~ 20,000 °C/min) such as the traditional French type straw (Fig. 2), open pulled straw, electron grid, and cryoloops (Fowler and Toner 2005; Gardner et al. 2007; Vajta and Nagy 2006; Yavin and Arav 2007). As a result, the amount of cryoprotectant required for vitrification can be reduced to around 4 M. To achieve an ultrafast cooling rate, two recent studies reported the use of a micro-fabricated oscillating heat pipe (OHP) device (Han et al. 2008; Jiao et al. 2006). Although their theoretical analysis shows that a cooling rate of ~ 10^6 °C/min could be achieved, testing of the device for low-CPA vitrification using living cells has not been reported to date. Another recent study reported that a cooling rate as high as ~ 200,000 °C/min can be achieved by plunging an ultra-thin walled (10 μm) quartz microcapillary (QMC, 180 μm inner diameter that is slightly bigger than a human oocyte, Fig. 2) into liquid nitrogen (He et al. 2008b). As a result, the CPA concentration required for vitrification of mouse ES cells and mouse oocytes can be reduced to as low as 2.5 M altogether (He et al. 2008b; Lee et al. 2010), which is close to the highest CPA concentration usually used for slow-freezing.

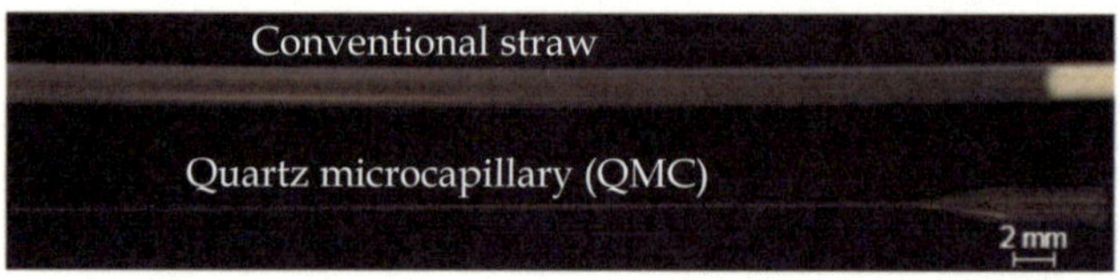

Fig. 2. (Adapted from (He et al. 2008b)) A comparison of the conventional French-type straw (top) used today for cell vitrification at an unusually high CPA concentration and the 200 μm (outer diameter), thin-walled (10 μm) quartz microcapillary (QMC, bottom) used to achieve ultrafast cooling to minimize the CPA concentration required for vitrification

Another way to improve cell vitrification is to confine cells in a small space such as sub-milimeter (in diameter) sized liquid droplets of aqueous cell suspension (Berejnov et al. 2006; Edd et al. 2008; Franks et al. 1983). A major disadvantage of using small liquid droplets to confine cells is that the droplets will merge with each other unless they are dispersed in an oil phase, which makes it difficult to retrieve cells from the droplets.

The hydrogel microcapsule (~250-1000 μm) of natural, biocompatible polymers such as alginate has been widely explored to confine or encapsulate a variety of living cells for transplantation and cell-based therapy (Chang 1996; Maguire et al. 2006; Magyar et al. 2001; Orive et al. 2003; Orive et al. 2004; Orive et al. 2006; Rohani et al. 2008; Torre et al. 2007; Wang et al. 2006a; Wang et al. 2006b). Recently, living cells have been encapsulated in even smaller (~100 μm) microcapsules for better cell survival and transplanation effcacy (Zhang and He 2009). These microcapsules can well retain their morphology for an extended period of time in physiologic solutions both *in vitro* and *in vivo*. However, cryopreservation of cells encapsulated in the large (≥ ~ 250 μm) microcapsules by slow-freezing has been challenging because the inevitable ice formation always results in significant damage to the microcapsules, which in turn can damage the encapsulated cells (Heng et al. 2004; Herrler et al. 2006; Stensvaag et al. 2004; Wu et al. 2007). Although the conventional vitrification approach can overcome this problem, the unusually high concentration CPA needed is detrimental to stress-sensitive cells such as the ES cells (Fahy et al. 1987; Fahy et al. 1984; Fahy et al. 2004b; Rall and Fahy 1985; Wu et al. 2007).

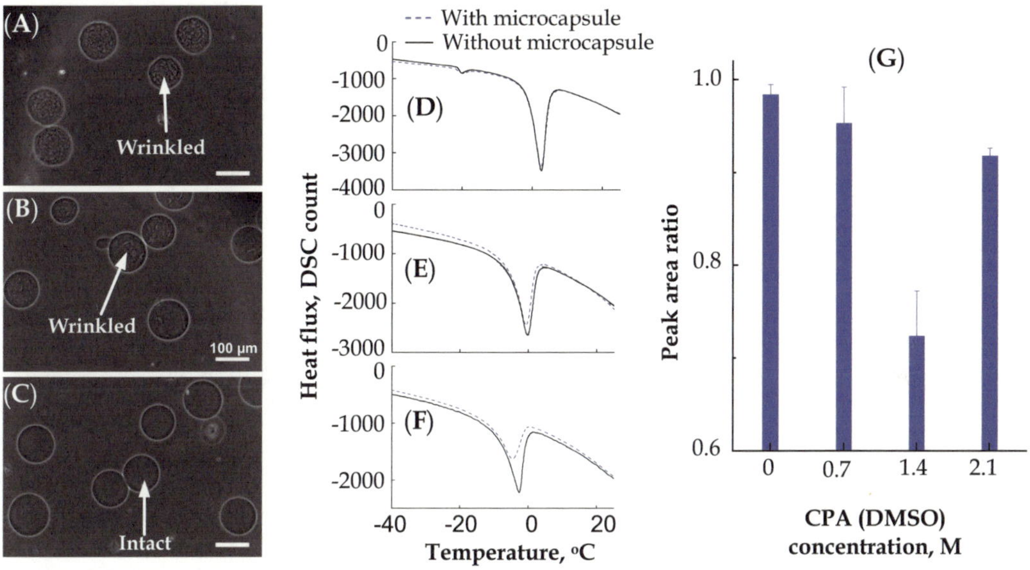

Fig. 3. (Adapted from (Zhang et al. 2010)) Typical morphology of alginate microcapsules after cryopreservation in physiological saline with (A) 0, (B) 0.7, and (C) 1.4 M DMSO (dimethylsulfoxide); typical scanning calorimetry data for samples with (D) 0, (E) 0.7, and (F) 1.4 M DMSO; and (G) the ratio of peak area on the scanning calorimetry data for samples with microcapsules (~ 30% in volume) to that of samples without microcapsules at various CPA concentrations: Cooling rate, 100 °C/min

By careful cryomicroscopy and scanning calorimetry studies, it was identified in a recent publication that water enclosed in ~100 μm (in diameter) alginate microcapsules can be preferentially vitrified over the bulk water (where the microcapsules are suspended) with only 1.4 M DMSO at a cooling rate of 100 °C/min (Zhang et al. 2010). Typical results from the cryomicroscopy studies are shown in Fig. 3 for microcapsules cryopreserved with (A) 0, (B) 0.7, and (C) 1.4 M DMSO. The microcapsules appeared intact post cryopreservation when the DMSO concentration was 1. 4 M (C) (or higher) while they were damaged (wrinkled) when ≤ 0.7 M DMSO was used (A and B). The wrinkled appearance of microcapsules in (A) and (B) presumably was a result of significant ice formation in the microcapsules during freezing. Since water in the bulk solution was frozen under all the conditions, the data suggest that water enclosed in the microcapsules was preferentially vitrified in the presence of 1.4 M DMSO resulting in the intact morphology in (C).

The calorimetry data are also shown in Fig. 3 for samples with (D) 0, (E) 0.7, and (F) 1.4 M DMSO either in the absence or presence of ~ 30% (by volume) alginate microcapsules. The area of the major peak on the heat flux curve for each sample is proportional to the amount of ice formed in the sample. Clearly, the amount of ice formed in samples with and without microcapsules was not significantly different when the DMSO concentration was ≤ 0.7 M (D and E), suggesting ice formed equally in the bulk solution and the microcapsules resulting in damage to the microcapsules shown in Fig. 3A and B. When the DMSO concentration was increased to 1.4 M, however, the peak area for samples with microcapsules was much smaller than that for samples without microcapsules (Fig. 3F), indicating much less ice formation in the samples with microcapsules. Presumably, the reduced ice formation in the samples with microcapsules was due to vitrification of water enclosed in the microcapsules resulting in the intact microcapsules shown in Fig. 3C.

The ratio of the peak area for samples with ~ 30% (by volume) microcapsules to that of samples without microcapsules was further quantified and is shown in Fig. 3G. The ratio was not different from 1, suggesting equal ice formation in the bulk solution and the microcapsule when the DMSO concentration was ≤ 0.7 M. When the DMSO concentration was 1.4 M, the ratio was ~ 0.7 suggesting that water encloased in the microcapsules (30% by volume) were preferentially vitrified under this condition. When further increasing DMSO to 2.1 M, the ratio increased back to 0.91, presumably due to the vitrification of more bulk water. Therefore, water enclosed in the microcapsules can be preferentially vitrified in the presence of 1.4 M DMSO while more than 2.1 M DMSO is required to vitrify the same amount of water in the bulk solution, indicating the capability of alginate microcapsules in enhancing vitrification of the enclosed water even at a cooling rate of 100 ^{o}C/min. The preferential vitrification of water enclosed in the microcapsule is due to its higher viscosity (Ahearne et al. 2005; Qin 2008; Zhang et al. 2006) and small volume (sub-nanoliter) and is expected to be much more significant at much higher cooling rates (e.g., > 10,000 ^{o}C/min) (Chen and Li 2008; Karlsson et al. 1994; Yang et al. 2009; Zhang et al. 2010; Zhao et al. 2006).

The preferential vitrification of water enclosed in small alginate microcapsules demonstrated in Fig. 3 should be able to enhance vitrification of living cells encapsulated in the microcapsules at high cooling rates (e.g., > 10,000 ^{o}C/min). This is because it can not only depress ice formation and growth in the microcapsule but also prevent ice (if any) propagation into cells from the bulk solution where ice is usually formed first (because of its much bigger volume) (Berejnov et al. 2006; Fahy et al. 1987; Franks et al. 1983; He et al. 2008b; Karlsson et al. 1994; Mazur et al. 2005a; Mazur et al. 2005b; Toner 1993; Toner et al. 1990; Yavin and Arav 2007). This hypothesis is confirmed by a recent study where the C3H10T1/2 mouse mesenchymal stem cells encapsulated in ~100 μm alginate microcapsules were vitrified using a 400 μm, thin-walled quartz microcapillary at a low-CPA concentration (1.4 M DMSO) (Zhang et al. 2010). Typical images of the cells are shown in Fig. 4 for both before (A-D) and after (E-H) the low-CPA vitrification procedure.

Before vitrification, both the non-encapsulated (A and B) and microencapsulated (C and D) cells remained alive, indicating that the microencapsulation process did not result in any significant damage to the cells. After vitrification, many of the non-microencapsulated cells appeared swollen with damaged plasma membrane (E) and were significantly injured (red, F) with a cell viability of 42.0 ± 4.4%. For the microencapsulated cells, however, most of them appeared intact (G) and viable (green, H) after vitrification. The viability of the microencapsulated cells post cryopreservation was determined to be 88.9 ± 2.9%, which is more than twice of that of the non-encapsulated cells and is only ~ 5% less than that before vitrification.

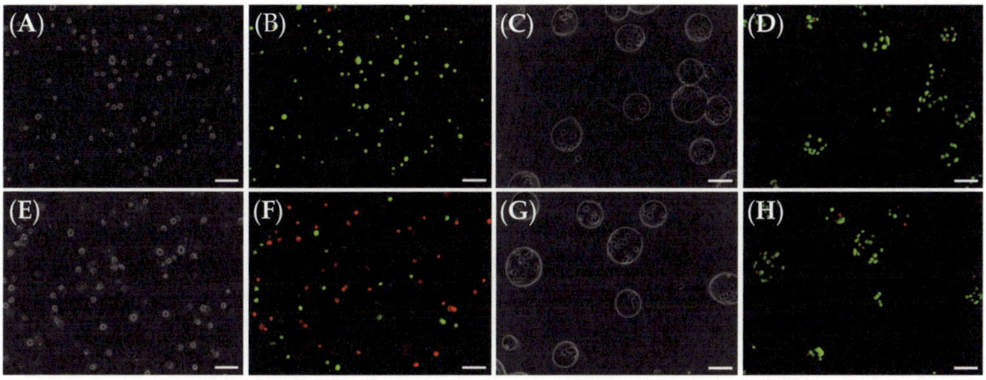

Fig. 4. (Adapted from (Zhang et al. 2010)) Typical phase and fluorescence images of non-encapsulated (A, B, E, and F) and microencapsulated (C, D, G, and H) cells before (A-D) and after (E-H) cryopreservation by low-CPA vitrification. In the fluorescence micrographs, live and dead cells were stained green and red, respectively. Scale bars: 100 µm

After liquefying the microcapsules, the collected cells were found to attach well with an attachment efficency (ratio of the number of cells attached in the cryopreserved samples to that in the control fresh samples at day 1, one day after seeding the cells) of ~ 85 and 37% for the microencapsulated and non-microencapsulated cells, respectively (Fig. 5). Moreover, the viable cells with low-CPA vitrification proliferate normally just like the control fresh cells (Fig. 5). These data clearly demonstrate the capability of the small alginate microcapsule in protecting cells from injury during low-CPA vitrification, presumably by minimizing ice formation (or enhancing vitrification) in the microencapsulated cells.

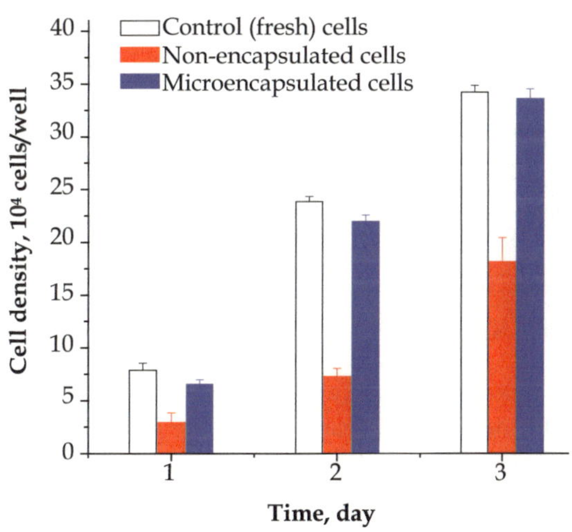

Fig. 5. Proliferation of the non-microencapsulated and microencapsulated mesenchymal stem cells in 3 days after vitrifcation using 1.4 M DMSO in the 400 µm quartz microcapillary together with control fresh cells without cryopreservation: The total number of cells seeded for each of the three conditions were the same

3. Lyopreservation at ambient temperature

The idea of dry or lyopreservation at ambient temperature is actually not new as many lower organisms, resurrection plants, and seeds can survive extreme drought in nature upon rehydration, a phenomenon called anhydrobiosis or life without water (Armstrong July, 1996; Browne et al. 2002; Clegg 2001; Crowe and Cooper 1971; Crowe and Crowe 2000; Crowe et al. 2004; Crowe et al. 1992; Perry 1999). A high concentration of small sugars (disaccharides typically sucrose for plants and trehalose for lower organisms) have been found in these organisms and plants when they are in the anhydrobiotic (or desiccated) state. Learning from nature, both sucrose and trehalose have been investigated as the protective agent (also called lyoprotectant) in protocols of lyopreservation (Crowe and Crowe 2000; Crowe et al. 2001; Crowe et al. 2005; Eroglu et al. 2000; Wolkers et al. 2002). Two methods of desiccation have been studied to dry cells in aqueous samples for lyopreservation: evaporative drying and freeze-drying (or lyophilization) which are illustrated in Fig. 1 as well.

3.1 Evaporative drying

During desiccation by evaporative drying (A→B→IV in Fig. 1), water in an aqueous sample with cells is removed by exposing the sample to a dry environment (e.g., dry air, inert gas such as nitrogen, or vacuum) without freezing (or ice formation) after loading with up to ~ 15 wt% lyoprotectants. Forced convection is usually used to increase the drying rate of natural convection. Desiccation by evaporative drying has been used to achieve lyopreservation of biomacromolecules such as proteins and lipids, pharmaceutical drugs,

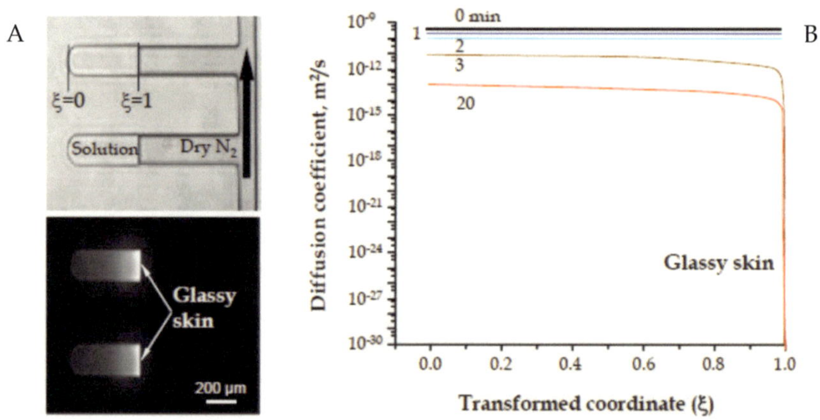

Fig. 6. (Adapted from (Aksan et al. 2006) and (He et al. 2008a)) (A) Evaporative drying of 0.2 M aqueous trehalose solution in micorchannels: The brightness in the solution in the lower panel indicate the viscosity of solution (the stronger the intensity, the higher the viscosity is in the solution), which clearly shows a glassy skin formed on the interface between the solution and the dry nitorgen gas (N_2) during drying and the heterogeneity of viscosity in the residual solution; and (B) predicted diffusivity in the trehalose sloution during evaporative drying at various times: The glassy skin forms after 3 minutes drying and has an extremely low diffusivity while the diffusivity in the rest of the solution is much higher, indicating a significant residual water in the dried residual solution

prokaryotic cells (microbials), and blood cells (red cells and platelets), but not eukaryotic mammalian cells.
A major engineering difficulty to dry the glass-forming disaccharide solution for cell lyopreservation by evaporative drying is that a thin glassy skin can easily form on the interface between the solution and the dry environment (Fig. 6A). The glassy skin has an extremely high viscosity and low diffusivity (Fig. 6B) leading to incomplete drying (up to 20% residual water after hours drying) and heterogeneity in the evaporatively dried sample. This problem might be alleviated by breaking up the solution into micron or submicron sized droplets or thin-films (He et al. 2008a).

3.2 Freeze-drying

A typical freeze-drying process is illustrated in Fig. 1 as A→B→C→D→G→ H→IV: (1), cells in an aqueous solution (state A) is first supplemented with lyoprotectants at a concentration of up to ~ 15 wt% (state B); (2), the sample is subcooled to usually between -2 and -7 °C (C) to seed extracellular ice in the sample by touching the sample with a deeply cooled object (C→D); (3), the sample is further cooled to between -30 and -50 °C slowly at a cooling rate usually less than 10 °C/min (D→G); (4), the ice formed in the sample during freezing is then sublimated (i.e., from solid ice crystal to vapor directly without going through the liquid water phase) by exposing the sample to a vacuum usually less than 10 Pa at the phase G (primary drying); and (5) a secondary drying process is then done by heating the sample in vacuum slowly to ambient temperature to further dehydrate the sample for additional hours to days (G→H→IV). The samples are then sealed and preserved in the dry phase (IV) at ambient temperature for future use. Freeze-drying has been used successfully in achieving lyopreservation of many biomacromolecules such as proteins and lipids and many pharmaceutical drugs. It has also been used for lyopreservation of prokaryotic cells such as microbials and blood cells including the red cells and platelets, but not eukaryotic cells at this time.
For freeze-drying, it is crucial to keep the temperature below the so-called collapse temperature (T_C) (Abdul-Fattah et al. 2007; Bellows and King 1972; Felix 2007; Fonseca et al. 2004a; Fonseca et al. 2004b; Gieseler et al. 2005; Kramer et al. 2009; MacKenzie 1966; Meister and Gieseler 2006; Meister et al. 2006; Nail et al. 2002; Pikal 1985; Pikal and Shah 1990; Pikal et al. 1983; Rey and May 1999)) during primary drying (at the phase G). Otherwise, the sample may collapse during primary drying (Fig. 7A) and blow up during secondary drying (Fig. 7B), resulting in incomplete drying and heterogeneity (Fig. 7C versus D) in the freeze-dried product as that in evaporatively dried sample. Consequently, the biostability of the freeze-dried biologicals could be significantly compromised (Hancock et al. 1995; He et al. 2008a; He et al. 2006b). More importantly, a recent study reported that the collapse temperature of cell culture medium-based trehalose solutions important for freeze-drying mammalian cells can be much lower than that of a simple binary trehalose-water solution (T_C = ~ -30 °C) and trehalose solutions used for freeze-drying pharmaceuticals and prokaryotes (Yang et al. 2010a), as shown in Fig. 7E.
Beside the engineering challenge to effectively dry the trehalose solutions, effective delivery of the small hydrophilic lyoprotectants (trehalose and sucrose) into mammalian cells has been challenging as the first step toward cell preservation at ambient temperature. This is because lyoprotectants such as trehalose must be present both intra and extracellularly to provide the maximum protection during drying, but mammalian cells lack a mechanism to synthesize trehalose endogenously and their plasma membrane is impermeable to the

sugars (Acker 2004; Chen et al. 2001b; Eroglu et al. 2002). Over the past decades, a number of approaches have been explored to introduce trehalose into living cells for preservation purposes. The most straightforward approach is to deliver exogenous trehalose into the cytosol of living cells by direct microinjection. This approach has been successfully used for intracellular delivery of trehalose to cryopreserve mammalian oocytes that have a large size (~ 100 μm in diameter) and are generally limited in quantity (less than a few hundred) (Bhowmick et al. 2002; Eroglu et al. 2005; Eroglu et al. 2003; Eroglu et al. 2002).

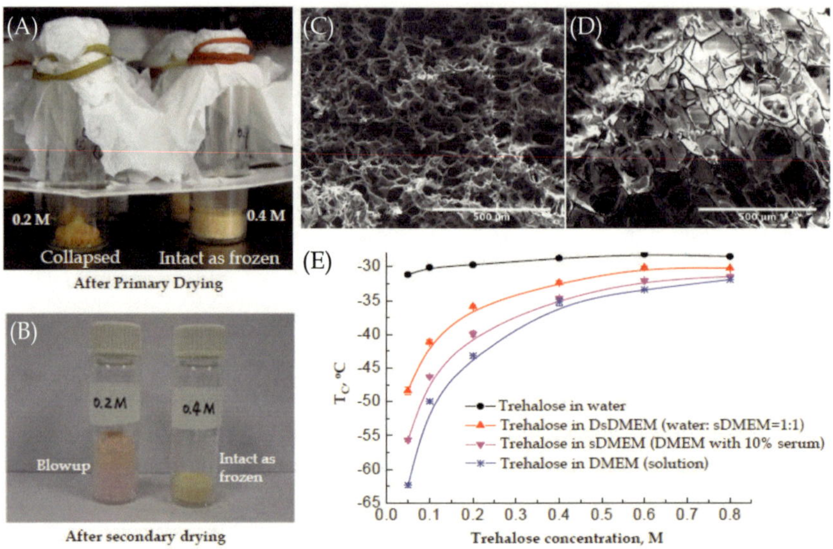

Fig. 7. (Adapted from (Yang et al. 2010a)) (A) Typical photograph showing collapsed vs. intact sample after primary drying (at -38 °C) of 0.2 versus 0.4 M trehalose in DMEM with 10% fetal bovine serum; (B) the collapse sample blew up during secondary drying (heating at 0.5 °C/min to room tempertaure) as a result of the evaporation of the significant residual water after primary drying while the non-collapsed sample appeared intact; SEM (scanning electron microscopy) images showing homogeneous microporous structure in the intact sample (C) and heterogeneous microstructure in the collapsed and blew-up sample (D) after secondary drying; and the collapse tempertaure (T_C) as a function of trehalose concentration in various solutions: DMEM, Dulbecco's modified eagle medium (aqueous) widely used for culturing mammalian cells

However, the microinjection approach is difficult (if not at all impossible) to apply for most living cells that are generally much smaller (< ~ 20 μm) than mammalian oocytes and usually present in a large quantity (millions). Small living cells have been genetically engineered to synthesize trehalose endogenously. This approach requires the constant production of adenoviral vectors that exhibit significant cytotoxicity, particularly at high multiplicities of infection (Gordon et al. 2001; Guo et al. 2000; Puhlev et al. 2001). Trehalose has also been introduced into mammalian cells or their organelles through engineered or native transmembrane pores (Acker et al. 2003; Chen et al. 2001b; Elliott et al. 2006; Eroglu et al. 2000; Liu et al. 2005), electroporation (Reuss et al. 2004; Shirakashi et al. 2002), fluid-phase endocytosis (He et al. 2006a; Oliver 2004; Wolkers et al. 2003), and lipid phase transition (Beattie et al. 1997; He et al. 2006a).

In spite of the various approaches being explored, a consistent report of cell preservation using trehalose for small eukaryotic living cells is still absent (Acker 2004; Crowe 2007; Crowe et al. 2005; Kanias and Acker 2006). This could be due to the inability to deliver a sufficient amount of intracellular trehalose ($\geq \sim 0.1$ M) for cell preservation using some of the approaches (e.g., fluid phase endocytosis). In addition, cells could be too severely compromised during the delivery step to withstand further freezing/dehydration stresses during preservation, considering the highly invasive nature of some of the approaches (e.g., electroporation). Recently, research has been sought to use nanoparticles (liposome and polymeric nanocapsules) as the intracellular delivery vehicles of the small hydrophilic lyoprotectants with promising outcomes (Holovati and Acker 2007; Holovati et al. 2009; Scott 2006; Zhang et al. 2009).

Besides the non-reducing disaccharides (trehalose and sucrose), small stress proteins particularly, the late embryogenesis abundant (LEA) proteins have been suggested to be an important part of the molecular repertoire that renders desiccation tolerance in anhydrobiotic organisms and are attracting more and more research attention (Browne et al. 2002; Clegg 2001; Crowe and Crowe 2000; Crowe et al. 2004; Crowe et al. 1992; de Castro et al. 2000; Hand et al. 2007; Huang and Tunnacliffe 2007; Iturriaga 2008; Li and He 2009; Perry 1999; Tunnacliffe et al. 2001).

4. Biophysics in cell preservation

As mentioned above, one or more protective agents (cryo and lyoprotectants for cryo and lyopreservation, respectively) are required to protect cells from being damaged during preservation. Although it has not been well established, the mechanism of protection provided by these agents is usually hypothesized to be three folds: 1) acting as water to form hydrogen bonds with proteins and lipids so that their functional conformations can be preserved during water deficit (Clegg et al. 1982; Crowe 1993a; Crowe 1993b; Crowe et al. 1998), 2) promoting preferential hydration of the biomacromolecules in cells during water loss (Cottone 2007; Cottone et al. 2005; D'Alfonso et al. 2003; Roche et al. 2006), and 3) forming a stable glassy matrix with extremely low molecular mobility to prevent the 3D intracellular structure from collapse and to suspend any degradative and metabolic reactions in response to water loss (Crowe et al. 1998; Crowe et al. 2001; He et al. 2006b; Sun et al. 1996).

At the cellular level, two biophysical events (cell dehydration and intracellular ice formation (IIF)) have been well established to be the major causes of cell injury. During slow-freezing, these two biophysical events result in the classical inverted U-curve of cell survival as a function of cooling rate during freezing with the cell survival being the highest at the optimal cooling rate (CR_{SF}), as demonstrated in Fig. 8. At a very slow cooling rate ($< CR_{SF}$), cell dehydration induced biochemical/biophysical alterations are the dominant mechanism of cell injury while at a not-so-high cooling rate ($< CR_V$), cells are mainly damaged by IIF. With the further increase of cooling rate to higher than CR_V, the kinetics of cooling is faster than that of both IIF and cell dehydration and cell injury due to both events is minimized. As a result, the cell survival increases with the increase of cooling rate till it reaches 100%. Both CR_{SF} and CR_V are dependent on the cell type, the CPA type (propylene glycol has been reported to be superior to ethylene glycol in terms of the capability of vitrification (He et al. 2008b)), and the CPA concentration. Of note, the damaging (both osmotic and metabolic) effect of an unusually high concentration of CPAs required by the conventional vitrification is not considered in the figure.

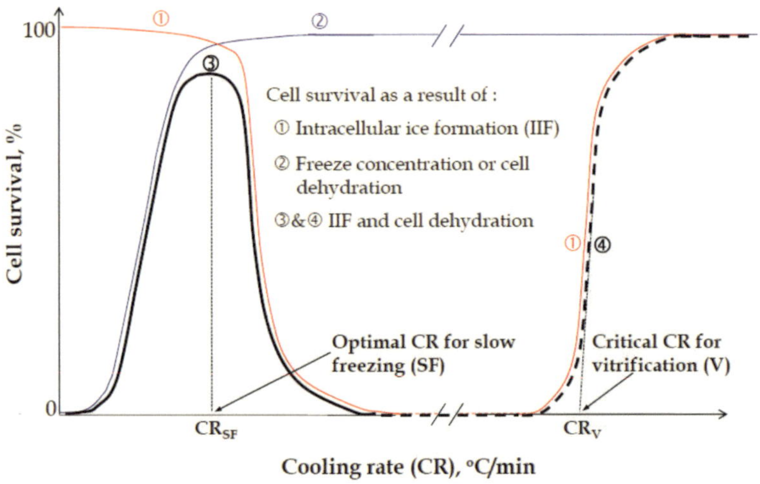

Fig. 8. Cell survival accounting for the effect of intracellular ice formation (IIF, ①), cell dehydration (②), the combination of IIF and cell dehydration at both low (for slow-freezing, ③) and high (for vitrification, ④) cooling rates. At a very slow cooling rate (< CR_{SF}), cell dehydration induced biochemical/biophysical alterations are the dominant mechanism of cell injury while at not-so-high cooling rates (< CR_V), cells are mainly damaged by IIF; with the further increase of cooling rates to higher than CR_V, cell injury due to both cell dehydration and IIF is negligible. Both CR_{SF} and CR_V are dependent on the cell type, the CPA type, and the CPA concentration. Of note, the damaging (both osmotic and metabolic) effect of an unusually high concentration of CPAs required by the conventional vitrification is not considered in the figure

Bothe cell dehydration and IIF can be quantified by modeling. IIF has been studied using both phenomenological and mechanistic models (Pitt 1990; Toner 1993). The mechanistic model has been widely used and delineates the ice formation process as two consecutive events: (1) Nucleation to form ice nuclei and (2) the subsequent growth of the nuclei (Hobbs 1974; Toner 1993). Nucleation of intracellular ice can be catalyzed by either a surface (surface catalyzed nucleation, SCN) such as the cell plasma membrane or a volume of subcooled solution (volume catalyzed nucleation, VCN) such as the cytoplasm (Hobbs 1974; Toner 1993). The rate of ice nucleation (I) due to either VCN or SCN can be estimated as follows (Toner et al. 1990; Toner et al. 1992):

$$I = \Omega_0 \frac{N}{N_0} \frac{\eta_0}{\eta} \left(\frac{T}{T_{f0}} \right)^{0.5} \exp\left[\frac{-\kappa_0 (T_f / T_{f0})^4}{(T - T_f)^2 T^3} \right] \tag{1}$$

where T_f is the equilibrium freezing point of the intracellular solution, N is the number of water molecules either in the cells (for VCN) or in contact with the cell plasma membrane (for SCN), η is viscosity, Ω and κ are two model parameters that are usually called the kinetic and thermodynamic model parameter, respectively, and the subscript 0 represents the isotonic solution state. The two model parameters (constants) under isotonic solution state (Ω_0 and κ_0) need to be determined *a priori* by experimental studies and have been reported in the literature for a number of cell types as summarized elsewhere (He and Bischof 2003; Toner 1993; Yang et al. 2010b). The cumulative probability of intracellular ice formation (P_{IIF}) can then be calculated as follows (Toner 1993; Toner et al. 1990):

$$P_{IIF}^{VCN} = 1 - \exp\left(-\int_0^t VI^{VCN} dt\right) \tag{2}$$

$$P_{IIF}^{SCN} = 1 - \exp\left(-\int_0^t AI^{SCN} dt\right) \tag{3}$$

$$P_{IIF}^{Tot} = P_{IIF}^{SCN} + (1 - P_{IIF}^{SCN})P_{IIF}^{VCN} \tag{4}$$

where V and A are the cell volume and surface area available for catalyzing the nucleation of intracellular ice, respectively. Significant IIF is usually manifested as darkening of the cell cytoplasm when observed under a bright field of light microscopy, which has been used widely to quantify the kinetics of IIF (Diller 2005; Toner et al. 1991; Yang et al. 2010b).

Of note, the above IIF model is valid only when the ice nucleation (the first step of IIF) is the rate-limiting step of IIF which is often true when freezing cells/tissues at not very high cooling rates (e.g., less than a few hundred Celsius per minute) and in the absence of a high concentration of CPA (e.g., less than 10 wt%) (Toner 1993). When cooling cells/tissue with much higher cooling rates and/or a high concentration of intracellular CPA (e.g., during vitrification and after significant cell dehydration during slow-freezing), the rate limiting-step of IIF is the growth of the ice nuclei. The IIF under these conditions is said to be diffusion-limited for which more complicated models are needed to predict the amount of intracellular ice (Chen and Li 2008; Karlsson et al. 1993; Karlsson et al. 1994; Yang et al. 2009; Zhao et al. 2006). To better predict the diffusion-limited ice nucleation and growth, an advanced model such as the free volume model that can account for the effect of glass transition on solution viscosity and diffusion coefficient might be necessary. Such free volume models for several cryo and lyoprotectants have been reported in (He et al. 2006b).

In order to predict the probability of IIF using the above model during slow-freezing where freeze concentration induced cell dehydration is significant, information on the cell volume, V (or cell surface area, A, related to the diameter of the cells when the cells assume a spherical geometry), during freezing is required. The following model has been commonly used to predict the cell volume change during freezing (Karlsson et al. 1994; Mazur 1963):

$$\frac{dV}{dt} = -\frac{L_p A R_g T}{\nu_w}\left[\ln\frac{V - V_b - V_s}{V - V_b - V_s + \phi_s n_s \nu_w} - \frac{\Delta h_f}{R_g}\left(\frac{1}{T_{ref}} - \frac{1}{T}\right)\right] \tag{5}$$

where ν_w is the partial molar volume of water, n represents amount (in mole), φ_s is the dissociation constant of solutes (e.g., 2 for NaCl), V_b is the osmotically inactive volume in cells, Δh_f is the latent heat of fusion of water, T_{ref} is a reference temperature (either ice-seeding temperature or the equilibrium melting point of intracellular solution), T is thermal history, the subscripts s and w represent solute (including CPAs) and water, respectively, and L_p is the cell plasma membrane permeability to water that can be calculated as follows (Levin et al. 1976):

$$L_p = L_{pg} \exp\left[-\frac{E_{Lp}}{R_g}\left(\frac{1}{T} - \frac{1}{T_{ref}} \right) \right] \tag{6}$$

where L_{pg} is the permeability of the cell membrane to water at the reference temperature (T_{ref}) and E_{Lp} is the activation energy for water transport across the cell plasma membrane. In the equation, L_{pg} and E_{Lp} are two model parameters (constants) that need to be determined *a priori* using experimental data. Cell dehydration during freezing can be measured using either a specialized cryostage mounted on a light microscope (Diller 2005; Toner et al. 1991) or differential scanning calorimetry (Bischof 2000; Devireddy et al. 2001; Devireddy et al. 1998; Luo et al. 2002). Many studies have been performed to determine the two model parameters for various cells, which have been reviewed and tabulated elsewhere (Bischof 2000; Han and Bischof 2004; He and Bischof 2003; Yang et al. 2010b).

Although cell dehydration and intracellular ice formation (IIF) can be predicted using the above models, a quantitative understanding of the mechanistic link between the two biophysical events and cell injury has not been well established despite some early efforts in this respect as reviewed in (He and Bischof 2003). The incidence of significant IIF (defined as darkening of cell cytoplasm) correlates strongly with cell death in many cell types (i.e., 50% of IIF in many cell populations yields 50% of dead cells) (Toner 1993). However, the exact amount or percentage of intracellular ice that is significant enough to result in irreversible cell death is still unclear. Some studies even suggest that a small amount of intracellular ice might be beneficial to cell survival (Acker and McGann 2002; Acker and McGann 2003). Therefore, further studies to establish mechanistic models capable of accounting for the effect of all the freezing induced biophysical events including IIF and freeze concentration (i.e., the so-called solute effect), and low temperatures *per se* is important to further our understanding of low temperature biology and its biomedical applications such as cryosurgery and cryopreservation.

5. Preservation of embryonic stem (ES) cells

Currently, the two most commonly used approaches for ES cell cryopreservation are slow-freezing and conventional vitrification (Hunt and Timmons 2007; Li et al. 2010; Martin-Ibanez et al. 2008). A summary of the major advantages and drawbacks of the two approaches is given in Table 1. Although a low, non-toxic CPA concentration (usually $\leq \sim$ 1.5 M) is used in slow-freezing, it is always associated with mechanical and physicochemical injury to cells due to ice formation and slow-freezing (usually ≤ 1 °C/min) induced cell dehydration (Bischof 2000; Gao and Critser 2000; Mazur 1984; Toner 1993). The conventional vitrification approach diminishes ice formation altogether to a harmless level (Fahy et al. 1987; Fahy et al. 1984; Fahy et al. 2004b; Rall and Fahy 1985; Wu et al. 2007). The unusually high (as high as 7 M) concentration of CPA required, however, can result in significant metabolic and osmotic injury to cells (Chen et al. 2000; Chen et al. 2001a; Fahy et al. 2004a; Fowler and Toner 2005; Heng et al. 2005; Hunt et al. 2006). Consequently, it is necessary to use multiple steps of CPA loading/dilution and maintain a short exposure time (within a few minutes) to high concentration CPA in each step to minimize injury (Reubinoff et al. 2001), which makes the procedure complicated, stressful, and particularly, difficult to control in that the time for the diffusion of CPAs into the cells to reach equilibrium usually takes at least 5-10 minutes (He et al. 2008b; Heng et al. 2005; Jain and Paulson 2006; Pedro et al. 2005). In addition, a cocktail of various CPAs rather than one CPA has been commonly used to reduce the CPA toxicity.

Slow-freezing	Low-CPA vitrification		Conventional vitrification
Drawbacks	Advantages	Advantages	Drawbacks
Cell injury due to ice formation and cell dehydration	Low, non-toxic CPA (≤ ~ 1.5 M)	Negligible ice formation and negligible cell dehydration	High CPA (4-7 M) induced metabolic and osmotic injury

Table 1. A summary of the major advantages and drawbacks of the commonly used slow-freezing and conventional vitrification approaches for cell cryopreservation today: The low-CPA vitrification approach combines all the advantages of the two commonly used approaches while avoiding their shortcomings

The inherent drawbacks associated with the two conventional approaches can result in damage that is (mild to many other types of cells though) sufficient to induce and/or accelerate apoptosis (programmed cell death or cell suicide) in dissociated ES cells considering that the ES cells are particularly susceptible to apoptosis (Heng et al. 2009; Heng et al. 2006; Martin-Ibanez et al. 2008). This may explain why adding ROCK (Rho-associated kinase) inhibitors in the cryopreservation medium to inhibit apoptosis can significantly improve the survival and function of human ES cells post cryopreservation (Baharvand et al. 2010; Claassen et al. 2009; Heng et al. 2007; Martin-Ibanez et al. 2008). Although keeping ES cells in aggregates (embryonic body or EB) can reduce apoptosis, it is even more difficult to cryopreserve the aggregates by either slow-freezing or conventional vitrification. Moreover, sub-optimal cryopreservation can induce epigenetic changes and impose a selection bias for their outgrowth (Baran and Ware 2007). Therefore, it is of great importance to achieve low-CPA vitrification of ES cells in that it combines all the advantages of the two conventional approaches while avoiding all their shortcomings, as demonstrated in Table 1.

A recent study has demonstrated that an ultrafast cooling rate (~200,000 °C/min) can be achieved by plunging a 200 μm (outer diameter), thin-walled (10 μm) quartz microcapillary (QMC, Fig. 2) into liquid nitrogen (He et al. 2008b). With this QMC ultrafast vitrification technique, R1 ES cells can be vitrified at a CPA concentration of as low as 2.5 M altogether (He et al. 2008b). Figure 9A shows the immediate (within 3 hr) and 1 day viability of the cells post cryopreservation using various CPAs. Only a small percentage of cells (~ 20%) can survive when using 2 M PROH (1,2-propanediol) alone as the CPA. When adding 0.5 M trehalose into the solution, however, the immediate cell viability increased to ~ 80%, even though trehalose could not permeate the cell membrane and was present only extracellularly. The immediate cell viability for cells cryopreserved using 0.5 M trehalose alone was ~ 65%.

Unlike the immediate viability, only few cells were able to survive at day 1 when using 0.5 M extracellular trehalose (< 2%) as the sole CPA (Fig. 9A). This result indicates the necessity of intracellular CPA to protect cells from within during cryopreservation. Similarly, only a minimal number of cells were able to survive at day 1 when using 2 M PROH (~12%) as the sole CPA. The 1 day viability, however, was much higher (~ 72%) when the cells were cryopreserved using the combination of 0.5 M extracellular trehalose and 2 M cell membrane permeable CPA (PROH). Therefore, PROH and extracellular trehalose appear to have a synergistic effect on protecting the ES cells from damage during vitrification. Such synergistic interaction between trehalose and PROH/DMSO has also been observed in other studies (Dash et al. 2008; Wusteman et al. 2003). The proliferation/growth of the attached ES cells post cryopreservation using the combination of 0.5 M trehalose and 2 M PROH was similar to that of the control (fresh) cells over a 3-day observation period (Fig. 9B).

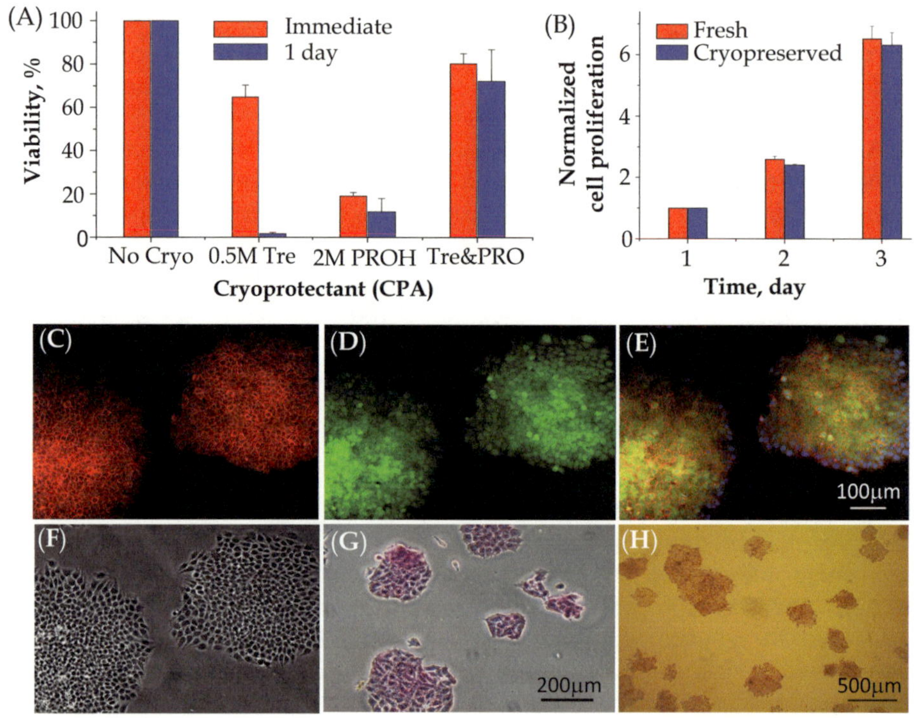

Fig. 9. (Adapted from (He et al. 2008b)) (A) Immediate and 1 day viability of fresh (No Cryo) cells and cells cryopreserved using 0.5 M trehalose (0.5 M Tre), 2 M PROH (1,2-propanediol), and the combination of 0.5 M trehalose and 2 M PROH (Tre&PROH); (B) normalized (to the data at day 1) proliferation of fresh and cryopreserved (using 0.5 M trehalose and 2 M PROH) cells over three days in culture; and micrographs showing undifferentiated properties of the ES cells post vitrification: (C) staining for the surface glycoprotein SSEA-1, (D) green fluorescence protein (GFP) expression denoting transcriptional activity, (E) merged view of SSEA-1, GFP and nuclei staining (in blue using DAPI), (F) phase contrast image of two ES cell colonies, and alkaline phosphatase expression viewed at both high (G) and low (H) magnifications

Typical Micrographs showing the undifferentiated properties of the ES cells post vitrification are given in Fig. 9C-H. Preservation of the undifferentiated properties were verified by the high level staining of the membrane surface glycoprotein SSEA-1 (C) and expression of GFP (green fluorescence protein) under the control of the transcription factor OCT-4 (D). The merged view (E) of the red (SSEA-1), green (GFP), and blue (DAPI to stain cell nuclei) channels indicates extensive co-expression of the two markers overlapping with the cell nuclei. The phase image (F) shows cells with high nuclei/cytoplasm ratios and compact colony formation typical of pluripotent mouse ES cells. The histochemical staining shows strong expression for alkaline phosphatase at high magnification (G) which was well distributed within each colony as observed at a lower magnification (H). These results suggest that the ES cells retained their undifferentiated properties post cryopreservation by ultrafast vitrification at the reduced CPA concentration (2.5 M altogether). Further studies to test the capability of the cryopreserved cells in differenriating into different types of cells *in vitro* and forming specific tissue *in vivo* are necessary to ultimately confirm preservation of the pluripotent properties of the cells after vitrification using the reduced CPA.

6. Outlook

Although the use of QMC can significantly reduce the required CPA concentration for ES cell vitrification from 4-7 M to 2.5 M, it is desired to further decrease the CPA concentration to ≤ 1.5 M which is usually used for slow-freezing. Therefore, it is of great interest and importance to further test the efficacy of the low-CPA vitrification technique in preserving the more stress sensitive ES cells by encapsulating the ES cells in small alginate microcapsules. The other advantage of microencapsulating the ES cells for vitrification is that the microencapsulated ES cells can be used directly for transplantation *in vivo*, provided that the wall permeability of the microcapsules is low enough to exclude immunoglobulin and other immunological factors from getting into the microcapsules to kill the encapsulated cells. The latter allows the use of non-autologous cells for the treatment of diseases, which significantly expands the capability of the ES cell-based medicine. Ultimately, it is important to achieve lyopreservation of ES cells at ambient temperature to allow convenient and wide distribution of the ES cell-based medicine to end users (just like what we are doing with lyophilized pharmaceutical drugs today), particulalry those in remote areas. However, no successful and consistent lyopreservation of mammalian cells (not to mention ES cells) has been reported in the literature. Hopefully, with the advances of modern nanotechnology for the intracellular deliver of small hydrophilic molecules (disaccharide such as trehalose) and our understanding on anhydrobiotism in nature and the biophysics of freeze-drying and evaporative drying, lyopreservation of ES cells can be realized in the near future.

7. Acknowledgements

This work was supported by a grant (CBET-1033426) from the National Science Foundation.

8. References

Abdul-Fattah AM, Kalonia DS, Pikal MJ. 2007. The challenge of drying method selection for protein pharmaceuticals: product quality implications. *J Pharm Sci* 96(8):1886-916.

Acker JP, Lu XM, Young V, Cheley S, Bayley H, Fowler A, Toner M. 2003. Measurement of trehalose loading of mammalian cells porated with a metal-actuated switchable pore. *Biotechnol Bioeng* 82(5):525-32.

Acker JP, McGann LE. 2002. Innocuous intracellular ice improves survival of frozen cells. *Cell Transplant* 11(6):563-71.

Acker JP, McGann LE. 2003. Protective effect of intracellular ice during freezing? *Cryobiology* 46:197-202.

Acker JP, T. Chen, A. Fowler, and M. Toner. 2004. Engineering desiccation tolerance in mammalian cells: tools and techniques. In: Fuller BJ, Lane, L., Benson, E.E., editor. *Life in the frozen state*. Boca Raton, FL: CRC Press LLC. p 563-581.

Ahearne M, Yang Y, El Haj AJ, Then KY, Liu KK. 2005. Characterizing the viscoelastic properties of thin hydrogel-based constructs for tissue engineering applications. *Journal of the Royal Society Interface* 2(5):455-463.

Aksan A, Irimia D, He XM, Toner M. 2006. Desiccation kinetics of biopreservation solutions in microchannels. *Journal of Applied Physics* 99(6):0647031-0647038.

Armstrong WP. July, 1996. The Power Of Plants. *Wayne's Word* (www.waynesword.net).

Baharvand H, Salekdeh GH, Taei A, Mollamohammadi S. 2010. An efficient and easy-to-use cryopreservation protocol for human ES and iPS cells. *Nat Protoc* 5(3):588-94.

Baran SW, Ware CB. 2007. Cryopreservation of rhesus macaque embryonic stem cells. *Stem Cells Dev* 16(2):339-44.

Beattie GM, Crowe JH, Lopez AD, Cirulli V, Ricordi C, Hayek A. 1997. Trehalose: a cryoprotectant that enhances recovery and preserves function of human pancreatic islets after long-term storage. *Diabetes* 46(3):519-23.

Bellows RJ, King CJ. 1972. Freeze-drying of aqueous solutions: Maximum allowable operating temperature. *Cryobiology* 9(6):559-61.

Berejnov V, Husseini NS, Alsaied OA, Thorne RE. 2006. Effects of cryoprotectant concentration and cooling rate on vitrification of aqueous solutions. *Journal of Applied Crystallography* 39:244-251.

Bhat SN, Sharma A, Bhat SV. 2005. Vitrification and glass transition of water: Insights from spin probe ESR. *Physical Review Letters* 95(23):2357021-2357024.

Bhowmick P, Eroglu A, Wright DL, Toner M, Toth TL. 2002. Osmometric behavior of mouse oocytes in the presence of different intracellular sugars. *Cryobiology* 45(2):183-7.

Bischof JC. 2000. Quantitative measurement and prediction of biophysical response during freezing in tissues. *Annu Rev Biomed Eng* 2:257-88.

Blow N. 2009. Biobanking: freezer burn. *Nature Methods* 6(2):173-7.

Boutron P. 1986. Comparison with the theory of the kinetics and extent of ice crystallization and of the glass-forming tendency in aqueous cryoprotective solutions. *Cryobiology* 23(1):88-102.

Browne J, Tunnacliffe A, Burnell A. 2002. Anhydrobiosis: plant desiccation gene found in a nematode. *Nature* 416(6876):38.

Bruggeller P, Mayer E. 1980. Complete Vitrification in Pure Liquid Water and Dilute Aqueous-Solutions. *Nature* 288(5791):569-571.

Chang PL. 1996. Microencapsulation - An alternative approach to gene therapy. *Transfusion Science* 17(1):35-43.

Chen C, Li WZ. 2008. Diffusion Controlled Ice Growth with Soft Impingement inside Biological Cells during Freezing. *Cryoletters* 29(5):371-381.

Chen SU, Lien YR, Chen HF, Chao KH, Ho HN, Yang YS. 2000. Open pulled straws for vitrification of mature mouse oocytes preserve patterns of meiotic spindles and chromosomes better than conventional straws. *Hum Reprod* 15(12):2598-603.

Chen SU, Lien YR, Cheng YY, Chen HF, Ho HN, Yang YS. 2001a. Vitrification of mouse oocytes using closed pulled straws (CPS) achieves a high survival and preserves good patterns of meiotic spindles, compared with conventional straws, open pulled straws (OPS) and grids. *Hum Reprod* 16(11):2350-6.

Chen T, Acker JP, Eroglu A, Cheley S, Bayley H, Fowler A, Toner M. 2001b. Beneficial effect of intracellular trehalose on the membrane integrity of dried mammalian cells. *Cryobiology* 43(2):168-81.

Claassen DA, Desler MM, Rizzino A. 2009. ROCK inhibition enhances the recovery and growth of cryopreserved human embryonic stem cells and human induced pluripotent stem cells. *Mol Reprod Dev* 76(8):722-32.

Clegg JS. 2001. Cryptobiosis--a peculiar state of biological organization. *Comp Biochem Physiol B Biochem Mol Biol* 128(4):613-24.

Clegg JS, Seitz P, Seitz W, Hazlewood CF. 1982. Cellular responses to extreme water loss: the water-replacement hypothesis. *Cryobiology* 19(3):306-16.

Coger R, Toner M. 2000. Preservation Techniques for Biomaterials. In: Bronzino JD, editor. *The Biomedical Engineering Handbook*: 2nd Edition. Boca Raton: CRC Press LLC.

Cottone G. 2007. A comparative study of carboxy myoglobin in saccharide-water systems by molecular dynamics simulation. *J Phys Chem* B 111(13):3563-9.

Cottone G, Giuffrida S, Ciccotti G, Cordone L. 2005. Molecular dynamics simulation of sucrose- and trehalose-coated carboxy-myoglobin. *Proteins* 59(2):291-302.

Crowe J, L. Crowe, and J. Carpenter. 1993a. Preserving dry biomaterials: The water replacement hypothesis, part II. *Biopharm* 6(4):40-43.

Crowe J, L. Crowe, and J. Carpenter 1993b. Preserving dry biomaterials: the water replacement hypothesis part I. *Biopharm* 6(3):28-33.

Crowe JH. 2007. Trehalose as a "chemical chaperone": fact and fantasy. *Adv Exp Med Biol* 594:143-58.

Crowe JH, Carpenter JF, Crowe LM. 1998. The role of vitrification in anhydrobiosis. *Annual Review of Physiology* 60:73-103.

Crowe JH, Cooper AF. 1971. Cryptobiosis. *Scientific American* 225(6):30-&.

Crowe JH, Crowe LM. 2000. Preservation of mammalian cells-learning nature's tricks. Nature *Biotechnology* 18(2):145-6.

Crowe JH, Crowe LM, Oliver AE, Tsvetkova N, Wolkers W, Tablin F. 2001. The trehalose myth revisited: introduction to a symposium on stabilization of cells in the dry state. *Cryobiology* 43(2):89-105.

Crowe JH, Crowe LM, Tablin F. 2004. Stabilization of dry mammalian cells: Lessons from nature. *Integrative and Comparative Biology* 44(6):542-542.

Crowe JH, Crowe LM, Wolkers WF, Oliver AE, Ma XC, Auh JH, Tang MK, Zhu SJ, Norris J, Tablin F. 2005. Stabilization of dry mammalian cells: *Lessons from nature. Integrative and Comparative Biology* 45(5):810-820.

Crowe JH, Hoekstra FA, Crowe LM. 1992. Anhydrobiosis. *Annual Review of Physiology* 54:579-599.

D'Alfonso L, Collini M, Baldini G. 2003. Trehalose influence on beta-lactoglobulin stability and hydration by time resolved fluorescence. *Eur J Biochem* 270(11):2497-504.

Dash SN, Routray P, Dash C, Guru BC, Swain P, Sarangi N. 2008. Use of the non-toxic cryoprotectant trehalose enhances recovery and function of fish embryonic stem cells following cryogenic storage. *Curr Stem Cell Res Ther* 3(4):277-87.

de Castro AG, Lapinski J, Tunnacliffe A. 2000. Anhydrobiotic engineering. *Nat Biotechnol* 18(5):473.

Deb K. 2009. A sweet potion to put embryonic stem cells to sleep. *ScientificWorldJournal* 9:236-49.

Devireddy RV, Coad JE, Bischof JC. 2001. Microscopic and calorimetric assessment of freezing processes in uterine fibroid tumor tissue. *Cryobiology* 42(4):225-43.

Devireddy RV, Raha D, Bischof JC. 1998. Measurement of water transport during freezing in cell suspensions using a differential scanning calorimeter. *Cryobiology* 36(2):124-55.

Diller KR. 2005. Bioheat and mass transfer as viewed through a microscope. *J Biomech Eng* 127(1):67-84.

Edd JF, Di Carlo D, Humphry KJ, Koster S, Irimia D, Weitz DA, Toner M. 2008. Controlled encapsulation of single-cells into monodisperse picolitre drops. *Lab on a Chip* 8(8):1262-1264.

Elliott GD, Liu XH, Cusick JL, Menze M, Vincent J, Witt T, Hand S, Toner M. 2006. Trehalose uptake through P2X7 purinergic channels provides dehydration protection. *Cryobiology* 52(1):114-27.

Eroglu A, Elliott G, Wright DL, Toner M, Toth TL. 2005. Progressive elimination of microinjected trehalose during mouse embryonic development. *Reprod Biomed Online* 10(4):503-10.

Eroglu A, Lawitts JA, Toner M, Toth TL. 2003. Quantitative microinjection of trehalose into mouse oocytes and zygotes, and its effect on development. *Cryobiology* 46(2):121-34.

Eroglu A, Russo MJ, Bieganski R, Fowler A, Cheley S, Bayley H, Toner M. 2000. Intracellular trehalose improves the survival of cryopreserved mammalian cells. *Nature Biotechnology* 18(2):163-167.

Eroglu A, Toner M, Toth TL. 2002. Beneficial effect of microinjected trehalose on the cryosurvival of human oocytes. *Fertility and Sterility* 77(1):152-8.

Fahy GM, Levy DI, Ali SE. 1987. Some emerging principles underlying the physical properties, biological actions, and utility of vitrification solutions. *Cryobiology* 24(3):196-213.

Fahy GM, MacFarlane DR, Angell CA, Meryman HT. 1984. Vitrification as an approach to cryopreservation. *Cryobiology* 21(4):407-26.

Fahy GM, Wowk B, Wu J, Paynter S. 2004a. Improved vitrification solutions based on the predictability of vitrification solution toxicity. Cryobiology 48(1):22-35.

Fahy GM, Wowk B, Wu J, Phan J, Rasch C, Chang A, Zendejas E. 2004b. Cryopreservation of organs by vitrification: perspectives and recent advances. *Cryobiology* 48(2):157-178.

Felix F. 2007. Freeze-drying of pharmaceuticals and biopharmaceuticals. Cambridge, UK: RSC Publishing.

Fonseca F, Passot S, Cunin O, Marin M. 2004a. Collapse temperature of freeze-dried Lactobacillus bulgaricus suspensions and protective media. *Biotechnol Prog* 20(1):229-38.

Fonseca F, Passot S, Lieben P, Marin M. 2004b. Collapse temperature of bacterial suspensions: the effect of cell type and concentration. *Cryo Letters* 25(6):425-34.

Fowler A, Toner M. 2005. Cryo-injury and biopreservation. *Ann N Y Acad Sci* 1066:119-35.

Franks F, Mathias SF, Galfre P, Webster SD, Brown D. 1983. Ice nucleation and freezing in undercooled cells. *Cryobiology* 20(3):298-309.

Fuller BJ. 2004. Cryoprotectants: The essential antifreezes to protect life in the frozen state. *Cryoletters* 25(6):375-388.

Gao D, Critser JK. 2000. Mechanisms of cryoinjury in living cells. *ILAR J* 41(4):187-96.

Gardner DK, Sheehan CB, Rienzi L, Katz-Jaffe M, Larman MG. 2007. Analysis of oocyte physiology to improve cryopreservation procedures. *Theriogenology* 67(1):64-72.

Gearhart J. 1998. New potential for human embryonic stem cells. *Science* 282(5391):1061-2.

Gieseler H, Lee H, Mulherkar B, Pikal MJ. 2005. Applicability of Manomentric Temperature Measurement (MTM) and SMART Freeze-Dryer Technology to Development of an Optimized Freeze-Drying Cycle: Prelim-inary Investigation of Two Amorphous Systems. *Proc. 1st European Congress on Life Science Process Technology*. Nuremberg, Germany.

Gordon SL, Oppenheimer SR, Mackay AM, Brunnabend J, Puhlev I, Levine F. 2001. Recovery of human mesenchymal stem cells following dehydration and rehydration. *Cryobiology* 43(2):182-7.

Guo N, Puhlev I, Brown DR, Mansbridge J, Levine F. 2000. Trehalose expression confers desiccation tolerance on human cells. *Nat Biotechnol* 18(2):168-71.

Han B, Bischof JC. 2004. Engineering Challenges in Tissue Preservation. *Cell Preservation Technology* 2(2):91-112.

Han X, Ma H, Jiao A, Critser JK. 2008. Investigations on the heat transport capability of a cryogenic oscillating heat pipe and its application in achieving ultra-fast cooling rates for cell vitrification cryopreservation. *Cryobiology* 56(3):195-203.

Hancock BC, Shamblin SL, Zografi G. 1995. Molecular mobility of amorphous pharmaceutical solids below their glass transition temperatures. *Pharm Res* 12(6):799-806.

Hand SC, Jones D, Menze MA, Witt TL. 2007. Life without water: Expression of plant LEA genes by an anhydrobiotic arthropod. *Journal of Experimental Zoology Part a-Ecological Genetics and Physiology* 307A(1):62-66.

He X, Bischof JC. 2003. Quantification of temperature and injury response in thermal therapy and cryosurgery. *Crit Rev Biomed Eng* 31(5-6):355-422.

He X, Fowler A, Menze M, Hand S, Toner M. 2008a. Desiccation kinetics and biothermodynamics of glass forming trehalose solutions in thin films. *Ann Biomed Eng* 36(8):1428-39.

He X, Park EY, Fowler A, Yarmush ML, Toner M. 2008b. Vitrification by ultra-fast cooling at a low concentration of cryoprotectants in a quartz micro-capillary: a study using murine embryonic stem cells. *Cryobiology* 56(3):223-32.

He XM, Amin AA, Fowler A, Toner M. 2006a. Thermally induced introduction of trehalose into primary rat hepatocytes. *Cell Preservation Technology* 4(3):178-187.

He XM, Fowler A, Toner M. 2006b. Water activity and mobility in solutions of glycerol and small molecular weight sugars: Implication for cryo- and lyopreservation. *Journal of Applied Physics* 100(7):074702 (11pp).

Heng BC, Clement MV, Cao T. 2007. Caspase inhibitor Z-VAD-FMK enhances the freeze-thaw survival rate of human embryonic stem cells. *Biosci Rep* 27(4-5):257-64.

Heng BC, Kuleshova LL, Bested SM, Liu H, Cao T. 2005. The cryopreservation of human embryonic stem cells. *Biotechnology and Applied Biochemistry* 41:97-104.

Heng BC, Richards M, Cao T. 2009. Are stem cells inherently more prone to cryopreservation-induced apoptosis compared to ordinary somatic cells? *Hum Reprod* 24(2):492; author reply 492-3.

Heng BC, Ye CP, Liu H, Toh WS, Rufaihah AJ, Cao T. 2006. Kinetics of cell death of frozen-thawed human embryonic stem cell colonies is reversibly slowed down by exposure to low temperature. *Zygote* 14(4):341-8.

Heng BC, Yu H, Ng SC. 2004. Strategies for the cryopreservation of microencapsulated cells. *Biotech Bioeng* 85(2):202-213.

Herrler A, Eisner S, Bach V, Weissenborn U, Beier HM. 2006. Cryopreservation of spermatozoa in alginic acid capsules. *Fertil Steril* 85(1):208-13.

Hobbs F. 1974. Ice physics. London: Oxford University Press.

Holovati JL, Acker JP. 2007. Spectrophotometric measurement of intraliposomal trehalose. *Cryobiology* 55(2):98-107.

Holovati JL, Gyongyossy-Issa MI, Acker JP. 2009. Effects of trehalose-loaded liposomes on red blood cell response to freezing and post-thaw membrane quality. *Cryobiology* 58(1):75-83.

Huang Z, Tunnacliffe A. 2007. Desiccation response of mammalian cells: anhydrosignaling. *Methods Enzymol* 428:269-77.

Hunt CJ, Pegg DE, Armitage SE. 2006. Optimising cryopreservation protocols for haematopoietic progenitor cells: A methodological approach for umbilical cord blood. *Cryoletters* 27(2):73-83.

Hunt CJ, Timmons PM. 2007. Cryopreservation of human embryonic stem cell lines. *Methods Mol Biol* 368:261-70.

Iturriaga G. 2008. The LEA proteins and trehalose loving couple: a step forward in anhydrobiotic engineering. *Biochem J* 410(2):e1-2.

Jain JK, Paulson RJ. 2006. Oocyte cryopreservation. *Fertil Steril* 86 Suppl 4:1037-46.

Jiao A, Han X, Critser JK, Ma H. 2006. Numerical investigations of transient heat transfer characteristics and vitrification tendencies in ultra-fast cell cooling processes. *Cryobiology* 52(3):386-92.

Kanias T, Acker JP. 2006. Mammalian cell desiccation: facing the challenges. *Cell Preservation Technology* 4(4):253-276.

Karlsson JO, Cravalho EG, Borel Rinkes IH, Tompkins RG, Yarmush ML, Toner M. 1993. Nucleation and growth of ice crystals inside cultured hepatocytes during freezing in the presence of dimethyl sulfoxide. *Biophys J* 65(6):2524-36.

Karlsson JOM, Cravalho EG, Toner M. 1994. A Model of Diffusion-Limited Ice Growth inside Biological Cells during Freezing. *Journal of Applied Physics* 75(9):4442-4445.

Kramer T, Kremer DM, Pikal MJ, Petre WJ, Shalaev EY, Gatlin LA. 2009. A procedure to optimize scale-up for the primary drying phase of lyophilization. *J Pharm Sci* 98(1):307-18.

Langer R, Vacanti JP. 1993. Tissue engineering. *Science* 260(5110):920-6.

Lee HJ, Elmoazzen H, Wright D, Biggers J, Rueda BR, Heo YS, Toner M, Toth TL. 2010. Ultra-rapid vitrification of mouse oocytes in low cryoprotectant concentrations. *Reprod Biomed Online* 20(2):201-208.

Levin RL, Cravalho EG, Huggins CE. 1976. A membrane model describing the effect of temperature on the water conductivity of erythrocyte membranes at subzero temperatures. *Cryobiology* 13(4):415-29.

Li D, He X. 2009. Desiccation induced structural alterations in a 66-amino acid fragment of an anhydrobiotic nematode late embryogenesis abundant (LEA) protein. *Biomacromolecules* 10(6):1469-77.

Li Y, Tan JC, Li LS. 2010. Comparison of three methods for cryopreservation of human embryonic stem cells. *Fertil Steril* 93(3):999-1005.

Liu XH, Aksan A, Menze MA, Hand SC, Toner M. 2005. Trehalose loading through the mitochondrial permeability transition pore enhances desiccation tolerance in rat liver mitochondria. *Biochim Biophys Acta* 1717(1):21-6.

Luo D, Han X, He L, Cui X, Cheng S, Lu C, Liu J, Gao D. 2002. A modified differential scanning calorimetry for determination of cell volumetric change during the freezing process. *Cryo Letters* 23(4):229-36.

MacKenzie AP. 1966. Basic principles of freeze-drying for pharmaceutics. Bull. *Parenteral Drug Assoc.* 20:101-129.

Maguire T, Novik E, Schloss R, Yarmush M. 2006. Alginate-PLL microencapsulation: effect on the differentiation of embryonic stem cells into hepatocytes. *Biotechnol Bioeng* 93(3):581-91.

Magyar JP, Nemir M, Ehler E, Suter N, Perriard JC, Eppenberger HM. 2001. Mass production of embryoid bodies in microbeads. *Ann N Y Acad Sci* 944:135-43.

Martin-Ibanez R, Unger C, Stromberg A, Baker D, Canals JM, Hovatta O. 2008. Novel cryopreservation method for dissociated human embryonic stem cells in the presence of a ROCK inhibitor. *Hum Reprod* 23(12):2744-54.

Mazur P. 1963. Kinetics of Water Loss from Cells at Subzero Temperatures and the Likelihood of Intracellular Freezing. *J Gen Physiol* 47:347-69.

Mazur P. 1984. Freezing of living cells: mechanisms and implications. *Am J Physiol* 247(3 Pt 1):C125-42.

Mazur P, Pinn IL, Seki S, Kleinhans FW, Edashige K. 2005a. Effects of hold time after extracellular ice formation on intracellular freezing of mouse oocytes. *Cryobiology* 51(2):235-9.

Mazur P, Seki S, Pinn IL, Kleinhans FW, Edashige K. 2005b. Extra- and intracellular ice formation in mouse oocytes. *Cryobiology* 51(1):29-53.

Meister E, Gieseler H. 2006. Evaluation of Collapse Temperatures by Freeze-Dry Microscopy: Impact of Excipient Concentration on Measured Transition and the Overall Dependance on Measurement Methodology. *Proc. 5th World Meeting on Pharmaceutics and Pharmaceutical Technology*. Geneva Switzerland.

Meister E, Marek E, Gieseler H. 2006. Collapse Temperature Measurement by Freeze-Dry Microscopy and Transferability to Freeze Drying Processes: Influence of Solute Concentration on Collapse Behavior and Effect on Cycle Design. *Proc. AAPS Annual Meeting and Exposition. San Antonio* (TX), USA.

Meyers SA. 2006. Dry storage of sperm: applications in primates and domestic animals. *Reprod Fertil Dev* 18(1-2):1-5.

Nail SL, Jiang S, Chongprasert S, Knopp SA. 2002. Fundamentals of freeze-drying. *Pharm Biotechnol* 14:281-360.

Oliver AE, K., Jamil, J.H.,Crowe, F., Tablin. 2004. Loading MSCs with trehalose by endocytosis. *Cell Preserv Technol* 2:35-49.

Orive G, Hernandez RM, Gascon AR, Calafiore R, Chang TM, De Vos P, Hortelano G, Hunkeler D, Lacik I, Shapiro AM and others. 2003. Cell encapsulation: promise and progress. *Nat Med* 9(1):104-7.

Orive G, Hernandez RM, Rodriguez Gascon A, Calafiore R, Chang TM, de Vos P, Hortelano G, Hunkeler D, Lacik I, Pedraz JL. 2004. History, challenges and perspectives of cell microencapsulation. *Trends Biotechnol* 22(2):87-92.

Orive G, Tam SK, Pedraz JL, Halle JP. 2006. Biocompatibility of alginate-poly-L-lysine microcapsules for cell therapy. *Biomaterials* 27(20):3691-700.

Pedro PB, Yokoyama E, Zhu SE, Yoshida N, Valdez DM, Jr., Tanaka M, Edashige K, Kasai M. 2005. Permeability of mouse oocytes and embryos at various developmental stages to five cryoprotectants. *J Reprod Dev* 51(2):235-46.

Perry RN. 1999. Desiccation survival of parasitic nematodes. *Parasitology* 119 Suppl:S19-30.

Pikal MJ. 1985. Use of laboratory data in freeze drying process design: heat and mass transfer coefficients and the computer simulation of freeze drying. *J Parenter Sci Technol* 39:115-138.

Pikal MJ, Shah S. 1990. The collapse temperature in freeze drying: Dependence on measurement methodology and rate of water removal from the glassy phase. *International Journal of Pharmaceutics* 62:165-186.

Pikal MJ, Shah S, Senior D, Lang JE. 1983. Physical chemistry of freeze-drying: measurement of sublimation rates for frozen aqueous solutions by a microbalance technique. *Journal of pharmaceutical sciences* 72(635-650).

Pitt RE. 1990. Cryobiological implications of diff erent methods of calculating the chemical potential of water in partially frozen suspending media. Cryo Letters 11:227-240.

Puhlev I, Guo N, Brown DR, Levine F. 2001. Desiccation tolerance in human cells. *Cryobiology* 42(3):207-17.

Qin Y. 2008. The preparation and characterization of fiber reinforced alginate hydrogel. *Journal of Applied Polymer Science* 108(5):2756-2761.

Rall WF, Fahy GM. 1985. Ice-Free Cryopreservation of Mouse Embryos at -196-Degrees-C by Vitrification. *Nature* 313(6003):573-575.

Reubinoff BE, Pera MF, Vajta G, Trounson AO. 2001. Effective cryopreservation of human embryonic stem cells by the open pulled straw vitrification method. *Hum Reprod* 16(10):2187-94.

Reuss R, Ludwig J, Shirakashi R, Ehrhart F, Zimmermann H, Schneider S, Weber MM, Zimmermann U, Schneider H, Sukhorukov VL. 2004. Intracellular delivery of carbohydrates into mammalian cells through swelling-activated pathways. *J Membr Biol* 200(2):67-81.

Rey L, May JC. 1999. Freeze-drying/lyophilization of pharmaceutical and biological products. New York: Marcel Dekker, Inc.

Roche CJ, Guo F, Friedman JM. 2006. Molecular level probing of preferential hydration and its modulation by osmolytes through the use of pyranine complexed to hemoglobin. *J Biol Chem* 281(50):38757-68.

Rohani L, Karbalaie K, Vahdati A, Hatami M, Nasr-Esfahani MH, Baharvand H. 2008. Embryonic stem cell sphere: a controlled method for production of mouse embryonic stem cell aggregates for differentiation. *Int J Artif Organs* 31(3):258-65.

Scott KL, Gyongyossy-Issa, M. I.C., Acker, J.P. 2006. Response of hemotopoietic progenitor cells to trehalose-loaded liposomes (Abstract). *Cryibiology* 53:380.

Shirakashi R, Kostner CM, Muller KJ, Kurschner M, Zimmermann U, Sukhorukov VL. 2002. Intracellular delivery of trehalose into mammalian cells by electropermeabilization. *J Membr Biol* 189(1):45-54.

Stensvaag V, Furmanek T, Lonning K, Terzis AJ, Bjerkvig R, Visted T. 2004. Cryopreservation of alginate-encapsulated recombinant cells for antiangiogenic therapy. *Cell Transplant* 13(1):35-44.

Sun WQ, Leopold AC, Crowe LM, Crowe JH. 1996. Stability of dry liposomes in sugar glasses. *Biophys J* 70(4):1769-76.

Toner M. 1993. Nucleation of ice crystals inside biological cells. *Advances in low-temperature biology* 2:1-51.

Toner M, Cravalho EG, Karel M. 1990. Thermodynamics and Kinetics of Intracellular Ice Formation during Freezing of Biological Cells. *Journal of Applied Physics* 67(3):1582-1593.

Toner M, Cravalho EG, Karel M, Armant DR. 1991. Cryomicroscopic analysis of intracellular ice formation during freezing of mouse oocytes without cryoadditives. *Cryobiology* 28(1):55-71.

Toner M, Kocsis J. 2002. Storage and translational issues in reparative medicine. *Ann N Y Acad Sci* 961:258-62.

Toner M, Tompkins RG, Cravalho EG, Yarmush ML. 1992. Transport Phenomena during Freezing of Isolated Hepatocytes. *Aiche Journal* 38(10):1512-1522.

Torre ML, Faustini M, Attilio KME, Vigo D. 2007. Cell Encapsulation in Mammal Reproduction. *Recent Patents on Drug Delivery & Formulation* 1:81-85.

Tunnacliffe A, Garcia de Castro A, Manzanera M. 2001. Anhydrobiotic engineering of bacterial and mammalian cells: is intracellular trehalose sufficient? *Cryobiology* 43(2):124-32.

Vajta G, Nagy ZP. 2006. Are programmable freezers still needed in the embryo laboratory? Review on vitrification. *Reprod Biomed Online* 12(6):779-96.

Wang W, Liu X, Xie Y, Zhang HA, Yu W, Xiong Y, Xie W, Ma X. 2006a. Microencapsulation using natural polysaccharides for drug delivery and cell implantation. *Journal of Materials Chemistry* 16:3252-3267.

Wang X, Wang W, Ma J, Guo X, Yu X, Ma X. 2006b. Proliferation and differentiation of mouse embryonic stem cells in APA microcapsule: A model for studying the interaction between stem cells and their niche. *Biotechnol Prog* 22(3):791-800.

Wolkers WF, Looper SA, Fontanilla RA, Tsvetkova NM, Tablin F, Crowe JH. 2003. Temperature dependence of fluid phase endocytosis coincides with membrane properties of pig platelets. *Biochim Biophys Acta* 1612(2):154-63.

Wolkers WF, Tablin F, Crowe JH. 2002. From anhydrobiosis to freeze-drying of eukaryotic cells. *Comp Biochem Physiol A Mol Integr Physiol* 131(3):535-43.

Wu Y, Yu H, Chang S, Magalhaes R, Kuleshova LL. 2007. Vitreous cryopreservation of cell-biomaterial constructs involving encapsulated hepatocytes. Tissue Eng 13(3):649-58.

Wusteman MC, Pegg DE, Wang LH, Robinson MP. 2003. Vitrification of ECV304 cell suspensions using solutions containing propane-1,2-diol and trehalose. *Cryobiology* 46(2):135-145.

Yang G, Gilstrap K, Zhang A, Xu LX, He X. 2010a. Collapse temperature of solutions important for lyopreservation of living cells at ambient temperature. *Biotechnol Bioeng* 106(2):247-259.

Yang G, Veres M, Szalai G, Zhang A, Xu LX, He X. 2010b. Biotransport phenomena in freezing mammalian oocytes. *Annals of Biomedical Engineering*(in press):DOI: 10.1007/s10439-010-0158-4.

Yang G, Zhang A, Xu LX, He X. 2009. Modeling the cell-type dependence of diffusion-limited intracellular ice nucleation and growth during both vitrification and slow freezing. *Journal of Applied Physics* 105(11):114701-11.

Yavin S, Arav A. 2007. Measurement of essential physical properties of vitrification solutions. *Theriogenology* 67(1):81-9.

Zhang H, Tumarkin E, Peerani R, Nie Z, Sullan RMA, Walker GC, Kumacheva E. 2006. Microfluidic production of biopolymer microcapsules with controlled morphology. *Journal of the American Chemical Society* 128(37):12205-12210.

Zhang W, He X. 2009. Encapsulation of Living Cells in Small (~ 100 mu m) Alginate Microcapsules by Electrostatic Spraying: A Parametric Study. *Journal of Biomechanical Engineering* 131(7):074515 (6pp).

Zhang W, Rong J, Wang Q, He X. 2009. The encapsulation and intracellular delivery of trehalose using a thermally responsive nanocapsule. *Nanotechnology* 20(27):275101 (14 pp).

Zhang W, Yang G, Zhang A, Xu LX, He X. 2010. Preferential vitrification of water in small alginate microcapsules significantly augments cell cryopreservation by vitrification. *Biomedical Microdevices* 12(1):89-96.

Zhao G, Luo D, Gao D. 2006. Universal model for intracellular ice formation and its growth. *AIChE Journal* 52(7):2596 - 2606.

8

Topographically and Chemically Modified Surfaces for Expansion or Differentiation of Embryonic Stem Cells

Lotte D'Andrea Markert, Jette Lovmand,
Mogens Duch and Finn Skou Pedersen
Department of Molecular Biology and The Interdisciplinary Nanoscience Center,
Aarhus University
Denmark

1. Introduction

Cells in the body respond to signals emanating from the interaction with neighbouring cells, from the surrounding extracellular matrix, and from soluble signalling molecules. These signals are perceived by the cell as topographical, mechanical, and chemical cues (Martínez et al., 2009). One example of a defined physical topography embedded in the extracellular matrix (ECM) is provided by the collagen structure. Collagen molecules are about 300 nm long and 1.5 nm wide and can form fibrils up to tens of microns in length and with a diameter of 260-410 nm (Bettinger et al., 2009). Furthermore, the spatial organisation and density of ECM is characteristic of individual tissue types (Martínez et al., 2009), and these natural structures serve to guide interacting cells in terms of cell morphology, migration, and function. Among mechanical cues, tissue elasticity has been found to vary from 0.1-0.3 kPa for embryonic stem (ES) cells and endoderm through increasing values for various differentiated cells to more than 30 kPa for demineralised bone (Reilly & Engler 2010). Such mechanical cues are also recognised in early development with cell-cortex tension being involved in germ-layer sorting (Krieg et al., 2008), indicating that mechanical signals have implications for the decision of stem cell fate. Chemical cues from the surroundings are provided from biochemical mixtures of soluble chemokines, cytokines, and growth factors, as well as insoluble receptor ligands and ECM molecules. Stem cells reside in three-dimensional tissue-specific stem-cell niches were the cells are exposed to a controlled microenvironment including both chemical, mechanical, and topographical cues from the surrounding matrix and cells (Reilly & Engler 2010).

For the development of cell-based therapies, where growth and differentiation of cells must be controlled in the laboratory or in the body, it is therefore a challenge to develop biomaterials that exploit these biological principles of guiding cells through specific interactions with their environment. Human embryonic stem (hES) cells are potentially valuable in cell-based therapies since they are able to differentiate into cells of all three germ layers as well as to self-renew and being expanded without loss of pluripotency. However, one prerequisite for such clinical use is that expansion and differentiation protocols must fulfil defined quality standards including xenofree culture conditions (Unger et al., 2008), maintenance of pluripotency during expansion, and uniform differentiation into a specific cell type.

Reproducible and defined ES cell-culturing systems must ensure cell expansion without compromising the quality of the cells as monitored by maintained chromosomal integrity and expression of pluripotency markers. For murine embryonic stem (mES) cells the ultimate quality control is the ability to enter the mouse germ line after blastocyst injection and give rise to viable offspring (Longo et al., 1997). For hES cells the strictest control of maintenance of an undifferentiated state is the ability to differentiate into three germ layers during teratoma formation after injection into immunodeficient mice (Thomson et al., 1998).
Murine and human ES cells share common characteristics such as high alkaline phosphatase (ALP) levels, a high nucleus to cytoplasm ratio, as well as the expression of a set of ES-cell-characteristic transcription factors including Nanog, Oct4, and Sox2. They exhibit distinct growth properties with mES cells growing in attached three-dimensionally rounded colonies and hES cells growing in flat colonies (Koestenbauer et al., 2006). While mES cells can be re-seeded as single cells after trypsination, the maintenance of hES cells requires a laborious step of manually cutting colonies for re-seeding, which is labour-intensive and comprises the means for clonal expansion. Differences with respect to the specific signals needed to maintain pluripotency of ES cells are reflected by the different soluble additives incorporated in the specialised culturing protocols developed for mES or hES cells. Additives to the culture media include biological signalling molecules like LIF (leukemia inhibitory factor) (Williams et al., 1988), BMP4 (bone morphogenetic protein 4) (Ying et al., 2003), or artificial inhibitors targeting specific cellular enzymes (Ying et al., 2008) in mES cell culture, and bFGF (basic fibroblast growth factor) and activin A (Akopian et al., 2010) in hES cell culture.
Both hES (Thomson et al., 1998) and mES (Evans & Kaufman 1981) cells were originally derived on mouse feeder cells, (i.e. a layer of fibroblasts) in serum containing culture medium. The layer of feeder cells contributes soluble molecules, ECM proteins (Horák & Fléchon 1998), as well as topographical and mechanical cues. However, the development of defined, xenofree culture conditions is a requirement for the use of hES cells in human cell therapy. Control and reproducibility of hES cell protocols have been improved by applying a gelatinous protein mixture secreted by a mouse sarcoma cell line manufactured under the product names Matrigel or Geltrex (Akopian et al., 2010) as culture surface. This ECM protein mixture is currently used as the standard reference surface in experiments concerning feeder-cell-free culture of hES cells, but during routine expansion feeder cells are still the preferred substrate in many laboratories in spite of the laborious preparation demands (Brafman et al., 2009; Braam et al., 2008; Mei et al., 2010; Nagaoka et al., 2010). Furthermore, neither Matrigel nor Geltrex are xenofree or chemically defined products and consequently incompatible with human cell therapy. With Matrigel and Geltrex being the generally applied feeder cell free standard surface there is a perspective for new smart biomaterials in trying to mimic the environmental demand of the ES cells, thereby contributing to better defined, xenofree, and reproducible expansion protocols.
Similar to the advantages of means for controlled expansion of undifferentiated ES cells there is also a great need for stringent protocols for controlled differentiation of ES cells for scientific and potential therapeutic purposes. Differentiation of mouse and human ES cells takes place in cell aggregates called embroid bodies or on two-dimensional surfaces. Both strategies result in a mixture of cell types even though populations are enriched for the desired cell type by means of soluble supplements (Keller 2005). A major hurdle towards the clinical use of hES cells is that the ability of pluripotent cells to induce teratomas must be lost from the batch of cells eventually used for therapy (Fong et al., 2010). It is therefore

desirable that protocols for hES cell differentiation eliminate any pluripotent cells with teratogenic properties. To identify culture conditions that lead to efficient and controlled differentiation of ES cells much effort has been put into the development of specific media supplements. However, it is now also recognised that biomaterials influence differentiation of ES cells not only through ECM-ligand presentation to integrin receptors but that also signals inherent in the synthetic biomaterials influence stem-cell state (Bakeine et al., 2009; Evans et al., 2009; Mahlstedt et al., 2010; Villa-Diaz et al., 2010). Consequently cues contributed by biomaterials are a promising and relatively unexplored tool which could increase the control of differentiation into desired cell types and help eliminate potentially teratogenic pluripotent cells from batches intended for therapeutic purposes.

The contemporary production of smart biomaterials takes advantage of several manufacturing techniques (Reviewed in (Schmidt & Healy 2009; von der Mark et al., 2010)). Development is accelerated by the screening of large biomaterial arrays, including combinations of patterning of proteins and signalling molecules (Brafman et al., 2009; Flaim et al., 2005), topography (Lovmand et al., 2009; Markert et al., 2009), modified self-assembling monolayers (Derda et al., 2007), or polymers (Anderson et al., 2004; Mei et al., 2010). In this chapter we review how stem cells respond to mechanical, chemical, and topographical cues and how such cues in synthetic two-dimensional biomaterials determine the cellular response. We refer to the vast amount of data generated by combinatorial screening approaches using two-dimensional polymer libraries derived by organic synthesis, and illustrate through our own work that the equally important input from surface topography can also be addressed by the systematic screening of libraries. To rationally design biomaterials with a predictable effect on stem cells we need to integrate optimised surface topography with optimal mechanical properties and surface chemistry. We expect that such synthetic designs based upon integrated random screening approaches will provide superior materials for use in stringent protocols that control stem-cell fate as well as novel insight into how environmental cues affect the state and signalling processes of ES cells.

2. Mechanical cues

In this section, mechanical cues affecting the programming of ES cells towards specific lineage differentiation or self-renewal are discussed. To illustrate fundamental principles we also refer to studies using multipotent stem cells.

The main mechanical link between the ECM and the intracellular actin skeleton is made up of the integrin family of membrane-spanning proteins. Integrins bind to RGD sequences in ECM proteins like for instance laminin, fibronectin, and vitronectin (Geiger et al., 2001). A key cellular response to binding of extracellular ligands to integrins is the recruitment of multiple structural proteins and signalling molecules into focal adhesions (Geiger et al., 2001). Such binding of extracellular ligands to integrins has been found to influence the differentiation state of ES cells (Burdon et al., 1999; Li et al., 2007; Na et al., 2010). E.g. a distinct type of laminin, which is present in the ES cell niche in vivo (Cooper & MacQueen 1983), supported the propagation of undifferentiated ES cells through activation of a specific integrin while other laminin subtypes did not have the same effect (Rodin et al., 2010). In mES cells focal adhesion assembly following integrin ligation was found to lead to activation of the kinases FAK, Rho, and ERK1/2 (Park et al., 2010), where ERK1/2 activation is known to link focal adhesion formation to the control of ES cell differentiation state (Burdon et al., 1999).

Externally applied force through integrin ECM interactions promotes the local growth of focal adhesions through an assembly process dependent upon the Rho kinase (Riveline et al., 2001). However, not only the size of focal adhesions but also their molecular composition changes in response to applied tensional stress (Sawada & Sheetz 2002). When integrins pull on the ECM, more tension will result from stiffer than from softer extracellular substrates, and it has been hypothesised that this might in turn lead to various conformational changes in mechanosensory proteins that connect the cytoskeleton and the ECM (Reilly & Engler 2010). Such changes may cause a change in binding-site exposure and focal-adhesion composition, thereby potentially transforming mechanical signals into a downstream intracellular response.
By using growth supports that mimicked different tissue-specific elasticities, multipotent stem cells were found to be stimulated towards the expression of markers characteristic of tissues with the elasticity in question (Engler et al., 2006). In the same study, cells grown on a stiffer matrix were stiffer and more tense than cells grown on a more elastic matrix. In the case of ES cells, such an effect of matrix stiffness on cellular differentiation is less clarified than in the case of multipotent stem cells, but mES cell morphology has been found to respond to changes in synthetic matrix stiffness, and mesendoderm marker expression to be upregulated on stiffer substrates compared to softer ones (Blin et al., 2010; Evans et al., 2009). In addition, cell softness regulated the spreading of mES cells in response to mechanical stress imposed via integrin ligands in a process dependent on myosin II, F-actin, and cdc42, but not Rac (Chowdhury et al., 2010). Myosin II activity was also critical for the previously mentioned elasticity-directed lineage-specific differentiation of multipotent stem cells (Engler et al., 2006) indicating that cytoskeletal tension is involved in the regulation of stem cell differentiation state in response to mechanical cues. This response is mediated by modulation of focal adhesion length and myosin II activity (Engler et al., 2006). The finding that disruption of cytoskeletal structures manipulated cell morphology and specifically directed the cells towards an adipogenic fate (Feng et al., 2010) also supports an involvement of cytoskeletal tension in regulation of hES cell differentiation state. Besides externally imposed mechanical cues like matrix stiffness also cyclic biaxial strain affects the differentiation state of hES cells (Saha et al., 2006). Although diverse, these studies establish that mechanical cues affect the differentiation state of ES cell by means of an interplay between cell shape, focal adhesion organisation, cell stiffness, myosin II activity, and cytoskeleton rearrangement.

3. Chemical cues

3.1 Immobilisation of signalling ligands

In vitro, many signals are imposed on cells through addition of soluble factors to the culture medium. This approach, however, is poorly compatible with transplantation based therapies. Hence, the immobilization of naturally soluble macromolecules to affect intracellular signalling poses an attractive alternative. Biomaterials with immobile growth factors can be used to modulate cell behaviour including attachment, proliferation, alignment, and migration (Mieszawska & Kaplan 2010), but have also recently been used to control the ES cell differentiation state (Alberti et al., 2008; Brafman et al., 2009; Chiang et al., 2010; Nagaoka et al., 2008).
Immobilised LIF activated the same intracellular pathways as soluble LIF and independently supported formation and growth of pluripotent mES cells (Alberti et al., 2008; Makino et al., 2004). The intensity of activation of intracellular signalling pathways by immobilised LIF was dose-dependent illustrating the possibility of fine tuning the cellular response to immobilised

signalling molecules (Alberti et al., 2008). Furthermore, lower amounts of immobilised LIF was necessary to maintain the pluripotent phenotype compared to soluble LIF (Nagaoka et al., 2008). The efficient cytokine induced signalling observed with immobilised LIF might be related to the fact that the soluble form induced LIF receptor internalisation whereas the immobilised form did not. This suggests that the high efficiency of immobilised LIF is due to prolonged signalling owing to the lack of receptor internalisation.

With respect to hES cells, formation of undifferentiated colonies was supported by immobilising bFGF in combination with ECM molecules (Brafman et al., 2009). The same approach was used to induce hES cell differentiation by co-immobilisation of BMP4 and retinoic acid (Brafman et al., 2009). These experiments proved the applicability of immobilising growth factors for differentiation purposes. Taking this a level further, a cutting-edge setup showed that 100 μm wide lanes of VEGF (vascular endothelial growth factor) and collagen IV stamped onto an otherwise non-adhesive surface directed site-specific differentiation of mES cell derived angiogenic progenitor cells into endothelial cells (Chiang et al., 2010). Taken together, the immobilisation of signalling molecules appears to be a versatile approach offering numerous possibilities of direction of site-specific differentiation in a dose-dependent manner.

Following a different line of thinking, non-adhesive surfaces with immobilised E-cadherin were used to mimic ES cell-cell interactions (Nagaoka et al., 2006; Nagaoka et al., 2010). Under these conditions, mES cells exhibited scattered growth of discrete cells with pseudopodial protrusions but still they maintained their germline competence (Nagaoka et al., 2006). Also, the morphology of hES cell colonies was affected by the non-adhesive surfaces with immobilised E-cadherin (Nagaoka et al., 2010). Interestingly, the cells proliferated at the same rate as cells cultured on Matrigel even though no integrin binding was involved in the initial attachment to the E-cadherin surfaces.

Immobilised signalling molecules thus provide an extremely powerful tool for mimicking the ES cell niche and thereby modulating ES cell differentiation and colony morphology in a site specific manner. This opens new perspectives for transplantation based therapies.

3.2 Modified organic polymers

Another promising approach for the development of chemically defined biomaterial surfaces for ES-cell culture involves the synthesis of organic polymers and takes advantage of screening of large combinatorial arrays. Acrylate polymers synthesised from different combinations of two monomers have been applied in high through-put screening of hES cells (Anderson et al., 2004; Mei et al., 2010). In such screenings the initial arrays identify hits based upon a small number of events whereas subsequent secondary arrays with a high number of replicates are used to decrease experimental error, thereby achieving an efficient and thorough analysis. Acrylate polymers possess both mechanical, chemical, and topographical cues (Mei et al., 2010). By means of different arrays, specific polymers were found to support proliferation and differentiation into cytokeratin-positive cells (Anderson et al., 2004) and some were found to support clonal propagation of undifferentiated hES cells (Mei et al., 2010). Especially the hydrophilicity and the chemical composition of the surface appeared to be important for the biological performance (Mei et al., 2010).

Extensive analysis of surface chemistry by secondary ion-mass spectrometry revealed that biological performance could in fact be predicted from the surface chemical composition with small hydrocarbon ions, oxygenated hydrocarbon ions derived from esters, as well as ions from cyclic structures being associated with favoured formation of undifferentiated colonies (Mei et al., 2010).

Polymers built from monomers with a high acrylate content supported the clonal expansion of undifferentiated ES cells (Mei et al., 2010). Further coating of di- or triacrylate polymers of moderate surface wettability with recombinant vitronectin supported clonal expansion in chemically defined media, thus providing a protocol for the clonal expansion of hES cells under fully defined conditions (Mei et al., 2010). Tissue-culture plastic coated with recombinant vitronectin was previously shown to support hES-cell expansion (Braam et al., 2008) but under these conditions clonal propagation led to differentiation (Mei et al., 2010), thereby emphasising the importance of the physicochemical characteristics of the coated material. This result might partly be related to the fact that surface chemistry alters the exposure of integrin binding sites of ECM proteins (Keselowsky et al., 2005). The importance of ECM conformation is emphasised by the finding that cyclic RGD strongly sustained hES cell expansion as compared to a linear RGD sequence (Kolhar et al., 2010). Melkoumian et al. applied an acrylate surface with covalently bound short (~ 15 aa long) peptides of vitronectin and bone sialoprotein including the RGD motif and found that this surface supported long-term self-renewal of hES cells (Melkoumian et al., 2010). This points to a fruitful avenue for enhancing the biological function of optimised polymer surfaces through covalent coupling of chemically synthesised minimal peptides recognised by specific cellular receptors.

Through analysis of the combinatorial arrays of acrylate polymers, surfaces of moderate hydrophilicity (i.e. a water contact angle ~ 70°) were found to robustly support the expansion of hES cells (Mei et al., 2010). However, another study reported that plasma etching of tissue-culture plast, which was found to reduce the water contact angle from the value 66° of non-etched plast to less than 10°, supported hES-cell expansion (Mahlstedt et al., 2010), indicating that hydrophilicity is not the sole determining parameter. This is also confirmed by studies including poly(α-hydroxy esters) (Harrison et al., 2004b), where the surface that best facilitated mES cell colonisation had a water-contact angle around 70° while other surfaces of the same hydrophilicity performed less well. Interestingly, surface hydrophilicity also affects embroid body formation (Valamehr et al., 2008). Embroid body size as well as the expression of differentiation markers could be tuned by varying surface hydrophilicity. Hence, hydrophilic surfaces resulted in smaller aggregates and hydrophobic surfaces in intermediate, regularly sized aggregates, which were found to exhibit sustained expression of germ-layer markers.

Studies of individual polymers or other chemically defined biomaterials are informative in their specific context (Harrison et al., 2004a; Horák et al., 2004; Kroupová et al., 2006; Li et al., 2006). However, the pioneering polymer-array study (Anderson et al., 2004) showed that differences among polymers with respect to their effect on proliferation of hES cells in some cases depend upon soluble media supplements. This re-enforces the value of systematic high through-put screening for biomaterial effects in various cellular protocols to avoid overlooking superior effects from specific combinations.

4. Topographical cues

4.1 The basis of topographical effects on cells

In their natural environments cells are exposed to both nano and microscale topographical cues provided by other cells and the surrounding ECM. For years it has been known that designed topographical cues influence important cellular processes such as adhesion, morphology, migration, proliferation (Curtis & Wilkinson 1997), and differentiation

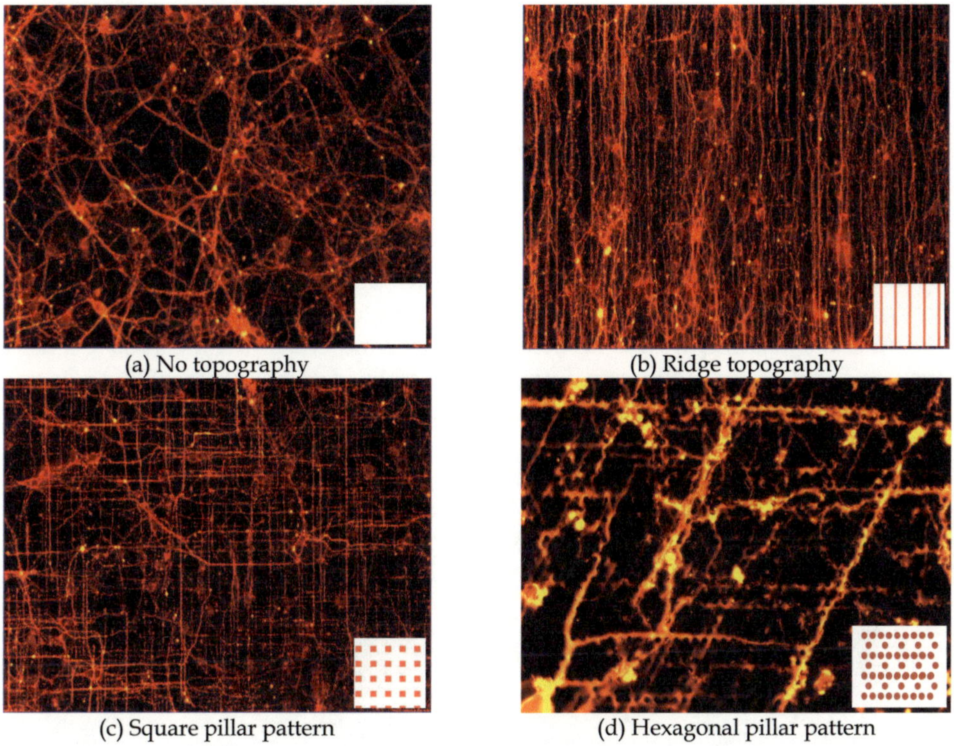

Fig. 1. Topographical alignment of neurites of fetal rat cortex cells on poly-D-lysine-coated structured surfaces. The topographical patterns are indicated by the inserts (not to scale) in the lower right corner of each panel. (a), (b), (c): Stained for Tau1 expression in axons. (d): Stained for β-tubulin expression in neurites. The images were kindly provided by Trine Elkjær Crovato.

(Dalby et al., 2007), and it has become clear that also ES cells are affected by topographical signals (Gerecht et al., 2007). The most commonly designed topographies applied in biomaterials are usually groove/ridge patterns, defined or random pillar patterns, pit patterns, or surface roughness (Bettinger et al., 2009). Figure 1 shows an illustrative example of the guidance of cell morphology by topography, namely the alignment of neurites according to topographical patterns. A variety of cell types have been applied to numerous nano and microscale variations of these different topographies and this scattered approach has brought about a lot of non-systematic information, which makes it difficult to draw general conclusions. There are, however, consistent ideas of how topography imposes an effect on mammalian cells. In this regard a distinction between micro and nanoscale topographies must be made.

Biomaterials with designed microscale topographies are aimed at controlling cell shape and position. The generally appreciated idea of how microscale topography induces a cellular response involves irregular tension imposed on the cell in the encounter with the geometric features. This cytoskeletal stress leads to a change in nucleus shape and possibly concomitant alterations in gene expression (Martínez et al., 2009).

Nanoscale topography, where feature sizes lie in the range of the dimensions of single proteins, modulate integrin binding and focal adhesion formation in that integrins require less than 73 nm separation between the ligands presented to them in order to cluster and facilitate cell spreading and attachment (Arnold et al., 2004). In addition, topography in the range 0-250 nm affects adsorption, conformation, and therefore biological functionality of ECM proteins (Martínez et al., 2009).

Whenever the effect of topography relies on focal adhesions it is relevant that both nano and microscale topographies contribute a spatial organisation of focal adhesion formation. Also, focal adhesion formation is modulated by the size of integrin ligand-presenting protein patches presented to the cell on planar surfaces (Malmström et al., 2010). Hence, 200 nm patches only supported formation of small and few focal complexes while 500 and 1000 nm patches facilitated increasingly more and large complexes with matching effects on cell attachment and spreading. Increasing pillar diameter from one to two micrometers has been observed to facilitate focal adhesion formation in embryonic fibroblasts (Figure 2). In fact, individual focal adhesions can adopt sizes up to the extent of several μm² (Geiger et al., 2001).

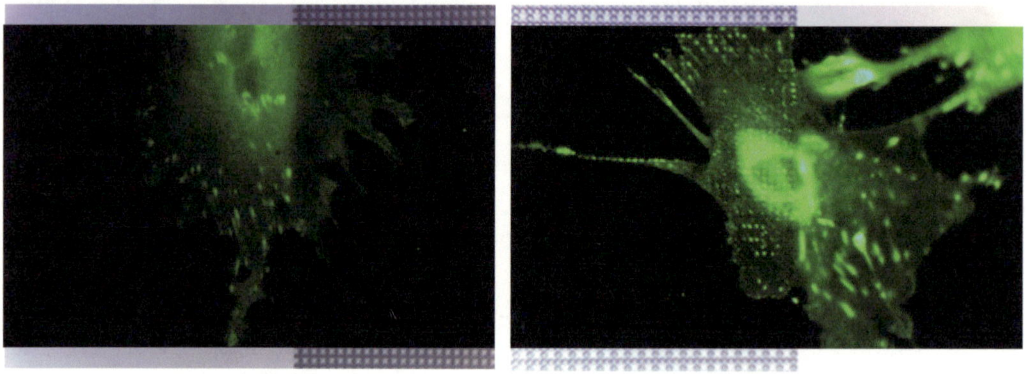

(a) Pillar diameter 1 μm (b) Pillar diameter 2 μm

Fig. 2. Embryonic fibroblasts on topographies with pillar distances of 1 μm and pillar diameters of 1 μm (a) and 2 μm (b) compared to planar surfaces. Cells were stained with anti-vinculin antibody. Notice the border between the topographical pattern and the planar surface. The images were kindly provided by Annette Füchtbauer.

Therefore, when designing topographical biomaterials it should be taken into consideration at which level the different topographical magnitudes affect the cells depending on which cue one intends to impose on the cells.

4.2 ES cell response to topographical surfaces

4.2.1 Morphology

Actin polymerisation is essential for cell morphology and attachment and is expected to be involved in the cellular response to topography by a mechanism involving filopodia formation during contact guidance (Bettinger et al., 2009). In this connection signalling through the Rho, Rac, and Cdc42 pathway, which is involved in the formation and organisation of the actin cytoskeleton, is a possibility. Clarification of this, however, awaits further investigation.

One of the most intensively studied topographical designs is line gratings to which ES cells morphologically respond in ways similar to somatic cells (Curtis & Wilkinson 1997). Hence, line gratings directed elongation of hES cells in an orientation parallel with the topographical features under culture conditions which conventionally favour self-renewal (Gerecht et al.,

2007) or neuronal differentiation (Lee et al., 2010). Under other conditions morphological aspects like cell spreading and protrusion formation of hES cell derived osteoprogenitor cells were changed in response to topography (Smith et al., 2010). Taken together, this implies that topography is able to cause versatile morphological responses of ES cells.

This morphological response was reflected in alteration of the organisation of the cytoskeletal components α-actin, tubulin, and vimentin (Gerecht et al., 2007), thus emphasising the effect of topography on the ES cell cytoskeleton.

4.2.2 Attachment and proliferation

Biomaterial topography also affects cell attachment and proliferation. When the topography of the biologically derived ECM material Matrigel was changed owing to adsorption onto different surfaces this affected both attachment and proliferation of hES cells (Kohen et al., 2010). The importance of surface topography for attachment was further emphasised by the finding that mES cells were not able to adhere to a planar polyamide surface in contrast to a fibrillar surface of the same material (Nur-E-Kamal et al., 2006). Proliferation of hES cells, mES-cell-derived cardiomyocytes, and ES-cell-derived neural precursors was reduced in response to defined topographical patterns as different as fibronectin-coated 600 nm groove/ridges (Gerecht et al., 2007), fibronectin coated pillars (diameter 5 μm) of 15 μm but not of 5 μm height (Biehl et al., 2009), and nano-roughness (Bakeine et al., 2009), respectively. More specifically, proliferation and/or attachment of mES cell derived neural precursors inversely correlated with increasing gold surface roughness in the range 0-30 nm (Bakeine et al., 2009). The reduction in proliferation caused by nano grooves/ridges was abolished by addition of actin skeleton-disrupting agents (Gerecht et al., 2007) and the inhibitory effect of microscale topography was reversed by blocking the actions of Rho or myosin light-chain kinase (Biehl et al., 2009), indicating that the actin skeleton organisation and tension play an important part in mediating the topographical cue that affects proliferation of ES cells. The fact that this breadth of topographical features produced from different materials inhibits cellular proliferation of ES cells in spite of differentiation state and species of origin suggests that topography in general inhibits proliferation of ES cells compared to planar surfaces. However, a single study showed that mES-cell proliferation was increased by the three-dimensionality of a nanofibrillar polyamide surface (Nur-E-Kamal et al., 2006). The increase in proliferation on this topographic surface correlated with an increase in PI3K (phosphoinositide 3-kinase) and Rac activity while Rho and Cdc42 activities were practically unchanged. PI3K activity has been linked to regulation of both proliferation and self-renewal of ES cells in other studies (Welham et al., 2007).

4.2.3 Self-renewal and differentiation

In recent years it has become clear that topography is also involved in the regulation of differentiation. The first convincing report in this field came from Dalby et al., who showed that mesenchymal stem cells were induced to differentiation into bone-mineralising cells in response to nothing but topography (Dalby et al., 2007). Lately also a few reports on the influence of defined topography on the differentiation of ES cells or ES-cell-derived progenitor cells have been published (Markert et al., 2009; Bakeine et al., 2009; McFarlin et al., 2009; Lee et al., 2010).

The few available studies concerning ES cells and topography have been carried out with different biomaterials and feature-size ranges which makes comparison difficult. In our laboratory we chose a systematic, array-based approach called BioSurface Structure Array (BSSA) (Lovmand et al., 2009), in order to achieve a more general insight into the connection between topography and ES-cell phenotype (Markert et al., 2009). We designed 10 defined topographical pattern series based on circular and/or square protrusions (Figure 3). Also a single pattern series with interrupted groove/ridge features inspired by shark-skin configuration was included. Iteration series of the patterns were generated by varying the horizontal pattern dimensions illustrated by X and Y in figure 3. Thus, protrusion-based patterns with varying feature (X=1, 2, 4, or 6 μm) and gap sizes (Y = 1, 2, 4, or 6 μm) were generated giving rise to 16x10 topographies with highly different protruding areas.

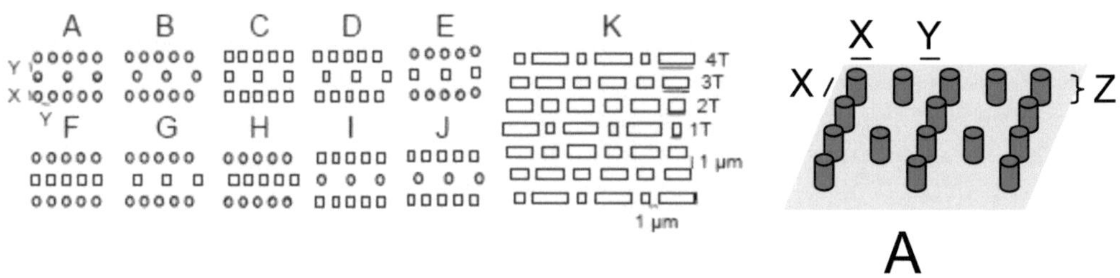

(a) 2-D projection of BSSA topographical pattern design (b) 3-D view of topography A

Fig. 3. Topography pattern designs of BSSA. (a) Two-dimensional projections of pattern A-K. Series A-J are each produced with 16 different combinations of feature size X and gap between features Y: (X,Y) =(1,1), (1,2), (1,4), (1,6); (2,1), (2,2), (2,4), (2,6); (4,1), (4,2), (4,4), (4,6); (6,1), (6,2), (6,4), (6,6). Iteration of the K series specified by T = 1,2,3,4,5,6,7, or 8 defines K patterns of varying length, but with constant gap and feature broadness (1 μm). (b) Three dimensional presentation of topography A where Z is the protrusion height; A(X,Y,Z). X, Y, Z and T are in μm.

Interrupted groove/ridge topographies were produced in different lengths by letting T adopt the values 1 through 8 μm resulting in 8 topographies (Figure 3). This total of 169 different topographies including one non-structured surface was produced as a tantalum BSSA for cell culturing. Furthermore, the topographical BSSA was produced in three different heights (Z = 0.6, 1.6, or 2.4 μm).

The surface of the BSSA was coated with a layer of tantalum (i.e. tantalum oxide) which has previously been successfully applied for cell culture and is a promising material for stem cell-based implantation procedures (Stiehler et al., 2008). To our knowledge it is, however, unprecedented to seed ES cells on this material.

When seeding two feeder-dependent mES cell lines on the topographical tantalum BSSA without feeder cells, we found that colony number and morphology varied among the different topographies. The examples shown in figure 4 are representative of the differences observed. It appears that colony number increased with structure height for both cell lines. Also both cell lines exhibited larger/more spread colony morphology on the F(X,Y)=F(4,1) topography than on the F(X,Y)=F(1,2) where the colonies formed were mainly compact and well defined as expected for undifferentiatied mES cells. On the K structures, here exemplified by K(T)=K(7), the colonies were elongated and tended to spread out. Even though the two cell lines behaved differently on the non-structured control surface, they exhibited similar responses to the topographies, thus emphasising the strength of the topographical cues.

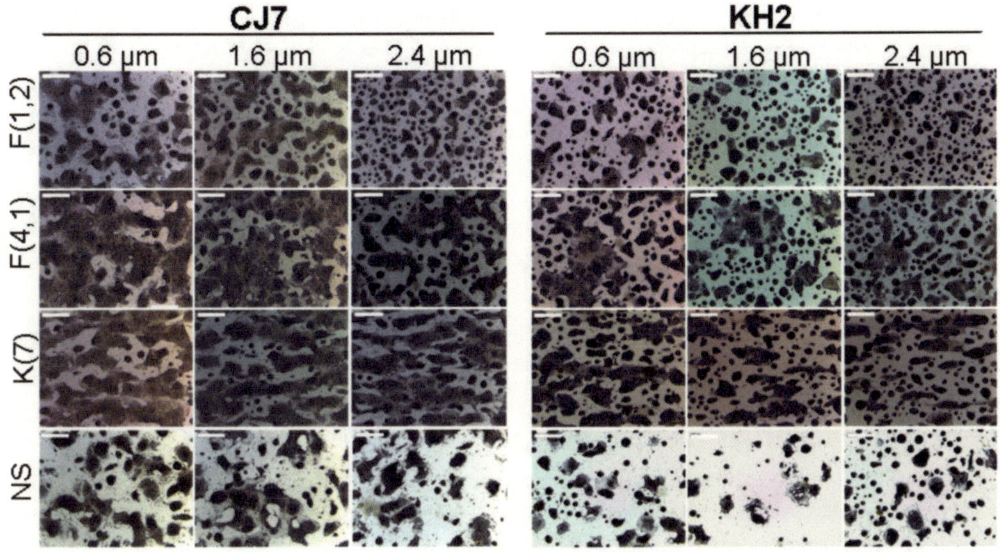

Fig. 4. Mouse ES cell colonies on selected tantalum BSSA topographies. CJ7 and KH2 mES cells were seeded at a density of 2.6×10^4 cells/cm^2 in conventional medium containing LIF. Colonies were fixed after three days and stained for alkaline phosphatase activity. Colony formation on the selected topographies F(1,2), F(4,1), and K(7) is shown at three different vertical feature heights (Z=0.6,1.6, or 2.4 µm). A non-structured (NS) control surface is also included. Scalebar: 500 µm.

To substantiate the topographically induced variations in colony number and colony spreading the findings were quantified using automated estimation procedures. It became obvious that the heights of the topographies were of major importance for formation of undifferentiated colonies and colony spreading while the shape of the geometric features (circle, square) in the A-J topography patterns did not contribute consistently to colony variations. More specifically, we found that increasing feature height, Z, from 0.6 µm to 2.4 µm heavily increased the number of undifferentiated colonies formed on the tantalum topographies while decreasing feature height resulted in colony spreading and colony areas of reduced alkaline phosphatase activity, which suggested that differentiation had commenced in spite of the presence of LIF. Others have reported that increasing surface roughness of acrylate polymers in the root mean-square range 0-110 nm had no major impact on the formation of undifferentiated hES-cell colonies (Mei et al., 2010). Our findings, however, indicate that an even higher number of undifferentiated colonies might be achieved by further increasing the height of our defined micro topographies. Moreover, we found that the lateral dimensions of the A-J topographies were decisive for the morphology and the number of colonies formed, in the sense that the largest numbers of undifferentiated mES-cell colonies were generated on topographies with feature width, X, 1 µm and a distance between adjacent features, Y, of 2 or 4 µm. At the same time colony spreading was most heavily induced on the topographical patterns with the smallest distances between features, Y=1. Combining the latter observation with unpublished results from our lab showing that mES cells lie on top of topographical features (Figure 5) suggests that cells spread more easily onto adjacent protrusions when the distance between them is reduced.

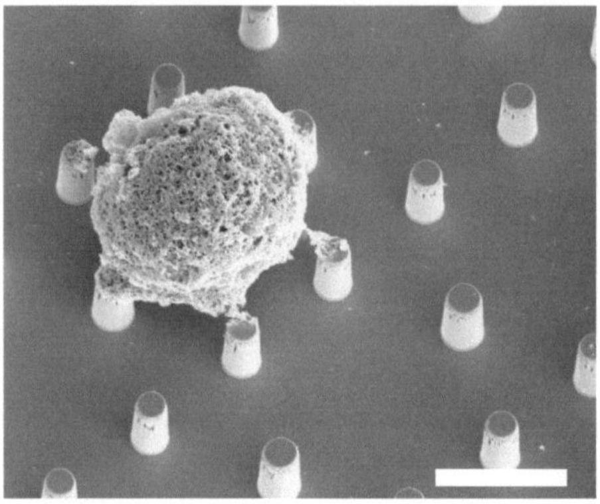

(a) Single mES cell on pillar topography

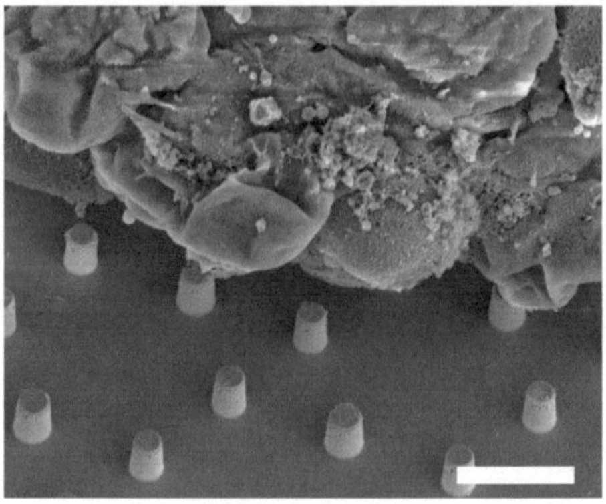

(b) Colony of mES cells on pillar topography

Fig. 5. Scanning-electron microscopy of CJ7 cells on B(X,Y,Z)= B(1,4,2.4) topography. (a) Single mES cell. (b) mES cell colony. We thank Jenny Malmström for kindly sharing her scanning electron microscopy expertise. Scalebar 5 µm.

The elongation of the mES-cell colonies induced by the interrupted groove/ridge K topography is in line with the elongation induced in individual cells in response to line gratings (Gerecht et al., 2007; Lee et al., 2010). The observed colony elongation was accompanied by increased colony spreading and loss of alkaline-staining intensity in the presence of LIF (Markert et al., 2009) - both characteristics of mES-cell differentiation. This makes topography a plausible cue for aiding the elimination of pluripotent cells which is critical for cell based therapies.

Colony morphology is an important pluripotency marker for mES cells. Therefore, we investigated the strength of the topographical cues on mES-cell-colony morphology in the absence of LIF. Surprisingly, it turned out that the variations in colony spreading induced by the iterated A-J topographies were also valid when LIF was not added to the culture (Markert et al., 2009). To further explore the sustained impact of the topographical cues we serially passaged mES cells 6 times on the F(X,Y,Z) = F(1,2,2.4) topography in conventional mES-cell-culture medium containing LIF. Under these conditions we found that the F(1,2,2.4) topography sustained the germ line competence of mES cells throughout 6 passages. It is an intriguing question to which extent the topographical cues provided by the various topographical patterns in the BSSA affect the differentiation state of mES cells. Equally exciting would it be to find out in more detail how the signals imposed by the defined topographies are translated into an intracellular response. However, our observations that topography has a positive effect on ES-cell self-renewal has been reported by others as well (McFarlin et al., 2009). The expression of the pluripotency marker Nanog in mES cells was found to be upregulated in response to a topographic polyamide surface (Nur-E-Kamal et al., 2006) and when hES cells were cultured on gelatin-coated polyurethane groove/ridge patterns of 300 nm feature height and a number of pitches in the range of 400-4000 nm, these were also found to affect the mES-cell differentiation state (McFarlin et al., 2009). It is, however, unclear exactly from which pitch ranges the published data have been obtained. Still, these authors found that topography reinforced the soluble cues in the sense that topography inhibited spontaneous differentiation under conditions which promoted self-renewal and stimulated differentiation under culture conditions which promoted differentiation. While it has been amply demonstrated that topography per se has an influence, indirect effects owing to altered access of the topographically elevated ES-cell colonies to the growth medium may also play a role in this case.

Topographical biomaterials have also been applied for directed differentiation. Neuronal differentiation of neuronal precursors generated from mES cells was increased by a gold surface roughness of 21 nm compared to a planar surface or a surface with 30 nm roughness (Bakeine et al., 2009). These authors combined the optimal (21 nm) nanoscale roughness with microscale groove/ridge topographies and found that neurons differentiating on this surface extended their neurites in parallel with the gratings. Moreover, unpublished results from our laboratory show that mES-cell-derived neuronal cells also can align in a characteristic pattern on pillar-structured surfaces (Figure 6). Also, neurons derived from hES cells showed extensions of neurites in parallel with the topographical features (Lee et al., 2010). Topography alone was able to stimulate the differentiation of hES cells into neurons when cells were cultured in media free from traditional neuronal inducers on ridge/groove (350 nm/350 nm) topography with ridge height 500 nm (Lee et al., 2010).

Altogether, these results emphasize the importance of considering topographical cues when designing biomaterials for both self-renewal and differentiation of ES cells. Our work highlights the important effects of defined micro-topography on ES-cell-colony formation and morphology. Since cell shape (Feng et al., 2010) and colony morphology (Lee et al., 2009) are linked to ES-cell differentiation, topography offers an extra level of possibilities for directing ES-cell fate through modulation of cell and colony morphology. Furthermore, the BSSA approach has been applied to other cell types as well, thereby facilitating immediate comparison between cell types over the same topographical patterns and ranges produced in the same material (Kolind et al., 2010; Lovmand et al., 2009).

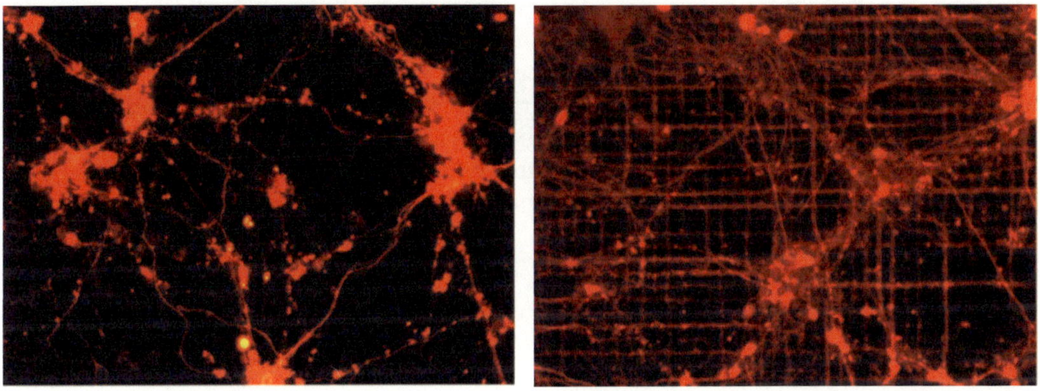

(a) Planar surface (b) Surface with pillar topography

Fig. 6. Neuronal cells derived from mES cells on (a) planar surface and (b) pillar-structured surface F(X,Y,Z)=F(6,4,2.4) from the BSSA library. The cells were stained for β-tubulin expression in neurites. The pictures were kindly provided by Trine Elkjær Crovato.

5. Conclusion

In this chapter we have highlighted the controlled self-renewal or directed differentiation of ES cells through an integrated input from chemical, topographical, and mechanical cues.

In general contemporary methodologies for self-renewal of ES cells a layer of feeder cells grown on standard tissue-culture plastic and the feeder-cell generated ECM provide topographical, and mechanical signals, whereas the chemical signals include specfic media supplements as well as products of the feeder-layer. Among the developments towards easy and stringent protocols for expansion or differentiation of ES cells under xeno-free and fully defined conditions, we emphasise development of novel biomaterials in which cell fate is partly or fully controlled by signals embedded in the surface of these materials.

The coupling of natural ligands such as specific peptides or recombinant proteins to surfaces have already led to sophisticated biomaterials with specific influence on cell behaviour. Immobilisation of soluble signalling molecules such as differentiation blockers or growth factors may enhance control and alleviate the need for specific soluble media additives. In another development integrin-ligands such as ECM proteins or artificial RGD-containing peptides may control not only cell-adhesion and morphology, but also the state of differentiation.

Array-based screening of surfaces generated by combinatorial chemistry of non-biological polymers have provided immense information on differences in cellular response to variations in surface chemistry. Remarkably, in such arrays a statistical analysis of the relationship between surface chemistry and biological read-out allowed the prediction of the cellular response to a given surface chemistry, even though the underlying mechanisms are far from understood (Mei et al., 2010; Yang et al., 2010). In a more recent development, polymer-based screenings have also been applied to screening of the impact of mechanical properties of the growth substrate on cellular behaviour. In fact, in such polymer-based screenings it may not be possible to fully discriminate between a chemical or a mechanical effect. Array-based screenings of the response of ES cells to chemical and mechanical cues currently contribute vast amounts of information that may aid the design of specialised biomaterials and provide new insight into regulatory mechanisms in ES cells.

It is well established that surface topography plays a role for the behaviour of ES cells, with nanosized features affecting interaction at the single protein level and microsized features affecting overall cell attachment and geometry. In this area, array-based screenings for the impact of topography have defined optimal structures for self-renewal or differentiation, but only on a given type of material and within a given window of analysis. However, we expect the overall potential of optimising topography for ES cell cultivation to be far from exhausted.

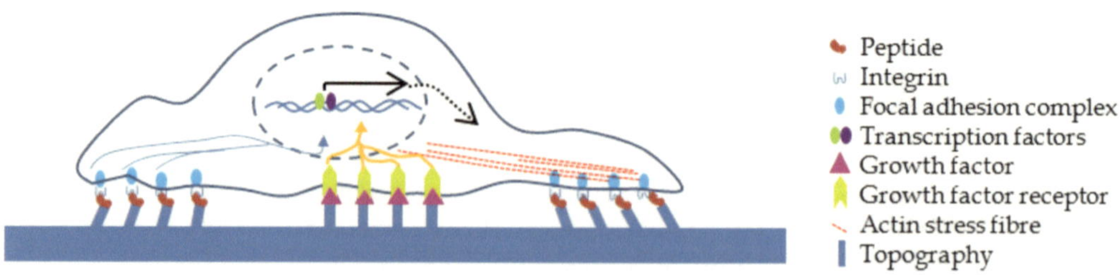

Fig. 7. Cellular reponse to topographical, chemical, and mechanical cues provided by synthetic biomaterials. The responding cell integrates signals from various external cues through complex regulatory circuits to determine the morphology and mechanical properties of a cell as well as its gene expression programme and state of differentiation.

Future specialized biomaterials may mimic the natural stem-cell niche by combining optimised surface topography with mechanical features, signals from specialised ECM-components and other chemical cues as illustrated in figure 7. Such materials may be fine-tuned to allow improved signalling in a concentration-dependent manner and optimisation of geometric patterning for site-specific attachment and differentiation to allow extended applications in regenerative medicine. The signals from mechanical, chemical, and topographic cues act in concert to determine cell fate through an influence on cell shape and cytoskeleton organisation as well as gene expression and differentiation state. Sophisticated biomaterials with cues that control stem cell fate will allow defined means for investigation of the integrated regulatory mechanisms that determine stem cell fate in response to signals in the environment.

6. References

Akopian V.; Andrews P.; Beil S.; Benvenisty N.; Brehm J.; Christie M.; Ford A.; Fox V.; Gokhale P.; Healy L.; Holm F.; Hovatta O.; Knowles B.; Ludwig T.; McKay R.; Miyazaki T.; Nakatsuji N.; Oh S.; Pera M.; Rossant J.; Stacey G. & Suemori H. (2010). Comparison of defined culture systems for feeder cell free propagation of human embryonic stem cells, *In Vitro Cellular & Developmental Biology - Animal,* 46(3): 247-58.

Alberti K.; Davey R. E.; Onishi K.; George S.; Salchert K.; Seib F. P.; Bornhauser M.; Pompe T.; Nagy A.; Werner C. & Zandstra P. W. (2008). Functional immobilization of signaling proteins enables control of stem cell fate, *Nature Methods,* 5(7): 645-50.

Anderson D. G.; Levenberg S. & Langer R. (2004). Nanoliter-scale synthesis of arrayed biomaterials and application to human embryonic stem cells, *Nature Biotechnology,* 22(7): 863-6.

Arnold M.; Cavalcanti-Adam E. A.; Glass R.; Blümmel J.; Eck W.; Kantlehner M.; Kessler H. & Spatz J. P. (2004). Activation of Integrin Function by Nanopatterned Adhesive Interfaces, *ChemPhysChem,* 5(3): 383-8.

Bakeine G. J.; Ban J.; Grenci G.; Pozzato A.; Zilio S. D.; Prasciolu M.; Businaro L.; Tormen M. & Ruaro M. E. (2009). Design, fabrication and evaluation of nanoscale surface topography as a tool in directing differentiation and organisation of embryonic stem-cell-derived neural precursors, *Microelectronic Engineering,* 86(4-6): 1435-8.

Bettinger C. J.; Langer R. & Borenstein J. T. (2009). Engineering Substrate Topography at the Micro- and Nanoscale to Control Cell Function, *Angewandte Chemie International Edition,* 48(30): 5406-15.

Biehl J. K.; Yamanaka S.; Desai T. A.; Boheler K. R. & Russell B. (2009). Proliferation of Mouse Embryonic Stem Cell Progeny and the Spontaneus Contractile Activity of Cardiomyocytes Are Affected by Microtopography, *Developmental Dynamics,* 238): 1964-73.

Blin G.; Lablack N.; Louis-Tisserand M.; Nicolas C.; Picart C. & Pucéat M. (2010). Nano-scale control of cellular environment to drive embryonic stem cells selfrenewal and fate, *Biomaterials,* 31(7): 1742-50.

Brafman D. A.; Shah K. D.; Fellner T.; Chien S. & Willert K. (2009). Defining Long-Term Maintenance Conditions of Human Embryonic Stem Cells With Arrayed Cellular Microenvironment Technology, *Stem Cells and Development,* 18(8): 1141-54.

Braam S. R.; Zeinstra L.; Litjens S.; Ward-van Oostwaard D.; van den Brink S.; van Laake L.; Lebrin F.; Kats P.; Hochstenbach R.; Passier R.; Sonnenberg A. & Mummery C. L. (2008). Recombinant Vitronectin Is a Functionally Defined Substrate That Supports

Human Embryonic Stem Cell Self-Renewal via alphaVbeta5 Integrin, *Stem Cells,* 26(9): 2257-65.

Burdon T.; Stracey C.; Chambers I.; Nichols J. & Smith A. (1999). Suppression of SHP-2 and ERK Signalling Promotes Self-Renewal of Mouse Embryonic Stem Cells, *Developmental Biology,* 210(1): 30-43.

Chiang C. K.; Mohammad Fahad C.; Rohin K. I.; William L. S. & Milica R. (2010). Engineering surfaces for site-specific vascular differentiation of mouse embryonic stem cells, *Acta biomaterialia,* 6(6): 1904-16.

Chowdhury F.; Na S.; Li D.; Poh Y.-C.; Tanaka T. S.; Wang F. & Wang N. (2010). Material properties of the cell dictate stress-induced spreading and differentiation in embryonic stem cells, *Nature Materials,* 9(1): 82-8.

Cooper A. R. & MacQueen H. A. (1983). Subunits of laminin are differentially synthesized in mouse eggs and early embryos, *Developmental Biology,* 96(2): 467-71.

Curtis A. & Wilkinson C. (1997). Topographical control of cells, *Biomaterials,* 18(24): 1573-83.

Dalby M. J.; Gadegaard N.; Tare R.; Andar A.; Riehle M. O.; Herzyk P.; Wilkinson C. D. W. & Oreffo R. O. C. (2007). The control of human mesenchymal cell differentiation using nanoscale symmetry and disorder, *Nature Materials,* 6(12): 997-1003.

Derda R.; Li L.; Orner B. P.; Lewis R. L.; Thomson J. A. & Kiessling L. L. (2007). Defined Substrates for Human Embryonic Stem Cell Growth Identified from Surface Arrays, *ACS Chemical Biology,* 2(5): 347-55.

Engler A. J.; Sen S.; Sweeney H. L. & Discher D. E. (2006). Matrix Elasticity Directs Stem Cell Lineage Specification, *Cell,* 126(4): 677-89.

Evans M. J. & Kaufman M. H. (1981). Establishment in culture of pluripotential cells from mouse embryos., *Nature,* 292(5819): 154-6.

Evans N. D.; Gentleman E.; Chen X.; Roberts C. J.; Polak J. M. & Stevens M. M. (2009). Extracellular matrix-mediated osteogenic differentiation of murine embryonic stem cells, *Biomaterials,* 31(12): 3244-52.

Feng T.; Szabo E.; Dziak E. & Opas M. (2010). Cytoskeletal Disassembly and Cell Rounding Promotes Adipogenesis from ES Cells, *Stem Cell Reviews and Reports,* 6(1): 74-85.

Flaim C. J.; Chien S. & Bhatia S. N. (2005). An extracellular matrix microarray for probing cellular differentiation, *Nature Materials,* 2(2): 119-25.

Fong C. Y.; Gauthaman K. & Bongso A. (2010). Teratomas from pluripotent stem cells: A clinical hurdle, *Journal of Cellular Biochemistry,* 111(4): 769-81

Geiger B.; Bershadsky A.; Pankov R. & Yamada K. M. (2001). Transmembrane crosstalk between the extracellular matrix and the cytoskeleton, *Nat Rev Mol Cell Biol,* 2(11): 793-805.

Gerecht S.; Bettinger C. J.; Zhang Z.; Borenstein J. T.; Vunjak-Novakovic G. & Langer R. (2007). The effect of actin disrupting agents on contact guidance of human embryonic stem cells, *Biomaterials,* 28(28): 4068-77.

Harrison J.; Melville A. J.; Forsythe J. S.; Muddle B. C.; Trounson A. O.; Gross K. A. & Mollard R. (2004a). Sintered hydroxyfluorapatites--IV: the effect of fluoride substitutions upon colonisation of hydroxyapatites by mouse embryonic stem cells, *Biomaterials,* 25(20): 4977-86.

Harrison J.; Pattanawong S.; Forsythe J. S.; Gross K. A.; Nisbet D. R.; Beh H.; Scott T. F.; Trounson A. O. & Mollard R. (2004b). Colonization and maintenance of murine embryonic stem cells on poly([alpha]-hydroxy esters), *Biomaterials,* 25(20): 4963-70.

Horák D.; Kroupová J.; Slouf M. & Dvorák P. (2004). Poly(2-hydroxyethyl methacrylate)-based slabs as a mouse embryonic stem cell support, *Biomaterials,* 25(22): 5249-60.

Horák V. & Fléchon J.-E. (1998). Immunocytochemical characterisation of rabbit and mouse embryonic fibroblasts, *Reproduction Nutrition Development,* 38(6): 683-95.

Keller G. (2005). Embryonic stem cell differentiation: emergence of a new era in biology and medicine., *Genes Development,* 19(10): 1129-55.

Keselowsky B. G.; Collard D. M. & Garcia A. J. (2005). Integrin binding specificity regulates biomaterial surface chemistry effects on cell differentiation., *Proceedings of the National Academy of Sciences of the United States of America,* 102(17): 5953-7.

Koestenbauer S.; Zech N. H.; Juch H.; Vanderzwalmen P.; Schoonjans L. & Dohr G. (2006). Embryonic Stem Cells: Similarities and Differences Between Human and Murine Embryonic Stem Cells, *American Journal of Reproductive Immunology,* 55(3): 169-80.

Kohen N. T.; Little L. E. & Healy K. E. (2010). Characterization of Matrigel interfaces during defined human embryonic stem cell culture, *Biointerphases,* 4(4): 69-79.

Kolhar P.; Kotamraju V. R.; Hikita S. T.; Clegg D. O. & Ruoslahti E. (2010). Synthetic surfaces for human embryonic stem cell culture, *Journal of Biotechnology,* 146(3): 143-6.

Kolind K.; Dolatshahi-Pirouz A.; Lovmand J.; Pedersen F. S.; Foss M. & Besenbacher F. (2010). A combinatorial screening of human fibroblast responses on micro-structured surfaces, *Biomaterials,* 31(35): 9182-91

Krieg M.; Arboleda-Estudillo Y.; Puech P. H.; Kafer J.; Graner F.; Muller D. J. & Heisenberg C. P. (2008). Tensile forces govern germ-layer organization in zebrafish, *Nature Cell Biology,* 10(4): 429-36.

Kroupová J.; Horák D.; Pacherník J.; Dvořák P. & Šlouf M. (2006). Functional polymer hydrogels for embryonic stem cell support, *Journal of Biomedical Materials Research Part B: Applied Biomaterials,* 76B(2): 315-25.

Lee L. H.; Peerani R.; Ungrin M.; Joshi C.; Kumacheva E. & Zandstra P. W. (2009). Micropatterning of human embryonic stem cells dissects the mesoderm and endoderm lineages, *Stem Cell Research,* 2(2): 155-62

Lee M. R.; Kwon K. W.; Jung H.; Kim H. N.; Suh K. Y.; Kim K. & Kim K.-S. (2010). Direct differentiation of human embryonic stem cells into selective neurons on nanoscale ridge/groove pattern arrays, *Biomaterials,* 31(15): 4360-6.

Li J.; Wang G.; Wang C.; Zhao Y.; Zhang H.; Tan Z.; Song Z.; Ding M. & Deng H. (2007). MEK/ERK signaling contributes to the maintenance of human embryonic stem cell self-renewal, *Differentiation,* 75(4): 299-307.

Li Y. J.; Chung E. H.; Rodriguez R. T.; Firpo M. T. & Healy K. E. (2006). Hydrogels as artificial matrices for human embryonic stem cell self-renewal, *Journal of Biomedical Materials Research Part A,* 79A(1): 1-5.

Longo L.; Bygrave A.; Grosveld F. G. & Pandolfi P. P. (1997). The chromosome make-up of mouse embryonic stem cells is predictive of somatic and germ cell chimaerism, *Transgenic Research,* 6(5): 321-8.

Lovmand J.; Justesen J.; Foss M.; Lauridsen R. H.; Lovmand M.; Modin C.; Besenbacher F.; Pedersen F. S. & Duch M. (2009). The use of combinatorial topographical libraries for the screening of enhanced osteogenic expression and mineralization, *Biomaterials,* 30(11): 2015-22.

Mahlstedt M. M.; Anderson D.; Sharp J. S.; McGilvray R.; Muñoz M. D. B.; Buttery L. D.; Alexander M. R.; Rose F. R. A. J. & Denning C. (2010). Maintenance of pluripotency in human embryonic stem cells cultured on a synthetic substrate in conditioned medium, *Biotechnology and Bioengineering,* 105(1): 130-40.

Makino H.; Hasuda H. & Ito Y. (2004). Immobilization of leukemia inhibitory factor (LIF) to culture murine embryonic stem cells, *Journal of Bioscience and Bioengineering,* 98(5): 374-9.

Malmström J.; Christensen B.; Jakobsen H. P.; Lovmand J.; Foldbjerg R.; Sørensen E. S. & Sutherland D. S. (2010). Large Area Protein Patterning Reveals Nanoscale Control of Focal Adhesion Development, *Nano Letters,* 10(2): 686-94.

Markert L. D. A.; Lovmand J.; Foss M.; Lauridsen R. H.; Lovmand M.; Füchtbauer E.-M.; Füchtbauer A.; Wertz K.; Besenbacher F.; Pedersen F. S. & Duch M. (2009). Identification of Distinct Topographical Surface Microstructures Favoring Either Undifferentiated Expansion or Differentiation of Murine Embryonic Stem Cells, *Stem Cells and Development,* 18(9): 1331-42.

Martínez E.; Lagunas A.; Mills C.; Rodríguez-Seguí S.; Estévez M.; Oberhansl S.; Comelles J. & Samitier J. (2009). Stem cell differentiation by functionalized micro- and nanostructured surfaces, *Nanomedicine,* 4(1): 65-82.

McFarlin D. R.; Finn K. J.; Nealey P. F. & Murphy C. J. (2009). Nanoscale Through Substratum Topographic Cues Modulate Human Embryonic Stem Cell Self-Renewal, *Journal of Biomimetics, Biomaterials and Tissue Engineering,* 2): 15-26.

Mei Y.; Saha K.; Bogatyrev S. R.; Yang J.; Hook A. L.; Kalcioglu Z. I.; Cho S.-W.; Mitalipova M.; Pyzocha N.; Rojas F.; Van Vliet K. J.; Davies M. C.; Alexander M. R.; Langer R.; Jaenisch R. & Anderson D. G. (2010). Combinatorial development of biomaterials for clonal growth of human pluripotent stem cells, *Nature Materials,,* 9(9): 768-78.

Melkoumian Z.; Weber J. L.; Weber D. M.; Fadeev A. G.; Zhou Y.; Dolley-Sonneville P.; Yang J.; Qiu L.; Priest C. A.; Shogbon C.; Martin A. W.; Nelson J.; West P.; Beltzer J. P.; Pal S. & Brandenberger R. (2010). Synthetic peptide-acrylate surfaces for long-term self-renewal and cardiomyocyte differentiation of human embryonic stem cells, *Nature Biotechnology,* 28(6): 606-10.

Mieszawska A. & Kaplan D. (2010). Smart biomaterials - regulating cell behavior through signaling molecules, *BioMed Central Biology,* 8(1): 59.

Na J.; Furue M. K. & Andrews P. W. (2010). Inhibition of ERK1/2 prevents neural and mesendodermal differentiation and promotes human embryonic stem cell self-renewal, *Stem cell Research,* 5(2): 157-69.

Nagaoka M.; Hagiwara Y.; Takemura K.; Murakami Y.; Li J.; Duncan S. A. & Akaike T. (2008). Design of the Artificial Acellular Feeder Layer for the Efficient Propagation of Mouse Embryonic Stem Cells, *The Journal of Biological Chemistry,* 283(39): 26468-76

Nagaoka M.; Koshimizu U.; Yuasa S.; Hattori F.; Chen H.; Tanaka T.; Okabe M.; Fukuda K. & Akaike T. (2006). E-Cadherin-Coated Plates Maintain Pluripotent ES Cells without Colony Formation, *PLoS ONE,* 1(1): e15.

Nagaoka M.; Si-Tayeb K.; Akaike T. & Duncan S. (2010). Culture of human pluripotent stem cells using completely defined conditions on a recombinant E-cadherin substratum, *BioMed Central Developmental Biology,* 10(1): 60.

Nur-E-Kamal A.; Ahmed I.; Kamal J.; Schindler M. & Meiners S. (2006). Three-Dimensional Nanofibrillar Surfaces Promote Self-Renewal in Mouse Embryonic Stem Cells, *Stem Cells,* 24(2): 426-33.

Park J. H.; Ryu J. M. & Han H. J. (2011). Involvement of caveolin-1 in fibronectin-induced mouse embryonic stem cell proliferation: Role of FAK, RhoA, PI3K/Akt, and ERK1/2 pathways, *Journal of Cellular Physiology*, 226(1): 267-75.

Reilly G. C. & Engler A. J. (2010). Intrinsic extracellular matrix properties regulate stem cell differentiation, *Journal of Biomechanics*, 43(1): 55-62.

Riveline D.; Zamir E.; Balaban N. Q.; Schwarz U. S.; Ishizaki T.; Narumiya S.; Kam Z.; Geiger B. & Bershadsky A. D. (2001). Focal contacts as mechanosensors: externally applied local mechanical force induces growth of focal contacts by an mDia1-dependent and ROCK-independent mechanism., *Journal of Cell Biology*, 153(6): 1175-86.

Rodin S.; Domogatskaya A.; Strom S.; Hansson E. M.; Chien K. R.; Inzunza J.; Hovatta O. & Tryggvason K. (2010). Long-term self-renewal of human pluripotent stem cells on human recombinant laminin-511, *Nature Biotechnology*, 28(6): 611-5.

Saha S.; Ji L.; de Pablo J. J. & Palecek S. P. (2006). Inhibition of human embryonic stem cell differentiation by mechanical strain, *Journal of Cellular Physiology*, 206(1): 126-37.

Sawada Y. & Sheetz M. P. (2002). Force transduction by Triton cytoskeletons., *Journal of Cell Biology*, 156(4): 609-15.

Schmidt R. C. & Healy K. E. (2009). Controlling biological interfaces on the nanometer length scale, *Journal of Biomedical Materials Research Part A*, 90A(4): 1252-61.

Smith L. A.; Liu X.; Hu J. & Ma P. X. (2010). The Enhancement of human embryonic stem cell osteogenic differentiation with nano-fibrous scaffolding, *Biomaterials*, 31(21): 5526-35.

Stiehler M.; Lind M.; Mygind T.; Baatrup A.; Dolatshahi-Pirouz A.; Li H.; Foss M.; Besenbacher F.; Kassem M. & Bünger C. (2008). Morphology, proliferation, and osteogenic differentiation of mesenchymal stem cells cultured on titanium, tantalum, and chromium surfaces, *Journal of Biomedical Materials Research Part A*, 86A(2): 448-58.

Thomson J. A.; Itskovitz-Eldor J.; Shapiro S. S.; Waknitz M. A.; Swiergiel J. J.; Marshall V. S. & Jones J. M. (1998). Embryonic Stem Cell Lines Derived from Human Blastocysts, *Science*, 282(5391): 1145-7.

Unger C.; Skottman H.; Blomberg P.; Sirac Dilber M. & Hovatta O. (2008). Good manufacturing practice and clinical-grade human embryonic stem cell lines, *Human Molecular Genetics*, 17(R1): R48-53.

Valamehr B.; Jonas S. J.; Polleux J.; Qiao R.; Guo S.; Gschweng E. H.; Stiles B.; Kam K.; Luo T. J.; Witte O. N.; Liu X.; Dunn B. & Wu H. (2008). Hydrophobic surfaces for enhanced differentiation of embryonic stem cell-derived embryoid bodies., *Proceedings of the National Academy of Sciences of the United States of America*, 105(38): 14459-64.

Villa-Diaz L. G.; Nandivada H.; Ding J.; Nogueira-de-Souza N. C.; Krebsbach P. H.; O'Shea K. S.; Lahann J. & Smith G. D. (2010). Synthetic polymer coatings for long-term growth of human embryonic stem cells, *Nature Biotechnology*, 28(6): 581-3.

von der Mark K.; Park J.; Bauer S. & Schmuki P. (2010). Nanoscale engineering of biomimetic surfaces: cues from the extracellular matrix, *Cell and Tissue Research*, 339(1): 131-53.

Welham M. J.; Storm M. P.; Kingham E. & Bone H. K. (2007). Phosphoinositide 3-kinases and regulation of embryonic stem cell fate, *Biochemical Society Transactions*, 35(Pt2): 225-8.

Williams R. L.; Hilton D. J.; Pease S.; Willson T. A.; Stewart C. L.; Gearing D. P.; Wagner E. F.; Metcalf D.; Nicola N. A. & Gough N. M. (1988). Myeloid leukaemia inhibitory factor maintains the developmental potential of embryonic stem cells, *Nature,* 336(6200): 684-7.

Yang J.; Mei Y.; Hook A. L.; Taylor M.; Urquhart A. J.; Bogatyrev S. R.; Langer R.; Anderson D. G.; Davies M. C. & Alexander M. R. (2010). Polymer surface functionalities that control human embryoid body cell adhesion revealed by high throughput surface characterization of combinatorial material microarrays, *Biomaterials,* 31(34): 8827-38

Ying Q.-L.; Nichols J.; Chambers I. & Smith A. (2003). BMP Induction of Id Proteins Suppresses Differentiation and Sustains Embryonic Stem Cell Self-Renewal in Collaboration with STAT3, *Cell,* 115(3): 281-92.

Ying Q.-L.; Wray J.; Nichols J.; Batlle-Morera L.; Doble B.; Woodgett J.; Cohen P. & Smith A. (2008). The ground state of embryonic stem cell self-renewal, *Nature,* 453(7194): 519-23.

9

Directed Differentiation of Human Embryonic Stem Cells in Combination of Biomaterials

Huihui Wang, Xiaojing Xu, Rong Ye and Huanxiang Zhang
Department of Cell Biology, Medical College of Soochow University, Suzhou 215123, China

1. Introduction

Human embryonic stem cells (hESC) are derived from the inner cell mass of preimplantation embryos and defined by their extensive self-renewal capacity and their potential to differentiate into any cell types of three germ layers (ectoderm, mesoderm and endoderm). Since the first hESC line was established in the year of 1998, much progress has been made on studies of hESC culture, cell line establishment, and directed differentiation.

2. Culture system of hESC

Culturing hESC in vitro requires not only keeping their multiplication and undifferentiated diploid condition but also maintaining their potential to differentiate into three germ layers. hESC lines are traditionally derived and maintained on mouse embryonic fibroblasts (MEF); even now using MEF as feeder is one widespread way that scientists adopt for hESC cultures. Because of the xenogeneic property of MEF and rapid senescence, people began to search for various kinds of human fibroblasts as feeders and the use of them has been suggested as a plausible alternative. At present, there have developed many kinds of human cell feeders and attempted feeder-free (and serum-free) culture system, non-conditioned medium culture system.

2.1 Feeder-containing culture system

As one of the earliest used feeders for hESC, MEF promotes hESC proliferation by secreting mitogenic factors like fibroblast growth factor (FGF) and inhibit hESC differentiation by secreting differentiation inhibiting factors like leukaemia inhibitory factor (LIF). However, not only MEF enter senescence rapidly, but also need prepare primary MEF cells uninterruptedly. Moreover the use of MEF and other components of animal origin in the culture media for hESC substantially elevates the risk of contaminating these cell lines with infectious agents of animal origin, thereby severely limiting their potential for clinical application[1]. As the therapeutic potential of hESC lies in the transplantation of differentiated cell types for various degenerative diseases, it is important to eliminate potential xenogeneic contamination[2]. Otherwise, hESC in prolonged culture in vitro may cause higher karyotype instability and uncontrollable spontaneous differentiation phenomenon, which also confine manipulation in clinical application[3]. Accordingly, many research groups attempt to use various kinds of human-source cells as feeders, such as human embryo fibroblast[4], adult fetal muscle fibroblast, adult skin and adult fallopian tube, human marrow stromal cells (hMSC)[5], human foreskin fibroblast (hFF)[6], human placenta-derived feeder (HPC)[7], human amniotic epithelial (HAE)[8], and post natal human fibroblast.

The number of colonies, number of cells produced, and cell survival rates, all of these indexes are significantly higher on MEF than on human feeder cells ($P < 0.01$) and the number of AP-positive colonies and cell quantity are also significantly higher on hMSC than on hFF ($P < 0.01$). It implies that hMSC is superior to hFF in supporting the proliferation of hESC[9]. SNUhES3 (hESC line) cultured on MEF for 50 passages display three chromosomal abnormalities: t(3:5) in the 10th, trisomy 12 in the 20th, and complex of t(3:5) and trisomy 12 in the 30th passage. In the HPC group, only one chromosomal abnormality is noted at the 30th passage: t(3:5)[7].

In another study, people derivatized FGF2 expressing germ layer derived fibroblast cells (GLDF) from hESC lines[2]. These feeder cells with fibroblast cells-like properties maintained the properties of hESC in prolonged culture over 30 passages. Furthermore, these GLDF cells could secrete FGF2 to maintain pluripotency of hESC cultures even in the absence of supplemental FGF2. Otherwise human embryonic lung fibroblasts (hELF)[10] overexpress LIF that result in the higher expressing of hESC multipotency genes Oct4 and Nanog.

Choosing human-derived cells as feeders solves the puzzle of biological safety in some degree, but need to prepare and identify the feeder uninterruptedly. Moreover, the growth cycle is finite, the manipulation is elaborate, all of this limite hESC large amplification and stable passage, bring disadvantages to large-scale culture of hESC[11]. To overcome this constraint, there has developed feeder-free culture system. It may not only offer chance for hESC large-scale culture but also accelerate the step of clinical application.

2.2 Feeder-free culture system

In order to simplify the culture of hESC, several research groups have engaged to establish the feeder-free culture system. Unlike feeder-based cultures, which require the simultaneous growth of feeder and stem cells, resulting in mixed cell populations, stem cells grown on feeder-free systems are easily separated from the surface, presenting a pure population of cells for downstream applications. The conditioned medium is prepared by adding proliferation promoting factors and differentiation inhibiting factors, or cells that secrete the above factors are used. Two main strategies are adopted: (1) using conditioned medium (CM) and culture dishes coated with extracellular matrix as matrigel, laminin or fibronectin; (2) feeder-free and serum-free culture system, in which hESC are cultured in special medium containing serum replacement (SR) and various kinds of growth factors promoting hESC self-renewal.

It has been successful to culture hESC using feeder-free system with matrigel or laminin, but MEF conditioned medium is required which may propagate infectious agents of animal origin to hESC[12]. Besides, matrigel is a mixture derived from murine tumors, containing many extracellular matrix and various kinds of growth factors. For this reason, the ideal feeder-free cultures should deploy xeno-free extracellular matrix and conditioned medium.

A novel feeder layer-free culture system for hESC, has been presented. The defined culture system is based on serum replacement, a combination of growth factors including transforming growth factor β1 (TGFβ1), LIF, bFGF and fibronectin matrix[13]. BMP-11/GDF-11 and Myostatin/GDF-8 are both members of the TGF-β superfamily that can activate SMAD2/3phosphorylation via the type I receptors ALK4, ALK5, or ALK7, ultimately maintain hESC undifferentiated, maintain POU5f1, NANOG, TRA-1-60, and SSEA4 normally expression[14]. BMP4 synergizing with LIF can maintain mouse ESC self-renewal, while BMPs alone induce hESC differentiation. hESC cultured in unconditioned medium (UM) are subject to high levels of BMP signaling activity. The BMP antagonist noggin synergizes with bFGF to repress BMP signaling and sustain undifferentiated proliferation of hESC in the absence of fibroblasts[15].

Genetic manipulation of hESC is important for both present research and future commercial applications. Under the feeder-free and serum-free conditions, plasmid transfection, virus infection, and siRNA transfection are highly effective. Stable genetically modified hESC lines can be generated with these genetic manipulations without loss of pluripotency or differentiation potential. The majority of lines generated in this system display a normal karyotype[16]. Immunofluorescence microscopy and quantitative PCR analyses of hESC under feeder-free conditions have illustrated with new results on cellular localization of transcriptional factors and components of the Hedgehog, Wnt, and PDGF signaling pathways to primary cilia in stem cell maintenance and differentiation[17].

3. Directed differentiation of hESC

Upon differentiation, hESC can give rise to a variety of cell types, including nerve cells, cardiac muscle cells, endothelial cells, hematopoietic cells, and insulin-producing beta cells, highlighting the importance of these cells in studying the developmental mechanisms and the potential in transplant therapy.

3.1 Neural cells

Studies indicated that neuroectodermal precursors in neural rosettes could be induced by replating EBs into the DMEM/F12 medium containing insulin, transferrin (TFN), corporin, putrescine, liquemine, FGF2. Immunostaining of neuroectodermal cells in rosettes for early neural markers Sox1 and nestin, and the neuronal marker TuJ1, showed that the radially organized columnar neuroepithelial cells coexpressed Sox1 and nestin[18]. All these hESC-derived neural progenitor cells could differentiate into neurons, astrocytes and oligodendrocytes.

Parkinson's disease (PD) is a neurodegenerative disorder characterized by progressive and selective loss of dopaminergic (DA) neurons in the midbrain substantia nigra[19]. The prevailing pharmacological strategy has side effects over time, therefore, much attention has been focused on the transplantation of DA-synthesizing cells. Studies showed that after differentiation of passage 2 cells, 30–50% of the total cells derived from hESC were Tuj1-positive neurons, among the Tuj1-positive neuronal population 64–79% of the cells expressed TH[20].

For the purpose of applying the hESC to PD, many researchers have tried to develop protocols increasing the purity of DA neurons. Myung Soo Cho, etc [21] introduce a method that hESC give rise to functional tyrosine hydroxylase-positive (TH^+) neurons up to nearly 86% of the total hESC-derived neurons. The most unique feature of this method is the generation of homogeneous spherical neural masses (SNMs) from the hESC-derived neural precursors, and it only takes two weeks to induce of DA neurons from SNMs.

hESC-derived DA nerons and neural progenitors raise Ca^{2+} from intra- and extracellular compartments in response to depolarization, glutamate, ATP and dopamine D_2 receptor activation, while cAMP is elevated in response to forskolin and 3-isobutyl-methylxanthine[22]. Analysis of hESC and hESC-derived neural stem cell nuclear extracts revealed an increased expression of Reptin52 in neurosphere nuclei[23] that serves a pivotal regulatory role in nuclear activities such as transcription regulation and histone modification.

The development of stem cell-based neural repair strategies requires detailed knowledge on the interaction of migrating donor cells with the host brain environment. hESC-derived glial precursors (ESGPs) transduced with a retrovirus encoding the polysialyltransferase STX exhibit overexpression of polysialic acid(PSA) which is a carbohydrate polymer attached to the neural cell adhesion molecule (NCAM) [24]. Chemotaxis assays show that overexpression of PSA results in an enhanced chemotactic migration of these cells toward gradients of a variety of chemoattractants, including FGF2, platelet-derived growth factor (PDGF), and brain-derived neurotrophic factor (BDNF), and this effect is mediated via the phosphatidylinositol 3'-kinase (PI3K) pathway.

3.2 Hematopoietic stem cells

The most common human cell-based therapy applied today is hematopoietic stem cell (HSC) transplantation[25]. Human bone marrow, mobilized peripheral blood, and umbilical cord blood are considered as the major sources of transplantable HSCs, but both compatibility between donor and recipient and required quantity limit their clinical application. In contrast, hESC have been shown to differentiate into the hematopoietic cell fate, providing an alternative source of transplantable blood cells. hESC-derived hematopoietic stem cells which emerge from a subset of embryonic endothelium expressing PECAM-1, Flk-1, and VE-Cadherin, but lacking CD45 (CD45negPFV) show similar clonogenic capacity and primitive phenotype to somatic sources of hematopoietic progenitors and possess limited in vivo repopulating capacity in immunodeficient mice. A subset of hESC-derived $CD45^+$ hematopoietic cells coexpress CD34 and show progenitor function in colony-forming units assays. However, compared to HSC from fetal blood (FB) or cord blood, these hESC-HSC display distinct functional properties, including poor repopulation ability, impaired differentiation and lack of homing[26]. Only 2.4% of differentially expressed transcripts were common for FB–HSCs and candidate hESC-HSCs[26], suggesting a completely different molecular signature for HSCs isolated from two different in utero ontogeny stages. Several key hematopoietic transcription factors, such as RUNX3, TAL1, VAV1, LMO2, AML-1, and c-mac, apoptosis and cycle regulators, such as CDC42, CDC27, CyclinD3 and CDK4, and cell aggregation and homing genes may contribute to explain the functional differences between hESC-HSCs and FB-HSCs. Importantly, the gene expression profiling study revealed that developmental signaling pathways such as Notch, Wnt, BMP, Shh are involved in HSC self-renewal and hematopoietic specification.

Several complementary methods are used to demonstrate canonical Wnt signaling that is important for development of hESC-derived cells with both hematopoietic and endothelial potential[27]. The development of cells with hematoendothelial potential decreases dramatically when treated with dickkopf1 which inhibits Wnt signaling. In addition, activation of the canonical Wnt signaling pathway in hESC by coculture with stromal cells that express Wnt1, but not use of noncanonical Wnt5-expressing stromal cells, results in an accelerated differentiation and higher percentage of $CD34^{bright}CD31^{+}Flk1^{+}$cells at earlier stages of differentiation.

3.3 Mesenchymal stromal cells

Mesenchymal stromal cells (MSCs), originally isolated from adult bone marrow (BM), have also been derived from hESC, either through coculturing with the OP9 murine BM stromal cell line or directly from ESCs cultured without feeder cells[28,29]. The simultaneous generation of $CD73^{+}$ MSCs along with $CD34^{+}$ hematopoietic cells from HESC has been shown when they are cocultured with OP9 murine stromal cells[28]. hESC–derived MSCs show spindle-shaped fibroblast-looking morphology, exhibit similar cell surface marker CD29, CD44, CD54, CD73, CD90 and CD105, but negative for CD34, CD45, and the endothelial marker CD31, characteristics when compared to BM-derived MSCs, and could differentiate into osteocytes, adipocytes, and chondrocytes.

3.4 Insulin-producing β cells

Type I diabetes is a disease in which β-cells of the pancreas are destroyed by an autoimmune mechanism[30]. Use of exogenous insulin to treat diabetes is life-saving; however, it does not truly mimic the body's natural response to blood glucose[31]. Islet cell transplantation is currently considered to be an important method of therapy[32]. But the scarcity of islets prevents this therapy. The isolation of hESC introduced a new prospect for obtaining a sufficient number of β cells for transplantation. Segev et al. present a method for forming immature islet-like clusters of insulin-producing cells derived from hESC[33]. Seven-day-old EBs were first cultured and plated in insulin-transferrin-selenium-fibronectin medium (ITSF) about a week, followed by medium supplemented with N2, B27, and bFGF. At the next stage, bFGF was removed and nicotinamide was added, and the total glucose concentration in the medium was reduced. After 4 days of culture, the formation of clusters exhibited higher insulin secretion and had longer durability than cells grown as monolayers. Gen-hong Mao[34] demonstrates a five-stage protocol with adding exendin-4 instead of nicotinamide and generates islet-like cells from human embryonic stem (ES) cells. Immunofluorescence analysis revealed that most c-peptide positive cells coexpressed inlusin, PDX-1 (pancreas duodenum homeobox-1), glucagon, somatostatin or pancreatic polypeptide. Insulin and other pancreatic β-cell-specific genes were all present in the differentiated cells.

Following the illumination of hESC differentiation mechanism and successfully culturing various kinds of terminal differentiated cells, tissues and organs, it may apply functioning cells of hESC-source in replacement therapies.

4. hESC coupled with biomaterials

Tissue engineering is the study of the growth of new connective tissues or organs, from cells and a collagenous scaffold to produce a fully functional organ for implantation back into the donor host. Scaffolding can be used as a cell support device upon which cells are seeded in vitro; cells are then encouraged to lay down matrix to produce the foundations of a tissue for transplantation. Tissue engineering scaffolds are designed to influence the physical, chemical and biological environment surrounding a cell population[35]. Many matrix materials that are in application of current research are mainly natural materials, synthetic polymer and biodegradable materials, ceramic materials and their complex formed between the composite materials. Natural biological materials, compared with synthetic materials, have many advantages, the abundant source, low cost, and with good biocompatibility, and that some materials have pore structure of natural systems. Currently, the mainly applied materials are collagen (type I collagen in the main), gelatin, chitin, chitosan, natural coral and its derivatives. Now the synthetic biodegradable polymer materials that extensively used are mainly polylactic acid (PLA), polyglycolic acid (PGA) and various copolymers between them. These materials are bulk-degrade by hydrolysis, providing a controllable drug release and degradation profile to match tissue in-growth. The main structures of PLA, PGA and their copolymers used in tissue engineering applications are in the form of fibrous, porous foam and tubular structure.

4.1 Electrospun fibrous scaffolds

Electrospun fibrous scaffolds can be prepared with high degree of control over their structure creating highly porous meshes of ultrafine fibers that mimic the natural three-dimensional environment of the in vivo extracellular matrix (ECM), and are amenable to various functional modifications targeted towards enhancing stem cell survival and proliferation, directing specific stem cell fates, or promoting tissue organization[36]. Evidence shows that of electrospun biodegradable polymers as scaffolds not only enhances the differentiation of mouse ES cells into neural lineages but also promotes and guides the neurite outgrowth[37]. Cultivation of hESC on electrospun fibrous polyurethane scaffolds also proved successful and neuronal differentiation was observed via standard immunocytochemistry[38]. Scanning electron micrographs confirmed neurite outgrowth and connection to adjacent cells, as well as cell attachment to individual fibers of the fibrous scaffold. Scaffold architecture has been shown to modify the response of cells and subsequent tissue formation, as demonstrated by the generation of mineralization fronts in specific regions of scaffolds. Nano to microscale topography has been demonstrated to affect cell behavior by modification of cytoskeleton arrangements.

Although complex structures are formed, the mechanism of adhesion of hESC to the surfaces of the scaffolds remains largely unknown. Steven Y. Gao[39] achieved the efficient adhesion of pluripotent hESC to 3D PLGA scaffolds. In the 2-D PLGA model, the scaffold surfaces were coated with laminin and flow cytometry analysis revealed that almost all of the pluripotent single cells expressed the integrin α6, with a small percentage also expressing α3β1, which facilitates adhesion to laminin. In the 3-D enviroment, the results were similar to those gained from the 2-D model, which suggests that the adhesion of pluripotent hESCs is mainly affected by the surface properties that they adhere onto rather than the geometric properties of the culture system.

4.2 Porous foam scaffolds

Porous biodegradable polymer scaffolds can be used to support ES cells because they represent a promising system for allowing formation of complex 3D tissues during differentiation[40]. The scaffolds which consist of a 50/50 blend of PLGA and PLLA could support hESC that remain viable for at least 2 weeks in vivo. Two weeks after transplantation of hESC-scaffold constructs into SCID mice, defined epithelial tubular structures and neural tube-like rosettes were observed. In addition, when cultured in vitro with both NGF and NT-3, enhanced numbers of neural structures and staining of nestin and βIII-tubulin were observed with hESC-seeded polymer scaffolds[41]. NGF and RA preferentially favor the differentiation toward ectodermal and mesodermal lineages[42]. Subcutaneous transplantation of PLGA scaffolds seeded with the hESC-derived islet-like cells or cell transplantation under kidney capsules for further differentiation in vivo could improve 6 h fasted blood glucose levels and diabetic phenotypes in streptozotocin-induced diabetic SCID mice[34]. More interestingly, blood vessels of host origin can be observed, invading the cell–scaffold complexes. This suggests scaffolds can serve as vehicles for islet-like cell transplantation.

Other study has certificated that 3D porous alginate scaffolds provide a conductive environment for generation of well-vascularized EBs from hESC[43]. EBs can be generated directly from hESC suspensions within 3D porous alginate scaffolds. The alginate scaffold pores which provide a confining environment enable efficient formation of round, small-sized EBs with a relatively high degree of cell proliferation and differentiation, and induce vasculogenesis in the forming EBs to a greater extent. Recent study demonstrated that 3D porous natural polymer scaffold, comprised of chitosan and alginate, could support sustained self-renewal of hESCs without the support of feeder cells or conditioned medium[44].

4.3 Other scaffolds

Scaffolds can be used as a growth factor/drug delivery device[35]. This strategy involves the scaffold being combined with growth factors, so upon implantation cells from the body are recruited to the scaffold site and form tissue upon and throughout the matrices. Recently, vascular progenitor cells as well as endothelial and smooth muscle cells have been isolated from hESC encapsulated in bioactive hydrogel-based scaffolds combined with immobilized regulatory factors: a tethered RGD peptide and microencapsulated VEGF165[45]. Others have investigated the functions of hESC-derived neural progenitor cells (hESC-NPC) after transplanted in collagen scaffolds supplemented with growth factors[46] and reveal that implantation of hESC-NPC into the spinal cord with collagen scaffold improved the recovery of hindlimb locomotor function and sensory responses in an adult rat model of SCI.

Tzu-I Chao, et al.[47] have reported that 2D thin film scaffolds composed of biocompatible polymer grafted carbon nanotubes (CNTs), can selectively differentiate hESC into neurons while maintaining excellent cell viability. According to fluorescence image analysis, neuron differentiation efficiency of polyacrylic acid grafted CNT thin films is greater than that on polyacrylic acid thin films. Other comparison has been done between silk and chitosan scaffolds in chondrogenesis of human cell sources[48]. After 4 weeks of cultivation, hESC-derived MSCs were promising for chondrogenesis, particularly in the silk scaffolds with BMP6. The polyurethane acrylate (PUA), fabricated into nanoscale ridge/groove pattern arrays with its surface treated in oxygen plasma for 60s Results showed that after alone[49] can effectively and rapidly induce the differentiation of hESC into a neuronal lineage without the addition of any biochemical or biological agents.

Future applications of those versatile scaffolds platform to human embryonic and induced pluripotent stem cells for functional tissue repair and regeneration will further expand its potential for regenerative therapies.

5. Prospect of HESC

Expecting to use hESC largely and safely in clinical, there are a lot of works to do. hESC differentiation is typically stimulated via generation of EBs and lineage commitment of individual cells depends upon numerous cues throughout the EB environment, including EB shape and size. Common EB formation protocols, however, produce a very heterogeneous size distribution, perhaps reducing efficiency of directed differentiation[50]. Some scholars have developed a 3-D microwell-based method to maintain undifferentiated hESC cultures for weeks without passaging using physical and extracellular matrix patterning constraints to limit colony growth. Microwell culture also permits formation of hESC colonies with a defined size, which can then be used to form monodisperse EBs. When cultured in this system, hESC retained pluripotency and self-renewal, and were able to be passaged to standard unconstrained culture conditions.

Successful clinical implementation of tissue engineering products generated from ESCs would require a controlled reproducible culture system for the expansion of the cells and differentiation into functional tissues. It is a marvelous scenario which involve combination of the bioreactor expansion and subsequent differentiation of the ESCs to generate the specialized cell types. N. I. zur Nieden[51] showed that suspension bioreactors which can be adapted for long term culture could be used in the regulated large-scale expansion of highly pluripotent murine ESCs. The ESC cultures remain highly undifferentiated, when serially passaged in suspension bioreactors for extended periods. Otherwise different growth factors have different effects on cell proliferation and differentiation[52]. BMPs have previously been shown to induce hESC differentiation, in apparent contrast to mouse embryonic stem (ES) cells, in which BMP4 synergizes with LIF to maintain self-renewal. The BMP antagonist noggin synergizes with bFGF to repress BMP signaling and sustain undifferentiated proliferation of hESC in the absence of fibroblasts or CM. These findings suggest a basic difference in the self-renewal mechanism between mouse and human ES cells.

Stem cells and human tissue engineering are two hot spots. Many scholars attempt to combine hESC with tissue engineering. Although transplantation of ES cells-derived neural progenitor cells has been demonstrated with some success for either spinal cord injury repair in small animal model, control of ES cell differentiation into complex, viable, higher ordered tissues is still challenging[37].

Silk fibroin (SF), the core structural protein of *Bombyx mori* silkworm silks, has widely been studied as biomaterials in tissue engineering and regenerative medicine due to the biocompatibility, impressive mechanical property, high yields through sericulture and controllability of degradability, as well as the versatility in processing, the availability of different morphologies and the ease of sterilization and surface modification of these SF-based materials. While a number of studies detailed the ability of SF to support the attachment and growth of a variety of human cell types and the promising applications of SF scaffolds in wound healing and in bone, cartilage, blood vessel, tendon, or ligament tissue engineering, there is little information about the interaction of astroglial and neuronal cells with SF-based scaffolds.

Among all the silk materials in different morphologies, of great interest are the electrospun nonwoven SF nanofibers or nets that mimic the nanostructured components of the extracellular matrix and are supposed to provide more surface area and rougher topography for cell attachment. Upon implantation these nanofibrous SF scaffolds might ameliorate the conditions at the injury sites via the formation of the oriented network of astroglial and neuronal cells by arranging these cells along the nanofibers and guiding axonal outgrowth, and finally prevent the formation of physical and molecular barriers to the CNS repair. Nonwoven SF nanofibers were studied with a variety of cells, but little information is hitherto available with neurons, astrocytes and their progenitors, the main cell types participating in the glial scaring.

Tussah silk is a type of wild silks, and the chemical structure, molecular conformation and physical properties of tussah silk fibroin (TSF) have been extensively studied. In contrast to *Bombyx mori* silkworm silk fibroin (SF), the amino acid composition of TSF is characterized by more Ala, Asp and Arg contents, and less Gly, and the presence of tripeptide sequence Arg-Gly-Asp (RGD), which is known to promote cell attachment and thus might make TSF an even more promising biomaterial for use in biomedical applications and tissue engineering.

In an attempt to develop strategies for fabricating nerve implants, using the regenerated SF, TSF, or blends of both at different proportions, that could be applied in vivo as the cell substrates to control the arrangement of astroglial cells and guide the axon growth, we prepared nonwoven nanofiber scaffolds by electrospinning and analyzed the behavior of astrocytes, neurons and neural progenitor cells, isolated from newborn rat cerebral cortex or derived from hESC, on these nonwoven fibroin nanofibers. In particular, we focused on the migration and spreading, in addition to the attachment, viability and proliferation of these cells. Our long-term goal is to identify the optimum conditions in the context of the properties of the nanofiber substrates that may allow for the manipulation of astroglial cells and ameliorate the nonpermissiveness of the CNS injury sites upon implantation.

6. Acknowledgments

This work was supported by National Natural Science Foundation of China (No.30870642, 30671041), Natural Science Foundation and Natural College Science Foundation of Jiangsu Province (BK2009119, 06KJA18025), and National Basic Research Program of China (973 Program No.2005CB623902) and China 973 Project (Reconstruction of spinal functions and bone tissue engineering after minimal invasive surgery).

Corresponding author: Huanxiang Zhang, Department of Cell Biology, Medical College of Soochow University, Suzhou 215123, China. Email: hzhang@suda.edu.cn

7. References

[1] Richards, M., et al., *Comparative evaluation of various human feeders for prolonged undifferentiated growth of human embryonic stem cells.* STEM CELLS, 2003. 21(5): p. 546-56.

[2] Saxena, S., et al., *FGF2 secreting human fibroblast feeder cells: a novel culture system for human embryonic stem cells.* Mol Reprod Dev, 2008. 75(10): p. 1523-32.

[3] Escobedo-Lucea, C. and M. Stojkovic, *Growth of human embryonic stem cells using derivates of human fibroblasts.* Methods Mol Biol, 2010. 584: p. 55-69.

[4] Heng, B.C., G.Q. Tong, and S.C. Ng, *Human embryonic stem cell-derived fibroblastic and epitheloid lineages as xeno-free support?* In Vitro Cell Dev Biol Anim, 2004. 40(5-6): p. 129-30.

[5] Cheng, L., et al., *Human adult marrow cells support prolonged expansion of human embryonic stem cells in culture.* STEM CELLS, 2003. 21(2): p. 131-42.

[6] Hovatta, O., et al., *A culture system using human foreskin fibroblasts as feeder cells allows production of human embryonic stem cells.* Hum Reprod, 2003. 18(7): p. 1404-9.

[7] Kim, S.J., et al., *Human placenta-derived feeders support prolonged undifferentiated propagation of a human embryonic stem cell line, SNUhES3: comparison with human bone marrow-derived feeders.* Stem Cells Dev, 2007. 16(3): p. 421-8.
[8] Miyamoto, K., et al., *Human placenta feeder layers support undifferentiated growth of primate embryonic stem cells.* STEM CELLS, 2004. 22(4): p. 433-40.
[9] Zhu, W.W., et al., *[Different types of feeder cells for maintenance of human embryonic stem cells].* Zhongguo Yi Xue Ke Xue Yuan Xue Bao, 2009. 31(4): p. 468-72.
[10] Li, F., et al., *Leukemia inhibitory factor-expressing human embryonic lung fibroblasts as feeder cells for human embryonic germ cells.* Cells Tissues Organs, 2007. 186(4): p. 221-8.
[11] Amit, M., *Feeder-layer free culture system for human embryonic stem cells.* Methods Mol Biol, 2007. 407: p. 11-20.
[12] Xu, C., et al., *Feeder-free growth of undifferentiated human embryonic stem cells.* Nat Biotechnol, 2001. 19(10): p. 971-4.
[13] Amit, M., et al., *Feeder layer- and serum-free culture of human embryonic stem cells.* Biol Reprod, 2004. 70(3): p. 837-45.
[14] Hannan, N.R., et al., *BMP-11 and myostatin support undifferentiated growth of human embryonic stem cells in feeder-free cultures.* Cloning Stem Cells, 2009. 11(3): p. 427-35.
[15] Xu, R.H., et al., *Basic FGF and suppression of BMP signaling sustain undifferentiated proliferation of human ES cells.* Nat Methods, 2005. 2(3): p. 185-90.
[16] Braam, S.R., C. Denning, and C.L. Mummery, *Genetic manipulation of human embryonic stem cells in serum and feeder-free media.* Methods Mol Biol, 2010. 584: p. 413-23.
[17] Awan, A., et al., *Immunoflourescence and mRNA analysis of human embryonic stem cells (hESCs) grown under feeder-free conditions.* Methods Mol Biol, 2010. 584: p. 195-210.
[18] Ma, W., et al., *Cell-extracellular matrix interactions regulate neural differentiation of human embryonic stem cells.* BMC Dev Biol, 2008. 8: p. 90.
[19] Noble, C., *Understanding Parkinson's disease.* Nurs Stand, 2007. 21(34): p. 48-56; quiz 58.
[20] Perrier, A.L., et al., *Derivation of midbrain dopamine neurons from human embryonic stem cells.* Proc Natl Acad Sci U S A, 2004. 101(34): p. 12543-8.
[21] differentiationCho, M.S., et al., *Highly efficient and large-scale generation of functional dopamine neurons from human embryonic stem cells.* Proc Natl Acad Sci U S A, 2008. 105(9): p. 3392-7.
[22] Malmersjo, S., et al., *Ca2+ and cAMP Signaling in Human Embryonic Stem Cell-Derived Dopamine Neurons.* Stem Cells Dev, 2009.
[23] Barthelery, M., et al., *Reptin52 expression during in vitro neural differentiation of human embryonic stem cells.* Neurosci Lett, 2009. 452(1): p. 47-51.
[24] Glaser, T., et al., *Neural cell adhesion molecule polysialylation enhances the sensitivity of embryonic stem cell-derived neural precursors to migration guidance cues.* STEM CELLS, 2007. 25(12): p. 3016-25.
[25] Wang, L., et al., *Hematopoietic development from human embryonic stem cell lines.* Exp Hematol, 2005. 33(9): p. 987-96.
[26] Shojaei, F. and P. Menendez, *Molecular profiling of candidate human hematopoietic stem cells derived from human embryonic stem cells.* Exp Hematol, 2008. 36(11): p. 1436-48.
[27] Woll, P.S., et al., *Wnt signaling promotes hematoendothelial cell development from human embryonic stem cells.* Blood, 2008. 111(1): p. 122-31.

[28] Trivedi, P. and P. Hematti, *Simultaneous generation of CD34+ primitive hematopoietic cells and CD73+ mesenchymal stem cells from human embryonic stem cells cocultured with murine OP9 stromal cells.* Exp Hematol, 2007. 35(1): p. 146-54.

[29] Trivedi, P. and P. Hematti, *Derivation and immunological characterization of mesenchymal stromal cells from human embryonic stem cells.* Exp Hematol, 2008. 36(3): p. 350-9.

[30] Eshpeter, A., et al., *In vivo characterization of transplanted human embryonic stem cell-derived pancreatic endocrine islet cells.* Cell Prolif, 2008. 41(6): p. 843-58.

[31] *The effect of intensive treatment of diabetes on the development and progression of long-term complications in insulin-dependent diabetes mellitus. The Diabetes Control and Complications Trial Research Group.* N Engl J Med, 1993. 329(14): p. 977-86.

[32] Shapiro, A.M., et al., *Islet transplantation in seven patients with type 1 diabetes mellitus using a glucocorticoid-free immunosuppressive regimen.* N Engl J Med, 2000. 343(4): p. 230-8.

[33] Segev, H., et al., *Differentiation of human embryonic stem cells into insulin-producing clusters.* STEM CELLS, 2004. 22(3): p. 265-74.

[34] Mao, G.H., et al., *The reversal of hyperglycaemia in diabetic mice using PLGA scaffolds seeded with islet-like cells derived from human embryonic stem cells.* Biomaterials, 2009. 30(9): p. 1706-14.

[35] Howard, D., et al., *Tissue engineering: strategies, stem cells and scaffolds.* J Anat, 2008. 213(1): p. 66-72.

[36] Lim, S.H. and H.Q. Mao, *Electrospun scaffolds for stem cell engineering.* Adv Drug Deliv Rev, 2009. 61(12): p. 1084-96.

[37] Xie, J., et al., *The differentiation of embryonic stem cells seeded on electrospun nanofibers into neural lineages.* Biomaterials, 2009. 30(3): p. 354-62.

[38] Carlberg, B., et al., *Electrospun polyurethane scaffolds for proliferation and neuronal differentiation of human embryonic stem cells.* Biomed Mater, 2009. 4(4): p. 45004.

[39] Gao, S.Y., et al., *Modeling the adhesion of human embryonic stem cells to poly(lactic-co-glycolic acid) surfaces in a 3D environment.* J Biomed Mater Res A, 2010. 92(2): p. 683-92.

[40] Levenberg, S., et al., *Differentiation of human embryonic stem cells on three-dimensional polymer scaffolds.* Proc Natl Acad Sci U S A, 2003. 100(22): p. 12741-6.

[41] Levenberg, S., et al., *Neurotrophin-induced differentiation of human embryonic stem cells on three-dimensional polymeric scaffolds.* Tissue Eng, 2005. 11(3-4): p. 506-12.

[42] Inanc, B., A.E. Elcin, and Y.M. Elcin, *Human Embryonic Stem Cell Differentiation on Tissue Engineering Scaffolds: Effects of NGF and Retinoic Acid Induction.* Tissue Eng Part A, 2008.

[43] Gerecht-Nir, S., et al., *Three-dimensional porous alginate scaffolds provide a conducive environment for generation of well-vascularized embryoid bodies from human embryonic stem cells.* Biotechnol Bioeng, 2004. 88(3): p. 313-20.

[44] Li, Z., et al., *Feeder-free self-renewal of human embryonic stem cells in 3D porous natural polymer scaffolds.* Biomaterials, 2010. 31(3): p. 404-12.

[45] Gerecht, S., L.S. Ferreira, and R. Langer, *Vascular differentiation of human embryonic stem cells in bioactive hydrogel-based scaffolds.* Methods Mol Biol, 2010. 584: p. 333-54.

[46] Hatami, M., et al., *Human embryonic stem cell-derived neural precursor transplants in collagen scaffolds promote recovery in injured rat spinal cord.* Cytotherapy, 2009. 11(5): p. 618-30.

[47] Chao, T.I., et al., *Carbon nanotubes promote neuron differentiation from human embryonic stem cells.* Biochem Biophys Res Commun, 2009. 384(4): p. 426-30.

[48] Seda Tigli, R., et al., *Comparative chondrogenesis of human cell sources in 3D scaffolds.* J Tissue Eng Regen Med, 2009. 3(5): p. 348-60.

[49] Lee, M.R., et al., *Direct differentiation of human embryonic stem cells into selective neurons on nanoscale ridge/groove pattern arrays.* Biomaterials, 2010. 31(15): p. 4360-6.

[50] Mohr, J.C., J.J. de Pablo, and S.P. Palecek, *3-D microwell culture of human embryonic stem cells.* Biomaterials, 2006. 27(36): p. 6032-42.

[51] zur Nieden, N.I., et al., *Embryonic stem cells remain highly pluripotent following long term expansion as aggregates in suspension bioreactors.* J Biotechnol, 2007. 129(3): p. 421-32.

[52] Willerth, S.M., et al., *The effects of soluble growth factors on embryonic stem cell differentiation inside of fibrin scaffolds.* STEM CELLS, 2007. 25(9): p. 2235-44.

Part 2

Methods of Imaging Embryonic Stem Cells

10

Non-invasive Imaging of Human Embryonic Stem Cells Derived Endothelial Cells Using Reporter Genes

Yizhou Zheng[1], Deling Kong[2] and Zongjin Li[2,3]
[1]*State Key Lab of Experimental Hematology, Institute of Hematology & Hospital of Blood Diseases, Chinese Academy of Medical Sciences, Tianjin,*
[2]*The Key Laboratory of Bioactive Materials, Ministry of Education, Nankai University, the College of Life Science, Tianjin,*
[3]*Department of Pathophysiology, Nankai University School of Medicine, Tianjin*
China

1. Introduction

Human embryonic stem (hES) cells are pluripotent stem cells capable of self-renewal and differentiation into virtually all cell types (Thomson *et al.*, 1998, Reubinoff *et al.*, 2000). Thus, they hold tremendous potential as cell sources for regenerative therapies. Using endothelial cells for therapeutic angiogenesis/vasculogenesis of ischemia diseases has led to exploring hES cells as a potential source for endothelial progenitor cells. Due to the limitation pluripotency of hES cells, hES cells derived endothelial cells are advantageous when compared with other endothelial cell origin (Li *et al.*, 2007, Li *et al.*, 2009b). Understanding the *in vivo* behavior of transplanted cells requires novel imaging techniques to longitudinally monitor hES cells localization, proliferation, differentiation, and viability. Molecular imaging has given investigators a high-throughput, inexpensive, and sensitive means for tracking *in vivo* cell proliferation over times (Li *et al.*, 2009a). This advancement has significantly increased the understanding of the spatiotemporal kinetics of hES cells engraftment in living animals and greatly expedited basic research prior to future clinical translation.

In this chapter, the specific methods needed for tracking hES cells proliferation with bioluminescence imaging and *in vitro* endothelial differentiation of hES cells will be described. We first established stable hES cell line constitutively expressing double fusion reporter genes (enhanced green fluorescent protein and firefly luciferase) using lentiviral transduction (Li *et al.*, 2009b). Unlike other methods of cell tracking, reporter genes are inherited genetically and can be used to monitor cell proliferation and survival of transplanted cells and daughter cells. Next, we introduced two-step procedures to increase endothelial differentiation efficiency of hES cells by subcultured embryoid bodies (EBs) in collagen. Using whole genome microarrays, we investigated the hES cells derived endothelial cells (hES-ECs) transcriptome that occur among sequenced hES cells differentiation processes and human umbilical vein endothelial cells (HUVECs). By intravital microscope, we demonstrated that hES-ECs can form function vessels with blood flow. To monitor cell survive and therapeutic potential of hES-ECs for ischemia diseases, we then employed bioluminescence imaging techniques that allow noninvasive monitoring of transplanted hES-ECs engraftment.

2. Molecular imaging methods for stem cell therapy

For stem cell research to make the next quantum leap, it is imperative to understand the dynamic processes of hES cells homing, migration, biodistribution, proliferation, and differentiation in the same subject over time. The development of noninvasive imaging techniques is essential for conducting detailed preclinical studies to optimize the delivery methods and strategies that can enhance cell survival. A number of methods are available to track stem cells by molecular imaging. In general, there are two methods to label the cells: (1) direct labeling method, which physically introduce marker(s) into the cells before transplant; (2) indirect labeling method, which genetically introduce reporter gene(s) into the cells before transplant (Wu et al., 2004).

2.1 Direct labeling methods

For direct labeling, detectable probes can be loaded into or attached to the cells during tracking. Examples include labeling with super paramagnetic iron oxide (SPIO) for magnetic resonance (MR) imaging, 18F-fluoro-deoxy-glucose ([^{18}F]-FDG) for positron emission tomography (PET) imaging, [^{111}In]oxine for single-photon emission computerized tomography (SPECT), and quantum dots for fluorescence imaging. These techniques have been used for adult and embryonic stem cell imaging (Zhang *et al.*, 2007). Although direct labeling can be used to evaluate the initial deposition of implanted stem cells, the imaging signals tend to diminish with cell division and proliferation or become undetectable after the decay of radioactive tracers. Another confounding factor with this type of labeling and imaging is that non-viable cells may still generate a robust MR signal, as shown in a recent study comparing iron labeling imaging (direct) versus reporter gene imaging (indirect)(Li *et al.*, 2008).

2.2 Indirect labeling methods

Indirect imaging such as the reporter gene approach involves inserting reporter gene(s) into stem cells for the purpose of tracking. Products of reporter gene expression generally can be divided into 3 categories: enzyme-based (e.g., herpes simple virus type 1 thymidine kinase [HSV1-tk] or firefly luciferase [Fluc]), receptor-based (e.g., dopamine type 2 receptor [D2R]), and transporter-based (e.g., sodium-iodide symporter [NIS]). Stable transfection or transduction with reporter genes is useful in assessing kinetic survival status of the implanted cells because the reporter genes can be expressed as long as the cells are alive; the inserted reporter gene(s) can be passed on to daughter cells upon cell division. However, a major disadvantage for future clinical applications of reporter gene approach in cell tracking is the requirement for genetic manipulations of the cells, which may lead to insertional mutagenesis. Thus, the recent advances in site-specific chromosomal integration mediated by phiC31 integrase may become a useful tool to overcome this obstacle (Thyagarajan *et al.*, 2008).

3. Endothelial differentiation of hES cells

Due to the limitation of postnatal cell sources and expanding efficiency, endothelial differentiation of hES cells provide an alternative source to generate a large supply of healthy, functional cells. hES-ECs can be generated by two main approaches, spontaneous differentiation of embryoid bodies (EBs), also called 3-dimensional differentiation (3D), and 2-dimensional differentiation (2D) (Wang *et al.*, 2007, Yamahara *et al.*, 2008, Kaufman *et al.*, 2001).

For the 3D differentiation, hES cells need to be cultured in low attachment dish for 9-13 days(Levenberg *et al.*, 2002, Li *et al.*, 2008, Wang *et al.*, 2007). Different mediums have been used with or without growth factors. Whole-mount immunostaining CD31 confirmed that hES cells cultivated as EBs spontaneously differentiated into endothelial cells and formed blood vessel like structure(Levenberg *et al.*, 2002, Li *et al.*, 2008). Subsequently, EBs were digested into single cells. Differentiated endothelial cells or hemangioblasts were sorted by fluorescence-ctivated cell sorting (FACS) or magnetic-activated cell separation (MACS). For the 2D differentiation, hES cells were cultured on various mouse fibroblast feeder layers, including mouse embryo fibroblasts (MEF) (Wang *et al.*, 2007), OP9 (Yamahara *et al.*, 2008, Vodyanik *et al.*, 2005), S17 (Kaufman *et al.*, 2001, Vodyanik *et al.*, 2005), MS-5 (Vodyanik *et al.*, 2005), or mouse endothelial cells (Kaufman *et al.*, 2001). hES cells also can be cultured on collagen IV-coated plates for endothelial differentiation (Gerecht-Nir *et al.*, 2003).

Both *in vitro* assays and *in vivo* transplantation have been used to characterize the hES derived endothelial cells. After *in vitro* subculturing, these cells express CD31, CD34, Flk-1, VE-cadherin and vWF. They are capable of Dil-ac-LDL uptake and can generate tube-like structure formation on Matrigel. For *in vivo* assay, hES-ECs can form a tube-like structure within Matrigel plug. Interestingly, postnatal hemangioblast cell markers such as KDR and CD133 are robustly expressed in undifferentiated hES cells (Chen *et al.*, 2007, Li *et al.*, 2008) and thus are *not* good markers for endothelial selection of hES cells differentiation.

A major challenge of hES-based therapy is the generation of sufficient numbers of differentiated endothelial cells. The efficiency of endothelial differentiation of the 3D EB system is typically low, ranging from 1%-3% (Levenberg *et al.*, 2002, Li *et al.*, 2008). Moreover, it is difficult to get single cells from EBs by enzyme digestion as cell viability is low after this harsh digestion step. Several investigators have introduced modified approaches to bypass the 3D EB formation or serial differentiation procedures to increase differentiation efficiency (Cho *et al.*, 2007, Lu *et al.*, 2007). Wang *et al.* demonstrated that ~10% CD34+ progenitor cells are present by monolayer culturing of hES cells on MEF for 10 days (Wang *et al.*, 2007).

To optimize hES cells differentiation to endothelial lineage, we designed a staged protocol that involved EB formation (stage 1) and expansion of endothelial lineage by subcultured EB in collagen (stage 2).

3.1 Maintenance and differentiation of human embryonic stem cells

Undifferentiated hES cells (H9 line from Wicell, passages 35 to 45) were grown on an inactivated mouse embryonic fibroblast (MEF) feeder layer as previously described (Chen *et al.*, 2007, Efroni *et al.*, 2008). Briefly, the cell was maintained at an undifferentiated stage on irradiated low-passage MEF feeder layers on 0.1% gelatin-coated plates. The medium was changed daily. The medium consisted of Dulbecco's modified Eagle's medium (DMEM)/F-12, 20% knockout serum replacement, 0.1 mM nonessential amino acids, 2 mM L-glutamine, 0.1 mM β-mercaptoethanol, and 4 ng/ml rhFGF-2 (R&D Systems Inc., Minneapolis). The undifferentiated hES cells were treated by 1 mg/ml collagenase type IV in DMEM/F12 and scraped mechanically on the day of passage. To deplete feeder cells before endothelial differentiation, hES cells were cultured on Matrigel coated plates with mTeSR™1 medium (Stem Cell Technologies Inc, Vancouver, Canada).

3.2 Two-step *In vitro* differentiation of hES cells

To induce hES cell differentiation, undifferentiated hES cells were cultured in differentiation medium containing Iscove's modified Dulbecco's medium (IMDM) and 15% defined fetal bovine serum (FBS) (Hyclone, Logan, UT), 1XBIT (BSA, insulin, transferring; Stem Cell Technologies), 0.1 mM nonessential amino acids, 2 mM L-glutamine, 450 μM monothioglycerol (Sigma, St. Louis, MO), 50 U/ml penicillin, and 50 μg/ml streptomycin, with 20ng/ml bFGF (R&D Systems Inc., Minneapolis) and 50ng/ml VEGF (R&D Systems Inc.), either in ultra-low attachment plates for the formation of suspended embryoid bodies (EBs) as previously described (Chen *et al.*, 2007, Levenberg *et al.*, 2002, Li *et al.*, 2008). Briefly, hES cells cultured on Matrigel coated plate with mTeSR™1 medium were treated by 2 mg/ml dispase (Invitrogen, Carlsbad, CA) for 15 minutes at 37°C to loosen the colonies. The colonies were then scraped off, and transferred into ultra low-attachment plates (Corning Incorporated, Corning, NY) for EB formation. EB sprouting differentiation in collagen type I was performed as described(Li *et al.*, 2005), with minor modifications. Briefly, 12 days-old EBs were harvested, and then suspended into rat tail collagen type I (Becton Dickinson, San Jose, CA) at a final concentration of 1.5 mg/ml collagen. After thoroughly mixing EBs into collagen, 1.5 ml/well of mixture were added into six-well-plate. The plates were incubated at 37°C for 30 min, allowing gel polymerization prior to addition of medium. After gel formation, each dish was supplemented with EGM-2 medium (Lonza, Basel, Switzerland) with 50ng/ml VEGF additional. The cultures were then incubated for 3 days without media change **(Figure 1)**.

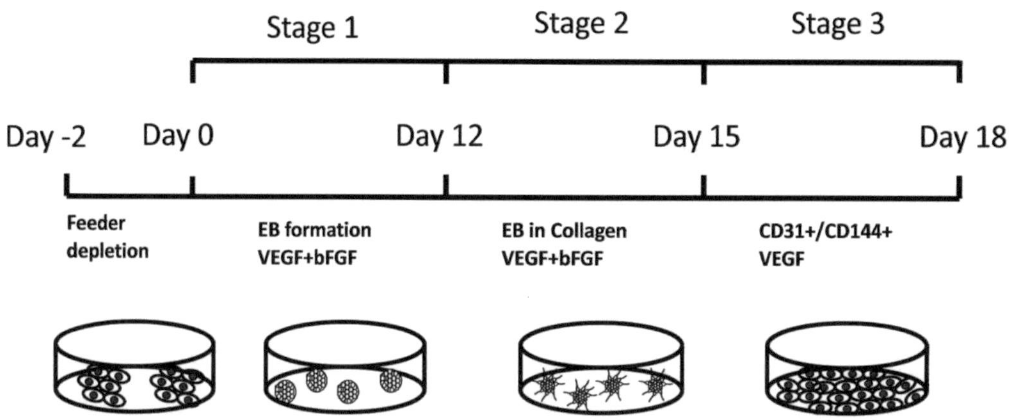

Fig. 1. Specification of the endothelial lineage differentiation from hES cells. An outline of the protocol used for the differentiation of hES cells to the endothelial lineage. Undifferentiated hES cells were grown to 60%–70% confluence on Matrigel and subcultured in low attachment dish with differentiation medium supplement VEGF and bFGF, dated as day 0. At day 12, hEBs were collected and digested and $CD31^+$/ $CD144^+$ cells were isolated by FACS and sub-cultured in EGM-2 medium to expand and induce endothelial maturation

After embedded into collagen, EBs rapidly developed into vascular sprouting in the presence of growth factors within 3-4 days. Whole-mount immunostaining confirmed that $CD31^+$ and $CD144^+$ cells were organized into channel-like structures within day-12 EBs (**Figure 2A**). These data demonstrate that some cells within EBs can spontaneously differentiate into endothelial cells that are then able to form blood vessel-like structures, confirming previous reports from our lab and others (Levenberg *et al.*, 2002, Li *et al.*, 2008,

Wang *et al.*, 2007). CD31/CD144 expression increased swiftly to 10-15% after subculture in collagen as confirmed by FACS analysis, as compared to 1-2% using the conventional EB culturing technique **(Figure 2B)**. After sorting, CD31⁺/CD144⁺ cells were further expanded as a nearly pure population (98%) and these cells were used for subsequent experiments.

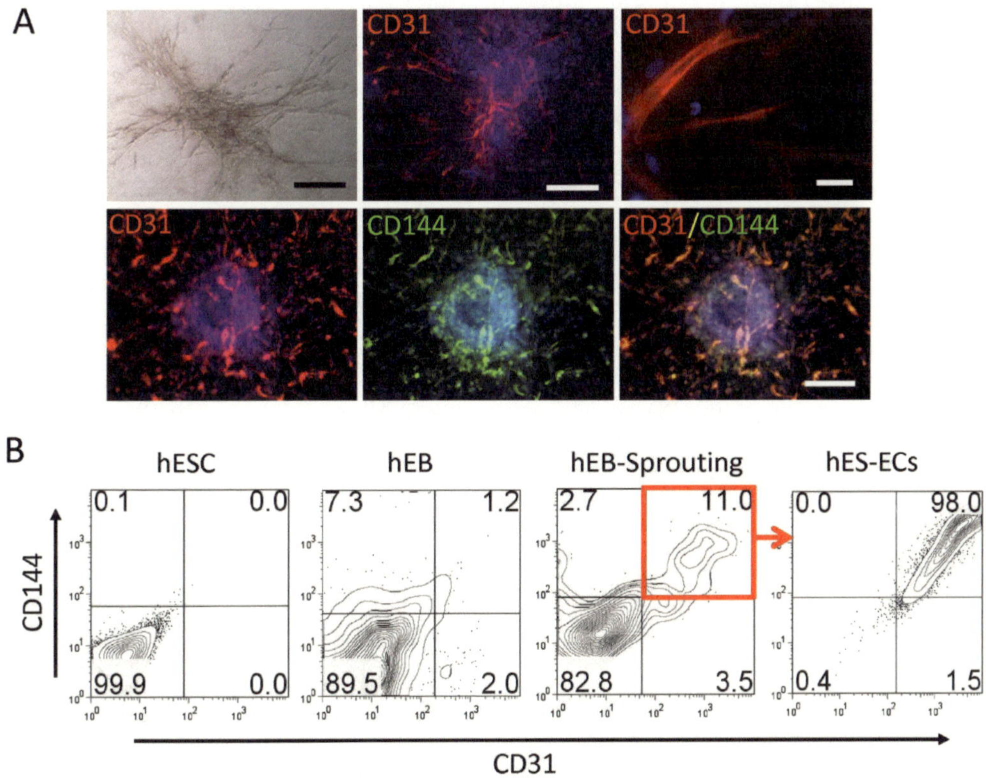

Fig. 2. Specification of the endothelial lineage differentiation from hES cells. (A) Endothelial differentiation of sprouting hEBs in collagen. Representative sprouting hEBs after cultured 3 days in collagen matrix (upper panel, left). Whole-mount immunostaining showed CD31 sprouting with channel-like vessel structures and CD31 cell clumps (upper panel). Also, those sproutings are CD31 and CD144 double positive. (B) Kinetic expression of CD31 and CD144 during two-step hES cells differentiation procedures. hES-ECs were enriched by CD31⁺/CD144⁺ isolation (right panel)

3.3 Flow cytometry sorting (FCS) of hES-ECs

Single cell suspension from hEB-sprouting was obtained by treatment with 0.25% collagenase I (Invitrogen, Carlsbad, CA) at 37°C for 20-30 min, then with 0.56 units/ml Liberase Blendzyme IV (Roche Diagnostics, Indianapolis) at 37 °C for 10-20 minutes. Cells were passed through a 40- µm cell strainer (BD Falcon, San Diego) (Xu *et al.*, 2006). Cells were incubated with mouse anti-human CD31 (BD) and CD144 (Abcam, Cambridge, MA). The CD31⁺/CD144⁺ cells were isolated using FACScan (Becton Dickinson). To generate hES-ECs, the isolated CD31⁺/CD144⁺ cells from hEB-sprouting were grown on 4 µg/cm² human fibronectin (Calbiochem, San Diego, CA) coated plates in EGM-2 (Lonza) with additional 5 ng/ml VEGF. The medium was changed every 2–3 days.

3.4 Biological characteristics of hES-ECs

Flow cytometry analysis, immunostaining, DiI-ac-LDL uptake assay, and Matrigel assay were used to confirm endothelial cell phenotype within these CD31⁺/CD144⁺ purified hES cells. Antibodies used for flow cytometry analysis were phycoerythrin (PE) conjugated anti-CD31, CD34 (BD Pharmingen) and Allophycocyanin (APC) conjugated anti-KDR, CD133 (R&D Systems), APC conjugated anti-mouse IgG2a, and rabbit anti-human CD144 (Abcam). The stained cells were analyzed using FACS Vantage (Becton-Dickinson, MA). Dead cells stained by propidium-iodide (PI) were excluded from the analysis. Isotype-identical antibodies served as controls (BD Pharmingen).

For immunostaining, the cells were fixed with with 4% paraformaldehyde in PBS at room temperature for 15 minutes. The fixed cells were permeated by 1% Triton 100 and incubated with 2% BSA for 30 minutes to block nonspecific binding, and stained for 1 hour with the primary antibodies: CD31, CD144 (BD Pharmingen) and vWF (Chemicon International Inc), respectively. The cells were then incubated for 30 minutes with either Alexa 593-conjugated donkey anti-mouse secondary antibody or Alexa 488-conjugated donkey anti-rabbit secondary antibody (Invitrogen), and counter stained with DAPI.

For DiI-ac-LDL uptake assay, hES-ECs were incubated with 10 μg/ml of DiI-Ac-LDL (Molecular Probes, Eugene, OR) at 37°C for 6 hours. After washing with PBS twice, cells were fixed and counterstained with DAPI (4, 6-diamidino-2-phenylindole) as described (Chen *et al.*, 2007). The formation of endothelial tubes was assessed by seeding cells in 24-well plates coated with Matrigel (BD Pharmingen) and incubating them at 37° for 12 hours as described(Li *et al.*, 2007).

hES-ECs morphologically resembled HUVECs, which were uniformly flat, adherent, and cobblestone-like in appearance. When CD31⁺/CD144⁺ cells were cultured in endothelial growth medium, the majority of cells were adherent and expressed endothelial markers (CD31, CD144) at the endothelial cell adherent junctions as well as von Willebrand factor (vWF) located within the cytoplasm **(Figure 3A)**. The CD31+/CD144+ cells also uptake DiI-acetylated low-density lipoprotein, and rapidly formed vascular network-like structures when placed on Matrigel **(Figure 3B)**. Taken together, these data confirm that these differentiated cells were of endothelial lineage.

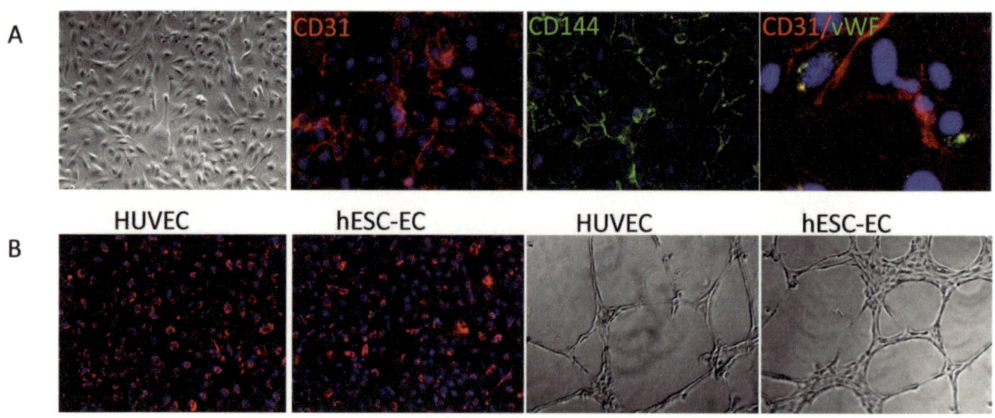

Fig. 3. *In vitro* characterization of hES-EC. (A) Morphogenesis shows the cells were cobble stone like; histology show CD31 and CD144 express on cell membranes, and vWF in the cytoplasm. (B) Compare to HUVEC, hES-ECs also can uptake ac-DiI-LDL and form tube like structure on Matrigel

3.5 Microarray hybridization and data acquisition

In order to define at a molecular level the changes occurring at each stage of hES cells differentiation to endothelial progeny, and to validate that these cells are similar to human umbilical vein endothelial cells (HUVECs), we also perform transcriptional profiling using whole human genome microarrays and real-time PCR arrays. Total RNA samples were isolated in Trizol (Invitrogen) followed by purification over a Qiagen RNeasy column (Qiagen) from undifferentiated hES cells at day 0, differentiated hEBs at day 12, hES-derived endothelial cells (after CD31/CD144 sort), and human umbilical endothelial cells (HUVEC). Four samples from each group (for a total of 16 unique samples) were harvested for RNA isolation. Using Agilent Low RNA Input Fluorescent Linear Amplification Kits, cDNA was reverse transcribed from each of 16 RNA samples representing four biological quadruplicates, as well as the pooled reference control, and cRNA was then transcribed and fluorescently labeled with Cy5/Cy3. cRNA was purified using an RNeasy kit (Qiagen, Valencia, CA, USA). 825 ng of Cy3- and Cy5- labeled and amplified cRNA was hybridized to Agilent 4x44K whole human genome microarrays (G4112F) and processed according to the manufacturer's instructions. The array was scanned using Agilent G2505B DNA microarray scanner. The image files were extracted using Agilent Feature Extraction software version 9.5.1 applying LOWESS background subtraction and dye-normalization.

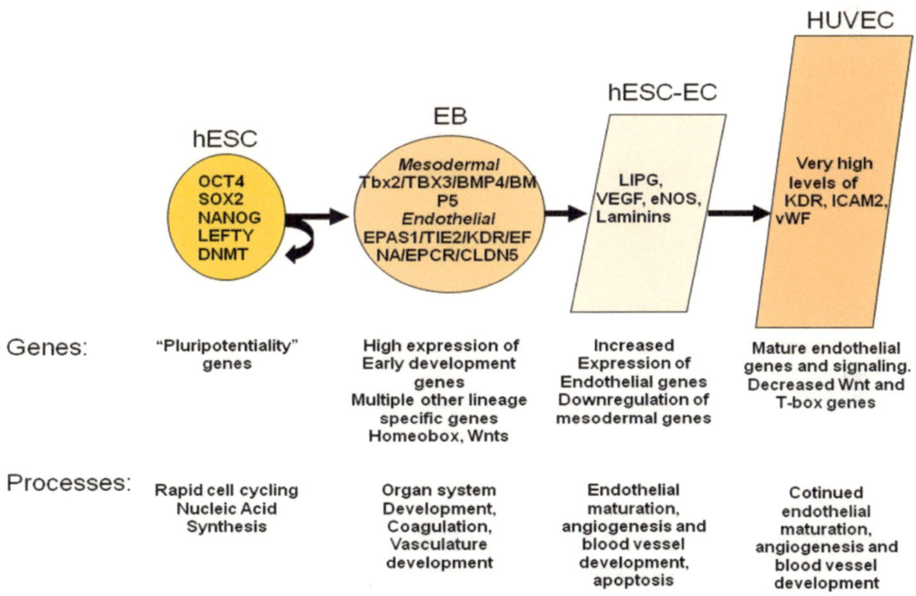

Fig. 4. Major themes in gene expression profiles at each stage of differentiation. (A) hES cells express high levels of pluripotency-associated genes including Oct4, Sox2, NANOG, Lefty, and DNMT. At the EB stage, the cells express high levels of mesodermal master regulators such as Tbx2, and BMP4 as well as very enriched levels of endothelial specific master regulators including EPAS1, TIE2, KDR, and EFNA. This population also expresses genes from other cell layers, and many developmental genes from Wnt and homeobox families. hES-ECs downregulate early mesodermal genes and express more endothelial specific genes, while HUVECs have the highest levels of mature endothelial gene expression with very few other developmental lineages represented

The resulting data were analyzed using GeneSpring GX 7.3.1 to identify genes which had changed expression significantly between stages. A summary of our major findings is shown in **Figure 4**. To obtain an overview of the transcriptional landscape, we looked at the data using principal components analysis (PCA), a dimensional reduction technique which identifies "principal components" or major trends in gene expression in the overall data (**Figure 5**). PCA demonstrates that each of the four replicates from each stage has very similar transcriptional profiles to one another, but distinctly different between stages, as expected.

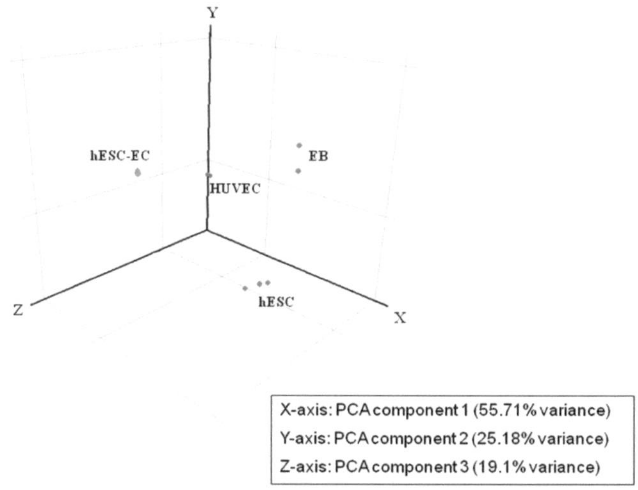

Fig. 5. Principal Components Analysis (PCA) shows that replicate experiments of each cell type are very similar while differentiation groups separate significantly along components 1 and 2

3.6 Real time RT-PCR (qRT-PCR)

To investigate changes in endothelial related genes during endothelial cell differentiation in collagen, we performed quantitative real-time PCR analysis using the Human Endothelial Cell Biology PCR array on undifferentiated hES cells, EBs, EB sprouting, and hES-ECs, and on HUVEC as a positive control. qRT-PCR assays were performed using the human endothelial cell biology RT^2 Profiler™ PCR Array (SuperArray Bioscience, Frederick, MD) on an ABI PRISM 7900 HT (Applied Biosystems, Foster City, CA). Data analysis is available at the company website (http://www.superarray.com/pcr/arrayanalysis.php). Briefly, total RNAs were isolated using RNeasy (Qiagen, Waltham, MA) from undifferentiated hES cells at day 0, differentiated hEBs at day 12, hEB-sprouting, hES-derived endothelial cells (after CD31/CD144 sort), and human umbilical endothelial cells (HUVEC) as positive control. First-strand cDNAs were generated using iScript Select cDNA Synthesis Kit (BioRad, Hercules, CA). For real-time PCR reaction, first-strand cDNAs were added to RT qPCR Master Mix (SuperArray Bioscience). Samples were heated for 10 min at 95°C and then subjected to 40 cycles of denaturation at 95°C for 15 sec and annealing and elongation at 60°C for 1 min.

Analysis of scatter plot demonstrated 28 genes were upregulated when day-12 EBs were embedded into collagen type I for an additional 3 days **(Figure 6A)**. Analysis of scatter plots also revealed that 30 genes were upregulated and 8 genes downregulated after undifferentiated hES cells differentiated into EBs, and 33 genes upregulated and 2 genes downregulated after hES cells differentiated into EB sprouting. Quantitative analysis showed after endothelial induction, ICAM1, CD31, VCAM1, and vWF were significantly upregulated, while FGF1, Flt1, KDR and VEGFA were expressed in undifferentiated cells and upregulated with endothelial differentiation of hES cells **(Figure 6B)**, which were also consistent with the microarray data. To explore the mechanism of promotion of endothelial differentiation in collagen, quantitative analysis was carried out and showed that endothelial cell activation genes were upregulated swiftly and endothelial differentiation was triggered after hEBs were embedded into collagen **(Figure 6C)**.

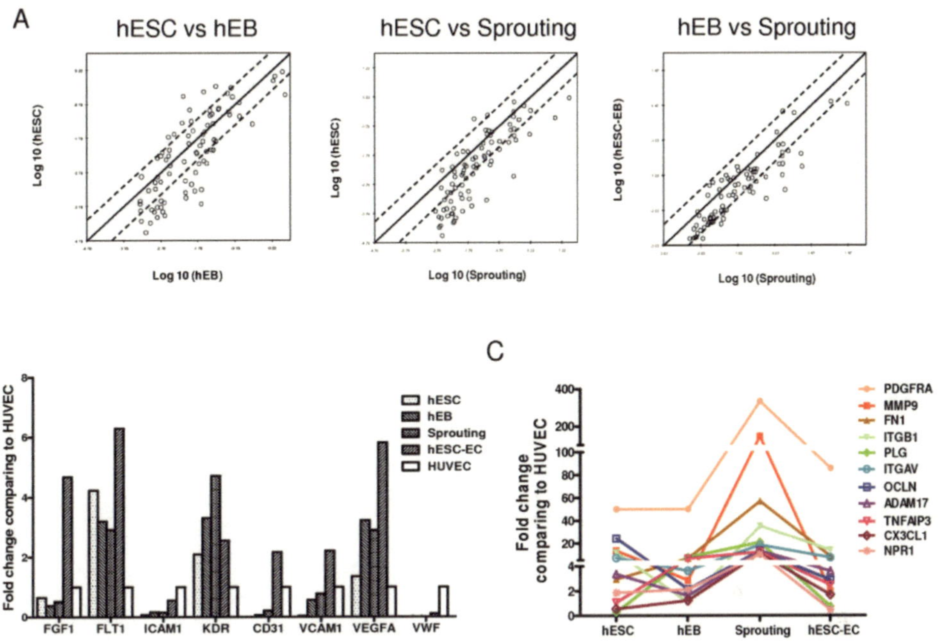

Fig. 6. Quantitative PCR analysis of endothelial gene in hES, hEB, hEB sproutin, hES-EC and HUVEC. (A) Scatter plots of the endothelial related gene-expression were compared between hES and hEB, hES and sprouting , and hEB and sprouting . Endothelial related genes expression was analyzed by Human Endothelial Cell Biology RT[2] Profiler PCR Array. The array includes 84 genes related to endothelial cell biology. The lines indicate the diagonal and 4-fold changes between the two samples. (B) Kinetic expression of selected endothelial gene on hES, hEB, sprouting and hES-EC between HUVEC. Compared to HUVEC, hES-ECs express abundant endothelial gene except adult endothelial gene vWF. (C) Cell matrix related gene increase swiftly in collagen regardless beginning level

4. Imaging fate of transplanted hES-ECs

Stem cell therapy is an exciting area of research that promises future treatment of many diseases (Chiu, 2003). However, to fully understand the beneficial effects of stem cell therapy, investigators must be able to track the biology and physiology of transplanted cells in *living* subjects over time. At present, most cell therapy protocols require histological analysis to determine viable engraftment of the transplanted cells. The development of sensitive, noninvasive technologies to monitor this fundamental engraftment parameter will greatly aid clinical implementation of cell therapy. Therefore, to validate the *in vivo* engraftment potential of our hES-ECs, we performed noninvasive monitoring of transplanted hES-ECs in a murine myocardial ischemia model by bioluminescence imaging (BLI).

Among the different reporter gene imaging techniques, bioluminescence imaging (BLI) is extremely useful because of its high sensitivity, high-throughput screening, and straightforward imaging procedures. Firefly luciferase (Fluc) is a 61-kDa monomeric protein, which reacts with its substrate D-luciferin in the presence of oxygen, Mg^{2+}, and ATP to emit luminescence. The light can be captured by high sensitive CCD camera, such as IVIS imaging system (Caliper Life Sciences, Mountain View, CA). This imaging modality is currently limited to small animal models, because bioluminescence imaging relies on low-energy photons (2-3 eV) that become attenuated within deep tissues. Recent work demonstrated that hES cells can be stably transduced with a lentiviral vector carrying a novel double-fusion reporter gene that consists of firefly luciferase and enhanced green fluorescence protein (Fluc-eGFP) (Li *et al.*, 2008). Reporter gene expression has shown *no* significant adverse effects on hES cells viability, proliferation, or differentiation into hES-ECs. After transplantation into the mouse hindlimb, hES-ECs can be tracked up to 4 week by BLI. Moreover, postmortem histology can be used to confirm engraftment of transplanted hES-ECs based on GFP immunostaining. Finally, the reporter gene approach may also permit longitudinal monitoring of endothelial differentiation process.

4.1 Lentiviral transduction of hES cells with double fusion (DF) reporter gene

In order to track transplanted cells *in vivo*, hES cells were transduced at multiplicity of infection (MOI) of 10 with self-inactivating (SIN) lentiviral vector carrying a human ubiquitin promoter driving firefly luciferase and enhanced green fluorescence protein (Fluc-eGFP) **(Figure 7A)**. Stable clones were isolated using FACS for eGFP expression. Afterwards, Fluc activity within different cell numbers was confirmed *ex vivo* using Xenogen IVIS 200 system (Xenogen, Alameda, CA) as described (Cao *et al.*, 2006). Non-transduced hES cells (control) and hES cells with DF reporter gene (hES-DF) were stained for Oct-4 (Chemicon, Temecula, CA). The undifferentiated hES cell colonies were fixed in 4% paraformaldehyde in PBS for 15 minutes. Nonspecific binding was blocked with 4% normal goat serum for 30 minutes, following which the colonies were stained with antibodies to Oct-4 and incubated with Alexa 594-conjugated rabbit anti-goat secondary antibodies (Invitrogen) for 30 minutes and nuclear counterstained with DAPI. Images were obtained with a Zeiss Axiovert microscopy (Sutter Instrument Co., USA). Subsequently, the processes for *in vitro* endothelial cell differentiation and characterization were the same as control non-transduced hES cells described earlier.

The efficiency of self-inactivating lentiviral vector for transducing hES cells was ~20% (data not show). Both control non-transduced hES cells and stably transduced hES cells showed similar expression patterns of stem cell markers Oct-4 on immunostaining, suggesting minimal side effects by reporter gene on maintaining stem cell state **(Figure 7B)**. Upon culturing onto 24-well plates, we also observed a strong correlation (r^2=0.99) between Fluc activity and cell numbers *ex vivo* using the Xenogen IVIS system **(Figure 7C)**. The cell proliferation and cell viability data are also similar between control hES cells and transduced hES cells (data not shown). Overall, these data are consistent with our previous studies showing minimal effects of reporter genes on mouse ES cell survival, proliferation, and differentiation (Cao *et al.*, 2006, Wu *et al.*, 2006a, Wu *et al.*, 2006b).

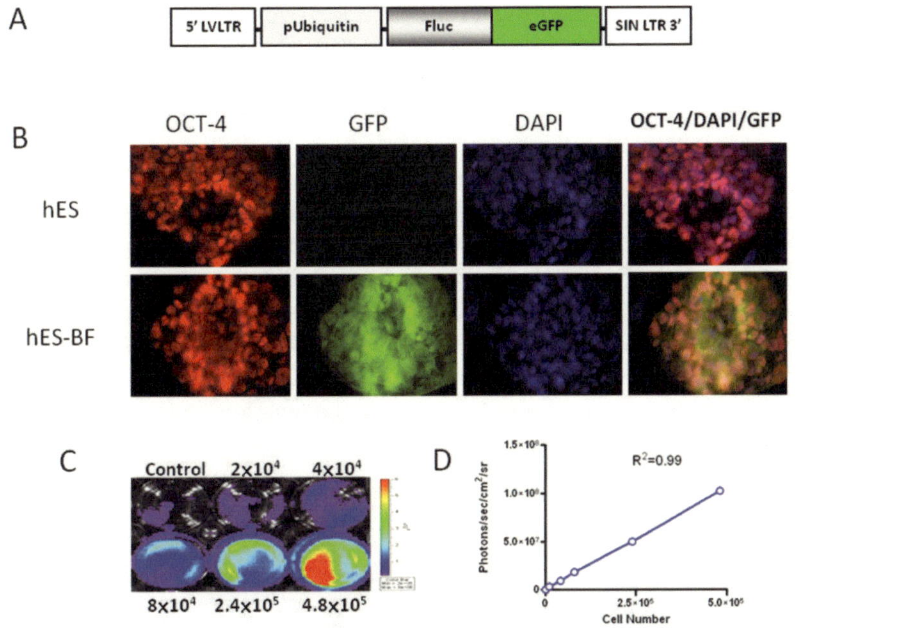

Fig. 7. Stable lentiviral transduction of hES cells with the double fusion reporter genes. (A) Schema of the double fusion reporter gene containing fusion of Fluc-sGFP. The double fusion reporter gene was cloned into a self-inactivating lentiviral vector downstream from the ubiquitin promoter. (B) Control hES cells and transduced hES cells showed similar expression pattern of Oct-4 under fluorescence microscopy. DAPI staining is used as a nuclear marker. (C) *Ex vivo* imaging analysis of stably transduced cells show increasing bioluminescence signals with cell numbers of hES cells (r^2=0.99). Data are representative of three independent experiments. Abbreviations: DAPI, 4′, 6-diamidino-2-phenylindole

4.2 Longitudinal bioluminescence imaging of hES cell survival in living animals

Previous studies have demonstrated that hES-derived cardiomyocyte transplantation can improve cardiac function after myocardial infarction (van Laake *et al.*, 2007, Laflamme *et al.*, 2007). To understand the therapeutic potential of hES-ECs for treatment of ischemia heart disease, we subjected SCID mice to LAD ischemia and reperfusion followed by injection with either $1x10^6$ cultured hES-ECs (n=28) or PBS (n=15). For sham operated animals (n=5), open thoracotomy was performed without ligation of the LAD and without injection of hES-ECs or PBS. All surgical procedures were performed on 8-10 week old female SCID Beige mice (Charles River Laboratories, Wilmington, MA) by a single experienced micro-surgeon. Protocols were approved by the Stanford University Animal Care and Use Committee guidelines. Following induction with inhaled isoflurane (2% to 3%), mice were intubated and ventilated and anesthesia was maintained with inhaled isoflurane (1% to 2.5%). A left thoracotomy was performed followed by ligation of the middle of left anterior descending (LAD) artery for 30 minutes followed by reperfusion. Infarction was visually confirmed by blanching of the anterolateral region of the left ventricle along with dyskinesis. After 30 minutes, 1×10^6 hES-ECs were injected intramyocardially into the peri-infarct zone at 20 μl of total volume. Control animals received PBS injection instead. For sham operated animals, suturing was performed without ligation. Post-operative analgesia was provided by a one-time, subcutaneous injection of buprenorphine (0.1 mg/kg body weight). Animals were recovered in a warmed, humidified chamber.

Cardiac bioluminescence imaging was performed using the Xenogen IVIS 200 system. After intraperitoneal injection of the reporter probe D-Luciferin (150 mg luciferin/kg), animals were imaged for 1-10 minutes. The same mice were imaged for up to 8 weeks. Bioluminescence signal was quantified in units of maximum photons per second per cm square per steridian (photons/sec/cm^2/sr). Injection of hES-ECs into infarcted myocardium resulted in a robust bioluminescence signal at day 2. However, serial imaging of the same animals out to 8 weeks demonstrated a significant decay in bioluminescence, suggesting acute donor cell death. When normalized to the signal on day 2 after injection, the percent bioluminescence signals were 36.8±15.5 at day 4, 6.7±1.5 at day 7, 3.5±0.6 at day 14, 2.3±0.5 at day 28, and 0.8±0.4 at day 56 **(Figure 8).** Thus, we believe less than 1% of transplanted hES-ECs survived beyond 8 weeks.

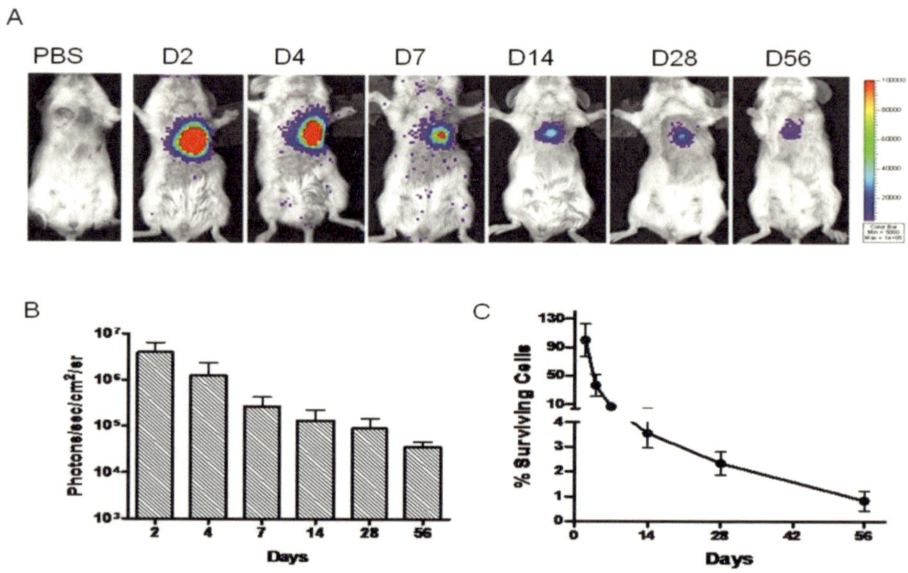

Fig. 8. Molecular imaging of hES-EC fate after transplantation. (A) A representative animal injected with $1\text{x}10^6$ hES-ECs shows significant bioluminescence activity at day 2, which decreases progressively over the following 8 weeks. A representative control animal injected with PBS shows no imaging signals as expected. (B) Detailed quantitative analysis of signals from all animals (n=28) transplanted with hES-ECs (signal activity is expressed as photons/sec/cm^2/sr). (C) Donor cell survival plotted as % signal activity from day 2 to week 8

4.3 Assessment of cardiac viability following hES-EC based therapy by PET imaging

Small animal PET imaging was carried out at day -7 (baseline) and days 2, 14, 28, and 56, post-operatively. Mice were fasted for 3 hours prior to radioisotope injection and then were injected with 140±17 μCi of [^{18}F]-FDG via the tail vein. At 45 to 60 minutes post-injection, animals were anesthetized with inhaled 2% isoflurane and imaged with a Vista microPET system (GE Health Care). Images were reconstructed by filtered back projection (FBP) and analyzed by image software Amide (http://amide.sourceforge.net/index.html). Three-dimensional regions of interest (ROIs) were drawn encompassing the heart. Counts/pixel/min were converted to counts/ml/min (assuming a tissue density of 1 g/ml) with a calibration constant derived from scanning a cylindrical phantom (Toyama *et al.*, 2004). For each ROI, counts/ml/min were then converted to counts/gram/min and divided by the injected dose to obtain the image ROI-derived [^{18}F]-FDG percentage injected dose per gram of heart (% ID/g). To measure the myocardium infarction size, [^{18}F]-FDG PET images were assembled into polar maps (Kudo *et al.*, 2002). The size of perfusion defects was measured as the myocardium with 50% of maximum activity and expressed as percent total myocardium polar map by GE Healthcare Advantage Workstation (GE Healthcare, UK). Myocardial metabolic activity and perfusion defect imaging can be detected using a microPET system and ^{18}F-fluorodeoxyglucose ([^{18}F]-FDG) radiotracer (Kudo *et al.*, 2002). This technique can be used to accurately detect changes in cardiac viability following infarctions in small animal models (Schelbert *et al.*, 2003). Therefore, we performed [^{18}F]-FDG PET scans at baseline, day 2, day 14, day 28, and day 56 post infarction for hES-EC treated and control animals. We observed a significant reduction in cardiac [^{18}F]-FDG uptake for both groups at day 2, confirming successful myocardial infarction. At day 14,

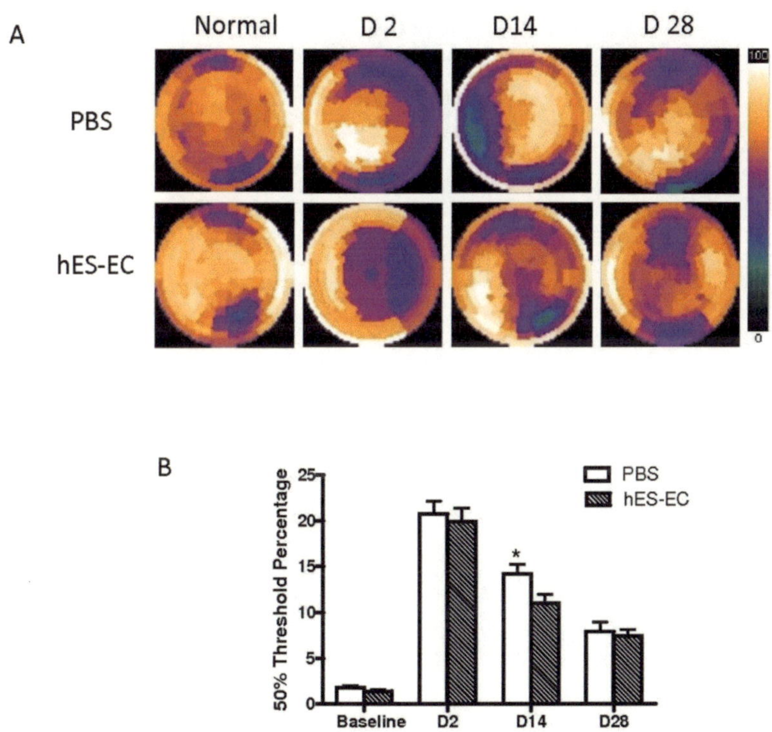

Fig. 9. [^{18}F]-FDG PET imaging of cardiac viability. (A) Representative polar map of the microPET images obtained from mice treated with PBS vs. hES-ECs. (B) Measurements are based on 50% thresholds. *P<0.05 vs. PBS group at day 14

there was a significant increase in cardiac viability (indicated by increased [^{18}F]-FDG uptake) in the hES-EC treated group compared to controls (26.3±1.8 vs. 22.4±3.4; $P<0.05$). Analysis of the [^{18}F]-FDG PET images using polar-map reconstruction revealed reduced infarction size on the hES-EC transplantation group compared to the PBS group, but there was no significant difference at any day other than day 14 (**Figure 9**). Overall, the trends for changes in functional contractility and metabolic activity were very similar based on the microPET imaging results.

4.4 Histologic assessment of hES-ECs engraftment

Histologic analysis of the myocardium was performed by examining thin sections of the gross specimen and via immunofluorescent microscopic examination. Microscopic examination showed the presence of GFP positive cells within myocardium **(Figure 10)**. At day 4, clumps of hES-ECs can be found in the interstitial spaces between cardiomyocytes, as seen with α-sarcomeric actin (α-SA) and GFP double staining **(Figure 10A)**. At day 28, GFP, α-SA, and mouse-specific CD31 triple staining revealed hES-ECs that had formed microvessel structures and integrated with host vasculature. However, the overall frequency of GFP+ cells was significantly decreased at day 28 **(Figure 10B, C)**, which is consistent with the decrease in bioluminescence signals measured over this same period of time and likely due to massive cell death. Moreover, immunostaining with mouse CD31 and α-SA at day 4

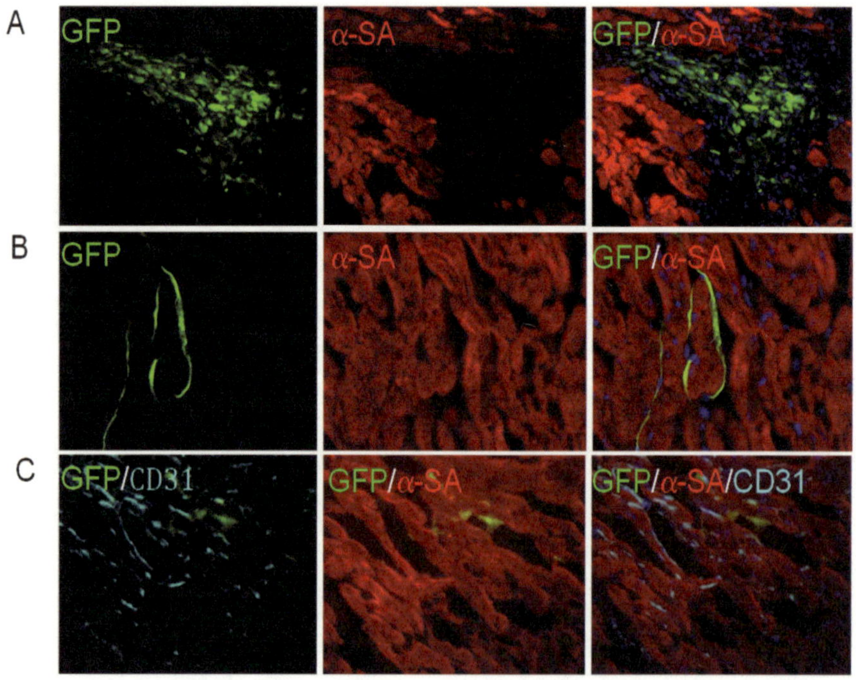

Fig. 10. Confirmation of engrafted hES-ECs by immunofluorescence. (A) hES-ECs within the recipient myocardium 4 days after injection showed clump formation. (B, C) At day 28, transplanted hES-ECs can differentiate into vasculature and integrate with host myocardium as confirmed by GFP, mouse CD31, and α-sarcomeric actin (α-SA) co-staining. However, this population became significantly rare compared to day 4

showed no host vasculature extension into the hES-EC clumps (data not shown), which suggests no direct nutrient transport to transplanted cells other than diffusion. Finally, examination of the explanted hearts showed a downward trend in infarct size and an upward trend in microvascular density (MVD) in the hES-EC treated group, though these differences with controls were not statistically significant (data not shown).

4.5 Functional imaging with intravital microscope

To understand whether hES-ECs also possess vasculogenic ability *in vivo*, we next evaluated the *in vivo* vascular formation potential of hES-ECs after transplantation in the model of mouse dorsal window chamber. Previous data showed that supporting fibroblast cells were needed to stabilize engineered blood vessels (Au *et al.*, 2008, Wang *et al.*, 2007). However, there are obvious drawbacks in using mouse support cells in study of vessel maturation of human cell. In this study, we introduced collagen type I and Matrigel, without support cells, and showed hES-ECs can form functional vasculature and are stable up to 2 months post transplantation. Briefly, one million hES-ECs were suspended in 1 ml solution of rat-tail type I collagen (at final concentration 1.5 mg/ml) (BD) and Matrigel (BD) (1:1) at 4°C. The cell suspension was pipetted into 12-well plates and warmed to 37°C for 30 minutes to allow polymerization of collagen and Matrigel and 1 ml EGM-2 medium was added after that (**Figure 11A**). After one day of culture in 5% CO_2, a skin puncher was applied to create circular disk-shape pieces of the construct (8-mm diameter), and they were implanted into

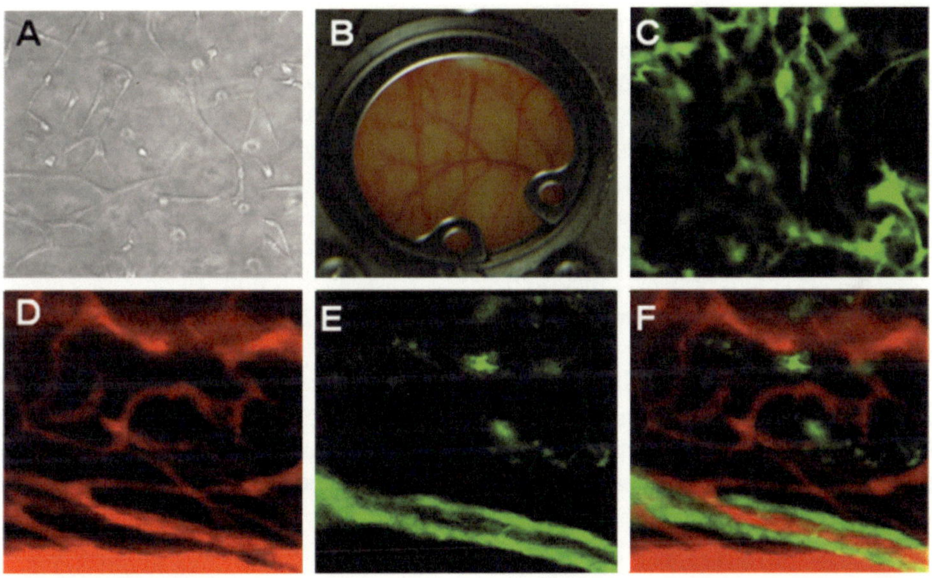

Fig. 11. Demonstration of functional vessels *in vivo* using Matrigel plug and dorsal window chamber. (A) Dorsal window chamber model in SCID mice. GFP+ hES-ECs were cultured in a mix of collagen and Matrigel for 1 day (left, upper panel), and implanted into dorsal windows in SCID mice (middle, upper panel). Images were taken at day 2, 14 and 21 after implantation. After 14 days, Angiosense 680 was injected by tail vein to highlight perfused vessels within the dorsal window. Green, hES-ECs expressing GFP; red, functional blood vessels with contrast enhanced by Angiosense 680

the dorsal skin fold windows in SCID mice. The titanium dorsal skin fold chamber (APJ Trading Co. Inc, Ventura, Ca) surgically mounted onto each mouse was described previously(Lehr *et al.*, 1993). Briefly, mice were anesthetized with inhaled isoflurane (2%–3%), and two symmetrical titanium frames were implanted to sandwich the extended double layer of the skin. One layer was removed in a 10-mm-diameter circular area. The remaining layer, consisting of epidermis, subcutaneous tissue, and striated skin muscle, was covered with a glass coverslip incorporated in one of the titanium frames **(Figure 11B)**. The animals were housed one per cage and had free access to water and food throughout the experiment. A recovery period of 2 days was allowed before the collagen/Matrigel patch was transplanted. Multiphoton laser-scanning intravital microscopy (IV-100, Olympus, Center Valley, PA) was used to visualize and quantify the morphological changes of EGFP-expressing hES-ECs.

To test whether the engineered vascular networks integrated with the host murine circulatory system, we next injected Angiosense 680 intravenously to enhance the contrast of perfused vessels. Importantly, several of the hES-EC derived vessels formed conduits that contained blood flow from day 12 to day 60 **(Figure 11, D, E & F)**. In summary, these data suggest that transplanted hES-ECs have angiogenic potential and are able to integrate into the host vasculature.

5. Conclusion

Taken together, the results showed that endothelial output efficiency was increased by a novel two-step differentiation of hES cells with collagen, and hES-ECs possess functional vasculogenic ability in vivo. However, the numbers surviving hES-ECs in infarcted tissue decreased significantly in the first few weeks and improved heart function at 2 weeks, but not to a significant degree later, demonstrating the requirement for long-term follow-up in all cardiac cell transplantation studies. Alternative transplantation protocols with larger numbers of hES-ECs, multiple grafts, or addition of matrix or prosurvival factors to prevent donor cell death after transplantation could eventually lead to the realization of further and sustained enhancement of heart function post myocardial infarction. These data suggest that prolonged heart function recovery may require more permanent graft survival of transplanted cells. In conclusion, two-step differentiation of hES cells can provide sufficient and functional endothelial cells for cell therapy, but exploring alternative transplant protocols that can improve cell survival and lead to sustained enhancement of heart function post infarction should be introduced in future. Furthermore, we believe molecular imaging will likely play a critical role in monitoring the localization and viability of these transplanted cells for cardiovascular diseases.

6. Prospect

Cardiovascular disease (CVD) accounts for approximately 30% of all deaths in the United States (Rosamond *et al.*, 2008). Since CVD is primarily caused by endothelial dysfunction, therapeutic angiogenesis/vasculogenesis holds great promise for a potential cure. The logic behind angiogenesis/vasculogenesis is to improve the spontaneous healing process by supplementation of vascular progenitor cells or growth factors(Li *et al.*, 2007). Endothelial progenitor cell transplantation can foster the formation of arterial collaterals and promote the regeneration of damaged tissues. Recently, human embryonic stem (hES) cells have generated much interest because of their capacity for self-renewal and pluripotency. In practical terms, hES cells can be cultured indefinitely *ex vivo*, and can differentiate into virtually any cell type in the adult body (Thomson *et al.*, 1998, Reubinoff *et al.*, 2000). hES cells are thus an attractive source for the derivation of large numbers of cells to be used in various tissue repair and cell replacement therapies. However, upon transplantation into living organisms, undifferentiated hES cells can spontaneously differentiate into rapidly proliferating teratomas, which are disordered amalgam of all three germs layers (Thomson *et al.*, 1998, Reubinoff *et al.*, 2000). Therefore, safely coaxing hES cells into *committed* progenitor lineages for therapeutic applications is an innovative and feasible strategy that can minimize the risk of cellular misbehavior and teratoma formation.

At present, hES cells are in the approval process for use in the treatment of neurodegenerative disease and Stargardt's Macular Dystrophy (SMD) by the Food and Drug Administration (FDA) of United States (Sun *et al.*, 2010). The isolation of hES-derived endothelial cells (hES-ECs) may also have potential therapeutic applications, including cell transplantation for repair of ischemic tissues and tissue-engineered vascular grafts. However, to fully understand the beneficial effects of stem cell therapy, investigators must be able to track the functional biology and physiology of transplanted cells in *living* subjects over time. At present, most cell therapy protocols are limited by their requirement for histological analysis to determine viable engraftment of the transplanted cells. The development of sensitive, noninvasive technologies to monitor this fundamental engraftment parameter will greatly aid clinical implementation of cell therapy. Moreover, recent research on hES-based therapy showed poor long-term engraftment of hES-ECs by serial bioluminescence imaging (Li *et al.*, 2008). Thus, to sustain long-term engraftment of hES-ECs and realize the full benefits of hES-EC therapies, alternative transplantation protocols with addition of matrix (Cao *et al.*, 2007) or pro-survival factors (Laflamme *et al.*, 2007) to prevent cells death after transplantation must be developed.

7. Acknowledgment

This work was supported by grants from the National Key Scientific Program of China (2011CB964903) and National Natural Science Foundation of China (31071308)

8. References

Au, P., Daheron, L.M., Duda, D.G., Cohen, K.S., Tyrrell, J.A., Lanning, R.M., *et al.* (2008). Differential in vivo potential of endothelial progenitor cells from human umbilical cord blood and adult peripheral blood to form functional long-lasting vessels. *Blood* 111, 1302-1305.

Cao, F., Lin, S., Xie, X., Ray, P., Patel, M., Zhang, X., *et al.* (2006). In vivo visualization of embryonic stem cell survival, proliferation, and migration after cardiac delivery. *Circulation* 113, 1005-1014.

Cao, F., Sadrzadeh Rafie, A.H., Abilez, O.J., Wang, H., Blundo, J.T., Pruitt, B., *et al.* (2007). In vivo imaging and evaluation of different biomatrices for improvement of stem cell survival. *Journal of tissue engineering and regenerative medicine* 1, 465-468.

Chen, T., Bai, H., Shao, Y., Arzigian, M., Janzen, V., Attar, E., *et al.* (2007). Stromal cell-derived factor-1/CXCR4 signaling modifies the capillary-like organization of human embryonic stem cell-derived endothelium in vitro. *Stem cells (Dayton, Ohio)* 25, 392-401.

Chiu, R.C. (2003). Bone-marrow stem cells as a source for cell therapy. *Heart failure reviews* 8, 247-251.

Cho, S.W., Moon, S.H., Lee, S.H., Kang, S.W., Kim, J., Lim, J.M., *et al.* (2007). Improvement of Postnatal Neovascularization by Human Embryonic Stem Cell Derived Endothelial-Like Cell Transplantation in a Mouse Model of Hindlimb Ischemia. *Circulation* 116, 2409-19.

Efroni, S., Duttagupta, R., Cheng, J., Dehghani, H., Hoeppner, D.J., Dash, C., *et al.* (2008). Global transcription in pluripotent embryonic stem cells. *Cell stem cell* 2, 437-447.

Gerecht-Nir, S., Ziskind, A., Cohen, S. and Itskovitz-Eldor, J. (2003). Human embryonic stem cells as an in vitro model for human vascular development and the induction of vascular differentiation. *Laboratory investigation* 83, 1811-1820.

Kaufman, D.S., Hanson, E.T., Lewis, R.L., Auerbach, R. and Thomson, J.A. (2001). Hematopoietic colony-forming cells derived from human embryonic stem cells. *Proceedings of the National Academy of Sciences of the United States of America* 98, 10716-10721.

Kudo, T., Fukuchi, K., Annala, A.J., Chatziioannou, A.F., Allada, V., Dahlbom, M., *et al.* (2002). Noninvasive measurement of myocardial activity concentrations and perfusion defect sizes in rats with a new small-animal positron emission tomograph. *Circulation* 106, 118-123.

Laflamme, M.A., Chen, K.Y., Naumova, A.V., Muskheli, V., Fugate, J.A., Dupras, S.K., *et al.* (2007). Cardiomyocytes derived from human embryonic stem cells in pro-survival factors enhance function of infarcted rat hearts. *Nature biotechnology* 25, 1015-1024.

Lehr, H.A., Leunig, M., Menger, M.D., Nolte, D. and Messmer, K. (1993). Dorsal skinfold chamber technique for intravital microscopy in nude mice. *The American journal of pathology* 143, 1055-1062.

Levenberg, S., Golub, J.S., Amit, M., Itskovitz-Eldor, J. and Langer, R. (2002). Endothelial cells derived from human embryonic stem cells. *Proceedings of the National Academy of Sciences of the United States of America* 99, 4391-4396.

Li, Z., Han, Z. and Wu, J.C. (2009a). Transplantation of human embryonic stem cell-derived endothelial cells for vascular diseases. *Journal of cellular biochemistry* 106, 194-199.

Li, Z., Suzuki, Y., Huang, M., Cao, F., Xie, X., Connolly, A.J., *et al.* (2008). Comparison of reporter gene and iron particle labeling for tracking fate of human embryonic stem cells and differentiated endothelial cells in living subjects. *Stem cells (Dayton, Ohio)* 26, 864-873.

Li, Z., Wilson, K.D., Smith, B., Kraft, D.L., Jia, F., Huang, M., *et al.* (2009b). Functional and Transcriptional Characterization of Human Embryonic Stem Cell-Derived Endothelial Cells for Treatment of Myocardial Infarction. *PLoS ONE* 4, e8443.

Li, Z., Wu, J.C., Sheikh, A.Y., Kraft, D., Cao, F., Xie, X., *et al.* (2007). Differentiation, Survival, and Function of Embryonic Stem Cell Derived Endothelial Cells for Ischemic Heart Disease. *Circulation* 116, I-46-I-54.

Li, Z.J., Wang, Z.Z., Zheng, Y.Z., Xu, B., Yang, R.C., Scadden, D.T. and Han, Z.C. (2005). Kinetic expression of platelet endothelial cell adhesion molecule-1 (PECAM-1/CD31) during embryonic stem cell differentiation. *J Cell Biochem* 95, 559-570.

Lu, S.J., Feng, Q., Caballero, S., Chen, Y., Moore, M.A., Grant, M.B. and Lanza, R. (2007). Generation of functional hemangioblasts from human embryonic stem cells. *Nat Methods* 4, 501-9.

Reubinoff, B.E., Pera, M.F., Fong, C.-Y., Trounson, A. and Bongso, A. (2000). Embryonic stem cell lines from human blastocysts: somatic differentiation in vitro. *Nat Biotech* 18, 399-404.

Rosamond, W., Flegal, K., Furie, K., Go, A., Greenlund, K., Haase, N., *et al.* (2008). Heart disease and stroke statistics--2008 update: a report from the American Heart Association Statistics Committee and Stroke Statistics Subcommittee. *Circulation* 117, e25-146.

Schelbert, H.R., Inubushi, M. and Ross, R.S. (2003). PET imaging in small animals. *J Nucl Cardiol* 10, 513-520.

Sun, N., Longaker, M.T. and Wu, J.C. (2010). Human iPS cell-based therapy: Considerations before clinical applications. *Cell Cycle* 9, 880-5.

Thomson, J.A., Itskovitz-Eldor, J., Shapiro, S.S., Waknitz, M.A., Swiergiel, J.J., Marshall, V.S. and Jones, J.M. (1998). Embryonic stem cell lines derived from human blastocysts. *Science* 282, 1145-1147.

Thyagarajan, B., Liu, Y., Shin, S., Lakshmipathy, U., Scheyhing, K., Xue, H., *et al.* (2008). Creation of engineered human embryonic stem cell lines using phiC31 integrase. *Stem cells (Dayton, Ohio)* 26, 119-126.

Toyama, H., Ichise, M., Liow, J.S., Vines, D.C., Seneca, N.M., Modell, K.J., *et al.* (2004). Evaluation of anesthesia effects on [18F]FDG uptake in mouse brain and heart using small animal PET. *Nuclear medicine and biology* 31, 251-256.

van Laake, L.W., Passier, R., Monshouwer-Kloots, J., Verkleij, A.J., Lips, D.J., Freund, C., *et al.* (2007). Human embryonic stem cell-derived cardiomyocytes survive and mature in the mouse heart and transiently improve function after myocardial infarction *Stem Cell Research* 1, 9-24.

Vodyanik, M.A., Bork, J.A., Thomson, J.A. and Slukvin, II (2005). Human embryonic stem cell-derived CD34+ cells: efficient production in the coculture with OP9 stromal cells and analysis of lymphohematopoietic potential. *Blood* 105, 617-626.

Wang, Z.Z., Au, P., Chen, T., Shao, Y., Daheron, L.M., Bai, H., *et al.* (2007). Endothelial cells derived from human embryonic stem cells form durable blood vessels in vivo. *Nature biotechnology* 25, 317-318.

Wu, J.C., Cao, F., Dutta, S., Xie, X., Kim, E., Chungfat, N., *et al.* (2006a). Proteomic analysis of reporter genes for molecular imaging of transplanted embryonic stem cells. *Proteomics* 6, 6234-6249.

Wu, J.C., Spin, J.M., Cao, F., Lin, S., Xie, X., Gheysens, O., *et al.* (2006b). Transcriptional profiling of reporter genes used for molecular imaging of embryonic stem cell transplantation. *Physiol Genomics* 25, 29-38.

Wu, J.C., Tseng, J.R. and Gambhir, S.S. (2004). Molecular imaging of cardiovascular gene products. *J Nucl Cardiol* 11, 491-505.

Xu, C., He, J.Q., Kamp, T.J., Police, S., Hao, X., O'Sullivan, C., *et al.* (2006). Human embryonic stem cell-derived cardiomyocytes can be maintained in defined medium without serum. *Stem cells and development* 15, 931-941.

Yamahara, K., Sone, M., Itoh, H., Yamashita, J.K., Yurugi-Kobayashi, T., Homma, K., *et al.* (2008). Augmentation of neovascularizaiton in hindlimb ischemia by combined transplantation of human embryonic stem cells-derived endothelial and mural cells. *PLoS ONE* 3, e1666.

Zhang, S.J. and Wu, J.C. (2007). Comparison of imaging techniques for tracking cardiac stem cell therapy. *J Nucl Med* 48, 1916-1919.

11

Fourier Transform Infrared Microspectroscopy as a Tool for Embryonic Stem Cell Studies

Diletta Ami[1,2], Paolo Mereghetti[3], Antonino Natalello[1] and Silvia M. Doglia[1]
[1]Department of Biotechnology and Biosciences, Univeristy of Milano Bicocca
[2]Fondazione IRCCS Policlinico San Matteo, Pavia
[3]Center for Modelling and Simulation in the Biosciences, University of Heidelberg and HITS gGmbH, Heidelberg
[1,2]Italy
[3]Germany

1. Introduction

Embryonic stem (ES) cells are self-renewing and pluripotent cells that arise from the inner cell mass of the mammalian blastocyst (Smith, 2001). Their unique capability of potentially generating every cell type continuously attracts the interest of different fields of research. Indeed, ES cells represent a powerful tool for the study of the molecular mechanisms of cell differentiation with important applications in cell therapies, tissue engineering, regenerative medicine and pharmaceutical screening (Trounson, 2006; Vats et al., 2005). All these applications require rapid and sensitive assays to evaluate the differentiation process through the identification of specific markers of the differentiation status. To date, ES cell differentiation is mainly monitored by biochemical methods such as immunohistochemistry, gene expression analysis, functional assays of the differentiating cells and flow cytometry, that - even if providing a comprehensive characterization of cells - are time consuming, expensive and often require a complex sample handling. For these reasons, the development of new approaches for stem cell studies is highly desirable.

In the last decades, optical spectroscopy approaches were applied to the study of intact cells and in particular vibrational spectroscopies revealed to be powerful techniques for the characterization of complex biological systems (Heraud & Tobin, 2009). In particular, Fourier transform infrared (FTIR) and Raman are non invasive and label-free vibrational (micro)spectroscopies that allow to obtain information on the molecular composition and structure of intact cells, tissues and whole organisms (Schulze et al., 2010 ; Tanthanuch et al., 2010; Chan & Lieu, 2009; Walsh et al., 2009; Ami et al., 2004; Choo et al., 1996), providing a unique molecular fingerprint within a single measurement. In this way, it is possible to characterize rapidly different processes that take place simultaneously in biological systems, a non easy task for the standard biochemical approaches. Thanks to the use of an infrared microscope coupled to a FTIR spectrometer, it becomes possible to collect the absorption spectrum from a selected sample area. Interestingly, these techniques, thank to their fast – time resolution, have been successfully used to snapshot and "freeze" molecular events in complex systems (Miller & Dumas, 2010; Hamm, 2009).

The potential of FTIR and Raman spectroscopies has been widely exploited to monitor biological processes in-situ, for instance in cancer diagnosis (Krafft et al., 2009; Baker et al., 2008; Wang et al., 2007), in protein aggregation (Diomede et al., 2010; Doglia et al., 2008; Ami et al., 2005; Choo et al., 1996) and in stem cell research (Schulze et al., 2010; Heraud & Tobin, 2009; Ami et al., 2008; Notingher et al., 2004a; Notingher et al., 2004b). Indeed, these vibrational approaches have been shown to be a promising tool for the characterization of cellular mechanisms, providing not only structural information, but also details on the dynamics of the structures (Miller & Dumas, 2010).
Concerning in particular the study of complex biological systems, it is important to underline that the multivariate statistical analysis is an essential support to fully understand the spectroscopic response. Among the different statistical approaches, the combined principal component - linear discriminant analysis (PCA-LDA) allows to find in the spectrum the wavenumbers that contribute to the largest inter-spectral variance, thus validating the identification of the marker bands obtained by the direct inspection of the spectral data (Ami et al., 2008; Walsh et al., 2007; Fearn, 2002). In Figure 1 we illustrated the procedure that should be followed to successfully tackle the FTIR characterization of complex systems.

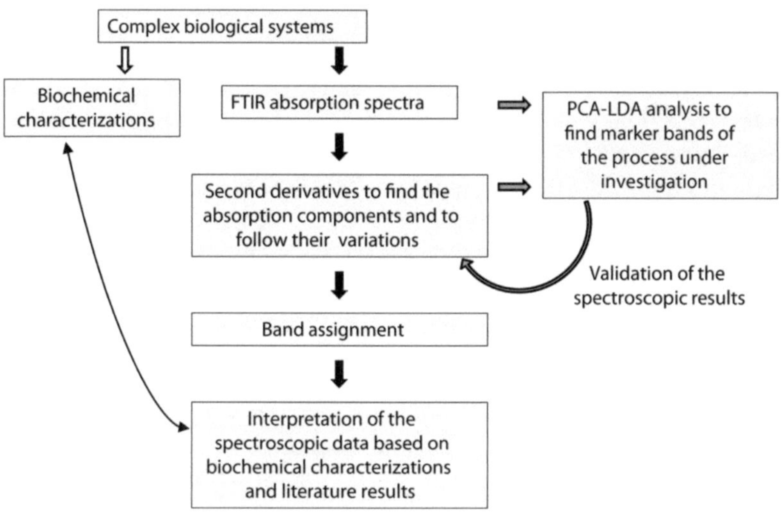

Fig. 1. Scheme of a FTIR approach to study complex biological systems. The measured absorption spectra are analyzed by resolution enhancement approaches, as second derivatives, to resolve the overlapped absorption components and to follow their variations during the process under investigation. To validate the spectroscopic results, a multivariate analysis - such as PCA-LDA - is required. The assignment of the identified marker bands to specific biomolecules involved in the process is the next crucial step. The interpretation of the spectroscopic data should be then confirmed by standard biochemical characterizations.

In this chapter we will first give an overview of FTIR spectroscopy of isolated biomolecules: proteins, nucleic acids, lipids, and carbohydrates. Then, we will illustrate in detail the basis of the multivariate analysis applied to the study of complex biological systems. Finally, we will extend the spectroscopic study on intact cells, focusing our interest on the characterization of embryonic stem cell differentiation.

2. FTIR spectroscopy of biomolecules: proteins, nucleic acids, lipids and carbohydrates

The application of FTIR (micro)spectroscopy to the study of biological systems is based on the knowledge of the band assignment of the infrared absorption due to the functional groups of the most important biomolecules. Indeed, proteins, nucleic acids, lipids and carbohydrates have specific absorptions in the mid infrared range, between 4000 and 400 cm^{-1}. To better illustrate this point, in Figure 2 the IR absorption spectra of model biomolecules are reported and compared with that of intact eukaryotic cells.

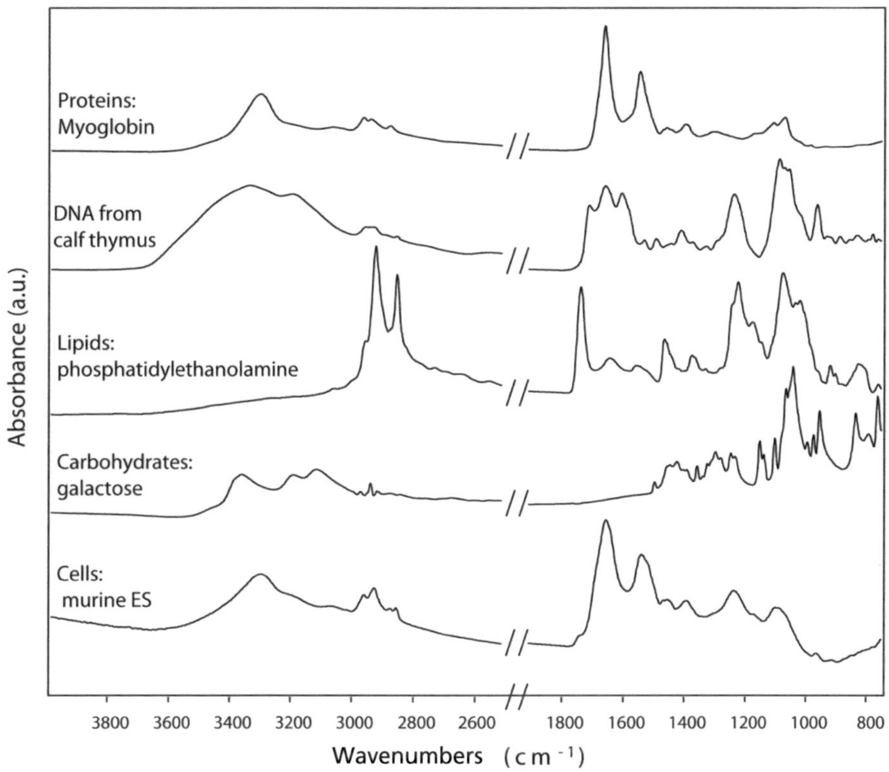

Fig. 2. FTIR absorptions of model biomolecules and intact eukaryotic cells. Myoglobin, calf thymus DNA, phosphatidylethanolamine and galactose are taken as models for protein, nucleic acid, lipid, and carbohydrate IR absorptions respectively. A representative FTIR spectrum of intact murine ES cells is also reported for comparison. Hydrated films of the isolated biomolecules were measured in attenuated total reflection (ATR), while murine embryonic stem (ES) cells were measured in transmission, after dry-fixing for about 30 minutes.

Several spectroscopic studies on complex biological systems are found in literature and became possible since IR spectroscopy is not limited by the physical state of the sample (liquid, solid, etc).

A limiting factor in the infrared characterization of biological molecules in their natural state was initially represented by the strong water absorption in the mid-IR range, where their internal vibrational modes occur. However, the development of high performing FTIR spectrometers that allow to obtain spectra with an excellent signal to noise ratio and baseline stability enables, nowadays, to subtract the solvent spectrum. Moreover, several strategies – as the use, for instance, of deuterated water – could help to overcome this problem.

For the FTIR study of secondary structures, stability, and aggregation of proteins, particularly useful are the Amide I and the Amide II bands that occur, respectively, in the 1700-1600 cm^{-1} and 1600-1500 cm^{-1} spectral regions. The Amide I band, the most used for protein analyses, is mainly due to the C=O stretching vibration of the peptide bond and it is sensitive to the protein secondary structures (Barth, 2007; Barth & Zscherp, 2002; Arrondo & Goni, 1999; Arrondo et al., 1993).

Since FTIR spectroscopy allows to examine also highly scattering samples, proteins can be studied in different environmental conditions, including solutions, hydrated film, and also within intact cells and tissues.

The Amide I of proteins and peptides usually appears as a broad band due to the overlapping of several spectral components arising from the peptide bond absorption in the different secondary structures. On the basis of computational analyses and experimental studies on model compounds - peptides and proteins with known three dimensional structures - it has been possible to assign these components to specific protein secondary structures according to their peak position (Barth, 2007; Barth & Zscherp, 2002; Arrondo & Goni, 1999; Arrondo et al., 1993). For this reason the first critical step on the FTIR analysis of proteins is the identification of the spectral components that contribute to the Amide I envelop, while the second step is the assignment of each component to a specific protein secondary structure. For the first step, two resolution enhancement procedures can be used: the second derivative analysis of the spectra (see below; Susi & Byler, 1986) and the Fourier self-deconvolution (FSD) method (not reported in this chapter, see Arrondo et al., 1993; Kauppinen et al., 1981). Several reviews discuss extensively the Amide I band assignment to protein secondary structures (for instance Barth, 2007; Barth & Zscherp, 2002; Arrondo & Goni, 1999; Arrondo et al., 1993) and here we report only a scheme for protein in not-deuterated solvent: alpha-helices (1660–1648 cm^{-1}), beta-sheets (1640–1623 cm^{-1} and 1695–1674 cm^{-1}), turns (1686–1662 cm^{-1}), random coils (1657–1642 cm^{-1}), and aggregation and protein-protein interactions (1630–1620 cm^{-1} and 1698-1692 cm^{-1}) .

Concerning nucleic acids, their IR absorption is very complex and covers a wide range of frequencies (Banyay et al., 2003; Zhizhina & Oleinik, 1972; Tsuboi, 1961). For simplicity, the range of absorption is conventionally divided in different spectral regions. Here, we will briefly illustrate the most studied ones.

The 1800-1500 cm^{-1} range is mainly due to nucleobase vibrations, sensitive to base stacking and base pairing interactions, while in the 1500-1250 cm^{-1} range marker bands sensitive to sugar puckering, glycosidic bond rotation and backbone conformation are found. Bands sensitive to nucleic acid backbone conformation occur also between 1250-1000 cm^{-1}, due to vibrations along the sugar-phosphate chain. This last spectral region is of particular interest for cell biology studies, since a number of marker bands of the different DNA conformations (A, B, and Z) can be found in the spectra, allowing to obtain important insights into the nucleic acid dynamics and functions. For instance, it is known that double stranded DNA exists in two main family forms, the A and the B geometries. In particular, it is known to assume the A form in low relative humidity conditions and in the hybrid with RNA, during transcription (Banyay et al., 2003 and references therein).

Also of particular interest is the spectral range between 1000 and 800 cm^{-1} where bands due to the different sugar puckering modes (S-N types) are found. Since these are sensitive to changes in the DNA sugar conformation induced by cytosine methylation (Banyay & Graslund, 2002), the analysis of this range could be relevant for biological studies considering the extent of DNA methylation in cells. Theophanides & Tajmir-Riahi (1985) studied the conformational changes of DNA, identifying IR marker bands of A and B DNA forms thanks to the absorption of their different sugar conformations.

Among the first nucleic acid FTIR studies, it should be mentioned the work of Tsuboi (1961), that studied the secondary structure of DNA in solution, native and denatured, enabling to detect the spectral changes due to the breakdown of its secondary structure.

Being the main components of biological membranes, also the infrared absorption of lipids has been widely characterized (Casal & Mantsch, 1984; Arrondo & Goni, 1998). It originates mainly from molecular vibrations of the hydrophylic head-group and of the hydrophobic hydrocarbon tail. The most studied lipid spectral range is that between 3100-2800 cm^{-1}, where the acyl chain vibrational modes occur with generally strong bands due to CH2 and CH3 stretching modes. The frequencies of these bands are conformation-sensitive and respond to temperature-induced changes of the trans/gauche ratio in acyl chains. In this way, it is possible to study lipid phase transitions and changes in lipid composition (Casal & Mantsch, 1984). For instance, through the analysis of the changes in the CH2 and CH3 absorption bands, the stress response induced by protein aggregation in bacterial cells has been monitored in situ (Ami et al., 2009). Furthermore, through the CH stretching region analysis of mouse oocyte FTIR spectra and supported by the ester carbonyl band around 1740 cm^{-1} (see below), Wood and colleagues (Wood et al., 2008) found that lipids - whose composition within the oocytes drastically changes during maturation stages - could be considered potential markers of oocyte developmental competence.

Interestingly, the methyl group vibrations occurring in the 3100-2800 cm^{-1} spectral range could also give information on DNA/histone methylation and/or histone acetylation, important issues for epigenetic studies (Ami et al., 2010; O'Connell, 2005).

Furthermore, also of particular interest for lipid studies is the so called interfacial region, between 1750-1700 cm^{-1}, where the stretching vibrations of the C=O group involved in ester bonds occur. The resulting absorption bands are sensitive to changes in their local environment, such as polarity or hydrogen bonding (Casal & Mantsh, 1984; Arrondo & Goni, 1998).

We should add that FTIR spectroscopy has been also widely applied in several fields of carbohydrate research as it allows, for instance, to study mono and oligosaccharide composition and conformation (Kacurakova & Wilson, 2001) and protein glycosylation (Natalello et al., 2005). To this aim, the most used spectral range is in the fingerprint region between 1200 and 750 cm^{-1}, whose band peak positions and intensities are specific for every polysaccharide. In particular, the IR response has been found to be highly sensitive to the carbohydrate conformation, to hydrogen bonding, to hydration, to the type of substituent, and to the linkage positions (Kacurakova & Wilson, 2001). Due to the complexity of the carbohydrate IR absorption, the band assignment could be not unequivocal, requiring accurate data analysis and validation of the results.

Noteworthy, of particular interest for cell biology applications is the infrared response of glycogen that in tissues and in single cells displays its spectral signature at specific wavenumbers: $\approx$1028, $\approx$1081, and $\approx$1153 cm^{-1} (Heraud et al., 2010; Ami et al., 2008; Walsh et al., 2007; Wang et al., 2007; Steller et al., 2006).

3. Infrared microspectroscopy applied to the study of intact cells: sample preparation

Coupling an infrared microscope, with all reflecting optics, with a FTIR spectrometer offers the opportunity to study selected areas within the sample under investigation.
Two main different types of infrared microscopy can be used: the first, conventional, allows the collection of the IR absorption spectra from a microvolume within the sample, with a spatial resolution not only due to the diffraction limit of Mid IR light (3-10 µm), but also to the level of the absorbed light. For these reasons, a spectrum of good quality can be collected by an area larger than 20 µm x 20 µm – when using a nitrogen cooled Mercury Cadmium Telluride (MCT) detector (Orsini et al., 2000). Thanks to the variable aperture of the microscope, it is possible to select a small area of few tens of microns within the sample, enabling the study of intact cells (Tanthanuc et al., 2010; Thumanu et al., 2009; Ami et al., 2008; Wood et al., 2008), tissues (Choo et al., 1996) and whole model organisms, as nematodes (Diomede et al., 2010; Ami et al., 2004).
Instead, more advanced IR microscopes employ a focal plane array (FPA) detection to collect the IR chemical imaging of the sample, with a spatial resolution improved compared to that of the conventional microscope. The image contrast is determined by the response of the different sample regions to the particular IR wavelengths selected by the user (Kazarian & Chan 2006; Levin & Bhargava 2005; Lewis et al., 1995).
We should mention that infrared measurements can be performed mainly in transmission or in attenuated total reflection (ATR) mode. Typically, measurements on biomolecules and on complex biological systems, such as cells and tissues, are carried out in transmission, employing suitable IR transparent supports, for instance of barium fluoride or zinc selenide. However, sometimes, it could be useful to work in ATR when the samples of interest are highly absorbing or when they cannot be easily transferred onto an opportune infrared support (Walsh et al., 2007; Orsini et al., 2000). Indeed, in ATR measurements the sample is placed in contact with the ATR element (diamond, germanium, etc) characterized by a refractive index higher than that of the sample. In this device, an evanescent wave generates and penetrates into the sample for a path length of the size of a micron (Tamm & Tatulian 1997).
In addition, the development of synchrotron light sources further improved the application of FTIR microspectroscopy to cell characterization. Indeed, due to its radiation source 100–1000 times brighter than that of a conventional thermal one, it is possible to collect an infrared absorption spectrum at a higher spatial resolution from a sample area of only few microns. In this way, a synchrotron IR source enables to collect high signal – to –noise ratio spectra of subcellular compartments, providing better insights useful for the study of biological processes within single cells (Miller & Dumas, 2010).
We should recall that - even if samples in different physical states can be examined by FTIR spectroscopy - the sample condition can strongly affect the FTIR spectra. This makes it necessary to standardize sample preparation and data acquisition procedures. Indeed, as discussed above for isolated biomolecules, water absorption – very high in the mid-IR - could represent a limit for FTIR analysis, as it makes difficult to perform measurements in vivo that require an aqueous environment. A successful strategy to overcome this problem is the dry fixing procedure: a cell suspension is deposited on an IR transparent support and then dried at room temperature for about 30 minutes, in order to remove excess water that would mask the IR response of the different biological components (Ami et al. 2008). Noteworthy, using Raman spectroscopy - a vibrational technique where the water signal is weak and does not affect the

Raman response of hydrated samples - it has been recently demonstrated that the rapid desiccation of cells doesn't affect their spectroscopic response (Konorov et al., 2010). In particular, using Raman microspectroscopy the authors measured different stages of differentiation of human embryonic stem cells live and dry-fixed, and compared the spectroscopic responses obtained in the two conditions. The relative intensities of the bands due to tryptophan in proteins and to nucleic acid backbone and base vibrations - used as differentiation markers – were found to be the same in living and in dry-fixed cells, allowing to monitor the same temporal patterns during differentiation in the two conditions.

These results strongly indicate that dry-fixing is a suitable method for the study of intact cells by FTIR microspectroscopy.

Furthermore, as recently pointed out by Zhao and colleagues (Zhao et al., 2010), changes due to aging of cells in culture could also interfere with the cell IR response. For this reason, the in-situ spectroscopic characterization of cell processes requires an accurate control of the stage of cell growth in culture, in order to obtain reliable and reproducible results.

4. FTIR second derivative analysis

Considering the complexity of the FTIR spectra of biological systems, it is often necessary to better resolve their absorption bands, often broad and overlapped one to the other, using the so called *resolution enhancement procedures*. To this aim, the second derivative analysis is widely applied to the measured spectra, as described and discussed by Susy & Byler (1986). In this way, the overlapping absorption components in the spectrum are identified as negative bands in the second derivative. This analysis requires spectra with high signal-to-noise ratio and free of vapour absorption, as second derivative band intensity is inversely proportional to the square of the original band half-width, leading to an enhancement of the relative contribution of sharp lines, such as due to noise and vapour.

Noteworthy, changes in the relative contributions of the different spectral components can be accurately monitored through the variations in intensity and peak position of the second derivative spectrum.

5. Multivariate statistical analysis

The multivariate statistical analysis (MVA) is an essential tool which allows to tackle the study of complex phenomena which are in general dependent by more than one statistical variable.

In general, the MVA allows the:

- simultaneous treatment of many variables and observations;
- discovery and visualization of complex associations;
- reduction of number of variables;
- construction of descriptive models;
- classification of data into groups.

Among the numerous potential applications of MVA, we are now stressing the aspects related to variable reduction, descriptive models and data classification, which are generally applied to the analysis of spectroscopic data.

In several cases, we are facing the question to find out which are the distinctive traits (if they exist) among samples of experiments done in different conditions, or at different times.

Each experiment can be repeated many times keeping fixed the experimental conditions. We then define a *group* as a collection of two or more replica of the same experiments. We also define the term *instance* or *observation* to refer to a specific experiment within one group.
On every instance, we perform one or more measurements, which we believe are able to capture the fundamental variation among our groups. In such a way, we can express every instance as a vector composed by all our measurements. The measurement can also be a single one, but intrinsically composed of many variables, for example, an IR spectra, where each wavenumber corresponds to a different variable within the same measurement.
In some cases it is also possible that we do not know, *a priori*, the distinction among different groups, but we would like only to determine if it is possible to classify our experiments into distinct groups.
A very broad range of techniques has been developed to address these issues; they span from the statistical analysis to the machine learning field.
For the purpose of this book, we will focus on those methodologies which are particularly successful in the field of spectroscopy, namely the principal component analysis (PCA) and the linear discriminant analysis (LDA).
In the first part, we will explain the basis of PCA with the role of dimensionality reduction combined with LDA as a method for descriptive analysis and classification.
In the second section, we will briefly describe some other fundamental methodologies frequently used in multivariate statistical analysis.

5.1 Principal component analysis (PCA)

We describe here the basis of PCA with the specific aim of reducing the number of variables of our problem.
As already mentioned in the introduction, we can express every observation as a vector composed by all our measurements. For example, suppose we have n observations, each one defined by a vector $\mathbf{y}_i$ composed of m variables, where $i=1,2,\dots,n$ stands for the i-th observation.
The matrix of the original data $\mathbf{Y}$ is then composed by n rows (the observations) and m columns (the variables).
We do not need, for the PCA, any information about the group membership of the observations, since no grouping of the observations or partitioning of the variables into subsets is assumed.
By using PCA, our intent is to develop a smaller number of artificial variables (called *principal components*) that will account for most of the variance in the observed variables.
We make the assumption that the original variables are redundant, which means that some variables are correlated to each other.
Considering the linear combination of the original data, $\mathbf{Z} = \mathbf{AY}$, we want to find the matrix $\mathbf{A}$ such that the new variables $\mathbf{Z}$ (the principal components) are uncorrelated. The correlation between variables can be measured using the covariance matrix. Given the sample mean of the m-dimensional vector $\mathbf{y}_i$, $\mathbf{y}_i$ $\langle \mathbf{y} \rangle = \frac{1}{n}\sum \mathbf{y}_i$, an unbiased estimator of the sample covariance matrix is

$$\mathbf{S} = \frac{1}{n-1}\Sigma(\mathbf{y}_i - \langle \mathbf{y} \rangle)(\mathbf{y}_i - \langle \mathbf{y} \rangle)^T \quad (1)$$

For uncorrelated variables, the off-diagonal values of the sample covariance matrix are zero, that is, $\mathbf{S}$ is diagonal. The covariance of linearly transformed variables $\mathbf{Z} = \mathbf{AY}$ is equal to $\mathbf{S}_z = \mathbf{ASA}^T$, where $\mathbf{S}$ is the sample covariance of the original data $\mathbf{Y}$ (Rencher, 2002).
Thus, we want to find $\mathbf{A}$ such that the covariance matrix of the transformed data, $\mathbf{S}_z$, is diagonal, which corresponds to find the eigenvectors of the covariance matrix and the corresponding eigenvalues.
The eigenvalues, which coincide with the matrix $\mathbf{S}_z$, are the sample variance of the principal components $\mathbf{Z}$ and are ranked according to their magnitude. The first principal component is then the linear combination with maximal variance (largest eigenvalue). The second principal component is the linear combination with the maximal variance along a direction orthogonal to the first component, and so on (Manly, 2004).
The number of eigenvalues is equal to the number of original variables; however, since the eigenvalues are equal to the variance of the principal components and they are sorted in a decreasing order, the first k eigenvalues explain a large portion of the variance of the data. Hence, to describe our original dataset we can use only the first k uncorrelated principal components, instead of the complete set of redundant m variables. In matrix notation this can be written as $\mathbf{Z}_k = \mathbf{A}_k\mathbf{Y}$ where $\mathbf{A}_k$ is the eigenvectors matrix truncated to the k-th eigenvector, and $\mathbf{Z}_k$ is the matrix of the first k principal components.
To choose how many principal components should be retained in order to summarize our data, we can use several strategies (Eriksson et al., 2006; Rencher, 2002). For example, one way commonly used is to retain sufficient components to explain a given total percentage of the variance, e.g 90% (Eriksson et al., 2006; Manly, 2004).
The principal components obtained in this way can be used as a non redundant input for another analysis.

5.2 Linear discriminant analysis (LDA)

LDA is mainly a supervised technique, that is, it requires the knowledge of the group membership of the observations. Contrary to PCA, we assume that our data are partitioned into k groups (Fearn, 2002; Rencher, 2002; Fukunaga, 1990).
LDA can have mainly two objectives. First, it can be a descriptive analysis used to describe and explain the differences among the groups. As we will see later, mathematically LDA finds the optimal hyperplane that separates the groups among each other. Or, in other words, it finds the optimal linear combination of the original variables that maximizes the distance among the groups. The transformed observations are called *discriminant functions.* The use of a linear combination implies that each original variable is weighted by a coefficient, which can be used to study the relative importance of the variable in the separation among the groups. A second possible role of LDA is to classify observations into groups. An observation, whose group membership is not known, is evaluated by a discriminant function (already calibrated) and it is assigned to one of the groups at which most likely it belongs (Eriksson et al., 2006; Manly, 2004; Rencher, 2002).
Firstly, we will explain the "several groups discriminant analysis" applied as a descriptive technique, then we will show how it can be used as a classifier.

5.2.1 Several groups descriptive discriminant analysis

The initial dataset is an ensemble of multivariate observations partitioned into k distinct groups (e.g. different experimental treatments, times or conditions). Each of the k groups contains n_i observations, where i runs from 1 to k and refers to the i-th group. The multivariate observation vectors can be written as $\mathbf{y}_{ij}$ where $i = 1,\ldots,k,\ j = 1,\ldots,n_i$ (i is the i-th group and j is the j-th observation). The vector has size m, which corresponds to the number of variables.

Our goal in LDA is to search for the linear combination that optimally separates our multivariate observation into k groups.

This can be visualized in the two group case in Figure 3. The new axis (the discriminant function) allows a better separation of the two clouds of points representing the two dimensional observations of two groups.

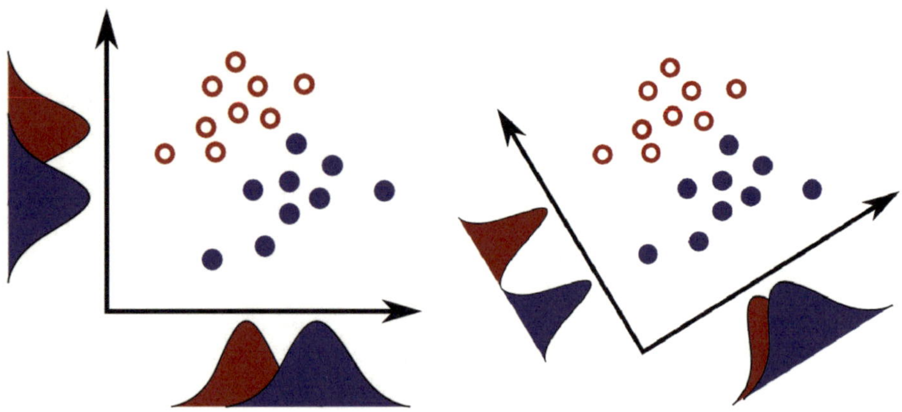

Fig. 3. LDA two group case separation. Rotation of the axis along the direction of the maximal separation between the two groups. After the rotation the two groups can be totally distinguished using only one axis.

The linear transformation of $\mathbf{y}_{ij}$ is written as $z_{ij} = \mathbf{w}^T \mathbf{y}_{ij}$

since z_{ij} is a linear transformation of $\mathbf{y}_{ij}$, the mean of the group i of the transformed data can be written as

$$\langle z_i \rangle = \mathbf{w}^T \langle \mathbf{y}_i \rangle \tag{2}$$

where $\langle \mathbf{y}_i \rangle$ is the mean of the original variables, obtained as $\langle \mathbf{y}_i \rangle = \sum \mathbf{y}_{ij} / n_i$

We now introduce the between-groups sum of squares **B** (measure of dispersion among the groups) and the within-group sum of squares **E** (measure of dispersion within one group). First, we define them for the unidimensional case relatively to the untransformed data.

$$\mathbf{B}(\mathbf{y}) = \sum_i n_i (\langle \mathbf{y}_i \rangle - \langle \mathbf{y} \rangle)^2$$

and

$$\mathbf{E}(\mathbf{y}) = \sum_i \sum_j (\mathbf{y}_{ij} - \langle \mathbf{y}_{ij} \rangle)^2$$

where $\langle \mathbf{y} \rangle = \frac{1}{k} \sum \frac{1}{n_i} \sum \mathbf{y}_{ij}$ is the total average of the data.

Analogously, in the multivariate case (where each observation is constituted by m variables) we have the two matrices

$$\mathbf{B}(\mathbf{y}) = \sum n_i (\langle \mathbf{y}_i \rangle - \langle \mathbf{y} \rangle)(\langle \mathbf{y}_i \rangle - \langle \mathbf{y} \rangle)^T \tag{3}$$

and

$$\mathbf{E}(\mathbf{y}) = \sum \sum (\langle \mathbf{y}_{ij} \rangle - \langle \mathbf{y}_i \rangle)(\langle \mathbf{y}_{ij} \rangle - \langle \mathbf{y}_i \rangle)^T \tag{4}$$

Finding the optimal linear combination that separates our multivariate observations into k groups means to find the vector $\mathbf{w}$ which maximizes the rate between the between-groups sum of squares over the within-groups sum of squares. Using the equation for the mean of the transformed data (eq. 2) into the equations 3 and 4 we can write

$$\lambda = \frac{\mathbf{w}^T \mathbf{B}(\mathbf{y}) \mathbf{w}}{\mathbf{w}^T \mathbf{E}(\mathbf{y}) \mathbf{w}} = \frac{\mathbf{B}(\mathbf{z})}{\mathbf{E}(\mathbf{z})} \tag{5}$$

We want to find $\mathbf{w}$ such that lambda is maximized.
Equation 5 can be rewritten in the form $\mathbf{w}^T(\mathbf{Bw} - \lambda \mathbf{Ew}) = \mathbf{0}$; then we search for all the non trivial ($\mathbf{w}^T = \mathbf{0}$ is excluded) solutions of this equation and we choose the one which gives the maximum value of lambda. This means to solve the eigenvalue problem $\mathbf{Bw} - \lambda \mathbf{Ew} = \mathbf{0}$ which can be written in the usual form

$$(\mathbf{A} - \lambda \mathbf{I})\mathbf{w} = \mathbf{0} \tag{6}$$

where $\mathbf{A} = \mathbf{E}^{-1}\mathbf{B}$
The solutions of equation 6 are the eigenvalues $\lambda_1, \lambda_2, ..., \lambda_m$ associated to the eigenvectors $\mathbf{w}_1, \mathbf{w}_2, ..., \mathbf{w}_m$. The solutions are ranked for the eigenvalues $\lambda_1 > \lambda_2 > ... > \lambda_m$. Hence, the first eigenvalue λ_1 corresponds to the maximum value of eq. 5.
The discriminant functions are then obtained considering only the first s positive eigenvalues and multiplying the original data by the eigenvectors

$$\mathbf{z}_1 = \mathbf{w}_1^T \mathbf{Y},\ \mathbf{z}_2 = \mathbf{w}_2{}^T \mathbf{Y}, ..., \mathbf{z}_s = \mathbf{w}_s^T \mathbf{Y}$$

Discriminant functions are uncorrelated but not orthogonal since the matrix $\mathbf{A} = \mathbf{E}^{-1}\mathbf{B}$ is not symmetric.
In many cases the first two or three discriminant functions account for most of $\lambda_1 + \lambda_2 + ... + \lambda_s$. This allows to represent the multivariate observations as 2 or 3 dimensional points which can be plotted on a scatter plot. These plots are particularly helpful to visualize the separation of our observations into the different groups. Moreover, we can deduce, looking at the scatter plot, the meaning of a given discriminant function, i.e. we can associate the discriminant function to a given property of the analyzed system.
The weighting vectors $\mathbf{w}_1, \mathbf{w}_2, ..., \mathbf{w}_s$ are called *unstandardized discriminant function coefficients* and give the weight associated to each variable on every discriminant function.
If the variables are on very different scales and with different variance, to assess the importance of each variable in the group separation the standardized discriminant functions

can be used. The standardization is done by multiplying the unstandardized coefficients by the square root of the diagonal element of the within-group covariance matrix.
Another way to assess the variable importance is to look at the correlation between each variable and the discriminant function. These correlations are called *structure* or *loading* coefficients. However, it has been shown (Rencher, 2002) that these parameters are intrinsically univariate, and they only show how a single variable contributes to the separation among groups, without taking into account the presence of the other variables.

5.2.2 LDA as classification method

The discriminant analysis can be applied to a given ensemble of data to produce a set of discriminant functions as described in the previous section. Afterwards, this model can be used to classify new observations into the most probable groups. From this point of view the linear discriminant analysis becomes a predictive tool, since it is able to classify observations whose group membership is unknown (Eriksson et al., 2006; Rencher, 2002). In the same way we can test the discrimination ability of our LDA model by a procedure called *"re-substitution"* (Rencher, 2002). This method consists of producing an LDA model using our dataset (i.e. finding the optimal w). Then, each observation vector is re-submitted to the classification function ($z_{ij} = \mathbf{w}^T\mathbf{y}_{ij}$) and assigned to a group. Since we know the group membership of the submitted vector, we can count the number of observations correctly classified and the number of observations misclassified.
Then, we can estimate the apparent classification rate as the number of correctly classified observations over the total number of observations. This is summarized in a classification table or confusion matrix. As an example, given N observations, n_1 belongs to the group 1 and n_2 belongs to the group 2. C_{11} is the total number of observations correctly classified in group 1 and C_{12} is the total number of data misclassified in group 2. Similarly, C_{22} is the total number of observations correctly classified in group 2 and C_{21} is the number of misclassified in group 1.
The confusion matrix becomes then:

Actual group	**Predicted group**	
	1	2
1	C_{11}	C_{12}
2	C_{21}	C_{22}

And the accuracy (the apparent classification rate (acr)) is computed as

$$acr = \frac{C_{11} + C_{22}}{n_1 + n_2}$$

In general, in evaluating the accuracy of a model, we have then to distinguish between two types of accuracy: the fitting accuracy and the prediction accuracy (Eriksson et al., 2006; Bishop, 1995).
The fitting accuracy is the ability to reproduce the data, namely, how the model is able to reproduce the data that were used to build the model. This corresponds to the apparent classification rate, and it is obtained using the *re-substitution* procedure. The data used to build the model are called *training set*.

The prediction accuracy is the ability to predict the value or the class of an observation, which was not included in the construction of the model. This kind of accuracy is often referred as the ability of the model to generalize. The data used to measure this accuracy are called *test set*. The prediction accuracy can be called actual classification rate. This is mainly used in settings where the goal is prediction, and one wants to estimate how accurately a predictive model will perform in practice.

To have an estimation of the actual classification rate, two main procedures can be applied: the hold-out and *cross-validation* (Eriksson et al., 2006).

In the hold-out, the dataset is divided into two partitions, one partition is used to develop the model (e.g the discriminant functions) and the second partition is given as input to the model. The first partition is usually called training set or calibration set, while the second partition is the validation set (Bishop, 1995).

When the number of observations is small, the *cross-validation* is usually preferred over the *hold-out*. The basic idea of the cross-validation procedure is to divide the entire dataset into L disjoint sets. L-1 sets are used to develop the model (i.e. this is the calibration set on which the discriminant functions are computed) and the omitted portion is used to test the model (i.e. the validation set given as input to the model). This is repeated on for all the L sets and an average result is obtained.

5.3 Principal component - linear discriminant analysis (PCA-LDA)

A powerful analysis tool is the combination of the principal component analysis with the linear discriminant analysis (Fearn, 2002). This is particularly helpful when the number of variables is large. In particular, if the number of observations (N) is less than the number of variables (m) - specifically $N\text{-}1<m$ - the covariance matrix is singular and can not be inverted. We then need to find a way to reduce the number of variables, for example by using the PCA (Rencher, 2006; Jonathan et al., 1996).

This procedure has been widely used for several problems in different fields (Ami et al., 2008; Rezzi et al., 2007; Skrobot et al., 2007; Walsh et al., 2007; Pereira et al., 2006; Héberger et al., 2003; Fearn, 2002).

In particular, a low rate of $(N\text{-}1)/m$ happened normally in spectroscopy, where the number of observations (N) is usually $<10^2$ and the number of variables (m) is typically within 10^2 to 10^3.

Let's take into account the same situation described for many group linear discriminant analysis. The original dataset is an ensemble of multivariate observations which is partitioned into k distinct groups. Again, we want to find the discriminant functions which optimally separate our multivariate observation into the k groups. Then, the discriminant functions can be used to identify the most important variables in terms of ability of distinguishing among the groups.

Thus, first the original dataset is submitted to PCA to reduce the number of variables, subsequently the reduced dataset is analyzed using LDA.

5.4 Other multivariate techniques

In the following section, we will briefly illustrate other multivariate statistical approaches, relevant for the spectroscopic studies reported in this chapter.

5.4.1 Multivariate Linear Regression (MLR)

MLR can be used to model a linear relationship among a numerical variable $\mathbf{z}$ and one or more independent variables $\mathbf{Y}$ (Manly, 2004). $\mathbf{Y}$ is the usual matrix already introduced,

composed by n rows corresponding to observations and m columns corresponding to independent variables. The MLR is based, as many other statistical techniques, on the generalized linear model $\mathbf{z} = \mathbf{Y}\boldsymbol{\beta} + \boldsymbol{\varepsilon}$, where $\boldsymbol{\beta}$ is a matrix containing the parameters to be estimated, $\boldsymbol{\varepsilon}$ is a matrix which models the errors or noise.
The coefficients beta are usually estimated using the ordinary least square, which consists of minimizing the sum of the squared differences of the n observed y's from their modeled values. Mathematically, the optimal values of beta are obtained by $\boldsymbol{\beta} = (\mathbf{Y}^{\mathrm{T}}\mathbf{Y})^{-1}\mathbf{Y}^{\mathrm{T}}\mathbf{z}$. To apply the least square method we must have $n - 1 > m$, otherwise the matrix $\mathbf{Y}^{\mathrm{T}}\mathbf{Y}$ is singular and can not be inverted. Moreover, none of the independent variables must be a linear combination of any other (muticollinearity) (Eriksson et al., 2006; Manly, 2004).

5.4.2 Partial Least Square (PLS)

The goal of PLS regression is to predict ***Z*** from ***Y*** and to describe their common structure. When the number of variables is large compared to the number of observations, ***Y*** is likely to be singular and the regression approach is no longer feasible (i.e., because of multicollinearity) (Eriksson et al., 2006). Several approaches have been developed to cope with this problem. One approach is to eliminate some predictors (e.g., using stepwise methods); another one, called *principal component regression*, is to perform a PCA of the ***Y*** matrix and then use the principal components (i.e., eigenvectors) of ***Y*** as regressors on ***Z***.
The problem is then of choosing an optimum subset of predictors that gives the best regression. One possibility is to choose the first k principal components; however, these components are obtained to best explain **Y** rather than **Z**, and so, nothing guarantees that they are also relevant for **Z**. In PLS we seek the components from ***Y*** that are relevant also for ***Z***. In particular, PLS regression performs a simultaneous decomposition of ***Y*** and ***Z*** into principal components with the constraint that the components explain as much as possible the covariance between ***Y*** and ***Z*** (Rencher, 2002).

5.4.3 Factor Analysis (FA)

Factor analysis is a statistical method used to discover if the observed variables can be explained in terms of a much smaller number of variables called factors. It is closely related to PCA in that they both try to reduce the redundancy among the variables by using a smaller number of factors (or principal components in PCA); however it has some important differences: i) in PCA the components are defined as linear combinations of the original variables while in FA the original variables are linear functions of the factors; ii) in PCA we seek to explain the total variance, while in FA we attempt to reproduce the covariance; iii) in PCA essentially no assumptions are required while in FA some fundamental assumptions are defined; iv) the principal components are unique, whereas the factors can be rotated. By rotating your factors you attempt to find a factor solution that is equal to that obtained in the initial extraction but which has the simplest interpretation. This last point is one of the main advantage of the FA over PCA, if our goal is to find and describe the underlying factors of the data. On the other hand, if we are simply searching for a smaller number of variables as input for another analysis, the PCA is preferred (Manly, 2004; Rencher, 2002; Bryant & Yarnold, 1994).

5.4.4 Cluster Analysis (CA)

Cluster analysis is a procedure used to partition the data into groups so that the most similar observations are assigned in the same cluster and clusters are dissimilar to each other (Manly, 2004).

CA is an unsupervised technique, that is, the group membership of the observations (and often the number of groups) is not known in advance.

Since we are trying to group similar observations, a measure of similarity or dissimilarity is required. The most common distance functions are: i) the Euclidean distance; ii) the Manatthan distance; iii) the Mahalanobis distance; iv) the maximum norm.

Several types of clustering algorithms have been developed. Based on the procedure they use, they can be divided into three main groups: hierarchical, partitional and density-based clustering.

Hierarchical clustering algorithms are sequential. They can be agglomerative or divisive. The agglomerative clustering starts with all observations placed in different clusters and in each step an observation or a cluster of observations are merged into another cluster. The divisive method starts with one single cluster containing all observations and then it divides the cluster into two sub-clusters at each step.

The partitional algorithm assigns the observations to a set of clusters without using hierarchical approaches. One of the most used non-hierarchical approach is the k-means clustering.

The density-based clustering seeks to search for region of high density without any assumption about the shape of the cluster.

5.4.5 Artificial Neural networks (ANN)

The artificial neural networks are mathematical models that were developed in analogy to a network of biological neurons (Krogh, 2008). In the brain, the highly interconnected network of neurons communicates sending electric pulses through the neural wiring of axons, synapses and dendrites. Mathematically, a neuron can be modeled as a switch that receives a series of values as input and produces an output consisting of a weighted sum of the input vectors eventually filtered by a function *f*. Many neurons can be combined to create more complex networks. Depending on the type of neurons and how the neurons are connected to each other, different kinds of neural networks can be created. The most common type of neural network is the feed-forward neural network, in which neurons are grouped into layers, each neuron of a layer is connected to all the neurons of the next layer and the information flows from the input to the output without loops. For a comprehensive description of neural networks and their applications see Haykin (1999) and Bishop (1995).

6. Applications of vibrational spectroscopies to the study of stem cell differentiation

Stem cells (SCs) are self-renewing cells characterized by the capacity to differentiate into a wide range of specialized cells. Two main types of SCs exist: embryonic and adult.

Embryonic stem cells (ESCs) are derived from cells of an embryo - the inner cell mass of blastocyst - and are considered the most versatile type of SCs because they have the unique ability to retain the developmental capacity of generating all functional adult cell types (Thomson et al., 1998; Evans & Kaufman, 1981).

Adult SCs are found among differentiated cells of a tissue or organ that can renew itself and can differentiate into specialized cell types of the same tissue or organ where they reside (the *stem cell niche*). Indeed, their major role is to maintain and repair the tissue in which they are found. Adult SCs have been identified in many organs and tissues, including bone marrow, brain, liver, skeletal muscle, and skin (Verfaillie, 2002; Peterson & Davidson, 2000).
As discussed in the introduction, Raman and FTIR microspectroscopies are successfully applied in stem cell research (Chan & Lieu, 2009; Heraud & Tobin, 2009). In the following pages, we will first report some studies that illustrate the potential of these vibrational tools to characterize embryonic stem cells. We will then illustrate a few research works on adult stem cells, that we believe could be relevant for a better understanding of stem cell biology.

6.1 Embryonic stem cells

FTIR and Raman microspectroscopies allow to detect rapidly and in a non invasive way biochemical changes during ES cell differentiation, providing unique markers for the identification in-situ of SCs differentiation status.
One of the earliest studies aimed at the spectroscopic characterization of embryonic stem (ES) cells is that of Notingher and colleagues, who applied Raman microspectroscopy to monitor the murine ES cell differentiation process, spontaneous and via embryoid body (EB) formation (Notingher et al., 2004 a and b). The authors showed that undifferentiated, spontaneously differentiated, and EB differentiated murine ES cells exhibit unique Raman markers that, in association with PCA, could be used to identify the differentiation state of the ES cells. In particular, it was found that the most significant differences could be attributed to cell RNA content that in undifferentiated cells was higher than in differentiated ones, a result that suggests to the authors that differentiating ES cells use the pool of dormant mRNA to produce new specific proteins of the new phenotype. Indeed, as the ES cells start to differentiate toward various phenotypes, the translation of mRNA increases, as indicated by the decrease in the ratio between the areas of the 813 cm^{-1} RNA peak and the phenylalanine peak at 1005 cm^{-1}, to reach values similar to those found in fully differentiated cells, after 16-20 days of differentiation. These results indicated that RNA and protein peaks in the Raman spectra of murine ES cells can be used as a differentiation marker, with important applications for the development of engineered tissues.
Raman microspectroscopy, coupled with the multivariate PCA-LDA analysis, was also applied to explore the possibility to discriminate between undifferentiated human ES cells and their cardiac derivatives (Chan et al, 2009). Indeed, unlike other cell lineages, cardiomyocytes lack specific surface markers required for their physical identification and separation, making desirable the development of new analytical tools. In this work, the authors were able to detect spectroscopic signatures of ES cells and of their cardiac derivatives, mainly involving RNA and protein content. In particular, the authors found that undifferentiated cells were characterized by a more elevated mRNA level than differentiated cells, resulting from their different active cell cycles, as suggested by the different intensity of the peak at 811 cm^{-1} (phosphodiester bond) observed in the two cases. Interestingly, their results were in agreement with those obtained by Notingher and colleagues on murine stem cells (Notingher et al., 2004 a and b) - as discussed above - and by Schulze and colleagues (2010) in a Raman study of the spontaneous differentiation process of human ES cells. Noteworthy, they also investigated the effect of laser exposure on cells, in order to verify the non-invasiveness of the spectroscopic method. Indeed, they demonstrated that the laser irradiation does not

compromise cell pluripotency, as it didn't affect the expression of the human ES cells transcription factor OCT4, required to sustain ES cell self renewal. Moreover, no effects on cell morphology and cell proliferation were detected.

Of great interest is the study of Heraud and colleagues (2010) that employed FTIR microspectroscopy with focal plane array detection to characterize human ES cell differentiation directed toward specific cell lineages, namely mesendoderm and ectoderm. Well defined spectral differences – confirmed also by partial least squares discriminant analysis (PLS-DA) and artificial neural network analysis (ANN) - were detected among the three different cell populations, mainly involving the lipid and the glycogen bands (respectively at 2920 cm^{-1} and at 1155 cm^{-1}), whose intensities were found to be higher in the undifferentiated than in the differentiated cell populations. The results demonstrated that FTIR signatures can be used to successfully discriminate between human stem cells and their differentiated progenies, even at early stages of differentiation.

Another application of FTIR microspectroscopy, coupled with PCA and unsupervised hierarchical cluster analysis (UHCA), was aimed at identifying specific marker bands of murine ES cell differentiation toward neural cell types (Tanthanuc et al., 2010). In particular, by applying focal plane array detection and synchrotron based FTIR microspectroscopy, the authors were able to find significant differences between undifferentiated and differentiated cells, mainly in spectral regions due to lipid and protein absorptions. In particular, they observed a dramatic increase of the acyl chain CH2 symmetric and asymmetric stretching modes - around 2850 cm^{-1} and 2920 cm^{-1} respectively - during the differentiation process, increment possibly related to changes in membrane lipids responsible for neural cell differentiation and signal transduction. This result has been also confirmed monitoring the lipid carbonyl band around 1740 cm^{-1}, whose peak position and intensity were observed to change during differentiation. Furthermore, important changes in protein secondary structures were detected and in particular the differentiated cells appeared to be characterized by a higher content of alpha-helix proteins than undifferentiated cells. The authors explained this result as due to the increased expression of alpha-helix rich proteins of the cytoskeleton, as tubulin and actin, important for the establishment of neural structure and function.

We applied FTIR microspectroscopy – supported by PCA-LDA analysis - to characterize in situ the early stages of murine ES cell spontaneous differentiation (Ami et al., 2010; Ami et al., 2008). We found that significant changes in nucleic acid, protein and lipid content occurred during the differentiation process.

In Figure 4 we reported the second derivative spectra of ES cells at different maturation stages (from undifferentiated to 14 days of differentiation).

As illustrated in the Figure, we first found that undifferentiated ES cells were characterized by a RNA content higher than differentiating cells, in agreement with what reported by Notingher and colleagues with Raman microspectroscopy (Notingher et al., 2004 a and b), as previously discussed. Moreover, we monitored the formation of the DNA/RNA hybrid through the simultaneous presence of the three components respectively around 954 cm^{-1} (CC stretching of DNA backbone), at 914 cm^{-1} (ribose ring) and at 899 cm^{-1} (deoxyribose), after 4-7 days of differentiation. These results, indicating that the transcription activity for the new phenotype was taking place in that temporal range, were further supported by changes in the secondary structures of the whole protein content, likely due to the emergence of the new phenotype. Indeed, again starting from 4-7 days of differentiation, we observed an increase of alpha-helix and beta-turn components, respectively at 1658 cm^{-1}

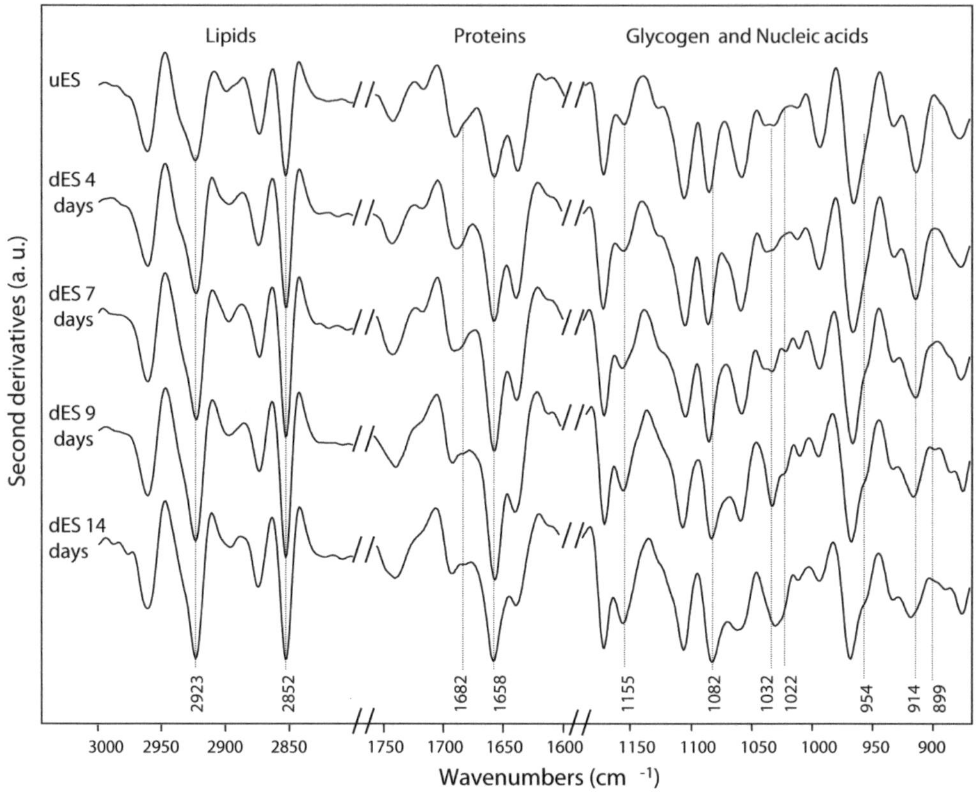

Fig. 4. Murine embryonic stem cell differentiation monitored by FTIR microspectroscopy. The second derivative of FTIR spectra of murine ES cells, undifferentiated (uES) and spontaneously differentiated (dES), are reported in three different spectral regions: i) 3000-2800 cm^{-1} mainly due to lipid acyl chains; ii) 1750-1600 cm^{-1} where protein amide I band occurs; iii) 1200-800 cm^{-1}, mainly due to glycogen and nucleic acid absorptions (see text). Spectra have been normalized at the tyrosine band around 1515 cm^{-1} and reported after magnification in each region, for the presentation of the data.

and 1682 cm^{-1}, that suggested that the expression of proteins typical of cardiomyocyte precursors was taking place. Indeed, it is known that these cells are rich in alpha-myosin, a protein belonging to alpha-helix fold and that they are characterized by the formation of gap junctions (Oyamada et al., 1996), whose main protein components are connexins, containing again alpha-helix structures and an important percentage of beta-turns. To support our hypothesis - confirmed by cytochemical analysis - after the "switch" of the new phenotype we also observed the emergence of IR bands due to glycogen at 1155 cm^{-1}, 1081 cm^{-1} and between 1035 and 1020 cm^{-1}, typical of cardiomyocytes (Pasumarthi & Field 2002).

Also dramatic changes in lipid absorption were detected during ES cell differentiation. In particular, an increase of the CH2 vibrational modes at 2923 cm^{-1} and at 2852 cm^{-1} was monitored, starting as soon as the differentiation process was taking place, and up to the end of our investigation (9-14 days). These results indicated that significant changes in lipid composition occurred, suggesting that the new phenotype was characterized by new membrane properties.

The spectroscopic results were then validated by PCA-LDA analysis that allowed to obtain an excellent segregation of the data into five separated clusters, each corresponding to a specific differentiation stage, as reported in Figure 5. Moreover, this analysis enabled us to identify in the spectrum the wavenumbers that contributed to the largest inter-spectral variance during the differentiation process, and - in agreement with the direct inspection of the spectral data - they were found to be due to protein and nucleic acid components.

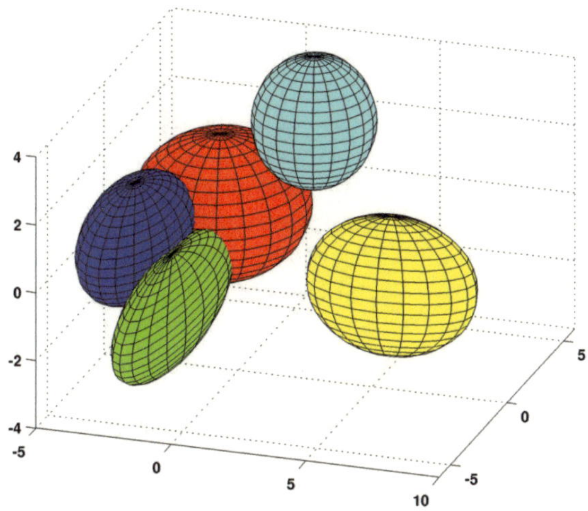

Fig. 5. PCA-LDA analysis of murine ES cell differentiation. The clustering of FTIR absorption spectra - from 1800 to 800 cm^{-1} - as 3D score plot is shown. Data for undifferentiated cells (red) and at 4 (blue), 7 (green), 9 (light blue), and 14 (yellow) days of differentiation have been analysed. The ellipsoid semi-axes correspond to two standard deviations of the data.

6.2 Adult stem cells

An interesting investigation performed by synchrotron based FTIR microspectroscopy - coupled with principal component analysis (PCA) - enabled to discriminate in bovine cornea among SCs, transit-amplifying (TA) and terminally differentiated (TD) cells (German et al., 2006). Measuring the absorption spectra of individual cells in cryosections, the authors found significant spectral differences among the three different cell types, with only a slight overlap between SC and TA cells. The most important differences mainly involved changes in spectral components due to nucleic acid absorptions, like the RNA band at 1120 cm^{-1} and the phosphate band around 1080 cm^{-1}. Moreover, the authors found that TD cells formed a well separated and homogeneous population with spectral features closer to TA cells than to SCs. As expected, the spectral response of the terminal differentiation state is characterized by important changes in nucleic acid and protein content, being associated with a loss of proliferative ability and the production of proteins associated with the new phenotype.

A FTIR characterization of human corneal epithelium performed by the same research group (Bentley et al., 2007) confirmed the previous results obtained on bovine cornea. Also in this case, the authors were able to discriminate among SC, TA and TD cells, finding again important changes mainly in nucleic acid and protein content. In particular, the main spectral differences between TA and TD cells were found to involve the protein secondary structures and the RNA expression, as discussed above. Noteworthy, in the two works, the authors detected small subpopulations of cells within the corneal epithelium SC niche with TA cell like characteristics, strongly suggesting that the TA cells are newly generated prior to their migration.

Furthermore, the entire tissue architecture was investigated using IR spectral imaging that enabled to localize and better characterize SCs (Nakamura et al., 2010). By this approach further details on the differences among SC, TA and TD cells were obtained, confirming that nucleic acid response - between 1425 and 900 cm^{-1} - accounts for the most significant differences among the three types of cell populations. Important changes in protein content, between 1800 and 1480 cm^{-1}, have been also detected in the examined cell types, as expected considering their different functions. Interestingly, the most discriminating spectral features of SCs were associated to DNA and RNA conformations, as indicated by the bands at 1225 cm^{-1} and at 1080 cm^{-1} respectively, whereas IR bands due to proteins and lipids, respectively at 1558 cm^{-1} and at 1728 cm^{-1}, allowed to discriminate between TA and TD cells.

Of particular relevance is also the work of Walsh and colleagues, where synchrotron FTIR microspectroscopy - supported by PCA-LDA analysis – was applied to characterize the different cell types, derived from stem cells, along the length of gastrointestinal tract, one of the most regenerative human tissue (Walsh et al., 2009). Through IR image maps with the related IR spectra collected from tissue sections at the single cell level, the authors detected spectral changes in the differentiation states along the gastrointestinal tract - with common features in related cell types – mainly involving DNA conformational changes, with one of the most important spectral marker at 1080 cm^{-1}, due to phosphate vibrational mode. These results were further confirmed by PCA-LDA multivariate analysis, whose crucial role has been critically highlighted in Walsh' work. Indeed, this analysis allowed to identify as the most contributory wavenumbers those due to the phosphate mode absorptions, partly associated to protein phosphorylation. Overall, these results suggested to the authors that DNA conformational changes could be considered a significant stemness markers in gastrointestinal crypts.

7. Conclusive remarks

The examples reported in this chapter highlight the great potential of spectroscopic approaches providing new insights in stem cell biology. In particular, FTIR microspectroscopy is a powerful tool that enables to obtain - in a non invasive way - a chemical fingerprint of the cell types, giving information on the overall changes in the macromolecular content occurring during a biological event. In this way, this approach allows to assess in-situ the differentiation status of the cells through the identification of specific marker bands. Moreover, the time evolution of these bands enabled to follow the progress of the process by the simultaneous monitoring of the most important cellular components, as nucleic acids and proteins.

We should underline that the successful application of the spectroscopic approach requires the use of an appropriate multivariate analysis to validate the spectral data and to identify the marker bands of the process under investigation. The integration with the established biochemical methods is, of course, an important requisite to understand the biological significance of the spectroscopic results.

As a final comment, we would also like to point out that FTIR and Raman spectroscopic approaches, indeed, might offer preliminary tools – rapid and inexpensive - to obtain useful information on complex systems, in order to design conclusive biological experiments. As discussed above, these techniques allow to characterize the temporal correlation of biological events that occur simultaneously in a complex system, a result not easily tackled by the standard biochemical methods.

8. Acknowledgements

D. A. is indebted to Fondazione IRCCS Policlinico San Matteo, Pavia (I) for the supporting scholarship. S.M. D. acknowledges the financial support of the FAR (Fondo di Ateneo per la Ricerca) of the University of Milano-Bicocca (I). The authors wish to thank Prof. Carlo Alberto Redi and its research group at the University of Pavia (I) for the research collaboration on embryonic stem cell differentiation. We are grateful to Dr Carla Smeraldi for the language revision of this chapter.

9. References

Ami D., Natalello, A., Zullini, A., Doglia, S.M. (2004). Fourier transform infrared microspectroscopy as a new tool for nematode studies. *FEBS Letters*, 576, 297–300, ISSN: 0014-5793.

Ami, D., Natalello, A., Gatti-Lafranconi, P., Lotti, M., Doglia, S.M. (2005). Kinetics of inclusion body formation studied in intact cells by FT-IR spectroscopy. *FEBS Letters*, 579, 3433–3436, ISSN: 0014-5793.

Ami, D., Neri, T., Natalello, A., Mereghetti, P., Doglia, S.M., Zanoni, M., Zuccotti, M., Garagna, S., Redi, C.A. (2008). Embryonic stem cell differentiation studied by FT-IR spectroscopy. *Biochimica et Biophysica Acta*, 1783, 98–106, ISSN: 0167-4889.

Ami, D., Natalello, A., Schultz, T., Gatti-Lafranconi, P., Lotti, M., Doglia, S.M., de Marco, A. (2009). Effects of recombinant protein misfolding and aggregation on bacterial membranes. *Biochimica et Biophysica Acta*, 1794, 263–269, ISSN: 1570-9639.

Ami, D., Natalello, A., Mereghetti, P., Neri, T., Zanoni, M., Monti, M., Doglia, S.M., Redi, C.A. (2010). FT-IR spectroscopy supported by PCA-LDA analysis for the study of embryonic stem cell differentiation. *Spectroscopy-Biomedical Applications*, 24, 89-97, ISSN: 0712-4813.

Arrondo, J.L.R., Muga, A., Castresana, J., Goni, F.M. (1993). Quantitative studies of the structure of proteins in solution by Fourier-transform infrared spectroscopy. *Progress in Biophysics and Molecular Biology*, 59, 23–56, ISSN: 0079-6107.

Arrondo, J.L.R. & Goni, F.M. (1998). Infrared studies of protein-induced perturbation of lipids in lipoproteins and membranes. *Chemistry and physics of lipids*, 96, 53-68, ISSN: 0009-3084.

Arrondo, J.L.R., & Goni, F.M. (1999). Structure and dynamics of membrane proteins as studied by infrared spectroscopy. *Progress in Biophysics and Molecular Biology*, 72, 367-405, ISSN: 0079-6107.

Baker, M.J., Gazi, E., Brown, M.D., Shanks, J.H., Gardner, P., Clarke, NW. (2008). FTIR-based spectroscopic analysis in the identification of clinically aggressive prostate cancer. *British Journal of Cancer*, 99, 1859 - 1866, ISSN: 0007-0920.

Banyay, M. & Graslund, A. (2002). Structural effects of cytosine methylation on DNA sugar pucker studied by FTIR. *Journal of Molecular Biology*, 324, 667-676, ISSN: 0022-2836.

Banyay, M., Sarkar, M., Graslund, A. (2003). A library of IR bands of nucleic acids in solution. *Biophysical Chemistry*, 104, 477–488, ISSN: 0301-4622.

Barth, A. & Zscherp, C. (2002). What vibrations tell us about proteins. *Quarterly Reviews of Biophysics*, 35, 369–430, ISSN: 0033-5835.

Barth, A. (2007). Infrared spectroscopy of proteins. *Biochimica et Biophysica Acta*, 1767, 1073–1101, ISSN: 0005-2728.

Bentley, A.J., Nakamura, T., Hammiche, A., Pollock, H.M., Martin, F.L., Kinoshita, S., Fullwood, N.J. (2007). Characterization of human corneal stem cells by synchrotron infrared micro-spectroscopy. *Molecular Vision*, 13, 237-242, ISSN: 1090-0535.

Bishop, C.M. (1995). *Neural Networks for Pattern Recognition*, Oxford University Press, ISBN:0198538642, New York, NY, USA.

Bryant, F.B. & Yarnold, P.R. (1994). Principal components analysis and exploratory and confirmatory factor analysis, In: *Reading and understanding multivariate analysis*, Grimm & Yarnold (Eds.), American Psychological Association Books, ISBN: 978-1-55798-273-5, Washington DC, USA.

Casal, H.L. & Mantsch, H.H. (1984). Polymorphic phase behaviour of phospholipid membranes studied by infrared spectroscopy. *Biochimica et Biophysica Acta*, 779, 381-401, ISSN: 03044157.

Chan, J.W., Lieu, D.K., Huser T., Li R.A. (2009). Label-free, non-invasive spectroscopic separation of human embryonic stem cells (hESCs) and their cardiac derivatives. *Analytical Chemistry*, 81, 1324-1331, ISSN: 0003-2700.

Chan, J.W. & Lieu, D.K. (2009). Label-free biochemical characterization of stem cells using vibrational spectroscopy. *Journal of Biophotonics*, 2, 656–668, ISSN: 1864-063X.

Choo, L.P., Wetzel, D.L., Halliday, W.C., Jackson, M., LeVine, S.M, Mantsch H.H. (1996). In situ characterization of beta-amyloid in Alzheimer's diseased tissue by synchrotron Fourier transform infrared microspectroscopy. *Biophysical Journal*, 71, 1672 - 1679, ISSN: 0006-3495.

Diomede, L., Cassata, G., Fiordaliso, F., Salio, M., Ami, D., Natalello, A., Doglia, S.M., De Luigi, A., Salmona, M. (2010). Tetracycline and its analogues protect Caenorhabditis elegans from β amyloid-induced toxicity by targeting oligomers. *Neurobiology of Disease*, 40, 424-431, ISSN: 0969-9961.

Doglia, S.M., Ami, D., Natalello, A., Gatti-Lafranconi, P., Lotti, M. (2008). Fourier transform infrared spectroscopy analysis of the conformational quality of recombinant proteins within inclusion bodies. *Biotechnology Journal*, 3, 193–201, ISSN: 1860-6768.

Eriksson, L., Johansson, E., Kettaneh-Wold, N., Trygg, J., Wikstrom, C., Wold, S. (2006). *Multivariate and Megavariate Data Analysis Basic Principles and Applications*, Umetrics Academy, ISBN-10: 91-973730-2-8, San Jose, CA, USA.

Evans, M.J. & Kaufman, M.H. (1981). Establishment in culture of pluripotential cells from mouse embryos. *Nature*, 292, 154–156, ISSN: 0028-0836.

Fearn, T. (2002). Discriminant analysis, In: *Handbook of Vibrational Spectroscopy*, Chalmers, J.M. & Griffiths, P.R. (eds.), 2086–2093, ISBN: 0-471-98847-2 , Wiley, New York, USA.

Fukunaga, K. (1990). *Introduction to Statistical Pattern Recognition*, Academic Press Professional Inc., ISBN: 0-12-269851-7, San Diego, CA, USA.

German, M.J., Pollock, H.M., Zhao, B., Tobin, M.J., Hammiche, A., Bentley, A., Cooper, L.J., Martin, F.L., Fullwood, N.J. (2006). Characterization of putative stem cell populations in the cornea using synchrotron infrared microspectroscopy. *Investigative Ophthalmology & Visual Science.*, 47, 2417–2421, ISSN: 0146-0404.

Hamm, P. (2009). For structural biology, try infrared instead. *Structure*, 17, 149-150, ISSN: 0969-2126.

Haykin, S. (1999). *Neural Networks: A Comprehensive Foundation*, Prentice Hall Inc., ISBN: 0132733501, Englewood Cliffs, New Jersey, USA.

Héberger, K.; Csomós, E. & Livia S. (2003). Principal Component and Linear Discriminant Analyses of Free Amino Acids and Biogenic Amines in Hungarian Wines. *Journal of agricultural and food chemistry*, 51, 27, 8055-8060, ISSN: 0021-8561.

Heraud, P. & Tobin, M.J. (2009). The emergence of biospectroscopy in stem cell research. *Stem cell research*, 3, 12-14, ISSN: 1873-5061.

Heraud, P., Ng, E.S., Caine, S., Yua, Q.C., Hirst, C., Mayberry, R., Bruce, A., Wood, B.R., McNaughton, D., Stanley, E.G., Elefanty, A.G. (2010). Fourier transform infrared microspectroscopy identifies early lineage commitment in differentiating human embryonic stem cells. *Stem Cell Research*, 4, 140–147, ISSN: 1873-5061.

Jonathan, P., McCarthy, W.V. & Roberts, M.I.A. (1996). Discriminant Analysis With Singular Covariance Matrices. A Method Incorporating Cross-Validation And Efficient Randomized Permutation Tests. *Journal Of Chemometrics*, 10, 4, 189-213, ISSN: 0886-9383.

Kacurakova, M. & Wilson, R.H. (2001). Developments in mid-infrared FT-IR spectroscopy of selected carbohydrates. *Carbohydrate Polymers*, 44, 291–303, ISSN: 0144-8617.

Kauppinen, J. K., Moffatt, D.J., Mantsch, H.H., Cameron, D.G. (1981). Fourier self-deconvolution—a method for resolving intrinsically overlapped bands. *Applied Spectroscopy*, 35, 271–276, ISSN: 0003-7028.

Kazarian, S.G. & Chan, K.L.A. (2006). Applications of ATR-FTIR spectroscopic imaging to biomedical samples. *Biochimica et Biophysica Acta*, 1758, 858–867, ISSN: 0005-2736.

Konorov, S.O., Schulze, H.G., Caron, N.J., Piret, J.M., Blades, M.W., Turner, R.F.B. (2010). Raman microspectroscopic evidence that dry-fixing preserves the temporal pattern of non-specific differentiation in live human embryonic stem cells. *Journal of Raman spectroscopy*, in press DOI 10.1002/jrs.2769, ISSN: 0377-0486.

Krafft, C., Steiner, G., Beleites, C., Salzer, R. (2009). Disease recognition by infrared and Raman spectroscopy. *Journal of Biophotonics*, 2, 13-28, ISSN: 1864-063X.

Krogh, A. (2008). What are artificial neural networks? *Nature Biotechnology*, 25, 2, 195-197, ISSN: 1087-0156.

Levin, W. & Bhargava, R. (2005). Fourier transform infrared vibrational spectroscopic imaging: Integrating Microscopy and Molecular Recognition. *Annual Review of Physical Chemistry*, 56, 429-474, ISSN: 0066-426X.

Lewis, E.N., Treado, P.J., Reedor, R.C., Stay, G.M., Dowrey, A.E., Marcott, C., Levint, I.W. (1995). Fourier Transform Spectroscopic Imaging Using an Infrared Focal-Plane Array Detector. *Analytical Chemistry*, 67, 3377-3381, ISSN: 0003-2700.

Manly, B.F.J. (2004). *Multivariate Statistical Methods*, Chapman & Hall/CRC press, ISBN: 9781584884149, London, UK.

Miller, L.M. & Dumas P. (2010). From structure to cellular mechanism with infrared microspectroscopy. *Current opinion in structural biology*, 20, 1-8, ISSN: 0959-440X.

Nakamura, T., Kelly, J.G., Trevisan, J., Cooper, L.J., Bentley, A.J., Carmichael, P.L., Scott, A.D., Cotte, M., Susini, J., Martin-Hirsch, P.L., Kinoshita, S., Fullwood, N.J.,

Martin, F.L. (2010). Microspectroscopy of spectral biomarkers associated with human corneal stem cells. *Molecular Vision*, 16, 359-368, ISSN: 1090-0535.

Natalello, A., Ami, D., Brocca, S., Lotti, M., Doglia, S.M. (2005). Secondary structure, conformational stability and glycosylation of a recombinant Candida rugosa lipase studied by Fourier-transform infrared spectroscopy. *Biochemical Journal*, 385, 511–517, ISSN: 0264-6021.

Notingher, I., Bisson, I., Polak, J.M., Hench, L.L. (2004a). In situ spectroscopic study of nucleic acids in differentiating embryonic stem cells. *Vibrational Sprectroscopy*, 35, 199–203, ISSN: 0924-2031.

Notingher, I., Bisson, I., Bishop, A.E., Randle, W.L., Polak, J.M., Hench, L.L. (2004b). In situ spectral monitoring of mRNA translation in embryonic stem cells during differentiation in vitro. *Analytical chemistry*, 76, 3185–3193, ISSN: 0003-2700.

O'Connell, D., Bertram, J., Giffin, M., Ebesu, J., Empey, C., Mulrooney, M., Wall, M. (2005). Detection of acetylated and non-acetylated histones in chromatin. *Microscopy And Microanalysis*, 11, 1004-1005, ISSN: 1431-9276.

Orsini, F., Ami, D., Villa, A.M., Sala, G., Bellotti, M.G., Doglia, S.M. (2000). FT-IR microspectroscopy for microbiological studies. *Journal of microbiological methods*, 42, 17–27, ISSN: 0167-7012.

Oyamada, Y., Komatsu, K., Kimura, H., Mori, M., Oyamada, M. (1996). Differential regulation of gap junction protein (connexin) genes during cardiomyocytic differentiation of mouse embryonic stem cells in vitro. *Experimental Cell Research*, 229, 318–326, ISSN: 0014-4827.

Pasumarthi, K.B.S. & Field, L.J. (2002). Cardiomyocyte enrichment in differentiating ES cell cultures: strategies and applications, In: *Embryonic Stem Cells Methods and Protocols*, Turksen, K. (Ed.), 157–168, Humana Press, ISBN: 978-0-89603-881-3, New Jersey, USA.

Pereira, R.C.C., Skrobot, V.L., Castro, E.V.R., Fortes, I.C.P., Pasa, V.M.D. (2006). Determination of Gasoline Adulteration by Principal Components Analysis-Linear Discriminant Analysis Applied to FTIR Spectra. *Energy & Fuels*, 20, 1097-1102, ISSN: 0887-0624.

Peterson, K.J. & Davidson, E.H. (2000). Regulatory evolution and the origin of the bilaterians. *Proceedings of the National Academy of Sciences*, 97, 4430-4433, ISSN: 0027-8424.

Rencher, A.C. (2002). *Methods of Multivariate Analysis*, Wiley, ISBN: 978-0-471-41889-4, Hoboken, New Jersey, USA.

Rezzi, S., Giani, I., Héberger, K., Axelson, D.E., Moretti, V.M., Reniero, F. & Claude Guillou. (2007). Classification of Gilthead Sea Bream (Sparus aurata) from 1H NMR Lipid Profiling Combined with Principal Component and Linear Discriminant Analysis. *Journal of agricultural and food chemistry*, 55, 24, 9963-9968, ISSN: 0021-8561.

Schulze, H.G., Konorov, S.O., Caron, N.J., Piret, J.M, Blades, M.W., Turner, R.F.B. (2010). Assessing Differentiation Status of Human Embryonic Stem Cells Noninvasively Using Raman Microspectroscopy. *Analytical Chemistry*, 82, 5020–5027, ISSN: 0003-2700.

Skrobot, L.S., Castro, V.R.E., Pereira, R.C.C., Pasa V.M.D. & Fortes I.C.P. (2007). Use of Principal Component Analysis (PCA) and Linear Discriminant Analysis (LDA) in

Gas Chromatographic (GC) Data in the Investigation of Gasoline Adulteration. *Energy & Fuels*, 21, 6, 3394-3400, ISSN: 0887-0624.

Smith, A.G. (2001). Embryo-derived stem cells: of mice and men. *Annual Review of Cell and Developmental Biology*, 17, 435–462, ISSN: 1081-0706.

Steller, W., Einenkel, J., Horn, L.C., Braumann, U.D., Binder, H., Salzer, R., Krafft, C. (2006). Delimitation of squamous cell cervical carcinoma using infrared microspectroscopic imaging. *Anaytical Bioanalytical Chemistry*, 384, 145–154, ISSN: 1618-2642.

Susi, H. & Byler, D.M. (1986). Resolution-enhanced Fourier transform infrared spectroscopy of enzymes. *Methods in Enzymology*, 130, 290–311, ISSN: 0076-6879.

Tamm, L.K. & Tatulian, S.A. (1997). Infrared spectroscopy of proteins and peptides in lipid bilayers. *Quarterly reviews of biophysics*, 30, 365-429, ISSN: 0033-5835

Tanthanuch, W., Thumanu, K., Lorthongpanich, C., Parnpai, R., Heraud, P. (2010). Neural differentiation of mouse embryonic stem cells studied by FTIR spectroscopy. *Journal of molecular Structure*, 967, 189–195, ISSN: 0022-2860.

Theophanides, T. & Tajmir-Riahi, H.A. (1985). FT-IR Spectroscopic Evidence of C2′-endo, anti, C3′-endo,anti Sugar Ring Pucker in 5′-GMP and 5′-IMP Nucleotides and their Metal-Adducts, In: *Structure and Motion: Membrane, Nucleic Acids and Proteins*, Clementi, E., Corongiu, G., Sarma, M.H., Sarma R.H. (Eds.), 521-530, Adenine Press, ISBN: 0-940030-12-8, New York, USA.

Thomson, J.A., Itskovitz-Eldor, J., Shapiro, S.S., Waknitz, M.A., Swiergiel, J.J., Marshall, V.S., Jones, J.M. (1998). Embryonic stem cell lines derived from human blastocysts. *Science*, 282, 1145–1147, ISSN: 0036-8075.

Thumanu, K., Tanthanuch, W., Lorthongpanich, C., Heraud, P., Parnpai, R. (2009). FTIR microspectroscopic imaging as a new tool to distinguish chemical composition of mouse blastocyst. *Journal of Molecular Structure*, 933, 104–111, ISSN: 0022-2860.

Trounson, A. (2006). The Production and Directed Differentiation of Human Embryonic Stem Cells. *Endocrine Reviews.*, 27, 208-219, ISSN: 0163-769X.

Tsuboi, M. (1961), Infrared spectra and secondary structure of deoxyribonucleic acid. *Progress of Theoretical Physics Supplement*, 17, 99-107, ISSN: 0375-9687.

Vats, A., Bielby, R.C., Tolley, N.S., Nerem, R., Polak, J.M. (2005). Stem cells. *Lancet*, 366, 592-602, ISSN: 0140-6736.

Verfaillie, C.M. (2002). Adult stem cells: assessing the case for pluripotency. *TRENDS in Cell Biology*, 12, 502-508, ISSN: 0962-8924.

Walsh, M.J., Hammiche, A., Fellous, T.G., Nicholson, J.M., Cotte, M., Susini, J., Fullwood, N.J., Martin-Hirsch, P.L., Alison, M.R., Martin, F.L. (2009). Tracking the cell hierarchy in the human intestine using biochemical signatures derived by mid-infrared microspectroscopy. *Stem Cell Research*, 3, 15-27, ISSN: 1873-5061.

Walsh, M.J., Singh, M.N., Pollock, H.M., Cooper, L.J., German, M.J., Stringfellow, H.F., Fullwood, N.J., Paraskevaidis, E., Martin-Hirsch, P.L., Martin, F.L. (2007). ATR microspectroscopy with multivariate analysis segregates grades of exfoliative cervical cytology. *Biochemical and Biophysical Research Communications*, 352, 213–219, ISSN: 0006-291X.

Wang, T.D., Triadafilopoulos, G., Crawford, J.M., Dixon, L.R., Bhandari, T., Sahbaie, P., Friedland, S., Soetikno, R., Contag, C.H. (2007). Detection of endogenous biomolecules in Barrett's esophagus by Fourier transform infrared spectroscopy. *Proceedings of the National Academy of* Sciences, 104, 15864–15869, ISSN: 0027-8424.

Wood, B.R., Chernenko, T., Matthäus, C., Diem, M., Chong, C., Bernhard, U., Jene, C., Brandli, A.A., McNaughton, D., Tobin, M.J., Trounson, A., Lacham-Kaplan, O. (2008). Shedding new light on the molecular architecture of oocytes using a combination of synchrotron Fourier transform-infrared and Raman spectroscopic mapping. *Analytical Chemistry*, 80, 9065-9072, ISSN: 0003-2700.

Zhao, R., Quaroni, L., Casson, A.G. (2010). Fourier transform infrared spectromicroscopic characterization of stem like cell populations in human esophageal normal and adenocarcinoma cell lines. *Analyst*, 135, 53-61, ISSN: 0003-2654.

Zhizhina, G.P. & Oleinik, E.F. (1972). Infrared spectroscopy of nucleic acids. *Russian Chemical Reviews*, 41, 258 -280, ISSN: 0036-021X.

Part 3

Methods for Reprogramming Cells

12

Reprogrammed Parthenogenetic ES Cells - New Choice for Regenerative Medicine

Takuro Horii and Izuho Hatada
Gunma University
Japan

1. Introduction

Regenerative medicine using pluripotent stem cells is indispensable for cell transplantation therapy for the patients; however, the production of human embryonic stem (hES) cells from fertilized embryos or cloned embryos often become the ethical concerns because it needs destruction of viable embryos. On the other hand, the production of induced pluripotent stem (iPS) cells do not need viable embryos; whereas, it is necessary to confirm safety for the use of iPS cells that containing viruses or expression plasmids. Here, we would like to propose parthenogenetic ES (PGES) cells as the 3rd pluripotent stem cells for cell transplantation therapy because the production of PGES cells from fertilization-failure or surplus oocytes overcome both problems (Fig. 1). PGES cells do not need destruction of viable embryos and viruses or expression plasmids for the establishment. Nevertheless, PGES cells have a big hurdle to overcome, namely genomic imprinting.

Mammalian parthenotes cannot develop to term. Mouse parthenogenetic embryos die by day 10 of gestation (Surani et al., 1984; Surani et al., 1986). Most notably, they fail in the trophectoderm and primitive endoderm, which results in failure of the extraembryonic tissues in the whole parthenogenetic conceptus (Surani et al., 1983). Alternatively, viable parthenogenetic chimeras can be produced by normal host embryo rescue, and parthenogenetic cells can give rise to a functional germline (Stevens et al., 1977; Stevens, 1978). In the somatic-lineages of chimeras, parthenogenetic cells are allocated initially randomly in the embryo proper (Clarke et al., 1988a; Clarke et al., 1988b; Thomson & Solter, 1989), but this is followed by a progressive elimination of parthenogenetic cells, most notably between days 13 and 15 of gestation (Fundele et al., 1990). In addition, PG chimeras often show the reduction of body weight. These can be explained by parent-specific epigenetic modification of the genome, genomic imprinting which leads to the altered expressions of imprinted genes in parthenogenetic cells. In general, gene expressions of imprinted genes are greatly dependent on the cytosine-guanine (CpG) methylation status in differentially methylated regions (DMRs) of imprinted genes (Fig. 2). In PG with the two maternal genomes, paternally expressed genes, *Peg1/Mest* (Kaneko-Ishino et al., 1995), *Peg3* (Kuroiwa et al., 1996), *Snrpn* (Barr et al., 1995) and *Igf2* (DeChiara et al., 1991), are silenced; whereas, maternally expressed genes, *Igf2r* (Barlow et al., 1991), *p57*^{kip2} (Hatada & Mukai, 1995) and *H19* (Bartolomei et al., 1991; Ferguson-Smith et al., 1991), are expressed excessively (Fig. 3). Thus, biallelic expression and repression of imprinted genes in parthenogenetic cells could restrict to produce transplantable tissues for regenerative

medicine; however, PGES cell chimeras are more normal in their tissue contribution of donor cells and body weight compared to PG chimeras (Allen et al., 1994). We expected that these phenomena were associated with the more normal epigenotype of PGES cells.
To elucidate the epigenetic mechanisms underlying this, we analyzed DNA methylation status and mRNA expression of imprinted genes in PG and PGES cell chimeras (Horii et al., 2008). Interestingly, the PGES cells showed reprogramming of maternal imprints and acquired more normal pluripotency. In this chapter, we propose that such reprogrammed PGES cells might be utilized for the regenerative medicine.

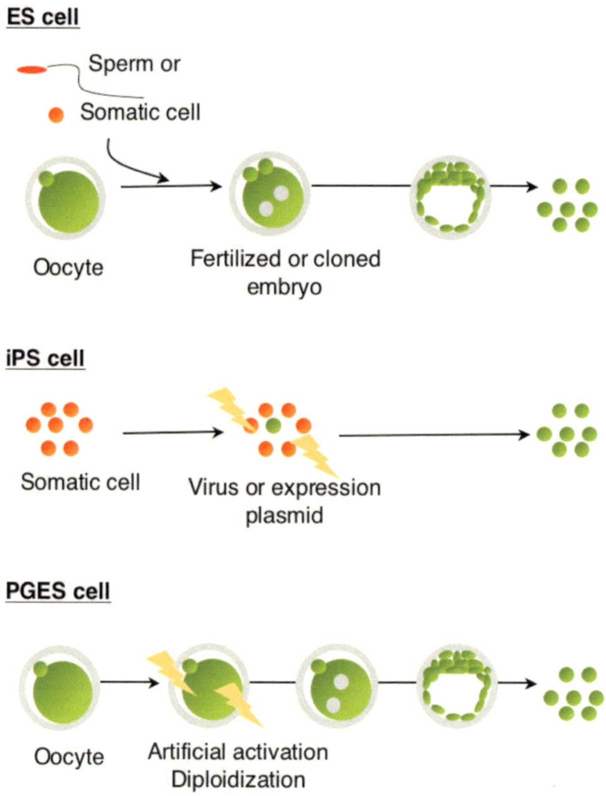

Fig. 1. Production of pluripotent stem cells using various methods.

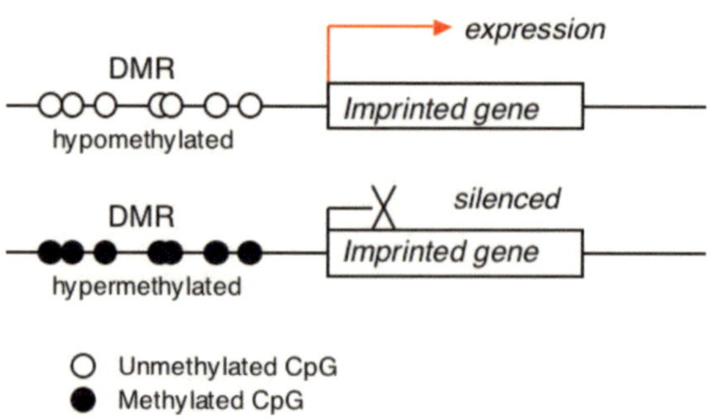

Fig. 2. Representative regulation of gene expression by CpG methylation.

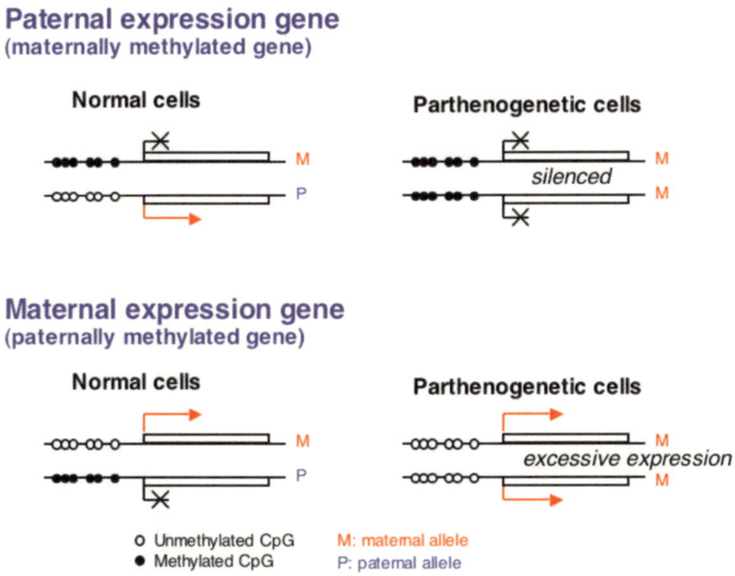

Fig. 3. Genomic imprinting in normal and parthenogenetic cells.

2. Epigenetic reprogramming occurred in PGES cells

2.1 Production of PG and PGES cells

To elucidate the epigenetic difference between PG and PGES cells, we analyzed DNA methylation status and mRNA expression of imprinted genes in PG and PGES cells. Diploid PG were produced as previously described (Horii et al., 2008). Briefly, oocytes at the metaphase stage of the second meiotic division (MII) were collected from the oviducts after human chorionic gonadotrophin (hCG) superovulation, and then cumulus cells were removed by digestion with hyaluronidase in M2 medium. Artificial activation was performed by brief exposure to SrCl2 and cytochalasin B in Ca^{2+}-free M16 embryo culture medium for 6 hours. After activation, the embryos were cultured in M16 medium until developing to the blastocyst stage.

For establishment of PGES cell lines, parthenogenetic blastocysts were cultured for 7 days in Serum-free ES medium, following standard procedures (Horii et al., 2003). After 7 days, ICM outgrowths were harvested in Trypsin/EDTA, disaggregated by mouth pipetting and plated onto feeder cells in ES medium. Clones resembling ES cells in morphology were then picked and disaggregated a second time. They were then expanded and passaged prior to freezing or use.

2.2 Epigenetic reprogramming occurred in PGES cells

The methylation status of PG and PGES cells was analyzed for the DMRs of maternally methylated imprinted genes *Peg1/Mest*, *Snrpn* and *Igf2r*. To identify methylated CpG sites, sodium bisulfite treatment, by which only unmethylated cytosine residues were changed to thymines, was carried out. PCR amplification for DMRs of *Peg1/Mest*, *Snrpn* and *Igf2r* was carried out on each set of isolated cells as described (Horii et al., 2008). PCR products were subcloned into the TA cloning vector, and positive clones in each sample were sequenced.

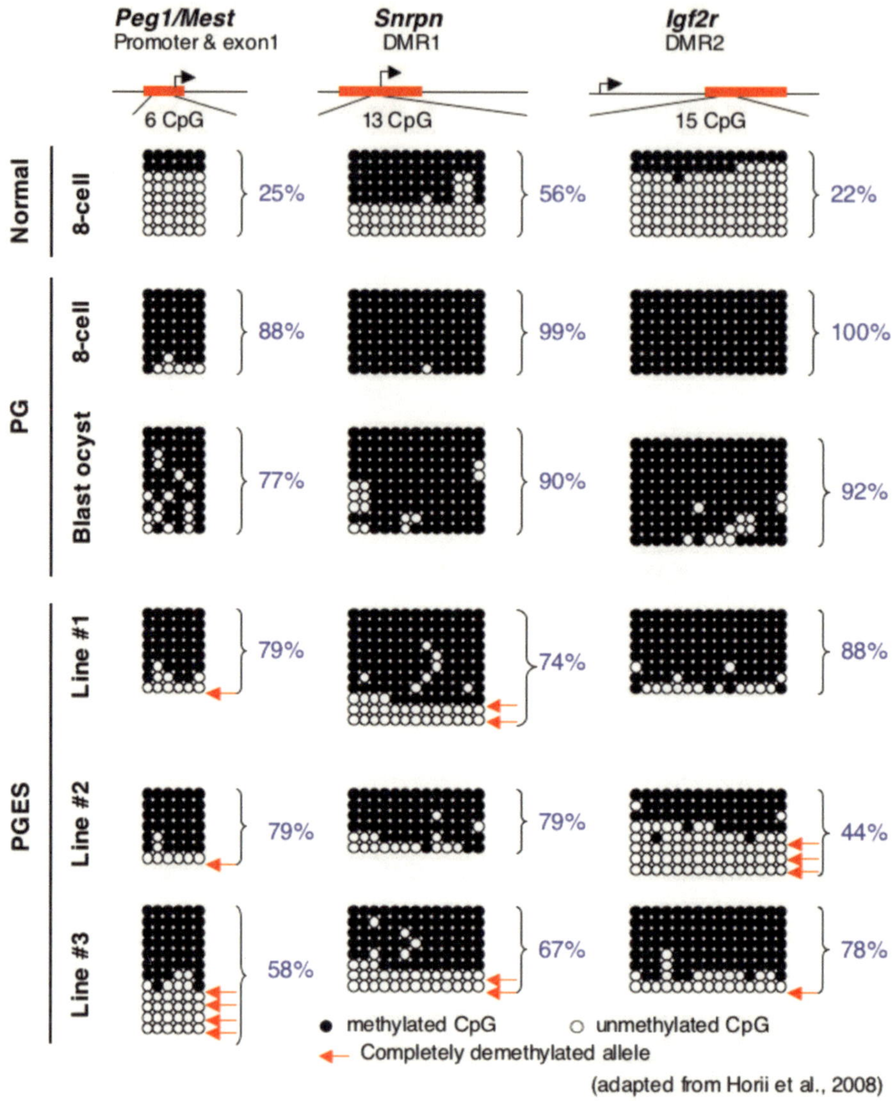

Fig. 4. Epigenetic reprogramming occurred in PGES cells. DNA methylation of normal embryos, PG and PGES cells was analyzed by bisulfite genomic sequencing.
The methylation status of maternally methylated imprinted genes, *Peg1/Mest*, *Snrpn* and *Igf2r* is shown schematically. Percentages of methylated CpGs are shown to the right of the sequences.

At the 8-cell stage, normal embryos had both methylated and unmethylated alleles of maternally methylated imprinted genes; whereas, almost all alleles were methylated in PG (Fig. 4). The loss of imprinting was observed at parthenogenetic blastocysts in more 2 days of culture. This partial demethylation also occurred in *in vitro* cultured normal blastocysts as reported previously (Doherty et al., 2000; Mann et al., 2004). Perhaps, parthenogenetic blastocysts also occurs demethylation as well as in normal blastocysts *in vitro*. Anyway, this demethylation was very partial and sparse. On the other hand, completely demethylated

alleles existed in almost all PGES cell samples (Fig. 4; arrows). At this point, loss of imprinting in PGES cells seems to be more progressive than that of PG.
We found that the methylation difference between PG and PGES cells have already exist before differentiation. In the following section, we elucidate whether such a difference affects the pluripotency of PG and PGES cells.

3. Improvement of pluripotency by epigenetic reprogramming

3.1 Production of PG and PGES cell chimeras

Chimeric mice were produced to examine developmental potential and epigenetic status of PG and PGES cells (Fig. 5). Briefly, PG chimeric embryos were produced by aggregating the 4-8 cell stage of GFP+ parthenogenetic embryos with the same stage of normal host embryos. PGES chimeric embryos were produced by introducing GFP+ PGES cells to host embryos. Then, 135 PG chimeric embryos and 338 PGES cell chimeric embryos were

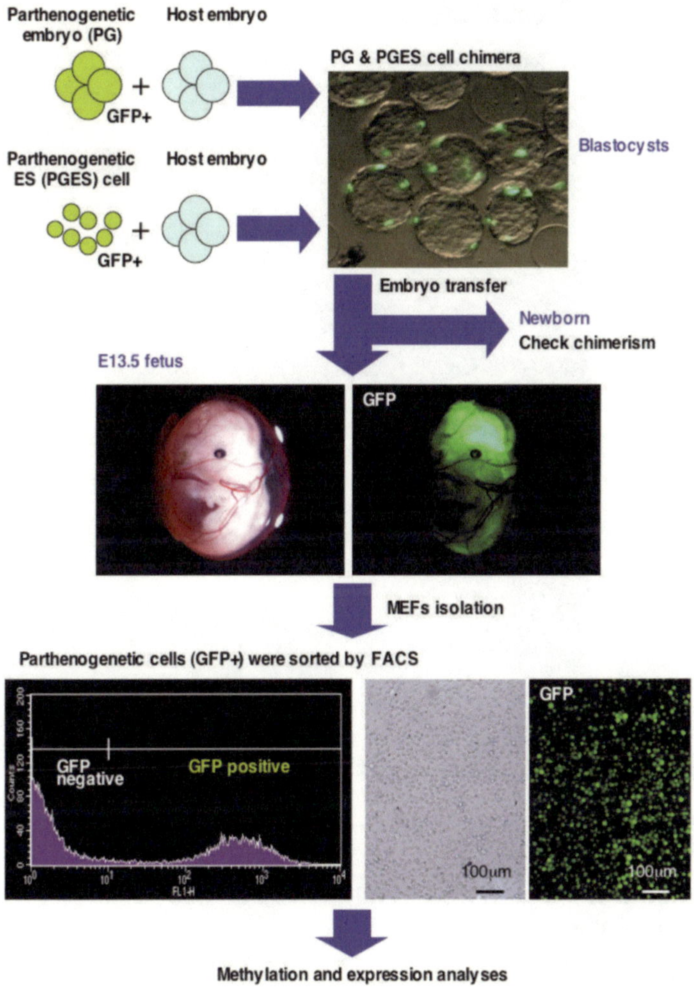

Fig. 5. Production of PG and PGES cell chimeras and analysis for their derivatives.

transferred to the uterine horns of pseudopregnant recipient females, and 18 PG chimeras and 56 PGES cell chimeras were obtained, respectively. In newborns, growth retardation was not observed in PGES chimeras (normal 1.48 +/- 0.21g vs chimera 1.48 +/- 0.37 g; P = 0.45), as previously reported (Allen et al., 1994). Contribution of PGES cells was found in all tissues tested (Fig. 6).

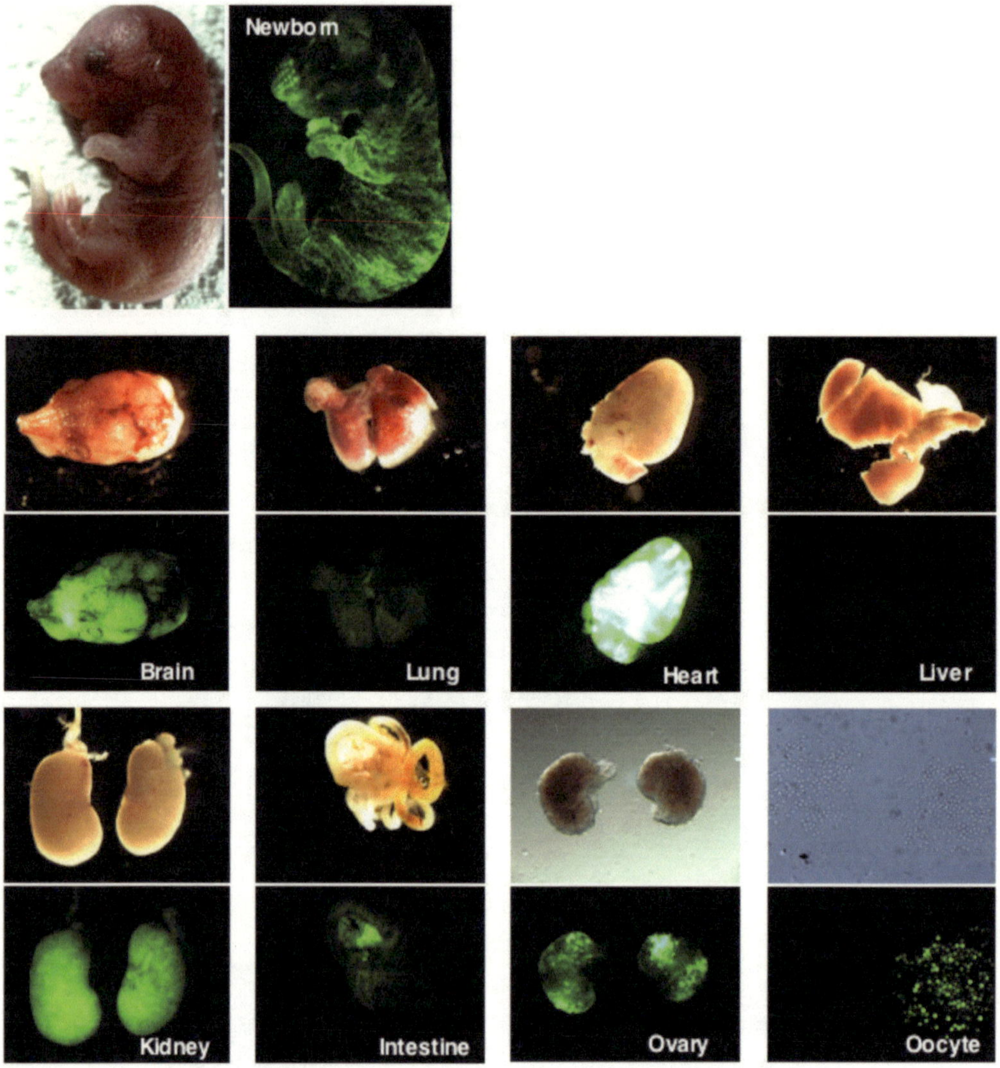

Fig. 6. Various tissue contributions of PGES cells (1days post partum).

3.2 Epigenetic status of PG- and PGES cell-derived somatic cells

To examine epigenetic status of PG- and PGES cell-derived cells in chimeras, primary mouse embryonic fibroblasts (MEFs) from E13.5 chimeras were isolated, and sorted by fluorescent-activated cell sorter (Fig. 5). In MEFs of PG chimeras, genomic imprinting of donor cells were almost totally maintained (Fig. 7, PG). In contrast, genomic imprinting of donor cells in

PGES cell chimeras were frequently reprogrammed (Fig. 7, PGES #1 and #2). Completely demethylated alleles were included in some PGES cell-derivatives.

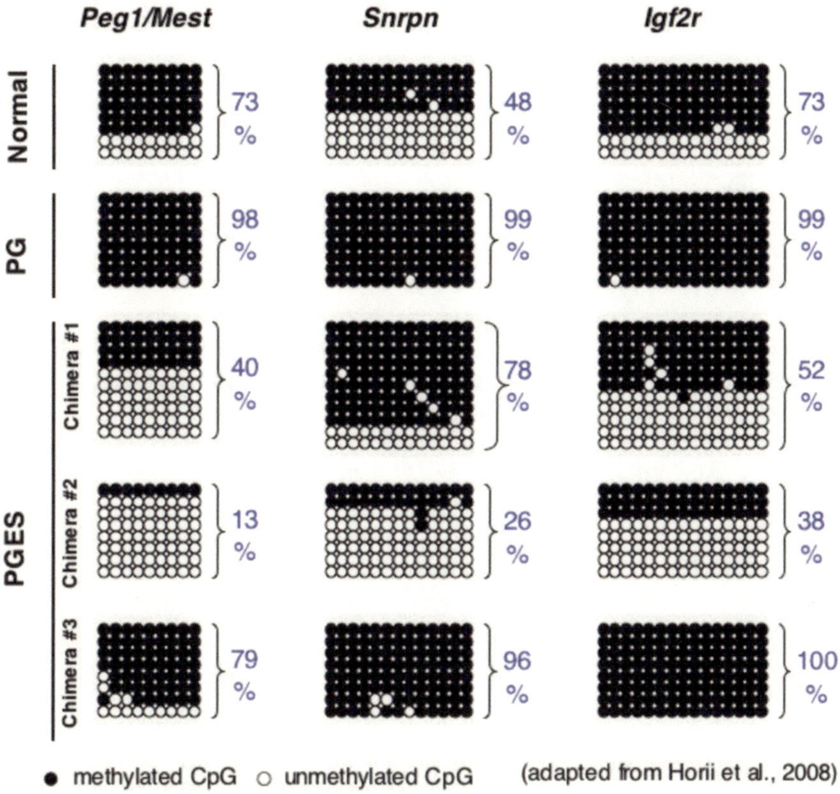

Fig. 7. Epigenetic reprogramming occurred in PGES cell-derived somatic tissues (E13.5 MEFs). DNA methylation of normal, PG-derived and PGES cell-derived MEFs were analyzed by bisulfite genomic sequencing. Percentages of methylated CpGs are shown to the right of the sequences.

Next, quantitative real-time RT-PCR was carried out to clarify whether demethylation is correlated to expression levels of imprinted genes. Maternally methylated imprinted genes, *Peg1/Mest* and *Snrpn* are expressed only from the paternal allele because these expressions are suppressed by DNA methylation in maternal allele. Therefore, in parthenogenetic cells, the loss of imprints leads to the upregulation of *Peg1/Mest* and *Snrpn*. In fact, the average expression level of each gene were upregulated in PGES cell-derivatives (Fig. 8). There were significant correlations ($P<0.05$) between the methylation status of DMRs and the gene expression level in these two genes (Fig. 9). On the other hand, paternally imprinted genes, *Igf2*, which are regulated by *H19* DMR methylation, were generally unmethylated in both PG and PGES cell-derived cells (data not shown), and the expression level of both genes did not differ between PG and PGES chimeras (Fig. 8). Then, correlations between the demethylation of imprinted genes and the tissue contribution of PGES cell-derived cells were examined. For E13.5 chimeras, there was low correlation between the percentage of methylation and the percentage of chimerism ($R^2 = 0.4719$); however, this correlation was much higher in newborn chimeras ($R^2 = 0.6981$; Fig. 10).

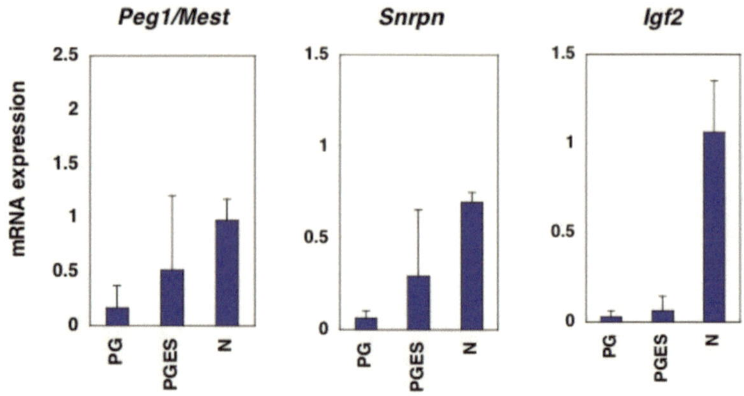

Fig. 8. Expression of imprinted genes in MEFs derived from PG (n = 6), PGES (n = 11) and normal biparental embryos (N; n = 2). Quantitative real-time PCR was performed for *Peg1/Mest*, *Snrpn* and *Igf2* genes. Standard deviations are indicated by bars.

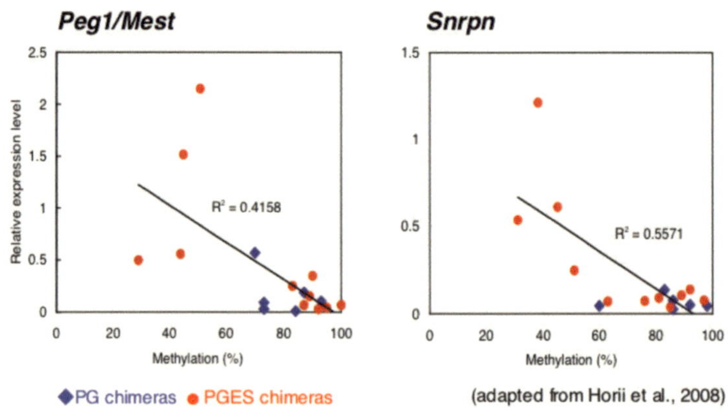

Fig. 9. Epigenetic reprogramming correlated to the gene expression level. The percentage of DMR methylation is plotted against the relative gene expression level for PG chimeras (filled circles; n = 6) and PGES cell chimeras (open circles; n = 11). Significant correlations (P<0.05) were found for these imprinted genes.

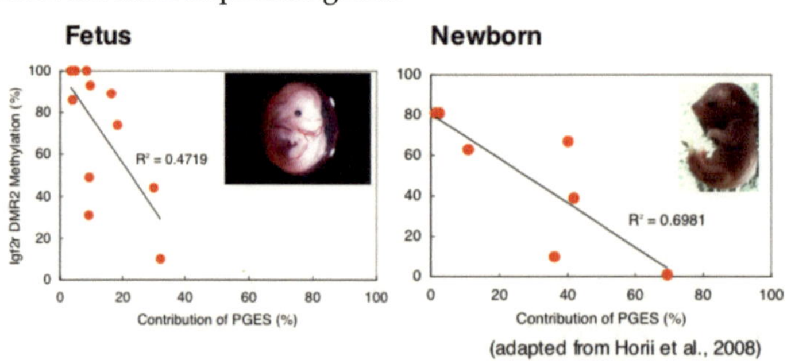

Fig. 10. Epigenetic reprogramming corelated to the pluripotency of PGES cells. Low correlation was observed in E13.5 fetuses (n = 11), whereas higher correlation was observed in newborns (n = 7).

3.3 Improvement of pluripotency by epigenetic reprogramming

In parthenogenetic chimeras, PGES cells resembled PG cells in their pluripotency. However, PGES chimeras are more normal in body weight and tissue contribution than PG chimeras. We postulated that this difference might be caused by the modified expressions of imprinted genes, due to loss of imprinting in PGES cells. To investigate the epigenetic status of parthenogenetic cells in somatic-lineages, we produced parthenogenetic chimeras using PG and PGES cells.

In general, higher overall levels of PGES cells are detected than PG cells in terms of tissue contribution. Furthermore, no significant growth retardation is apparent in PGES chimeras, irrespective of their degree of chimerism or the PGES cell lines used (Allen et al., 1994). Also in our study, growth retardation was not found in PGES chimeras. A phenotypic difference is expected to be caused by the difference in expressions of imprinted genes due to the loss of imprints in PGES derivatives. In some PGES chimeras, the loss of imprints was observed in *Peg1/Mest* and *Snrpn* genes. Especially, the *Peg1/Mest* gene is related to embryonic growth (Lefebvre et al., 1998). Therefore, there is no doubt that these alterations of gene expressions improve the tissue contribution of PGES cells in chimeras. In PG chimeras, progressive elimination of PG cells occurs after day 13 of gestation (Fundele et al., 1990). We found higher positive correlations between demethylation and chimerism in newborns than in E13.5 fetuses, suggesting that demethylated parthenogenetic cells evaded progressive elimination from tissues after day 13 of gestation. Summary of results are shown in Fig. 11.

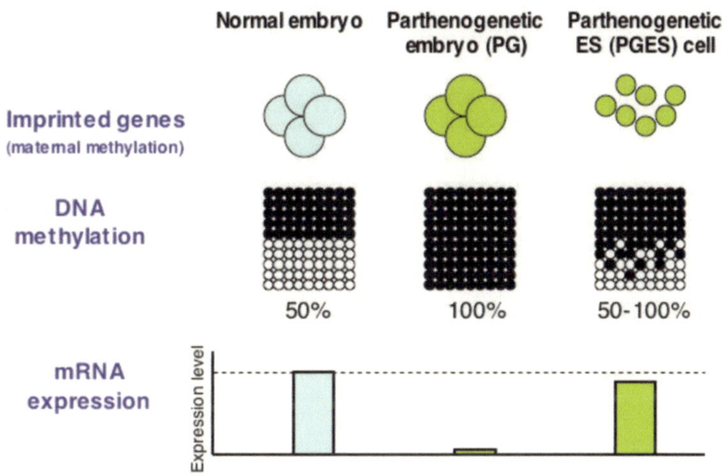

Fig. 11. Summary of epigenetic status of normal embryo, PG and PGES cell.

4. Why epigenetic reprogramming occurred in PGES Cells?

The partial or complete loss of imprints was observed in undifferentiated PGES cells and its derivatives. Why did loss of imprints occur mostly in PGES cells?

A first consideration is that the culture conditions of preimplantation embryos and ES cells sometimes influence the methylation status of genomic imprinting. For example, a sub-optimal culture medium can cause aberrant genomic imprinting of the *Snrpn* and *H19* gene, whereas embryos cultured in potassium simplex optimized medium with added amino acids (KSOMAA) show global gene expression, genomic imprinting and embryo development resembling that found in *in vivo* developed embryos (Doherty et al., 2000; Mann et al., 2004). Furthermore, long term culture of ES cells also affects the methylation status of imprinted genes and their totipotency (Dean et al., 1998; Horii et al., 2010).

A second consideration is that XX ES cells including PGES cells are more susceptible to demethylation than XO and XY ES cells (Zvetkova et al., 2005). XO and XY ES cells are able to restore the methylation imprints; whereas, XX ES cells are not able to restore them enough. PGES cells have generally two X chromosomes, so that demethylated PGES cells could be demethylated, or progress the demethylation of imprinted genes in chimeras. Then, why XX ES cells including PGES cells show demethylation of imprints? There is a speculation that X chromosome encodes a modifier locus whose product represses *de novo* methyltransferases. The *de novo* methyltransferase, Dnmt3a and Dnmt3b, are known to play a critical role for the restoration of methylation post implantation (Okano et al., 1998; Okano et al., 1999). Cells with two active X chromosomes will overexpress the modifier and therefore have reduced levels of the enzymes. In evidence, the forced expression of Dnmt3a or Dnmt3b restores the DNA methylation of XX ES cells (Zvetkova et al., 2005), suggesting that the expression levels of Dnmt3a and/or Dnmt3b are not sufficient for PGES cells. On the other hand, once methylation of imprinted genes is completely lost in XX ES cells, the loss is not restored by the forced expression of Dnmt3a or Dnmt3b *in vitro* (Zvetkova et al., 2005). According to the bisulfite genomic sequencing of PG and PGES cells, demethylation was sparse in parthenogenetic blastocysts; whereas, completely demethylated alleles existed in undifferentiated PGES cells (Fig. 4), suggesting that completely demethylated alleles in PGES cells were not able to be remethylated post implantation.

5. Comparison among ES cell, iPS cell and PGES cell.

PGES cells have been proposed as a source of patient-derived therapeutic materials (Cibelli et al., 2002). In addition to normal ES cells and iPS cells, parthenogenesis is another tool for creating pluripotent stem cells. Human PGES cells have already been isolated from human parthenogenetic blastocysts (Mai et al., 2007; Revazova et al., 2007). As described in the introduction, the advantage using PGES cells is that PGES cells do not need destruction of viable biparental embryos like normal ES cells. In addition, PGES cells do not need viruses or expression plasmids for the establishment like iPS cells. The genomic imprinting with uniparental genome sets is the biggest problem for PGES cells; however, we and others clarified that PGES cells partially lost maternally methylated imprints and obtained more normal imprint patterns (Jiang et al., 2007; Horii et al., 2008; Li et al., 2009). In this study, we clarified correlation of expression and methylation of imprinted genes with pluripotency of PGES cells. Even more surprisingly, live parthenogenetic pups were recently produced from reprogrammed PGES cells through tetraploid embryo complementation (Chen et al., 2009). These reports suggest that PGES cells have more normal pluripotency than PG. In contrast, even normal biparental ES cells sometimes obtain abnormal imprinting which influences pluripotency during long-term culture (Dean et al., 1998; Horii et al., 2010). In addition, iPS cells occasionally show aberrant silencing of imprinted genes on chromosome 12qF1 (Stadtfeld et al., 2010). Thus, reprogrammed PGES cells have pluripotency nearly equivarent to normal biparental ES cells or iPS cells. Besides these merits, there are other advantages to use PGES cells for cell transplantation therapy.

For example, normal biparental ES cells derived from fertilized embryos are genetically divergent from any patient requiring tissue transplantation and bring an immune response resulting in rejection (Drukker & Benvenisty, 2004). The hES cell bank that contains sufficient cell lines with diverse human leukocyte antigen (HLA) genotypes (Taylor et al., 2005; , Nakajima et al., 2007) might serve most of the patients in a region; however, this system need numerous numbers of human embryos and might cause ethical issues. On the other hand, PGES cell lines which are homozygous for HLA loci significantly reduce the number of cell lines required for the repository because tissues derived from homozygous PGES cells express only one set of histocompatibility antigens and are more readily matched to patients with less risk of immunological rejection (Fig. 12). Thus, homozygous PGES cells have the potential for cell-based therapy in a significant number of individuals. Athough recombination events occur between paired chromosomes in meiosis I (Kim et al., 2007), the successful derivation of stable homozygous PGES cell lines was reported (Lin et al., 2007; Revazova et al., 2008).

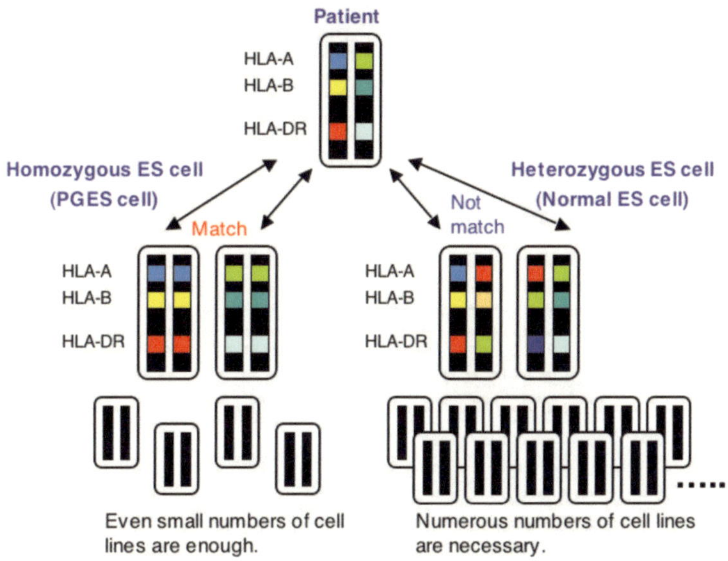

Fig. 12. HLA genotypes in homozygous and hetero zygous ES cells. The most important HLA molecules to match for are the HLA class I molecules HLA-A and HLA-B, and the class II molecule HLA-DR.

PGES cells also have a merit in viewpoint of tumorigenicity. Transplanted stem cell-derived tissues occasionally forming tumors becomes a serious problem (Brickman et al., 2002). In many cases, such tumors are teratomas or teratocarcinomas arising from undifferentiated stem cells residing in the differentiated cell population that have not completed the differentiation process. A variety of approaches, such as selective pluripotent apoptotic agents (Bieberich et al., 2004), magnetic and fluorescent activated cell sorting (MACS and FACS; Shibata et al., 2006; Fong et al., 2009) and antibodies against undifferentiated stem cells (Choo et al., 2008; Tan et al., 2009), have been reported to help eliminate tumorigenesis; however, the final obstacle of teratoma formation has not been adequately addressed and remains a major safety hurdle that has to be overcome before tissue transplantations. Interestingly, primary MEFs, whose entire genome is either exclusively paternal (androgenetic) or maternal (parthenogenetic), exhibit dramatically contrasting patterns of growth and tumorigenesis (Hernandez et al., 2003). Parthenogenetic MEFs reach a lower saturation density and senesce; whereas, androgenetic and biparental MEFs increased saturation density, spontaneous

transformation, and formation of tumors. Analysis of individual imprinted genes revealed that *Igf2* regulates transformation and functions as a potent oncogene, converting primary fibroblasts into forming rapidly growing tumors. In our study, high *Igf2* expression was not detected in PGES cell-derived MEFs as well as PG-derived MEFs (Fig. 8), indicating that PGES cells rarely cause tumorigenesis. In addition, *H19*, *p57^{kip2}* and *Igf2r*, which show excessive expression in parthenogenetic cells, are candidate tumor supressor genes (Hao et al., 1993; Matsuoka et al., 1995; De souza et al., 1995; Yoshimizu et al., 2008). Therefore, expression patterns of oncogenes and tumor supressor genes in PGES cell-derived tissue could be one of the advantages for cell transplantation therapy (Fig. 13).
Last, summary of comparison among PGES cells and other pluripotent stem cells are shown in Fig. 14.

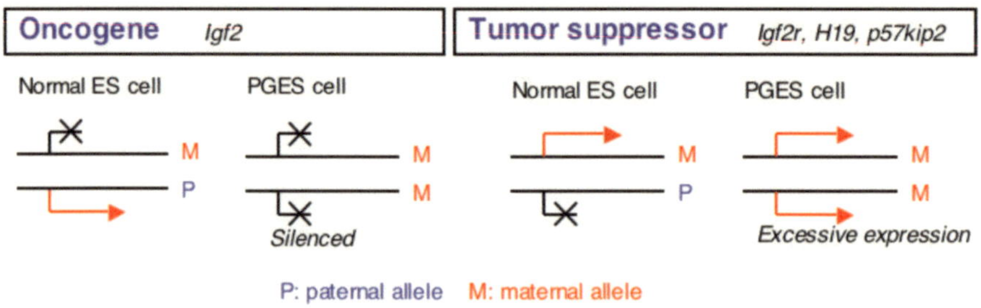

Fig. 13. Gene expression patterns of tumorigenesis related genes in PGES cells.

	PGES cell	ES cell	iPS cell
Ethics	It is **not necessary** to destroy viable biparental embryos.	It is **necessary** to destroy viable biparental embryos for establishment.	It is **not necessary** to destroy viable biparental embryos.
Safety	Viruses or expression plasmids is **not necessary** for establishment.	Viruses or expression plasmids is **not necessary** for establishment.	Viruses or expression plasmids is **necessary** for establishment.
Epigenetics	Genomic imprinting is only maternal origin. → However, PGES cells obtained **more normal** imprint pattern.	Genomic imprinting is generally **normal**. → However, aberrant methylation imprints occurred occasionally.	Genomic imprinting is generally **normal**. → However, aberrant methylation imprints occurred occasionally.
HLA matching	HLA homozygous cell lines are **easily** obtained.	HLA completely matched cell lines are **rarely** obtained.	HLA completely matched cell lines are **always** obtained.
Tumori-genicity	**Low** frequency of tumorigenesis?	**High** frequency of tumorigenesis.	**High** frequency of tumorigenesis.

Fig. 14. Comparison among PGES cells and other pluripotent stem cells.

6. Conclusion

In this study, we clarified epigenetic reprogramming occurred in PGES cells, resulted in nearly normal expression patterns of imprinted genes. These reprogrammed PGES cells could be used for regenerative medicine as the 3rd pluripotent stem cells. In mice, it is reported that uniparental ES cells can differentiate into transplantable hematopoietic progenitors *in vitro* that contribute to long-term hematopoiesis in recipients (Eckardt et al., 2007). The human PGES cells might be utilized for cell transplantation therapy in the near future.

7. References

Allen, N.D.; Barton, S.C.; Hilton, K.; Norris, M.L. & Surani, M.A. (1994). A functional analysis of imprinting in parthenogenetic embryonic stem cells. *Development*, Vol. 120, No. 6, (Jun 1994) pp. 1473-1482, ISSN 0950-1991

Barlow, D.P.; Stoger, R.; Herrmann, B.G.; Saito, K. & Schweifer, N. (1991). The mouse insulin-like growth factor type-2 receptor is imprinted and closely linked to the Tme locus. *Nature*, Vol. 349, No. 6304, (January 1991) pp. 84-87, ISSN 0028-0836

Barr, J.A.; Jones, J.; Glenister, P.H. & Cattanach, B.M. (1995). Ubiquitous expression and imprinting of Snrpn in the mouse. *Mamm Genome*, Vol. 6, No. 6, (June 1995) pp. 405-407, ISSN 0938-8990

Bartolomei, M.S.; Zemel, S. & Tilghman, S.M. (1991). Parental imprinting of mouse H19 gene. *Nature*, Vol. 351, No. 6322, (May 1991) pp. 153-155, ISSN 0028-0836

Bieberich, E.; Silva, J.; Wang, G.; Krishnamurthy, K. & Condie, B.G. (2004). Selective apoptosis of pluripotent mouse and human stem cells by novel ceramide analogues prevents teratoma formation and enriches for neural precursors in ES cell-derived neural transplants. *J Cell Biol*, Vol. 167, No. 4, (November 2004) pp. 723-734, ISSN 0021-9525

Brickman, J.M. & Burdon, T.G. (2002). Pluripotency and tumorigenicity. *Nat Genet*, Vol. 32, No. 4, (December 2002) pp. 557-558, ISSN 1061-4036

Chen, Z.; Liu, Z.; Huang, J.; Amano, T.; Li, C.; Cao, S.; Wu, C.; Liu, B.; Zhou, L.; Carter, M.G.; Keefe, D.L.; Yang, X. & Liu, L. (2009). Birth of parthenote mice directly from parthenogenetic embryonic stem cells. *Stem Cells*, Vol. 27, No. 9, (September 2009) pp. 2136-2145, ISSN 1066-5099

Choo, A.B.; Tan, H.L.; Ang, S.N.; Fong, W.J.; Chin, A.; Lo, J.; Zheng, L.; Hentze, H.; Philp, R.J.; Oh, S.K. & Yap, M. (2008). Selection against undifferentiated human embryonic stem cells by a cytotoxic antibody recognizing podocalyxin-like protein-1. *Stem Cells*, Vol. 26, No. 6, (June 2008) pp. 1454-1463, ISSN 1066-5099

Cibelli, J.B.; Grant, K.A.; Chapman, K.B.; Cunniff, K.; Worst, T.; Green, H.L.; Walker, S.J.; Gutin, P.H.; Vilner, L.; Tabar, V.; Dominko, T.; Kane, J.; Wettstein, P.J.; Lanza, R.P.; Studer, L.; Vrana, K.E. & West, M.D. (2002). Parthenogenetic stem cells in nonhuman primates. *Science*, Vol. 295, No. 5556, (February 2002) pp. 819, ISSN 0036-8075

Clarke, H.J.; Rossant, J. & Masui, Y. (1988a). Suppression of chromosome condensation during meiotic maturation induces parthenogenetic development of mouse oocytes. *Development*, Vol. 104, No. 1, (September 1988) pp. 97-103, ISSN 0950-1991

Clarke, H.J.; Varmuza, S.; Prideaux, V.R. and Rossant, J. (1988b). The developmental potential of parthenogenetically derived cells in chimeric mouse embryos:

implications for the actions of imprinted genes. *Development*, Vol. 104, No. 1, (September 1988) pp. 175-182, ISSN 0950-1991

Dean, W.; Bowden, L.; Aitchison, A.; Klose, J.; Moore, T.; Meneses, J.J.; Reik, W. & Feil, R. (1998). Altered imprinted gene methylation and expression in completely ES cell-derived mouse fetuses: association with aberrant phenotypes. *Development*, Vol. 125, No. 12, (June 1998) pp. 2273-2282, ISSN 0950-1991

DeChiara, T.M.; Robertson, E.J. & Efstratiadis, A. (1991). Paternal imprinting of the mouse insulin-like growth factor II gene. *Cell*, Vol. 64, No. 4, (February 1991) pp. 849-959, ISSN 0092-8674

De Souza, A.T.; Hankins, G.R.; Washington, M.K.; Orton, T.C. & Jirtle, R.L. (1995). M6P/IGF2R gene is mutated in human hepatocellular carcinomas with loss of heterozygosity. *Nat Genet*, Vol. 11, No. 4, (December 1995) pp. 447-449, ISSN 1061-4036

Doherty, A.S.; Mann, M.R.; Tremblay, K.D.; Bartolomei, M.S. & Schultz, R.M. (2000). Differential effects of culture on imprinted *H19* expression in the preimplantation mouse embryo. *Biol Reprod*, Vol. 62, No. 6, (June 2000) pp. 1526-1535, ISSN 0006-3363

Drukker, M. & Benvenisty, N. (2004). The immunogenicity of human embryonic stem-derived cells. *Trends Biotechnol*, Vol. 22, No. 3, (March 2004) pp. 136-141, ISSN 0167-7799

Eckardt, S.; Leu, N.A.; Bradley, H.L.; Kato, H.; Bunting, K.D. & McLaughlin, K.J. (2007). Hematopoietic reconstitution with androgenetic and gynogenetic stem cells. *Gene Dev*, Vol. 21, No. 4, (February 2007) pp. 409-419, ISSN 0890-9369

Ferguson-Smith, A.C.; Cattanach, B.M.; Barton, S.C.; Beechey, C.V. & Surani MA. (1991). Embryological and molecular investigations of parental imprinting on mouse chromosome 7. *Nature*, Vol. 351, No. 6328, (June 1991) pp. 667-670, ISSN 0028-0836

Fong, C.Y.; Peh, G.S.; Gauthaman, K. & Bongso, A. (2009). Separation of SSEA-4 and TRA-1-60 labelled undifferentiated human embryonic stem cells from a heterogeneous cell population using magnetic-activated cell sorting (MACS) and fluorescence-activated cell sorting (FACS). *Stem Cell Rev*, Vol. 5, No. 1, (March 2009) pp. 72-80, ISSN 1550-8943

Fundele, R.H.; Norris, M.L.; Barton, S.C.; Fehlau, M.; Howlett, S.K.; Mills, W.E. & Surani, M.A. (1990). Temporal and spatial selection against parthenogenetic cells during development in fetal chimeras. *Development*, Vol. 108, No. 1, (January 1990) pp. 203-211, ISSN 0950-1991

Hao, Y.; Crenshaw, T.; Moulton, T.; Newcomb, E. & Tycko, B. (1993). Tumour-suppressor activity of *H19* RNA. *Nature*, Vol. 365, No. 6448, (October 1993) pp. 764-767, ISSN 0028-0836

Hatada, I. & Mukai, T. (1995). Genomic imprinting of $p57^{KIP2}$, a cyclindependent kinase inhibitor, in mouse. *Nat Genet*, Vol. 11, No. 2, (October 1995) pp. 204-206, ISSN 1061-4036

Hernandez, L.; Kozlov, S.; Piras, G. & Stewart, C.L. (2003). Paternal and maternal genomes confer opposite effects on proliferation, cell-cycle length, senescence, and tumor formation. *Proc Natl Acad Sci U S A*, Vol. 100, No. 23, (November 2003) pp. 13344-13349, ISSN 0027-8424

Horii, T.; Nagao, Y.; Tokunaga, T. & Imai, H. (2003). Serum-free culture of murine primordial germ cells and embryonic germ cells. *Theriogenology*, Vol. 59, No. 5-6, (March 2003) pp. 1257-1264, ISSN 0093-691X

Horii, T.; Kimura, M.; Morita, S.; Nagao, Y. & Hatada, I. (2008). Loss of genomic imprinting in mouse parthenogenetic embryonic stem cells. *Stem Cells*, Vol. 26, No. 1, (January 2008) pp. 79-88, ISSN 1066-5099

Horii, T.; Yanagisawa, E.; Kimura, M.; Morita, S. & Hatada, I. (2010). Epigenetic differences between embryonic stem cells generated from blastocysts developed *in vitro* and *in vivo*. *Cloning Stem Cells* (in press), ISSN 1536-2302

Jiang, H.; Sun, B.; Wang, W.; Zhang, Z.; Gao, F.; Shi, G.; Cui, B.; Kong, X.; He, Z.; Ding, X.; Kuang, Y.; Fei, J.; Sun, Y.J.; Feng, Y. & Jin, Y. (2007). Activation of paternally expressed imprinted genes in newly derived germline-competent mouse parthenogenetic embryonic stem cell lines. *Cell Res*, Vo. 17, No. 9, (September 2007) pp. 792-803, ISSN 1001-0602

Kaneko-Ishino, T.; Kuroiwa, Y.; Miyoshi, N.; Kohda, T.; Suzuki, R.; Yokoyama, M.; Viville, S.; Barton, S.C.; Ishino, F. & Surani, M.A. (1995). Peg1/Mest imprinted gene on chromosome 6 identified by cDNA subtraction hybridization. *Nat Genet*, Vol. 11, No. 1, (September 1995) pp. 52-59, ISSN 1061-4036

Kim, K.; Lerou, P.; Yabuuchi, A.; Lengerke, C.; Ng, K.; West, J.; Kirby, A.; Daly, M.J. & Daley, G.Q. (2007). Histocompatible embryonic stem cells by parthenogenesis. *Science*, Vol. 315, No. 5811, (January 2007) pp. 482-486, ISSN 0036-8075

Kuroiwa, Y.; Kaneko-Ishino, T.; Kagitani, F.; Kohda, T.; Li, L.L.; Tada, M.; Suzuki, R.; Yokoyama, M.; Shiroishi, T.; Wakana, S.; Barton, S.C.; Ishino, F. & Surani, M.A. (1996). Peg3 imprinted gene on proximal chromosome 7 encodes for a zinc finger. *Nat Genet*, Vol. 12, No. 2, (February 1996) pp. 186-190, ISSN 1061-4036

Lefebvre, L.; Viville, S.; Barton, S.C.; Ishino, F.; Keverne, E.B. & Surani, M.A. (1998). Abnormal maternal behaviour and growth retardation associated with loss of the imprinted gene Mest. *Nat Genet*, Vol. 20, No. 2, (October 1998) pp. 163-199, ISSN 1061-4036

Li, C.; Chen, Z.; Liu, Z.; Huang, J.; Zhang, W.; Zhou, L.; Keefe, D.L. & Liu, L. (2009). Correlation of expression and methylation of imprinted genes with pluripotency of parthenogenetic embryonic stem cells. *Hum Mol Genet*, Vol. 18, No. 12, (June 2009) pp. 2177-2187, ISSN 0964-6906

Lin, G.; OuYang, Q.; Zhou, X.; Gu, Y.; Yuan, D.; Li, W.; Liu, G.; Liu, T. & Lu, G. (2007). A highly homozygous and parthenogenetic human embryonic stem cell line derived from a one-pronuclear oocyte following in vitro fertilization procedure. *Cell Res*, Vol. 17, No. 12, (December 2007) pp. 999-1007, ISSN 1001-0602

Mai, Q.; Yu, Y.; Li, T.; Wang, L.; Chen, M.J.; Huang, S.Z.; Zhou, C. & Zhou, Q. (2007). Derivation of human embryonic stem cell lines from parthenogenetic blastocysts. *Cell Res*, Vol. 17, No. 12, (December 2007) pp. 1008-1019, ISSN 1001-0602

Mann, M.R.; Lee, S.S.; Doherty, A.S.; Verona, R.I.; Nolen, L.D.; Schultz, R.M. & Bartolomei, M.S. (2004). Selective loss of imprinting in the placenta following preimplantation development in culture. *Development*, Vol. 131, No. 15, (August 2004) pp. 3727-3235, ISSN 0950-1991

Matsuoka, S., Edwards, M.C., Bai, C., Parker, S., Zhang, P., Baldini, A., Harper, J.W. & Elledge, S.J. (1995). $p57^{KIP2}$, a structurally distinct member of the $p21^{CIP1}$ Cdk inhibitor family, is a candidate tumor suppressor gene. *Genes Dev*, Vol. 9, No. 6, (March 1995) pp. 650-662, ISSN 0890-9369

Nakajima, F.; Tokunaga, K. & Nakatsuji, N. (2007). Human leukocyte antigen matching estimations in a hypothetical bank of human embryonic stem cell lines in the Japanese population for use in cell transplantation therapy. *Stem Cells*, Vol. 25, No. 4, (April 2007) pp. 983-985, ISSN 1066-5099

Okano, M.; Xie, S. & Li, E. (1998). Cloning and characterization of a family of novel mammalian DNA (cytosine-5) methyltransferases. *Nat Genet*, Vol. 19, No. 3, (July 1998) pp. 219-220, ISSN 1061-4036

Okano, M.; Bell, D.W.; Haber, D.A. & Li, E. (1999). DNA methyltransferases Dnmt3a and Dnmt3b are essential for de novo methylation and mammalian development. *Cell*, Vol. 99, No. 3, (October 1999) pp. 247-257, ISSN 0092-8674

Revazova, E.S.; Turovets, N.A.; Kochetkova, O.D.; Kindarova, L.B.; Kuzmichev, L.N.; Janus, J.D. & Pryzhkova, M.V. (2007). Patient-specific stem cell lines derived from human parthenogenetic blastocysts. *Cloning Stem Cells*, Vol. 9, No. 3, (Fall 2007) pp. 432-449, ISSN 1536-2302

Shibata, H.; Ageyama, N.; Tanaka, Y.; Kishi, Y.; Sasaki, K.; Nakamura, S.; Muramatsu, S.; Hayashi, S.; Kitano, Y.; Terao, K. and Hanazono, Y. (2006). Improved safety of hematopoietic transplantation with monkey embryonic stem cells in the allogeneic setting. *Stem Cells*, Vol. 24, No. 6, (June 2006) pp. 1450-1457, ISSN 1066-5099

Stadtfeld, M.; Apostolou, E.; Akutsu, H.; Fukuda, A.; Follett, P.; Natesan, S.; Kono, T.; Shioda, T. & Hochedlinger, K. (2010). Aberrant silencing of imprinted genes on chromosome 12qF1 in mouse induced pluripotent stem cells. *Nature*, Vol. 465, No. 7295, (May 2010) pp. 175-181, ISSN 0028-0836

Stevens, L.C.; Varnum, D.S. & Eicher, E.M. (1977). Viable chimaeras produced from normal and parthenogenetic mouse embryos. *Nature*, Vol. 269, No. 5628, (October 1977) pp. 515-517, ISSN 0028-0836

Stevens, L.C. (1978). Totipotent cells of parthenogenetic origin in a chimaeric mouse. *Nature*, Vol. 276, No. 5685, (November 1978) pp. 266-267, ISSN 0028-0836

Surani, M.A. & Barton, S.C. (1983). Development of gynogenetic eggs in the mouse: implications for parthenogenetic embryos. *Science*, Vol. 222, No. 4627, (Dec 1983) pp. 1034-1036, ISSN 0036-8075

Surani, M.A.; Barton, S.C, & Norris, M.L. (1984). Development of reconstituted mouse eggs suggests imprinting of the genome during gametogenesis. *Nature*, Vol. 308, No. 5959, (April 1984) pp. 548-550, ISSN 0028-0836

Surani, M.A.; Barton, S.C, & Norris, M.L. (1986). Nuclear transplantation in the mouse: heritable differences between parental genomes after activation of the embryonic genome. *Cell*, Vol. 45, No. 1, (April 1986) pp. 127-136, ISSN 0092-8674

Tan, H.L.; Fong, W.J.; Lee, E.H.; Yap, M. & Choo, A. (2009). mAb 84, a cytotoxic antibody that kills undifferentiated human embryonic stem cells via oncosis. *Stem Cells*, Vol. 27, No. 8, (August 2009) pp. 1792-1801, ISSN 1066-5099

Taylor, C.J.; Bolton, E.M.; Pocock, S.; Sharples, L.D.; Pedersen, R.A. & Bradley, J.A. (2005). Banking on human embryonic stem cells: estimating the number of donor cell lines needed for HLA matching. *Lancet*, Vol. 366, No. 9502, (December 2005) pp. 2019-2025, ISSN 0140-6736

Thomson, J.A. & Solter, D. (1989). Chimeras between parthenogenetic or androgenetic blastomeres and normal embryos: allocation to the inner cell mass and trophectoderm. *Dev Biol*, Vol, 131, No. 2, (February 1989) pp. 580-583, ISSN 0012-1606

Yoshimizu, T.; Miroglio, A.; Ripoche, M.A.; Gabory, A.; Vernucci, M.; Riccio, A.; Colnot, S.; Godard, C.; Terris, B.; Jammes, H. & Dandolo, L. (2008). The *H19* locus acts *in vivo* as a tumor suppressor. *Proc Natl Acad Sci U S A*, Vol. 105, No. 34, (August 2008) pp. 12417-12422, ISSN 0027-8424

Zvetkova, I.; Apedaile, A.; Ramsahoye, B.; Mermoud, J.E.; Crompton, L.A.; John, R.; Feil, R. & Brockdorff, N. (2005). Global hypomethylation of the genome in XX embryonic stem cells. *Nat Genet*, Vol, 37, No. 11, (November 2005) pp. 1274-1279, ISSN 1061-4036

13

Reprogramming Somatic Cells by Fusion with Embryonic Stem Cells: Present Status and Prospects in Regenerative Medicine

Jinnuo Han and Kuldip S. Sidhu
University of New South Wales
Australia

1. Introduction

Both embryonic and somatic stem cells have attracted many scientists' interest as they appear to be ideal candidates for cell therapy to treat various degenerative diseases and tissue injury for which there are no cures available. However, during the last few years, extensive amount of publications have focused on somatic cell reprogramming, a reversal of cell fate from a mature differentiated state to an undifferentiated state or directly to that of progenitors or mature cells of a different cell type. The advantage of that strategy is to obtain an autologous source of cells, which when differentiated and/or transplanted will not be rejected by the recipients. Several strategies have been applied for reprogramming purpose, including somatic cell nuclear transfer (SCNT), somatic-stem cell fusion, cell-free extract treatment and induced pluripotent stem cell (iPSC) generation. As a model for studying nuclear reprogramming, cell-cell fusion has been studied since the 1970's. Recent observations demonstrate that transplanted stem cells can differentiate into multiple cell types *in vivo* and have brought investigators to a new stage for applying cell fusion to regenerative therapies, cherishing the idea that fusion-based reprogramming may possess more potential than the original belief. In this chapter, we have briefly discussed the current progress of somatic cell reprogramming induced by cell fusion. In our lab, using human embryonic stem cells (hESCs) and human fetal fibroblasts (HFFs) as fusion candidates, we have compared the fusion efficiency *in vitro* using different methods and successfully generated hybrid cells. Research publications to date suggest that cell fusion between stem cells and somatic cells may restore regenerative capacity of terminally differentiated cells and can be applied for transplantation and cell therapy. However, certain aspects of disadvantages cannot be neglected. The limitations of cell fusion based reprogramming includes the presence of stem cell nuclei in the hybrid cells, the possibility of causing immune-rejection and the genetic instability of hybrid cells. Insights of both pros and cons have been described in this chapter and we conclude that reprogramming induced by cell fusion can be achieved at different efficiencies when using various fusion methods *in vitro*. The hybrid cells show ESC-like properties with double nuclear content. More detailed studies will be needed to remove ESC nuclei before these cells can be used for clinical purposes.

2. Stem cells and regenerative medicine

Stem cells have been identified as clonal cells that have the capacity to self-renew as well as the ability to generate more than one type of specialized cells (Weissman, 2000; Cowan and Melton, 2006; Sidhu and Ryan *et al.*, 2010). These properties make them important in wound healing and in the processes of regeneration throughout life. Regenerative medicine is an emerging field with the aim of repairing tissue/organs caused by injury, disease or aging by restoring the function of cells, tissues and organs faster and better. The techniques being investigated and applied include manipulation of stem cell behaviour by genetic (gene therapy) or non-genetic means, followed by transplantation of stem cells (cell therapy) and *in vitro*-grown cells/tissues/organs taking the dynamics and microenvironment into account (tissue engineering).

The traditional approach to reduce disability and improve health of patients has been organ transplantation, which started with the successful transplantation of the kidney between identical twins (Guild and Harrison *et al.*, 1955). With the development and usage of immunosuppressant drugs, improved survival rate of patients have been reported (Chkhotua and Klein *et al.*, 2003; Kim and Kwon *et al.*, 2004) and solid organ types have expanded to the heart, liver, pancreas and lung (Hariharan and Johnson *et al.*, 2000; Sayegh and Remuzzi, 2007). However, despite these positive aspects, there is shortage of transplantable organs. According to the official U.S. Government information on organ and tissue donation and transplantation (http://www.organdonor.gov/), by June 2010 more than 108,000 people are waiting for transplantation surgeries and this number continues to rise by approximately 300 people per month due to scarcity of available donors.

Regenerative medicine is a broad phrase used to describe the field of medicine that covers a diverse range of research including cell therapy, tissue engineering and transplantation. The use of stem cells in regenerative medicine provides unlimited cell sources for clinical application that overcome the shortage of tissue/organs. There are three classes of stem cells according to their origin: embryonic, germinal and adult stem cells (also known as somatic stem cells). Depending on their capacity to differentiate into other cell types, stem cells may be classified into totipotential, pluripotential and multipotential stem cells (Rao and Mattson, 2001). Despite the fact that more and more research are showing evidence for the potential use of stem cells in regenerative medicine (Bajada and Mazakova *et al.*, 2008), it is not clear which type of stem cell provides the best approach for cell therapy.

2.1 Adult stem cells in regenerative medicine

There are different types of adult stem cells, some of these include hematopoietic stem cells (HSCs), mesenchymal stem cells (MSCs), adipose tissue-derived stem cells (ADSCs), cardiac stem cells (CSCs), neural stem cells (NSCs), pancreatic stem cells (PSCs), hepatic oval cells (HOCs) and bronchioalveolar stem cells (BASCs). Adult stem cells are located within a stem cell niche and upon injury, proliferate to maintain the stable stem cell number and generate differentiated cells to replace damaged tissue cells. Adult stem cells have had applications in the treatment of many diseases including nervous system disorders, heart disorders, diabetes, muscular disorders, vascular disorders and interstitial lung diseases (Mimeault and Batra, 2008). An easy and straightforward approach for applying stem cells in regenerative medicine is to transplant/inject adult stem cells or differentiated cells into the targeted diseased or injured site, whereas the approach of tissue engineering is designed to generate solid tissue or organ by using bioreactors, scaffolds and other biomaterials *in vitro* before transplantation.

Bone marrow (BM) has long been known to contain at least three types of stem cells: HSCs, MSCs and endothelial progenitor cells (EPCs) (Alison and Islam, 2009). HSCs are able to reconstitute the hematopoietic system by differentiation into all types of blood cells and MSCs are able to differentiate into skeletal tissues (Bonnet, 2003). Currently, BM or BM-derived stem cell transplantation is the best-known and well-established stem cell-based therapy in regenerative medicine. When HSCs are transplanted for either allogenic or autologous therapy, function of bone marrow that was damaged can be re-established through engraftment and continuous generation of a new hematopoietic system (Armitage, 1994; Shizuru and Negrin *et al.*, 2005). HSC transplantation method has been established and has become a standard therapy for the treatment of many hematopoietic diseases including Hodgkin's and non-Hodgkin's lymphoma, acute and chronic myelogenousleukemia and myelodysplastic syndromes (Armitage, 1994; Copelan, 2006).

A concept that has brought much attention recently in the field of adult stem cell research is plasticity, which describes the phenomenon that restrictions in cell fates are flexible and the capacity of stem cells to differentiate into various cell types, including those not of their lineage of origin (Raff, 2003). Being the most extensively studied adult stem cells, HSCs have received much more attention as a result of recent transplantation studies, showing their plasticity to give rise to many non-hematopoietic cells both *in vivo* and *in vitro* (Quesenberry and Dooner *et al.*, 2010). This will be discussed in further details in Section 3 of this chapter.

2.1.1 Skeletal muscle

Being the largest tissue in the body, skeletal muscle contains a population of stem cells known as satellite cells (SCs). SCs are normally quiescent but can be activated to proliferation upon injury or disruption of the basal lamina resulting in the generation of multinucleated myofibers for muscle regeneration (Shi and Garry, 2006). Many regulatory factors, including SC niche, stimulatory and inhibitory growth factors are involved in muscle regeneration. Allogenic SCs have been transplanted for therapeutic purposes. However, limited migration ability and poor survival of the injected cells have hampered their application in regenerative medicine (Broek and Grefte *et al.*, 2010). Transgenic strategy to promote proliferation of SCs and derivation of other precursor cells involved in skeletal muscle regeneration have been investigated, showing great potential for the treatment of skeletal muscle injury and diseases.

2.1.2 Heart

The heart of human adults has been shown to contain resident adult stem cells with differentiation and regeneration capacity (Leri and Kajstura *et al.*, 2005; Lyngbæk and Schneider *et al.*, 2007). As precursors of cardiac muscle, transplantation of autologous myoblast cells and other cell types, for example, cardiomyocytes and BM-derived cells have also been suggested to be potential therapy for the treatment of myocardial dysfunction (Rubart and Field, 2006).

Apart from cell therapy, scientists have been attempted to create three dimensional scaffolds and engineering heart tissue. A recent improvement in this aspect was the establishment of a bio-artificial heart with a perfusion-decellularized matrix obtained from real heart (Ott and Matthiesen *et al.*, 2008).

2.1.3 Liver, kidney and pancreas

Both whole liver and hepatocytes have been transplanted to treat various liver diseases. However, due to limited number of available liver and hepatocytes, stem cells (especially BM-derived stem cells) have been considered a valuable source for obtaining sufficient number of cells for transplantation. Yet, the existence or absence of a stem cell population in adult liver is still not clear. Studies on liver development suggested the presence of progenitors in fetal liver (Schmelzer and Zhang *et al.*, 2007) and another side population (SP) cells have also been found to contribute to hematopoietic and epithelial lineages (Terrace and Hay *et al.*, 2009), representing the existence of a second potential hepatic progenitor cells.

The adult kidney consists at least 26 types of different cells with low proliferative potential. Thus, kidney is considered a highly terminally differentiated organ. However, the kidney has regenerative and repair potential upon injury. Much effort has been made in looking for renal stem/progenitor cells residing in the adult kidney. Renal epithelial stem cells have been reported to exist and SP cells have been isolated from adult kidneys (Iwatani and Ito *et al.*, 2004; Benigni and Morigi *et al.*, 2010). In addition, there has been controversy as to whether the adult pancreas contains stem cells. In normal adult pancreas, β-cells are known to renew at a low rate, but the mechanism of this cell renewal is not clear and has been suggested to be either replication of differentiated β-cells or neogenesis of pancreatic progenitor cells. The presence of pancreatic stem/progenitor cells has been proposed in the adult pancreatic exocrine, pancreatic duct and islets (Efrat, 2008). However, further characterization of these cells is essential and to date, whether there is a true renal and pancreatic stem cell population still remains elusive.

In addition, BM-derived stem cells have been reported to transdifferentiate to repair kidney and to generate insulin producing cells (Efrat, 2008; Iwatani and Imai, 2010).

2.1.4 Cartilage and Bone

When damaged, bone repair occurs within a compressed time frame and in a precise location, whereas damaged cartilage has shown apparent lack of self-renewal. Current research on cell-based skeletal tissue regeneration focuses on identification of an ideal cell type for tissue repair. Bone has been found to contain stem cells as well as osteoprogenitor cells that produce bone and cartilage (Nuttall and Patton *et al.*, 1998; Gronthos and Zannettino *et al.*, 1999). At the same time, BM-derived MSCs have been widely investigated for their potential to differentiate into bone and cartilage tissue (Arinzeh, 2005).

2.1.5 Nervous system

It has been suggested that diseases of the nervous system including spinal cord injury and degenerative diseases can be treated or at least improved by replacing dysfunctional cells with new ones derived from NSCs. The existence of NSCs in the central nervous system (CNS) was supported by the fact that the CNS can regenerate neuronal axons, replenish lost neural cells as well as recovery of neuronal functions. More specifically, NSCs reside in the subventricular zone of the forebrain and the dentate gyrus of the hippocampus, where they consistently generate new neural cells (McKay, 1997; Temple, 2001). To repair the nervous system, scientists have applied two main methods：transplantation of stem cells and allow them to differentiate into neurons/glial cells *in vivo* or induce differentiation *in vitro* before transplantation; alternatively, growth factors that are involved in brain development have been used to stimulate stem cells within the brain to repair damage (Okano, 2006). NSCs have been considered an important source of neurons for transplantation in treating Parkinson's disease. Other adult stem cells, including BM-derived stem cells have also been shown to differentiate towards neuronal lineage and possess the potential to be used for the relief from Parkinson's symptoms (Morizane and Li *et al.*, 2008).

The major advantage of using adult stem cells is that these cells can be autologous (isolated from the patients themselves), lessening the possibility of rejection by the immune system. However, it is difficult to isolate pure population of adult stem cells due to the extremely low numbers existing in the adult tissue of interest (approximately 1 in 10,000 cells) (Marshak and Gardner *et al.*, 2001). Another problem that hinders research on adult stem cells has been maintaining and expanding long-term cultures of these cells *in vitro*.

2.2 Pluripotent stem cells in regenerative medicine

Embryonic stem cells (ESCs) derived from spare embryos can be cultured in large numbers on feeder layers or extracellular matrices in defined medium (Yu and Thomson, 2008). These cells are pluripotent in that they are able to form all three germ layers when directly and spontaneously differentiated. A number of cell types, including neural progenitors, dopaminergic (DA) neurons, insulin-producing cells, cardiomyocytes, endothelial cells, hematopoietic cells and many others, have been generated from ESCs (Choumerianou and Dimitriou *et al.*, 2008; Sidhu, 2008). Yet, destruction of embryos to isolate human embryonic stem cells (hESCs) is considered a major ethical concern regarding the use of these cells. Another limitation is the allogenic resource of these cells, which may result in immune rejection when transplantation is carried out. New technologies such as somatic cell nuclear transfer (SCNT) and somatic cell reprogramming (cell fusion and induced pluripotent stem cell, iPCS generation) offer opportunities to overcome transplantation issues (Figure 1). However, SCNT technique raises other ethical issues regarding the use of human eggs while the latter is being investigated extensively. The past couple of years have witnessed the progress of reprogramming and re-differentiation of these reprogrammed cells, although research on reprogramming is still in its early stages and much more work needs to be done before somatic cell reprogramming can be used in clinical therapy.

One of the most exciting report in reprogramming was the generation of iPSCs from terminally differentiated somatic cells by transduction of four transcription factors (*OCT4, SOX2, KLF4* and *c-MYC*) into fibroblasts (Takahashi and Yamanaka, 2006; Takahashi and Tanabe *et al.*, 2007). Later, it was shown that the four factors can be substituted with different combinations or with small molecules and a variety of somatic cells were successfully reprogrammed (Amabile and Meissner, 2009; Cox and Rizzino, 2010). The iPSCs closely resemble ESCs and are pluripotent (Robbins and Prasain *et al.*, 2010). The approach of generating patient-specific and disease-specific iPSCs lines has already been achieved (Dimos and Rodolfa *et al.*, 2008; Park and Arora *et al.*, 2008; Ebert and Yu *et al.*, 2009; Lee and Papapetrou *et al.*, 2009). The use of these cells could pave the way for regenerative medicine without immune response that limits allogenic cell therapy or the ethics with human embryos destruction when isolating hESCs.

However, one of the most important questions in iPSC generation is to identify the starting cells amongst somatic tissue that give rise to iPSCs, and this remains elusive. In addition, although new techniques using non-integrating viruses, transient plasmid transfection or small molecules have been developed to avoid changes to the somatic cell genome, the results need to be further confirmed. Most recently, research revealed a gene expression signature in

iPSCs that is different from ESCs, suggesting that iPSCs are not identical to ESCs; rather, they could be a unique subtype of pluripotent cell (Chin and Mason *et al.*, 2009).
Another strategy being explored to induce reprogramming of differentiated cells is through fusion with pluripotent cells (Figure 1). Fusion of human fibroblasts with hESCs generates tetraploid hybrid cells with gene expression patterns, epigenetic status and differentiation ability characteristic of hESCs (Cowan and Atienza *et al.*, 2005). A reprogramming study based on hESC claimed re-expression of enhanced green fluorescent protein (EGFP) in OCT4-EGFP knock-in ESC-derived myeloid precursor cells after cell fusion (Yu and Vodyanik *et al.*, 2006). Most recently, it was reported that iPSCs generated from fibroblasts also possess the ability to reprogram other somatic cells after fusion (Sumer and Jones *et al.*, 2010). In the following sections of this chapter, we will discuss fusion between different cell types and prospect future challenges and application of fusion-based reprogramming in regenerative medicine.

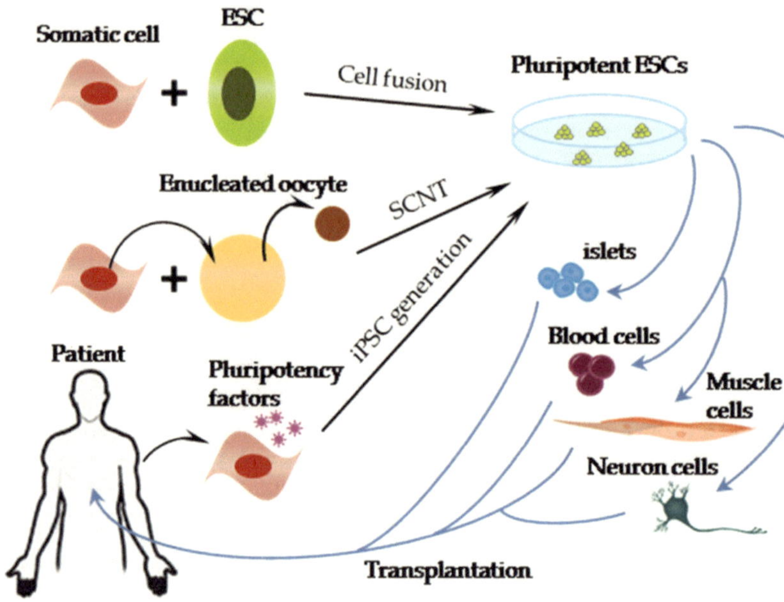

Fig. 1. Approaches to reprogram somatic cells and differentiation of the resulting pluripotent cells for regenerative medicine.

3. Cell-cell fusion

3.1 Cell fusion during development *in vivo*

Cell-cell fusion is a process that occurs in a range of normal development and infection conditions *in vivo*. The beginning of a new life and the first step of development is the event of fusion between the sperm and the egg, known as fertilization. While sperm-egg fusion has been a subject of intense investigation, the mechanism and factors that are essential for cell fusion are still not clear. Knockout studies on mouse models have shown that CD9 on the egg plasma membrane is essential for fusion (Boucheix, 2000; Kaji and Oda *et al.*, 2000; Miyado and Yamada *et al.*, 2000). Later, a sperm-specific glycoprotein Izumo, which is only detectable after the acrosome reaction, was found to be required for sperm-egg fusion

(Inoue and Ikawa *et al.*, 2005). Later, during normal development, the formation of syncytia, where multiple distinct nuclei exist, was observed in placenta, skeletal muscle and bone formation.

Trophoblast cells of mammals are known to fuse with neighbouring cells to form the syncytiotrophoblast. These syncytiotrophoblasts are maintained by further fusion with cytotrophoblasts and they serve as a good candidate for transportation of nutrients and hormones across the maternal and fetal blood vessels (Huppertz and Frank *et al.*, 1998; Pötgens and Schmitz *et al.*, 2002).

Skeletal muscle is composed of multinucleate muscle fibres, which are the products of multistep fusion process during development. Fusion starts with mononucleatemyoblasts fusing with each other to form nascent myotubes containing a small number of nuclei. This is then followed by additional fusion between the multinucleate myotubes and myoblasts, leading to the formation of large and mature myotubes (Horsley and Pavlath, 2004). At a later stage of development, SCs can fuse with muscle fibres for regeneration and maintenance of the skeletal muscle. The mechanism and regulation of myoblast fusion is reviewed in detail recently (Pavlath 2010).

Another example of cell fusion that occurs *in vivo* in mammals is macrophage fusion, which results in the formation of multinucleate osteoclasts and giant cells (Vignery, 2000). During osteoclastogenesis, macrophages fuse with each other to form multinucleate osteoclasts in the bone, or giant cells in chronic inflammatory sites. As a result, macrophages increase in size and consequently endow them enhanced capacity of resorption, which is illustrated by the difference in gene expression patterns of mono-/multi-nucleated macrophages (Teitelbaum and Ross, 2003). This resoption capacity has been shown to be important for bone remodelling and failure in fusion induces thick and brittle bone formation in osteopetrosis.

3.2 Cell-cell fusion in tissue regeneration

It has been reported that different types of cells can fuse spontaneously when co-cultured *in vitro* (Weiss and Green, 1967; Ying and Nichols *et al.*, 2002). More recently, cell fusion has also been suggested to be involved in transdifferentiation and tissue regeneration. Transdifferentiation is a concept describing the process of conversion of cells from a specific lineage to that not of their lineage of origin and has been demonstrated by the plasticity of adult stem cells. BM-derived HSCs and MSCs are the most extensively studied adult stem cells, which have also been considered to possess the most plasticity and can give rise to non-hematopoietic cells. After BM transplantation, cells carrying markers indicating their origin can be found in different tissues, including the liver, lung, pancreas and muscle (Quesenberry and Dooner *et al.*, 2010). Some scientists proposed that this is the consequence of fusion between transplanted cells and local somatic/precursor cells, rather than real transdifferentiation.

An early transplantation study aimed to treat a muscle defective disease, Duchenne muscular dystrophy (DMD), suggested that fusion occurred between implanted cells and host myoblasts/muscle fibres. Later, transplantation of BM-derived progenitors suggested possible fusion between the progenitors and myofibers and their contribution in muscle regeneration (Gibson and Karasinski *et al.*, 1995; Ferrari and Cusella-De Angelis *et al.*, 1998). In humans, it was shown that exogenous BM cells fused with myofibres *in vitro* after BM transplantation contributed to tissue repair (Gussoni and Bennett *et al.*, 2002). Additional data also showed that repair of muscle injury was not the result of HSC transdifferentiation, but was the consequence of fusion between transplanted HSCs and muscle fibres, further confirming the importance of cell fusion during tissue repair at the site of damage (Camargo and Green *et al.*, 2003; Corbel and Lee *et al.*, 2003).

Results from two groups using a hepatic lethal murine models showed regain of normal liver function after BM transplantation. The newly generated hepatocytes express genes of both donor and host, indicating fusion of donor and host cells. Southern blot and karyotyping analysis further confirmed that the hepatocytes derived, arise from BM and somatic cell fusion and not by transdifferentiation of HSCs (Vassilopoulos and Wang *et al.*, 2003; Wang and Willenbring *et al.*, 2003). Together with the studies of muscle repair, it is accepted that in a damaged tissue model, BM can fuse with somatic cells at the site of injury and regenerate organ function.

It was then indicated that this transdifferentiation induced by cell fusion is injury-independent and could occur under normal conditions. Analysis of female brains that had male BM transplantation, revealed the existence of tetraploid Purkinje neurons (XXXY), indicating fusion between HSCs and existing Purkinje neurons. Similarly, after transplantation of green fluorescent protein (GFP)-positive BM into lethally irradiated wild-type mice, GFP-positive Purkinje-BM cell heterokaryons containing two nuclei were observed, suggesting fusion had occurred (Weimann and Charlton *et al.*, 2003; Weimann and Johansson *et al.*, 2003). BM-derived MSCs have also been transplanted into Niemann-Pick mice. The transplanted MSCs were shown to fuse with Purkinje neurons and develop into functional neurons to promote brain function (Bae and Furuya *et al.*, 2005).

BM cell fusion was also investigated using transgenic mouse models carrying either Cre-recombinase or LacZ reporter gene, which is under the control of LoxP-flanked stop cassette mediated by Cre recombination. Spontaneous fusion was observed in the brain, liver and heart of irradiated healthy mice (Alvarez-Dolado and Pardal *et al.*, 2003). More recently, it was suggested that fusion between BM cells and Purkinje cells and the formation of heterokaryon occurs at a much higher efficiency upon tissue injury than under normal conditions (Johansson and Youssef *et al.*, 2008). Then another group sought to address fusion under strictly physiological conditions using the Cre-LoxP system. They claimed absence of binucleated cell generation after BM transplantation and favour the idea that fusion between HSCs and Purkinje neurons is only a transient event and under non-invasive conditions, no stable heterokaryons are formed (Nern and Wolff *et al.*, 2009).

Other studies on BM transplantation suggested that cell fusion could also be the mechanism of tissue repair for heart (Nygren and Jovinge *et al.*, 2004), kidney (Fang and Alison *et al.*, 2005) and intestinal epithelium (Rizvi and Swain *et al.*, 2006). However, there are also studies reporting HSC plasticity with no cell fusion observations (Newsome and Johannessen *et al.*, 2003; Bailey and Jiang *et al.*, 2004; Harris and Herzog *et al.*, 2004).

While some may ask whether stem cells regenerate tissue by fusing with other cells or by transdifferentiation, it will be fair to say that cell fusion and transdifferentiation do not exclude each other. In transplantation and transdifferentiation studies mentioned above, each group used a different cell population, including whole BM cells, HSCs and MSCs isolated based on expression of different cell markers. The diversity thus makes it hard to compare the results and conclude. Transdifferentiation of BM cells have been shown *in vitro*; however, cell fusion does occur during tissue repair as well as in normal tissues although at a low frequency. It is most possible that the mechanism of plasticity involves both fusion and transdifferentiation. For one thing, upon fusion of BM cells with another cell type, the latter may induce differentiation of BM cells; for another, fusion may result in transient hybrid cells with plasticity, which can differentiate into different cell types. Further research will be needed to provide more information on stem cell plasticity and tissue regeneration.

3.3 Fusion-induced reprogramming

3.3.1 Fusion between pluripotent and somatic cells

It has long been known that under co-culturing system, different cell types can fuse together, although at rather low rates. Somatic cells were shown to fuse spontaneously with ESCs *in vitro* and were reprogrammed (Pells and Di Domenico *et al.*, 2002).

The pioneer study was designed to fuse pluripotent embryonic carcinoma cells (ECCs) with primary thymocytes with the resultant hybrid cells showing similarities to the parent ECCs in their multipotent differentiation abilities (Miller and Ruddle, 1976). This indicated that pluripotency was not lost by the introduction of a somatic cell; on the contrary, pluripotency prevailed the cell fate and overwhelmed the differentiation pathways.

More recent research has shown that pluripotency can be re-established in somatic cells by fusion with ECCs, embryonic germ cells (EGCs) and ESCs, suggesting that these pluripotent cells possess the same or similar components that mediates reprogramming (Han and Sidhu, 2008). Compared to ECCs and EGCs, ESCs may represent a better source for this fusion-based reprogramming and have been paid much attention. The programming abilities of mouse ESCs (mESCs) were first illustrated by showing that mouse spleen cells obtained pluripotency after fusion with ESCs (Matveeva and Shilov *et al.*, 1998). After fusion with mESCs, the Oct4 gene was reactivated 48 hours later in mouse lymphocytes, which suggest pluripotent properties of the hybrid cells. This was also testified by the hybrids' contribution to all three germ layers as well as the epigenetic status as assessed by DNA methylation pattern of imprinted and non-imprinted genes (Tada and Takahama *et al.*, 2001). Similar to mESCs, hESCs can also reprogram differentiated somatic cells to an embryonic state by fusion (Figure 2). This was first achieved by Eggan's group in which hESCs were fused with human fibroblasts, resulting in tetraploid hybrid cells with gene expression patterns, epigenetic status and differentiation ability characteristic of hESCs (Cowan and Atienza *et al.*, 2005). Another fusion-induced reprogramming based on hESCs claimed re-expression of EGFP in OCT4-EGFP knock-in ESC-derived myeloid precursor cells after fusion and ESC-specific marker gene expression in the hybrid cells. Furthermore, pluripotency of the hybrids were demonstrated by the formation of three germ layers and extra-embryonic tissues (Yu and Vodyanik *et al.*, 2006). These results provide an alternative pathway to SCNT and iPSCs generation for reprogramming somatic cells and especially for studying the mechanisms of nuclear reprogramming.

Fusion between same or similar cells results in synkaryons or homokaryons (Figure 2). The former are single nucleated cells in which chromosome loss or re-sorting has occurred to homokaryons, an intermediate cell type in the process of synkaryon formation. Similarly, the cells formed by the fusion of different types of cells are either synkaryons or heterokaryons. Heterkaryons contain multiple nuclei and each nucleus remains separate and stable over time. Heterokaryon formation has been used as a tool to overcome chromosome loss and rearrangement after fusion and as a model to study the role of cytoplasmic factors on gene expressions. When nuclear fusion occurs, the fused nucleus initially contains chromosomal content of both fusion partners, but then chromosome loss or re-sorting leads to the formation of snykaryons.

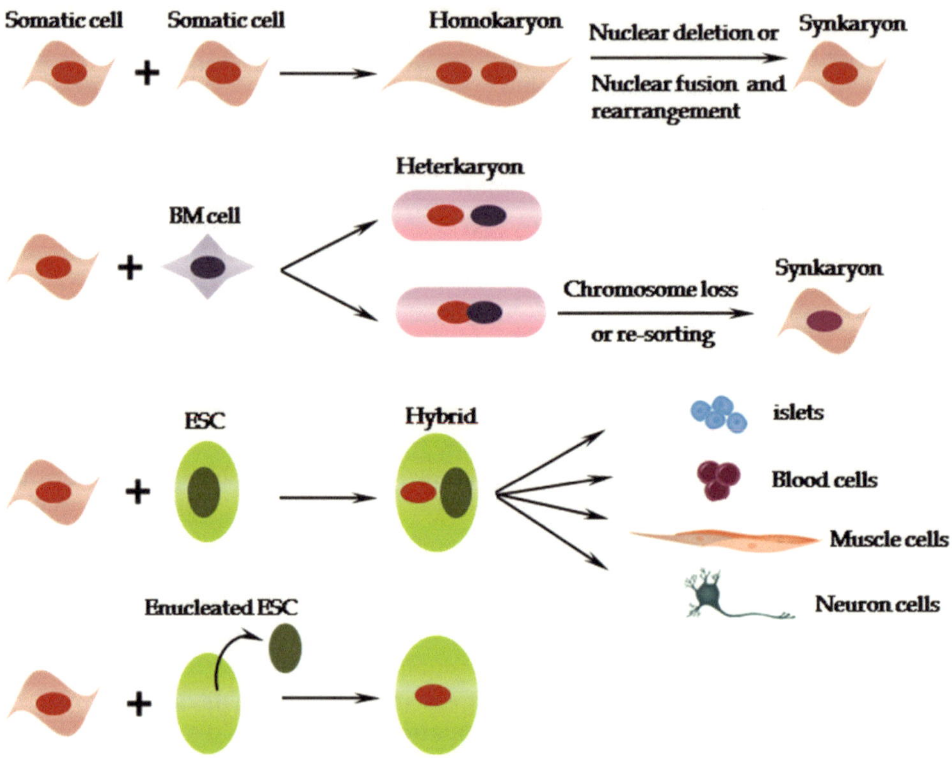

Fig. 2. Fusion between same or different types of cells and the resulting cell types.

3.3.2 Comparison of different fusion methods

Reactivation of previously silenced genes in the somatic cells after fusion with another cell type was first reported in heterokaryons of muscle and amniotic cells, indicating plasticity of differentiated cells can be changed by fusion (Blau and Chiu *et al.*, 1983). Subsequent studies fusing different cell types demonstrated that silenced genes in differentiated cell types can be reactivated after fusion and the differentiated state is not fixed. Instead, the differentiated state of cells can be reset and regulated in the heterokaryons (Yamanaka and Blau, 2010). In the case of somatic-stem cell fusion, reprogramming of somatic cells have been reported as mentioned in this chapter and shown in Figure 2.

Three approaches have been developed and applied for fusion-induced reprogramming: electro-fusion, polyethylene glycol (PEG)-induced fusion and Sendai virus-induced fusion. Electro-fusion, which involves delivery of an electric pulse to the fusing cells, was developed in the 1980s. It allows the two types of cells in a low conductivity medium suspension, be brought into close contact by dielectrophoresis, which uses high frequency alternating current and aligns the cells in a chain. Then, the high voltage electrical field pulses applied brings about an abrupt change in the ionic conductivity and membrane permeability, which results in temporary reversible membrane breakdown, pore formation on the membranes and finally combining of the membranes. The cells are held in place transiently by an alternating voltage to allow fused membranes to mature and components of the cell cytoplasm to mix. A number of studies reported high fusion rate and reproducibility when using electro-fusion (Radomska and Eckhardt, 1995). Electro-fusion has been applied to induce fusion between mouse EGCs/ESCs and somatic cells for reprogramming (Tada and Tada *et al.*, 1997; Tada and Takahama *et al.*, 2001).

PEG is a chemical that has long been used to induce cell-cell fusion (Ahkong and Howell *et al.*, 1975). Several mechanisms were suggested on how PEG mediates cell fusion. Being a linear polymer of ethylene oxides and hydroxyl terminals, PEG has the ability of binding to water molecules and form compounds. It is now widely accepted that PEG aggregates cell membranes as well as dehydrating in the areas of contact (MacDonald, 1985). At the same time, PEG also promotes fusion via positive osmotic pressure that likely helps stabilize fusion intermediates (Lentz, 2007). Fusion experiments mediated by PEG between both mouse and human ESCs have demonstrated that 50% PEG can be applied to reprogram somatic cells although at a rather low efficiency (Cowan and Atienza *et al.*, 2005; Do and Schöler, 2005).

Sendai virus (Hemagglutinating Virus of Japan, HVJ) is a type of enveloped virus that can bind to cell surface receptors on the target membrane and fuse with cell plasma membrane at neutral pH with the mediation of viral envelope glycoproteins (Robert Blumenthal, 1991). Cell fusion induced by HVJ was first reported to induce fusion among tumor cells, resulting in the formation of giant polynuclear structures (Okada, 1962). Then, cell fusion between mouse and human was induced by virus in 1965 (Harris and Watkins *et al.*, 1965). The HVJ-Envelope (HVJ-E) is inactivated and purified HVJ, with only the cell membrane-fusing capability retained. When HVJ-E are added to the cells at a concentration of several hundred per cell, they are absorbed on the cell surface immediately by the receptor acetyl type sialic acid (at the terminal of sugar chains), which is recognized by haemagglutinin-neuraminidase (HN) protein at 4°C. This leads to distortion of the cell membranes due to the penetration of an F protein of the envelope into the lipid layer of the cell membrane, which allows an inflow of ions. At this stage, the cell/HVJ-E complex is brought to a 37°C environment to induce temporary alteration of the cell membrane structure and cell fusion (Hoekstra and Klappe *et al.*, 1985; Henis and Gutman, 1987; Loyter and Chejanovsky *et al.*, 1989; Düzgüneş and Shavnin, 1992). Because the genomic RNA of the Sendai virus has been inactivated, there is no potential infective or proliferative harm to the cells.

However, the fusion efficiency of these methods in reprogramming has not been compared. It is in fact hard to compare the results from the above mentioned studies directly. First of all, in the reports of different groups, different somatic cell types and different ESC lines were used, which may affect fusion efficiency. Secondly, even for the same fusion approach, conditions applied by each group differ slightly.

We modified protocols of electro-fusion, virus induced fusion and PEG-induced fusion to suit two hESC lines and a human fetal fibroblast (HFF) cell line that have not been used for fusion-based reprogramming with the aim of determining their effectiveness of reprogramming human somatic cells. The results indicated that by optimizing fusion methods, hybrid cells between hESCs and HFFs could be generated at different efficiencies. These hybrid cells formed stem cell-like colonies and expressed stem cell specific markers.

To identify the two types of cells, an auto-fluorescent hESC line Envy and CMFDA- or SNARF-labelled cells (MEL1 and HFF10T, respectively) were used. An additional drug selection process was applied for hybrid selection, which enabled screening of stem cells that fused with puromycin-resistant HFFs (HFF10T). Envy cells contain a GFP sequence and the constitutive expression of GFP makes them easily visible and distinguishable from feeder HFFs (Costa and Dottori *et al.*, 2005). MEL1 hESCs were labelled with CMFDA to express a similar level of fluorescence as Envy under the microscope and flow cytometry conditions, whereas HFF10T cells were labelled fluorescent red with SNARF. As shown in Figure 3A, cells that were dual fluorescent represent the fused hybrid cells. After induced fusion of hESCs and puromycin-resistant HFFs, cells were cultured in normal hESC growth medium with puromycin supplemented from day 2 to 14 days post-fusion to screen for hybrid cells. Morphological changes during this period were assessed at different time points and shown in Figure 3B (day 5 and passage 2 day 7). Puromycin is toxic to wild type hESCs and the cell death was observed within 24 h. Only hESCs that had fused with

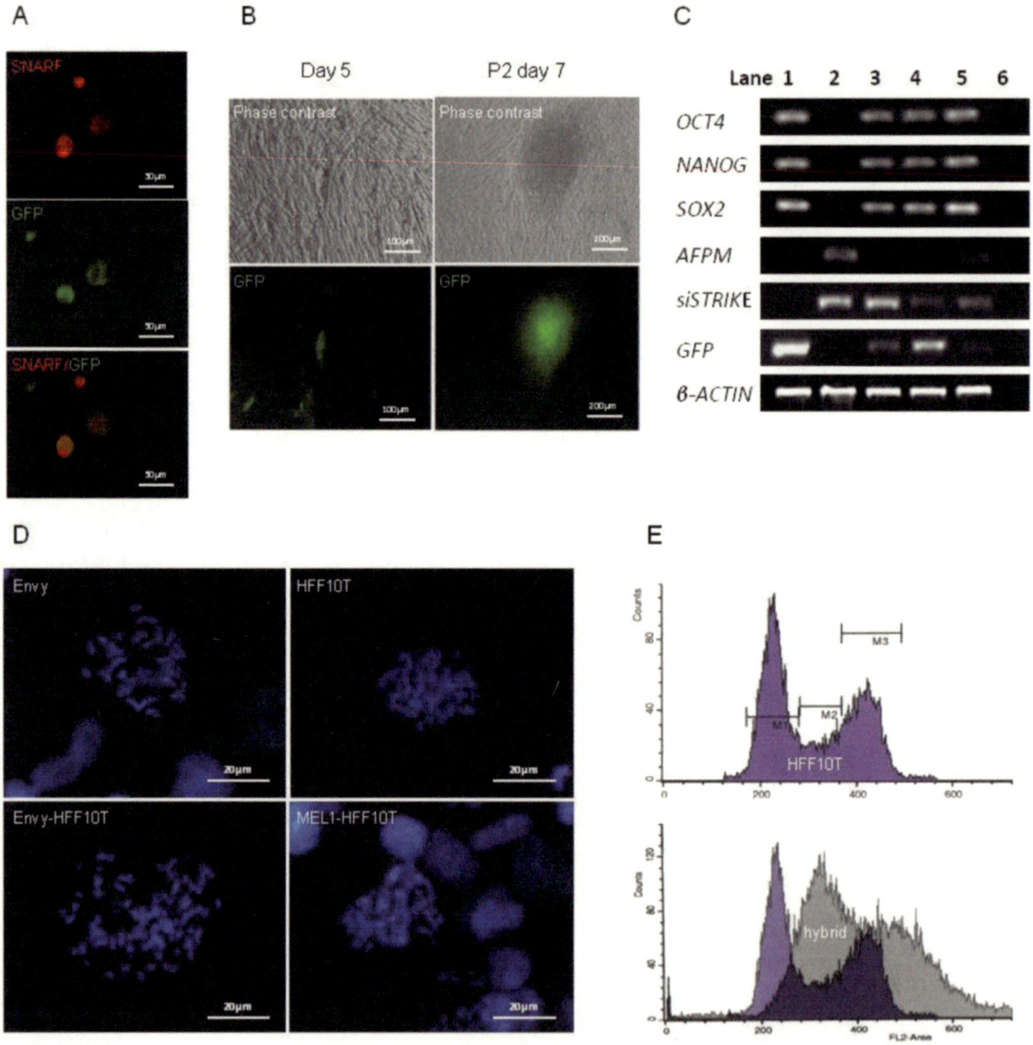

Fig. 3. hESC-HFF fusion. (A) A typical example of hybrid generation induced by cell fusion. (B) Formation of hybrid colony after fusion between Envy and HFF10T induced by PEG. (C) RT-PCR analysis of gene expression. Lanes 1 to 6 were loaded with samples (1) Envy, (2) HFF10T, (3) hybrid P1, (4) hybrid P2, (5) hybrid P3 and (6) water. (D) Chromosome numbers in hESCs, HFFs and hybrid cells. (D) Analysis of DNA content in HFF10T cells and hybrids (Han, unpublished data).

HFF10T cells that bear the puromycin-resistant plasmid sequence can survive the antibiotic selection system. Stem cell-like colonies started to form on day 4 and after 2 weeks, the colonies can be manually dissected for sub-culture and they illustrated the compact, phase-bright cell cluster feature of hESCs (Figure 3B).

RT-PCR (Figure 3C) shows that through the culturing process, pluripotency markers such as *OCT4*, *NANOG* and *SOX2* were expressed in the putative hybrids at a comparable level to that in the hESC control, indicating their ESC properties. In contrast, AFPM, a somatic cell marker, was strongly expressed in HFF10T cells but not in hESCs. In the hybrid cells at P1 and P2, this AFPM expression was removed, most probably due to reprogramming. However, a weak expression of AFPM was observed from P3 hybrid cells, suggesting re-differentiation. Primers that specifically amplify GFP sequence that was inserted into Envy cells' DNA and the puromycin-resistant sequence of the plasmid (siSTRIK) that was transferred into HFF10T cells were used to examine the origin of the hybrid cells.

DNA content analysis and chromosome number counting reflects the ploidity of the cells. As shown in Figure 3D and 3E, a normal human diploid cell has 46 chromosomes (44 autosomes and 2 sex chromosomes), whereas in a typical hybrid cell generated from PEG induced fusion between Envy/MEL1 and HFF10T cell, 92 chromosomes were contained within a single nucleus. The DNA content of hybrids showed an obvious loss of S phase cells and a trend of increased DNA content.

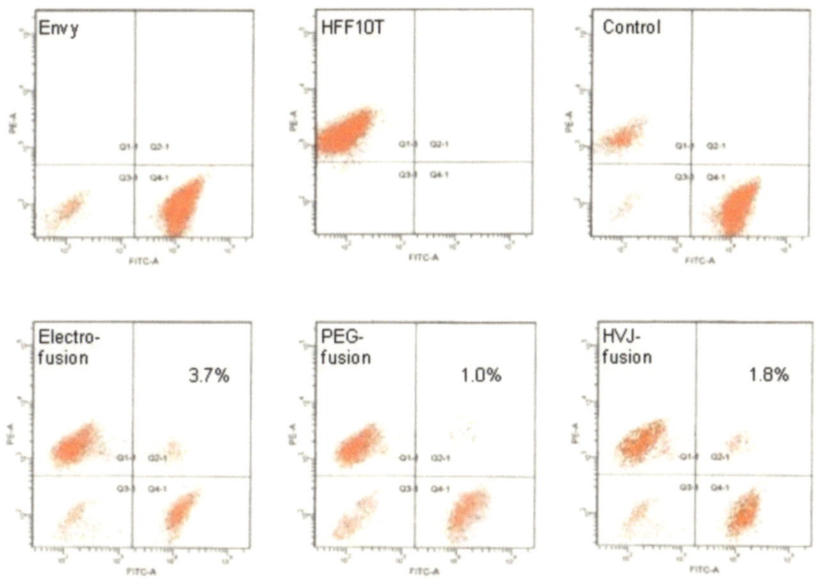

Fig. 4. Quantification of fusion efficiency by flow cytometry between Envy and HFF10T cells induced by different methods (Han, unpublished data).

When flow cytometer was applied, as shown in Figure 4, the GFP positive Envy and the SNARF-labelled HFF populations are separate and distinct. The fusion products appeared as a dual fluorescing population with magnitudes that are approximately equal to those of the stained fusion partners. The fusion efficiencies of the three methods (electro-fusion, virus-induced fusion and PEG-induced fusion) were compared using two different hESC lines, Envy and MEL1. There were no considerable differences in fusion results between the two cell lines (data not shown). However, when fusion strategies are considered, it was found that in electro-fusion, voltage applied for generating pores on the membranes of cells caused cell lysis and resulted in less survival of hybrid cells or generation of stem cell-like colonies. Furthermore, PEG 1500 was able to generate the highest number of colonies with typical stem cell characteristics despite its cytotoxicity and low percentage of double-stained hybrid cells (Figure 4). Sendai virus fusion was shown to have a moderate fusion ability as determined by hybrid generation efficiency and number of colony formation. In the study, it was concluded that fusion using PEG was still superior compared to other fusion methods used. Our results showed a fusion rate of 1.0 ± 1.2% and 1.2 ± 1.0 % for two different hESC lines (Envy and MEL1 respectively). When colony number was counted, only 0.0003 ± 0.0001% of the cell population was observed to form stem cell-like colonies. This was not improved by optimization attempts which involved changing the PEG concentration or modifying the incubation time. Nor did the density of the cells in the PEG solution have any effect on the fusion rate. This could be because PEG mediated fusion is affected by many factors that are hard to normalize, such as the size and shape of the cell pellet and the speed and vigor with which PEG is stirred into the resuspended cell pellet (Radomska and Eckhardt, 1995).

4. Prospects of cell fusion in future regenerative medicine

After injury or under pathological conditions, stem cells can proliferate and differentiate in order to regenerate and repair damaged tissues. During regeneration process, adult stem cells normally differentiate to cells that reside in the damaged site. As discussed earlier, in the organs of muscle, brain and liver, cell fusion has been demonstrated during tissue regeneration, despite of controversial reports. Stem cells from the BM origin have been of particular interest because they have shown to migrate to alternative locations and contribute to functional cells. Transdifferentiation was postulated as the mechanism of generation of the differentiated cell types after HSC and MSC transplantation. However, the finding that HSCs and MSCs can fuse with other cell types *in vivo* indicated that cell fusion could be the mechanism to explain the low number of newly generated functional cells in tissue repair after injury. Currently, it is still not clear whether the improvement of organ function after damage was a result of fusion-mediated recovery or production of growth factors and cytokines of the residue stem cells. Furthermore, it is possible that fusion occurs in more organs than we know and at a higher frequency than we expected. In addition, the function of cell fusion may be more important than what we have known in the process of organ or tissue repair.

In vitro cell fusion studies have shown that fusion between different cell types can change cell fate and gene expression profiles. Fusion reverses the differentiated state of mature cells towards a more immature or even embryonic-like state (Tada and Takahama *et al.*, 2001; Cowan and Atienza *et al.*, 2005).

Despite successful reprogramming, the strategy of inducing reprogramming by cell fusion has been problematic in the following aspects. Firstly, although the somatic cell genome is reprogrammed, the existence of ESC or other pluripotent cell genome in the tetraploid hybrids is a technical hurdle for therapeutic application. When transplanted or injected into somatic donors, immune rejection caused by the ESC genome could happen at a high frequency. Secondly, after fusion of two cells, both heterokaryons and synkaryons can be formed. Chromosome loss or spitting a nucleus of a heterokaryon results in the formation of synkaryons and chromosomal instability and genetic instability have been reported (Vasilkova and Kizilova *et al.*, 2007; Nowak-Imialek and Kues *et al.*, 2010). This needs to be taken into consideration when clinical transplantation is the aim of generating these hybrids. Thirdly, the efficiency of fusion induced reprogramming has been low. Since the first study reporting the success of reprogramming human somatic cells after fusion with hESCs, efforts have been made in order to increase reprogramming efficiency in the system. Over expression of certain transcription factors, such as Nanog and Sall4, as well as manipulation of histonemethylase Jhdm2a and methyltransferease G9a, have been claimed to significantly enhance the ability to generate reprogrammed hybrids from fusion (Silva and Chambers *et al.*, 2006; Ma and Chiang *et al.*, 2008; Wong and Gaspar-Maia *et al.*, 2008). However, these approaches need to be further tested in different cell lines and cell types. Lastly, fusion induced reprogramming was evidenced by morphological changes, gene expressions and epigenetic states, yet only limited functional reprogramming has been demonstrated (Yu and Vodyanik *et al.*, 2006).

To fully explore the mechanisms of fusion induced reprogramming for cell therapy and regenerative medical application, there are important questions yet to be answered. For example, the subcellular location of reprogramming factors has always been a subject of interest. When the nucleus and the cytoplasm of mESCs were fused with neurosphere cells separately, only karyoplasts could reactivate Oct4 in the somatic genome (Do and Scholer, 2004). A conflicting observation was made later by Strelchenko et al., showing that hybrid cells generated by fusion of human somatic cells with enucleated ESCs express pluripotency markers OCT4 and TRA-2-39 (Strelchenko and Kukharenko *et al.*, 2006). Most recently, it was reported that hESC cytoplast fusion can only initiate but unable to complete reprogramming, indicating the importance of the ESC nucleus in fusion-induced reprogramming (Hasegawa and Zhang *et al.*, 2010) (Figure 2). Another issue is the possibility of immune rejection caused by the existence of the allogenic genome in the hybrid cells when they are transplanted into the somatic donor. In this case, removal of the pluripotent genome would be necessary. However, this would be hindered if the major elements responsible for reprogramming within ESCs are in the nucleus. Although a remedial approach has been developed in the mouse by allowing enough time for ESC nucleus to reprogram the somatic genome before removing the ESC nucleus (Pralong and Mrozik *et al.*, 2005), the reprogramming ability of those nuclei-removed ESCs needs to be confirmed and it is not known if this approach could also be applied to hESCs. Targeted elimination of ESC chromosomes from mouse somatic cell-ESC hybrids is introduced as an alternative method to expel the effect of ESC genome (Matsumura and Tada *et al.*, 2007). It was also reported that human B-lymphocytes can be reprogrammed towards a multipotent state in transient heterokaryons before nuclear fusion (Pereira and Terranova *et al.*, 2008). In that case, it may be possible to remove ESC genome after reprogramming has occurred although this needs further investigation. Finally, the lacking of effective approaches to monitor early reprogramming events and being unaware of the molecules that mediate reprogramming in the ESCs, makes it challenging to achieve efficient fusion and reprogramming. However, considering that cell fusion is technically simple and reprogramming occurs fast after fusion, it is widely accepted that it is an ideal way to investigate the regulatory mechanisms of reprogramming. For example, it was demonstrated by heterokaryoyic cell fusion between mESCs and human fibroblasts that activation-induced cytidinedeaminase (AID), a DNA demethylation enzyme, is required for reprogramming initiation (Bhutani and Brady *et al.*, 2010).

5. Conclusion

In vivo cell fusion is a tightly regulated process that occurs in normal development or under pathological conditions. There are indications that induced *in vitro* cell fusion could be potentially applied for regeneration when stem cells are fused with somatic cells. A better understanding of the process will enable us to use cell fusion as an approach for cell therapy. Fusion between ESCs and somatic cells is considered a way of reprogramming somatic cells, with the generation of pluripotent cells that can be further differentiated into desired cell types. Cell-cell fusion may be faster and more efficient than iPSC technique as hESCs provide all the necessary factors for reprogramming. However, enucleation of ESC nucleus must be achieved before the hybrid cells can be used clinically. In addition, cell fusion is a useful tool for studying the molecular mechanism that controls somatic cell reprogramming.

6. References

Ahkong and Howell, *et al.* (1975). "Fusion of hen erythrocytes with yeast protoplasts induced by polyethylene glycol." Nature 255(5503): 66-67.

Alison and Islam (2009). "Attributes of adult stem cells." The Journal of Pathology 217(2): 144-160.

Alvarez-Dolado and Pardal, *et al.* (2003). "Fusion of bone-marrow-derived cells with purkinje neurons, cardiomyocytes and hepatocytes." Nature 425(6961): 968-973.

Amabile and Meissner (2009). "Induced pluripotent stem cells: Current progress and potential for regenerative medicine." Trends in Molecular Medicine 15(2): 59-68.

Arinzeh (2005). "Mesenchymal stem cells for bone repair: Preclinical studies and potential orthopedic applications." Foot and Ankle Clinics of North America 10(4): 651-665.

Armitage (1994). "Bone marrow transplantation." New England Journal of Medicine 330(12): 827-838.

Bae and Furuya, *et al.* (2005). "Neurodegeneration augments the ability of bone marrow-derived mesenchymal stem cells to fuse with purkinje neurons in niemann-pick type c mice." Human Gene Therapy 16(8): 1006-1011.

Bailey and Jiang, *et al.* (2004). "Transplanted adult hematopoietic stems cells differentiate into functional endothelial cells." Blood 103(1): 13-19.

Bajada and Mazakova, *et al.* (2008). "Updates on stem cells and their applications in regenerative medicine." Journal of Tissue Engineering and Regenerative Medicine 2(4): 169-183.

Benigni and Morigi, *et al.* (2010). "Kidney regeneration." The Lancet 375(9722): 1310-1317.

Bhutani and Brady, *et al.* (2010). "Reprogramming towards pluripotency requires aid-dependent DNA demethylation." Nature 463(7284): 1042-1047.

Blau and Chiu, *et al.* (1983). "Cytoplasmic activation of human nuclear genes in stable heterocaryons." Cell 32(4): 1171-1180.

Bonnet (2003). "Biology of human bone marrow stem cells." Clinical and Experimental Medicine 3(3): 140-149.

Boucheix (2000). "Severely reduced female fertility in cd9-deficient mice." Science 287(5451): 319-321.

Broek and Grefte, *et al.* (2010). "Regulatory factors and cell populations involved in skeletal muscle regeneration." Journal of Cellular Physiology 224(1): 7-16.

Camargo and Green, *et al.* (2003). "Single hematopoietic stem cells generate skeletal muscle through myeloid intermediates." Nature Medicine 9(12): 1520-1527.
Chin and Mason, *et al.* (2009). "Induced pluripotent stem cells and embryonic stem cells are distinguished by gene expression signatures." Cell Stem Cell 5(1): 111-123.
Chkhotua and Klein, *et al.* (2003). "Kidney transplantation from living-unrelated donors: Comparison of outcome with living-related and cadaveric transplants under current immunosuppressive protocols." Urology 62(6): 1002-1006.
Choumerianou and Dimitriou, *et al.* (2008). "Stem cells: Promises versus limitations." Tissue engineering. Part B, Reviews 14(1): 53-60.
Copelan (2006). "Hematopoietic stem-cell transplantation." New England Journal of Medicine 354(17): 1813-1826.
Corbel and Lee, *et al.* (2003). "Contribution of hematopoietic stem cells to skeletal muscle." Nature Medicine 9(12): 1528-1532.
Costa and Dottori, *et al.* (2005). "The hesc line envy expresses high levels of gfp in all differentiated progeny." Nat Meth 2(4): 259-260.
Cowan and Atienza, *et al.* (2005). "Nuclear reprogramming of somatic cells after fusion with human embryonic stem cells." Science 309(5739): 1369-1373.
Cowan and Melton (2006). 'stemness': Definitions, criteria, and standards. Essentials of stem cell biology, Elsevier Inc.: xxv-xxxi.
Cox and Rizzino (2010). "Induced pluripotent stem cells: What lies beyond the paradigm shift." Exp. Biol. Med. 235(2): 148-158.
Dimos and Rodolfa, *et al.* (2008). "Induced pluripotent stem cells generated from patients with als can be differentiated into motor neurons." Science 321(5893): 1218-1221.
Do and Scholer (2004). "Nuclei of embryonic stem cells reprogram somatic cells." Stem Cells 22(6): 941-949.
Do and Schöler (2005). "Comparison of neurosphere cells with cumulus cells after fusion with embryonic stem cells: Reprogramming potential." Reproduction, Fertility and Development 17(1-2): 143-149.
Düzgüneş and Shavnin (1992). "Membrane destabilization by n-terminal peptides of viral envelope proteins." Journal of Membrane Biology 128(1): 71-80.
Ebert and Yu, *et al.* (2009). "Induced pluripotent stem cells from a spinal muscular atrophy patient." Nature 457(7227): 277-280.
Efrat (2008). "Beta-cell replacement for insulin-dependent diabetes mellitus." Advanced Drug Delivery Reviews 60(2): 114-123.
Fang and Alison, *et al.* (2005). "Proliferation of bone marrow-derived cells contributes to regeneration after folic acid-induced acute tubular injury." J Am Soc Nephrol 16(6): 1723-1732.
Ferrari and Cusella-De Angelis, *et al.* (1998). "Muscle regeneration by bone marrow-derived myogenic progenitors." Science 279(5356): 1528-1530.
Gibson and Karasinski, *et al.* (1995). "Dermal fibroblasts convert to a myogenic lineage in mdx mouse muscle." J Cell Sci 108(1): 207-214.
Gronthos and Zannettino, *et al.* (1999). "Differential cell surface expression of the stro-1 and alkaline phosphatase antigens on discrete developmental stages in primary cultures of human bone cells." Journal of Bone and Mineral Research 14(1): 47-56.

Guild and Harrison, *et al.* (1955). "Successful homotransplantation of the kidney in an identical twin." Transactions of the American Clinical and Climatological Association 67: 167-173.

Gussoni and Bennett, *et al.* (2002). "Long-term persistence of donor nuclei in a duchenne muscular dystrophy patient receiving bone marrow transplantation." Journal of Clinical Investigation 110(6): 807-814.

Han and Sidhu (2008). "Current concepts in reprogramming somatic cells to pluripotent state." Current Stem Cell Research and Therapy 3(1): 66-74.

Hariharan and Johnson, *et al.* (2000). "Improved graft survival after renal transplantation in the united states, 1988 to 1996." New England Journal of Medicine 342(9): 605-612.

Harris and Herzog, *et al.* (2004). "Lack of a fusion requirement for development of bone marrow-derived epithelia." Science 305(5680): 90-93.

Harris and Watkins, *et al.* (1965). "Mitosis in hybrid cells derived from mouse and man." Nature 207(4997): 606-608.

Hasegawa and Zhang, *et al.* (2010). "Comparison of reprogramming efficiency between transduction of reprogramming factors, cell–cell fusion, and cytoplast fusion." Stem Cells 28(8): 1338-1348.

Henis and Gutman (1987). "Lateral mobility of reconstituted sendai virus envelope glycoproteins on human erythrocytes: Correlation with cell-cell fusion." Biochemistry 26(3): 812-819.

Hoekstra and Klappe, *et al.* (1985). "Characterization of the fusogenic properties of sendai virus: Kinetics of fusion with erythrocyte membranes." Biochemistry 24(18): 4739-4745.

Horsley and Pavlath (2004). "Forming a multinucleated cell: Molecules that regulate myoblast fusion." Cells Tissues Organs 176(1-3): 67-78.

Huppertz and Frank, *et al.* (1998). "Villous cytotrophoblast regulation of the syncytial apoptotic cascade in the human placenta." Histochemistry and Cell Biology 110(5): 495-508.

Inoue and Ikawa, *et al.* (2005). "The immunoglobulin superfamily protein izumo is required for sperm to fuse with eggs." Nature 434(7030): 234-238.

Iwatani and Imai (2010). "Kidney repair using stem cells: Myth or reality as a therapeutic option?" Journal of Nephrology 23(2): 143-146.

Iwatani and Ito, *et al.* (2004). "Hematopoietic and nonhematopoietic potentials of hoechstlow//side population cells isolated from adult rat kidney." Kidney Int 65(5): 1604-1614.

Johansson and Youssef, *et al.* (2008). "Extensive fusion of haematopoietic cells with purkinje neurons in response to chronic inflammation." Nat Cell Biol 10(5): 575-583.

Kaji and Oda, *et al.* (2000). "The gamete fusion process is defective in eggs of cd9-deficient mice." Nature Genetics 24(3): 279-282.

Kim and Kwon, *et al.* (2004). "Experience with cyclosporine in adult living donor kidney transplantation: From 1984 to 2002 at yonsei university." Transplantation Proceedings 36(2, Supplement 1): S186-S192.

Lee and Papapetrou, *et al.* (2009). "Modelling pathogenesis and treatment of familial dysautonomia using patient-specific ipscs." Nature 461(7262): 402-406.

Lentz (2007). "Peg as a tool to gain insight into membrane fusion." European Biophysics Journal 36(4-5): 315-326.

Leri and Kajstura, *et al.* (2005). "Cardiac stem cells and mechanisms of myocardial regeneration." Physiol. Rev. 85(4): 1373-1416.

Loyter and Chejanovsky, *et al.* (1989). Implantation of isolated carriers and receptors into living cells by sendai virus envelope-mediated fusion. Methods in enzymology. F. Sidney and F. Becca, Academic Press. Volume 171: 829-850.

Lyngbæk and Schneider, *et al.* (2007). "Cardiac regeneration by resident stem and progenitor cells in the adult heart." Basic Research in Cardiology 102(2): 101-114.

Ma and Chiang, *et al.* (2008). "G9a and jhdm2a regulate embryonic stem cell fusion-induced reprogramming of adult neural stem cells." Stem Cells: 2008-0388.

MacDonald (1985). "Membrane fusion due to dehydration by polyethylene glycol, dextran, or sucrose." Biochemistry 24(15): 4058-4066.

Marshak and Gardner, *et al.* (2001). Stem cell biology, Cold Spring Harbor:Cold Spring Harbor Laboratory Press.

Matsumura and Tada, *et al.* (2007). "Targeted chromosome elimination from es-somatic hybrid cells." Nat Meth 4(1): 23-25.

Matveeva and Shilov, *et al.* (1998). "In vitro and in vivo study of pluripotency in intraspecific hybrid cells obtained by fusion of murine embryonic stem cells with splenocytes." Molecular Reproduction and Development 50(2): 128-138.

McKay (1997). "Stem cells in the central nervous system." Science 276(5309): 66-71.

Miller and Ruddle (1976). "Pluripotent teratocarcinoma thymus somatic cell hybrids." Cell 9(1): 45-55.

Mimeault and Batra (2008). "Recent progress on tissue-resident adult stem cell biology and their therapeutic implications." Stem Cell Reviews 4(1): 27-49.

Miyado and Yamada, *et al.* (2000). "Requirement of cd9 on the egg plasma membrane for fertilization." Science 287(5451): 321-324.

Morizane and Li, *et al.* (2008). "From bench to bed: The potential of stem cells for the treatment of parkinson's disease." Cell and Tissue Research 331(1): 323-336.

Nern and Wolff, *et al.* (2009). "Fusion of hematopoietic cells with purkinje neurons does not lead to stable heterokaryon formation under noninvasive conditions." J. Neurosci. 29(12): 3799-3807.

Newsome and Johannessen, *et al.* (2003). "Human cord blood-derived cells can differentiate into hepatocytes in the mouse liver with no evidence of cellular fusion." Gastroenterology 124(7): 1891-1900.

Nowak-Imialek and Kues, *et al.* (2010). "Preferential loss of porcine chromosomes in reprogrammed interspecies cell hybrids." Cellular reprogramming 12(1): 55-65.

Nuttall and Patton, *et al.* (1998). "Human trabecular bone cells are able to express both osteoblastic and adipocytic phenotype: Implications for osteopenic disorders." Journal of Bone and Mineral Research 13(3): 371-382.

Nygren and Jovinge, *et al.* (2004). "Bone marrow-derived hematopoietic cells generate cardiomyocytes at a low frequency through cell fusion, but not transdifferentiation." Nature Medicine 10(5): 494-501.

Okada (1962). "Analysis of giant polynuclear cell formation caused by hvj virus from ehrlich's ascites tumor cells. Iii. Relationship between cell condition and fusion reaction or cell degeneration reaction." Experimental Cell Research 26(1): 119-128.

Okano (2006). Adult neural stem cells and central nervous system repair. Stem cells in reproduction and in the brain. J. Morser, S. I. Nishikawa and H. R. Schöler, Springer Berlin Heidelberg. 60: 215-228.

Ott and Matthiesen, *et al.* (2008). "Perfusion-decellularized matrix: Using nature's platform to engineer a bioartificial heart." Nat Med 14(2): 213-221.

Park and Arora, *et al.* (2008). "Disease-specific induced pluripotent stem cells." Cell 134(5): 877-886.

Pavlath, G. K. (2010). "Spatial and functional restriction of regulatory molecules during mammalian myoblast fusion." Experimental Cell Research 316(18): 3067-3072.

Pells and Di Domenico, *et al.* (2002). "Multipotentiality of neuronal cells after spontaneous fusion with embryonic stem cells and nuclear reprogramming in vitro." Cloning and Stem Cells 4(4): 331-338.

Pereira and Terranova, *et al.* (2008). "Heterokaryon-based reprogramming of human b lymphocytes for pluripotency requires oct4 but not sox2." PLoS Genet 4(9): e1000170.

Pötgens and Schmitz, *et al.* (2002). "Mechanisms of syncytial fusion: A review." Placenta 23(SUPPL. 1).

Pralong and Mrozik, *et al.* (2005). "A novel method for somatic cell nuclear transfer to mouse embryonic stem cells." Cloning and Stem Cells 7(4): 265-271.

Quesenberry and Dooner, *et al.* (2010). "Stem cell plasticity revisited: The continuum marrow model and phenotypic changes mediated by microvesicles." Experimental Hematology 38(7): 581-592.

Radomska and Eckhardt (1995). "Mammalian cell fusion in an electroporation device." Journal of Immunological Methods 188(2): 209-217.

Raff (2003). "Adult stem cell plasticity: Fact or artifact?" Annual Review of Cell and Developmental Biology 19(1): 1-22.

Rao and Mattson (2001). "Stem cells and aging: Expanding the possibilities." Mechanisms of Ageing and Development 122(7): 713-734.

Rizvi and Swain, *et al.* (2006). "Bone marrow-derived cells fuse with normal and transformed intestinal stem cells." Proceedings of the National Academy of Sciences of the United States of America 103(16): 6321-6325.

Robbins and Prasain, *et al.* (2010). "Inducible pluripotent stem cells: Not quite ready for prime time?" Current Opinion in Organ Transplantation 15(1): 61-67 10.1097/MOT.1090b1013e3283337196.

Robert Blumenthal (1991). "A dissection of steps leading to viral envelope protein-mediated membrane fusion." Annals of the New York Academy of Sciences 635(Calcium Entry and Action at the Presynaptic Nerve Terminal): 285-296.

Rubart and Field (2006). "Cardiac regeneration: Repopulating the heart." Annual Review of Physiology 68(1): 29-49.

Sayegh and Remuzzi (2007). "Clinical update: Immunosuppression minimisation." Lancet 369(9574): 1676-1678.

Schmelzer and Zhang, *et al.* (2007). "Human hepatic stem cells from fetal and postnatal donors." Journal of Experimental Medicine 204(8): 1973-1987.

Shi and Garry (2006). "Muscle stem cells in development, regeneration, and disease." Genes & Development 20(13): 1692-1708.

Shizuru and Negrin, *et al.* (2005). "Hematopoietic stem and progenitor cells: Clinical and preclinical regeneration of the hematolymphoid system." Annual Review of Medicine 56(1): 509-538.

Sidhu (2008). Motoneurons from human embryonic stem cells: Present status and future strategies for their use in regenerative medicine. Neurovascular medicine pursuing cellular longevity for healthy aging. K. Maiese, Oxford University Press: 231-254.

Sidhu and Ryan, *et al.* (2010). "Derivation of a new human embryonic stem cell line, endeavour-2, and its characterization." In Vitro Cellular & Developmental Biology - Animal 46(3): 269-275.

Silva and Chambers, *et al.* (2006). "Nanog promotes transfer of pluripotency after cell fusion." Nature 441(7096): 997-1001.

Strelchenko and Kukharenko, *et al.* (2006). "Reprogramming of human somatic cells by embryonic stem cell cytoplast." Reproductive BioMedicine Online 12(1): 107-111.

Sumer and Jones, *et al.* (2010). "Reprogramming of somatic cells after fusion with induced pluripotent stem cells and nuclear transfer embryonic stem cells." Stem Cells and Development 19(2): 239-246.

Tada and Tada, *et al.* (1997). "Embryonic germ cells induce epigenetic reprogramming of somatic nucleus in hybrid cells." EMBO Journal 16(21): 6510-6520.

Tada and Takahama, *et al.* (2001). "Nuclear reprogramming of somatic cells by in vitro hybridization with es cells." Current Biology 11(19): 1553-1558.

Takahashi and Tanabe, *et al.* (2007). "Induction of pluripotent stem cells from adult human fibroblasts by defined factors." Cell 131(5): 861-872.

Takahashi and Yamanaka (2006). "Induction of pluripotent stem cells from mouse embryonic and adult fibroblast cultures by defined factors." Cell 126(4): 663-676.

Teitelbaum and Ross (2003). "Genetic regulation of osteoclast development and function." Nature Reviews Genetics 4(8): 638-649.

Temple (2001). "The development of neural stem cells." Nature 414(6859): 112-117.

Terrace and Hay, *et al.* (2009). "Side population cells in developing human liver are primarily haematopoietic progenitor cells." Experimental Cell Research 315(13): 2141-2153.

Vasilkova and Kizilova, *et al.* (2007). "Dominant manifestation of pluripotency in embryonic stem cell hybrids with various numbers of somatic chromosomes." Molecular Reproduction and Development 74(8): 941-951.

Vassilopoulos and Wang, *et al.* (2003). "Transplanted bone marrow regenerates liver by cell fusion." Nature 422(6934): 901-904.

Vignery (2000). "Osteoclasts and giant cells: Macrophage-macrophage fusion mechanism." International Journal of Experimental Pathology 81(5): 291-304.

Wang and Willenbring, *et al.* (2003). "Cell fusion is the principal source of bone-marrow-derived hepatocytes." Nature 422(6934): 897-901.

Weimann and Charlton, *et al.* (2003). "Contribution of transplanted bone marrow cells to purkinje neurons in human adult brains." Proceedings of the National Academy of Sciences of the United States of America 100(4): 2088-2093.

Weimann and Johansson, *et al.* (2003). "Stable reprogrammed heterokaryons form spontaneously in purkinje neurons after bone marrow transplant." Nat Cell Biol 5(11): 959-966.

Weiss and Green (1967). "Human-mouse hybrid cell lines containing partial complements of human chromosomes and functioning human genes." Proceedings of the National Academy of Sciences of the United States of America 58(3): 1104-1111.

Weissman (2000). "Stem cells: Units of development, units of regeneration, and units in evolution." Cell 100(1): 157-168.

Wong and Gaspar-Maia, *et al.* (2008). "High-efficiency stem cell fusion-mediated assay reveals sall4 as an enhancer of reprogramming." PLoS ONE 3(4): e1955.

Yamanaka and Blau (2010). "Nuclear reprogramming to a pluripotent state by three approaches." Nature 465(7299): 704-712.

Ying and Nichols, *et al.* (2002). "Changing potency by spontaneous fusion." Nature 416(6880): 545-548.

Yu and Thomson (2008). "Pluripotent stem cell lines." Genes & Development 22(15): 1987-1997.

Yu and Vodyanik, *et al.* (2006). "Human embryonic stem cells reprogram myeloid precursors following cell-cell fusion." Stem Cells 24(1): 168-176.

14

Use of Transposon-Transposase Systems for Stable Genetic Modification of Embryonic Stem Cells

Tamás I. Orbán[1], Ágota Apáti[1], Zsuzsanna Izsvák[2,3], Zoltán Ivics[2,3] and Balázs Sarkadi[1]

[1]*Membrane Research Group of the Hungarian Academy of Sciences, Semmelweis University and National Blood Center, Budapest*

[2]*Mobile DNA Group, Max-Delbrück Center for Molecular Medicine, Berlin*

[3]*University of Debrecen, Debrecen*

[1,3]*Hungary,*

[2]*Germany*

1. Introduction

Human stem cells provide new hopes in the clinical treatment of a number of diseases, are excellent models for tissue and cell differentiation, and serve as the basis of new screening systems for drug development and toxicity. Regenerative medicine makes use of cells that can grow and differentiate to replace a damaged tissue. Hematopoietic stem/progenitor cells are successfully applied for bone marrow transplantation in otherwise lethal clinical conditions, while many other cell-based treatments are still experimental.

Based on their basic features, we distinguish two major kinds of stem cells. Pluripotent stem cells, capable of differentiating to all types of the cells of human body, were first derived from early human embryos (HuES) and could be grown to provide cell lines with preserved pluripotent characteristics (Thomson et al., 1998). A recently discovered method of generating induced pluripotent stem (iPS) cells from differentiated cell types (Takahashi & Yamanaka, 2006) provides a potential to obtain autologous human stem cell lines without using embryonic tissues. In both cases a major concern, regarding therapeutic applications, is the formation of teratomas, consisting of numerous types of partially differentiated tissues (Reubinoff et al., 2000).

Another major source of human stem cells is our own body. Although in a relatively small number, the so called "tissue-derived stem cells", sometimes referred to as "adult stem cells", are present throughout our life in various tissues and organs. The "stemness" of these cells, that is their actual stage of differentiation, depends on age, tissue origin and many other still unrecognized conditions. We know that the bone marrow or the cord blood of the neonate contains mostly hematopoietic progenitor cells, the skin, the liver or the intestine has a large number of repopulating epithelial progenitors, and even the muscle or the brain have dormant cells capable of tissue regeneration. Whether our body also contains early cell types similar to the embryonic stem cells, capable of differentiating into any tissue, is still an unanswered question (Vieyra et al., 2005; Bussolati & Camussi, 2006; Bussolati & Camussi, 2007; Satija et al., 2007).

Under laboratory conditions undifferentiated cells from HuES or iPS cell lines grow in small clumps, but retain their undifferentiated form only under special conditions. They require media which contain a specific combination of growth factors, and in most cases the presence of appropriate feeder cells. The passage of the pluripotent cells can be performed by mild protease treatment and each cell type may require individual culturing and passaging conditions.

The differentiation of the HUES or iPS cells is initiated spontaneously, as soon as these cells are removed from the special media and/or feeder cells, and placed into normal tissue culture plates or flasks. If the attachment of the cells is prevented, pluripotent stem cells generate so called embryoid bodies (EBs), which are complex, teratoma-like tissue structures with highly variable forms and tissue elements. If these EBs are placed onto tissue culture plates, they attach to the surface and start a further spontaneous differentiation process, by forming in many cases well recognizable tissue-types. Under these conditions, the formation of endothelial, epithelial, and neuronal cells, as well as of fibroblasts or cardiomyocytes can be observed.

Tissue formation greatly depends on the culture conditions, which can be relatively well adjusted to obtain a specific enrichment of a desired tissue type. By applying protease-based cell separation, specific protein and nucleic acid extractions, or even by studying the formed tissues in situ, the pattern of protein expression can be followed during the stages of differentiation of stem cells.

Well characterized human pluripotent stem cells thus indeed represent a great new tool for developmental studies, drug screening as well as cell- and gene-therapy applications. However, all these approaches usually require the development of efficient, stable gene delivery, and proper progenitor cell and tissue separation methods. Following of cell fate of differentiation by stably expressed marker proteins, or the introduction of new or corrected genes into stem cells, greatly facilitate and expand their research and therapeutic potential.

Currently, the most widely applied methods for gene delivery into stem cells are based on the use of viral vector constructs. There are numerous efficient retrovirus- or lentivirus-based methods which allow stable genomic incorporation of the foreign DNAs with high gene product expression levels. However, virus-based gene therapy technologies also have serious drawbacks, including safety concerns of virus production, and the preferential incorporation of foreign genes into active host gene loci, which may cause uncontrolled proliferation of the gene-modified stem cells (Schroder et al., 2002; VandenDriessche et al., 2003). Non-viral gene delivery techniques are usually considered to be less efficient, however, with the emergence and refinement of the transposon based methods, they represent a valid alternative to viral applications (Ivics & Izsvak, 2004; Izsvak & Ivics, 2004; Ivics & Izsvak, 2006). In the following sections we detail the basic features, advantages and concerns of using transposon-based gene delivery into human stem cells.

2. Transposon systems as genetic tools

Transposons are "selfish" genetic elements that can move from one DNA locus to another either by a replicative or a non-replicative manner. They are widely present in the genome of all organisms and are also believed to be important driving forces for evolution (Kazazian, 2004; Hedges & Batzer, 2005; Feschotte & Pritham, 2007). The human genome carries a significant proportion of transposable elements: it is estimated that ca. 45% of our genetic material is made up of transposons (Biemont & Vieira, 2006; Mills et al., 2006; Wicker et al., 2007; Goodier & Kazazian, 2008). The majority of these belong to the Class I retrotransposons which contain currently active elements moving in our genome by the replicative "copy and paste" mechanism (Mills et al., 2007). Although applications using certain retrotransposons as genetic tools exist (Uren et al., 2005; Ostertag et al., 2007), their obvious disadvantages (the potential remobilization of the delivered transgene and the high mutational rate resulting from reverse transcription) make them less favorable as genetic delivery vehicles.

As opposed to retrotransposons, the Class II DNA transposons make up a relatively small portion (~3%) of the human mobile elements (Feschotte & Pritham, 2007) and currently none of them have been shown to be active in our genome (Collier & Largaespada, 2007; Izsvak et al., 2010). They move by the non-replicative "cut and paste" mechanism and their active and modified forms have been widely used in genetic analysis of lower (mostly invertebrate) model organisms including *Drosophila* species and *Caenorhabditis elegans* (Mates et al., 2007). Various gene trapping or insertional mutagenesis experiments proved that transposon based gene delivery is an efficient system, however, lacking similar genetic methods in mammalian cells was an obvious disadvantage.

The first breakthrough in this field was the resurrection of *Sleeping Beauty* (SB), an artificial Tc1/Mariner-type transposon system "remastered" from old fish transposon fossils, and the

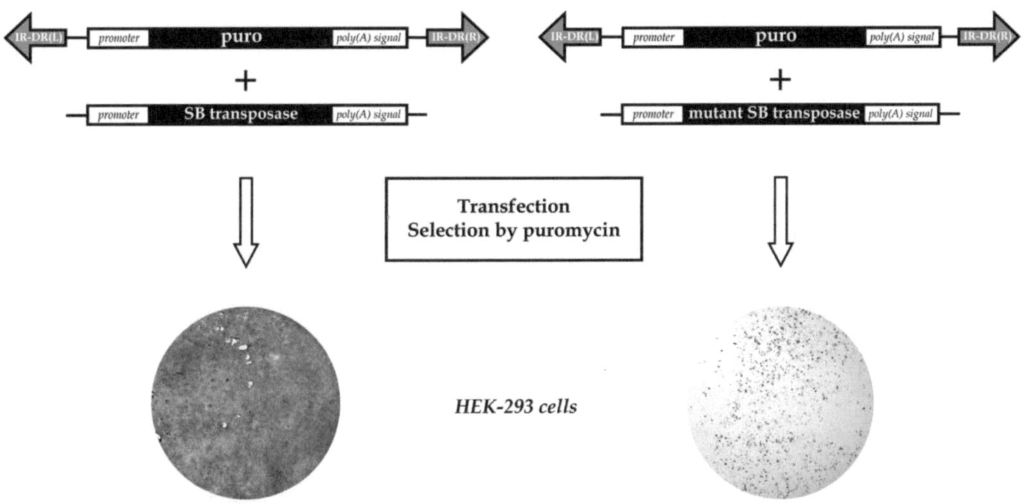

Fig. 1. The principle of gene delivery by the *Sleeping Beauty* transposon system.
The gene of interest in a transcription unit (here a puromycin resistance gene, "puro") is placed between the transposon sequences (IR-DR(L) and-(R): inverted repeat - direct repeat Left and Right sequences). Such an engineered transposon vector is co-transfected with a transposase expressing plasmid into the cells (left panel) and selected by puromycin for 10 days. Following selection, live cells are visualized by Giemsa-staining. The control reaction is carried out using a mutant transposase (right panel) to detect the level of random integration without transposition. As shown in this experiment, transposon based gene delivery is several orders of magnitude higher than the background random integration.

proof of principle that it was active in mammalian cells (Ivics et al., 1997). Its structure is relatively simple (**Figure 1**) and an elegant way of creating a non-autonomous version by separating the transposase from its targets (the terminal inverted repeat sequences) made it an easily controllable system and therefore an attractive tool for functional genetics (Izsvak et al., 2000). Nevertheless, its efficiency in gene delivery at that stage was still behind that of viral vectors, the canonical genetic vehicles used in human applications.

Following the reconstruction of SB, other DNA transposons were shown to be active in mammalian cells, including transposons isolated from other species (such as *piggyBac* from the insect *Trichoplusia ni* (Ding et al., 2005) and *Tol2* from medaka fish (Balciunas et al., 2006)), and another "awakened" Tc1/Mariner transposon called *Frog Prince* (Miskey et al., 2003). However, transposons are naturally not selected for maximal activity in order to minimize the insertional mutagenesis in the host genome and the reconstructed ones were not expected to be the most active forms either. It was then shown that it is possible to create hyperactive versions of these transposons by molecular engineering (Zayed et al., 2004; Baus et al., 2005; Pledger & Coates, 2005), and such variants would likely represent more efficient genetic vehicles.

Recently, a 100 times more active form of SB (SB100x) was created, and its activity was already comparable to the most efficient viral vectors (Mates et al., 2009). The system could also tolerate practically any inserted sequence and the cargo size capacity is less limited than that of viral vectors: it can efficiently move inserts of >8 kb, although transposition efficiency decreases with larger cargo size (Izsvak et al., 2010). Unexpectedly, the amount of the transposase seemed to be a more critical issue: the transposition efficiency paradoxically decreases when the amount of transposase raises beyond a certain level, a phenomenon called overproduction inhibition (Lohe & Hartl, 1996). All of the used DNA transposons seem to share this feature but the careful titration to set up the optimal transposase level provided evidence that among them, SB was the most efficient system in conditions when the amount of transposon DNA is limiting (Grabundzija et al., 2010). This finding made the SB system attractive to many applications where transgene delivery into hard-to-transfect cell types is required, including embryonic stem cells (see **Part 3**). In addition, the lack of endogenous copies in vertebrate (particularly in human) genomes represents an important safety issue as it ensures that the integrated transgenes are not being remobilized (Ivics et al., 1997; Ivics et al., 2004). Such attractive characteristics prognosticated that the SB100x transposon version would likely represent a method of choice when carrying out gene delivery into mammalian cells.

The "technology transfer" from invertebrates to use transposons for genetic manipulations had an immediate effect on mammalian forward genetic screens: SB transposon was successfully used in cancer genetics in order to identify genes involved in certain malignant phenotypes (Carlson et al., 2005; Collier et al., 2005; Collier & Largaespada, 2005; Starr et al., 2009). In addition, various insertional mutagenesis screens could be set up, often exploiting the phenomenon of "local hopping" when the transposon preferentially moves in the vicinity of the donor chromosomal locus (Dupuy et al., 2005; Lu et al., 2007; Takeda et al., 2007; Takeda et al., 2008). This characteristic seems to be a common feature of "cut and paste" transposons and could be very useful for saturation mutagenesis of particular chromosomal regions (Luo et al., 1998; Fischer et al., 2001; Carlson et al., 2003; Carlson & Largaespada, 2005; Keng et al., 2005). For SB, however, this "local hopping" does not appear to be very stringent since the transposition intervals are higher than that of other DNA transposons (Carlson et al., 2003). Nevertheless, the SB transposon was successfully

established as an efficient genetic tool for forward genetics in mammals, similarly to the P element based applications in *Drosophila* (Ryder & Russell, 2003; Miskey et al., 2005).
When considering gene therapy applications in human, however, efficiency is only one of the important issues that must be addressed before a genetic system becomes approved. For example, an important drawback of the efficient viral methods are their non-random integration profile: they integrate favorably into transcription units, often preferably into 5′ regions of active genes (Schroder et al., 2002; Bushman, 2003; VandenDriessche et al., 2003; Wu et al., 2003; Narezkina et al., 2004). To a lesser extent, this adverse site preference is also the characteristic of some transposons, including *piggyBac* (Wilson et al., 2007) and *Tol2* (Grabundzija et al., 2010). On the other hand, *Sleeping Beauty* seems to be a favorable system from this point of view: the integration profile was revealed to be very close to random on the genomic level (Vigdal et al., 2002; Liu et al., 2005; Yant et al., 2005). This important feature significantly lowers the risk of insertional mutagenesis which is beneficial for gene therapy applications.
Other issues include the potential silencing of the transgene which could hinder the applicability of viral vectors (Ellis, 2005). Embryonic stem cells, for instance, have clear molecular defense mechanisms against viral promoter sequences (Meilinger et al., 2009; Rowe et al., 2010). When addressing this question for the SB transposon, it was revealed that the effect of silencing depends rather on the cargo sequence and not on the vector itself (Garrison et al., 2007; Zhu et al., 2010). All these encouraging characteristics further supported the use of SB as a tool for gene therapy and provided the basis for the first clinical trial initiated by a non-viral vector: SB is used in the treatment of a B-lymphoid malignancy by ex vivo genetically modified autologous T-cells (Williams, 2008). The outcome of the trial will apparently provide valuable information on the efficiency and biosafety of transposon based gene delivery and could potentially set new standards in gene therapy application, especially since the use of the SB transposon system was recently shown to be applicable also in embryonic stem cells (Wilber et al., 2007; Orban et al., 2009). This promising scenario undoubtedly played a role in that the SB100x transposase version was recently nominated as the "Molecule of the Year" in 2009 (http://www.biotechniques.com/news/Sleeping-Beauty-named-Molecule-of-the-Year/biotechniques-187068.html?autnID¼191663).

3. Selecting transgene expressing stem cells after transposition

The basis of any successful gene therapy applications is undeniably an efficient and stable gene delivery into stem cells. For this purpose, viral based applications were traditionally favored as viruses have been selected through evolution to efficiently deliver genetic material (DNA or RNA) into their host cells. However, two negative aspects of their usage made non-viral approaches favorable despite their lower efficiency: the biased integration profile of viral vectors, and the limitation of the cargo size due to the packaging constraint of the virus particles. The latter one is an important issue when considering relatively large cDNAs of certain human genes, or the need for simultaneous delivery of more cDNAs into one sample. Although potentially overcoming these problems, the uses of first generation non-viral vectors were clearly several magnitudes less efficient than their peer viral counterparts. With the emerging hyperactive transposons, however, gene delivery efficiency was partly resolved. Nevertheless, the use of any transposon system requires the transfection of DNA into the host cells, and several applications involve cell types that are generally difficult to transfect, including embryonic stem cells. Albeit various transfection protocols are currently available, it is infrequent to achieve higher than 50-80% transfection efficiency without severely affecting cell survival, therefore efficient and preferably non-invasive selection protocols should always be worked out to establish homogenous transgene expressing stem cells following transfection and transposition.

One widely used selection method is to apply chemical selection (e.g. antibiotics) to enrich for transgene expression. This approach usually serves well if cell source is not limiting and when the chemical selection does not significantly perturb cell physiology. Embryonic stem cells, however, represent much more sensitive cell types and if later clinical applications are taken into account, it has to be borne in mind that the precious cells on which the therapy can begin with will most likely come from a limited source. Therefore, a selection must be efficient and at the same time, the least invasive procedure. The commonly used drugs, however, could induce undesired gene expression profiles, or initiate partial differentiation of the stem cells. If chemical selection is inescapable, the cells must always be examined carefully whether they retained their pluripotency status at least by immunostaining for accepted surface markers or preferably also by scrutinizing their differentiation potential (Duan et al., 2007; Tomescot et al., 2007; Orban et al., 2009). Moreover, since the use of a marker gene (eg. an antibiotic resistance gene) evidently means a use of a larger cargo, the overall gene delivery efficiency will decrease. An elegant way of reducing this problem is the use of viral linker peptides between cDNA sequences instead of independent transcription units (see **Part 4**), but the potentially altered genetic profiles still disfavors such selection approaches, if possible.

Another unexpected problem of chemical selection originates from the multidrug resistance phenotype, more precisely, from the presence of MDR-ABC transporters, especially the ABCG2 protein. It is now well established that this multidrug transporter is present in embryonic stem cells and is responsible for the so called "side population" phenotype of a wide variety of tissue-derived stem cells (Zhou et al., 2001; Sarkadi et al., 2010). The exact function of ABCG2 in these cell types is not fully elucidated yet but numerous evidence points to its role to protect these valuable sanctuaries against various noxae by extruding undesired drugs out of the cells (**Figure 2**). Clearly, the presence of such defense mechanism can work against chemical selection since the increase of endogenous ABCG2 expression could work against the enrichment of transgene expressing stem cells, e.g. by pumping out puromycin from stem cells (Takenaka et al., 2007). Also, antibiotic selection may greatly increase MDR-ABC protein expression as a stress-related response (Theile et al., 2010).

Another approach for the enrichment of transgene expression is the use of fluorescent proteins as markers. The genetically modified, fluorescent stem cells can be separated by FACS (Fluorescent Activated Cell Sorting) analysis or positive clumps can be sequestered in sterile conditions using fluorescent microscopy. In our laboratory, we have routinely established different transgene expressing human embryonic stem cell clones, usually combining both methods (**Figure 3**). Although less invasive than chemical selection, this procedure can be more laborious and time consuming. Moreover, as strongly expressing cells have a higher chance of being selected, this method inherently favors stem cells with higher copy numbers which should be considered if a certain application requires low copy number (or even single copy number) clones. Nevertheless, this approach is still less invasive as a chemical selection, even though the problem of delivering another transgene additionally to the selection marker again places a burden by the increase of the cargo size, therefore lowering the efficiency.

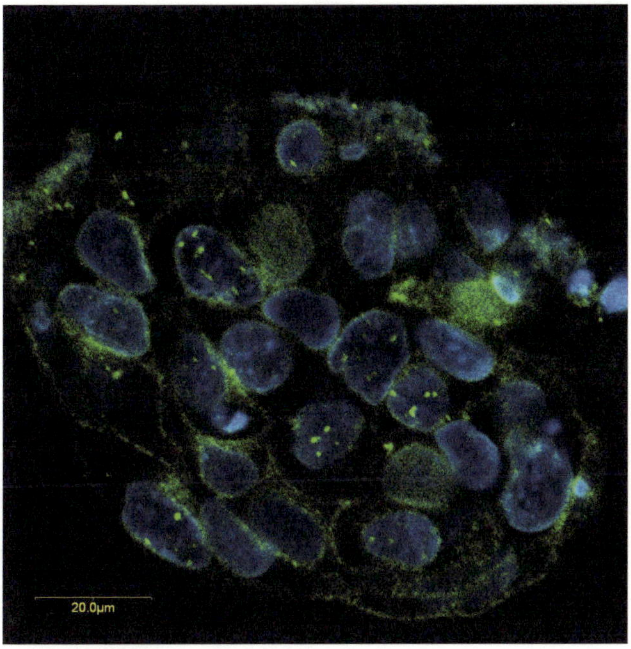

Fig. 2. Expression of the ABCG2 membrane transporter in a human embryonic stem cell clump.
HUES9 cells were immunostained with the 5D3 anti-ABCG2 antibody (green); the Hoechst 33342 dye (blue) was used to visualize cell nuclei. The confocal microscopy image clearly shows that ABCG2 is localized in the plasma membrane of all examined cells. Scale bar represents 20 μm.

To combine the advantages of viral vectors and transposons, several groups attempted to create genetic chimera vehicles, using non-integrating virus forms with the SB system (Bowers et al., 2006; Staunstrup et al., 2009; Vink et al., 2009; de Silva et al., 2010b). This approach can overcome the inefficient delivery often associated with transfection of DNA into certain cell types, and presents the favorable transgene integration profile provided by the SB transposase. Such hybrid vectors certainly open new vistas in gene therapy, although rigorous testing should still be carried out to carefully examine the safety and efficiency of these methods. There is one study claiming that the chimera vector of a Herpes simplex virus and a hyperactive version of SB loses its transposition "hyperactivity" in vivo (de Silva et al., 2010a). However, as this study used an earlier version of the transposase, the results should be carefully repeated with the new SB100x system which provides a far more robust gene delivery in vivo than any previous transposons, therefore could potentially overcome this negative side effect.

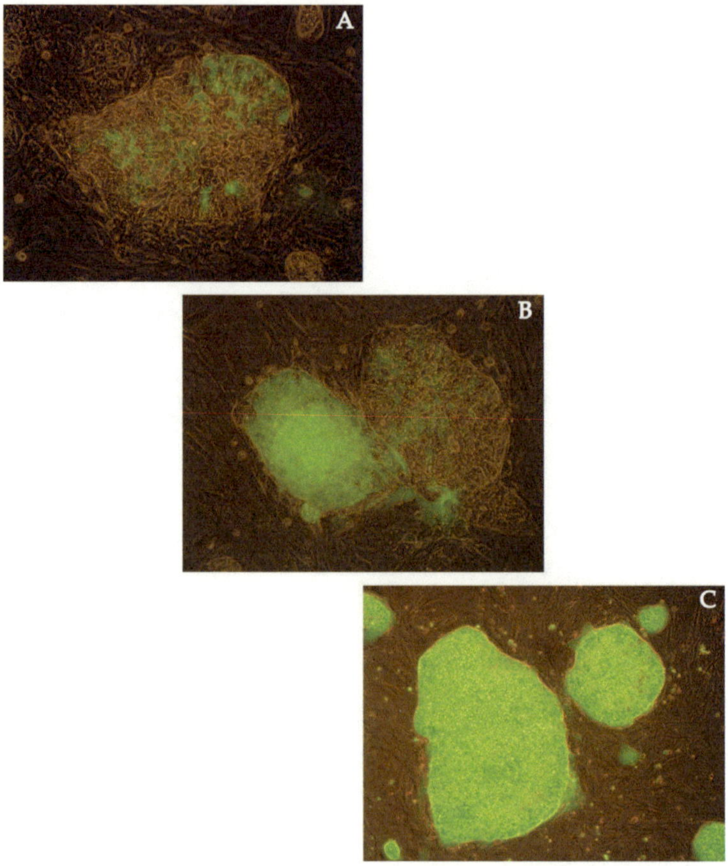

Fig. 3. Cloning of the transgene expressing embryonic stem cells following transposon mediated gene delivery.
(A) GFP expression cassette is delivered into HUES9 cells by the SB transposon system. The image is taken 48 hours post-transfection, showing a heterogeneous cell population of a clump on mouse embryonic feeder cells. (B) Enrichment of transgene expressing cells by manual sequestration of GFP positive cells using fluorescent microscopy. (C) The results of cloning GFP expressing stem cells by FACS from clumps in stage (B). Phase contrast fluorescence microscopy images, x40 magnification.

4. Applications of transposon based gene delivery in embryonic stem cells

HuES cells represent excellent models for cell and tissue differentiation, however, directing the genetic program toward a certain lineage is often challenging. For various applications, such as pharmacological screening models, the aim is to achieve a reasonable enrichment of a given tissue type among the various progeny of cell types. Current methods often apply endogenous morphogenic proteins or invasive chemicals to obtain the tissue(s) of interest, however, the use of such artificial chemical cocktails could have serious side effects, including undesired gene expression profiles and/or distorted differentiation pathways. On the other hand, the alternatively used spontaneous differentiation of HuES cells (e.g. via the embryoid body pathway) is a stochastic process and the efficiency of obtaining a particular cell type is often very low. In addition, some tissue types are difficult to recognize solely by morphological signs, and other - often invasive - molecular identification methods are necessary to apply.

In our laboratory, we are examining cardiovascular differentiation, with an obvious advantage of having a clearly recognizable phenotype at the end of differentiation. Such contracting cell populations can be found following spontaneous differentiation, however, their number can vary significantly. To increase the efficiency of cardiac cell detection, we have developed a method using a specific "double-feature" promoter (Orban et al., 2009). It is based on the unprecedented characteristic of a CAG promoter variant: as being a constitutive promoter, it is expressed in all tissue types which was the reason why it was chosen to drive the expression of a fluorescent reporter gene. On the other hand, the transcriptional activity of this CAG variant becomes extremely high in differentiated cardiomyocytes, providing an excellent platform of selecting these cell types based on the intensity of the fluorescent signal. This behavior is also very useful when a transposon based gene delivery is applied: the "double-feature" promoter is used to identify the transgene expressing undifferentiated cells after transfection and later on, it offers the possibility to select for differentiated cardiomyocytes (**Figure 4**). This approach represents a great advantage because up to our knowledge, no commercially available antibodies against cardiac specific cell surface markers exist which would allow gentle separation of these cell types. Moreover, as this promoter is less prone for silencing (Chung et al., 2002; Xia et al., 2007), the loss of cells containing inactive transgene copies is also significantly reduced. The background of this "double-feature" phenomenon is still under investigations, and deciphering its structural basis could lead to a promising scenario of creating promoters with different tissue specificities. Such achievement would represent a great technological

SB-CAG-GFP **SB-EF1α-GFP**

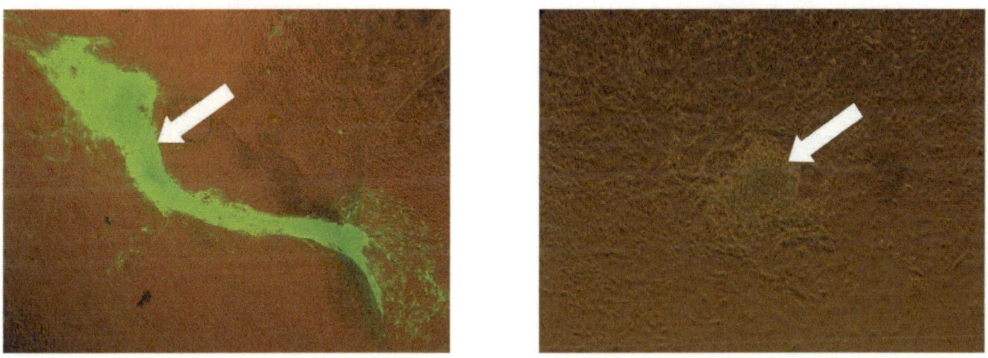

Fig. 4. Using the "double-feature" CAG promoter to visualize cardiomyocytes.

White arrows depict contracting cardiomyocytes arising from spontaneously differentiated HUES9 cell clones expressing either CAG promoter- (SB-CAG-GFP) or EF1α promoter-driven (SB-EF1α-GFP) transgene. Note that in SB-EF1α-GFP cells, the entire population expresses GFP at low level almost uniformly, whereas in SB-CAG-GFP cells, GFP expression in cardiomyocytes is extremely high as compared to the surrounding tissues; see further details in text. Phase contrast fluorescence microscopy images, x40 magnification.

breakthrough as tissues lacking easily recognizable morphological signs could be separated without invasive identification protocols; and the SB transposon based transgene delivery would ensure the lowest possible risk of mutagenesis by its random integration profile.
The technology of generating iPS cells is an obvious example where the use of transposons as genetic vehicles is clearly beneficial due to the large cargo size. For efficient reprogramming of fibroblasts, at least 4 transcription factors need to be overexpressed (Takahashi & Yamanaka, 2006) which places a heavy burden on the otherwise also not very efficient method itself. The use of independent transcription units are clearly represent too large cargos; the use of IRES (Internal Ribosome Entry Site) sequences instead also faces the problem of inefficient and non-equimolar expression of several cDNAs. An elegant way of overcoming this issue is the use of the 2A viral linker peptides: it basically allows to establish a polycistronic mRNA from which separate peptide chains can be translated equimolarly in eukaryotic systems (Szymczak et al., 2004). However, when considering an additional selection marker, the length of such a cargo (~7 kb or higher) still pushes the packaging limits of most viral vectors so the need to deliver such a long transgene calls for the use of transposons. Indeed, the *piggyBac* system was applied to successfully establish pluripotent iPS cell lines (Kaji et al., 2009; Woltjen et al., 2009). Nevertheless, some disadvantages of the *piggyBac* system, e.g. the non-random integration profile (Wilson et al., 2007) or the presence of endogenous elements potentially capable of remobilizing the transgene (Newman et al., 2008) clearly awaits for the use of another, technically safer system such as the SB transposon. In addition, for any gene therapy application constrained by the size of a large human transgene cDNA calls for the application of transposons, preferably the SB system.

5. Conclusion

Embryonic stem cells represent promising new tools in the clinical treatment of various diseases, and in the meantime, they provide emerging new systems in modeling tissue differentiation and pharmacological screens for drug development and toxicity. With the development of laboratory protocols for the maintenance of these cell types, it is also important to work out efficient and biologically safe methods for gene delivery as it often represents the "take-off" point of any successful work with HuES cells. Among the non-viral gene delivery techniques, the hyperactive *Sleeping Beauty* transposon-transposase system represents a particularly attractive method with several advantages. It is a powerful gene delivery methodology with the least currently known genotoxic effects mainly due to its random integration profile at the genomic level. Also, it is a favorable genetic vehicle in terms of cargo capacity, tolerating relatively long transgene sequences. Moreover, as opposed to most viral vectors, the transposon sequences are less prone to epigenetic silencing which also provides the background for a long-term stable transgene expression. The last but not the least argument for the use of this transposon system is the cheaper and easier production of vector DNA, especially when clinical-grade applications are considered. Taken together, all these favorable characteristics definitely make the hyperactive SB system an attractive alternative for any gene therapy purposes, although, as it is the case with any other newly developed techniques, further investigations are still necessary to validate its biological safety in clinical applications.

6. Acknowledgements

Tamás I. Orbán is a recipient of the János Bolyai Scholarship of the Hungarian Academy of Sciences. Scientific research in our laboratory was supported by grants from OTKA (NK72057), ETT (213-09), ES2Heart Jedlik (OM00203/2007), STEMKILL Jedlik (OM00108/2008), KMOP-1.1.2-07/1-2008-0003 and TÁMOP-4.2.2-08/1-2008-0015.

7. References

Balciunas, D.; Wangensteen, K. J.; Wilber, A.; Bell, J.; Geurts, A.; Sivasubbu, S.; Wang, X.; Hackett, P. B.; Largaespada, D. A.; McIvor, R. S. & Ekker, S. C. (2006). Harnessing a high cargo-capacity transposon for genetic applications in vertebrates. *PLoS Genet,* 2(11):e169

Baus, J.; Liu, L.; Heggestad, A. D.; Sanz, S. & Fletcher, B. S. (2005). Hyperactive transposase mutants of the Sleeping Beauty transposon. *Mol Ther,* 12(6):1148-1156

Biemont, C. & Vieira, C. (2006). Genetics: junk DNA as an evolutionary force. *Nature,* 443(7111):521-524

Bowers, W. J.; Mastrangelo, M. A.; Howard, D. F.; Southerland, H. A.; Maguire-Zeiss, K. A. & Federoff, H. J. (2006). Neuronal precursor-restricted transduction via in utero CNS gene delivery of a novel bipartite HSV amplicon/transposase hybrid vector. *Mol Ther,* 13(3):580-588

Bushman, F. D. (2003). Targeting survival: integration site selection by retroviruses and LTR-retrotransposons. *Cell,* 115(2):135-138

Bussolati, B. & Camussi, G. (2006). Adult stem cells and renal repair. *J Nephrol,* 19(6):706-709

Bussolati, B. & Camussi, G. (2007). Stem cells in acute kidney injury. *Contrib Nephrol,* 156:250-258

Carlson, C. M.; Dupuy, A. J.; Fritz, S.; Roberg-Perez, K. J.; Fletcher, C. F. & Largaespada, D. A. (2003). Transposon mutagenesis of the mouse germline. *Genetics,* 165(1):243-256

Carlson, C. M.; Frandsen, J. L.; Kirchhof, N.; McIvor, R. S. & Largaespada, D. A. (2005). Somatic integration of an oncogene-harboring Sleeping Beauty transposon models liver tumor development in the mouse. *Proc Natl Acad Sci U S A,* 102(47):17059-17064

Carlson, C. M. & Largaespada, D. A. (2005). Insertional mutagenesis in mice: new perspectives and tools. *Nat Rev Genet,* 6(7):568-580

Chung, S.; Andersson, T.; Sonntag, K. C.; Bjorklund, L.; Isacson, O. & Kim, K. S. (2002). Analysis of different promoter systems for efficient transgene expression in mouse embryonic stem cell lines. *Stem Cells,* 20(2):139-145

Collier, L. S.; Carlson, C. M.; Ravimohan, S.; Dupuy, A. J. & Largaespada, D. A. (2005). Cancer gene discovery in solid tumours using transposon-based somatic mutagenesis in the mouse. *Nature,* 436(7048):272-276

Collier, L. S. & Largaespada, D. A. (2005). Hopping around the tumor genome: transposons for cancer gene discovery. *Cancer Res,* 65(21):9607-9610

Collier, L. S. & Largaespada, D. A. (2007). Transposable elements and the dynamic somatic genome. *Genome Biol,* 8 Suppl 1:S5

de Silva, S.; Mastrangelo, M. A.; Lotta Jr, L. T.; Burris, C. A.; Izsvak, Z.; Ivics, Z. & Bowers, W. J. (2010a). HSV/Sleeping Beauty Vector-based Embryonic Gene Transfer Using the HSB5 Mutant: Loss of Apparent Transposition "Hyperactivity" In Vivo. *Hum Gene Ther*, 21(11):1603-1313

de Silva, S.; Mastrangelo, M. A.; Lotta, L. T., Jr.; Burris, C. A.; Federoff, H. J. & Bowers, W. J. (2010b). Extending the transposable payload limit of Sleeping Beauty (SB) using the Herpes Simplex Virus (HSV)/SB amplicon-vector platform. *Gene Ther,* 17(3):424-431

Ding, S.; Wu, X.; Li, G.; Han, M.; Zhuang, Y. & Xu, T. (2005). Efficient transposition of the piggyBac (PB) transposon in mammalian cells and mice. *Cell,* 122(3):473-483

Duan, Y.; Catana, A.; Meng, Y.; Yamamoto, N.; He, S.; Gupta, S.; Gambhir, S. S. & Zern, M. A. (2007). Differentiation and enrichment of hepatocyte-like cells from human embryonic stem cells in vitro and in vivo. *Stem Cells,* 25(12):3058-3068

Dupuy, A. J.; Akagi, K.; Largaespada, D. A.; Copeland, N. G. & Jenkins, N. A. (2005). Mammalian mutagenesis using a highly mobile somatic Sleeping Beauty transposon system. *Nature,* 436(7048):221-226

Ellis, J. (2005). Silencing and variegation of gammaretrovirus and lentivirus vectors. *Hum Gene Ther,* 16(11):1241-1246

Feschotte, C. & Pritham, E. J. (2007). DNA transposons and the evolution of eukaryotic genomes. *Annu Rev Genet,* 41:331-368

Fischer, S. E.; Wienholds, E. & Plasterk, R. H. (2001). Regulated transposition of a fish transposon in the mouse germ line. *Proc Natl Acad Sci U S A,* 98(12):6759-6764

Garrison, B. S.; Yant, S. R.; Mikkelsen, J. G. & Kay, M. A. (2007). Postintegrative gene silencing within the Sleeping Beauty transposition system. *Mol Cell Biol,* 27(24):8824-8833

Goodier, J. L. & Kazazian, H. H., Jr. (2008). Retrotransposons revisited: the restraint and rehabilitation of parasites. *Cell,* 135(1):23-35

Grabundzija, I.; Irgang, M.; Mates, L.; Belay, E.; Matrai, J.; Gogol-Doring, A.; Kawakami, K.; Chen, W.; Ruiz, P.; Chuah, M. K.; VandenDriessche, T.; Izsvak, Z. & Ivics, Z. (2010). Comparative analysis of transposable element vector systems in human cells. *Mol Ther,* 18(6):1200-1209

Hedges, D. J. & Batzer, M. A. (2005). From the margins of the genome: mobile elements shape primate evolution. *Bioessays,* 27(8):785-794

Ivics, Z.; Hackett, P. B.; Plasterk, R. H. & Izsvak, Z. (1997). Molecular reconstruction of Sleeping Beauty, a Tc1-like transposon from fish, and its transposition in human cells. *Cell,* 91(4):501-510

Ivics, Z. & Izsvak, Z. (2004). Transposable elements for transgenesis and insertional mutagenesis in vertebrates: a contemporary review of experimental strategies. *Methods Mol Biol,* 260:255-276

Ivics, Z. & Izsvak, Z. (2006). Transposons for gene therapy! *Curr Gene Ther,* 6(5):593-607

Ivics, Z.; Kaufman, C. D.; Zayed, H.; Miskey, C.; Walisko, O. & Izsvak, Z. (2004). The Sleeping Beauty transposable element: evolution, regulation and genetic applications. *Curr Issues Mol Biol,* 6(1):43-55

Izsvak, Z.; Hackett, P. B.; Cooper, L. J. & Ivics, Z. (2010). Translating Sleeping Beauty transposition into cellular therapies: victories and challenges. *Bioessays,* 32(9):756-767

Izsvak, Z. & Ivics, Z. (2004). Sleeping beauty transposition: biology and applications for molecular therapy. *Mol Ther,* 9(2):147-156

Izsvak, Z.; Ivics, Z. & Plasterk, R. H. (2000). Sleeping Beauty, a wide host-range transposon vector for genetic transformation in vertebrates. *J Mol Biol,* 302(1):93-102

Kaji, K.; Norrby, K.; Paca, A.; Mileikovsky, M.; Mohseni, P. & Woltjen, K. (2009). Virus-free induction of pluripotency and subsequent excision of reprogramming factors. *Nature,* 458(7239):771-775

Kazazian, H. H., Jr. (2004). Mobile elements: drivers of genome evolution. *Science,* 303(5664):1626-1632

Keng, V. W.; Yae, K.; Hayakawa, T.; Mizuno, S.; Uno, Y.; Yusa, K.; Kokubu, C.; Kinoshita, T.; Akagi, K.; Jenkins, N. A.; Copeland, N. G.; Horie, K. & Takeda, J. (2005). Region-specific saturation germline mutagenesis in mice using the Sleeping Beauty transposon system. *Nat Methods,* 2(10):763-769

Liu, G.; Geurts, A. M.; Yae, K.; Srinivasan, A. R.; Fahrenkrug, S. C.; Largaespada, D. A.; Takeda, J.; Horie, K.; Olson, W. K. & Hackett, P. B. (2005). Target-site preferences of Sleeping Beauty transposons. *J Mol Biol,* 346(1):161-173

Lohe, A. R. & Hartl, D. L. (1996). Autoregulation of mariner transposase activity by overproduction and dominant-negative complementation. *Mol Biol Evol,* 13(4):549-555

Lu, B.; Geurts, A. M.; Poirier, C.; Petit, D. C.; Harrison, W.; Overbeek, P. A. & Bishop, C. E. (2007). Generation of rat mutants using a coat color-tagged Sleeping Beauty transposon system. *Mamm Genome,* 18(5):338-346

Luo, G.; Ivics, Z.; Izsvak, Z. & Bradley, A. (1998). Chromosomal transposition of a Tc1/mariner-like element in mouse embryonic stem cells. *Proc Natl Acad Sci U S A,* 95(18):10769-10773

Mates, L.; Chuah, M. K.; Belay, E.; Jerchow, B.; Manoj, N.; Acosta-Sanchez, A.; Grzela, D. P.; Schmitt, A.; Becker, K.; Matrai, J.; Ma, L.; Samara-Kuko, E.; Gysemans, C.; Pryputniewicz, D.; Miskey, C.; Fletcher, B.; Vandendriessche, T.; Ivics, Z. & Izsvak, Z. (2009). Molecular evolution of a novel hyperactive Sleeping Beauty transposase enables robust stable gene transfer in vertebrates. *Nat Genet,* 41(6):753-761

Mates, L.; Izsvak, Z. & Ivics, Z. (2007). Technology transfer from worms and flies to vertebrates: transposition-based genome manipulations and their future perspectives. *Genome Biol,* 8 Suppl 1:S1

Meilinger, D.; Fellinger, K.; Bultmann, S.; Rothbauer, U.; Bonapace, I. M.; Klinkert, W. E.; Spada, F. & Leonhardt, H. (2009). Np95 interacts with de novo DNA methyltransferases, Dnmt3a and Dnmt3b, and mediates epigenetic silencing of the viral CMV promoter in embryonic stem cells. *EMBO Rep,* 10(11):1259-1264

Mills, R. E.; Bennett, E. A.; Iskow, R. C. & Devine, S. E. (2007). Which transposable elements are active in the human genome? *Trends Genet,* 23(4):183-191

Mills, R. E.; Bennett, E. A.; Iskow, R. C.; Luttig, C. T.; Tsui, C.; Pittard, W. S. & Devine, S. E. (2006). Recently mobilized transposons in the human and chimpanzee genomes. *Am J Hum Genet,* 78(4):671-679

Miskey, C.; Izsvak, Z.; Kawakami, K. & Ivics, Z. (2005). DNA transposons in vertebrate functional genomics. *Cell Mol Life Sci,* 62(6):629-641

Miskey, C.; Izsvak, Z.; Plasterk, R. H. & Ivics, Z. (2003). The Frog Prince: a reconstructed transposon from Rana pipiens with high transpositional activity in vertebrate cells. *Nucleic Acids Res,* 31(23):6873-6881

Narezkina, A.; Taganov, K. D.; Litwin, S.; Stoyanova, R.; Hayashi, J.; Seeger, C.; Skalka, A. M. & Katz, R. A. (2004). Genome-wide analyses of avian sarcoma virus integration sites. *J Virol,* 78(21):11656-11663

Newman, J. C.; Bailey, A. D.; Fan, H. Y.; Pavelitz, T. & Weiner, A. M. (2008). An abundant evolutionarily conserved CSB-PiggyBac fusion protein expressed in Cockayne syndrome. *PLoS Genet,* 4(3):e1000031

Orban, T. I.; Apati, A.; Nemeth, A.; Varga, N.; Krizsik, V.; Schamberger, A.; Szebenyi, K.; Erdei, Z.; Varady, G.; Karaszi, E.; Homolya, L.; Nemet, K.; Gocza, E.; Miskey, C.; Mates, L.; Ivics, Z.; Izsvak, Z. & Sarkadi, B. (2009). Applying a "double-feature" promoter to identify cardiomyocytes differentiated from human embryonic stem cells following transposon-based gene delivery. *Stem Cells,* 27(5):1077-1087

Ostertag, E. M.; Madison, B. B. & Kano, H. (2007). Mutagenesis in rodents using the L1 retrotransposon. *Genome Biol,* 8 Suppl 1:S16

Pledger, D. W. & Coates, C. J. (2005). Mutant Mos1 mariner transposons are hyperactive in Aedes aegypti. *Insect Biochem Mol Biol,* 35(10):1199-1207

Reubinoff, B. E.; Pera, M. F.; Fong, C. Y.; Trounson, A. & Bongso, A. (2000). Embryonic stem cell lines from human blastocysts: somatic differentiation in vitro. *Nat Biotechnol,* 18(4):399-404

Rowe, H. M.; Jakobsson, J.; Mesnard, D.; Rougemont, J.; Reynard, S.; Aktas, T.; Maillard, P. V.; Layard-Liesching, H.; Verp, S.; Marquis, J.; Spitz, F.; Constam, D. B. & Trono, D. (2010). KAP1 controls endogenous retroviruses in embryonic stem cells. *Nature,* 463(7278):237-240

Ryder, E. & Russell, S. (2003). Transposable elements as tools for genomics and genetics in Drosophila. *Brief Funct Genomic Proteomic,* 2(1):57-71

Sarkadi, B.; Orban, T. I.; Szakacs, G.; Varady, G.; Schamberger, A.; Erdei, Z.; Szebenyi, K.; Homolya, L. & Apati, A. (2010). Evaluation of ABCG2 expression in human embryonic stem cells: crossing the same river twice? *Stem Cells,* 28(1):174-176

Satija, N. K.; Gurudutta, G. U.; Sharma, S.; Afrin, F.; Gupta, P.; Verma, Y. K.; Singh, V. K. & Tripathi, R. P. (2007). Mesenchymal stem cells: molecular targets for tissue engineering. *Stem Cells Dev,* 16(1):7-23

Schroder, A. R.; Shinn, P.; Chen, H.; Berry, C.; Ecker, J. R. & Bushman, F. (2002). HIV-1 integration in the human genome favors active genes and local hotspots. *Cell,* 110(4):521-529

Starr, T. K.; Allaei, R.; Silverstein, K. A.; Staggs, R. A.; Sarver, A. L.; Bergemann, T. L.; Gupta, M.; O'Sullivan, M. G.; Matise, I.; Dupuy, A. J.; Collier, L. S.; Powers, S.; Oberg, A. L.; Asmann, Y. W.; Thibodeau, S. N.; Tessarollo, L.; Copeland, N. G.; Jenkins, N. A.; Cormier, R. T. & Largaespada, D. A. (2009). A transposon-based genetic screen in mice identifies genes altered in colorectal cancer. *Science,* 323(5922):1747-1750

Staunstrup, N. H.; Moldt, B.; Mates, L.; Villesen, P.; Jakobsen, M.; Ivics, Z.; Izsvak, Z. & Mikkelsen, J. G. (2009). Hybrid lentivirus-transposon vectors with a random integration profile in human cells. *Mol Ther,* 17(7):1205-1214

Szymczak, A. L.; Workman, C. J.; Wang, Y.; Vignali, K. M.; Dilioglou, S.; Vanin, E. F. & Vignali, D. A. (2004). Correction of multi-gene deficiency in vivo using a single 'self-cleaving' 2A peptide-based retroviral vector. *Nat Biotechnol,* 22(5):589-594

Takahashi, K. & Yamanaka, S. (2006). Induction of pluripotent stem cells from mouse embryonic and adult fibroblast cultures by defined factors. *Cell,* 126(4):663-676

Takeda, J.; Izsvak, Z. & Ivics, Z. (2008). Insertional mutagenesis of the mouse germline with Sleeping Beauty transposition. *Methods Mol Biol,* 435:109-125

Takeda, J.; Keng, V. W. & Horie, K. (2007). Germline mutagenesis mediated by Sleeping Beauty transposon system in mice. *Genome Biol,* 8 Suppl 1:S14

Takenaka, K.; Morgan, J. A.; Scheffer, G. L.; Adachi, M.; Stewart, C. F.; Sun, D.; Leggas, M.; Ejendal, K. F.; Hrycyna, C. A. & Schuetz, J. D. (2007). Substrate overlap between Mrp4 and Abcg2/Bcrp affects purine analogue drug cytotoxicity and tissue distribution. *Cancer Res,* 67(14):6965-6972

Theile, D.; Staffen, B. & Weiss, J. (2010). ATP-binding cassette transporters as pitfalls in selection of transgenic cells. *Anal Biochem,* 399(2):246-250

Thomson, J. A.; Itskovitz-Eldor, J.; Shapiro, S. S.; Waknitz, M. A.; Swiergiel, J. J.; Marshall, V. S. & Jones, J. M. (1998). Embryonic stem cell lines derived from human blastocysts. *Science,* 282(5391):1145-1147

Tomescot, A.; Leschik, J.; Bellamy, V.; Dubois, G.; Messas, E.; Bruneval, P.; Desnos, M.; Hagege, A. A.; Amit, M.; Itskovitz, J.; Menasche, P. & Puceat, M. (2007). Differentiation in vivo of cardiac committed human embryonic stem cells in postmyocardial infarcted rats. *Stem Cells,* 25(9):2200-2205

Uren, A. G.; Kool, J.; Berns, A. & van Lohuizen, M. (2005). Retroviral insertional mutagenesis: past, present and future. *Oncogene,* 24(52):7656-7672

VandenDriessche, T.; Collen, D. & Chuah, M. K. (2003). Biosafety of onco-retroviral vectors. *Curr Gene Ther,* 3(6):501-515

Vieyra, D. S.; Jackson, K. A. & Goodell, M. A. (2005). Plasticity and tissue regenerative potential of bone marrow-derived cells. *Stem Cell Rev,* 1(1):65-69

Vigdal, T. J.; Kaufman, C. D.; Izsvak, Z.; Voytas, D. F. & Ivics, Z. (2002). Common physical properties of DNA affecting target site selection of sleeping beauty and other Tc1/mariner transposable elements. *J Mol Biol,* 323(3):441-452

Vink, C. A.; Gaspar, H. B.; Gabriel, R.; Schmidt, M.; McIvor, R. S.; Thrasher, A. J. & Qasim, W. (2009). Sleeping beauty transposition from nonintegrating lentivirus. *Mol Ther,* 17(7):1197-1204

Wicker, T.; Sabot, F.; Hua-Van, A.; Bennetzen, J. L.; Capy, P.; Chalhoub, B.; Flavell, A.; Leroy, P.; Morgante, M.; Panaud, O.; Paux, E.; SanMiguel, P. & Schulman, A. H. (2007). A unified classification system for eukaryotic transposable elements. *Nat Rev Genet,* 8(12):973-982

Wilber, A.; Linehan, J. L.; Tian, X.; Woll, P. S.; Morris, J. K.; Belur, L. R.; McIvor, R. S. & Kaufman, D. S. (2007). Efficient and stable transgene expression in human embryonic stem cells using transposon-mediated gene transfer. *Stem Cells,* 25(11):2919-2927

Williams, D. A. (2008). Sleeping beauty vector system moves toward human trials in the United States. *Mol Ther,* 16(9):1515-1516

Wilson, M. H.; Coates, C. J. & George, A. L., Jr. (2007). PiggyBac transposon-mediated gene transfer in human cells. *Mol Ther,* 15(1):139-145

Woltjen, K.; Michael, I. P.; Mohseni, P.; Desai, R.; Mileikovsky, M.; Hamalainen, R.; Cowling, R.; Wang, W.; Liu, P.; Gertsenstein, M.; Kaji, K.; Sung, H. K. & Nagy, A. (2009). piggyBac transposition reprograms fibroblasts to induced pluripotent stem cells. *Nature,* 458(7239):766-770

Wu, X.; Li, Y.; Crise, B. & Burgess, S. M. (2003). Transcription start regions in the human genome are favored targets for MLV integration. *Science,* 300(5626):1749-1751

Xia, X.; Zhang, Y.; Zieth, C. R. & Zhang, S. C. (2007). Transgenes delivered by lentiviral vector are suppressed in human embryonic stem cells in a promoter-dependent manner. *Stem Cells Dev,* 16(1):167-176

Yant, S. R.; Wu, X.; Huang, Y.; Garrison, B.; Burgess, S. M. & Kay, M. A. (2005). High-resolution genome-wide mapping of transposon integration in mammals. *Mol Cell Biol,* 25(6):2085-2094

Zayed, H.; Izsvak, Z.; Walisko, O. & Ivics, Z. (2004). Development of hyperactive sleeping beauty transposon vectors by mutational analysis. *Mol Ther,* 9(2):292-304

Zhou, S.; Schuetz, J. D.; Bunting, K. D.; Colapietro, A. M.; Sampath, J.; Morris, J. J.; Lagutina, I.; Grosveld, G. C.; Osawa, M.; Nakauchi, H. & Sorrentino, B. P. (2001). The ABC transporter Bcrp1/ABCG2 is expressed in a wide variety of stem cells and is a molecular determinant of the side-population phenotype. *Nat Med,* 7(9):1028-1034

Zhu, J.; Park, C. W.; Sjeklocha, L.; Kren, B. T. & Steer, C. J. (2010). High-level genomic integration, epigenetic changes, and expression of sleeping beauty transgene. *Biochemistry,* 49(7):1507-1521

Part 4

Methods for Genetic Delivery, Manipulation and Assessment in Embryonic Stem Cells

15

Gene Transduction Approaches in Human Embryonic Stem Cells

David Brafman and Karl Willert
University of California, San Diego
La Jolla, California,
USA

1. Introduction

Human pluripotent stem cells (hPSCs, which include both human embryonic and induced pluripotent stem cells [hESCs and hiPSCs]) provide a unique model system to study early human development, derive functionally mature cell populations, and hold great promise to advance medical treatments for currently incurable diseases. Modulating gene function in these cells is of critical importance to gain insight into the molecular underpinnings of the pluripotent state and the process of differentiation. Furthermore, efficient transgenesis in hPSC is essential in generating cells carrying specific disease-associated alleles, thereby enabling the study of human diseases at the cellular level.

In contrast to murine ESCs, hESCs have proven quite recalcitrant to gene modification using many of the traditional methods. The reason(s) for this difference between mESCs and hESCs are poorly understood, but it has been postulated that hESCs represent a slightly different stage with distinct developmental potential than mESCs and that this difference accounts for the differences observed in transgenesis and homologous recombination (Buecker et al., 2010).

DNA can be introduced into hPSCs using viral and non-viral methods, yielding both transient and stable transduction. Stable integration into the genome can occur at random sites, thus potentially disrupting endogenous gene expression or function. In addition, transduced cell lines can be selected to carry targeted gene insertions through site-specific recombination or homologous recombination. In this chapter, we describe several methods to transduce genetic elements into hPSCs and discuss their strengths and weaknesses.

2. Random transgenesis

Several methods have been used successfully to randomly integrate DNA into the genome of hPSCs and thereby generating marker lines that express reporter genes in a cell- or tissue-specific manner. The most commonly used methods to transduce DNA into cells, either transiently or stably, involve chemical-based transfection reagents, electroporation, or viral infection. The advantages and limitations of each transgenic method will be discussed in the following section. In general, transgenic approaches to create reporter lines are limited by the fact that regulatory elements controlling cell-specific gene expression are often either poorly defined or located at great distances from the gene. To faithfully reproduce a gene expression pattern may require DNA fragment lengths that exceed conventional recombinant DNA methods. In addition, stably transduced reporter genes may be under control of endogenous regulatory elements near the sight of integration. For example, genetic integration into certain compact regions of the genome can lead to silencing or lower levels of expression due to reduced accessibility of transcription factors. Another caveat of using random insertional transgenesis is that in rare instances gene insertion may interrupt expression of other genes that influence cell phenotype. As a result of these positional effects, screening and analysis of multiple clones for reporter gene expression patterns over time and during differentiation must be performed. Unfortunately, generation of single-cell clones is often difficult given the low cloning efficiency of hESCs (Amit et al., 2000). Recently, this challenge has been partially mitigated by the use of neurotrophins (such as NT3 and NT4) (Pyle et al., 2006) or selective inhibitors of Rho-associated kinase ROCK (e.g. Y-27632) (Damoiseaux et al., 2009; Watanabe et al., 2007), both of which increase survival of dissociated hESCs.

The methods described in this section have been used to introduce transgenes encoding fluorescent proteins or drug resistance markers under the transcriptional control of cell-specific promoters into hPSCs. Specific sub-populations of cells expressing the reporter gene can be enriched and isolated using drug selection or flow cytometry-based cell sorting.

2.1 Electroporation

Electroporation involves the use of electric pulses of varying voltage, length, and number to transiently increase the permeability of the cell membrane and permit entry of DNA. Several groups have reported the use of electroporation for the transfection of hESCs (Eiges et al., 2001; Lakshmipathy et al., 2004; Mohr et al., 2006; Zwaka and Thomson, 2003). Using a plasmid containing a gene encoding green fluorescent protein (GFP) under control of a cytomegalovirus immediate-early gene 1 (CMV) promoter/enhancer, Mohr et al. investigated the electroporation parameters of pulse, voltage, duration, and number on hESC viability and transfection efficiency (Mohr et al., 2006). A single 300V, 0.5 millisecond pulse achieved an optimum survival rate of 50% and a transient transfection rate of 10%. These values were consistent with other reported transfection efficiencies using electroporation (Lakshmipathy et al., 2004). It is important to note that this study focused on the transient transfection of plasmid DNA. In contrast, stable transfection efficiencies are much lower because of the low rate of DNA integration into the genome. Typically, transient transgene expression in hESCs is not retained for longer than 5 passages (Vallier et al., 2004). By comparison, the yield of stably transfected clones using electroporation has been reported to be on the order of 1 in 10^{-5} (Adewumi et al., 2007; Zwaka and Thomson, 2003). Moreover, less than 1% of electroporated hESCs in single-cell suspension survive (Zwaka and Thomson, 2003). However, electroporation of hESCs in clumps suspended in standard cell culture medium greatly improved cell survival (Zwaka and Thomson, 2003).

2.2 Nucleofection

Nucleofection is a recently developed and commercialized technology (Amaxa® Nucleofector® Technology, Lonza) that expands on the principles of electroporation. Specifically, nucleofection is a customized procedure in which buffer solutions and electric pulses are optimized for specific cell types (Siemen et al., 2005; Siemen et al., 2008). In contrast to electroporation, in which efficiency is dependent on cell cycle because DNA only enters the nucleus when the nuclear envelope is fragmented during cell division, nucleofection leads to a direct transfer of the DNA into the cell nucleus (Hamm et al., 2002). Nucleofection of a GFP reporter construct into hESCs yielded survival rates of >70% and a transient transgene expression of 66% in surviving cells. Furthermore, transfected cells maintained expression of markers of pluripotency, such as TRA-1-60 and OCT4, and could be expanded to stably expressing clones. Additionally, the small amount of cells and DNA required for nucleofection has allowed it to be a valuable tool for high-throughput genetic manipulation of hESCs (Moore et al., 2010).

2.3 Chemical-based transfection methods

As an alternative to electroporation, chemical transfection reagents (lipid and polymer-based reagents) have been used in attempts to achieve substantial levels of transgene expression in hESCs. The commonly used calcium-phosphate (Ca/PO_4) mediated transfection methods have had limited success in hESCs, due to the cytotoxicity of the Ca/PO_4 precipitate.

An early report compared the efficiency of three different lipofection-based reagents, Lipofectamine (Invitrogen), Fugene (Boehringer Mannnheim) and ExGen 500 (Fermentas), in hESCs by using an expression construct in which GFP was under control of the elongation factor I (EF1α) promoter (Eiges et al., 2001). Transient transfection with ExGen 500 resulted in transgene expression an order of magnitude higher than Lipofectamine or Fugene. However, GFP transgene expression occurred in only 10% of the ExGene500 transfected cells, a transfection efficiency comparable to electroporation. This report conflicts with that of others (Siemen et al., 2005), which state that ExGen 500 was less effective than other reagents such as Lipofectamine and Fugene. One study compared several chemical transfection reagents and confirmed the poor efficiency of ExGen 500 and determined that GeneJammer was the most effective method (Anderson et al., 2007). A recent report describes the development of a class of polymers, poly(β-amino esters), which have higher gene delivery efficiencies because of their ability to condense DNA into nanoparticles, thus enhancing cellular uptake (Green et al., 2008). These nanoparticles have gene delivery efficiencies 4 times higher than other transfection reagents such as Lipofectamine. It is important to note that the reported differences between the efficacies of these reagents could be attributed to differences in the culture systems and individual hESC lines tested. Therefore, some reagents may be better suited for use in specific hESC lines.

In order to achieve stable transfection into hESCs using chemical-based transfection methods, a gene encoding antibiotic-resistance or fluorescent protein must be present in the construct so that stable transfectants can be selected. HESC lines developed employing such methods have been utilized to monitor the differentiation status of hESCs. Eiges et al. used lipofection to transfect hESCs with GFP under the control of a murine Rex1 promoter, a gene that is rapidly downregulated upon differentiation of hESCs (Eiges et al., 2001). The transfected hESCs could be sorted to homogeneity using fluorescent-activated cell sorting (FACS). The purified cell population showed high levels of GFP expression when in an undifferentiated state. As the cells differentiated this expression was significantly reduced. Along similar lines, lipofection was used to generate a hESC line with GFP driven by an OCT4 promoter. Such OCT4-GFP lines have proven useful in monitoring the exit from the undifferentiated state and acquisition of differentiation markers.

Chemical-based transfection methods have also been used to tag and purify specific hESC derivatives. For example, chemical transgenic methods were used to enrich cardiomyocytes using a reporter system in which the human α-myosin promoter drove expression of puromycin–N-acetyltransferase (PAC)(Anderson et al., 2007). Positive selection with puromycin of hESCs differentiated towards cardiomyocytes resulted in a cardiomyocytes enrichment of 14.5 fold (change from 7% to 92% cardiomyocytes). Enriched cardiomyocytes expressed high levels of cardiac specific markers and displayed cardiac-specific action potentials, demonstrating that functionality was not compromised by drug selection. In another example, lipofection was used to transfect hESCs with plasmids encoding GFP placed under the transcriptional control of a motor neuron specific enhancer within the 5'-regulatory region of the gene encoding the transcription factor Hb9. Motor neurons induced *in vitro* with the growth factor SHH and retinoic acid activated Hb9-driven GFP expression, allowing their isolation by FACS. Sorted cells achieved physiological and functional maturation *in vitro*, demonstrating the feasibility of promoter/enhancer-based FACS for the isolation of specific derivatives from hESCs. Chemical transfection has also been used to generate hESC lines with fluorescent reporter transgenes under transcriptional control of murine *albumin* (Lavon and Benvenisty, 2005; Lavon et al., 2004) and *Pdx1* (Lavon et al., 2006) genes in order to identify hESC differentiating toward the hepatic and pancreatic lineages, respectively.

2.4 Lentiviral systems

Lentiviral vectors offer another strategy to stably introduce DNA into hESCs. Pfeifer et al. first reported the use of lenti-viral vectors for high levels of transduction efficiency without transgene silencing in hESCs (Pfeifer et al., 2002). Specifically, almost 100% efficiency was reported at a multiplicity of infection of 50, and transgene expression was maintained over several passages. Several additional groups have reported high-level sustained transgene expression in hESCs and hESC derivatives using similar viral vectors. For example, Gropp et al. used lentiviral vectors to stably transduce a GFP encoding transgene under the control of an EF1α promoter (Gropp et al., 2003). Transduction of hESCs with these lentiviral vectors allowed for stable transgene expression for long-term (38 weeks) undifferentiated culture. Furthermore, transgene expression was not silenced upon differentiation as demonstrated by continuous GFP expression throughout differentiation both *in vitro* (EB formation) and *in vivo* (teratoma formation).

Lentiviral vectors have also been used to identify and select hESC derivative cell populations. For example, Huber et al. generated stable transgenic hESC lines using lentiviral vectors and isolated single-cell clones that expressed a GFP transgene under the transcriptional control of a cardiac specific promoter, human myosin light-chain-2V promoter (Huber et al., 2007). As hESCs differentiated, GFP expressing cells were isolated and purified to near homogeneity using FACS. The GFP-positive cells stained for cardiac specific markers, expressed cardiac-specific genes, displayed cardiac-specific action-potentials, and demonstrated long-term engraftment in rat hearts. A recent report expanded on these methods to engineer hESCs with lentiviral vectors that combined blasticidin, neomycin, and puromycin resistance based drug selection of pure populations of stem cells and cardiomyocytes with constitutive or lineage-specific promoters that control expression of fluorescent proteins. This allowed for cardiomyocytes and their progenitors to be visualized and tracked (Kita-Matsuo et al., 2009). Specifically, hESCs were infected with lenti vectors carrying Brachyury (T) and α-myosin heavy chain (αMHC) promoters driving expression of fluorescent or drug-resistance proteins. HESCs differentiated to early mesoderm and cardiomyocytes were enriched using cell sorting or drug selection. Moreover, a drug selection differentiation protocol yielded 96% pure cardiomyocytes that could be cultured for over 4 months. Additionally, these drug-selected cardiomyocytes exhibited a gene expression profile similar to that of adult human cardiomyocytes and generated force and action potentials consistent with normal fetal cardiomyocytes. Lentiviral approaches have also been used to engineer hESCs with glial fibrillary acidic protein (GFAP) (Dhara et al., 2009) and alpha-fetoprotein (AFP) (Chiao et al., 2008) promoters driving expression of genes encoding fluorescent proteins to enrich for hESCs differentiating towards the neural and hepatic lineages, respectively.

2.5 Comparison of random transgenic methods

Very few studies have directly compared the non-viral and viral gene delivery methods discussed above. A recent study compared the transduction efficiencies and effect on cell viability of each method in several hESC lines using a fusion construct with an ubiquitin promoter driving GFP and firefly luciferase along with a neomyocin selection marker (Cao et al., 2009). Lentiviral transduction demonstrated the highest efficiency (range: 22.4-25.3%) with >95% cell viability. Nucleofection demonstrated significantly lower efficiency (range: 5.8-16.1%) with moderate cell survival (range: 70-75%). Minimal transfection efficiency was observed with electroporation (range: 1.9-2.1%) and lipofection (range: 1.3-1.5%). Electroporation resulted in the lowest cell survival (range: 38-58%) while cell survival with lipofection (>90%) was comparable to that of lentiviral infection. Moreover, lentiviral transduction resulted in the greatest number of hESCs stably expressing the fusion reporter gene.

Even though lentiviral transduction results in the highest efficiency and cell survival of all transgenic approaches, it has several drawbacks that may make use of other transgenic methods more attractive. One major weakness of lentiviral methods is that constructs larger than 6-8 kb package poorly, thereby reducing the efficiency of transduction and limiting the size of DNA that can be used (Kumar et al., 2001). Another disadvantage of lentiviral vectors is that they integrate randomly and thus have the ability to activate or inactivate nearby genes (Kohn et al., 2003). A final concern that may limit the use of lentiviral vectors is that viral transgenes may be silenced in the infected cells. Several studies have shown that gene silencing can occur during propagation of undifferentiated ESCs (Cherry et al., 2000) as well as during differentiation (Laker et al., 1998). Such silencing is likely due to epigenetic modifications of the viral DNA following its integration. Furthermore, studies have demonstrated that transgenes delivered by lentiviral vectors are suppressed in ESCs in a promoter-dependent manner (Hong et al., 2007; Xia et al., 2007). For example, Xia et al. examined the silencing of four ubiquitous promoters in lentiviral vectors driving expression of GFP in hESCs: CMV, hybrid CMV enhancer/chicken β-actin (CAG), phosphoglycerate kinase (PGK) and EF1α (Xia et al., 2007). By calculating the ratio between the percentage of nonfluorescent cells and the GFP copy number per cell, the authors showed that more than 95% of the GFP copies driven by the CMV or CAG promoter and 75% of the GFP copies driven by the EF1α promoter were inactive. Meanwhile, GFP driven by the PGK promoter showed the least extent of suppression with ~55% inactive. Results such as these clearly demonstrate that genes delivered through the use of lentivirus are highly susceptible to gene silencing and that the choice of promoter is critical for long term ubiquitous expression of transgenes. Whether similar gene silencing occurs with cell- and tissue-specific promoters remains to be explored.

3. Site-specific recombination

One way to avoid the problems caused by random integration of transgenes into the genome is the creation of hESCs lines where large, complex genetic elements can be stably introduced into cells at defined chromosomal locations in a single copy. Several studies have identified sites on chromosomes 2, 6, 10, 12, 13, 17, and 21 that are transcriptionally active in undifferentiated hESCs and remain transcriptionally active upon differentiation into all tested mature cell populations (Costa et al., 2005; Irion et al., 2007; Thyagarajan et al., 2008). Two strategies that have been used for site-specific recombination to introduce transgenes into hESCs are the Cre/loxP and phiC3I recombination systems.

3.1 Cre/lox

The Cre protein is a site-specific DNA recombinase that catalyzes the recombination of DNA between *loxP* sequences which contain binding sites for Cre. When cells that have *loxP* sites in their genome express Cre, a recombination event occurs between the *loxP* sites (Bouhassira et al., 1997). Depending on the orientation of the *loxP* sites relative to each other, the recombination reaction can result in an insertion, deletion (direct repeat of two *loxP* sites), inversion (two inverted *loxP* on the same chromosome), or translocation (two *loxP* sties on different chromosomes). Nolden et al. first reported the Cre-mediated recombination of a chromosomally integrated *loxP*-modified allele in hESCs and hESC derived neural progenitors (Nolden et al., 2006). In this system, a constitutively active CAG promoter drove transcription of the *HcRed* (an enhanced version of red fluorescent protein [RFP]) while a neomycin-resistance gene (Neo) was under transcriptional control of the PGK promoter, thus allowing for selection of stable clones. Cre-mediated recombination caused deletion of *HcRed* and *Neo* resulting in the expression of a second reporter GFP. In another study, Cre-mediated recombination was used to insert genes at the human homolog of the mouse *Rosa26* locus (*hROSA26*) (Irion et al., 2007). Specifically, a vector containing a *loxP*-flanked promoterless Neo-resistance gene followed by an inverted RFP variant, tandem dimer RFP (tdRFP), flanked by mutant *lox*P2272 was targeted to the *hROSA26* locus using homologous recombination (see below for further discussion of homologous recombination). The *loxP* and mutant loxP2272 sites were positioned so that after Cre expression, the neomycin resistance cassette was deleted and the tdRFP is inverted, placing it under control of the endogenous *hROSA26* promoter. The use of wildtype and mutant *loxP* allowed for the exchange of the tdRFP cassette with any cDNA of interst at the *hROSA26* locus using Cre recombinase-mediated cassette exchange.

One drawback of the Cre-*loxP* system is that its efficiency depends on the level of Cre expression, which is typically achieved by either transfection or viral methods. As a result, recombination only occurs in a low percentage of the cell population (10-15%). As an alternative strategy to introduce Cre, several groups have reported the use of cell-permeable versions of Cre recombinase that induce recombination by direct protein delivery and lead to higher levels of recombination (>90%) (Jo et al., 2001; Joshi et al., 2002; Will et al., 2002).

3.2 phiC3I

Another method to carry out site-specific recombination in hESCs is the use of bacteriophage phiC3I integrase. Unlike Cre recombinase, the phiC3I catalyzes recombination between non-identical sites which makes recombination unidirectional, ensuring that integrated constructs do not act as substrates for the reverse excision reaction. PhiC3I carries out site-specific recombination between its attachment site (*attB*) and an attachment site (*attP*) in host DNA. Moreover, phiC3I integrases target a small number of loci (pseudo-*attP* sties) in transcriptionally active regions ("hot spots") of the human genome.

Recently, hESC lines were developed in which phiC31 integrase was used to place a target site for the R4 integrase into a pseudo *attP* site, and the R4 integrase was used to place genes of interest into the specific R4 target site (Figure 1) (Liu et al., 2009; Thyagarajan et al., 2008). Specifically, hESCs were co-transfected with a plasmid with a CMV promoter driving expression of phiC3I integrase and a plasmid containing the R4 *attP* target site upstream of a promoter-less Zeocin-resistance gene, as well as a hygromycin-resistance marker and a phiC3I *attB* site (pJTI/Zeo). Individual colonies carrying a R4 *attP* site upstream of the Zeocin selectable marker was achieved with selection with hygromyocin and screened for insertion into one of the genome's pseudo-*att* sites (e.g. chromosomal location 13q32 represents a hot spot). Retargeting these "R4" cell lines was achieved by co-transfecting a plasmid expressing the R4 integrase with a plasmid that had the human EF1α promoter upstream of the complementary R4 *attB* recognition site (pER4B-EG and pER4B-hOG). Recombination between the R4 *attP* and *attB* sites positioned the EF1α promoter upstream of

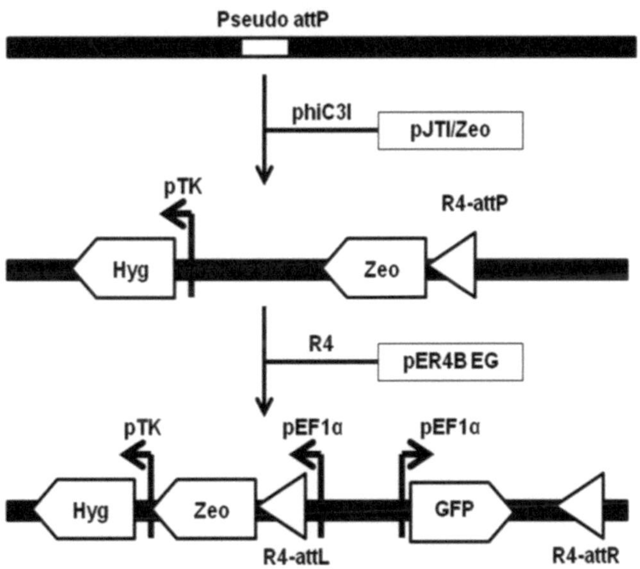

Fig. 1. Site-specific recombination strategy using phiC3I integrase. In the first step a "platform line" is created using phiC3I integrase, which mediates site-specific recombination of the pJTI/Zeo plasmid into one of the host cell genome's pseudo-*att* sites. In a second "retargeting" step, the R4 integrase mediates site-specific recombination of the pER4B vector with the previously inserted *att* site. Proper R4 recombination events lead to the transcriptional activation of the Zeocin-resistance gene. Figure and legend adapted from Liu et al., 2009.

the Zeocin-resistance gene. Thus, only clones that had undergone a site-specific recombination event were resistant to Zeocin. This method has been used to place cassettes with GFP driven by promoters such as EF1α and OCT4 in specific loci in hESCs. Retargeting was extremely efficient with an average of 17 colonies per 10^6 transfected cells with all Zeocin resistant clones containing a single integration event occurring at the chromosome 13q23 hot spot.

A major disadvantage of the phiC3I site-specific integration techniques is that it requires the generation of "platform" hESC lines (e.g. the "R4" cell line). Although retargeting efficiency is extremely high (100% after Zeomycin selection), initial targeting to one of the pseudo-*att* sites is inefficient and requires the screening of hundreds of clones. Given that hESC lines vary in their behavior and propensity to differentiate into various lineages (Osafune et al., 2008), multiple "platform" lines would need to be generated.

4. Homologous recombination

Homologous recombination (HR) has long been a powerful "reverse genetics" approach to interrogate gene function in mice. Methods for HR exploit the endogenous DNA repair and recombination machinery to alter specific sequences within the genome by aligning the engineered transduced DNA sequences with genomic sequences. The outcome can include single base pair mutations, deletion of entire exons or open reading frames, and insertion of sequences that encode protein tags or reporter proteins, such as fluorescent proteins. In contrast to other genetic modifications described above, HR faithfully reproduces expression patterns of genes with largely insignificant effects on expression of neighboring genes.

There are three main reasons to target genes by HR in hESCs. First, hESC lines can be designed to express a reporter gene under the control of an endogenous and developmentally regulated promoter. Such marker lines are useful to monitor and interrogate the developmental progression from undifferentiated to mature cell states and to optimize protocols for efficient and directed differentiation. By using cell isolation and purification methods, such as FACS, even extremely rare and specialized cell types can be isolated and separated from other cell types, especially from undifferentiated hESC that carry the potential to generate tumors. Such purified cell populations are essential for cell transplantation studies and eventual cell replacement therapies. Second, efficient HR can be used to attach tags onto genes so that the function of their protein products can be studied *in vitro* and *in vivo*. By modifying genes with tags that permit detection and purification, endogenous protein function can be explored without relying on current overexpression approaches that are prone to yield artifactual results. Finally, HR can be employed to introduce specific mutations in disease-associated genes to develop models of human disease. Currently, the majority of research to study human diseases in cell culture utilize hiPSCs derived from patients with specific genetic defects or predispositions. However, methods for iPSC generation are still wrought with potential complications, e.g. the various transgenes required for reprogramming may influence disease progression. Efficient HR in hPSCs would provide a powerful alternative method to insert mutations in disease associated genes and model human diseases.

In comparison to murine ESCs, a limited, yet gradually expanding number of genes has been modified by HR in hESCs (see Table 1 for examples). One reason for the limited success of HR is that hPSCs, in contrast to murine ESCs, grow poorly as single cells, a necessary condition to isolate the cells carrying the rare HR events. This problem can be partially overcome by treating dissociated cells with ROCK inhibitor, which has been shown to increase survival of dissociated hESCs (Damoiseaux et al., 2009; Watanabe et al., 2007). Most reports of HR in hESCs have targeted the X-chromosome linked *HPRT1* (*Hypoxanthine-guanine phosphoribosyltransferase*) locus, a gene which is inherently easier to target since deletion of the single copy of HPRT in male cells yields cells that are resistant to the drug 6-Thioguinine (6-TG). While these studies provide compelling proof-of-principle, methods for HR in hESC are far from optimized. In this section, we discuss four methods that show great promise in generating HR in hESCs.

4.1 Conventional HR method

The conventional HR method in mESCs utilizes standard transduction methods, such as electroporation (see descriptions above), to introduce linearized DNA constructs comprised of homology arms of 3 to 5 kilo bases (kb) flanking drug selection cassettes, such as Neomycin or Hygromycin resistance genes. These drug resistance genes are placed between the homology arms to disrupt or delete one or more exons of the targeted gene. Positive drug selection alone yields both targeted and random insertions into the genome. To increase the frequency of targeted insertions, many strategies incorporate a negative selectable marker, generally Herpes simplex virus thymidine kinase (HSV-TK), which confers sensitivity to Gancyclovir (Ganc), inserted distally to one of the homology arms (Mansour et al., 1988). Cells in which a random insertion has occurred will have retained HSV-TK and will consequently be eliminated in the presence of Ganc. Cells that are resistant to Ganc have not integrated the HSV-TK gene, thus potentially representing a HR event. HR rates, which vary between <0.1-5%, are generally indicated as a percentage of cells carrying a targeted insertion (as determined by Southern blot or PCR) among all Neomycin resistant clones. To further increase targeting efficiencies it is critical to utilize isogenic DNA, i.e. the DNA used to create the homology arms in the targeting vector matches the DNA of the cells used for targeting. While this is possible with the various in-bred mouse strains, the large number of single nucleotide polymorphisms (SNP) in the human population (one every 1,000 base pairs) rule out an approach that utilizes isogenic DNA. However, it is noteworthy that two reports found that HR in humans is efficient even with non-isogenic DNA (Sedivy et al., 1999; Urbach et al., 2004). The need for isogenic DNA is overcome as the length of the homology arms increases, however, plasmids of greater than 15 kb are difficult to engineer and propagate. This size restriction is overcome with the use of bacterial artificial chromosomes (see below). Due to the low targeting frequencies observed for many loci, this conventional HR method rarely succeeds at disrupting both copies of a gene. To date only a small number of genes have been successfully targeted in hESCs using positive-negative selection methods (see Table 1).

Gene	hESC line	HR Method	Efficiency	Reference
ATM	HUES9, H9	BAC	21% (3/14) for first allele, 27% (10/37) for second allele	(Song et al., 2010)
FEZF2	HUES9	Conventional positive (Neo)-negative (HSV-TK) selection	1.5% (2/130 clones)	(Ruby and Zheng, 2009)
HMGA1	BG01, BG02, H1	AAV	96% (24 of 25 Hyg^R clones)	(Khan et al., 2010)

HPRT1	H1	Conventional positive (Neo)-negative (6TG) selection	2% (7 6-TG^R out of 350 G418^R)	(Zwaka and Thomson, 2003)
	H13	Conventional positive (Hyg)-negative (6TG) selection	~1-2%	(Urbach et al., 2004)
	H1	Conventional positive (Neo)-negative (6TG) selection	2.3% (6 6TG^R out of 260 G418^R)	(Di Domenico et al., 2008)
	KhES-1	HDAdV	45% (14 of 31 Ganc^R and G418^R clones)	(Suzuki et al., 2008)
	HUES3, HUES 8	BAC	HUES3: 1 of 3 Puro^R clones HUES8: 1of 6 Puro^R clones	(Song et al., 2010)
	BG01, iPSCs	AAV	BG01: 26% (8 of 31 G418^R clones) iPSCs: 19-29%	(Khan et al., 2010)
IL2RG	HUES1, HUES3	ZFN and IDLV	13-39%	(Lombardo et al., 2007)
MIXL1	HES3	Conventional positive (Neo) selection	Not indicated	(Davis et al., 2008)
OLIG2	BG01	Conventional positive (Neo)-negative (HSV-TK) selection	5.7% (6/106 clones)	(Xue et al., 2009)
p53	HUES9, H9	BACs	3.3% (2/60) for first allele, 22% (7/32) for second allele	(Song et al., 2010)
PIG-A	H1, H9, iPSC lines	ZFN	50% (6 out 12 Hyg^R clones)	(Zou et al., 2009)
PITX3	BG01, iPSCs	ZFN	11%	(Hockemeyer et al., 2009)
POU5F1 (OCT4)	H1	Conventional positive (Neo) selection	27% (28/103 G418^R clones) 39% (22/56 G418^R clones)	(Zwaka and Thomson, 2003)
	BG01	ZFN	39-100% depending on ZFN pair	(Hockemeyer et al., 2009)
ROSA26	HES2	Conventional positive (Neo)-negative (D-TA) selection	2.3% (2/88 G418^R clones)	(Irion et al., 2007)

Table 1. Examples of genes targeted by homologous recombination in hPSCs. Abbreviations: 6-TG^R = resistance to drug 6-thioguanine; BAC = Bacterial artificial chromosome; D-TA = diphtheria toxin A-fragment; G418^R = resistance to drug G418; Ganc^R = resistance to drug ganciclovir; HSV-TK = herpes simplex virus thymidine kinase; Hyg^R = resistance to drug Hygromycin; IDLV = integration defective lenti virus; iPSC = induced pluripotent stem cell; Neo = Neomycin; Puro^R = resistance to drug Puromycin; ZNF = Zinc finger nuclease.

4.2 BAC-based HR

A major reason for the low HR efficiency observed for conventional methods in both mouse and human ESCs is related to the limited length of homology arms. It has been documented that with an increase in homology arm lengths from 1.3 to 6.8 kb, gene targeting frequencies dramatically increase (Hasty et al., 1991), and further lengthening homology regions increases targeted gene insertion frequencies. The upper limit for the length of conventional targeting vectors is approximately 20 kb, with standard targeting vectors carrying approximately 10 kb of homologous DNA (e.g. a short and long arm of 4 and 6 kb, respectively). Bacterial artificial chromosomes (BACs) with usual insert sizes of 150-350 kb (and up to 700 kb) provide a unique tool to overcome the length constraint of conventional targeting vectors. Manipulation of these large DNA constructs requires recombineering technologies (Copeland et al., 2001), which exploits the HR system of Escherichia coli to generate large recombinant DNA constructs. Importantly, the large homology regions overcome the need for isogenic DNA, thus permitting HR in hESCs of various genetic backgrounds.

A recent report described successful BAC-based HR in hESCs at three loci, p53, ATM and HPRT, with targeting efficiencies of 21 to 27% (Song et al., 2010). The linearized BAC-targeting vectors were transduced into hESCs, both HUES9 and H9, by electroporation. Using the same targeting vectors with an alternative drug selection cassette, the authors were able to generate homozygous mutants in ATM and p53. One frequently cited drawback of BAC-based gene targeting is the difficulty in distinguishing HR from random integration. Song et al. overcame this challenge by shortening one homology arm to less than 10 kb while retaining the other at 80 to 100 kb. This allowed confirmation of HR by Southern blotting and a ligation-mediated PCR.

4.3 Zinc finger nucleases mediated HR

It has been well documented that double-stranded DNA breaks significantly enhance HR in human cells (Porteus and Carroll, 2005; Porteus et al., 2003). Such double-stranded breaks (DSB) are then repaired either by nonhomologous end joining (NHEJ), which is highly error prone thereby disrupting the targeted gene, or by homology-directed repair (HDR), where the sequence of a homologous strand of DNA (generally the intact homologous chromosome) is accurately copied. However, introduction of a single DSB in the genome is impossible with conventional restriction enzymes.

Zinc-finger nucleases (ZFN) offer a unique technology for cutting at single sites within the genome. In this approach a Zinc-finger DNA binding domain of the C2H2 class is engineered to recognize a single DNA sequence of interest and fused to an endo-nuclease, generally Fok1. Since Fok1 needs to dimerize to cleave DNA, two Zinc-finger DNA binding proteins with oppositely oriented binding sites are engineered. These two ZFNs are designed to be highly specific so that the Fok1 nuclease introduces a single DSB in the entire genome. One study demonstrated that design of the ZFN can greatly affect targeting efficiency, with rates varying between 39 to 100% for three distinct ZFN pairs designed to the OCT4 locus (Hockemeyer et al., 2009). In addition to the difficult and time-consuming nature of ZFN design and construction, it should be noted that site-specificity is difficult to control and off-target cleavage events have been observed (Radecke et al., 2010). As the ZFN introduces a single DSB, the DNA repair machinery will either introduce a deletion at the break site by NHEJ, or, if an exogenous DNA fragment is provided, a genetic modification near the DSB by HDR.

The ZFN technology has been successfully applied in multiple organisms, including Zebrafish, and in mouse and human cell culture, including hESC and hiPSCs (Davis and Stokoe, 2010; Hockemeyer et al., 2009; Lee et al., 2010a, b; Lombardo et al., 2007; Porteus and Carroll, 2005; Zou et al., 2009). A major obstacle to successfully applying this method is that two ZFNs and the homologous DNA have to be co-delivered into the cells of interest. Lombardo et al. addressed this issue by utilizing integration defective lentiviral vectors (IDLV) to co-deliver the ZFNs and the donor sequence. While this method yielded HR at the targeted site, the IL-2 receptor common γ-chain gene (IL2RG), in hESCs (Lombardo et al., 2007), such IDLVs are potentially capable of integrating randomly into the genome, thereby potentially disrupting critical genes.

An alternative virus free application of ZFNs in hESCs and hiPSCs was recently reported to target the PIG-A locus and a chromosomally integrated GFP gene (Zou et al., 2009). In this study, the donor DNA (2 kb homology arm) was transduced with or without the DNA encoding the ZFNs, into hESCs under optimized conditions that included the use of an immortalized cell line (W3R) that expresses Wnt3a, a growth factor that promotes hESC growth (Cai et al., 2007). The Amaxa Nucleofector® Technology was employed to achieve 50% transfection efficiency and high stable integration rates (10^{-5} cells). In the presence of the ZFNs, HR frequencies were increased 200 to 2,000 fold.

4.4 Adenoviral vector- and adeno associated viral-mediated HR.

Adenoviral vectors (AdVs) have been used extensivley to transduce a broad range of cell types, including hPSCs, and are used in pre-clinical studies involving gene therapy. AdVs have been modified to remove all viral genes from the viral genome, thereby producing helper-dependent AdVs (HDAdVs) that are less cytotoxic than their wildtype counterparts. Suzuki et al. used HDAdV to successfully target the *HPRT1* locus in hESCs at high efficiency (Suzuki et al., 2008).

Adeno-associated virus (AAV) is a single-stranded DNA virus, which derives its name from the fact that it is often found in cells that are simultaneously infected with adenovirus. However, in contrast to adenovirus, AAV does not elicit a host immune response or stimulate inflammation and can infect non-dividing cells. Wildtype AAV predominately integrates into one location on human chromosome 19. Like adenovirus, AAV can be engineered to remove all viral genes, including those needed for integration, and has been utilized for HR in mammalian cell culture (Hirata et al., 2002; Porteus et al., 2003; Russell and Hirata, 1998; Vasileva and Jessberger, 2005; Vasileva et al., 2006), including hESCs and hiPSCs (Khan et al., 2010).

Upon endocytosis, a wildtype AAV particle moves to the nucleus via motor proteins in the cytoplasm and nuclear tubular structures (Seisenberger et al., 2001), thereby "protecting" the single stranded genome on its journey to the nucleus. Once within the nucleus, "the linear single-stranded monomers mimic DNA damage and induce DNA-repair pathways" (Vasileva and Jessberger, 2005). In Escherichia coli it is well established that single stranded DNA triggers DNA repair via the SOS system. Consequently, it has been postulated that in eukaryotic cells, single stranded DNA would trigger a similar DNA repair process. The various methods described above involve donor DNA that is double stranded, which only efficiently triggers a DNA response pathway and homology-based repair when double stranded breaks exist in the genome. Thus, with its single stranded DNA genome, recombinant AAV may represent an ideal substrate for HR.

With a genome of 4.8 kb, AAV provides limited capacity of targeting vector design. At the same time, this limited genome size greatly simplifies cloning strategies relative to the coventional methods or BAC recombineering. Targeting vectors can be designed to replace all viral sequences except two essential palindromic inverted terminal repeats (ITR) flanking the donor DNA. In contrast to the conventional HR methods and BAC-mediated HR, left and right homology arms of 1 kilobase length have been shown to be sufficient to promote high HR rates using AAV. To permit selection of infected cells, the recombinant AAV can be constructed to contain drug selection genes under control of a constitutive promoter. HESCs can be efficiently infected with AAV serotype 2, 4 and 5 (Smith-Arica et al., 2003), and among drug selected clones HR frequencies approaching 100% have been reported (Hirata et al., 2002; Khan et al., 2010; Porteus et al., 2003; Suzuki et al., 2008; Vasileva et al., 2006).

5. Concluding remarks

Manipulation of gene expression using one or more of the various transgenesis methods described here are critical in advancing hPSC science and accelerating applications of hPSC in regenerative medicine. While many methods of transgenesis developed in other cell culture systems are available, it is clear that hPSCs exhibit unique properties and charcateristics so that all methods require significant refinement and optimization. The current differences in gene transduction efficiencies among the various hPSC lines will likely evaporate as culture conditions for hPSC improve and become standardized. Recent advances and improvements in culture substrates and media formulations specifically designed for hPSCs will greatly augment and enhance the current methods to modify the genome of hPSCs. Additionally, further development of technologies for single cell isolation and purification, such as flow cytometry and cell sorting, will yield transduced and targeted cell populations in sufficiently large numbers to permit further analysis of transgenic cell lines and to enable transplantation studies.

6. References

Adewumi, O., Aflatoonian, B., Ahrlund-Richter, L., Amit, M., Andrews, P.W., Beighton, G., Bello, P.A., Benvenisty, N., Berry, L.S., Bevan, S., *et al.* (2007). Characterization of human embryonic stem cell lines by the International Stem Cell Initiative. *Nat Biotechnol* 25, 803-816.

Amit, M., Carpenter, M.K., Inokuma, M.S., Chiu, C.P., Harris, C.P., Waknitz, M.A., Itskovitz-Eldor, J., and Thomson, J.A. (2000). Clonally derived human embryonic stem cell lines maintain pluripotency and proliferative potential for prolonged periods of culture. *Dev Biol* 227, 271-278.

Anderson, D., Self, T., Mellor, I.R., Goh, G., Hill, S.J., and Denning, C. (2007). Transgenic enrichment of cardiomyocytes from human embryonic stem cells. *Mol Ther* 15, 2027-2036.

Bouhassira, E.E., Westerman, K., and Leboulch, P. (1997). Transcriptional behavior of LCR enhancer elements integrated at the same chromosomal locus by recombinase-mediated cassette exchange. *Blood* 90, 3332-3344.

Buecker, C., Chen, H.H., Polo, J.M., Daheron, L., Bu, L., Barakat, T.S., Okwieka, P., Porter, A., Gribnau, J., Hochedlinger, K., *et al.* (2010). A murine ESC-like state facilitates transgenesis and homologous recombination in human pluripotent stem cells. *Cell Stem Cell* 6, 535-546.

Cai, L., Ye, Z., Zhou, B.Y., Mali, P., Zhou, C., and Cheng, L. (2007). Promoting human embryonic stem cell renewal or differentiation by modulating Wnt signal and culture conditions. *Cell Res* 17, 62-72.

Cao, F., Xie, X., Gollan, T., Zhao, L., Narsinh, K., Lee, R.J., and Wu, J.C. (2009). Comparison of gene-transfer efficiency in human embryonic stem cells. *Mol Imaging Biol* 12, 15-24.

Cherry, S.R., Biniszkiewicz, D., van Parijs, L., Baltimore, D., and Jaenisch, R. (2000). Retroviral expression in embryonic stem cells and hematopoietic stem cells. *Mol Cell Biol* 20, 7419-7426.

Chiao, E., Elazar, M., Xing, Y., Xiong, A., Kmet, M., Millan, M.T., Glenn, J.S., Wong, W.H., and Baker, J. (2008). Isolation and transcriptional profiling of purified hepatic cells derived from human embryonic stem cells. *Stem Cells* 26, 2032-2041.

Copeland, N.G., Jenkins, N.A., and Court, D.L. (2001). Recombineering: a powerful new tool for mouse functional genomics. *Nat Rev Genet* 2, 769-779.

Costa, M., Dottori, M., Ng, E., Hawes, S.M., Sourris, K., Jamshidi, P., Pera, M.F., Elefanty, A.G., and Stanley, E.G. (2005). The hESC line Envy expresses high levels of GFP in all differentiated progeny. *Nat Methods* 2, 259-260.

Damoiseaux, R., Sherman, S.P., Alva, J.A., Peterson, C., and Pyle, A.D. (2009). Integrated chemical genomics reveals modifiers of survival in human embryonic stem cells. *Stem Cells* 27, 533-542.

Davis, D., and Stokoe, D. (2010). Zinc finger nucleases as tools to understand and treat human diseases. *BMC Med* 8, 42.

Davis, R.P., Ng, E.S., Costa, M., Mossman, A.K., Sourris, K., Elefanty, A.G., and Stanley, E.G. (2008). Targeting a GFP reporter gene to the MIXL1 locus of human embryonic stem cells identifies human primitive streak-like cells and enables isolation of primitive hematopoietic precursors. *Blood* 111, 1876-1884.

Dhara, S.K., Gerwe, B.A., Majumder, A., Dodla, M.C., Boyd, N.L., Machacek, D.W., Hasneen, K., and Stice, S.L. (2009). Genetic manipulation of neural progenitors derived from human embryonic stem cells. *Tissue Eng Part A* 15, 3621-3634.

Di Domenico, A.I., Christodoulou, I., Pells, S.C., McWhir, J., and Thomson, A.J. (2008). Sequential genetic modification of the hprt locus in human ESCs combining gene targeting and recombinase-mediated cassette exchange. *Cloning Stem Cells* 10, 217-230.

Eiges, R., Schuldiner, M., Drukker, M., Yanuka, O., Itskovitz-Eldor, J., and Benvenisty, N. (2001). Establishment of human embryonic stem cell-transfected clones carrying a marker for undifferentiated cells. *Curr Biol* 11, 514-518.

Green, J.J., Zhou, B.Y., Mitalipova, M.M., Beard, C., Langer, R., Jaenisch, R., and Anderson, D.G. (2008). Nanoparticles for gene transfer to human embryonic stem cell colonies. *Nano Lett* 8, 3126-3130.

Gropp, M., Itsykson, P., Singer, O., Ben-Hur, T., Reinhartz, E., Galun, E., and Reubinoff, B.E. (2003). Stable genetic modification of human embryonic stem cells by lentiviral vectors. *Mol Ther* 7, 281-287.

Hamm, A., Krott, N., Breibach, I., Blindt, R., and Bosserhoff, A.K. (2002). Efficient transfection method for primary cells. *Tissue Eng* 8, 235-245.

Hasty, P., Rivera-Perez, J., and Bradley, A. (1991). The length of homology required for gene targeting in embryonic stem cells. *Mol Cell Biol* 11, 5586-5591.

Hirata, R., Chamberlain, J., Dong, R., and Russell, D.W. (2002). Targeted transgene insertion into human chromosomes by adeno-associated virus vectors. *Nat Biotechnol* 20, 735-738.

Hockemeyer, D., Soldner, F., Beard, C., Gao, Q., Mitalipova, M., DeKelver, R.C., Katibah, G.E., Amora, R., Boydston, E.A., Zeitler, B., *et al.* (2009). Efficient targeting of expressed and silent genes in human ESCs and iPSCs using zinc-finger nucleases. *Nat Biotechnol* 27, 851-857.

Hong, S., Hwang, D.Y., Yoon, S., Isacson, O., Ramezani, A., Hawley, R.G., and Kim, K.S. (2007). Functional analysis of various promoters in lentiviral vectors at different stages of in vitro differentiation of mouse embryonic stem cells. *Mol Ther* 15, 1630-1639.

Huber, I., Itzhaki, I., Caspi, O., Arbel, G., Tzukerman, M., Gepstein, A., Habib, M., Yankelson, L., Kehat, I., and Gepstein, L. (2007). Identification and selection of cardiomyocytes during human embryonic stem cell differentiation. *FASEB J* 21, 2551-2563.

Irion, S., Luche, H., Gadue, P., Fehling, H.J., Kennedy, M., and Keller, G. (2007). Identification and targeting of the ROSA26 locus in human embryonic stem cells. *Nat Biotechnol* 25, 1477-1482.

Jo, D., Nashabi, A., Doxsee, C., Lin, Q., Unutmaz, D., Chen, J., and Ruley, H.E. (2001). Epigenetic regulation of gene structure and function with a cell-permeable Cre recombinase. *Nat Biotechnol* 19, 929-933.

Joshi, S.K., Hashimoto, K., and Koni, P.A. (2002). Induced DNA recombination by Cre recombinase protein transduction. *Genesis* 33, 48-54.

Khan, I.F., Hirata, R.K., Wang, P.R., Li, Y., Kho, J., Nelson, A., Huo, Y., Zavaljevski, M., Ware, C., and Russell, D.W. (2010). Engineering of human pluripotent stem cells by AAV-mediated gene targeting. *Mol Ther* 18, 1192-1199.

Kita-Matsuo, H., Barcova, M., Prigozhina, N., Salomonis, N., Wei, K., Jacot, J.G., Nelson, B., Spiering, S., Haverslag, R., Kim, C., *et al.* (2009). Lentiviral vectors and protocols for creation of stable hESC lines for fluorescent tracking and drug resistance selection of cardiomyocytes. *PLoS One* 4, e5046.

Kohn, D.B., Sadelain, M., and Glorioso, J.C. (2003). Occurrence of leukaemia following gene therapy of X-linked SCID. *Nat Rev Cancer* 3, 477-488.

Kumar, M., Keller, B., Makalou, N., and Sutton, R.E. (2001). Systematic determination of the packaging limit of lentiviral vectors. *Hum Gene Ther* 12, 1893-1905.

Laker, C., Meyer, J., Schopen, A., Friel, J., Heberlein, C., Ostertag, W., and Stocking, C. (1998). Host cis-mediated extinction of a retrovirus permissive for expression in embryonal stem cells during differentiation. *J Virol* 72, 339-348.

Lakshmipathy, U., Pelacho, B., Sudo, K., Linehan, J.L., Coucouvanis, E., Kaufman, D.S., and Verfaillie, C.M. (2004). Efficient transfection of embryonic and adult stem cells. *Stem Cells* 22, 531-543.

Lavon, N., and Benvenisty, N. (2005). Study of hepatocyte differentiation using embryonic stem cells. *J Cell Biochem* 96, 1193-1202.

Lavon, N., Yanuka, O., and Benvenisty, N. (2004). Differentiation and isolation of hepatic-like cells from human embryonic stem cells. *Differentiation* 72, 230-238.

Lavon, N., Yanuka, O., and Benvenisty, N. (2006). The effect of overexpression of Pdx1 and Foxa2 on the differentiation of human embryonic stem cells into pancreatic cells. *Stem Cells* 24, 1923-1930.

Lee, H.J., Kim, E., and Kim, J.S. (2010a). Site-specific DNA excision via engineered zinc finger nucleases. *Trends Biotechnol.*

Lee, H.J., Kim, E., and Kim, J.S. (2010b). Targeted chromosomal deletions in human cells using zinc finger nucleases. *Genome Res* 20, 81-89.

Liu, Y., Thyagarajan, B., Lakshmipathy, U., Xue, H., Lieu, P., Fontes, A., MacArthur, C.C., Scheyhing, K., Rao, M.S., and Chesnut, J.D. (2009). Generation of platform human embryonic stem cell lines that allow efficient targeting at a predetermined genomic location. *Stem Cells Dev* 18, 1459-1472.

Lombardo, A., Genovese, P., Beausejour, C.M., Colleoni, S., Lee, Y.L., Kim, K.A., Ando, D., Urnov, F.D., Galli, C., Gregory, P.D., *et al.* (2007). Gene editing in human stem cells using zinc finger nucleases and integrase-defective lentiviral vector delivery. *Nat Biotechnol* 25, 1298-1306.

Mansour, S.L., Thomas, K.R., and Capecchi, M.R. (1988). Disruption of the proto-oncogene int-2 in mouse embryo-derived stem cells: a general strategy for targeting mutations to non-selectable genes. *Nature* 336, 348-352.

Mohr, J.C., de Pablo, J.J., and Palecek, S.P. (2006). Electroporation of human embryonic stem cells: Small and macromolecule loading and DNA transfection. *Biotechnol Prog* 22, 825-834.

Moore, J.C., Atze, K., Yeung, P.L., Toro-Ramos, A.J., Camarillo, C., Thompson, K., Ricupero, C.L., Brenneman, M.A., Cohen, R.I., and Hart, R.P. (2010). Efficient, high-throughput transfection of human embryonic stem cells. *Stem Cell Res Ther* 1, 23.

Nolden, L., Edenhofer, F., Haupt, S., Koch, P., Wunderlich, F.T., Siemen, H., and Brustle, O. (2006). Site-specific recombination in human embryonic stem cells induced by cell-permeant Cre recombinase. *Nat Methods* 3, 461-467.

Osafune, K., Caron, L., Borowiak, M., Martinez, R.J., Fitz-Gerald, C.S., Sato, Y., Cowan, C.A., Chien, K.R., and Melton, D.A. (2008). Marked differences in differentiation propensity among human embryonic stem cell lines. *Nat Biotechnol* 26, 313-315.

Pfeifer, A., Ikawa, M., Dayn, Y., and Verma, I.M. (2002). Transgenesis by lentiviral vectors: lack of gene silencing in mammalian embryonic stem cells and preimplantation embryos. *Proc Natl Acad Sci U S A* 99, 2140-2145.

Porteus, M.H., and Carroll, D. (2005). Gene targeting using zinc finger nucleases. *Nat Biotechnol* 23, 967-973.

Porteus, M.H., Cathomen, T., Weitzman, M.D., and Baltimore, D. (2003). Efficient gene targeting mediated by adeno-associated virus and DNA double-strand breaks. *Mol Cell Biol* 23, 3558-3565.

Pyle, A.D., Lock, L.F., and Donovan, P.J. (2006). Neurotrophins mediate human embryonic stem cell survival. *Nat Biotechnol* 24, 344-350.

Radecke, S., Radecke, F., Cathomen, T., and Schwarz, K. (2010). Zinc-finger nuclease-induced gene repair with oligodeoxynucleotides: wanted and unwanted target locus modifications. *Mol Ther* 18, 743-753.

Ruby, K.M., and Zheng, B. (2009). Gene targeting in a HUES line of human embryonic stem cells via electroporation. *Stem Cells* 27, 1496-1506.

Russell, D.W., and Hirata, R.K. (1998). Human gene targeting by viral vectors. *Nat Genet* 18, 325-330.

Sedivy, J.M., Vogelstein, B., Liber, H.L., Hendrickson, E.A., and Rosmarin, A. (1999). Gene Targeting in Human Cells Without Isogenic DNA. *Science* 283, 9.

Seisenberger, G., Ried, M.U., Endress, T., Buning, H., Hallek, M., and Brauchle, C. (2001). Real-time single-molecule imaging of the infection pathway of an adeno-associated virus. *Science* 294, 1929-1932.

Siemen, H., Nix, M., Endl, E., Koch, P., Itskovitz-Eldor, J., and Brustle, O. (2005). Nucleofection of human embryonic stem cells. *Stem Cells Dev* 14, 378-383.

Siemen, H., Nolden, L., Terstegge, S., Koch, P., and Brustle, O. (2008). Nucleofection of human embryonic stem cells. *Methods Mol Biol* 423, 131-138.

Smith-Arica, J.R., Thomson, A.J., Ansell, R., Chiorini, J., Davidson, B., and McWhir, J. (2003). Infection efficiency of human and mouse embryonic stem cells using adenoviral and adeno-associated viral vectors. *Cloning Stem Cells* 5, 51-62.

Song, H., Chung, S.K., and Xu, Y. (2010). Modeling disease in human ESCs using an efficient BAC-based homologous recombination system. *Cell Stem Cell* 6, 80-89.

Suzuki, K., Mitsui, K., Aizawa, E., Hasegawa, K., Kawase, E., Yamagishi, T., Shimizu, Y., Suemori, H., Nakatsuji, N., and Mitani, K. (2008). Highly efficient transient gene expression and gene targeting in primate embryonic stem cells with helper-dependent adenoviral vectors. *Proc Natl Acad Sci U S A* 105, 13781-13786.

Thyagarajan, B., Liu, Y., Shin, S., Lakshmipathy, U., Scheyhing, K., Xue, H., Ellerstrom, C., Strehl, R., Hyllner, J., Rao, M.S., *et al.* (2008). Creation of engineered human embryonic stem cell lines using phiC31 integrase. *Stem Cells* 26, 119-126.

Urbach, A., Schuldiner, M., and Benvenisty, N. (2004). Modeling for Lesch-Nyhan disease by gene targeting in human embryonic stem cells. *Stem Cells* 22, 635-641.

Vallier, L., Rugg-Gunn, P.J., Bouhon, I.A., Andersson, F.K., Sadler, A.J., and Pedersen, R.A. (2004). Enhancing and diminishing gene function in human embryonic stem cells. *Stem Cells* 22, 2-11.

Vasileva, A., and Jessberger, R. (2005). Precise hit: adeno-associated virus in gene targeting. *Nat Rev Microbiol* 3, 837-847.

Vasileva, A., Linden, R.M., and Jessberger, R. (2006). Homologous recombination is required for AAV-mediated gene targeting. *Nucleic Acids Res* 34, 3345-3360.

Watanabe, K., Ueno, M., Kamiya, D., Nishiyama, A., Matsumura, M., Wataya, T., Takahashi, J.B., Nishikawa, S., Muguruma, K., and Sasai, Y. (2007). A ROCK inhibitor permits survival of dissociated human embryonic stem cells. *Nat Biotechnol* 25, 681-686.

Will, E., Klump, H., Heffner, N., Schwieger, M., Schiedlmeier, B., Ostertag, W., Baum, C., and Stocking, C. (2002). Unmodified Cre recombinase crosses the membrane. *Nucleic Acids Res* 30, e59.

Xia, X., Zhang, Y., Zieth, C.R., and Zhang, S.C. (2007). Transgenes delivered by lentiviral vector are suppressed in human embryonic stem cells in a promoter-dependent manner. *Stem Cells Dev* 16, 167-176.

Xue, H., Wu, S., Papadeas, S.T., Spusta, S., Swistowska, A.M., MacArthur, C.C., Mattson, M.P., Maragakis, N.J., Capecchi, M.R., Rao, M.S., *et al.* (2009). A targeted neuroglial reporter line generated by homologous recombination in human embryonic stem cells. *Stem Cells* 27, 1836-1846.

Zou, J., Maeder, M.L., Mali, P., Pruett-Miller, S.M., Thibodeau-Beganny, S., Chou, B.K., Chen, G., Ye, Z., Park, I.H., Daley, G.Q., *et al.* (2009). Gene targeting of a disease-related gene in human induced pluripotent stem and embryonic stem cells. *Cell Stem Cell* 5, 97-110.

Zwaka, T.P., and Thomson, J.A. (2003). Homologous recombination in human embryonic stem cells. *Nat Biotechnol* 21, 319-321.

16

Pluripotent Stem Cells Induced from Testicular Tissue of a Man with Klinefelter Syndrome (47, XXY) by Four Transcription Factors (*OCT4, SOX2, KLF4,* and *C-MYC*)

Hideyuki Kobayashi
Department of Urology, Toho University School of Medicine
Japan

1. Introduction

Klinefelter syndrome (KS) is the most common chromosomal aberration in men, with an estimated frequency of 1:500 to 1:1000 among live deliveries (Lanfranco et al., 2004). KS is characterized by an X-chromosome polysomy, with X disomy being the most common variant (47, XXY). Although the classic description of men with KS has been that they are tall with eunuchoid body proportions, low testosterone, sparse facial and pubic hair, small, hard testicles, micropenis, sterility, and mild to moderate cognitive deficits, it is now well known that this description is not accurate. Rather, men with KS exhibit a broad spectrum of phenotypes, and they are represented in a variety of professions and can be found at all socioeconomic levels (Lanfranco et al., 2004).

The cardinal problems in men with KS—progressive testicular failure (and thus azoospermia or cryptozoospermia), small testes (5 to 7 cm^3), and low testosterone—are found in most men with KS, and men with these symptoms should undergo prompt cytogenic evaluation. Most men with KS are diagnosed as adults, when they present with infertility or hypogonadism (Graham et al., 1988). It is strongly recommended that men with KS undergo genetic counseling, because they have an increased medical risk for various diseases, including diabetes, cardiovascular disease, and cancer, and their offspring have an increased risk for chromosomal abnormalities (Lanfranco et al., 2004).

It is predicted that the aberrant expression of X chromosome-linked genes plays a role in the failure of spermatogenesis seen in men with 47, XXY. However, the mechanisms underlying the infertility of men with KS is still poorly understood, and its treatment is both difficult and rarely successful. The elucidation of the molecular mechanisms of X chromosome inactivation might therefore allow us to better predict the extent of reproductive failure in KS patients and to design novel therapies.

It was recently reported that introducing a set of transcription factors related to pluripotency can directly reprogram human somatic cells to produce induced pluripotent stem (iPS) cells (Lowry et al., 2008; Park et al., 2008; Takahashi et al., 2007; Yu et al., 2007). We reasoned that iPS cells derived from KS patients could be useful as a tool for studying the mechanism of X chromosome inactivation. In addition, iPS cells derived from testicular tissue might be a potential source for cell-based therapies to treat some cases of male infertility.

In this study, we successfully obtained iPS cells derived from the testicular tissue of men with KS, by introducing four transcription factors (*OCT4, SOX2, KLF4*, and *C-MYC*), using the lentiviral vector system. In the future, such iPS cells might be useful for uncovering the mechanisms of X-chromosome inactivation, and could lead to novel treatments for infertility.

2. Materials and methods

2.1 Human subjects

In accordance with the regulations set forth by the Human Investigations Committee of Toho University School of Medicine, written informed consent was obtained from male patients being treated for infertility, who visited the Reproduction Center of Toho Medical Center Omori Hospital. Testicular tissues were obtained from one patient with KS by testicular sperm extraction.

2.2 Adult human testis tissues with KS

First, we confirmed that there were no mature sperm in the testicular tissues from KS patients. These tissues included the rete testis, Leydig cells, Sertoli cells, and fibroblasts. The tissues were used fresh. Cell suspensions from the testis tissues were prepared by enzymatic digestion (Ogawa et al., 1997). The dissociated testis cell suspension was plated onto 10-cm culture dishes coated with 0.1% gelatin (Sigma) in a standard culture medium (Dulbecco's modified Eagle's medium (DMEM) (Invitrogen) containing 7% fetal bovine serum (FBS), 2 mM glutamine (Sigma), and antibiotics (50 U/ml penicillin and 50 μg/ml streptomycin; Sigma)). These cells were incubated at 37°C in a humidified 5% CO_2/95% air atmosphere for 17 days.

2.3 Cell culture

293FT cells were purchased (Invitrogen). The 293FT cells were maintained in the standard culture medium. Mitomycin C-treated MEF feeder cells were purchased (ReproCELL). The MEF feeder cells were plated in 0.1% gelatin-coated 10-cm culture dishes in the standard culture medium. iPS cells were generated and maintained in Primate ES medium (ReproCELL) supplemented with 4 ng/ml recombinant human basic fibroblast growth factor (bFGF) (R&D Systems). To grow the iPS cells, the Primate ES cell medium was supplemented with 4 ng/ml bFGF and 10 μM Y27632 (Wako) and changed every second day. For passaging, the iPS cells were rinsed with Hank's balanced salt solution (HBSS) (Invitrogen) once and incubated in dissociation medium (ReproCELL) at 37°C. All cultures were maintained at 37°C in a humidified 5% CO_2/95% air atmosphere.

2.4 Plasmid construction

The open reading frames of human *OCT4, SOX2, KLF4*, and *C-MYC* were amplified by RT-PCR and subcloned into pLenti 6.3/V5-TOPO (Invitrogen). A stop codon was inserted to avoid expressing the V5 protein. All the plasmids were constructed by Invitrogen.

2.5 Lentivirus production and infection

The ViraPower™ HiPerfome™ Lentiviral Expression Kit (Invitrogen) was used to produce the lentiviruses. 293FT cells were plated at 6×10^6 cells per 10-cm dish and incubated overnight. We prepared four dishes of 293FT cells to produce four lentiviruses (encoding of

OCT4, SOX2, KLF4, and *C-MYC*). The next day, the 293FT cells were transfected with 3 μg pLenti 6.3/V5-TOPO and 9 μg ViraPower packaging mix with the Fugene 6 transfection reagent (Roche), according to the manufacturer's instructions. The medium was collected 24 and 48 hours after the transduction and replaced with new medium. The collected supernatants were designated as the first and second virus-containing supernatants. Medium was also collected 72 hours after transduction, as the third (final) virus-containing supernatant. The virus-producing cells were then discarded. The virus-containing supernatants were filtered through a 0.45-μm pore filter (Millipore). An equal amount of the supernatants containing each of the four lentiviruses was mixed, transferred to the culture dish containing testis cells, and incubated at 37°C.

2.6 iPS cell generation

The virus-containing medium was used to replace the standard medium in the testis cell cultures, 24 hours after the third virus-mediated transduction of medium from the virus-producing cells. The medium was changed every second day. Six days later, the cells were harvested by trypsinization, and 5 X 10^4 cells were placed on MEF feeder cells (10-cm dish) in Primate ES cell medium supplemented with 4 ng/ml bFGF. The medium was changed every second day. We monitored these cells daily for morphological changes.

2.7 Immunofluorescence microscopy and immunostaining

The Human Embryonic Stem Cell Marker Antibody Panel was used (R&D Systems). Cells were washed twice with phosphate-buffered saline, fixed with 4% (w/v) paraformaldehyde for 20 min, permeabilized for 60 min with phosphate-buffered saline containing 0.1% (v/v) Triton X-100, and then blocked for 3 h with phosphate-buffered saline containing 20% donkey serum. For immunostaining, the fixed samples were incubated with anti-human alkaline phosphatase monoclonal, anti-human NANOG polyclonal, anti-human OCT4 polyclonal, anti-human SSEA-1 monoclonal, and anti-human SSEA-4 monoclonal antibodies (all from R&D Systems) as indicated, washed three times with PBS containing 0.1% (v/v) Triton X-100, and probed with the appropriate secondary antibodies (anti-goat IgG antibody conjugated with Alexa 488 or anti-mouse IgG antibody conjugated with Alexa 488) (Molecular Probes). Nucleic acid was detected using SYTOX^{R} Orange Nucleic Acid Stain (Molecular Probes).

2.8 Reverse-Transcription (RT)-PCR

Total RNA was prepared using the PureLink™ Micro-to-Midi Total RNA Purification System (Invitrogen). RT-PCR was performed using the SuperScript™III One-Step RT-PCR System with Platinum^{R} Taq DNA Polymerase (Invitrogen) for human SSC markers, with the following conditions: 55°C for 30 min for reverse transcription; 94°C for 2 min to inactivate the reverse transcriptase and activate the polymerase; 40 cycles of 94°C for 15 s, 55°C for 30 s, 68°C for 1 min; and 68°C for 5 min, for the final extension. The products were as follows: for *OCT4*, 315 bp; *NANOG*, 285 bp; *GAPDH*, 513 bp; *STELLAR*, 174 bp; *GDF3*, 150 bp; *DAZL*, 178 bp; and *VASA*, 199 bp, (Klimanskaya et al., 2006) (Ezeh et al., 2005).

2.9 Short tandem repeat analysis and karyotyping

DNA fingerprinting analyses with short tandem repeat (STR) markers were performed at SRL Laboratory, Japan. Chromosomal G-band analyses were performed at Nihon Gene Research Laboratories, Japan.

2.10 In vitro assay (embryoid body formation)

Confluent iPS cells in a 10-cm dish were harvested by trypsinization. The iPS cells were then transferred to a Poly (hydroxyethyl methacrylate-co-methyl methacrylate; HEMA-MMA)-coated dish in Primate ES cell medium. The medium was changed every other day, and the cells were maintained in floating culture for 8 days.

2.11 In vivo assay (teratoma formation)

Confluent iPS cells in a 6-well dish were harvested by trypsin treatment, collected into tubes, and spun, and the pellets were suspended in ES medium. The cells were then injected into the testes of SCID mice (8-weeks old) (Charles River). Twelve weeks after injection, the tumors were dissected and fixed with 10% formalin. Pathological analyses were performed at the Tokyo Central Pathology Laboratory, Japan.

3. Results

3.1 Generation of ES-like cell colonies from the testicular tissue of an infertile man with KS

Testicular tissue was obtained from an infertile man with KS (Fig. 1a). Colonies with a human ES cell-like morphology first became visible 15-16 days after their transduction with transcription-factor-expressing viruses under human ES cell-supporting conditions. The morphology of these colonies was similar to that of human ES cells. We obtained about 20 ES cell-like colonies from 5 X 10^4 testicular cells in this experiment (data not shown). The colonies were collected on days 20-25, and transferred into 24-well plates on MEF feeder cells. Colonies that continued to expand and maintained an ES cell-like cell morphology were further passaged, while those that failed to expand or produced aberrant colonies were discarded. Ultimately, only one ES cell-like cell line was obtained (Fig. 1b).

a

Sample	Disease	Doner cell	Age	After enzymatic digestion
A. T	Klinefelter syndrome	testicular tissue	32	8.7x10^6

b

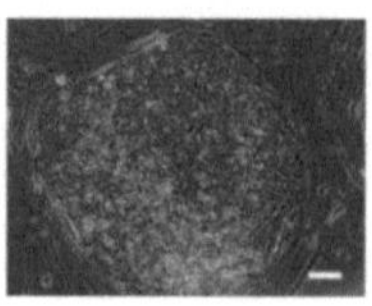

Locus		testis		colony	
STR	vWA	17	19	17	19
	D16S539	9	12	9	12
	TH01	7	-	7	-
	D5S818	11	12	11	12
	D13S317	11	14	11	14
	D7S820	11	-	11	-
	TPOX	8	11	8	11
	CSF1PO	10	12	10	12
	Amelogenin	X	Y	X	Y

Fig. 1. Generation of ES-like cells from human testicular tissue from an adult with KS by four transcription factors (*OCT4, SOX2, KLF4,* and *C-MYC*). a, Overview of the human testicular sample. b, Morphology and STR analysis of the ES-like cells. Bar = 60 μm

We then performed evaluations to confirm that this ES-like cell line was derived from testicular tissue. The pattern of 15 short tandem repeat (STR) loci and the Amelogenin locus was a perfect match between the ES cell-like colonies and one testicular tissue sample, as assessed using the PowerPlex^R 16 System (Promega). The results for 8 STRs and Amelogenin are shown in Figure 1b. In this analysis, Amelogenin only indicated the presence of the X or Y chromosome, and could not distinguish polysomies of the X or Y chromosome. These findings indicated that the ES cell-like colonies were generated from the testicular tissue and were not a result of cross-contamination.

3.2 Human testicular cells and ES-like cells derived from a KS patient expressed *OCT4, NANOG, STELLAR,* and *GDF3,* but not *DAZL* or *VASA*

RT-PCR showed that the ES-like cells derived from the KS patient expressed many marker genes for undifferentiated ES cells, including *OCT4, NANOG, STELLAR,* and *GDF3* (Fig. 2). We also confirmed that the testicular tissues from the same KS patient expressed *OCT4, NANOG, STELLAR,* and *GDF3* (Fig. 2). Neither the ES-like cells nor the original testes tissues expressed the germ-cell marker *DAZL* or *VASA* (Fig. 2). We could not find any differences between the ES cells and the testicular tissue derived from the KS patient by RT-PCR.

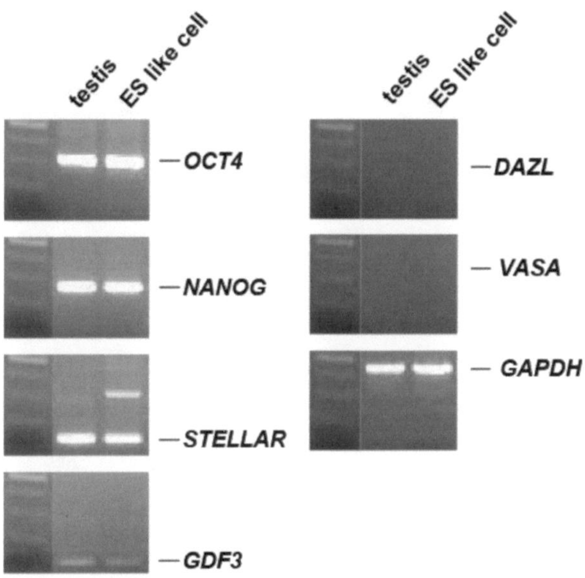

Fig. 2. Total gene expression in the ES-like cells and testis cells derived from a KS patient by RT-PCR. RT-PCR analysis for the expression of *OCT4, NANOG, STELLAR, GDF3, DAZL, VASA,* and *GAPDH* in ES-like cells and testis cells. The upper band in the RT-PCR results for *STELLAR* is nonspecific. *GAPDH* was used as a loading control.

3.3 Confirmation by chromosomal analysis that the iPS cells derived from the KS patient had a karyotype of 47, XXY

Karyotyping is the gold standard for diagnosing KS. To examine the karyotype of the iPS cells, we subjected twenty cells to chromosomal G-band analysis. The results confirmed that the iPS cells had the karyotype 47, XXY, which the same as that of the donor (Fig. 3).

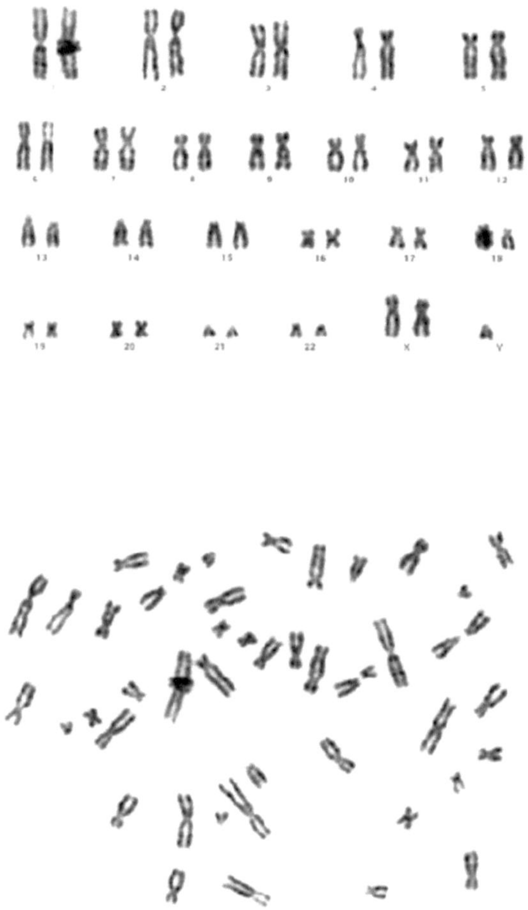

Fig. 3. G-banding chromosome analysis of the ES-like cells. A karyotype analysis of the ES-like cells derived from a patient with KS

3.4 iPS cells derived from a KS patient expressed human ES-cell markers by immunostaining

We characterized the iPS cells derived from the KS patient by immunostaining, and found that they expressed human ES-cell-specific surface antigens, including alkaline phosphatase (AP), NANOG, OCT4, and SSEA-4, but not SSEA-1 (Fig. 4a-e).

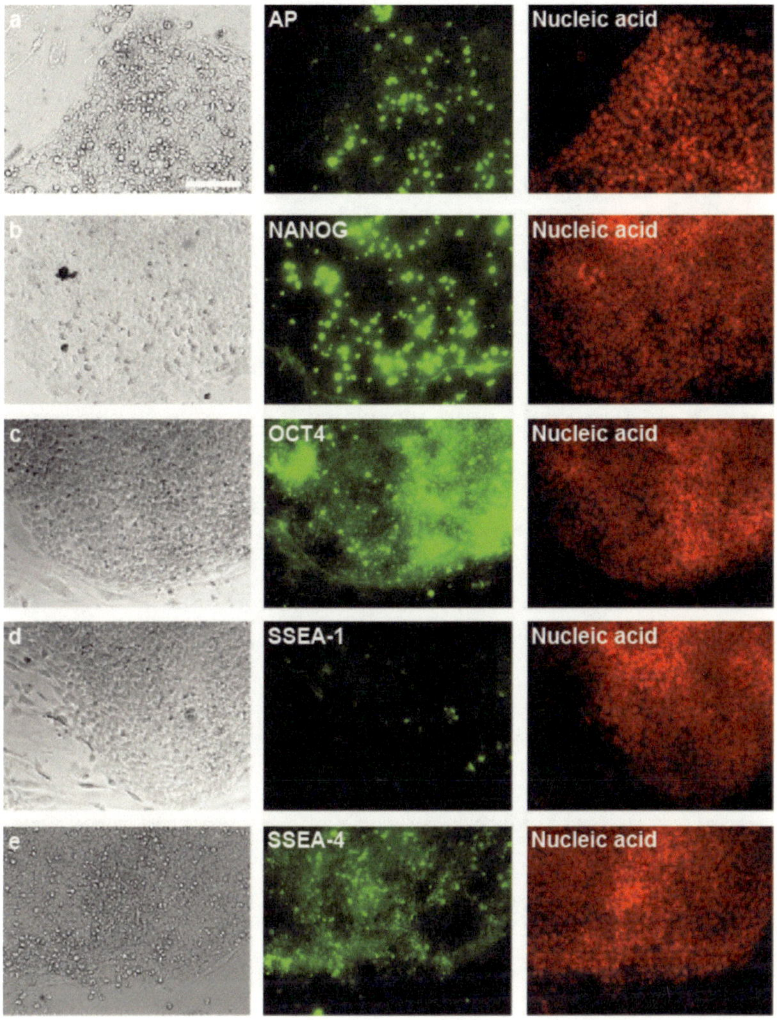

Fig. 4. Characterization of ES-like cells derived from a KS patient by immunostaining. a-e, Immunostaining of ES-like cells for alkaline phosphatase (AP) (a), NANOG (b), OCT4 (c), SSEA-1 (d), and SSEA-4 (e). Nuclei were stained with SYTOX^{R} Orange. Bar=100 μm.

3.5 iPS cells derived from a KS patient have the potential for multipotency in vitro and in vivo

To assess the pluripotency of the iPS cells in vitro, we examined their ability to form embryoid bodies. After 8 days in floating culture, embryoid bodies derived from the iPS cell were identified (Fig. 5a).

We next investigated whether the iPS cells derived from a KS patient were pluripotent in vivo, by transplanting them into the testis of SCID mice. We observed tumor formation in the testis 12 weeks after the injection. Histological inspection showed that the cells had differentiated into representatives of all three germ layers in vivo (Fig. 5b).

a

b

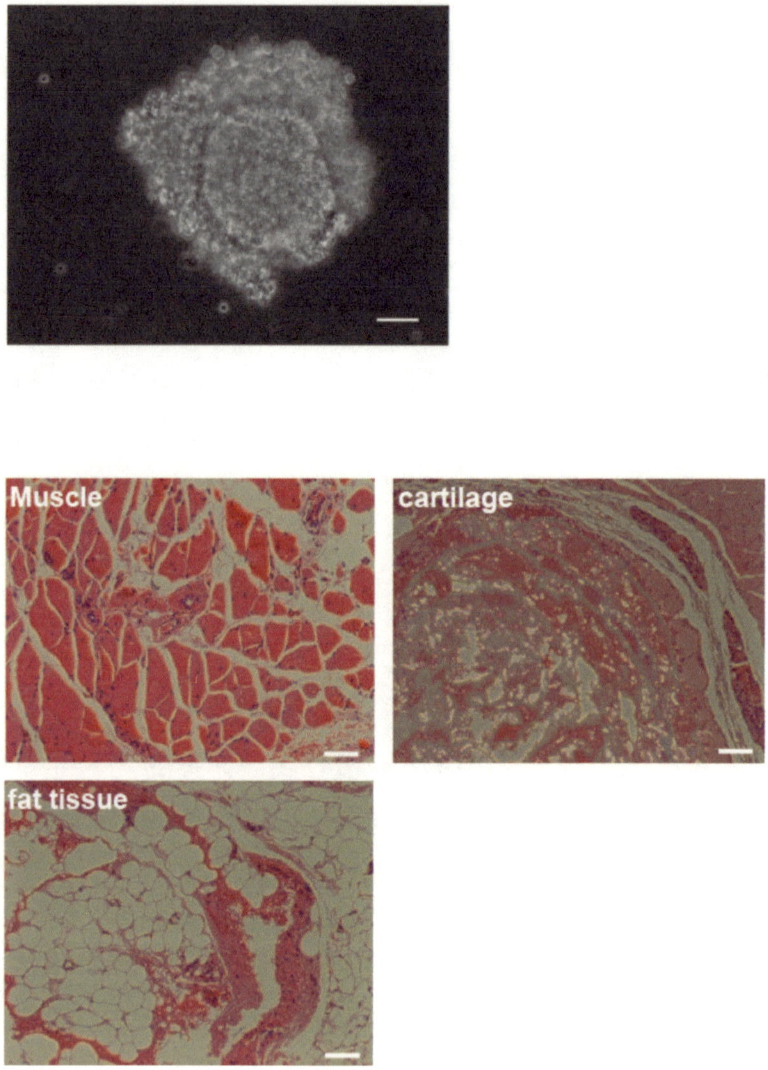

Fig. 5. Embryoid body formation and teratoma formation from the ES-like cells derived from a KS patient. a, Embryoid body derived from the iPS cells in vitro. Bars = 30 μm. **b,** Hematoxylin and eosin staining of sections of tumors generated from the ES-like cells derived from a KS patient. Bars = 60 μm.

4. Discussion

Male infertility has increasingly become a major health and social concern worldwide. At the same time, the development of molecular genetics and stem cell biology methods has led to major advances in genomic medicine. However, the genetic abnormalities that cause male infertility are still largely unknown.

Many of the most severe cases of nonobstructive azoospermia are in men with KS, which is the most common known genetic cause of azoospermia. In 90% of these cases, KS results from non-mosaic X chromosomal aneuploidy, in which the men carry an extra X chromosome (47, XXY). The other 10% of patients are mosaics who carry a combination of XXY/XY chromosomes (Graham et al., 1988). Approximately half of the Klinefelter's cases are thought to be paternally derived, and recent evidence suggests that KS may be related to advancing paternal age, although this finding is controversial (Jacobs et al., 1988; Lowe et al., 2001).

It is predicted that the aberrant expression of X chromosome-linked genes plays a role in the spermatogenic failure seen in men with 47, XXY (Vawter et al., 2007). However, the abnormal inactivation of the X chromosome in KS patients has not been demonstrated. Therefore, it is important to investigate the molecular mechanisms of X chromosome inactivation in KS, both to predict the extent of reproductive failure and to offer some treatment in the future.

The condition of nonobstructive azoospermia is typically characterized by small-volume testes and elevated follicle-stimulating hormone (FSH). Men with nonobstructive azoospermia have severely deficient spermatogenesis, resulting in so little sperm being produced in the testis that the sperm reach the ejaculate. Typically, a man with nonobstructive azoospermia cannot biologically father a child. Recently, however, microdissection testicular sperm extraction (micro-TESE) has shown promise for enabling nonobstructive azoospermia patients, including those with KS, to father a child (Schlegel, 1999; Schlegel, 2009). However, very little sperm is retrieved from the testis of nonobstructive azoospermia patients using this technique. We are therefore looking for new approaches to treat male infertility, including that due to nonobstructive azoospermia.

The reprogramming of human somatic cell nuclei towards pluripotency was recently achieved by the introduction of four defined factors, generating iPS cells (Lowry et al., 2008; Park et al., 2008; Takahashi et al., 2007; Yu et al., 2007). This remarkable technique enables the generation of patient-specific cells that can be used to model human diseases in tissue culture. We hypothesized that the generation of iPS cells derived from the testicular tissues of infertile men might provide a cell source for individual cell-based therapy that also avoids the ethical concerns associated with the use of stem cells. In support of this possibility, the establishment of pluripotent stem cells from the adult human testis under specific culture conditions was reported (Conrad et al., 2008) (Kossack et al., 2009). However, a later report indicated that the pluripotent stem cells of Conrad et al have a gene-expression profile that is similar to fibroblasts, and not similar to human ES cells (Ko et al., 2010). Therefore, the pluripotency of Conrad and colleagues' stem cells has been called into question. In addition, fibroblasts can be easily established from human testicular cultures (Chen et al., 1975).

Here we successfully generated iPS cells from the testicular tissue of KS patients by expressing four transcription factors in them (*OCT4, SOX2, KLF4,* and *C-MYC*). The morphology, proliferation, surface markers, and gene expression of these cells were similar to those of human ES cells. In the future, the investigation of iPS cells derived from KS patients may elucidate the mechanisms of infertility in KS patients and lead to a breakthrough in its treatment.

In particular, the X chromosome is the only chromosome in humans where one sex (female) has double the amount of genetic material as the other sex. In animals and hybridoma models, the presence of two active X chromosomes is lethal; in females, one of the X chromosomes undergoes random X chromosome inactivation (XIC) during embryogenesis, which is executed by a noncoding RNA called X chromosome inactivating transcript (XIST) (Hong et al., 2000). Therefore, since KS is not lethal, it is unlikely that both X chromosomes are active in men with KS, and XIC must occur. Such an abnormal inactivation of the X chromosome, which ordinarily would never happen in men, might be associated with some of the reproductive and cognitive sequelae associated with KS.

Data from several groups led to the assumption that most men with KS are born with spermatogonia (Lin et al., 2004) (Yamamoto et al., 2002) (Wikstrom et al., 2004). However, during early puberty, most likely after the initiation of spermatogenesis, the spermatogonia undergo massive apoptosis due to increased FSH levels in boys with KS (Wikstrom et al., 2004). It was later found that boys with KS have an adequate number of sperm in the ejaculate during early puberty (Paduch et al., 2009). The mechanism by which spermatogenesis is lost from puberty to adulthood is unknown. Three possible mechanisms have been suggested. The first is an intratesticular hormonal imbalance that results in hypersensitivity to the increasing intratesticular testosterone and estradiol concentrations (Paduch et al., 2009). The second is Sertoli cell dysfunction (Paduch et al., 2009). The third suggests that there are defects in spermatogonial stem cell renewal (Paduch et al., 2009). In particular, the abnormal pairing of X and Y chromosomes may lead to the loss of spermatocytes during meiosis (Paduch et al., 2009).

In the future, an understanding of the molecular mechanism governing X chromosome inactivation should help us resolve how male infertility develops and how its treatment might be improved.

5. Acknowledgements

This work was supported by a research grant from the Suzuki Urological Foundation and a Grant-in-Aid for Young Scientists (B) of the Japan Society for the Promotion of Science (JSPS).

6. References

Chen, AT., Fu, YS. & Reidy, JA. (1975). Human testicular cultures. II. Sertoli cells. *In Vitro* 11: 313-321.

Conrad, S., Renninger, M., Hennenlotter, J., Wiesner, T., Just, L., et al. (2008). Generation of pluripotent stem cells from adult human testis. *Nature* 456:344-349.

Ezeh, UI., Turek, PJ., Reijo, RA. & Clark, AT. (2005). Human embryonic stem cell genes OCT4, NANOG, STELLAR, and GDF3 are expressed in both seminoma and breast carcinoma. *Cancer* 104: 2255-2265.

Graham, JM, Jr., Bashir, AS., Stark, RE., Silbert, A. & Walzer, S. (1988). Oral and written language abilities of XXY boys: implications for anticipatory guidance. *Pediatrics* 81: 795-806.

Hong, YK., Ontiveros, SD. & Strauss, WM. (2000). A revision of the human XIST gene organization and structural comparison with mouse Xist. *Mamm Genome* 11: 220-224.

Jacobs, PA., Hassold, TJ., Whittington, E., Butler, G., Collyer, S., et al. (1988). Klinefelter's syndrome: an analysis of the origin of the additional sex chromosome using molecular probes. *Ann Hum Genet* 52: 93-109.

Kossack, N., Meneses, J., Shefi, S., Nguyen, HN., Chavez S., et al. (2009) Isolation and characterization of pluripotent human spermatogonial stem cell-derived cells. *Stem Cells.* 27:138-149.

Lanfranco, F., Kamischke, A., Zitzmann, M. & Nieschlag, E. (2004). Klinefelter's syndrome. *Lancet* 364: 273-283.

Lin, YM., Huang, WJ., Lin, JS. & Kuo, PL. (2004). Progressive depletion of germ cells in a man with nonmosaic Klinefelter's syndrome: optimal time for sperm recovery. *Urology* 63: 380-381.

Lowe, X., Eskenazi, B., Nelson, DO., Kidd, S., Alme, A., et al. (2001). Frequency of XY sperm increases with age in fathers of boys with Klinefelter syndrome. *Am J Hum Genet* 69: 1046-1054.

Lowry, WE., Richter, L., Yachechko, R., Pyle, AD., Tchieu, J., et al. (2008). Generation of human induced pluripotent stem cells from dermal fibroblasts. *Proc Natl Acad Sci U S A* 105: 2883-2888.

Klimanskaya, I., Chung, Y., Becker, S., Lu, SJ. & Lanza, R. (2006). Human embryonic stem cell lines derived from single blastomeres. *Nature* 444: 481-485.

Ko, K., Arauzo-Bravo, MJ., Tapia, N., Kim, J., Lin, Q., et al. (2010).Human adult germline stem cells in question. *Nature* 465: E1; discussion E3.

Ogawa, T., Arechaga, JM., Avarbock, MR. & Brinster, RL. (1997). Transplantation of testis germinal cells into mouse seminiferous tubules. *Int J Dev Biol* 41: 111-122.

Paduch, DA., Bolyakov, A., Cohen, P. & Travis, A. (2009). Reproduction in men with Klinefelter syndrome: the past, the present, and the future. *Semin Reprod Med* 27: 137-148.

Park, IH., Zhao, R., West, JA., Yabuuchi, A., Huo, H., et al. (2008). Reprogramming of human somatic cells to pluripotency with defined factors. *Nature* 451: 141-146.

Schlegel, PN. (1999). Testicular sperm extraction: microdissection improves sperm yield with minimal tissue excision. *Hum Reprod* 14: 131-135.

Schlegel, PN. (2009). Nonobstructive azoospermia: a revolutionary surgical approach and results. *Semin Reprod Med* 27: 165-170.

Takahashi, K., Tanabe, K., Ohnuki, M., Narita, M., Ichisaka, T., et al. (2007). Induction of pluripotent stem cells from adult human fibroblasts by defined factors. *Cell* 131: 861-872.

Vawter, MP., Harvey, PD. & DeLisi, LE. (2007). Dysregulation of X-linked gene expression in Klinefelter's syndrome and association with verbal cognition. *Am J Med Genet B Neuropsychiatr Genet* 144B: 728-734.

Wikstrom, AM., Raivio, T., Hadziselimovic, F., Wikstrom, S., Tuuri, T., et al. (2004). Klinefelter syndrome in adolescence: onset of puberty is associated with accelerated germ cell depletion. *J Clin Endocrinol Metab* 89: 2263-2270.

Yamamoto, Y., Sofikitis, N., Mio, Y., Loutradis, D., Kaponis, A., et al. (2002). Morphometric and cytogenetic characteristics of testicular germ cells and Sertoli cell secretory function in men with non-mosaic Klinefelter's syndrome. *Hum Reprod* 17: 886-896.

Yu, J., Vodyanik, MA., Smuga-Otto, K., Antosiewicz-Bourget, J., Frane, JL., et al. (2007). Induced pluripotent stem cell lines derived from human somatic cells. *Science* 318: 1917-1920.

17

Application of Magnet-based Nanofection in Embryonic Stem Cell Research

Ssang-Goo Cho, Sachin Honguntikar and Hyun Joo Lee
Department of Animal Biotechnology, Konkuk University, Seoul
Korea

1. Introduction

Embryonic stem (ES) cells are derived from the epiblast of the inner cell mass (ICM) of a blastocyst. These are pluripotent cells and can give rise to all derivatives of the three primary germ layers - ectoderm, endoderm, and mesoderm. ES cells require specific signals for specific lineage of differentiation - if injected directly into another body, ES cells will differentiate into many different types of cells, causing a teratoma (Wu et al., 2007).

The techniques for culturing mouse embryonic stem (ES) cells from the inner cell mass of the preimplantation blastocyst were first done in 1981 (Martin, 1981), and versions of these standard procedures are used today in laboratories throughout the world. The first successful derivation of human ES (hES) cells was reported by Thomson et al. (Thomson et al., 1998). They isolated and plated the cells onto mitotically inactivated MEF (mouse embryonic fibroblast) cells. In 2000, Reubinoff et al. confirmed that hES cells could be efficiently derived from surplus embryos and possess the differentiation potential under *in vitro* conditions. Since then, there has been rapid progress made and numerous studies have described the derivation of new hES cell lines including methods of growing both undifferentiated hES cells and their differentiated progeny. In last 6 years, there has been exponential increase in methods to improve culture conditions, differentiation patterns to produce human cells for transplantation and drug testing (Trounson, 2006, Gepstein, 2002) and genetic manipulation (Draper et al., 2004, Zwaka and Thomson, 2003).

Generating cultures of mouse or human ES cells that remain in a proliferating and undifferentiated state is multistep process. Typically, the inner cell mass of a preimplantation blastocyst is removed from the trophectoderm that surrounds it and cultured in the small plastic culture dishes containing growth medium supplemented with fetal calf serum. The culture dishes are sometimes coated with a "feeder" layer of non-dividing cells, which are often MEF cells that have been chemically inactivated so they will not divide. Mouse ES cells can be grown *in vitro* without feeder layers if the cytokine leukemia inhibitory factor (LIF) is added to the culture medium, but human ES cells do not respond to LIF.

The process of generating an embryonic stem cell line is somewhat difficult, so lines are not produced every time from the preimplantation-stage embryo maintained in a culture dish. However, if the plated cells survive, divide and multiply enough to crowd the dish, they are removed, and plated into several fresh culture dishes. The process of re-plating or subculturing of the cells is repeated many times and maintained for many months. Once the cell line is established, the original cells yield millions of embryonic stem cells. Embryonic stem cells that have proliferated in cell culture for six or more months without differentiating, are pluripotent, and appear genetically normal and then referred to as an embryonic stem cell line.

As long as the embryonic stem cells in culture are grown under appropriate conditions, they can remain undifferentiated condition. But, if cells are allowed to clump together, known as embryoid bodies and they start to differentiate spontaneously. They can form any types of cells like muscle cells, nerve cells, and many other cell types. Although spontaneous differentiation is a good indication that a culture of embryonic stem cells is healthy, it is not a proficient way to produce cultures of specific cell types. To generate cultures of particular types of differentiated cells like heart muscle cells, blood cells, or nerve cells, researchers tried to control the differentiation of embryonic stem cells. They changed the chemical composition of the culture medium, altered the surface of the culture dish, or modified the cells by inserting specific genes. Through years of experimentation, scientists have established some basic protocols or "recipes" for the directed differentiation of embryonic stem cells into some targeted cell types.

- Derived from the inner cell mass/epiblast of the blastocyst.
- Capable of undergoing an unlimited number of symmetrical divisions without differentiating (long- term self-renewal).
- Exhibit and maintain a stable, full (diploid), normal complement of chromosomes (karyotype).
- Give rise to differentiated cell types that are derived from all three primary germ layers of the embryo (endoderm, mesoderm, and ectoderm).
- Capable of integrating into all fetal tissues during development (Mouse ES cells maintained in culture for long periods can still generate any tissue when they are reintroduced into an embryo to generate a chimeric animal).
- Capable of colonizing the germ line and giving rise to egg or sperm cells.
- Clonogenic, which is a single ES cell can give rise to a colony of genetically identical cells, or clones, which have the same properties as the original cell.
- Express the transcription factor Oct-4, which then activates or inhibits a host of target genes and maintains ES cells in a proliferative, undifferentiating state.
- Can be induced to continue proliferating or to differentiate.
- Lacks the G1 checkpoint in the cell cycle. ES cells spend most of their time in the S phase of the cell cycle, during which they synthesize DNA.
- Unlike differentiated somatic cells, ES cells do not require any external stimulus to initiate DNA replication.
- Do not show X chromosome inactivation. In every somatic cell of a female mammal, one of the two X chromosomes becomes permanently inactivated. X chromosome inactivation does not occur in undifferentiated ES cells.

Table 1. Characteristics of embryonic stem cells

The ability of ES cells to develop into all cell types of the body has fascinated many scientists in recent years. The transcription factor Oct4 has been used as a key marker for ES cells and for the pluripotent cells of the intact embryo and its expression must be maintained at a high level for ES cells to remain undifferentiated (Mallon et al., 2006). However, the Oct4 protein itself is insufficient to maintain ES cells in the undifferentiated state (Matsuda et al., 1999). Two groups also identified another transcription factor, Nanog, which is essential for the maintenance of the undifferentiated state of mouse ES cells (Sato et al., 2004, Mitsui et al., 2003). By comparing gene expression patterns between different ES cell lines, other cell types such as adult stem cells and differentiated cells, genes that are enriched in the ES cells have been identified.

To understand the mechanism behind the undifferentiated property of ES cells and genes that play a vital role in the maintenance of the pluripotency of ES cells, development of efficient gene delivery system is necessary for embryonic stem cell research.

1.1 Significance of Gene delivery systems in embryonic stem cells

ES cells have generated great hope as a potential resource for cell and transplantion therapy, due to their ability to differentiate into many cell types (Odorico et al.,2001). However, such potential depends on efficient and reliable methods for specific tissue differentiation using several strategies, such as manipulation of the microenvironment by use of different cell culture conditions, and the use of growth factors or small molecules that can efficiently regulate cell differentiation at specific cell stages (Boheler et al., 2002, Thomson et al., 1998). Matrices designed for formation of specific tissues and genetic manipulation of stem cells include transient expression from exogenous constructs, expression of episomal vectors, stable expression from randomly integrated exogenous constructs, or site specific targeting by homologous recombination (Bethke and Sauer1997). The use of genetic manipulation in ES cells has provided a valuable tool in the research for gene function study (Dhara and Benvenisty, 2004).

For analyzing the molecular mechanism of the specific genes during proliferation and differentiation of ES cells, gene delivery into ES cells is very important technique. There are arrays of transfection reagents and methods which aim to achieve high efficiency in specific gene delivery. Determination of the best gene delivery approach for any particular experiment requires consideration of a number of different factors. Even with all the newest and currently available technologies, ES cells have very low efficiency in transduction during conventional transfection procedures, with their undifferentiated phenotype often being compromised. Many gene delivery systems are reported till today, like viral transfection, lipofection, and nucleofection, but all the methods have some or the other disadvantages. Here, we tried to explain several gene delivery systems used in ES cell research. Especially, we focus on magnet-based nanofection.

2. Gene delivery systems

In this chapter, we will mainly discuss about magnet-based nanofection and its benefits compared to the other viral and non viral gene delivery systems, but here we briefly explain several gene delivery techniques, such as calcium phosphate- (Jarcho, 1981), DEAE-dextran- (Gauss and Lieber, 1992), or liposome-mediated transfection (Schenborn and Oler, 2000), electroporation (Chu et al.,1987), nucleofection (Siemen et al., 2005), or viral vector-mediated transduction (Kaplitt et al., 1994). Under standard conditions, mammalian cells take up and express the externally applied DNA with very low efficiency. Various methods have been proposed to overcome this problem.

- Calcium phosphate-mediated transfection; Calcium phosphate relies on precipitates of plasmid DNA formed by its interaction with calcium ions. It is cheap and has been routinely used for both transient and stable transfection in a variety of cell types and easy technique to perform. However, this method is prone to high variability and is not suited for *in vivo* gene transfer.
- DEAE-dextran-mediated transfection; Diethylaminoethyl-dextran (DEAE–dextran) is a cationic polymer that tightly associates with negatively charged nucleic acids. This method successfully delivers nucleic acids into cells for transient gene expression, all of which seek to maximize the uptake of DNA and to minimize the cytotoxic effects. In this method, cells are exposed briefly to a high concentration of DEAE-dextran-DNA and then to chloroquine diphosphate, which acts as a facilitator for transfection. However, this technique is not generally useful for long-term transfection studies (stable transfection) that rely upon integration of transferred DNA into the chromosome.
- Liposome–mediated transfection; Liposomes are lipid bilayers entrapping a fraction of aqueous fluid. DNA is spontaneously associated to the external surface of cationic liposomes (by virtue of its negative charge) and these liposomes will interact with the cell membrane. However the disadvantages of this technique are poor target selectivity and decreased efficiency compared viral vectors and transient gene expression.
- Electroporation; Electroporation (or electropermeabilization) is a mechanical method which is used to introduce polar molecules into a host cell through the cell membrane. In this technique there is a significant increase in the electrical conductivity and permeability of the plasma membrane caused by an externally applied electrical field. It is generally used in molecular biology for introducing any new molecule into a cell, such as loading it with a molecular probe, a drug that can change the cell's function, or a piece of coding DNA.
- Nucleofection; Nucleofection is a gene delivery system, in which, there is a combination of electrical parameters, generated by a device called nucleofector, with cell-type specific reagents. The substrate is directly transferred into the cell nucleus and to the cytoplasm. The major concern of this technique involves safety risks, lack of reliability, and high costs.
- Viral vector-mediated transduction; Viruses are obligate intra-cellular parasites, designed through the course of evolution to infect cells, often with great specificity to a particular cell type. They tend to be very efficient at transfecting their own DNA into the host cell, which is expressed to produced new viral particles. By replacing genes that are needed for the replication phase of their life cycle (the non-essential genes) with foreign genes of interest, the recombinant viral vectors can transduce the specific cell type. Although a number of viruses have been developed, interest has centred on four types; retroviruses (including lentiviruses), adenoviruses, adeno -associated viruses, and herpes simplex virus type 1. However, the major disadvantages of viral vectors are that all viral vectors induce an immunological response to some degree and may have safety risks (such as insertional mutagenesis), and toxicity problems.

3. Magnet-based nanofection

In recent years, nanotechnology has reached such a level that allows nanoparticles with magnetic properties to be functionalized with specific ligands for many applications, such as site specific drug or gene delivery, magnetic resonance imaging, hyperthermic treatment for cancer cells and tumour targeting (Latorre and Rinaldi, 2009). The term nanoparticle refers to particles less than 1 mm in size, typically less than 200 nm. Thus they are of a similar length scale to biomolecules, making nanoparticles ideal for combining with biomolecules for medical applications (Jain, 2007, Freitas, 2006).

Several research groups have independently developed magnet-based nanofection methods. This method of transfection was inspired by the concept of magnetic drug targeting (Plank, Anton et al., 2003), which dates back to the late 1970s (Hafeli, 2004), where drugs are bound covalently or noncovalently to magnetic particles and suitable magnetic gradient fields are used to accumulate such 'magnetic' drugs at target sites and/or to retain them there.

Magnet-based nanofection is a gene delivery system that utilizes nanotechnology. It is achieved by the application of a magnetic field to superparamagnetic iron peroxide particles which are associated with gene vectors (Lee et al., 2008). In this technique, the cationic polymer polyethyleneimine (PEI) is coated with superparamagnetic nanoparticles (tsMAG-PEI) are complexed to plasmid DNA. Apart from suitable magnetic nanoparticles, magnet-based nanofection also requires suitable magnetic fields, which are provided by the Magnetofactor plate, especially designed for magnet-based nanofection. This plate will produce a pattern of higher and lower densities of transfected cells according to the geometry of the magnetic field lines. As a reporter gene, enhanced GFP (green fluorescent protein) was transfected in D3 embryonic stem cells by magnet-based nanofection and gave better result (45 %), when compared to Fugene 6- and liposome-mediated transfection (15 %) (Plank, Schillinger et al., 2003).

There are several different types of commercially available magnet-based nanofection systems, which can be purchased from the OZ biosciences (France) or chemicell (Germany). PolyMag is a universally applicable magnetic particle preparation for high efficiency gene delivery. It can be used to deliver DNA into cells with a single step procedure and has been used successfully with plasmid DNA, antisense oligonucleotides, and siRNAs. While, CombiMag can be combined with any commercially available transfection reagent like poly cations and lipids which can be associated with plasmids DNA or siRNA. Recently, PolyMag, a type of magnet-based nanofection, has been reported to provide highly efficient transfection in numerous cell types (Gersting et al., 2004).

The coupling of magnetic nanoparticles to gene vectors in presence of static magnetic feild has been shown to result in dramatic increase in transfection efficiency of reporter genes when compared with other gene delivery methods (Plank, Anton et al., 2003, Plank, Schillinger et al., 2003). Magnet-based nanofection has an improved dose dependent response and enhances transfection efficiency in both permanent and primary airway epithelial cells and also it leads to significant transgene expression at very short incubation times in an ex vivo airway epithelium organ model (Gersting et al., 2004). Magnet-based nanofection in human umbilical vein endothelial cells (HUVEC) had an increased transfection efficiency of a luciferase reporter gene up to 360-fold in comparison with various conventional transfection systems. Apart from the efficicency, there was only an up to 1.6-fold increase in toxicity caused by magnet-based nanofection, suggesting the advantages of this method (Krotz et al.,2003).

Here, we briefly explain six types of commercially available magnet-based nanofection reagents.

1. PolyMag has been optimized to be used with all types of nucleic acids and can be applied to any magnetic particle preparation for increased efficiency of nucleic acid

delivery. This reagent has been used successfully with plasmid DNA, antisense oligonucleotides, and siRNA (Kamau et al., 2006, Namiki et al., 2009).

2. PolyMag Neo represents the latest development in magnet-based nanofection reagent and can be applied to wide variety of nucleic acid delivery, especially in the primary cell lines, where it is hard to transfect. Its universality and strong formulation were reported to allow in achieving higher gene expression level (Kim et al., 2009).
3. CombiMag has been used successfully with plasmid DNA, antisense oligonucleotides, mRNA, siRNA, and viruses. This reagent is designed that it can combine with any commercially available transfection reagent such as cationic polymers and lipids (Kadota et al.,2005).
4. SilenceMag provides more efficient method for delivery of siRNA even at low doses. This formulation was reported to provide reliable high gene silencing efficiencies in primary cell lines and introduce large quantities of siRNA duplexes into cells leading to exceptional gene knockdown effects (Li et al.,2008).
5. ViroMag is reported to be applicable on a various cell lines and primary cells (hard to transfect cell lines) and can be combined with any viruses (specifically designed for retrovirus and lentivirus). It is reported that combination with adenovirus allows up to 500-fold improvement of gene expression compared with other commercial available method (Kamau et al., 2006).
6. NeuroMag is dedicated to be a magnet-based nanofection reagent for neurons. It is perfect for primary neurons and can also be used for glial cells. This reagent is also called magnetoencephalogram (MEG), an analogoue of electroencephalogram (EEG). However, in MEG, the magnetic fields are produced by ion flow and are associated with neuronal activity rather than the electric potentials measured in the EEG. By this system magnetic feilds can be recorded by neuronal activity occurring in the brain (Jensen and Tesche, 2002).

Overall, magnet-based nanofection is easy to use and unique solution for transfection in various cell lines. It has many advantages than that of other gene delivery systems, as follows.

1. High transgene expression.
2. Lead to transfect non-permissive cells (ability of transfect in primary cells which are usually hard to transfect).
3. Rapidly accumulate all virus doses on the cells/ culture medium.
4. Synchronize cell adsorption/infection without alteration to the virus or its genome.
5. Significantly progress virus infectivity.
6. Confined to target specific area of transduction.
7. Can be used for various types of nucleic acids including plasmid DNA, siRNA, oligonucleotides, linearized DNA, double stranded RNA, mRNA, shRNA, virus, and protein.
8. Simple, easy to handle, and non-toxic.
9. Compatible with and without serum-containing medium.

4. Methodology of magnet-based nanofection

4.1 Optimization of Magnet-based nanofection

4.1.1 Magnet-based nanofection reagents

Magnet-based nanofection reagents are universally applicable magnetic particle preparation for high efficiency gene delivery. The gene of interest to be transfected is mixed with magnetic particles in a single step. Magnet-based nanofection reagents have been used successfully with plasmid DNA, antisense oligonucleotides, and siRNA.

4.1.2 Magnetofactor plate

Apart from suitable magnetic nanoparticles, magnet-based nanofection reagent requires appropriate magnetic fields. A magnetofactor plate (Chemicell, Germany) is specially designed for magnet-based nanofection reagent.

Due to its special geometry, it produces powerful magnetic fields under different plate formats (T-75 flasks, 60 -100 -mm dishes, 6, 12, 24, and 96-well plates). The magnetofactor plate is designed in such way that it exerts a heterogeneous magnetic field that magnetizes the nanoparticles in solution, forms a very strong gradient, and covers all the surface of the plate.

4.1.3 Generalized protocol of magnet-based nanofection

For the better efficiency of transfection, it is always recommended to seed the cells on the previous day of the transfection. The suitable cell density will depend on the growth rate and the cells conditions. Cells should be 60-90 % confluent during magnet-based nanofection. Immediately preceeding transfection, the medium can be replaced with fresh medium (optionally without serum) if necessary.

It is always better to incubate at least 24 hours before exposing the transfected cells to selection media. Vectors are prepared in medium without serum or in PBS. Along with the standard magnet-based nanofection protocol, the serum- and supplement-free vector cocktail is added to the cells that are cultured in medium. Therefore, the addition of the transfection reagent will result in the further dilution of standard culture medium. For most cell types, change of media is not required after magnet-based nanofection. However, it may be necessary for cell lines that are sensitive to serum/supplement concentration. Alternatively, the cells may be kept in serum-free medium during magnet-based nanofection. In this case, a changing of media may be required after magnet-based nanofection.

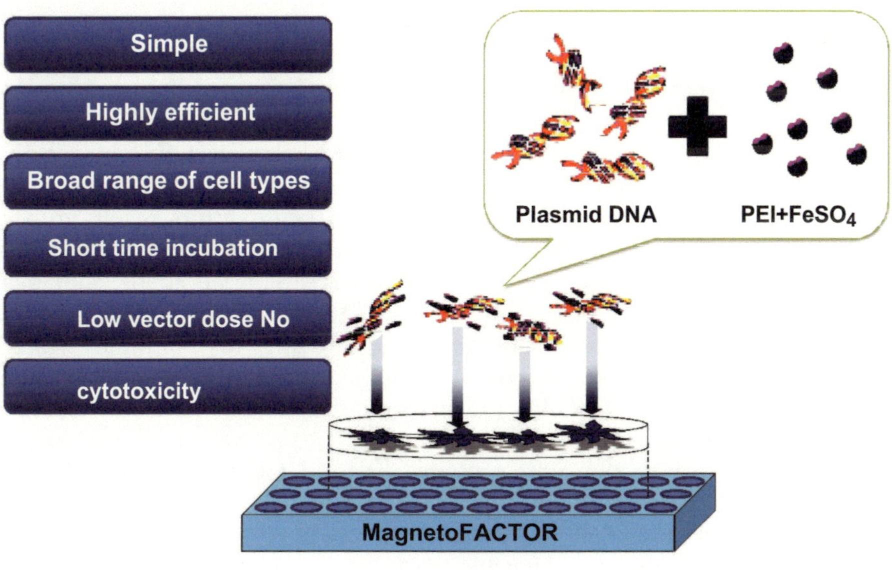

Fig. 1. Generalized view of magnet-based nanofection

4.1.4 Time

Usually, transfection reagents must be in contact with cells for a period of time, then additional medium is added or the medium is replaced to assist decrease toxic effects of the reagent. The optimal transfection time depends on the cell lines, transfection reagent, and nucleic acid used. These reagents usually require significantly shorter time when compared to other transfection reagents and can be decreased to as little as 30 minutes with certain cell lines. This shortened transfection time may help in considerably decreased risk of cell death.

4.1.5 Serum condition

Transfections by magnet-based nanofection protocols often require serum-free conditions for optimal performance, because serum can interfere with many commercially available transfection reagents. It has to be noted that the best results are obtained when variability is minimized among lot of serum.

4.1.6 DNA delivery mechanism of magnet-based nanofection

The magnetic nanoparticles are majorly made up of iron oxide particles, which are fully biodegradable, coated with specific cationic property of the molecules depending upon the applications. Their association with the gene vectors (DNA, siRNA, ODN, virus, etc.) is achieved by salt-induced colloidal aggregation or electrostatic interaction. The magnetic particles are then concentrated towards the target cells by the force of an external magnetic field generated by magnets. The cellular uptake of the genetic material is accomplished by endocytosis and pinocytosis, two biologically natural processes. Consequently, membrane architecture and structure are reported to stay intact, in contrast to other physical transfection methods that damage the cell membrane (Kamau et al., 2006). The nucleic acids are then released into the cytoplasm. However, the releasing of nucleic acid may be by different mechanisms depending upon the formulation used:

1. The proton sponge effect caused by cationic polymers coated on the nanoparticles promotes endosome osmotic swelling, disruption of the endosome membrane, and intracellular release of DNA.
2. The destabilization of endosome by cationic lipids coated on the particles that release the nucleic acid into cells by flip-flop of cell negative lipids and charge neutralization
3. The usual viral infection mechanism when virus is used in magnet-based nanofection works for primary cells and the cells which are hard to transfect, that are not dividing or slowly dividing, meaning that the genetic materials can go to the cell nucleus without cell division. Coupling magnetic nanoparticles to gene vectors results in a dramatic increase of the uptake of these vectors and consequently high transfection efficiency.

The biodegradable cationic magnetic nanoparticles are reported to be non toxic at the recommended doses and also even at higher doses (Mykhaylyk et al., 2010). Complexes of gene vectors and magnetic nanoparticles are seen in cells after 10–15 minutes which is much faster than any other transfection method. After 24, 48, or 72 hours, most of the particles are localized in the cytoplasm, vacuoles (membranes surrounded structure into cells) and occasionally inside the nucleus.

Thus, magnet-based nanofection uses a simple protocol for high efficient transfection in most of cell lines, up to several thousand fold increased levels of transgene expression upon short-term incubation (about 15~20 minutes). Moreover, great transfection rates and transgene expression levels are achievable with extremely low vector concentrations and without intense cytotoxicity.

5. Application of Magnet-based nanofection

5.1 Cell lines

The introduction of nucleic acids into cells is relevant for therapeutic (gene therapy) purposes. Many scientists have reported that high gene delivery can be obtained by magnet-based nanofection in various cell lines. Magnetic drug targeting was used to enhance nonviral transgene expression in airway epithelial cells and tracheal airway epithelium ex vivo (Gersting et al., 2004). They also found that efficiency of transfection by magnet-based nanofection increased compared to other non viral gene delivery methods like polyfection and lipofection by 2–3 orders of magnitude.

Transfection efficiency of a luciferase reporter gene was increased up to 360-fold by magnet-based nanofection in umbilical vein endothelial cells (HUVEC) compared to various conventional methods (Ido et al.,2002). In contrast, there was only an up to 1.6-fold increase in toxicity caused by magnet-based nanofection, signifying that the advantages of magnet-based nanofection outbalanced the increase in toxicity. It is quite simple, effective, and cheap improvement of plasmid gene delivery that results in hundred fold increased efficiency in gene expression in target cells (Plank, Anton et al., 2003). In this respect, they established a powerful tool in molecular physiological research and at the same time demonstrated its potential in gene therapies for cardiovascular diseases. As part of this study, they also showed the potency of magnet-based nanofection for the delivery of siRNA (small interfering RNA).

Among the many strategies that have been proposed, immuno gene therapy has been one of the most of the common approaches. The authors have introduced cytokine genes into tumor cells which when re-administered after irradiation to a patient, are supposed to elicit an immune response against tumor antigens by the immunostimulatory effect of the expressed cytokine in general.

In another study, the gene coding for human GM-CSF (granulocyte macrophage colony stimulating factor) under the influence of the CMV promoter in magnetic formulation was administered twice in a one week interval prior to surgery into the biologically active margins of the fibrosarcoma. Equal volume of plasmid DNA and magnetic particles (trans-MAG-PEI, chemicell, Berlin, Germany) was mixed in physiological saline. A neodymium-iron–boron permanent magnet was fixed on the tumor adjacent to the injection site during vector injection. They found that expression of the magnetofected cytokine gene was increased in tumor (Gersting et al., 2004).

5.2 Embryonic stem cells and magnet-based nanofection

5.2.1 The Magnet-based nanofection method: non-viral gene transfection technologies for mammalian cells

Although magnet-based nanofection was studied in various cell lines, Lee et al. originally used this technology for gene delivery in embryonic stem cells in 2008 (Lee et al., 2008). In this method, superparamagnetic nanoparticles (tsMAG) coated with cationic polymer polyethyleneimine (tsMAG-PEI) were combined with plasmid DNA. The tsMAG-PEI and DNA complex was generated by salt-induced aggregation and used for magnet-based nanofection.

In this study, we compared the transfection efficiency of different DNA delivery methods such as PolyMag-mediated magnet-based nanofection, FuGENE 6-mediated transfection, liposome–mediated transfection, nucleofection, and calcium phosphate-mediated transfection both in NIH3T3 and embryonic stem cells.

		Magnet-based nanofection	FuGENE 6-mediated transfection	Liposome–mediated transfection	Nucleofection (6well)	Calcium phosphate-mediated transfection
PlasmidDNA µg per 24-well plate		0.5-2	0.2-0.4	0.8 -1.2	1-5	1.1
Reagent µl per 24-well plate		0.5-2	0.6-1.8	2	100	1X HBS = 18, 2.5M CaCl2 =1.1
Cells per well (24-well plate)		70-80 %	70-80 %	70-80 %	70-80 %	70-80 %
Transfection efficiency	NIH3T3	> 90 %	> 90 %	> 80~90 %	> 80~90 %	20 ~100 %
	ES cells	> 70~80 %	> 40~50 %	> 40~50 %	-	-

Table 2. Comparison of transfection efficiency of different DNA delivery systems

Under standard conditions, mammalian cells take up and express externally applied DNA with very low efficiency. This is commonly due to the lipid bilayer of the eukaryotic cell membrane, which oppose the entry of charged molecules into a cell. All transfection methods cannot be applied to all types of cells or experiments. The different methods vary greatly with respect to the level of gene expression that can be achieved.

Liposome–mediated transfection is a cationic liposome based reagent that provides high transfection efficiency and increased levels of the transgene expression in a range of mammalian cell lines *in vitro* using simple protocol (Dalby et al., 2004). However, this transfection reagent had highest efficiency and also highest toxicity. These results indicate that liposome–mediated transfection cannot be a widely accepted gene delivery method in ES Cells.

FuGene 6-mediated transfection is one of the non liposomal transfection methods and has been used since 1997. Since that time, its popularity has increased due to its ease of use, minimal to no cytotoxicity, and the high level of transfection in various cell lines. This reagent can be easily used to transfect cells, very efficient and little numbers of cells are required for quantifiable response by following the standard protocol. High levels of transfection can be obtained by this method, compared to the traditional reagents (Mykhaylyk et al.,2007, Namiki et al., 2009, Plank, Schillinger et al., 2003). It was also reported that the efficiency of transfection by Fugene 6 in stem cells was very low (15 %) (Siemen et al.,2005). These results indicate that FuGENE 6 cannot be feasible method for ES cells.

Recently, it has been reported that magnet-based nanofection gave a significantly higher efficiency (45 %) of gene delivery in stem cells than did the FuGENE 6-mediated transfection method (15 %) and found that these cells maintained their usual un-differentiated characteristics of self-renewal and pluripotency for a long time (> 50 passages) (Lee et al., 2008).

5.2.2 Application of magnet-based nanofection in ES cell research

Many non viral systems have come up recently for gene delivery in embryonic stem cells. But due to low efficiency levels, scientists started focusing on other alternative methods. Considering that these non viral gene delivery methods do not significantly affect basic properties of ES cells.

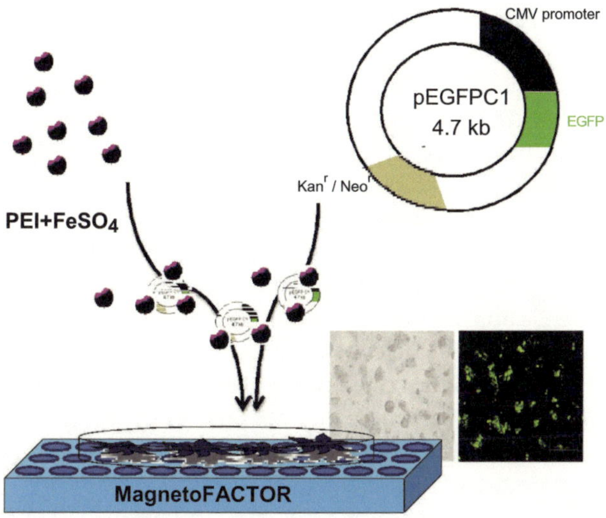

Fig. 2. Applicaton of the magnet-based nanofection method in ES cells

In this chapter, we discuss about magnet-based nanofection in detail. In this technique, plastic tissue culture dishes were pre-treated with 0.1 % gelatin in 24-well plate on previous day of transfection. Undifferentiated mouse embryonic stem cells (D3) were cultured in Dulbecco's modified eagle's medium (DMEM) and transfected with a CMV promoter-driven enhanced green fluorescent protein (pEGFP) expression vector through magnet-based nanofection. For the transfection method, 1 µg pEGFP vector was suspended in 98 µl serum free buffer and 1 µg PolyMag was added to the mixture and incubated at room temperature for 15~20 minutes. The complex was added to the serum-free DMEM medium in each of the 24 wells. The plates were placed on the magnetofactor plate device for 15~20 minutes at 37^0 C in 5 % CO_2 incubator. After selection by using G418 antibiotic for 2 weeks, the positive colonies were subcultured separately in a new cell culture dishes which were pretreated with 0.1 % gelatin. To confirm that the cells are in undifferentiated state, some of the basic confirmatory tests were performed.

1. AP staining protocol:
- Cells were cultured on sterile glass cover slips or slides overnight at 37 °C.
- Add fixative solution for 30 seconds.
- Rinse gently in deionised water for 45 seconds.
- Add to alkaline-dye mixture and incubate at 18° C ~ 26° C for 15 minutes in dark.
- Wash the excess dye with distilled water and observe under microscope.

2. Immunocytochemistry protocol:
- Fix the cells in 4 % PFA (paraformaldehyde) for 20~30 minutes at 37° C.
- Permeabilization of cells using 0.1 % triton X

- Incubate the cells in 10 % goat serum for 1 hour at room temperature.
- Incubate the cells with the primary antibody at 4° C overnight.
- Rinse the cells in 1 % goat serum for 10 minutes.
- Incubate the cells with secondary antibody.
- Mount coverslips and visualize under microscope.

3. FACS (*Fluorescence-activated cell sorting*) analysis protocol:

- Harvest the cells using 0.25 % trypsin and resuspend the cells in PBS (Phosphate buffer saline).
- Fix the cells in 0.5 ~ 1 ml of cold 70 % ethanol and incubate for 20~30 minutes at 4° C.
- Add 0.5 ~ 1 ml of cold PI (propidium iodide) solution to the cell pellet.
- Suspend the cells immediately and acquire using flow cytometry.

4. EB (Embryonic body) formation protocol:

- Harvest the cells using 0.25 % tryPLE.
- Decant the supernatant and resuspend the cells in ES medium.
- Gently mix the cells to form a single cell suspension.
- Spin down the cells and remove the supernatant and resuspend the cells in ES medium.
- Count the cells and add ES medium without LIF (Leukemia inhibitory factor).
- Incubate the cells for 24 hours at 37 ° C incubator.

5. Spontaneous differentiation protocol:

- EB (Embryonic body) formation by the 8 day induction method ($4^-/4^+$).
- EB's were plated onto 0.1 % gelatin-coated plates.
- Cultured for 15 days on N2 medium with 20 ng/ml bFGF (basic fibroblast growth factor).

6. Teratoma formation protocol:

- Embryonic stem cells are harvested.
- Centrifuge the cells for 5 minutes at 1, 200 rpm.
- Inject cells into the leg muscle of 4-week-old SCID-beige mice (male).
- After 10 weeks, examine for the teratoma formation.

7. Methylation analysis protocol:

- Isolation of genomic DNA and restriction enzyme digestion.
- Alkaline lysis and denaturation of DNA at 37° C for 15 minutes.
- Add 2 x volumes of 2 % low melting agarose to the DNA solution and mix thoroughly.
- Prepare the agarose beads by pipetting 10 µl aliquots of DNA/agarose mixture in cold mineral oil.
- Transfer beads in the tube containing 1 ml of modifying solution.
- Incubate the tubes for 4 hours at 50°C in dark.
- Wash the beads 6 times for 15 minutes with tris EDTA (ethylene diamine tetraacetic acid).
- Incubate for 15 minutes in 0.2N NaOH (sodium hydroxide) and wash the beads for 15 minutes with double distilled water.
- Methylation levels are checked by PCR (Polymerase chain reaction).

8. Chimera & germline transmission:

- Construct the homologous recombinant gene and transfect in cultured mouse embryonic stem cells by electroporation.
- Positive cells are selected by antibiotic mediated resistance marker provided in insertion cassette.
- Targeted gene is inserted into diploid mouse blastocyst.

- The injected blastocysts were implanted into pregnant female surrogate mouse, which directs the embryos to full term development and give birth to a mouse whose germline is derived from the donor mouse's ES cells.

So, the magnet-based nanofection is considered as extremely easy, less time- consuming, non cytotoxic and did not affect properties of ES cells like proliferation or differentiation. Magnet-based nanofection system may be the tool of choice for gene delivery and expression in ES cells.

5.3 Induced pluripotent stem (iPS) cells and magnet-based nanofection

5.3.1 Introduction of iPS cells

Most of the research on ES cells was on pluripotent stem cells, which are derived from the inner cell mass (ICM) of embryos in the blastocyst stage of development (Martin, 1981). Self-renewal and pluripotency are the two most essential characteristics of ES cells (Martin, 1981, Brimble et al., 2007) These cells owe the potential to differentiate into any cell type and provides a promising application in regenerative medicine (Mimeault et al.,2007, Keller, 2005). ES cells are the acknowledgment as a cell culture model for researching developmental mechanisms and their therapeutic modulations. The proliferative and developmental capacity of ES cells promises an essentially unlimited supply of specific cell type for basic research and transplantation therapies. However, one of the major concerns in ES cell research has been the ethical implications in using stem cells which are derived from embryo and also the therapeutic applications of stem cells depends on the availability that are limited by technical, ethical, or immunological considerations (Lowry et al., 2008, Wernig et al., 2007). Somatic cells reprogramming can be achieved by viral-mediated transduction using defined transcription factors (Feng et al.,2009). The iPS cells showed ES cell-like properties, such as morphology, gene expression profiles, differentiation into three germ layer lineages, formation of teratomas, and epigenetic status of several pluripotency markers (Lee et al., 2008). Their therapeutic potential is thought to be similar to that of ES cells in several disease models (Martin, 1980). Various reports from many labs gave contributed to a growing list of reprogramming factors used for iPS cells generation, such as Oct4, Sox2, Klf4, c-Myc, Lin28, Tcl1, Esrrb, Sall4, miRNA, SV40LT antigen, and telomerase reverse transcriptase (hTERT). Several small molecules also have been reported that enhances ES cell-like colony formimg efficiency such as supplementation with AZA (Azacytidine), VPA (Valproic acid), TSA (Tricholoro acetic acid), SAHA (suberoylanilide hydroxamic acid), BIX (Histone methyl transferase inhibitor), BayK (Calcium channel agonist), RG108 (DNA methyl transferase inhibitor), hypoxia condition, protein extract treatment, and knockdown of p53, p21, mdm2, and p16 (Plank, Anton et al., 2003, Plank, Schillinger et al., 2003).

5.3.2 Current trend of iPS cells research

Many researchers have discussed different methods for generating iPS cells. In this section, we will discuss the current techniques that are employed to introduce the reprogramming factors required for iPS cells generation (Zhao et al., 2009). Retroviral vectors were used for the initial derivation of iPS cells by Takahashi et al. But retroviral vectors may lead to many obstacles like oncogenesis and teratoma development in therapeutic applications. To avoid or reduce the potential oncogenic effect of genomic integration of viral vectors, non viral gene delivery methods was introduced (Wiles and Johansson, 1999, Lakshmipathy et al., 2004, Zeng et al., 2003, Plank, Schillinger et al., 2003). These non-viral gene delivery systems appear to have numerous advantages, including ease to manipulate, increased stability, low cost, safety, and high flexibility regarding the size of transgene delivery (Kircheis, Wightman et al.,2001, Nimesh et al., 2006). Several researchers reported non-viral gene delivery methods, such as adenoviral vectors (D. Kim et al., 2009), non-integrating plasmids (Papapetrou et al., 2009), Cre recombinase/loxP system (Wang et al., 2007), piggyBac transposon-based system (Yusa et al., 2009), RNA transfection (Fusaki et al.,2009), and reprogramming by recombinant proteins (D. Kim et al., 2009). However, the efficiency of non-viral gene delivery systems was less compared to that of viral gene delivery methods.

Sl.no.	Method of Gene delivery	Factors for Pluripotency	Cell lines	Efficiency	Reference
1	Virus mediated Transfection	Oct4, Sox2, Klf4, c-Myc	Fibroblast	0.0001 %	Takahashi and Yamanaka, et al.,2006
2	Virus mediated transfection	Oct4, Sox2, Nanog and Lin28	fibroblast	> 0.01 %	Yu et al., 2007.
3	Lentiviral vectors	Pax5 shRNA, C/EBPa	Mouse B cell	> 0.001 %	Hanna et al., 2008
4	Retroviral vectors	Oct4, Sox2, Klf4	Neural stem cells	0.1-1.0 %	Kim et al., 2008
5	Retroviral vectors	Oct4, Sox2, Klf4	hepatocytes	0.5-1 %	Aoi et al., 2008
6	Retroviral vectors	Oct4, Sox2, Klf4	MEF	0.001-0.010 %.	Nakagawa et al., 2008
7	Lentiviral vectors using Cre recombinase /loxP system	Oct4, Sox2, Klf4, c-Myc	human fibroblast	0.001-0.002 %	Soldner et al., 2009
8	Piggy BAC transposons	Oct4, Sox2, Klf4, c-Myc	fibroblast	2.5 %	Yusa K et al., 2009
9	Episomal Vectors	OCT4, SOX2, NANOG, LIN28, c-Myc, KLF4, and SV40LT	Human foreskin fibroblasts	~ 0.1%	Yu et al., 2009
10	Lipofection	OCT4, SOX2, NANOG, LIN28, c-Myc, and KLF4	Mouse fibroblasts	> 0.01 %	Okita et al., 2008

Table 3. Different methods and sources for generating iPS cells

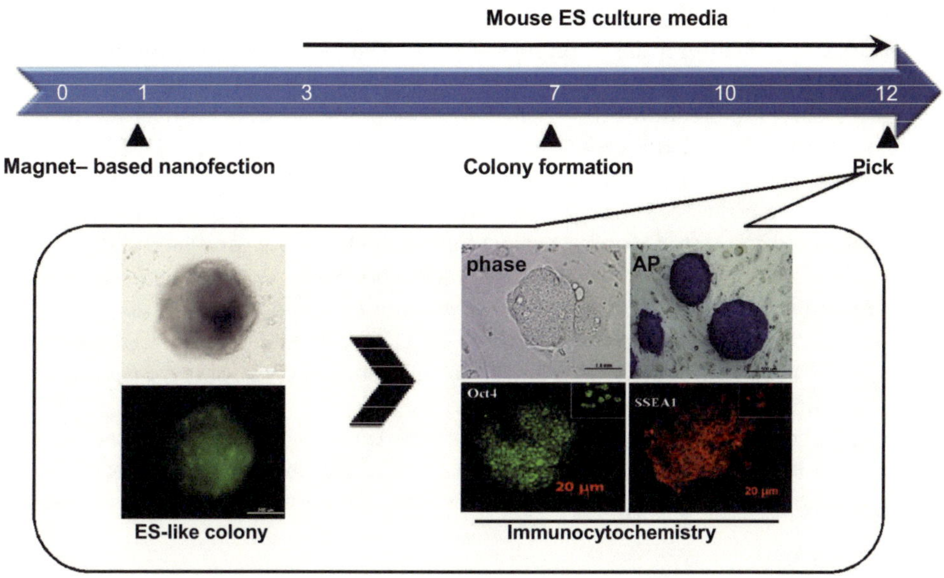

Fig. 3. Schematic representation of generation of iPS cells from MEF and characterization of ES-like colony with analysis of pluripotent marker expression

5.3.3 Generation of iPS cells using magnet-based nanofection

Since the initial derivation of iPS cells by Shinya Yamanaka's group, the field has advanced at a rapid pace. Main progress has been made in identifying new strategies to enhance the reprogramming efficiency and methods to improve clinical safety (Plank, Schillinger et al., 2003). In a recent study, generation of iPS cells using magnet-based nanofection was proved to be a very simple, highly efficient, cost -effective, and minimal to no cytotoxicity, implying that this system may be beneficial for generating safe iPS cells, compared to the other methods described previously.

Induced pluripotent stem cells were first generated by transfecting four factors in MEF cells (Takahashi and Yamanaka, 2006). For transfection, MEF cells were seeded (70~80 %) into each well of 96-well plates on the previous day of transfection and then four factors (Oct4, Sox2, Klf4, and c-Myc) was transfected into MEF by the magnet-based nanofection and ES cell media was changed alternative day. Some colonies of ES-like morphology were observed on seven days after transfection, and the iPS cell colonies generated by transduction of four factors were picked on day 12. The iPS cells were cultured in ES cell media on STO feeder cells, using standard embryonic stem cell culture protocol (Thomson et al., 1998, Xiao et al.,2006).

Magnet-based nanofection-derived iPS cells showed ES cell-like characteristics, such as expression of high levels of akaline phosphatase, expression of undifferentiated ES cell-specific cell surface markers (such as SSEA-1 and Oct4), and expression of endogenous pluripotency genes (Oct4, Nanog, Sox2, Klf4, and c-Myc). These cell lines also differentiated into various cell types through the embryid body (EB) formation. The three germ layer cell differentiation markers was observed through immunocytochemistry; endoderm (Tie2 and AFP), mesoderm (Brachury and Tn1) and ectoderm (Tuj1 and A2B5). Of particular interest, transcriptome, and proteome expression profiles of iPS cells confirmed that the iPS cells more similar to ES cells, but different from MEF. Magnet-based nanofection-derived iPS cells identified 2,102 differentially expressed genes (DEGs) between MEF and ES cells (J1 and R1). Clustering analysis was performed for the expression of DEGs for MEF, ES cells, and magnet-based nanofection-derived iPS cells. Gene expression profiles for the magnet-based nanofection derived iPS cells was highly similar to that of ES cells (J1 and R1) and clearly distinct from that of MEF. As expected, 80-90 % of proteins found in magnet-based nanofection derived iPS cell lines were similar to those in ES cell line (D3). When transcriptome was compared to proteomic analysis, it showed a more significant difference in protein expression patterns between iPS cells and MEF. According to gene ontology analysis, functional classification appeared to be similar between cell lines with stemness (magnet-based nanofection derived iPS cells and ES cells) and MEF in most biological processes, such as protein metabolism and modification, cell cycle, transport, immunity, cell proliferation, differentiation, and apoptosis etc.

Recent study showed that primate ES cells can be derived by magnet-based nanofection from mouse embryonic fibroblast cells. In near future, this novel gene delivery system may open the door to the possibility of generating mouse /human cells for "patient-specific" iPS cells.

4. Summary

Mouse embryonic stem cells, which are derived from the epiblast of inner cell mass of the blastocyst have the ablity to self-renewate and differentiate into any cell types. The therapeutic use of these cells depends on the availability of pluripotent cells that are limited by technical, ethical, or immunological considerations.One of major technical drawback of gene delivery in ES cells is method of viral gene delivery systems, which may cause insertional mutagenesis, and also may induce cytotoxicity problems. Apart from the viral gene delivery systems, there are many non viral methods of gene delivery systems like calcium phosphate-, DEAE -dextran-, Fugene 6-, or liposome-mediated transfection and nucleofection. However, due to low efficiency, high toxicity, and economical problems, these methods are not widely accepted.

In this review, we have described application of magnet-based nanofection in various cell lines including ES and iPS cells. Among all the non viral gene delivery methods, magnet-based nanofection is an appropriate tool to overcome the strong barriers like low efficiency, slow vector accumulation, low vector concentration at target tissues, and toxicity problems. It is very simple, less time-consuming, cost-effective, and highly efficient even with low doses of nucleic acids. In this methodology magnetic field is applied to superparamagnetic iron peroxide particles which are associated with gene vectors. Gene vectors and magnetic nanoparticles complexes are seen inside the cells after 15-20 minutes, which is much faster than any other transfection method. After 24, 48, or 72 hours, most of the particles are localized in the cytoplasm, vacuoles, and frequently in the nucleus. Importantly, this type of design is universally applicable to all types of gene vectors. The concept of magnet-based nanofection greatly benefits from the fact that the individual modules of the system can be optimized independently and variants can be assembled in a combinatorial manner, thus facilitating optimization towards specific applications. The size and surface chemistry of magnetic particles can be tailored to meet specific demands on physical and biological characteristics, and the linkage between vector and magnetic particle can be designed accordingly.

Magnet-based nanofection is used as gene delivery tool in several cell lines and also has been used in ES cells. Recently, iPS cells were generated by using magnet-based nanofection technique and there are several perspectives to the future use of magnet-based nanofection. The three important features of magnet-based nanofection are:

1. The drastically lowered vector dose.
2. The reduced incubation time required to achieve high efficiency.
3. The possibility of gene delivery in non dividing cells.

In conclusion, the magnet-based nanofection will be an ideal research tool where the available vector dose, the required process time, and the sustainable costs of the procedure are limiting factors. In near future, magnet-based nanofection may become a strong choice for the clinical applications. Obvious target diseases are cancer, cardiovascular, or neurological diseases, and potentially also genetic diseases. Combined with existing advanced concepts of gene delivery, magnet-based nanofection may provide additional specificity and efficiency which are required in many gene therapy approaches.

5. References

Aoi, T., Yae, K., Nakagawa, M., Ichisaka, T., Okita, K., Takahashi, K., et al. (2008). Generation of pluripotent stem cells from adult mouse liver and stomach cells. *Science, 321*(5889), 699-702.

Bethke, B., & Sauer, B. (1997). Segmental genomic replacement by Cre-mediated recombination: genotoxic stress activation of the p53 promoter in single-copy transformants. *Nucleic Acids Res, 25*(14), 2828-2834.

Boheler, K. R., Czyz, J., Tweedie, D., Yang, H. T., Anisimov, S. V., & Wobus, A. M. (2002). Differentiation of pluripotent embryonic stem cells into cardiomyocytes. *Circ Res, 91*(3), 189-201.

Brimble, S. N., Sherrer, E. S., Uhl, E. W., Wang, E., Kelly, S., Merrill, A. H., Jr., et al. (2007). The cell surface glycosphingolipids SSEA-3 and SSEA-4 are not essential for human ESC pluri

Chu, G., Hayakawa, H., & Berg, P. (1987). Electroporation for the efficient transfection of mammalian cells with DNA. *Nucleic Acids Res, 15*(3), 1311-1326.

Dalby, B., Cates, S., Harris, A., Ohki, E. C., Tilkins, M. L., Price, P. J., et al. (2004). Advanced transfection with Lipofectamine 2000 reagent: primary neurons, siRNA, and high-throughput applications. *Methods, 33*(2), 95-103.

Dhara, S. K., & Benvenisty, N. (2004). Gene trap as a tool for genome annotation and analysis of X chromosome inactivation in human embryonic stem cells. *Nucleic Acids Res, 32*(13), 3995-4002. potency. *Stem Cells, 25*(1), 54-62.

Draper, J. S., Smith, K., Gokhale, P., Moore, H. D., Maltby, E., Johnson, J., et al. (2004). Recurrent gain of chromosomes 17q and 12 in cultured human embryonic stem cells. *Nat Biotechnol, 22*(1), 53-54.

Feng, B., Ng, J. H., Heng, J. C., & Ng, H. H. (2009). Molecules that promote or enhance reprogramming of somatic cells to induced pluripotent stem cells. *Cell Stem Cell, 4*(4), 301-312.

Freitas, R. A., Jr. (2006). Pharmacytes: an ideal vehicle for targeted drug delivery. *J Nanosci Nanotechnol, 6*(9-10), 2769-2775.

Fusaki, N., Ban, H., Nishiyama, A., Saeki, K., & Hasegawa, M. (2009). Efficient induction of transgene-free human pluripotent stem cells using a vector based on Sendai virus, an RNA virus that does not integrate into the host genome. *Proc Jpn Acad Ser B Phys Biol Sci, 85*(8), 348-362.

Gauss, G. H., & Lieber, M. R. (1992). DEAE-dextran enhances electroporation of mammalian cells. *Nucleic Acids Res, 20*(24), 6739-6740.

Gepstein, L. (2002). Derivation and potential applications of human embryonic stem cells. *Circ Res, 91*(10), 866-876.

Gersting, S. W., Schillinger, U., Lausier, J., Nicklaus, P., Rudolph, C., Plank, C., et al. (2004). Gene delivery to respiratory epithelial cells by magnetofection. *J Gene Med, 6*(8), 913-922.

Hafeli, U. O. (2004). Magnetically modulated therapeutic systems. *Int J Pharm, 277*(1-2), 19-24.

Ido, Y., Carling, D., & Ruderman, N. (2002). Hyperglycemia-induced apoptosis in human umbilical vein endothelial cells: inhibition by the AMP-activated protein kinase activation. *Diabetes, 51*(1), 159-167.

Jain, K. K. (2007). Applications of nanobiotechnology in clinical diagnostics. *Clin Chem, 53*(11), 2002-2009.

Jarcho, M. (1981). Calcium phosphate ceramics as hard tissue prosthetics. *Clin Orthop Relat Res*(157), 259-278.

Jensen, O., & Tesche, C. D. (2002). Frontal theta activity in humans increases with memory load in a working memory task. *Eur J Neurosci, 15*(8), 1395-1399.

Kadota, S., Kanayama, T., Miyajima, N., Takeuchi, K., & Nagata, K. (2005). Enhancing of measles virus infection by magnetofection. *J Virol Methods, 128*(1-2), 61-66.

Kamau, S. W., Hassa, P. O., Steitz, B., Petri-Fink, A., Hofmann, H., Hofmann-Amtenbrink, M., et al. (2006). Enhancement of the efficiency of non-viral gene delivery by application of pulsed magnetic field. *Nucleic Acids Res, 34*(5), e40.

Kaplitt, M. G., Leone, P., Samulski, R. J., Xiao, X., Pfaff, D. W., O'Malley, K. L., et al.(1994). Long-term gene expression and phenotypic correction using adeno-associated virus vectors in the mammalian brain. *Nat Genet, 8*(2), 148-154.

Keller, G. (2005). Embryonic stem cell differentiation: emergence of a new era in biology and medicine. *Genes Dev, 19*(10), 1129-1155.

Kim, D., Kim, C. H., Moon, J. I., Chung, Y. G., Chang, M. Y., Han, B. S., et al. (2009). Generation of human induced pluripotent stem cells by direct delivery of reprogramming proteins. *Cell Stem Cell, 4*(6), 472-476.

Kim, J. B., Zaehres, H., Wu, G., Gentile, L., Ko, K., Sebastiano, V., et al. (2008). Pluripotent stem cells induced from adult neural stem cells by reprogramming with two factors. *Nature, 454*(7204), 646-650.

Kim, T. S., Lee, S. H., Gang, G. T., Lee, Y. S., Kim, S. U., Koo, D. B., et al. (2009). Exogenous DNA Uptake of Boar Spermatozoa by a Magnetic Nanoparticle Vector System. *Reprod Domest Anim.*

Kircheis, R., Wightman, L., & Wagner, E. (2001). Design and gene delivery activity of modified polyethylenimines. *Adv Drug Deliv Rev, 53*(3), 341-358.

Krotz, F., Sohn, H. Y., Gloe, T., Plank, C., & Pohl, U. (2003). Magnetofection potentiates gene delivery to cultured endothelial cells. *J Vasc Res, 40*(5), 425-434.

Lakshmipathy, U., Pelacho, B., Sudo, K., Linehan, J. L., Coucouvanis, E., Kaufman, D. S., et al.(2004). Efficient transfection of embryonic and adult stem cells. *Stem Cells, 22*(4), 531-543.

Latorre, M., & Rinaldi, C. (2009). Applications of magnetic nanoparticles in medicine: magnetic fluid hyperthermia. *P R Health Sci J, 28*(3), 227-238.

Lee, C. H., Kim, E. Y., Jeon, K., Tae, J. C., Lee, K. S., Kim, Y. O., et al. (2008). Simple, efficient, and reproducible gene transfection of mouse embryonic stem cells by magnetofection. *Stem Cells Dev, 17*(1), 133-141.

Li, M., Zacharia, J., Sun, X., & Wier, W. G. (2008). Effects of siRNA knock-down of TRPC6 and InsP(3)R1 in vasopressin-induced Ca(2+) oscillations of A7r5 vascular smooth muscle cells. *Pharmacol Res, 58*(5-6), 308-315.

Lowry, W. E., Richter, L., Yachechko, R., Pyle, A. D., Tchieu, J., Sridharan, R., et al. (2008). Generation of human induced pluripotent stem cells from dermal fibroblasts. *Proc Natl Acad Sci U S A, 105*(8), 2883-2888.

Mallon, B. S., Park, K. Y., Chen, K. G., Hamilton, R. S., & McKay, R. D. (2006). Toward xeno - free culture of human embryonic stem cells. *Int J Biochem Cell Biol, 38*(7), 1063-1075.

Martin, G. R. (1980). Teratocarcinomas and mammalian embryogenesis. *Science, 209*(4458), 768-776.

Martin, G. R. (1981). Isolation of a pluripotent cell line from early mouse embryos cultured in medium conditioned by teratocarcinoma stem cells. *Proc Natl Acad Sci U S A, 78*(12), 7634-7638.

Matsuda, T., Nakamura, T., Nakao, K., Arai, T., Katsuki, M., Heike, T., et al. (1999). STAT3 activation is sufficient to maintain an undifferentiated state of mouse embryonic stem cells. *Embo J, 18*(15), 4261-4269.

Mimeault, M., Hauke, R., & Batra, S. K. (2007). Stem cells: a revolution in therapeutics-recent advances in stem cell biology and their therapeutic applications in regenerative medicine and cancer therapies. *Clin Pharmacol Ther, 82*(3), 252-264.

Mitsui, K., Tokuzawa, Y., Itoh, H., Segawa, K., Murakami, M., Takahashi, K., et al. (2003). The homeoprotein Nanog is required for maintenance of pluripotency in mouse epiblast and ES cells. *Cell, 113*(5), 631-642.

Mykhaylyk, O., Antequera, Y. S., Vlaskou, D., & Plank, C. (2007). Generation of magnetic nonviral gene transfer agents and magnetofection *in vitro*. *Nat Protoc, 2*(10), 2391-2411.

Mykhaylyk, O., Sanchez-Antequera, Y., Vlaskou, D., Hammerschmid, E., Anton, M., Zelphati, O., et al. Liposomal magnetofection. *Methods Mol Biol, 605*, 487-525.

Nakagawa, M., Koyanagi, M., Tanabe, K., Takahashi, K., Ichisaka, T., Aoi, T., et al. (2008). Generation of induced pluripotent stem cells without Myc from mouse and human fibroblasts. *Nat Biotechnol, 26*(1), 101-106.

Namiki, Y., Namiki, T., Yoshida, H., Ishii, Y., Tsubota, A., Koido, S., et al. (2009). A novel magnetic crystal-lipid nanostructure for magnetically guided *in vivo* gene delivery. *Nat Nanotechnol4*(9), 598-606.

Nimesh, S., Goyal, A., Pawar, V., Jayaraman, S., Kumar, P., Chandra, R., et al. (2006). Polyethylenimine nanoparticles as efficient transfecting agents for mammalian cells. *J Control Release, 110*(2), 457-468.

Odorico, J. S., Kaufman, D. S., & Thomson, J. A. (2001). Multilineage differentiation from human embryonic stem cell lines. *Stem Cells, 19*(3), 193-204.

Papapetrou, E. P., Tomishima, M. J., Chambers, S. M., Mica, Y., Reed, E., Menon, J., et al. (2009). Stoichiometric and temporal requirements of Oct4, Sox2, Klf4, and c -Myc expression for efficient human iPSC induction and differentiation. *Proc Natl Acad Sci U S A, 106*(31), 12759-12764.

Plank, C., Anton, M., Rudolph, C., Rosenecker, J., & Krotz, F. (2003). Enhancing and targeting nucleic acid delivery by magnetic force. *Expert Opin Biol Ther, 3*(5), 745-758.

Plank, C., Schillinger, U., Scherer, F., Bergemann, C., Remy, J. S., Krotz, F., et al. (2003). The magnetofection method: using magnetic force to enhance gene delivery. *Biol Chem, 384*(5), 737-747.

Sato, N., Meijer, L., Skaltsounis, L., Greengard, P., & Brivanlou, A. H. (2004). Maintenance of pluripotency in human and mouse embryonic stem cells through activation of Wnt signaling by a pharmacological GSK-3-specific inhibitor. *Nat Med, 10*(1), 55-63.

Schenborn, E. T., & Oler, J. (2000). Liposome-mediated transfection of mammalian cells. *Methods Mol Biol, 130*, 155-164.

Siemen, H., Nix, M., Endl, E., Koch, P., Itskovitz-Eldor, J., & Brustle, O. (2005). Nucleofection of human embryonic stem cells. *Stem Cells Dev, 14*(4), 378-383.

Takahashi, K., & Yamanaka, S. (2006). Induction of pluripotent stem cells from mouse embryonic and adult fibroblast cultures by defined factors. *Cell, 126*(4), 663-676.

Thomson, J. A., Itskovitz-Eldor, J., Shapiro, S. S., Waknitz, M. A., Swiergiel, J. J., Marshall, V. S., et al. (1998). Embryonic stem cell lines derived from human blastocysts. *Science, 282*(5391), 1145-1147.

Trounson, A. (2006). The production and directed differentiation of human embryonic stem cells. *Endocr Rev, 27*(2), 208-219.

Wang, Y., Greenwood, J. S., Calcagnotto, M. E., Kirsch, H. E., Barbaro, N. M., & Baraban, S. C. (2007). Neocortical hyperexcitability in a human case of tuberous sclerosis complex and mice lacking neuronal expression of TSC1. *Ann Neurol, 61*(2), 139-152.

Wernig, M., Meissner, A., Foreman, R., Brambrink, T., Ku, M., Hochedlinger, K., et al. (2007). *In vitro* reprogramming of fibroblasts into a pluripotent ES-cell-like state. *Nature, 448*(7151), 318-324.

Wiles, M. V., & Johansson, B. M. (1999). Embryonic stem cell development in a chemically defined medium. *Exp Cell Res, 247*(1), 241-248.

Wu, D. C., Boyd, A. S., & Wood, K. J. (2007). Embryonic stem cell transplantation: potential applicability in cell replacement therapy and regenerative medicine. *Front Biosci, 12*, 4525-4535.

Xiao, L., Yuan, X., & Sharkis, S. J. (2006). Activin A maintains self-renewal and regulates fibroblast growth factor, Wnt, and bone morphogenic protein pathways in human embryonic stem cells. *Stem Cells, 24*(6), 1476-1486.

Yusa, K., Rad, R., Takeda, J., & Bradley, A. (2009). Generation of transgene-free induced pluripotent mouse stem cells by the piggyBac transposon. *Nat Methods, 6*(5), 363-369.

Zeng, X., Chen, J., Sanchez, J. F., Coggiano, M., Dillon-Carter, O., Petersen, J., et al. (2003). Stable expression of hrGFP by mouse embryonic stem cells: promoter activity in the undifferentiated state and during dopaminergic neural differentiation. *Stem Cells, 21*(6), 647-653.

Zhao, X. Y., Li, W., Lv, Z., Liu, L., Tong, M., Hai, T., et al. (2009). iPS cells produce viable mice through tetraploid complementation. *Nature, 461*(7260), 86-90.

Zwaka, T. P., & Thomson, J. A. (2003). Homologous recombination in human embryonic stem cells. *Nat Biotechnol, 21*(3), 319-321

18

Chromosome Engineering in Mouse Embryonic Stem Cells: Addition and Elimination of Targeted Chromosomes

Masako Tada
Division of Chromosome Engineering & Therapeutics, Chromosome Research Center, Tottori University, Nishi-cho 86, Yonago, Tottori 683-8503 Japan

1. Introduction

Human pluripotential stem cells, including both embryonic stem cells (ES cells) and induced pluripotent stem cells (iPS cells), possess self-renewing potency or the ability to differentiate into virtually any type of somatic cell. These features make them particularly advantageous as sources from which to generate specific types of human tissue cells *in vitro* for use in drug development and regenerative medicine. In most cases, however, human pluripotential cell lines and especially human ES cells can only be used in cell-based applications because of ethical issues. Animal models are therefore sought as an alternative to using human cells. An increasingly popular non-human primate model is the common marmoset (*Callithrix jacchus*). The recent successful creation of lentivirus-mediated transgenic marmosets provides a new animal model for human disease offering the powerful advantage of a close genetic relationship with humans (Sasaki et al., 2009), though this technique is not yet sufficiently developed for common use, and the numbers of monkeys available for experiments are limited. Mouse ES cells, meanwhile, still have great value as a research tool, even after the development of human pluripotent cells, as they can be used to create chimeric mice, achieve germline transmission, and generate normal offspring.

The available genetic engineering technologies often employ an embryonic manipulation approach in mice, using mouse ES cells to examine the gain-of-function or loss-of-function effects associated with certain chromosomal regions *in vivo*. Recently, we have developed chromosome elimination cassettes (CEC) using a Cre-inverted loxP system that was first used in mouse ES cells. In this system, transient *cre* expression can initiate immediate chromosomal loss over the course of a few cell cycles in the recombinant cells. This technology was developed to clarify chromosomal function through observing loss-of-function at the chromosomal level. In mammalian cells, chromosome composition and gene dosage are kept stable because large chromosome-wide deletions are usually fatal. Accordingly, we first applied the Cre-inverted loxP system to tetraploid cells composed of mouse ES cells fused with mouse somatic cells to generate conditions conducive to large-scale chromosomal imbalance. The CEC-tagged chromosomes could be targets of Cre-dependent chromosome elimination. In addition, we have demonstrated that the Cre-inverted loxP system enhances cohesion between a loxP site and an adjacent inverted loxP, which induces a remarkable degree of spontaneous mitotic recombination around loxP sites. These characteristics may also help us to resolve the mechanisms involved in spontaneous mitotic recombination events leading to chromosomal deletion during tumorigenesis. This review discusses the development of the chromosome elimination method over the last few years and provides examples of the emerging practical use of mouse ES cell lines containing CEC-tagged chromosomes as a chromosome engineering technology.

Recent advances in the opposite direction, that is, toward the insertion of chromosomes into cells as a chromosome engineering technology, are also discussed here. Human artificial chromosome (HAC) vectors have been used to transport intact foreign chromosomal segments. The Cre-loxP system is frequently employed in this form of chromosome transfer technology as well. HAC contains an acceptor loxP that promotes insertion of the chromosomal region through chromosome translocation with the donor loxP-tagged chromosome (Smith et al., 1995). If undesired chromosomal regions exist at the ends of the loxP-tagged donor chromosome, the regions can be trimmed using telomere-directed truncation technology; for the efficient application of this technique, chicken DT40 cells are often used as high homologous recombination-proficient cells. At the end of this process, the desired region is introduced into the final host cells (Dieken et al., 1996; Kuroiwa et al., 1998). In any step of this process toward the production of the final transformants, an HAC can be sequentially transferred from one type of cell to the other by means of the microcell-mediated chromosome transfer (MMCT) technique (Kuroiwa et al., 2000). MMCT requires a high level of skill, but can be used to produce 1:1 cell fusion products between micro-cells, which contain a recombinant chromosome, and host cells at high frequencies. Though some trials have been started, this remains a significant problem in chromosome transfer technology (Katoh et al., 2010). As a consequence, trans-chromosomal mice generated by germline transmission of an HAC have been created *via* the formation of chimeric mice with mouse ES cells that stably possess a defined chromosomal region on an HAC (Kuroiwa et al., 2002). Using this chromosome transfer technology with pluripotential cells offers advantages supporting the identification of the functions of given chromosome segments on the basis of gain-of-function and the repair of impaired genome-wide functions in the host cells. By observing the functional differences between wild-type cells and transformants, researchers can identify the responses to given stimuli and address these effects via chromosome transfer. These cells are not expected to over express any particular molecule on the transferred chromosome and are thought to exhibit more-or-less physiological responses to transcriptional regulation *in-vivo*. Accordingly, chromosome-transferred cells might sometimes constitute a more useful assay system than single-gene recombinant cells that highly express a single target molecule. Moreover, parallel inheritance of a human chromosome by cells from another animal species after elimination of the parallel chromosomal region makes it possible to create humanized animals.

The physiological relevance of biological analyses using CEC or HAC technology is actually demonstrated by the epigenetic reprogramming activity that occurs in host pluripotent cells. Reprogramming factors operating in pluripotent cells can impose epigenetic modifications on foreign somatic chromosomes that have been introduced by cell fusion, making their cells equivalent to host pluripotent cells. Without this activity, introduced chromosome segments would never be able to behave like their homologues or orthologues in the given cells *in vitro* and *in vivo*. Our previous important findings on the epigenetic reprogramming potencies working in mouse ES cells are discussed first.

2. Epigenetic reprogramming potencies in mouse ES cells

In keeping with the classic epigenetic landscape model, it was long believed that epigenetic marks were placed into the genome step by step during a developmental program and that the epigenetics of somatic nuclei would never revert to a pluripotential, multipotential, or bipotential state. The successful creation of adult normal frogs by nuclear transfer from highly specialized tadpole intestinal cells into ultraviolet-light-irradiated oocytes clearly demonstrated that developmental programs can be completely reversed (Gurdon, 1962). The first successfully cloned mammal was made by fusing a G0 cell with an enucleated unfertilized oocyte (Wilmut et al., 1997). As technology improved, it became possible to replace a mouse oocyte nucleus with a mouse somatic nucleus through enucleation; this led to the successful creation of cloned mice (Wakayama et al., 1998). This process is called somatic-cell nuclear transfer, and is now widely used. Complete epigenetic reprogramming *in vivo* through nuclear transfer creates another means by which the specific genotype of an animal can be propagated in the absence of mating.

Animal cloning suggested that epigenetic restoration was occurring in the somatic nuclei during embryonic development, but the these epigenetic changes were first analytically visualized through cell fusion between somatic cells and undifferentiated cells. Every female mammalian somatic cell possesses an inactivated X chromosome as a gene dosage compensation mechanism between XX female cells and XY male cells. To maintain X chromosome inactivation, numerous molecules, including the non-coding RNA *Xist*, are accumulated on the inactive X chromosome *in cis*. Moreover, DNA cytosine methylation of the *Xist* gene is kept at a low level on the inactivated somatic X chromosome. The somatic X chromosome is converted from an inactive state to an active state through cell fusion with mouse embryonic carcinoma cells (EC cells), in which case the fused cells acquire undifferentiated-cell morphology (Takagi et al., 1983).

As a next step, we demonstrated that *de novo* DNA cytosine methylation of the *Xist* gene might silence the *Xist* allele on the previously inactivated somatic X chromosome and initiate conversion of the X chromosome from an inactive state to an active state in somatic hybrid cells between certain kinds of EC cells and female somatic cells (Mise et al., 1996). By that time, however, we were asking whether the mouse EC cells had lost an important factor required for the maintenance of the X inactivation state through carcinogenesis or possessed dominant factors that converted the somatic nucleus to a host EC cell-like nucleus. To resolve this question, we designed a cell fusion experiment involving mouse embryonic germ cells (EG cells) and mouse somatic cells (Tada et al., 1997). The mouse EG cells, derived from primordial germ cells from embryonic-day (e)11.5 to e12.5 female embryos, had lost the epigenetic differences, also known as genomic imprinting, between the homologous chromosomes derived from the two parents. Inactivation marks on the X-chromosome and genome-wide CpG methylation had already been eliminated from the mouse gonadal EG cells. These cells could contribute to mouse development, but the resulting chimeric embryos were abnormal because of the deletion of the previous genomic imprints (Tada et al., 1998). Thus, unlike mouse EC cells, mouse EG cells are not derived through carcinogenesis, and they maintain a set of the epigenetic properties of the original cells. Several important characteristics of the epigenetic reprogramming activities working in embryonic pluripotent cells were first analytically demonstrated at the molecular level through cell fusion experiments using gonadal mouse EG cells and mouse somatic cells. The mouse EG cells contribute pluripotency to the resulting somatic nuclei, which in turn grants them several other EG cell-like characteristics, including the following: (1) reactivation of inactivated X chromosome derived from female somatic cells, (2) competence for embryonic development, (3) deletion of genome-wide DNA cytosine methylation, (4) deletion of methylation marks from several imprinted and non-imprinted genes, and (5) demonstrable reactivation of the repressed paternally imprinted allele of *Peg1/Mest*.

We then showed that somatic cells can acquire a pluripotent state after being fused with mouse ES cells (Tada et al., 2001). Using thymocytes from female mice that contained a GFP reporter transgene driven by the promoter of mouse *Oct4,* we monitored *Oct4* reactivation by an *Oct4–gfp* reporter. In contrast to germ cells, mouse ES cells do not possess demethylation activity for imprinted genes in fused tetraploid cells, as far as we could determine. Interestingly, methylation imprints on mouse ES cell chromosomes were erased in the hybrid cells made from mouse gonadal EG cells. This evidence shows that mouse EG cells possess dominant factors leading to the erasure of methylation imprints. The latter fusion experiments included our first use of intersubspecies hybrid cells, made from *Mus musculus domesticus* ES cells and *Mus musculus molossinus* thymocytes (Tada et al., 2003; Kimura et al., 2004; Kimura et al., 2002; Hatano et al., 2005). Frequent DNA sequence polymorphisms between these two subspecies allowed us to monitor the origin of the RNA and DNA derived from the somatic nuclei of the tetraploid hybrid cells. Using this mouse ES cell-somatic cell fusion system, we demonstrated (6) reactivation of some pluripotency-associated genes derived from somatic genomes (*i.e., Oct4, Nanog,* and *Tsix*) and (7) conversion of histone modification of somatic cell-derived chromatin to a pluripotent state. The reprogrammed somatic genomes in the hybrid cells made from mouse ES cells became hyperacetylated at histones H4 and H3 and globally dimethylated and trimethylated, with respect to the lysine residue K4, at H3. Such epigenetics are known as typical modifications for transcriptionally active regions. Later, another group demonstrated that overexpression of *Nanog* substantially enhanced fusion-based nuclear reprogramming (Silva et al., 2006). *Nanog* is known to be an important transcriptional factor involved in maintaining pluripotency (Chambers et al., 2003; Mitsui et al., 2003). We have previously shown that *Nanog* expression is controlled by *Oct4* and *Sox2* (Kuroda et al., 2006), which regulate the pluripotency of mouse ES cells in a dose-dependent manner (Hatano et al., 2005). Thus, we expected that two copies of endogenous *Nanog* mRNA might not be sufficient to make the tetraploid cells pluripotent during the initial full cell cycles after cell fusion with somatic cells when extrinsic *Nanog* was fully reprogrammed. We addressed this question by attempting the complete elimination of two copies of mouse ES cell-derived *Nanog*-bearing chromosome 6s in ES cell-somatic cell hybrid cells (Matsumura et al., 2007). The results clearly demonstrated that *Nanog* expression from reprogrammed somatic genomes could efficiently maintain pluripotency in mouse ES cell-somatic cell hybrid cells, but that at least three out of four copies of chromosome 6s were required to keep tetraploid cells pluripotent. This will be discussed in greater detail below.

ES cell-specific epigenetic profiles are predominantly regulated by intrinsic factors, so that the host cell's epigenotypes can be exposed to foreign DNAs and histones in ES cell-somatic cell hybrids. Several reprogramming core factors have already been identified, including *Oct4, Sox2, c-Myc,* and *Klf4* (OSCK). Overexpression of OSCK or OSK via virus- or nonvirus-mediated gene transformation can reprogram mouse and human somatic cells into pluripotential cells; cells reprogrammed through this means are known as iPS cells (Nakagawa et al., 2008; Takahashi et al., 2007a; Takahashi et al., 2007b; Takahashi & Yamanaka, 2006). The use of this *in-vitro* somatic reprogramming technology to produce human iPS cells is expected to be the long-awaited breakthrough that will allow researchers to produce syngenic tissue cells from personalized stem cells obtained from patients who require tissue-cell transplants. In the clinical setting, the propensity of iPS cells to form tumors is the main problem remaining to be solved, but functional somatic cells derived from human iPS cells could contribute to *in-vitro* analysis. Human iPS cells are currently regarded as a promising tool, satisfying the pharmacological industry's need for scalability and physiological relevance. Many researchers are planning to acquire patient-specific human iPS cell lines from somatic cells carrying genetic dysfunctions in order to develop medical compounds or gene therapies that reduce the symptoms associated with these genetic dysfunctions. Primary cultured somatic cells derived from patients are now becoming available from cell banks, helping us to produce models of gene therapy or chromosome-mediated therapy based on human iPS cells. Some kinds of biological analysis require accurate assessment of the phenotypic and physiological differences between normal cells and genetically affected cells in an identical genetic background. Even if normal cells were available from a patient with a genetic dysfunction, it might be difficult to compare the iPS cell line established from the affected tissue with that established from neighboring unaffected tissue because of the wide clonal variation among the iPS clones established through factor-mediated reprogramming. Thus genetically manipulated pluripotential cells created through knock-out, knock-in, transgene, or another molecular engineering technology might still provide some advantage over iPS cells because normal controls for the manipulated cells are always available.

3. General chromosome manipulation tools

3.1 Cre-loxP system

The Cre-loxP system is particularly useful for creating new chromosomes carrying site-specific deletions, duplications, inversions and translocations. A single loxP site contains two 13-bp inverted repeats (5′-ATAACTTCGTATA-3′ and 5′-TATACGAAGTTAT-3′) flanking an asymmetric 8-bp core sequence (5′-GCATACAT-3′). The central core sequence defines the orientation of the loxP site. Thus the core sequence in inverted-loxP can be described as 5′-ATGTATGC-3′. One Cre (causes recombination) recombinase monomer binds each inverted repeat. The synaptic complex is first formed through dimerization of the loxP-bound Cre molecules, and then Cre catalyzes DNA strand exchange between the homologous core regions *via* a Holliday intermediate. Only when two sites are placed as direct repeats on the same DNA strand does Cre induce restricted deletion of the DNA placed between the two loxP sites. Using this system, the targeting of a specific undesired DNA sequence can be Cre-dependently induced in a specific tissue type or cell type both *in vivo* and *in vitro* (Branda & Dymecki, 2004; Mills & Bradley, 2001). The Cre-loxP system also enables the induction of chromosome-wide recombination *in cis,* leading to megabase deletion, and can induce interchromosomal exchange between any loxP-tagged chromosomes, resulting in targeted chromosomal reciprocal translocation. Such *in-trans* recombination is induced to allow transfected exogenous sequences to integrate into the host genome through mitotic recombination at loxP sites (Figure 1A).

In addition from the experimental machinery, a native cellular mechanism for mitotic recombination exists in mammalian cells. For spontaneous mitotic recombination to occur, two DNA strands possessing high similarity first align spontaneously in a nucleus and then are exchanged through the repair of spontaneous DNA double-strand breaks between them.

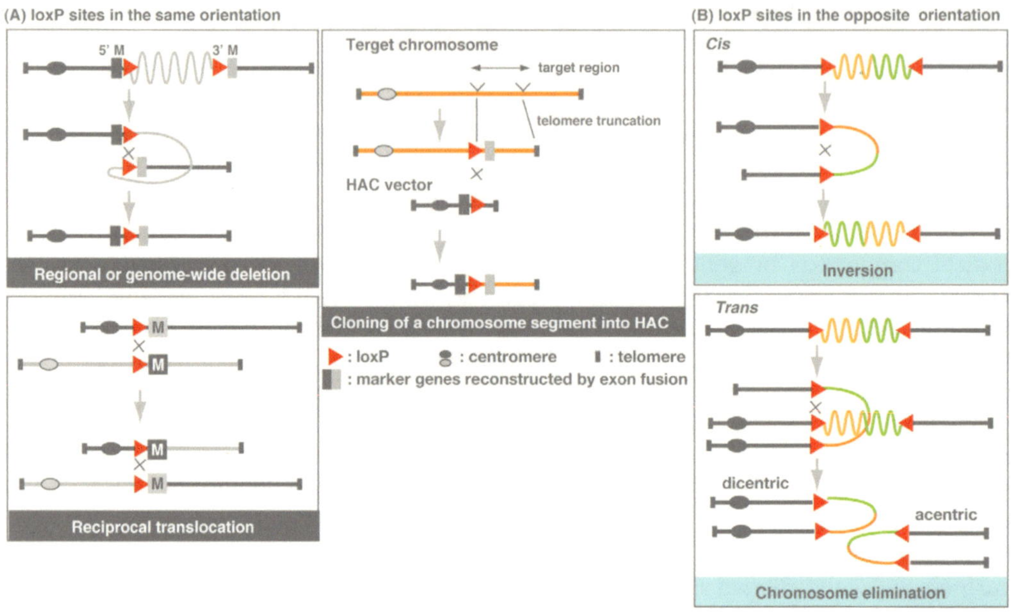

Fig. 1. Chromosome engineering technologies by means of Cre-loxP system. (A) Targeted insertion of loxP sites is practically used to induce a chromosome-wide deletion or a reciprocal translocation. Using sequential gene targeting, two loxP sites are introduced into two regions on the same or different chromosomes, resulting in chromosome-wide deletion or chromosome translocation, respectively. A loxP are often combined with 5' portion of a marker gene, while the other one are combined with the remaining portion of the marker gene. The marker genes reconstructed by exon fusion permit cell growth of the Cre recombinants. The Cre-loxP system is also applied to induce chromosome translocation between a desired chromosomal segment and an HAC vector. (B) Cre-inverted loxP system is used to induce elimination of the targeted chromosomes in the living cells. This system often induces inversion of the DNA region that has been placed between two loxP sites.

When two different chromosomes are targets of mitotic recombination, the daughter cell received a new recombinant created through balanced reciprocal translocation. In other cases, mitotic recombination causes the production of daughter cells carrying uniparental disomy. Even if recombinant cells do not contain any genetic imbalance, uniparental disomy sometimes causes detrimental effects because of the genomic imprinting phenomenon in mammals. Genomic imprinting, also known as epigenetic marking, leads to exclusive uniparental expression of the imprinted genes. Thus, uniparental disomy of imprinted regions causes either overexpression or nonexpression of the imprinted genes.

3.2 Cre-inverted loxP system

In the Cre-inverted loxP system, on the other hand, where the core sequence of the second loxP site is inverted, the DNA array placed between two loxP sites is often inverted through parallel synapses and recombination at the loxP sites located on the same DNA strand (Spitz et al., 2005). This inversion event yields intact recombined loxP sites, which will become targets of the second recombination. Moreover, Cre can initiate recombination between the loxP sites located on a pair of sister chromatids, resulting in derivation of a dicentric chromosome and an acentric chromosome. The subsequent cell divisions yield an accumulation of cells exhibiting targeted chromosomal loss (Lewandoski & Martin, 1997) (Figure 1B). We have developed CEC plasmid vectors containing a pair of loxP sites in an inverted orientation, which reproducibly induced hemizygous or homozyous loss of CEC-tagged chromosomes (Matsumura et al., 2007; Otsuji et al., 2008). To isolate transformants of CEC, a ubiquitously expressed drug-resistant gene and a green fluorescent protein (GFP) reporter gene were inserted in the into the Cre-inverted loxP sequence.

4. Chromosome elimination in mouse pluripotential cells

In mammalian cells, chromosome composition and gene dosage are kept stable, and large-scale chromosome-wide deletions are usually fatal. To create a non-fatal large-scale chromosomal imbalance, we applied the Cre-inverted-loxP system to tetraploid hybrid cells derived from mouse ES cells and somatic cells. As a first step, we prepared multiple stable CEC transformants of mouse ES cells, in which CEC-tagged chromosomes become targets of selective elimination. In the Cre-inverted loxP system, where the core sequence of the second loxP site is inverted, the inverted loxP sites are able to form parallel synapses between sister-chromatids after DNA synthesis, which produce a dicentric chromosome and an acentric chromosome as Cre-induced recombination products (CRPs). The CRPs are visible in many metaphases one to three days after Cre treatment. As a consequence, CRPs are eliminated in the course of the cell divisions that follow (Figure 1B). Such targeted chromosomal loss has been induced in progeny carrying a Y-transgene and containing an accidentally inverted loxP within a multi-copy array of directly repeated loxP sites (Lewandoski & Martin, 1997). Thus conditional elimination of targeted chromosomes might be technically inducible in mice, though viable offspring cannot be expected if any one of the intact autosomal chromosomes is eliminated.

4.1 Chromosome elimination cassette (CEC)

The CEC contains a ubiquitously expressed GFP-encoding gene and a drug-resistant gene between a loxP site and an inverted loxP site. The reporter genes were intended to allow researchers to isolate CEC-transformants as GFP-positive and drug-resistant cells, while Cre recombinants missing CEC-tagged chromosomes can be selectively obtained as GFP-negative cells through fluorescence activated cell sorting (FACS). Each sorted cell can be clonally expanded as an independent Cre-recombination product for further analysis.

4.2 Methods

4.2.1 CEC-tagging

Two kinds of CECs and their transgenic ES cell lines have been created: pCEC-CAG-*gfp*/IRES.puro-pA (CECpuro) and pCEC-Pgk-*neo*/IRES.*gfp*-pA (CECneo). The linealized DNAs of CECpuro and CECneo were electroporated into mouse ES cells to allow for the isolation of transformants through purimycin and G418, respectively. The transformants of linealized CEC vector DNA were isolated from HM1 mouse ES cells (129/Ola: *Mus musculus domesticus*) deficient in the *Hprt* gene. In such cases, CEC can be introduced into the desired part of the chromosome by homologous recombination. Thus fused cells created from mouse ES cells and wild-type somatic cells were isolated by means of HAT selection medium. All of the stable transformants were analyzed for their CEC-integration sites using the fluorescence-*in-situ*-hybridization (FISH) method. The location of FISH signals was determined on the basis of the G-banded pattern of the chromosomes. By now, CEC integration sites have been observed on 13 chromosomes out of the 19 pairs of autosomes and sex chromosomes of laboratory mice. These mapping data and chromosome identifications have been reported previously (Tada et al., 2009). The CEC-tagged chromosomes could be future targets for creating monosomy or uniparental disomy in a Cre-dependent manner.

4.2.2 Whole cell fusion

ES cell-like hybrid cells derived from mouse ES cells and mouse somatic cells are obtained by means of electric fusion followed by HAT selection. The electric fusion protocol has been described previously (Tada & Tada, 2006a, b). Only *Hprt*-positive cells can grow in HAT selection medium, so its use results in selective growth of the fused cells derived from *Hprt*-negative ES cells and wild-type somatic cells. Hybrid cells derived from CECneo-transgenic ES cells and CECpuro-transgenic ES cells can be isolated through co-treatment with G418 and puromycine after electric cell fusion. JF1 mice possessing a *Mus musculus molossinus* genetic background were often used as somatic cell donors so that the origin of the chromosomes derived from somatic cells in the tetraploid cells could be determined.

4.2.3 Cre treatment

The Cre expression vector pCMV-*cre* (Gibco-BRL, Invitrogen, Carlsbad, CA, USA) was modified into pCAG-*cre* and then transfected into CEC-tagged mouse ES cell hybrid cells by means of Lipofectamine™ 2000 (Invitrogen) or Nucleofector® (Amaxa, Basel, Switzerland). The best program and DNA content for nucleofection were A13 and 5 mg of DNA to $2x10^6$ cells. A high transfection rate was obtained: more than 60% of surviving cells showed transgene expression within one day after nucleofection. Consistent with this data, efficient induction of CRPs of CEC chromosomes was recorded one day after nucleofection. Further, multi-day culture yielded an accumulation of cells missing CEC-tagged chromosomes.

4.3 Whole chromosome elimination

First, we attempted single chromosome elimination in CECpuro transformants of mouse ES cell hybrid cells with mouse somatic cells (Matsumura et al., 2007). After conventional lipofection of the Cre expression vector, GFP-negative cells were isolated by FACS. All clones proliferated from a single GFP-negative cell were used for FISH analysis. The results clearly showed that CEC-tagged chromosomes were selectively eliminated in a Cre-dependent manner. Next we attempted to eliminate two chromosomes at once (Otsuji et al., 2008). We generated four sets of fusion clones derived from CECpuro-tagged mouse ES cells and CECneo-tagged mouse ES cells through cell selection with G418 and puromycin. In the case of the fused mES cells containing both CECpuro-tagged chromosome 12 and CECneo-tagged chromosome 17 (CEC12/17), flowcytometric analysis after conventional lipofection of the Cre expression vector showed only a 5% increase in GFP-negative cells in 7-day-old cultures. After nucleofection of the Cre expression vector, in contrast, nearly 50% of Cre-treated cells were GFP-negative hybrid cells. FACS-mediated isolation of GFP-negative cells was successively achieved by recovering the 78,XXYY, -12, -17 cell clones as Cre-recombinants with 96% purity from 80,XXYY tetraploid cells containing four chromosome 12s and four chromosome 17s. Similarly, we detected the loss of one set of chromosomes each from CEC11/17, CEC6/11 and CEC6/12 through FISH chromosome painting. In these hybrid cells, asynchronous chromosomal loss was also observed, even though two CEC regions were exposed to Cre activity in the same nucleus. This asynchrony gradually decreased over 5 days, however. In any case, following the DNA replication stages will provide Cre enzymes with access to loxP sites. As a next step, we need to precisely identify the kinds of autosomes to which chromosome elimination technology can be applied.

4.4 Homozygotic chromosome elimination

CEC6 mouse ES cells were created by means of the homologous recombination technology known as knock-in, through which CECneo vector was inserted into the *Rosa26* locus on mouse chromosome 6 (Matsumura et al., 2007). We prepared mouse ES cells that were homozyous for a CEC-tagged region through either mitotic recombination between homologous chromosomes or a combination of loss of wild-type chromosome and duplication of CEC-tagged chromosomes. Because there are two expressed copies of *neo* in a CECneo transformant, the homozygotes created from this transformant can be grown in the presence of high doses of G418. Using this system, we tried to demonstrate that the reprogrammed chromosome 6s derived from JF1 somatic cells could maintain an undifferentiated state without mouse ES cell-derived chromosome 6s as mentioned above. Cre recombinants missing both copies of chromosome 6s derived from mouse ES cells only expressed *Nanog* mRNA from JF1-derived chromosome 6s, and maintained their pluripotency. This fact clearly demonstrates that reprogrammed somatic chromosomes can functionally replace the ES chromosomes in hybrid cells derived from mouse ES cells and somatic cells. For tetraploid cells to survive and maintain their self-renewing potency, however, three copies of chromosome 6s were likely to be required. We found that every Cre-transformant possessed three chromosome 6s due to spontaneous duplication of somatic chromosome 6. Thus epigenetic reprogramming of the somatic cell–derived *Nanog* gene is sufficient, but the group of genes encoding chromosome 6 including *Nanog* might be sensitive to gene-dosage regulation.

4.5 Partial chromosome deletion

Theoretically, in the Cre-inverted loxP system, partial chromosomal deletions are not expected. Nevertheless, more than 30% of CEC6 mouse ES cells were observed to have turned into GFP-negative cells in 5-day-old Cre recombinants due to partial deletion, whereas whole chromosomal loss occurred at low frequencies (around 5%). It is therefore evident that the frequencies of regional mitotic recombination between short identical sequences in the genomic region of chromosome 6s are enhanced *via* a scaffold formation through Cre-mediated antiparallel cohesion between loxP sites (Otsuji et al., 2008). Thus a small deletion is less likely to be fatal; in fact, some diploid Cre recombinants isolated from CEC6 mouse ES cells maintained their self-renewing potencies and expressed both *Nanog* and *Oct4*. The *Rosa26* locus occurs close to the mouse von Hippel-Lindau tumor suppressor gene *Vh1h*, an orthologue of human *VHL*, which is located on human chromosome 3p25.5. Null mutation coupled with germline mutation and chromosomal deletion of the *VHL* gene have been reported in familial VHL syndrome, which predisposes affected persons to malignant or benign tumors (Maher & Kaelin, 1997; McGrath et al., 1992; Latif et al., 1993). Based on these cytogenetic reports, VHL syndrome is also known as 3p- syndrome (Sherr, 2004). In mice, heterozygotic loss of *Vhl1* is sufficient to predispose affected mice to vascular tumors, while homozygous loss causes embryonic lethality (Gnarra et al., 1997; Haase et al., 2001). Therefore CEC-tagged mouse ES cell clones may provide materials with which to create a loss of heterozygosity (LOH) model *in vitro* or *in vivo* by a combination of genetic mutation and inducible chromosomal deletion. We have already isolated several CEC-tagged mouse ES cell clones in which CEC-tagged regions are located near tumor suppressor genes, including *BRACA1* or *TP57* in CEC11Dpuro, *BRACA2* in CEC5puro, *P16* and *TP73* in CEC4neo, *TP53* in CEC7neo, and *VHL* in CEC6neo (Tada et al., 2009). In large chromosomal imbalances, the regions of insertion and deletion disrupt the complex interactions of many genes, not only within the chromosomal domains but also in other regions (Stallings, 2007). Thus chromosome-wide deletion technology might provide more information than simple gene disruption does as a model of deletion-mediated diseases in humans.

4.6 Future research

It has previously been shown that the frequency of mitotic recombination is about 100 times lower in mouse ES cells than it is in adult somatic cells or in isogenic mouse embryonic fibroblasts (Cervantes et al., 2002). Homozygous loss also takes place due to uniparental disomy induced either by sequential events of mitotic recombination and X segregation or by loss of functional chromosomes and duplication of the affected chromosomes. In LOH events in somatic cells, mitotic recombination predominates over uniparental disomy initiated by loss-and-duplication, for which a high level of nucleotide sequence homology is required. Despite the importance of chromosomal recombination in the LOH diseases, little is known about the properties of the junctions involved in the chromosomal rearrangements or about the responsible enzymes, because of the difficulty of inducing chromosomal deletion *in vitro* and *in vivo*. In this Cre-inverted loxP system, however, Cre significantly enhanced local rates of recombination at CEC-tagged regions. This may help to resolve the correlation between intra-chromosomal deletion events and either the flexibility of chromatin or the accumulation of junction sequences responsible for mitotic recombination. The application of CEC technology to diploid cells could help to isolate recombinants as GFP-negative clones passing through mitotic recombination more frequently rather than undergoing spontaneous recombination. This might promote better understanding of the modulator sequences responsible for the tangle structure formation and its solution mechanism that induces mitotic recombination leading to chromosomal deletions. It is clear that Cre-mediated cohesion enhances the rate of intra-chromosomal recombination in this region, but Cre is probably not involved in DNA strand exchange. Identifying the cellular factors that catalyze these recombination events will help resolve the mechanisms underlying the progression of cancers through chromosomal rearrangements.

5. Chromosome transfer into pluripotential cells

In early gene manipulation techniques, mRNA coding regions were introduced and ubiquitously expressed in a given set of cells. As a next step, endogenous transcriptional regulatory elements were connected to the open reading frames of the genes and used for developmental or tissue-specific expression. Now, to mimic the physiological expression profile of a certain gene, 100 kilobasepairs of genomic DNA are introduced into mammalian cells using bacterial artificial chromosome (BAC), P1-bacteriophage artificial chromosome (PAC), or Yeast artificial chromosome (YAC). Recently, there has arisen a need to realize the development of larger-capacity carriers for genes or gene clusters spanning a genomic region as large as a megabase. In one example, a chromosomal region comprises a major part of the gene function shared by many genes classified as a gene family. In another example, splicing forms are regulated tissue-specifically, so that a single form may be functionally divergent when expressed in different tissues. In immunoglobin genes, moreover, VDJ segments are recombined from a vast variety of the choices aligned in tandem over a wide chromosomal region. To induce VDJ recombination against a specific immunogen under physiological conditions, this whole chromosomal region should be introduced to the host cells to make them suitable for further use. Thus the development of a chromosome-wide gene transfer system would bring researchers a new genetic tool to analyze a proper function as seen *in vivo*. Human chromosome segments have been introduced not only into human host cells but also into mouse, avian, and other types of host cells. It is evident that human chromosome segments function in the trans-chromosome mice that are created via trans-chromosome mouse ES cells. There have recently been preliminary reports claiming the creation of transchromosomic animals producing a diverse repertoire of human immunoglobulin (Kuroiwa et al., 2002; Tomizuka et al., 2000; Tomizuka et al., 1997). Theoretically, depending on the size and region of the chromosome segment in question, an extra chromosome can contribute to half of the progeny through meiosis.

5.1 Chromosome manipulation

Centromeres, together with telomeres, are essential for segregation during cell division in any eukaryotic chromosome. Telomere function is also required in each of the ends to ensure chromosomal stability. Thus, chromosomal vectors have been created as "mini-chromosomes", each containing a short chromosome segment containing a loxP site, a human chromosome-derived centromere, and two telomere ends enclosing them. A telomere consists of an array of short tandem repeats, $(TTAGGG)_n$ in humans, which form a closed loop and protect chromosome ends. Through targeted insertion of telomere repeats, a new chromosome end can be created at a desired position in the chromosome. Native centromeres from human chromosomes that are involved in vectors are likely to behave as endogenous chromosomes. Vectors containing such chromosomes are called human artificial chromosome (HAC) vectors.

5.2 Human artificial chromosome (HAC) vector

A loxP site has been introduced into the HAC to allow for Cre-mediated site-specific insertion of circular DNAs at the beginning of the HAC (Dieken et al., 1996; Kuroiwa et al., 2000). Transgenes can therefore be introduced into these loxP sites by Cre-mediated lateral recombination with a loxP site located on a plasmid-, BAC-, or PAC-vector. The acceptor loxP site on the HAC can later be used for various purposes, such as a platform for megabase-level chromosomal segments. To insert the desired chromosomal segment into the HAC, one end of each defined chromosomal region is truncated by telomere insertion, while the other end is tagged by a donor loxP site. The loxP-tagged intact or truncated chromosome is introduced into the HAC-containing host cells. Transient Cre expression is then able to combine an HAC and a donor chromosomal segment at a loxP site, leading to the creation of an HAC containing an extra-chromosomal segment. Recombinants can be selectively grown through culturing with drugs to create a complete drug-resistant gene by exon fusion technology. HACs can be introduced into mouse ES cells through MMCT

(Fournier & Ruddle, 1977; Koi et al., 1989) as a cell fusion product (Figure 1A). For further *in-vitro* use, HAC transformants could be created using human ES cells and iPS cells.
5.3 Microcell-mediated chromosome transfer
First, to manipulate the donor chromosomes, normal mouse or human cells are fused with DT40 cells, on which targeting-mediated genetic manipulation is performed. Next, the manipulated donor chromosome is transferred from the DT40 hybrids to the HAC-containing CHO cells by MMCT technology (Kuroiwa et al., 2000). Usually, HAC is maintained in the CHO cell line, which is a more capable microcell donor. When the HAC uses a centromere of human chromosome 21, the HAC vector is originally constructed in the DT40 hybrid cells, in which intact human chromosome 21 is introduced from human fibroblasts through cell fusion (Katoh et al., 2004). The HAC vector recombined through Cre treatment is maintained in CHO cells. Again, the constructed HAC vectors containing donor chromosomal segments are transferred from CHO cells to other recipient cells (Kugoh et al., 1990) such as mouse A9 cells, mouse ES cells, or the other multipotent or pluripotent cells through MMCT using standard procedures.

5.4 Trisomy 21

Trisomy 21, the most common live-born human aneuploidy, causes Down's syndrome (DS), which encompasses many clinical phenotypes including a reduced incidence of solid tumors. An extra copy of human chromosome 21 has been introduced into mouse ES cells through MMCT (Shinohara et al., 2001). The authors showed that the progeny of these chimeric mice were able to contain human chromosome 21-containing cells after small-scale deletion. Chimeric mice showed various degrees of mosaicism as regards the retention of human chromosome 21, and there was a high correlation between the retention rate of human chromosome 21 in the brain and impairment in learning or emotional behaviors. Hypoplastic thymus and cardiac defects were also reported in a considerable number of chimeric mouse fetuses, with a high contribution of human chromosome 21 in mouse somatic cells. Recently, another group has reported a trans-chromosomic mouse model of DS in which tumor angiogenesis is significantly repressed; in particular, *in-vitro* and *in-vivo* angiogenic responses to vascular endothelial growth factor (VEGF) are inhibited by overdose of a set of identified factors transcribed from the three-copy alleles (Reynolds et al., 2010). The evidence shows that some human genes are able to reproduce their physiological functions in mouse cells.

5.5 Functional compensation through additional chromosome segments

As mentioned above, Oshimura and colleagues have developed a large number of HAC applications for biological analysis in the fields of cancer research, DNA repair, telomere research, genomic imprinting, and others. Moreover, they have shown that HAC vector could be used for gene therapy to correct insulin deficiency in mice (Suda et al., 2006) and Duchenne muscular dystrophy (DMD) in mice and in human immortalized mesenchymal stem cells (Hoshiya et al., 2009). The DMD-HAC also functionally replaced the affected allele in a mouse DMD model and in patient iPS cells (Kazuki et al., 2010). Therefore, the combination of patient-specific iPS cells and an HAC containing the responsible genes represents a powerful tool for gene and cell therapies. As previous cases have demonstrated, HAC technology may enable us to create animal models using native human chromosomes.

6. Future research

In the recent successful applications of this technology, human chromosomal regions were transmitted stably to mouse progeny, resulting in the creation of humanized mice for desired chromosomal regions, as mentioned above. An important point to consider is that horizontal chromosome transfer can realize the exchange of genetic material between different strains and species without producing offspring. Combined with chromosome-wide elimination by means of the Cre-inverted loxP system and chromosome transfer technology, the creation of consomic mice derived from different strains might be possible.

7. References

Branda, C.S. & Dymecki, S.M. (2004). Talking about a revolution: The impact of site-specific recombinases on genetic analyses in mice. *Dev Cell* 6, 7-28.

Cervantes, R.B.; Stringer, J.R.; Shao, C.; Tischfield, J.A. & Stambrook, P.J. (2002). Embryonic stem cells and somatic cells differ in mutation frequency and type. *Proc Natl Acad Sci U S A* 99, 3586-3590.

Chambers, I.; Colby, D.; Robertson, M.; Nichols, J.; Lee, S.; Tweedie, S. & Smith, A. (2003). Functional expression cloning of Nanog, a pluripotency sustaining factor in embryonic stem cells. *Cell* 113, 643-655.

Dieken, E.S.; Epner, E.M.; Fiering, S.; Fournier, R.E. & Groudine, M. (1996). Efficient modification of human chromosomal alleles using recombination-proficient chicken/human microcell hybrids. *Nat Genet* 12, 174-182.

Fournier, R.E. & Ruddle, F.H. (1977). Stable association of the human transgenome and host murine chromosomes demonstrated with trispecific microcell hybrids. *Proc Natl Acad Sci U S A* 74, 3937-3941.

Gnarra, J.R.; Ward, J.M.; Porter, F.D.; Wagner, J.R.; Devor, D.E.; Grinberg, A.; Emmert-Buck, M.R.; Westphal, H.; Klausner, R.D. & Linehan, W.M. (1997). Defective placental vasculogenesis causes embryonic lethality in VHL-deficient mice. *Proc Natl Acad Sci U S A* 94, 9102-9107.

Gurdon, J. (1962). The developmental capacity of nuclei taken from intestinal epithelial cells of feeding tadpoles. *J Embryol Exp Morphol* 10, 622-640.

Haase, V.H.; Glickman, J.N.; Socolovsky, M. & Jaenisch, R. (2001). Vascular tumors in livers with targeted inactivation of the von Hippel-Lindau tumor suppressor. *Proc Natl Acad Sci U S A* 98, 1583-1588.

Hatano, S.Y.; Tada, M.; Kimura, H.; Yamaguchi, S.; Kono, T.; Nakano, T.; Suemori, H.; Nakatsuji, N. & Tada, T. (2005). Pluripotential competence of cells associated with Nanog activity. *Mech Dev* 122, 67-79.

Hoshiya, H.; Kazuki, Y.; Abe, S.; Takiguchi, M.; Kajitani, N.; Watanabe, Y.; Yoshino, T.; Shirayoshi, Y.; Higaki, K.; Messina, G.; Cossu, G. & Oshimura, M. (2009). A highly stable and nonintegrated human artificial chromosome (HAC) containing the 2.4 Mb entire human dystrophin gene. *Mol Ther* 17, 309-317.

Katoh, M.; Ayabe, F.; Norikane, S.; Okada, T.; Masumoto, H.; Horike, S.; Shirayoshi, Y. & Oshimura, M. (2004). Construction of a novel human artificial chromosome vector for gene delivery. *Biochem Biophys Res Commun* 321, 280-290.

Katoh, M.; Kazuki, Y.; Kazuki, K.; Kajitani, N.; Takiguchi, M.; Nakayama, Y.; Nakamura, T. & Oshimura, M. (2010). Exploitation of the interaction of measles virus fusogenic envelope proteins with the surface receptor CD46 on human cells for microcell-mediated chromosome transfer. *BMC Biotechnol* 10, 37.

Kazuki, Y.; Hiratsuka, M.; Takiguchi, M.; Osaki, M.; Kajitani, N.; Hoshiya, H.; Hiramatsu, K.; Yoshino, T.; Kazuki, K.; Ishihara, C.; Takehara, S.; Higaki, K.; Nakagawa, M.; Takahashi, K.; Yamanaka, S. & Oshimura, M. (2010). Complete genetic correction of ips cells from Duchenne muscular dystrophy. *Mol Ther* 18, 386-393.

Kimura, H.; Tada, M.; Hatano, S.; Yamazaki, M.; Nakatsuji, N. & Tada, T. (2002). Chromatin reprogramming of male somatic cell-derived XIST and TSIX in ES hybrid cells. *Cytogenet Genome Res* 99, 106-114.

Kimura, H.; Tada, M.; Nakatsuji, N. & Tada, T. (2004). Histone code modifications on pluripotential nuclei of reprogrammed somatic cells. *Mol Cell Biol* 24, 5710-5720.

Koi, M.; Shimizu, M.; Morita, H.; Yamada, H. & Oshimura, M. (1989). Construction of mouse A9 clones containing a single human chromosome tagged with neomycin-resistance gene via microcell fusion. *Jpn J Cancer Res* 80, 413-418.

Kugoh, H.M.; Hashiba, H.; Shimizu, M. & Oshimura, M. (1990). Suggestive evidence for functionally distinct, tumor-suppressor genes on chromosomes 1 and 11 for a human fibrosarcoma cell line, HT1080. *Oncogene* 5, 1637-1644.

Kuroda, T.; Tada, M.; Kubota, H.; Kimura, H.; Hatano, S.; Suemori, H.; Nakatsuji N. & Tada, T. (2005) Octamer and Sox elements are required for transcriptional cis-regulation of Nanog gene expression. *Mol Cell Biol*, 25, 2475-2485.

Kuroiwa, Y.; Shinohara, T.; Notsu, T.; Tomizuka, K.; Yoshida, H.; Takeda, S.; Oshimura, M. & Ishida, I. (1998). Efficient modification of a human chromosome by telomere-directed truncation in high homologous recombination-proficient chicken DT40 cells. *Nucleic Acids Res* 26, 3447-3448.

Kuroiwa, Y.; Tomizuka, K.; Shinohara, T.; Kazuki, Y.; Yoshida, H.; Ohguma, A.; Yamamoto, T.; Tanaka, S.; Oshimura, M. & Ishida, I. (2000). Manipulation of human minichromosomes to carry greater than megabase-sized chromosome inserts. *Nat Biotechnol* 18, 1086-1090.

Kuroiwa, Y.; Yoshida, H.; Ohshima, T.; Shinohara, T.; Ohguma, A.; Kazuki, Y.; Oshimura, M.; Ishida, I. & Tomizuka, K. (2002). The use of chromosome-based vectors for animal transgenesis. *Gene Ther* 9, 708-712.

Latif, F.; Tory, K.; Gnarra, J.; Yao, M.; Duh, F.M.; Orcutt, M.L.; Stackhouse, T.; Kuzmin, I.; Modi, W.; Geil, L. *et al.* (1993). Identification of the von Hippel-Lindau disease tumor suppressor gene. *Science* 260, 1317-1320.

Lewandoski, M. & Martin, G.R. (1997). Cre-mediated chromosome loss in mice. *Nat Genet* 17, 223-225.

Maher, E.R. & Kaelin, W.G., Jr. (1997). von Hippel-Lindau disease. *Medicine (Baltimore)* 76, 381-391.

Matsumura, H.; Tada, M.; Otsuji, T.; Yasuchika, K.; Nakatsuji, N.; Surani, A. & Tada, T. (2007). Targeted chromosome elimination from ES-somatic hybrid cells. *Nat Methods* 4, 23-25.

McGrath, F.P.; Gibney, R.G.; Morris, D.C.; Owen, D.A. & Erb, S.R. (1992). Case report: multiple hepatic and pulmonary haemangioblastomas--a new manifestation of von Hippel-Lindau disease. *Clin Radiol* 45, 37-39.

Mills, A.A. & Bradley, A. (2001). From mouse to man: generating megabase chromosome rearrangements. *Trends Genet* 17, 331-339.

Mise, N.; Sado, T.; Tada, N.; Takada, S. & Takagi, N. (1996). Activation of the inactive X chromosome induced by cell fusion between a murine EC and female somatic cell accompanies reproducible changes in the methylation pattern of the *Xist* gene. *Exp Cell Res* 223, 193-202.

Mitsui, K.; Tokuzawa, Y.; Itoh, H.; Segawa, K.; Murakami, M.; Takahashi, K.; Maruyama, M.; Maeda, M. & Yamanaka, S. (2003). The homeoprotein Nanog is required for maintenance of pluripotency in mouse epiblast and ES cells. *Cell* 113, 631-642.

Nakagawa, M.; Koyanagi, M.; Tanabe, K.; Takahashi, K.; Ichisaka, T.; Aoi, T.; Okita, K.; Mochiduki, Y.; Takizawa, N. & Yamanaka, S. (2008). Generation of induced pluripotent stem cells without Myc from mouse and human fibroblasts. *Nat Biotechnol* 26, 101-106.

Otsuji, T.; Matsumura, H.; Suzuki, T.; Nakatsuji, N.; Tada, T. & Tada, M. (2008). Rapid induction of large chromosomal deletions by a Cre/inverted loxP system in mouse ES Cell hybrids. *J Mol Biol* 378, 328-336.

Reynolds, L.E.; Watson, A.R.; Baker, M.; Jones, T.A.; D'Amico, G.; Robinson, S.D.; Joffre, C.; Garrido-Urbani, S.; Rodriguez-Manzaneque, J.C.; Martino-Echarri, E.; Aurrand-Lions, M.; Sheer, D.; Dagna-Bricarelli, F.; Nizetic, D.; McCabe, C.J.; Turnell, A.S.; Kermorgant, S.; Imhof, B.A.; Adams, R.; Fisher, E.M.; Tybulewicz, V.L.; Hart, I.R. & Hodivala-Dilke, K.M. (2010). Tumour angiogenesis is reduced in the Tc1 mouse model of Down's syndrome. *Nature* 465, 813-817.

Sasaki, E.; Suemizu, H.; Shimada, A.; Hanazawa, K.; Oiwa, R.; Kamioka, M.; Tomioka, I.; Sotomaru, Y.; Hirakawa, R.; Eto, T.; Shiozawa, S.; Maeda, T.; Ito, M.; Ito, R.; Kito, C.; Yagihashi, C.; Kawai, K.; Miyoshi, H.; Tanioka, Y.; Tamaoki, N.; Habu, S.; Okano, H. & Nomura, T. (2009). Generation of transgenic non-human primates with germline transmission. *Nature* 459, 523-527.

Sherr, C.J. (2004). Principles of tumor suppression. *Cell* 116, 235-246.

Shinohara, T.; Tomizuka, K.; Miyabara, S.; Takehara, S.; Kazuki, Y.; Inoue, J.; Katoh, M.; Nakane, H.; Iino, A.; Ohguma, A.; Ikegami, S.; Inokuchi, K.; Ishida, I.; Reeves, R.H. & Oshimura, M. (2001). Mice containing a human chromosome 21 model behavioral impairment and cardiac anomalies of Down's syndrome. *Hum Mol Genet* 10, 1163-1175.

Silva, J.; Chambers, I.; Pollard, S. & Smith, A. (2006). Nanog promotes transfer of pluripotency after cell fusion. *Nature* 441, 997-1001.

Smith, A.J.; De Sousa, M.A.; Kwabi-Addo, B.; Heppell-Parton, A.; Impey, H. & Rabbitts, P. (1995). A site-directed chromosomal translocation induced in embryonic stem cells by Cre-loxP recombination. *Nat Genet* 9, 376-385.

Spitz, F.; Herkenne, C.; Morris, M.A. & Duboule, D. (2005). Inversion-induced disruption of the Hoxd cluster leads to the partition of regulatory landscapes. *Nat Genet* 37, 889-893.

Stallings, R.L. (2007). Are chromosomal imbalances important in cancer? *Trends Genet* 23, 278-283.

Suda, T.; Katoh, M.; Hiratsuka, M.; Takiguchi, M.; Kazuki, Y.; Inoue, T. & Oshimura, M. (2006). Heat-regulated production and secretion of insulin from a human artificial chromosome vector. *Biochem Biophys Res Commun* 340, 1053-1061.

Tada, M.; Matsumura, H.; Kurse, Y.; Nakatsuji, N. & Tada, T. (2009). Target chromosomes of inducible deletion by a Cre/inverted loxP system in mouse embryonic stem cells. *Chromosome Research* 17, 443-450.

Tada, M.; Morizane, A.; Kimura, H.; Kawasaki, H.; Ainscough, J.F.; Sasai, Y.; Nakatsuji, N. & Tada, T. (2003). Pluripotency of reprogrammed somatic genomes in embryonic stem hybrid cells. *Dev Dyn* 227, 504-510.

Tada, M. & Tada, T. (2006a). Epigenetic reprogramming of somatic genomes by electrofusion with embryonic stem cells. *Methods Mol Biol* 325, 67-79.

Tada, M. & Tada, T. (2006b). Nuclear reprogramming of somatic nucleus hybridized with embryonic stem cells by electrofusion. *Methods Mol Biol* 329, 411-420.

Tada, M.; Tada, T.; Lefebvre, L.; Barton, S.C. & Surani, M.A. (1997). Embryonic germ cells induce epigenetic reprogramming of somatic nucleus in hybrid cells. *Embo J* 16, 6510-6520.

Tada, M.; Takahama, Y.; Abe, K.; Nakatsuji, N. & Tada, T. (2001). Nuclear reprogramming of somatic cells by in vitro hybridization with ES cells. *Curr Biol* 11, 1553-1558.

Tada, T.; Tada, M.; Hilton, K.; Barton, S.C.; Sado, T.; Takagi, N. & Surani, M.A. (1998). Epigenotype switching of imprintable loci in embryonic germ cells. *Dev Genes Evol* 207, 551-561.

Takagi, N.; Yoshida, M.A.; Sugawara, O. & Sasaki, M. (1983). Reversal of X-inactivation in female mouse somatic cells hybridized with murine teratocarcinoma stem cells *in vitro*. *Cell* 34, 1053-1062.

Takahashi, K.; Okita, K.; Nakagawa, M. & Yamanaka, S. (2007a). Induction of pluripotent stem cells from fibroblast cultures. *Nat Protoc* 2, 3081-3089.

Takahashi, K.; Tanabe, K.; Ohnuki, M.; Narita, M.; Ichisaka, T.; Tomoda, K. & Yamanaka, S. (2007b). Induction of pluripotent stem cells from adult human fibroblasts by defined factors. *Cell* 131, 861-872.

Takahashi, K. & Yamanaka, S. (2006). Induction of pluripotent stem cells from mouse embryonic and adult fibroblast cultures by defined factors. *Cell* 126, 663-676.

Tomizuka, K.; Shinohara, T.; Yoshida, H.; Uejima, H.; Ohguma, A.; Tanaka, S.; Sato, K.; Oshimura, M. & Ishida, I. (2000). Double trans-chromosomic mice: maintenance of two individual human chromosome fragments containing Ig heavy and kappa loci and expression of fully human antibodies. *Proc Natl Acad Sci U S A* 97, 722-727.

Tomizuka, K.; Yoshida, H.; Uejima, H.; Kugoh, H.; Sato, K.; Ohguma, A.; Hayasaka, M.; Hanaoka, K.; Oshimura, M. & Ishida, I. (1997). Functional expression and germline transmission of a human chromosome fragment in chimaeric mice. *Nat Genet* 16, 133-143.

Wakayama, T.; Perry, A.F.C.; Zuccotti, M.; Johnson, K.R. & Yanagimachi, R. (1998). Full-term development of mice from enucleated oocytes injected with cumulus cell nuclei. *Nature* 394, 369-374.

Wilmut, I.; Schnieke, A.E.; McWhir, J.; Kind, A.J. & Campbell, K.H.S. (1997). Viable offspring derived from fetal and adult mammalian cells. *Nature* 385, 810-813.

19

Microarray Analysis of Undifferentiated and Differentiated Human Pluripotent Stem Cells

Jane Synnergren[1,2] and Peter Sartipy[3]
[1]*University of Skövde,*
[2]*Sahlgrenska University Hospital,*
[3]*Cellartis AB*
Sweden

1. Introduction

Over the last decade, a tremendous progress has been made regarding our understanding of the molecular program involved in early human development. The main reason behind these advancements can be ascribed to the successful isolation of human embryonic stem cell (hESC) lines in the late 1990's. Based on their fundamental properties of pluripotency and unlimited proliferation, these unique cells have provided the possibility to study early human developmental processes in vitro. However, there are many obstacles to overcome before the potential of these cells can be fully realized. One important issue is to increase the understanding about the gene regulatory mechanisms that control the differentiation of hESCs. A wide variety of tools and technologies have been used to manipulate and study basic hESC characteristics and functions. Furthermore, the parallel analysis of functional derivatives of hESCs has provided important insights into the mechanisms that govern their differentiation into specific cell lineages. Global transcriptional changes in cells and tissues can be studied using molecular techniques such as DNA microarray, EST-enumeration, MPSS profiling, and SAGE. The results from such experiments provide a snapshot of the status of the cells under study. This approach has proven well suited for characterization of the "stemness" state of hESCs, but also for the identification of crucial pathways involved in their differentiation. Large scale gene expression databases have been generated using various hESC lines and technical platforms, and subsequently the information has been analyzed using different bioinformatic approaches. A discrete set of genes has been identified which are highly expressed in hESCs, and these genes are considered to be involved in preserving the pluripotency and self-renewal capacity of the undifferentiated cells. Furthermore, several studies have focused on characterizing the molecular signature of specific differentiation processes. Again, global expression analysis has proven to be a very suitable tool since novel important mechanisms can be revealed in the context of such experiments. More recently, the possibility to analyze also the global expression profile of microRNA (miRNA) has been realized, and microarray based platforms designed specifically for the detection of miRNA species are now available. The concurrent analysis of the global mRNA and miRNA expression profiles of hESCs and their differentiated progenies are anticipated to provide additional insights into the regulatory pathways which are active in the cells in the undifferentiated and differentiated states. In the present chapter, we will discuss issues regarding the platforms used and the bioinformatic tools applied to interpret the results and also review some important studies on transcriptional profiling of hESCs.

2. Gene transcription and protein translation

Gene expression is the process by which information from a gene is copied from the gene to an mRNA sequence, which is then used in the synthesis of a functional gene product, a protein. The properties of the expression products give rise to the phenotype of an organism. When conducting gene expression studies, it is important to understand the basic concepts behind these processes, for a proper interpretation of the data. The genetic code is mediated by the gene expression, and the process from transcription of a gene to a functional protein involves several steps, such as transcription of the gene in the nucleus, and transport of the mRNA to the cytoplasm where translation to a protein is carried out aided by ribosomes. What is usually measured with microarrays is the amount of mRNA in the cell. However, in recent years, the microarray technology has also been successfully used for assessment of miRNA expression and moreover, protein arrays have begun to enter the market. By means of gene regulation the cell has control over its structure and function, and this is the basis for cellular differentiation, morphogenesis, and the versatility and adaptability of any organism. Transcriptional regulation is also essential for evolutionary changes, since control of the timing, location, and amount of gene expression often have profound effects on the functions of the gene in a cell.

2.1 Transcriptional regulation

The transcription of genes involves intricate dynamic dependencies which makes it challenging to study. Several mechanisms have been shown to be critical for the initiation of transcription, the rate of transcription, and the subsequent processing of the mRNA. These regulatory mechanisms control when the transcription occurs and the amount of mRNA produced. The transcription of a gene is carried out by RNA polymerase and the process is regulated by several components (Alberts et al., 2004):

- *Specificity factors* control the ability for the RNA polymerase to bind to a specific promoter or set of promoters.
- *Repressors* bind to non-coding regions, close to or overlapping with the promoter for a gene, and impede the RNA polymerase's progress along the DNA strand, thus hampering the transcription of the gene.
- *General transcription factors* aid in positioning the RNA polymerase at the start of a protein coding sequence.
- *Activators* enhance the interaction between the RNA polymerase and the specific promoter.
- *Enhancers* are sites on the DNA helix that are bound to by activators in order to loop the DNA and bring a specific promoter to the initiation complex.

As microarrays provide the opportunity to globally monitor the transcriptome and the expression of interacting molecules, they provide a powerful tool for studies of transcriptional regulation.

2.2 Splicing of mRNAs

Splicing is a modification of an RNA sequence post-transcription, in which introns are removed and exons are joined together (Fig. 1.) Hence, after transcription of a gene, the pre-mRNA is spliced to mRNA, which is often performed in a series of reactions. RNA splicing allow for packing of more information into every gene as the transcripts from one single gene can be spliced in various ways to produce different mRNAs, depending on the cell type in which the gene is being expressed or the stage of the development of the organism (Alberts et al., 2004). As a consequence, different proteins can be produced by the same gene and it is estimated that 60% of the human genes undergo such alternative splicing (Alberts et al., 2004). Thus, RNA splicing increases the already enormous coding potential of eukaryotic genomes, at the same time as it complicates the studies of gene transcription. This is because the complexity increases dramatically when there, as in many cases, are several different transcripts transcribed by one single gene.

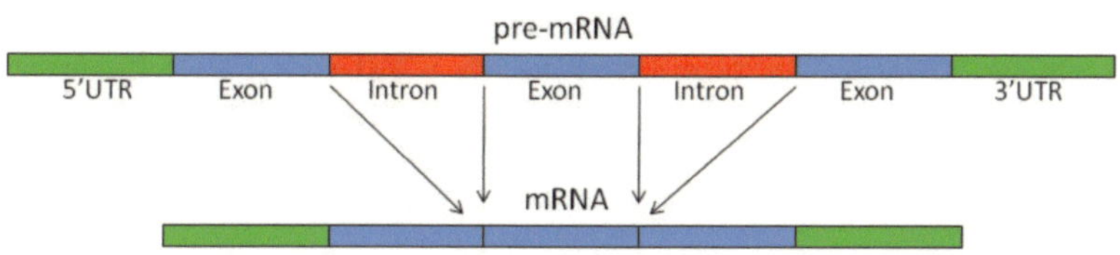

Fig. 1. Splicing of pre-mRNA where introns are removed before formation of the mRNA sequence. Different sets of exons can be selected to form the mRNA, which means that one pre-mRNA can give rise to several variants of mRNA sequences.

2.3 Translation to protein

After the splicing, the mRNA transcript is transported from the nucleus to the cytoplasm where the translation occurs by means of ribosomes. The same mRNA sequence can be translated many times, and therefore the period of time that a mature mRNA molecule persists in the cell, influences the amount of protein that is produced. The lifetime of mRNAs differs considerably and cannot be assessed by using single time point microarray analyses, as these only gives a snap-shot of the amount of mRNA at a specific time point. The mRNA lifetime is dependent on a multitude of factors, such as the nucleotide sequence of the mRNA itself, as well as the type of cell in which the mRNA is produced. The typical lifetime for mRNA molecules in eukaryotic cells ranges from 30 minutes up to 10 hours (Alberts et al., 2004). In the ribosome the nucleotide sequence is translated into an amino acid sequence by means of the *genetic code*. The sequence of nucleotides in the mRNA is read in groups of three, *codons*, and each codon specifies one amino acid. Depending on where in the sequence the de-coding begins, each mRNA sequence can be translated in three different, non-overlapping, reading frames but only one of these is the correct one (Alberts et al., 2004). To control that correct reading frame are used, the translation of an mRNA begins with a specific start codon (AUG), and is then performed in the direction 5′ cap to 3′ end. The end of a protein coding mRNA is indicated by the presence of one of three stop codons (UAA, UAG, UGA) which signal to the ribosome to stop the translation.

3. MicroRNAs

An additional level of cellular regulation involves a family of tiny molecules, known as miRNAs. These are 19-25 nucleotides non-coding RNAs that bind to the 3′ untranslated region of target mRNAs through imperfect matching. In mammalian genomes miRNAs are predicted to regulate the expression of approximately 30% of the protein-coding genes

(Bartel, 2004). Knowledge about the biological functions of most miRNAs identified thus far is still lacking, but it has been shown that they play important roles in embryo development, determination of cell fate, cell proliferation, and cell differentiation (Sartipy et al., 2009, Wang et al., 2009). MicroRNAs are derived from approximately 70 nucleotide long precursors, encoded by introns or intergenic regions, and are expressed in most organisms ranging from plants to humans. Many miRNAs appear to be expressed at different levels in various tissues, and the maturation and function of the tissues seem to be influenced by their presence. Interestingly, results from recent studies have indicated important roles for miRNAs in the control of diverse aspects of heart formation and cardiac function (Ivey et al., 2008, van Rooij & Olson, 2007). It is also known that miRNAs are involved in various types of cancer by targeting tumor suppressing genes (Lu et al., 2005, Zhu et al., 2008). MicroRNAs bind to their target mRNAs and negatively regulate their expression, either by repression of translation or by degradation of the mRNA (Bartel, 2004). Increased expression levels of miRNAs can also result in up-regulation of previously suppressed target genes either directly, by decreasing the expression of inhibitory proteins and/or transcription factors, or indirectly, by inhibiting the expression levels of inhibitory miRNAs (Gregory et al., 2008). Depending on the state of the cell, miRNAs have also been observed to affect the translation of target mRNAs by regulation of their stability (Gregory et al., 2008, Vasudevan et al., 2007). Moreover, it has been shown that combinatorial regulation by miRNAs is common, which enables complex regulatory programs that are exceptionally challenging to dissect (Zhou et al., 2007).

4. Global transcriptional profiling techniques

There are several high throughput techniques for measuring gene expression at the large scale, such as expressed sequence tags (EST)-enumeration, Serial Analysis of Gene Expression (SAGE), Massively Parallel Signature Sequencing (MPSS) and different types of microarrays (described in more detail below). In EST-enumeration the expression levels are assessed by counting the number of ESTs for a particular gene, randomly selected from a cDNA library derived from the sample. The ESTs are clustered into groups of sequences originating from the same transcript, and subsequently assembled to achieve a longer consensus sequence, which is then aligned to the genome to find the matching gene sequence. Both SAGE and MPSS are sequencing based techniques that use tags to identify and count the mRNAs, but the biochemical manipulation and the sequencing approaches differ substantially between these techniques. Both methods are based on the principle that a short sequence tag contains sufficient information to uniquely identify a transcript, provided that the tag is obtained from a specific position within each transcript. In SAGE, short tags, usually 9-10 base pairs in length, are extracted from each mRNA, at a defined position. These tags are then linked together to form long serial molecules that can be cloned and sequenced. The quantification is performed by counting the number of times a specific tag is observed in the sequenced molecule. Finally, the tags are matched to the corresponding genes.

In MPSS, the extracted signatures are longer, 17-20 base pairs. Every one of these signatures is cloned into a vector and labeled with a unique 32 base pair oligonucleotide tag. The tag is then attached to one of millions of microbeads, by hybridization of the tag to a complementary sequence on the bead. The signatures on the microbeads are then sequenced and matched to the corresponding genes, and subsequently quantified by counting the number of beads. The longer tag sequences, used in MPSS, provide higher specificity compared to SAGE. Another advantage of MPSS is the larger library size. One disadvantage that applies to both SAGE and MPSS is the loss of certain transcripts due to lack of restriction enzyme recognition sites, and ambiguity in tag annotation. Compared to microarray techniques, sequencing techniques, which are not based on hybridizations, give on the other hand a more exact quantitative value. This is because the number of transcripts is counted directly, instead of quantifying spot intensities which is prone to background noise. Another advantage is that the mRNA sequences do not need to be known beforehand, and therefore also previously unknown transcripts can be detected. Nevertheless, microarray experiments are much cheaper to perform and are therefore usually used in large scale experiments.

4.1 Microarray technology

The microarray technology has been around since the early 1990s, and during the last two decades the precision of the technology has increased considerably and, at the same time, the cost has decreased. Microarrays render the possibility to monitor the expression of thousands of genes simultaneously, which make them exceptionally useful for transcriptional studies on the global scale. Investigators are using the microarray technology to try to understand fundamental aspects of growth and development as well as to explore the underlying genetic causes of many human diseases. By monitoring the cells at various time points during a biological process or at specific biological conditions, one can take snapshots of the global transcriptional profile at different stages. The principle behind the microarray technology is base pairing of DNA/RNA. When two complementary sequences come together, such as the immobilized probe on the array and the mobile target in the sample, they will lock together (hybridize). The microarray consists of a surface on which millions of probes are immobilized. The surface is divided into features (locations), and each feature on the microarray has a superfluous number of probes that correspond to a specific transcript. When labeled target transcripts are hybridized onto the microarray, these bind complementary to their probes. The general procedure for performing a microarray experiment (which varies somewhat depending on the type of system) includes a series of steps. Initially, the RNA is reverse transcribed, usually to cDNA, and labeled with a fluorophore, and then the solution is hybridized onto the array. After the hybridization, the arrays are thoroughly washed, rinsed, and dried to remove non-hybridized transcripts from the surface. They are subsequently scanned to measure the fluorescence intensity for each spot on the array and these intensities are then translated into expression values. The spot intensities are directly proportional to the number of transcripts corresponding to each gene, and thus to the expression level of the gene.

4.2 Different types of microarrays

There are many different types of microarrays and the broadest distinction is whether the probes are spatially arranged on a slide made of glass, silicon or plastic or, if they are coded on microscopic polystyrene beads. They can be fabricated using different techniques, where the most common ones are robotic printing of the spots on the array (spotted arrays) or synthesis of the probes in situ using techniques such as photolithography. Moreover, the arrays vary in the way the signals are detected, and they are designed for hybridization of either one or two samples on the same array (one- or two-channel arrays) (Fig. 2).

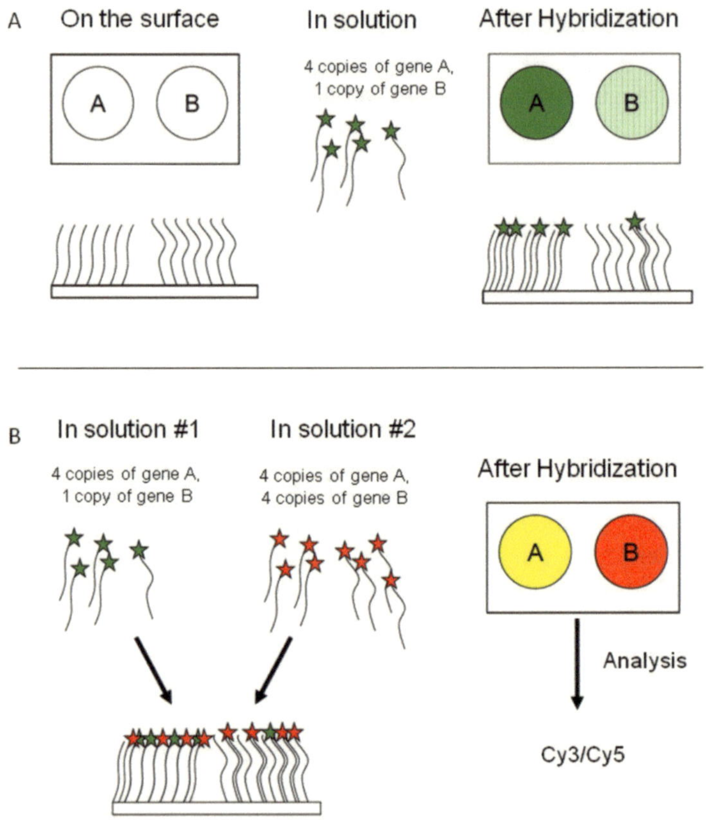

Fig. 2. Schematic picture of one-channel hybridization (A) and two-channel hybridization (B). Round-shaped features contain superfluous identical probes that hybridize with labeled targets from the samples. The shape of the features may vary between different microarray platforms. The intensity of the color is proportional to the number of probes that are hybridized to that feature. Yellow color means equal amounts of red and green labeled targets.

On one-channel arrays (also called oligonucleotide arrays) only one sample can be hybridized on each array, and the intensity levels are measured rather than the ratio between two intensities. Therefore, comparison of two conditions requires two separate single-dye hybridizations. On two-channel arrays two samples are labeled with two different fluorophores, typically Cy3 and Cy5, which have different fluorescence emission wavelengths. The two Cy-labeled cDNA samples are mixed and hybridized to a single microarray. Since the fluorophores have different excitation wavelengths it is possible to split the two signals during the scanning and calculate the intensities of each fluorophore, and use this in ratio-based analysis to identify up- and down-regulated genes. One benefit of one-channel arrays is that the data is more easily compared to data from different experiments, as long as batch effects have been accounted for. However, using the one-color system may require twice as many microarrays to compare samples within an experiment than with the two-color system. Depending on which system is used, the experimental design, and the generated data, the subsequent data analysis may differ.

4.3 Affymetrix microarrays

The Affymetrix platform is the most widely used commercial platform, providing a whole range of different types of arrays and covering various species. Affymetrix arrays are in situ synthesized, applying the photolithography technology to synthesize thousands to millions of 25-mer cDNA oligonucleotides in parallel. By using light-sensitive masking agents, a sequence is "built" one nucleotide at a time across the entire array. Typical for Affymetrix arrays are the multiple (11–20) probe pairs for each transcript. A new type of Affymetrix arrays which recently have entered the market is the Whole Transcript arrays, including both Gene ST 1.0 and Exon ST 1.0 arrays (Pradervand et al., 2008). The characteristics of these arrays are that they have an increased number of probes targeting exons along the whole transcript and not only in the 3' end. The Gene ST 1.0 array has 1-2 probes per exon and the more comprehensive Exon ST 1.0 has four probes per exon.

4.4 Reliability and reproducibility of microarray data

The microarray technology has had tremendous impact on gene expression analysis during the last decade. However, publications of studies with dissimilar or even contradictory results have raised concerns regarding the reliability of this technology (Draghici et al., 2006, Kuo et al., 2002, Tan et al., 2003). For example, several global gene expression studies of stem cells have shown poor overlap (Fortunel et al., 2003, Ivanova et al., 2002, Ramalho-Santos et al., 2002). To address these and other concerns, such as performance and data analysis issues, the MicroArray Quality Control project (Chen et al., 2007, Shi et al., 2006) was initiated by the US Food and Drug Administration. Using an impressive number of laboratories, this comprehensive study showed both intra-platform consistencies across laboratories and a high level of inter-platform concordance in terms of genes identified as differentially expressed genes. Nevertheless, there are several issues to be aware of when using this technology, and which can introduce substantial biases in the final results. Examples of such issues to consider are:

- Cross-hybridization: There is a risk that some mRNAs may cross-hybridize probes on the array that are supposed to detect other mRNAs.
- Fold change compression: Due to various technical limitations, such as limited dynamic range and signal saturation, a certain level of FC compression is expected for array data compared to e.g. RT-PCR data (Wang et al., 2006, Yuen et al., 2002).
- Poor sensitivity for low expressed transcripts: Problems with relatively poor sensitivity in detecting small FCs have been reported for several microarray platforms (Wang et al., 2006).
- Cross-platform inconsistency: Inconsistent probe annotations across platforms, which leads to difficulties to ascertain that probes on various platforms aimed at the same gene do in fact quantify the same mRNA transcript (Draghici et al., 2006).
- Dye-biases: In two-channel systems the fluorescent dyes usually have different dynamic ranges and quantum yields, which is partially adjusted for by appropriate normalization, but may not be completely eliminated.
- Non-biological variations: There is always a risk that variations may be introduced during the experimental procedure (e.g. different persons performing the experiment, minor variations in temperature or duration) (Frantz, 2005) and these sometimes add substantial noise to the system. However, this source of variation is not unique to microarray experiments but is also an issue in other reverse transcription reactions (Wang et al., 2006).

5. Bioinformatic and statistical analysis

Large scale gene expression experiments generate enormous datasets that are computationally demanding to analyze. Therefore, gene expression analysis is one research area where bioinformatic methods have had important impact. These datasets are often challenging to analyze because of complex dependencies of interacting molecules and the data is often fragmented, incomplete, and noisy. Today, there are a lot of tools and software available, both commercially and open source, for solving various bioinformatic problems, such as identification of differentially expressed genes, clustering of data, and identification of interaction networks.

5.1 Data analysis of microarray data

The raw data from microarray experiments need to be pre-processed in several steps, before conducting any high level data analysis. Depending on the array type and the platform, these pre-processing steps vary, but basically involve subtraction of background and normalization for removal of non-biological variations. The data is also typically log_2-transformed to achieve roughly normally distributed data, and potential outliers are excluded before performing the high level analysis. Due to the large amounts of data generated in microarray experiments, advanced bioinformatic algorithms (described below) are required for efficient interpretation of the data into valuable biological information. In the area of gene expression analysis there are e.g. algorithms for:

- identification of differentially expressed genes
- clustering of gene expression data
- pathway analysis
- derivation of protein interaction networks
- functional annotation of regulated genes

Although requiring some programming skills, the freely available R software environment[1] is highly recommended for various analyses of microarray data. This software has packages for normalization/standardization and statistical computing, as well as graphics. R can be used as a powerful standalone programming language, but the most prominent advantages are indeed all the implemented functions that are ready to use, and which make the R environment both flexible and extendible.

5.2 Identification of differentially expressed genes

There are several approaches on how to determine which genes that are differentially expressed in large gene expression datasets. Traditionally, researchers have applied different statistical tests and used the p-values as selection criteria. However, when dealing with microarray data one has to consider the *multiple testing problem*, as these datasets usually contain thousands of genes and the statistical test is applied for each of these genes. In other words, the multiple testing problem means that thousands of hypotheses are tested simultaneously, which leads to an increased chance of false positives. Individual p-values of a specified significance threshold no longer correspond to significant findings and thus, there is a need to adjust for multiple testing when assessing the statistical significance of genes in large microarray datasets. One such example of adjustment is Bonferroni

---

[1] http://www.r-project.org

correction, where the significance threshold is divided by the number of tests performed. However, this is often too strict criteria to be useful, and therefore other methods that are more adopted for analysis of microarray data have been developed. Examples of such methods are Significance Analysis of Microarray Data (SAM) (Tusher et al., 2001) and Empirical Bayes Analysis of Microarrays (EBAM) (Efron & Tibshirani, 2002), both included in the Siggenes package[2]. SAM and EBAM are statistical methods that controls for the false discovery rate (FDR), which is the fraction of false positives in the total set of genes that are selected as differentially expressed. The FDR is determined by using permutations of the repeated measurements to estimate the percentage of genes identified by chance. Although statistical methods usually are preferable since they provide a significance measure, the simple Fold Change (FC) method which calculates the ratio between two samples, are also frequently used. However, since FC provides no statistics regarding the significance of the results, combinations of FC and statistical methods are sometimes applied.

5.3 Clustering of gene expression data

To reduce the dimensionality and facilitate interpretation of microarray data one can apply different clustering techniques, such as hierarchical clustering, K-means (Świniarski et al., 1998), principle component analysis (PCA) (Jolliffe, 1986) or self-organizing maps (SOMs) (Tamayo et al., 1999), to group transcripts with similar transcriptional profiles. The purpose of clustering is to identify co-regulated and functionally related genes in large data sets. Clustering can also be used as a quality control tool to check the reproducibility of replicated arrays. For this purpose the agglomerative hierarchical clustering approach is most common. It starts with clusters containing a single item, and iteratively links and merges the two closest clusters together based on a distance measure. After each step, all the distances between the newly formed clusters are re-calculated. The output is a relationship tree (dendrogram) where the branches represent similarity and where the replicated arrays are expected to be grouped tightly together.

5.4 Pathway analysis

There are two main approaches for identification of pathways that are differentially expressed across various experimental conditions. These are Individual Gene Analysis (IGA) methods and Gene Set Analysis (GSA) methods (Nam & Kim, 2008). IGA is the most widely used approach and evaluates the significance of individual genes between two groups of compared samples. Methods using this approach typically yield a list of differentially expressed genes from a cutoff threshold, and evaluate this list for the enrichment of genes participating in specific pathways from a pathway database. A limitation with IGA approaches is that the final result is significantly affected by the selected threshold, which is often arbitrarily chosen. Notably, many genes with moderate, but biologically meaningful expression differences, are discarded by a strict cutoff threshold, which implies a reduction in statistical power. The GSA approach directly scores pre-defined pathways or gene sets based on differential expression, and specifically aims to identify pathways with subtle but coordinated expression changes that cannot be detected by IGA methods (Mootha et al., 2003, Nam & Kim, 2008). It is based on the principle that even weak expression changes for groups of related genes can have important effects. From

---

[2] http://www.bioconductor.org

a biological perspective, GSA methods are promising because functionally related genes often display coordinated expression (Nam & Kim, 2008). Various bioinformatic resources are available for carrying out pathway analysis such as WebGestalt[3] and DAVID[4] for the IGA approach, and GSEA[5] for the GSA approach.

5.5 Protein interaction networks

Protein–protein interactions are of central importance for virtually every biological process in a living cell. Typically, signal transduction, where mechanical/chemical stimuli to a cell are converted into specific cellular responses, plays a fundamental role in many biological processes and in many diseases. To investigate the putative interactions among proteins from the significantly up- or down-regulated genes identified from a biological experiment, protein interaction networks can be computationally generated by combining the experimental data with information from interaction databases. Several tools to aid derivation of protein interaction networks are available and currently one of the most comprehensive is STRING[6], which is a freely available database and web resource for experimentally determined and predicted protein-protein interactions (Jensen et al., 2009, von Mering et al., 2007). STRING includes both physical and functional interactions, and it weights and integrates information from numerous sources, including experimental repositories, computational prediction methods and public text collections. Thus, STRING is acting as a meta-database that maps all interaction evidence into a common set, which is then graphically visualized in a protein interaction network (Fig. 3).

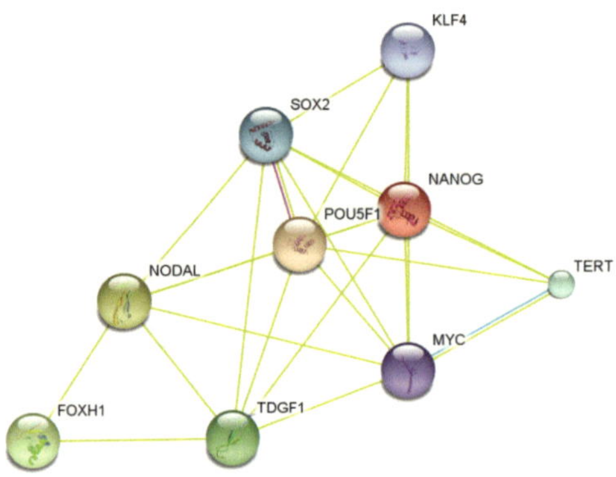

Fig. 3. Example of a protein interaction network generated by STRING, using pluripotency marker genes as input.

5.6 Functional annotation of differentially expressed genes

To increase our understanding of the biological properties of differentially expressed genes and further explore their functional properties, one can also use annotation information from Gene Ontology (Ashburner et al., 2000) which has annotation terms describing the

---

[3] http://bioinfo.vanderbilt.edu/webgestalt/

[4] http://david.abcc.ncifcrf.gov/

[5] http://www.broadinstitute.org/gsea/

[6] http://string-db.org/

genes or gene products. Gene Ontology consists of three categories of annotation terms, Biological Processes, Molecular Functions and Cellular Components. By comparing with a reference list, overrepresentation of annotations among sets of genes can be calculated by dividing the observed number of genes holding a specific annotation with the expected number of genes with that annotation. A common tool for Gene Ontology enrichment analysis is AmiGO[7], but also e.g. WebGestalt[8] and DAVID[9] provide similar functions.

6. Problems to address with transcriptional profiling

Presently, one of the main bottlenecks in stem cell research is insufficient yield and purity of the final cell preparations and immature phenotypes of the differentiated cells. There are substantial gaps in our understanding of the molecular programs that govern early cellular differentiation and maturation, which limits the possibilities to improve the culturing protocols. In order to fully realize the potential of hESCs, a better understanding of the regulatory mechanisms that control their differentiation towards specific lineages is needed. An extensive characterization of hESCs and their derivates is also urgently required. To address these challenges, information is needed about which genes that are differentially expressed at specific stages during differentiation. Some of these are likely candidate genes in regulatory mechanisms, important for transitions of cells between different developmental stages. We also need precise characterization methods to determine the identity of cells during the differentiation process, and therefore more reliable marker genes for specific developmental stages are required. In next chapter we describe how microarray technology can been applied to analyze global gene expression patterns in hESCs and their differentiated progenies. Below are some examples studies, with focus on cardiac- and hepatic lineages, described in more detail.

7. Examples of transcriptional profiling of hESCs and their derivatives

Several global gene expression studies have been conducted on hESCs and differentiated progenies thereof. Depending on the research questions addressed, these studies have had different experimental designs and they have used different analysis approaches.

7.1 Transcriptional profiling of hESC-derived cardiomyocytes

The first example is a study where hESC-derived cardiomyocyte clusters (CMCs) were characterized at the gene expression level and their global transcriptional pattern was investigated (Synnergren et al., 2008). This required only a rather simple design with not more than two groups to compare, undifferentiated (UD) hESCs and hESC-derived CMCs. The material consisted of one pooled sample of UD hESCs and two different biological replicates of pooled hESC-derived CMCs, harvested at a number of time points up to 22 days after initiation of differentiation (Fig. 4).

---

[7] http://amigo.geneontology.org/cgi-bin/amigo/go.cgi

[8] http://bioinfo.vanderbilt.edu/webgestalt/

[9] http://david.abcc.ncifcrf.gov/

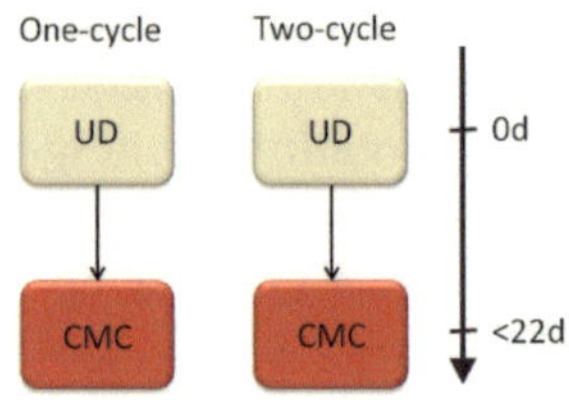

Fig. 4. Experimental design where two different groups (UD and CMC) were included and the experiment was repeated twice using one- and two-cycle amplification respectively.

The hESC line SA002 (Cellartis AB, Göteborg) was used in this experiment. Due to technical issues, two separate sets of microarray experiments were conducted. In the first, one-cycle amplified RNA was used, while in the second set of experiments two-cycle amplified RNA was used due to the limited amount of available RNA. Even though no obvious differences between the two data sets could be observed, all subsequent calculations between samples were conducted within each experiment separately. The quality of the RNA and cRNA, labeled by in vitro transcription, was tested and the fragmented cRNA was then hybridized to the microarrays. Each sample was hybridized to duplicate arrays from the Affymetrix microarray platform (GeneChip 133 Plus 2.0) (Affymetrix, Santa Clara, CA). Extraction of expression values and scaling of data was performed using the MAS5 algorithm and transcripts flagged as 'Absent' on all arrays were filtered before the data analysis. The SAM statistical algorithm was used to identify significantly up- and down-regulated genes between the CMCs compared to UD cells. In total 530 genes were identified as up-regulated and 40 genes were down-regulated in the CMCs (Synnergren et al., 2008). These sets of regulated genes were further analyzed using various bioinformatic tools. To further explore the biology of the significantly up-regulated genes in hESC-derived CMCs, Gene Ontology annotations were used to group the genes according to biological process, molecular function, and cellular component. To investigate possible interactions among proteins from the significantly up-regulated genes in hESC-derived CMCs, the search tool STRING was applied to derive protein interaction networks. These networks were used to identify hub-proteins, with many interactions to other genes. Moreover, differentially expressed pathways were assessed in the CMCs using the WebGestalt tool.

7.2 Investigation of putative correlation between mRNA and miRNA expression

This second example describes how transcriptional profiling can be applied to further our understanding of the regulatory mechanisms of transcription and translation, by investigation of putative correlation between mRNA and miRNA expression. Thus, mRNA and miRNA microarray experiments were designed where matched samples from hESCs and hESC-derived CMCs were collected for global mRNA and miRNA profiling. Using cell line SA002 (Cellartis AB, Göteborg) total RNA was extracted with a method which preserves small molecules. The RNA was split into two aliquots, and microarray experiments were conducted in parallel to measure both miRNA and mRNA expression of paired samples.

As illustrated in Fig. 5, the material consisted of samples of UD cells and hESC-derived CMCs, cultured for 3 (CMC3w) and 7 weeks (CMC7w) after onset of differentiation. In addition, samples from fetal heart (FH) and adult heart (AH) were included as reference material. Significantly up- or down-regulated miRNAs and mRNAs were identified using the SAM statistical algorithm. The differentially expressed miRNAs were grouped

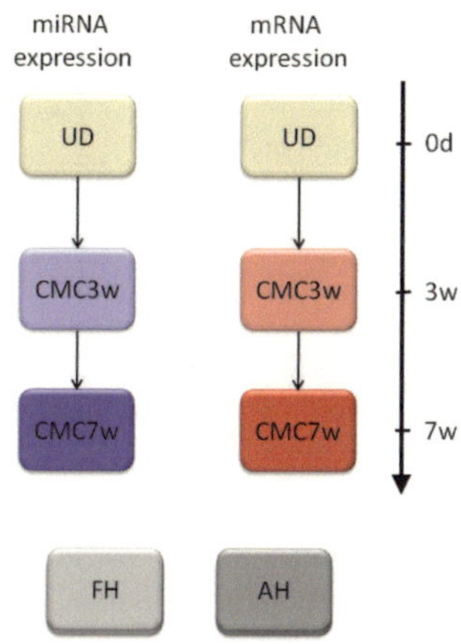

Fig. 5. Experimental design of a parallel miRNA and mRNA study. Three time points (UD, CMC 3 weeks, and CMC 7 weeks) were analyzed and fetal heart (FH) and adult heart (AH) were included as reference samples.

according to their expression profiles using hierarchical clustering. Putative target genes of up- and down-regulated miRNAs were predicted using the tool microT[10], and to investigate if the predicted target genes are present in the set of differentially expressed mRNAs, the overlap between predicted target genes and differentially expressed mRNAs were calculated. To further explore the molecular biology of the regulated target genes these were also analyzed for enrichment of Gene Ontology annotations related to cardiac development.

7.3 Transcriptional comparison between two endoderm differentiation protocols

The final example describes a transcriptional comparison between hESCs differentiated through the endoderm, either definitive endoderm (DE) or primitive endoderm (PrE), as well as a global transcriptional characterization of endoderm, hepatocyte progenitors, and hepatocyte-like cells (Synnergren et al., 2010a). A comprehensive experimental design was applied in this work including three cell lines (SA002, SA167, and SA461) and four time points, as well as the two separate differentiation protocols (Fig. 6).

The hepatocellular carcinoma cell line (HepG2) was included as a reference sample in the experiment. Each sample was cultured and harvested in biological duplicates. The RNA was extracted and assessed for quality before generation of cRNA, and subsequently hybridized to the arrays. The raw data was extracted and normalized using MAS5 and filtered and log_2 transformed before subsequent data analysis. In this experiment, the FC method was applied to identify differentially expressed genes at various stages. At the early time points these cells showed relatively large variations in the magnitude of up- or down-regulation of genes across the cell lines. However, the trend of the regulation was consistent in all three cell lines and therefore an interesting observation. In such situation, the statistical methods have problems to select genes of real biological interest, and consequently FC may be a preferable alternative. The DE differentiation pathway to generate hepatocytes mimics the development of hepatocytes in the embryo and the expression profiles in these samples

[10] http://diana.cslab.ece.ntua.gr/microT/

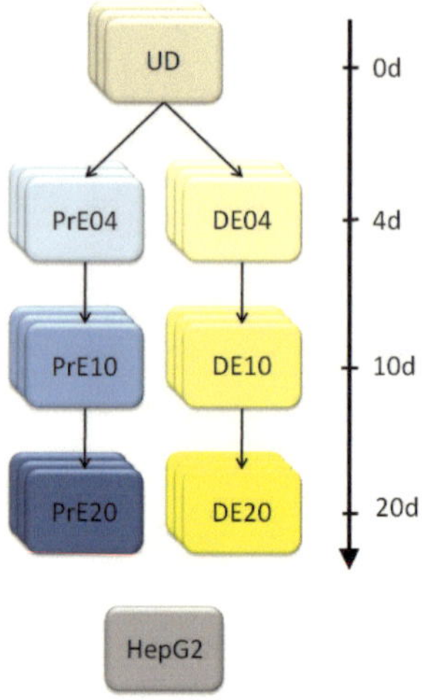

Fig. 6. Experimental design of the protocol comparison experiment. Four time points (UD, 4 days, 10 days, and 20 days) and two differentiation protocols (PrE and DE) were included in the experiment which was run in duplicates and repeated using three different cell lines. HepG2 was included as a reference sample in the study.

were therefore characterized in more detail. The genes that showed up-regulation in DE20 samples compared to UD samples were further analyzed for enrichment of Gene Ontology annotations and differentially expressed pathways were also identified using the DAVID bioinformatic resource. Moreover, protein interaction networks were derived using the STRING tool and based in these, hub proteins were identified among the up-regulated genes in DE20 samples.

8. Results and discussion

Extensive characterization of hESC-derived functional cell types, such as cardiomyocytes and hepatocytes has been performed by several investigators, with the purpose to explore the transcriptional programs that are activated during differentiation along these specific lineages. Results from these studies have identified large sets of genes that showed differential expression at specific developmental stages. The sets of regulated genes have been further explored by various bioinformatic analyses, to learn more about the molecular functions of the differentially expressed genes and understand their biology. Here we discuss results from studies on cardiomyocyte and hepatocyte differentiation, performed by us and others, and report on genes and pathways that showed up- or down-regulation particularly during these processes.

8.1 Molecular signature of hESC-derived cardiomyocyte clusters

Despite the substantial progress made by different investigators during recent years, the understanding of the molecular signature of hESC-derived cardiomyocytes (CMs) and the factors that induce cardiogenesis during embryonic development still remains limited. However, important knowledge about the transcriptional program that is activated during CM differentiation has been gained through transcriptional profiling of these cells (experimental set-up described in section 7.1). Selected colonies of hESC-derived contracting clusters of CMs were manually dissected, and pooled for subsequent microarray analysis. These samples were compared with samples from UD hESCs. In total 530 up-regulated and 40 down-regulated genes were identified in the CMCs (Synnergren et al., 2008). Among the up-regulated genes, there were several that have been used before to characterize hESC-derived CMs e.g., MYH6, MYH7, PLN, TNNT2, NPPA, GATA4, and MEF2C (Kehat et al., 2001, McDevitt et al., 2005). The functional properties of the up-regulated genes in the hESC-derived CMCs were further investigated, using available Gene Ontology annotations. Among the enriched annotations were 'muscle contraction', 'development of mesoderm and muscle', 'cellular differentiation', 'calcium ion binding', and 'tropomyosin binding'. Moreover, several induced cellular pathways where identified, that may be important for cardiogenic induction of hESCs but also for sustaining the CM phenotype. Results reported by other investigators show high overlap with our data, despite the fact that direct comparisons of results between different microarray studies are sometimes difficult to make. Different experiments often have major discrepancies in differentiation models, microarray platforms, cell lines used, and experimental set-ups, which partly may explain observed problems with poor overlap between published results from different stem cell studies (Fortunel et al., 2003, Ivanova et al., 2002, Ramalho-Santos et al., 2002).

To overcome some of these problems one should preferably re-analyze the data from the raw data files from each experiment, using a consistent data mining approach for all the data sets, when comparing results from multiple microarray studies. However, as an alternative, we have compared published lists of significantly enriched genes from similar studies, to explore the overlap of differentially expressed genes during CM differentiation. Importantly, in addition to the above mentioned challenges when comparing microarray data from different experiments, the final cell populations that have been analyzed in these studies differ in their composition (Cao et al., 2008, Xu et al., 2009). Nevertheless, and in contrast to previous findings (Fortunel et al., 2003, Ivanova et al., 2002, Ramalho-Santos et al., 2002), notable similarities were identified across results from the four global expression studies that so far have been published on hESC-derived CM-like cells (Beqqali et al., 2006, Cao et al., 2008, Synnergren et al., 2008, Xu et al., 2009). Importantly, this strengthens the reliability of the microarray technology and verifies that hESC-derived CMs express a uniform transcriptional profile, despite different cell lines and major differences in how these cells are derived.

Comparing results from our data (Synnergren et al., 2008) with the study performed by Beqqali et al. (Beqqali et al., 2006), where hESC-derived CMs were generated by co-culture with END-2 cells (Passier et al., 2005), 15 genes were reported as enriched in their hESC-derived CMs and in fetal heart tissue. Notably, eight (53%) of these genes are also up-regulated in our hESC-derived CMs (e.g. TNNT2, PLN, and MYL7). Another study published on hESC-derived CMs (Cao et al., 2008) report on analyses made on material from hESCs, hESC-derived beating embryoid bodies (EBs), hESC-derived CMs which were percol purified to 40-45% CMs, and purified CMs from fetal heart (FH) tissue samples. Notably, their study focused on transitions from one stage to the next one, and consequently they compared hESCs-EBs, EBs-CMs, and CMs-FH. In our work we compared hESCs with hESC-derived CM clusters and the corresponding direct comparison of CMs and hESCs was not done by Cao et al. (Cao et al., 2008) which hampers the comparison of our results with theirs. Nevertheless, we found that 33% of our up-regulated genes in the CMCs were in their study identified as enriched already at the EB stage. Five of our up-regulated genes in the CMCs (EPAS1, ITGB3, PLD1, MSRB3, EMP1) were significantly enriched in FH compared to the CM-sample, and six of our genes that were enriched in the CMCs (CLIC5, RUNX1, COL8A1, LONRF2, MSRB3, CAV2) were up-regulated in CMs compared to EBs. A similar comparison was made regarding the repressed genes across these two studies and 17 (43%) of the 40 significantly down-regulated genes in our data were already repressed at the EB stage in Cao et al., and one gene was among the significantly down-regulated genes between FH and CMs. Again, no comparison was made between CMs and hESCs (Cao et al., 2008) regarding down-regulated genes, but such a comparison is anticipated to generate a higher overlap with our list of genes that were down-regulated in CMCs.

The most recent work on global gene expression of hESC-derived CMs used a transgenic cell line with a construct comprising the CM-restricted α-myosin heavy chain (α-MHC) promoter (Xu et al., 2009). They applied antibiotic selection to purify their population of hESC-derived CMs and achieved a 99% pure population. Fetal and adult heart tissue were used as reference samples but notably, these samples were not purified, but contained a mixture of the cell types present in heart tissue. Despite substantial differences, such as different cell lines, differentiation protocols, purity of CMs, sampling day etc, a prominent overlap was observed between our data and the data from Xu and colleagues (Xu et al., 2009). In total 147 (27%) of the 540 genes that were up-regulated in our data were also identified as significantly up-regulated in their population of CMs, when compared to UD and EB samples. Remarkably, 115 (78%) of these 147 genes also show up-regulation in the FH and AH samples in data from Xu et al. Strikingly, a subset of 57 genes that show up-regulation in our hESC-derived CM clusters is also overlapping with the up-regulated genes both in Cao et al. (Cao et al., 2008) and in Xu et al (Xu et al., 2009). All of these 57 genes also show significantly up-regulation in FH and AH. Furthermore, three (RBM24, TCEA3, and FHOD3) of the four novel candidate cardiac markers, which by Xu and co-workers were validated by *in situ* hybridization during early mouse development, were indeed significantly up-regulated in our study of hESC-derived CMCs (Synnergren et al., 2008). The fourth one (C15orf52) was not present on the arrays we used. Interestingly, TCEA3 is also among the 57 genes that overlapped across all three studies (Cao et al., 2008, Synnergren et al., 2008, Xu et al., 2009). Taken together, this suggests that there are substantial similarities between the CM cell populations obtained from hESCs, independent of differentiation protocols and cell lines used. The results summarized here provide valuable information about the molecular program that is active in hESC-derived CMs.

However, to further analyze the regulatory mechanisms that may control CM differentiation, an additional level of gene regulation have also been explored (experimental set-up described in section 7.2) and hESC-derived CMCs have been characterized with respect to their miRNA expression. Global microarrays were employed to measure the expression of both miRNA and mRNA in parallel in samples of CMC, harvested at two different time points, 3 weeks and 7 weeks after onset of differentiation, as well as in UD cells and in fetal and adult heart tissue samples. Differentially expressed miRNAs and mRNAs were identified in these datasets, by using UD cells as control sample. Notably there were more than twice as many up-regulated than down-regulated miRNAs in the samples of CMCs, indicating the importance of increased expression of specific miRNAs during cardiac development (Synnergren et al., 2010b). Furthermore, we also identified more differentially expressed miRNAs (both up- and down-regulated) in the CMC samples than in the fetal and adult heart tissue samples. To define a set of miRNAs of putative importance in cardiac-like cells, differentially expressed miRNAs in samples from CMC and in samples from fetal and adult heart were compared, and an overlap of regulated miRNAs in all four samples was identified. In total 61 up-regulated and 24 down-regulated miRNAs were identified when investigating all four samples (Synnergren et al., 2010b). Moreover, possible correlations between differentially expressed miRNAs and mRNAs were investigated, by first conducting computational predictions for the differentially expressed miRNAs, and then determining putative concordance in miRNA expression and mRNA levels of the predicted target genes. Interestingly, a correlation between the global miRNA expression and corresponding target mRNA expression was observed. To further explore the biology of the predicted target genes of the differentially expressed miRNAs, enrichment of Gene Ontology annotations was determined and strikingly, several of the over-represented annotations relate to cardiac function and cardiac development. A number of induced cellular pathways were also identified among the predicted target genes, and several of these have been demonstrated to be important in cardiac development or functions e.g. 'NFAT and Hypertrophy of the heart', 'Wnt signaling pathway' and 'Calcium signaling pathway'. Results from this analysis provide an excellent starting point for further studies regarding the functional properties of the differentially expressed miRNAs in the context of cardiogenesis and regeneration of cardiac tissue.

Several other studies have also identified a number of miRNAs that are likely to play key roles during heart development and in cardiac function (Cordes & Srivastava, 2009, Divakaran & Mann, 2008, Sartipy et al., 2009, Thum et al., 2008, van Rooij & Olson, 2007, Zhang, 2008). It was recently demonstrated that miR-1 reinforces the expression of one of the earliest cardiac markers, NKX2.5, in both murine and human ESC lines and that it increases the fraction of contracting CMs compared to control samples (Ivey et al., 2008). Another group of miRNAs that are expressed during the stem cell state and progressively declines during differentiation are the nearly identical miRNAs miR-302a-d, collectively referred to as miR-302 (Rosa et al., 2009). By controlling the germ layer specification and promote mesendodermal fate specification while inhibiting neuroectoderm formation, it is suggested that miR-302 has a crucial role in embryogenesis (Rosa et al., 2009). In line with these reports, our data confirms that several variants of miR-302 are highly expressed in UD cells, and a substantial down-regulation is also observed in differentiated progenies as well as in fetal and adult heart tissue. Moreover, we identified miR-208a/b and miR-499 as significantly induced in all cardiac-like samples. Interestingly, these miRNAs are also reported as enriched in cardiac tissue by others (Adachi et al., 2010, Ji et al., 2009, Sluijter et al., 2010). Both miRNA-499 and miRNA-1 are suggested to regulate the proliferation of human CM progenitors and their further differentiation into CMs (Sluijter et al., 2010). In addition, miR-499 is also proposed as a marker for acute myocardial infarction in humans (Adachi et al., 2010). Moreover, miR-208 is suggested as a marker for myocardial injury in rat (Ji et al., 2009), and as a regulator of cardiac hypertrophy in mice (Callis et al., 2009).

Together with our results, these reports emphasize the importance of miRNAs for cardiac development as well as potentially useful markers in clinical applications. In the future, some of these miRNAs may also serve as prospective drug targets in various cardiac injuries. There is accumulating evidence supporting the importance of miRNAs for hESC self-renewal, pluripotency, and differentiation. Determining miRNAs that are associated with re-programming will yield significant insight into the specific miRNA expression patterns that are required for pluripotency. To further investigate which miRNAs that are associated with re-programming, investigators have now started to characterize the miRNA expression during the re-programming of iPS cells (Wilson et al., 2009). Interestingly, result shows that miR-302 is up-regulated in both hESCs and in iPS cells.

8.2 Transcriptional profiling of hESCs differentiating to definitive and primitive endoderm and further towards the hepatic lineage

The endoderm lineage can be subdivided into the DE which further develops into liver, pancreas and lung, and the PrE which develops into the yolksack, where it forms the placenta. To explore transcriptional differences between these two subtypes, global gene expression data from DE and PrE differentiation was analysed and the transcriptional patterns in these two cell lineages were compared with UD cells, as well as with control samples from HepG2 (experimental set-up described in section 7.3). Interesting differences and similarities were identified between these two endodermal subtypes. We also thoroughly characterized the DE-derivatives, by identifying up- and down-regulated genes in each of the three differentiation time points 4 days (DE), 10 days (DE-Prog) and 20 days (DE-Hep). In total, we identified 167, 439, and 921 transcripts which were significantly up-regulated in DE, DE-Prog, and DE-Hep, respectively, when compared to UD samples. None of these transcripts were significantly enriched in the PrE derivatives. Well-known markers for DE, such as SOX17, CXCR4, CER1 and GSC, showed a distinct peak of expression in the DE time point in all the three investigated cell lines. At the final time point, when using the DE differentiation protocol, several genes known to be expressed in mature hepatocytes, e.g. ALB, DPP4, SERPINA7, TF, TM4SF1 and UBD (Chiao et al., 2008, Hay et al., 2008, Kon et al., 2006), showed increased mRNA levels. Notably, also CD44, known to be expressed in hepatocyte progenitors (Kon et al., 2006), showed high expression at 20 days, which indicates that the DE-Hep have an immature phenotype or also contain a fraction of hepatocyte progenitors. Interestingly, ALB, which is a well-known marker for mature hepatocytes (Duncan, 2003), showed 3-11 times higher expression in DE-Hep than in the corresponding PrE-derivatives, and the expression of ALB was about 1,000-fold higher in the DE-Hep than in the UD samples. Of special note from our results are the less reported genes TM4SF1 and UBD, which demonstrate highly interesting expression patterns. Their expression patterns were consistent across all three cell lines, indicating a putative importance of controlled expression during hepatocyte differentiation. Both these genes demonstrate increasing expression levels during the differentiation towards hepatocytes, with the peak expression in the most mature samples. Interestingly, and consistent with our observations, these genes have also previously been reported as reliable hepatocyte markers (D´Amour et al., 2007). Another observation from our results is that the global transcriptional activity increases dramatically as the cells differentiate, with a larger number of differentially expressed genes in the more mature samples.

To the best of our knowledge, besides our work (Synnergren et al., 2010a), global transcriptional profiling of hESC-derived hepatocytes has only been reported once (Chiao et al., 2008), in which AFP^+ cells were selected for gene expression analysis. However, the data analysis approach used in that study differs from ours and instead of generating lists of up- vs. down-regulated genes during hepatocyte differentiation, they applied Gene Set Enrichment Analysis (Mootha et al., 2003) to identify affected sets of genes during hepatic specification, e.g. molecular pathways etc. An overlap comparison of lists of enriched genes in different stages is therefore not possible. However, when investigating the expression of 30 marker genes, which specifically were reported as enriched in their AFP^+ population, we observed important similarities between the data sets. Examples of genes that are enriched in the early DE stage in both studies are typical DE-markers such as SOX17, FOXA2 and MIXL1. Moreover, AFP, which is expressed during hepatocyte specification but also at very early stages of PrE, is up-regulated accordingly in our data. Additionally, CDH17 and KRT7, which are reported as expressed during hepatocyte specification (Chiao et al., 2008) are induced in our hepatocyte-like cells, which might indicate that this population also contains more immature cells. Also other genes such as ALB, NTN4, MET, and CEBPA, which are known to be expressed in mature hepatocytes, show induced expression patterns in both studies.

9. Conclusion and future perspectives

Human ESCs have a tremendous potential in many different applications such as drug development, regenerative medicine, and as a model system in basic research. However, to fully utilize the potential of these cells we need to better understand the regulatory mechanisms that control the differentiation of hESC into various functional cell types. Transcriptional profiling is a powerful approach to learn more about the global transcriptional pattern in differentiating hESC. In this chapter we have reviewed studies on global transcriptional profiling in hESCs and differentiated progenies of the cardiac- and hepatic lineages. The comprehensive datasets generated from these studies are especially useful for characterization purposes, identification of differentially expressed genes, and investigation of gene regulation. We, and others, have performed transcriptional profiling of hESC-derived CMs and hepatocytes, and sets of significantly up- and down-regulated genes in different stages during the differentiation processes have been identified. Interesting transcriptional patterns have been revealed, which confirms the expression of known marker genes as well as identifies large sets of novel genes that are differentially expressed at particular stages during the differentiation. The expected expression patterns of known marker genes is important information to confirm the efficiency of the differentiation, but most interesting are the novel sets of genes that not previously have been associated with differentiation or developmental processes. It is hypothesized that many of these genes have the potential to serve as novel markers and provide important information for optimization of the differentiation protocols, and further validation and investigation of their functional properties is therefore suggested.

Less studied than mRNA expression is the miRNA expression, which has shown to be important in stem cell specification. These tiny molecules provide an additional level of gene regulation by fine-tuning of the mRNA expression. An interesting approach to identify putative correlations between these interacting molecules is to explore the miRNA and mRNA expression in parallel, which has revealed interesting result when applied on data from hESC-derived CMs. MicroRNA expression may also provide a novel characterization of hESC-derivatives and results from studies on CMs show that specific miRNAs may serve as important complementary markers for this specific cell type. However, additional studies on the miRNA expression in other derivatives are needed to be able to compare the expression in different cell types, and ideally identify lineage specific miRNA patterns.

During the last decade the microarray technology has improved considerably and the recent development of whole transcript arrays has rendered the possibility to also explore different splicing variant genes. This is an important step forward since it is anticipated that more than half of all genes undergo alternative splicing. However, the data analysis of exon data sets are demanding and tools for proper interpretation of such data is still rare and

insufficient. Nevertheless, splicing variations are believed to be important in many disease and their function in stem cell differentiation remains to be investigated. The most recent progress in the field of human pluripotent stem cells is successful re-programming of differentiated somatic cells into induced pluripotent stem (iPS) cells (Takahashi et al., 2007). Human iPS-cells are generated from somatic cells by over-expression of specific factors and these cells share many characteristics of hESCs, including multi-lineage differentiation potential and infinitive proliferation capabilities *in vitro*. However, extensive characterization of the iPS cells and their differentiated progenies at the transcriptional level is still lacking, but it is nevertheless required to be able to assess their similarity to hESCs.
Taken together, transcriptional profiling of hESC and their derivatives offers great possibilities to explore global expression patterns that are activated during hESC differentiation, and provide a foundation for further dissection of the molecular mechanisms that control the stem cell specification.

10. References

Adachi, T., Nakanishi, M., Otsuka, Y., Nishimura, K., Hirokawa, G., Goto, Y., Nonogi, H. & Iwai, N. (2010) Plasma microrna 499 as a biomarker of acute myocardial infarction. *Clin Chem,* 56, 7, 1183-5.

Alberts, B., Bray, D., Hopkin, K., Johnson, A., Lewis, J., Raff, M., Roberts, K. & P, W. (2004) *Essential cell biology,* (Second edn) (Garland Science, Taylor & Francis Group, New York).

Ashburner, M., Ball, C. A., Blake, J. A., Botstein, D., Butler, H., Cherry, J. M., Davis, A. P., Dolinski, K., Dwight, S. S., Eppig, J. T., Harris, M. A., Hill, D. P., Issel-Tarver, L., Kasarskis, A., Lewis, S., Matese, J. C., Richardson, J. E., Ringwald, M., Rubin, G. M. & Sherlock, G. (2000) Gene ontology: Tool for the unification of biology. The gene ontology consortium. *Nat Genet,* 25, 1, 25-9.

Bartel, D. P. (2004) Micrornas: Genomics, biogenesis, mechanism, and function. *Cell,* 116, 2, 281-97.

Beqqali, A., Kloots, J., Ward-Van Oostwaard, D., Mummery, C. & Passier, R. (2006) Genome-wide transcriptional profiling of human embryonic stem cells differentiating to cardiomyocytes. *Stem Cells,* 24, 8, 1956-67.

Callis, T. E., Pandya, K., Seok, H. Y., Tang, R. H., Tatsuguchi, M., Huang, Z. P., Chen, J. F., Deng, Z., Gunn, B., Shumate, J., Willis, M. S., Selzman, C. H. & Wang, D. Z. (2009) Microrna-208a is a regulator of cardiac hypertrophy and conduction in mice. *J Clin Invest,* 119, 9, 2772-86.

Cao, F., Wagner, R. A., Wilson, K. D., Xie, X., Fu, J. D., Drukker, M., Lee, A., Li, R. A., Gambhir, S. S., Weissman, I. L., Robbins, R. C. & Wu, J. C. (2008) Transcriptional and functional profiling of human embryonic stem cell-derived cardiomyocytes. *PLoS ONE,* 3, 10, e3474.

Chen, J. J., Hsueh, H. M., Delongchamp, R. R., Lin, C. J. & Tsai, C. A. (2007) Reproducibility of microarray data: A further analysis of microarray quality control (maqc) data. *BMC Bioinformatics,* 8, 412.

Chiao, E., Elazar, M., Xing, Y., Xiong, A., Kmet, M., Millan, M. T., Glenn, J. S., Wong, W. H. & Baker, J. (2008) Isolation and transcriptional profiling of purified hepatic cells derived from human embryonic stem cells. *Stem Cells,* 26, 8, 2032-41.

Cordes, K. R. & Srivastava, D. (2009) Microrna regulation of cardiovascular development. *Circ Res,* 104, 6, 724-32.

D´Amour, K., Baetge, Emmanuel (2007) Hepatocytes lineage cells. US Patent WO 07127454.

Divakaran, V. & Mann, D. L. (2008) The emerging role of micrornas in cardiac remodeling and heart failure. *Circ Res,* 103, 10, 1072-83.

Draghici, S., Khatri, P., Eklund, A. C. & Szallasi, Z. (2006) Reliability and reproducibility issues in DNA microarray measurements. *Trends Genet,* 22, 2, 101-9.

Duncan, S. A. (2003) Mechanisms controlling early development of the liver. *Mech Dev,* 120, 1, 19-33.

Efron, B. & Tibshirani, R. (2002) Empirical bayes methods and false discovery rates for microarrays. *Genet Epidemiol,* 23, 1, 70-86.

Fortunel, N. O., Otu, H. H., Ng, H. H., Chen, J., Mu, X., Chevassut, T., Li, X., Joseph, M., Bailey, C., Hatzfeld, J. A., Hatzfeld, A., Usta, F., Vega, V. B., Long, P. M., Libermann, T. A. & Lim, B. (2003) Comment on " 'stemness': Transcriptional profiling of embryonic and adult stem cells" And "A stem cell molecular signature". *Science,* 302, 5644, 393; author reply 393.

Frantz, S. (2005) An array of problems. *Nat Rev Drug Discov,* 4, 5, 362-3.

Gregory, P. A., Bert, A. G., Paterson, E. L., Barry, S. C., Tsykin, A., Farshid, G., Vadas, M. A., Khew-Goodall, Y. & Goodall, G. J. (2008) The mir-200 family and mir-205 regulate epithelial to mesenchymal transition by targeting zeb1 and sip1. *Nat Cell Biol,* 10, 5, 593-601.

Hay, D. C., Fletcher, J., Payne, C., Terrace, J. D., Gallagher, R. C., Snoeys, J., Black, J. R., Wojtacha, D., Samuel, K., Hannoun, Z., Pryde, A., Filippi, C., Currie, I. S., Forbes, S. J., Ross, J. A., Newsome, P. N. & Iredale, J. P. (2008) Highly efficient differentiation of hescs to functional hepatic endoderm requires activina and wnt3a signaling. *Proc Natl Acad Sci U S A,* 105, 34, 12301-6.

Ivanova, N. B., Dimos, J. T., Schaniel, C., Hackney, J. A., Moore, K. A. & Lemischka, I. R. (2002) A stem cell molecular signature. *Science,* 298, 5593, 601-4.

Ivey, K. N., Muth, A., Arnold, J., King, F. W., Yeh, R. F., Fish, J. E., Hsiao, E. C., Schwartz, R. J., Conklin, B. R., Bernstein, H. S. & Srivastava, D. (2008) Microrna regulation of cell lineages in mouse and human embryonic stem cells. *Cell Stem Cell,* 2, 3, 219-29.

Jensen, L. J., Kuhn, M., Stark, M., Chaffron, S., Creevey, C., Muller, J., Doerks, T., Julien, P., Roth, A., Simonovic, M., Bork, P. & Von Mering, C. (2009) String 8--a global view on proteins and their functional interactions in 630 organisms. *Nucleic Acids Res,* 37, Database issue, D412-6.

Ji, X., Takahashi, R., Hiura, Y., Hirokawa, G., Fukushima, Y. & Iwai, N. (2009) Plasma mir-208 as a biomarker of myocardial injury. *Clin Chem,* 55, 11, 1944-9.

Jolliffe, I. T. (1986) *Principal component analysis,* (Springer-Verlag).

Kehat, I., Kenyagin-Karsenti, D., Snir, M., Segev, H., Amit, M., Gepstein, A., Livne, E., Binah, O., Itskovitz-Eldor, J. & Gepstein, L. (2001) Human embryonic stem cells can differentiate into myocytes with structural and functional properties of cardiomyocytes. *J Clin Invest,* 108, 3, 407-14.

Kon, J., Ooe, H., Oshima, H., Kikkawa, Y. & Mitaka, T. (2006) Expression of cd44 in rat hepatic progenitor cells. *J Hepatol,* 45, 1, 90-8.

Kuo, W. P., Jenssen, T. K., Butte, A. J., Ohno-Machado, L. & Kohane, I. S. (2002) Analysis of matched mrna measurements from two different microarray technologies. *Bioinformatics,* 18, 3, 405-12.

Lu, J., Getz, G., Miska, E. A., Alvarez-Saavedra, E., Lamb, J., Peck, D., Sweet-Cordero, A., Ebert, B. L., Mak, R. H., Ferrando, A. A., Downing, J. R., Jacks, T., Horvitz, H. R. & Golub, T. R. (2005) Microrna expression profiles classify human cancers. *Nature,* 435, 7043, 834-8.

Mcdevitt, T. C., Laflamme, M. A. & Murry, C. E. (2005) Proliferation of cardiomyocytes derived from human embryonic stem cells is mediated via the igf/pi 3-kinase/akt signaling pathway. *J Mol Cell Cardiol,* 39, 6, 865-73.

Mootha, V. K., Lindgren, C. M., Eriksson, K. F., Subramanian, A., Sihag, S., Lehar, J., Puigserver, P., Carlsson, E., Ridderstrale, M., Laurila, E., Houstis, N., Daly, M. J., Patterson, N., Mesirov, J. P., Golub, T. R., Tamayo, P., Spiegelman, B., Lander, E. S., Hirschhorn, J. N., Altshuler, D. & Groop, L. C. (2003) Pgc-1alpha-responsive genes involved in oxidative phosphorylation are coordinately downregulated in human diabetes. *Nat Genet,* 34, 3, 267-73.

Nam, D. & Kim, S. Y. (2008) Gene-set approach for expression pattern analysis. *Brief Bioinform,* 9, 3, 189-97.

Passier, R., Oostwaard, D. W., Snapper, J., Kloots, J., Hassink, R. J., Kuijk, E., Roelen, B., De La Riviere, A. B. & Mummery, C. (2005) Increased cardiomyocyte differentiation from human embryonic stem cells in serum-free cultures. *Stem Cells,* 23, 6, 772-80.

Pradervand, S., Paillusson, A., Thomas, J., Weber, J., Wirapati, P., Hagenbuchle, O. & Harshman, K. (2008) Affymetrix whole-transcript human gene 1.0 st array is highly concordant with standard 3' expression arrays. *Biotechniques,* 44, 6, 759-62.

Ramalho-Santos, M., Yoon, S., Matsuzaki, Y., Mulligan, R. C. & Melton, D. A. (2002) "Stemness": Transcriptional profiling of embryonic and adult stem cells. *Science,* 298, 5593, 597-600.

Rosa, A., Spagnoli, F. M. & Brivanlou, A. H. (2009) The mir-430/427/302 family controls mesendodermal fate specification via species-specific target selection. *Dev Cell,* 16, 4, 517-27.

Sartipy, P., Olsson, B., Hyllner, J. & Synnergren, J. (2009) Regulation of 'stemness' and stem cell differentiation by micrornas. *IDrugs,* 12, 8, 492-6.

Shi, L., Reid, L. H., Jones, W. D., Shippy, R., Warrington, J. A., Baker, S. C., Collins, P. J., De Longueville, F., Kawasaki, E. S., Lee, K. Y., Luo, Y., Sun, Y. A., Willey, J. C., Setterquist, R. A., Fischer, G. M., Tong, W., Dragan, Y. P., Dix, D. J., Frueh, F. W., Goodsaid, F. M., Herman, D., Jensen, R. V., Johnson, C. D., Lobenhofer, E. K., Puri, R. K., Schrf, U., Thierry-Mieg, J., Wang, C., Wilson, M., Wolber, P. K., et al. (2006) The microarray quality control (maqc) project shows inter- and intraplatform reproducibility of gene expression measurements. *Nat Biotechnol,* 24, 9, 1151-61.

Sluijter, J. P., Van Mil, A., Van Vliet, P., Metz, C. H., Liu, J., Doevendans, P. A. & Goumans, M. J. (2010) Microrna-1 regulate differentiation and proliferation in human-derived cardiomyocyte progenitor cells. *Arterioscler Thromb Vasc Biol,* 30, 4, 859-68.

Świniarski, R., Cios, K. J. & Pedrycz, W. (1998) *Data mining methods for knowledge discovery,* (Kluwer Academic).

Synnergren, J., Akesson, K., Dahlenborg, K., Vidarsson, H., Ameen, C., Steel, D., Lindahl, A., Olsson, B. & Sartipy, P. (2008) Molecular signature of cardiomyocyte clusters derived from human embryonic stem cells. *Stem Cells,* 26, 7, 1831-40.

Synnergren, J., Heins, N., Brolen, G., Eriksson, G., Lindahl, A., Hyllner, J., Olsson, B., Sartipy, P. & Bjorquist, P. (2010) Transcriptional profiling of human embryonic stem cells differentiating to definitive and primitive endoderm and further toward the hepatic lineage. *Stem Cells Dev,* 19, 7, 961-78.

Synnergren, J., Améen, C., Lindahl, A., Olsson, B., Sartipy, B. (2010) Molecular signature of cardiomyocyte clusters. Physiol Genomics, 2010 Sep 14. [Epub ahead of print].

Takahashi, K., Tanabe, K., Ohnuki, M., Narita, M., Ichisaka, T., Tomoda, K. & Yamanaka, S. (2007) Induction of pluripotent stem cells from adult human fibroblasts by defined factors. *Cell,* 131, 5, 861-72.

Tamayo, P., Slonim, D., Mesirov, J., Zhu, Q., Kitareewan, S., Dmitrovsky, E., Lander, E. S. & Golub, T. R. (1999) Interpreting patterns of gene expression with self-organizing maps: Methods and application to hematopoietic differentiation. *Proc Natl Acad Sci U S A,* 96, 6, 2907-12.

Tan, P. K., Downey, T. J., Spitznagel, E. L., Jr., Xu, P., Fu, D., Dimitrov, D. S., Lempicki, R. A., Raaka, B. M. & Cam, M. C. (2003) Evaluation of gene expression measurements from commercial microarray platforms. *Nucleic Acids Res,* 31, 19, 5676-84.

Thum, T., Catalucci, D. & Bauersachs, J. (2008) Micrornas: Novel regulators in cardiac development and disease. *Cardiovasc Res,* 79, 4, 562-70.

Tusher, V. G., Tibshirani, R. & Chu, G. (2001) Significance analysis of microarrays applied to the ionizing radiation response. *Proc Natl Acad Sci U S A,* 98, 9, 5116-21.

Van Rooij, E. & Olson, E. N. (2007) Micrornas: Powerful new regulators of heart disease and provocative therapeutic targets. *J Clin Invest,* 117, 9, 2369-76.

Wang, Y., Barbacioru, C., Hyland, F., Xiao, W., Hunkapiller, K. L., Blake, J., Chan, F., Gonzalez, C., Zhang, L. & Samaha, R. R. (2006) Large scale real-time pcr validation on gene expression measurements from two commercial long-oligonucleotide microarrays. *BMC Genomics,* 7, 59.

Wang, Y., Keys, D. N., Au-Young, J. K. & Chen, C. (2009) Micrornas in embryonic stem cells. *J Cell Physiol,* 218, 2, 251-5.

Vasudevan, S., Tong, Y. & Steitz, J. A. (2007) Switching from repression to activation: Micrornas can up-regulate translation. *Science,* 318, 5858, 1931-4.

Wilson, K. D., Venkatasubrahmanyam, S., Jia, F., Sun, N., Butte, A. J. & Wu, J. C. (2009) Microrna profiling of human-induced pluripotent stem cells. *Stem Cells Dev,* 18, 5, 749-58.

Von Mering, C., Jensen, L. J., Kuhn, M., Chaffron, S., Doerks, T., Kruger, B., Snel, B. & Bork, P. (2007) String 7--recent developments in the integration and prediction of protein interactions. *Nucleic Acids Res,* 35, Database issue, D358-62.

Xu, X. Q., Soo, S. Y., Sun, W. & Zweigerdt, R. (2009) Global expression profile of highly enriched cardiomyocytes derived from human embryonic stem cells. *Stem Cells,* 27, 9, 2163-74.

Yuen, T., Wurmbach, E., Pfeffer, R. L., Ebersole, B. J. & Sealfon, S. C. (2002) Accuracy and calibration of commercial oligonucleotide and custom cdna microarrays. *Nucleic Acids Res,* 30, 10, e48.

Zhang, C. (2008) Micrornas: Role in cardiovascular biology and disease. *Clin Sci (Lond),* 114, 12, 699-706.

Zhou, Y., Ferguson, J., Chang, J. T. & Kluger, Y. (2007) Inter- and intra-combinatorial regulation by transcription factors and micrornas. *BMC Genomics,* 8, 396.

Zhu, S., Wu, H., Wu, F., Nie, D., Sheng, S. & Mo, Y. Y. (2008) Microrna-21 targets tumor suppressor genes in invasion and metastasis. *Cell Res,* 18, 3, 350-9.

20

Tracking the Structure of Protein Interaction Network via Multiple Genetic Perturbations on Mouse Embryonic Stem Cells — Implementation of the Entropy Maximization Principle

Lei Mao[1], Rossella De Cegli[2], Mario Lauria[2], Grit Nebrich[1], Jean Maurice Delabar[3], Yann Herault[4], Gilda Cobellis[2] and Joachim Klose[1]
[1]*Institute for Medical Genetics, Charité – Universitätsmedizin, Berlin,*
[2]*Telethon Institute of Genetics and Medicine, Napoli,*
[3]*Université Paris, Paris,*
[4]*Institut de Génétique Biologie Moléculaire et Cellulaire, IGBMC, Institut clinique de la Souris, ICS, CNRS, INSERM, Université de Strasbourg, UMR7104, UMR964, Illkirch*
[1]*Germany*
[2]*Italy*
[3,4]*France*

1. Introduction

The Human Genome Project (HUGO) opened a new era for biomedical research. However, a decade after the completion of HUGO, we are still standing at the very beginning regarding the understanding of cellular dynamics under development, disease and aging. To bridge between genotype and phenotype, it is imperative to track the underlying genetic interaction network, and perhaps more importantly, the protein-protein interaction (PPI) network, which forms the basal layer of the majority of fundamental cellular processes [Antal et al., 2009].

Current approaches for protein-protein interaction network structure detection include gene fusion experiments such as yeast two-hybrids, co-occurrence evidences via immuno-coprecipitation, as well as literature or orthology-based knowledge extrapolation [Stelzl and Wanker, 2006]. Albeit powerful and foreseeing incremental technical improvements, these methods either rely heavily on experimental detection thresholds, or could be biased as they mirror the focus of individual fields of research. Hence, it is vital to path-find the underlying cellular PPI network with hypothesis-free (*de novo*) methods [Janes and Yaffe, 2006].

A biological organism can be perceived as a complex system with a set of interacting elements. Basically, when a meta-stable system consisting of steady number of interacting elements is disturbed systematically, one is able to use the system outcome to puzzle back the underlying network structure. This is commonly known as "reverse engineering" in system sciences [Bansal et al., 2007].

Previously, the use of such reverse engineering approaches on the exploration of cellular PPI network has been hampered by limited amount of high confident expression profiling data.

Nowadays, however, techniques for the probing of gene expression profile have been witnessing swirling advances. mRNA microarray, mRNA-Seq, protein antibody chips, as well as high resolution protein electrophoresis combined with mass spectrometry have been catching-up rapidly to become high throughput and high confident methodologies. These solid technical platforms promote the use of experimental data in sophisticated *de novo* inferring of cellular interaction networks. Recently, such approach has been applied on diverse complex systems [Shmulevich et al., 2002]. In a pioneering study of di Bernardo et. al., the reverse engineering approach was successfully applied on yeast to prioritize the gene products and cellular pathways involved in drug responses [di Bernardo et al., 2005].

It should be noted that all current reverse engineering approaches have practical limitations. For example, Bayesian methods need a sufficiently large number of experiments comparable to the number of genes. Undersampled datasets will not be adequate to generate confident outcome. Multiple regression methods work well only when the identity of the perturbed genes is known, or can be reliably inferred from the data [Lauria et al., 2009]. Fortunately, this methodological constraint of gene-by-gene perturbation can be partially relaxed if entropy maximization principle is invited into play. The entropy maximization principle was based on the information theory of Shannon [Shannon, 1948]. By this mean, the most parsimonious PPI interaction network structure that is able to give rise to experimental expression profiles can be probed out despite non-complete system perturbation. This principle of entropy maximization has been proved useful in inferring the genetic and signaling networks of peripheral nerve development and yeast chemostat [Dhadialla et al., 2009; Lezon et al., 2006].

Nevertheless, in multicellular organism and especially in higher eukaryotes, each differentiated cellular system can exhibit alternative PPI networks. This has been prohibiting the probing of mammalian interaction network in a large extent. Embryonic stem (ES) cells can be considered as the starting point of multicellular life. Apart from the high relevance to biomedical research, the ES cells are homogeneous self-maintaining systems. Upon a genetic mutation, either an ES cell will balance the effect of genetic insult and remain in their pluripotent state, or else undergo cell death or differentiation should the impact of the genetic mutation surpasses a certain threshold [Amit et al., 2000; Mao et al., 2007]. Such "all or none" cell faith decision constitutes a solid genomic and proteomic stability, hence put forward the ES cells as a brilliant model system for *de novo* exploration of the PPI network structure.

In this study, we applied a hybrid approach of multiple genetic perturbations and entropy maximization principle on the mouse embryonic stem cells to probe the underlying cellular protein-protein interaction network structure. Our results suggest the essential role of antioxidative response, transcriptional regulation and protein degradation pathways in the rebalancing of the proteomic network. Moreover, our predicted protein interaction data indicates that overexpression of Ripk4, a protein kinase originally located on human chromosome 21, could severely perturb the cellular cholesterol syntheses pathways via an adenosyl-homocystein mediated mechanism. Our study demonstrates the value of mouse ES cells in protein network research in the context of system biology.

2. Methods

Mouse embryonic stem (ES) cells were genetically perturbed with a pallet of 14 single genes overexpressed individually in different subclones. We studied the impacts of these genetic system stimuli on the PPI network at both transcriptomic and proteomic levels. The workflow of the current study is summarized in figure 1.

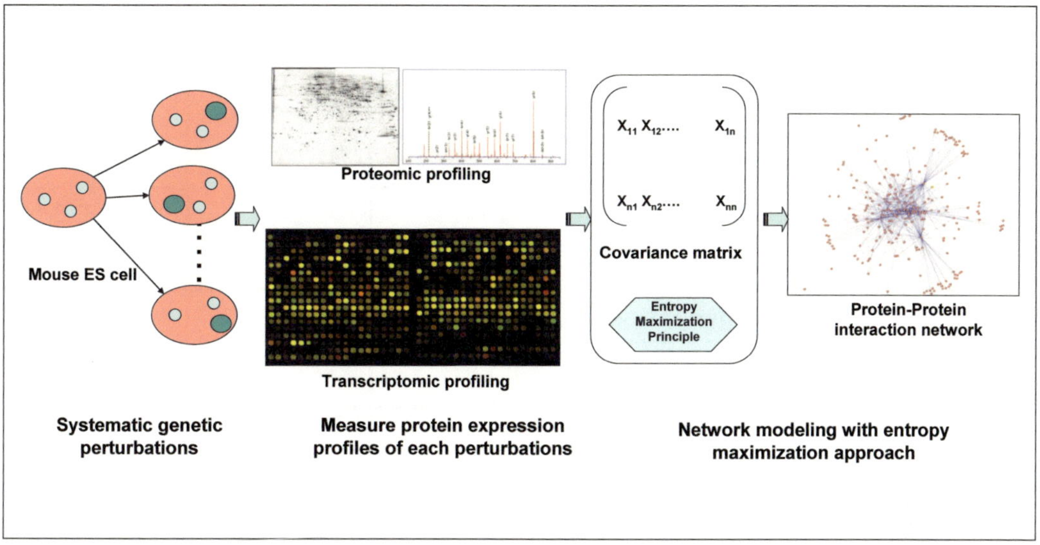

Fig. 1. Workflow of the current study. Mouse embryonic stem cells were genetically perturbed with a pallet of 14 single gene overexpressions. The impact of the system stimuli on the ES cell PPI network were readout with proteomic and mRNA microarray analyses. Gene expression profiling data undergo entropy maximization approach to reverse engineer the underlying PPI network structure (picture of microarray from Wikipedia)

2.1 Systematic genetic perturbations on mouse ES cells

This is a meta-study on a pellet of 14 transgenic mouse ES cell lines constructed in the frame of our European Research Grant "AnEUploidy". Please refer to our previous publications for detailed methodology for the generation of transgenic cell lines and expression profiling [De Cegli et al., 2010; Mao et al., 2007]. Except for Snca-overexpression cell line, each transgenic ES cell line overexpresses one of the human chromosome 21 genes (Table 1).

The impacts of these genetic system perturbations on the ES cells were examined at the proteomic and transcriptomic levels: All cell lines were measured by large-gel two-dimensional protein electrophoresis [Klose, 1999], whereas 8 of the cell lines were measured by mRNA microarray (Affymetrix GeneChip Mouse Genome 430_2 array) [De Cegli et al., 2010].

2.2 Theoretical framework of entropy maximum principle

We applied the principle of entropy maximization to identify the protein interaction network structure with the highest probability of giving rise to our experimentally observed proteomic and transcriptomic data [Lezon et al., 2006]. This relies on the Boltzmann's concept of entropy maximization. In effect, as the number of ways of realizing a given macroscopic state can very widely, the most likely state of the system is the one that corresponds to the largest number of microscopic states, or biggest entropy. This idea of entropy maximization provides a natural mechanism for revealing dominant protein-protein interaction structure inside of a quasi-stable system.

Gene Symbol	Protein name	Category	Original gene location
Aire	Autoimmune regulator	Transcription factor	21q22.3
Erg	Avian erythroblastosis virus E-26 (v-ets) oncogene related		21q22.3
Nrip1	Nuclear receptor interacting protein 1		21q11.2
Olig2	Oligodendrocyte transcription factor 2		21q22.11
Runx1	Runt related transcription factor 1		21q22.3
Sim2	Single-minded homolog 2		21q22.13
Pdxk	Pyridoxal (pyridoxine, vitamin B6) kinase	Protein kinase	21q22.3
Ripk4	Receptor-interacting serine-threonine kinase 4		21q22.3
Dyrk1A	Dual-specificity tyrosine-(Y)-phosphorylation regulated kinase 1A		21q22.13
Hunk	Hormonally up-regulated Neu-associated kinase		21q22.1
Dopey2	Dopey family member 2		21q22.2
App	Amyloid beta (A4) precursor protein		21q21.3
Snca	Synuclein, alpha		4q21
MirLet7Cdel 99a	---	Micro RNA	21q21.1
Mir99adelLet 7C	---		21q21.1
Mir802	---		21q22.12

Table 1. Characteristics of the transgenes used to genetically perturb the mouse ES cell system

2.3 Data-driven *de novo* PPI network exploration

We consider each genetic modification as a distinct stimulus on the ES cell protein interaction network. In this sense, the protein expression profile of each transgenic cell line is a sampling on the perturbed system. Given a collection of **n** genes, their expression profiles in a pellet of **k** perturbation experiments can be represented by a **k** × **n** expression matrix. Here, each row vector of this expression matrix is the expression alteration (ratio of transgenic to control) of a give gene under different perturbations. We generated the expression matrixes from the proteomic and microarray data, respectively.

In the second step, we determined the **n** × **n** expression covariance matrix **M**, whose entry M_{ij} is the expression covariance of gene Xi and Xj according to the following equation:

$$M_{ij} = Cov(X_i, X_j) = \sum_{m=1}^{k} \frac{(X_{i,m} - \overline{X_i}) \cdot (X_{j,m} - \overline{X_j})}{k} \tag{1}$$

where **m** stands for index of genetic perturbation experiments (in our context: measurement of each transgenic cell line, and **k** denotes the total number of perturbations (**k**=14 for proteomic data, **k**=8 for transcriptomic data).

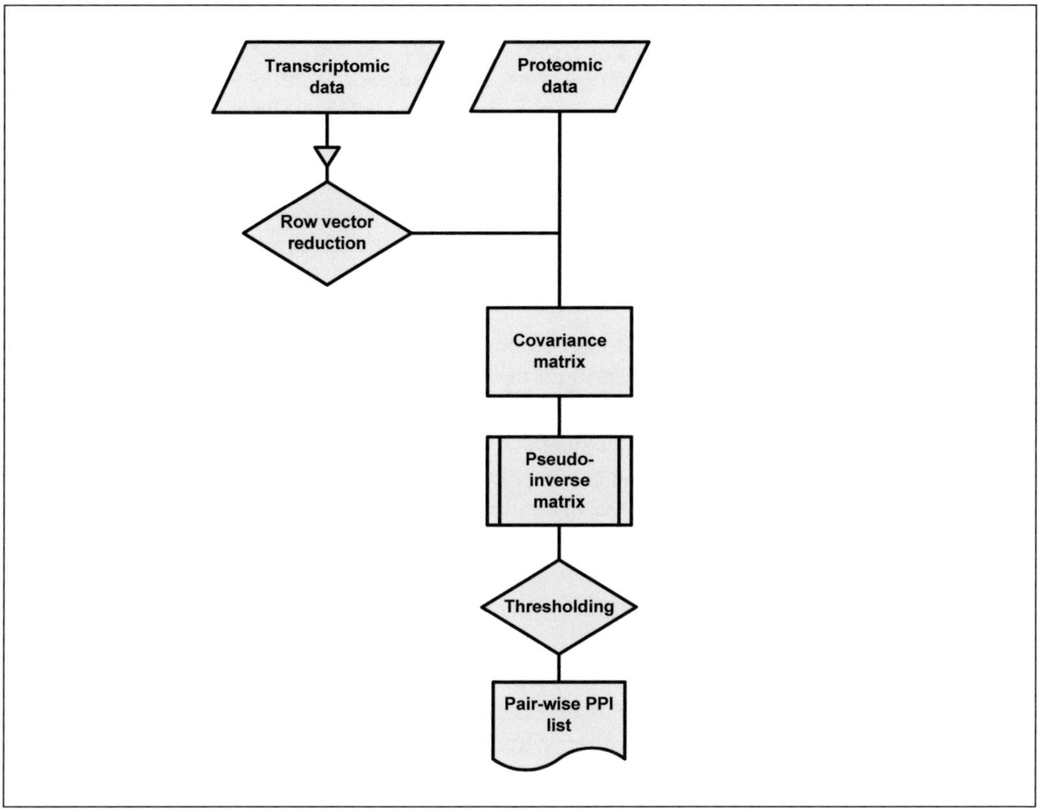

Fig. 2. Programming algorithms for the de novo inferring of protein-protein interaction information from protein expression data of multiple genetic perturbation experiments on mouse ES cells

According to the maximum entropy hypothesis, the corresponding pseudo-inverse matrix of the expression covariance matrix represents the best fit of the underlying protein-protein interaction network, while maximizing the entropy of the system. The strength of interactions is expressed by the value of M_{ij}^{-1}. Positive interaction values indicate the cis-action of the two genes. For instance, an up-regulation of gene A will lead to up-regulation of gene B, and *vice versa*. In contrast, negative values indicate the gene-gene trans-action.

Subsequently, the pair-wise protein interaction information was extracted from the pseudo-inverse matrix of the covariance matrix. By thresholding this PPI information, we obtained the dominant PPI network information from the experimental data solely, without any prior information of the genes.

The algorithm described above was implemented in a Java application. This Java program takes the input data of the expression matrix, and process it with three consecutive modules: In module one, the expression matrix was reduced in terms of redundant gene symbols. Gene expression profiles bearing the same gene symbol, but of different probeset identity, were averaged. In order to increase fidelity, only genes with multiple probesets were retained for further data evaluation. In module two, the covariance matrix was calculated according to equation 1. Subsequently, the pseudoinverse matrix of the covariance matrix was calculated using the BlueBit (http://www.bluebit.gr/matrix-calculator/) or ARARCNE software (for microarray data) [Margolin et al., 2006]. Finally, the last module extracts the PPI information in form of a list of pair-wise interaction: Gene A - Gene B, together with their corresponding edge weight obtained from the covariance values. We set the threshold for interaction to consider only those protein interactions with weight over 0.5 as significant. The overview of the programming structure is given in figure 2.

2.4 Investigation on the obtained PPI network

The protein interaction outcome from our entropy maximization approach was compared to the publicly available PPI databases via three online meta-databases:

- String (http://string-db.org),
- UniHI (http://theoderich.fb3.mdc-berlin.de:8080/unihi) and
- ConsensusPathDB (http://cpdb.molgen.mpg.de).

Together, they encompass a total of 29 individual PPI interaction databases. Network visualization was realized using the Cytospace software (www.cytoscape.org). The online graph analysis tool CFinder (cfinder.org) was used to grab the community structures (closely interlinked sub-graphs) in our predicted PPI network. This method first locates all cliques of the network and then identifies the communities by carrying out component analysis of the clique-clique overlap matrix [Adamcsek et al., 2006]. Gene Ontology functional enrichment analysis was performed using Webgestallt (bioinfo. vanderbilt.edu/webgesta).

2.5 Label-free mass spectrometric relative protein quantification

The global protein expression profile of the Ripk4-overexpressing ES cells was additionally monitored by label-free mass spectrometry quantification [Ishihama et al., 2005]. For this purpose, 200 μg of total protein extract from Ripk4-overexpressing and control ES cells (as control we consider the same mouse ES clone in which the expression of the protein Ripk4 is as basal expression level) were first separated by SDS-PAGE in a gel format of 13 x 25 cm. Subsequently, the gel strip was cut into 16 homogeneous gel slides. These gel slides were subjected to in-gel trypsin digestion as previously described [Mao et al., 2010]. LC/ESI-MS/MS was performed on a LCQ Deca XP ion trap instrument (Thermo Finnigan, Waltham, MA, USA). The eluting gradient is formed by 0.1 % (v/v) formic acid (FA) in water as solvent A and 0.1 % (v/v) FA in acetonitrile (ACN) as solvent B and run at a flow rate of 200 nL per minute. The gradient is linear starting with 5 % B increasing to 70 % B in 180 minutes and additional 10 min to 95 % B. ESI-MS data acquisition is performed throughout the LC run. The raw data were extracted by TurboSEQUEST algorithm, trypsin autolytic fragments and known keratin peptides were filtered out. These files were searched using our in-house licensed Mascot Version 2.1 (Matrix Sciences, London, UK). The MS/MS ion searches are performed with the following set of parameters: database = Swiss-Prot, taxonomy = Mus musculus, Proteolytic enzyme = trypsin, the maximum of accepted missed cleavages =1, mass value = monoisotopic, peptide mass tolerance = ±0.8 Da, fragment mass tolerance = ±0.8 Da and as variable modifications oxidation of methionine and acrylamide adducts (propionamide) on cysteine are expected. Only protein identifications with over 4 spectral counts were retained for further analysis.

The "Experimentally Modified Protein Abundance Index", or "emPAI", is a indexing value on the relative quantification of the proteins in a protein mixture [Ishihama et al., 2005]. The of emPAI ratios of the identified proteins from Ripk4-overexpression to that of control were used as a relative indication of protein expression alteration in ES cells bearing Ripk4 overexpression. We used the threshold of emPAI ratio >2 or emPAI ratio < 0.5 as an arbitrary cut-off for up-regulation or down-regulation of proteins, respectively.

3. Results

3.1 De novo protein interaction network attained from experimental data

We performed the entropy maximization based PPI exploration approach on our experimental data. For this purpose, we combined the proteomic and transcriptomic expression data of the transgenic mouse ES cell lines, whose protein expression profile have been investigated previously in our laboratories. This built up a 14 × 690 expression matrix at the proteomic data level (8 × 8383 for transcriptomic data), where each row vector of the

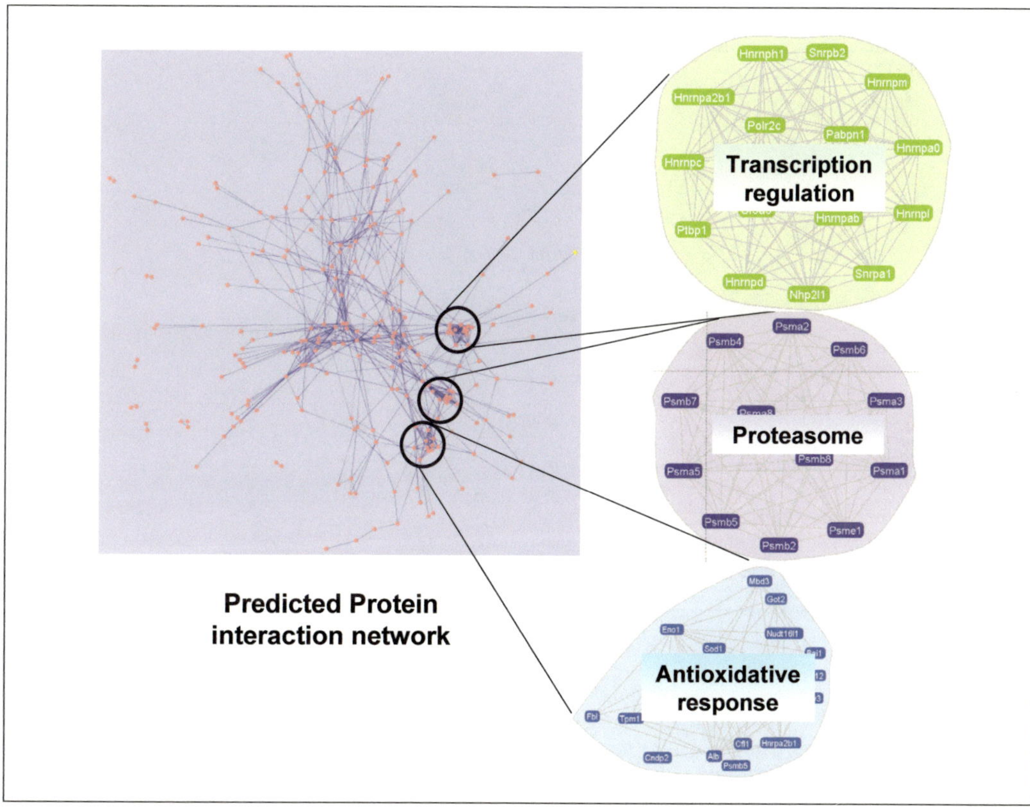

Fig. 3. Global network of protein interactions predicted from our experimental data using the entropy maximization approach. Three major community structures could be detected, which correspond to transctipitonal regulation, proteasome, and antioxidative response processes. Nodes represent proteins, edges represent predicted functional interaction. Illustrated using Cytospace

expression matrix represents the expression ratio (transgne vs. parental control lines) of a distinct protein under divergent single gene overexpressions.

In order to obtain the protein co-regulation data, we calculated the covariance matrix of these genes using the self-implemented Java program. Following equation 1, the program returns a 690x690 (8383 x 8383 for transcriptomic data) symmetrical matrix, with each element representing the covariance of gene Xi and Xj. This matrix contains information of the co-action of genes under genetic system perturbation. Here, the covariance values were normalized to dimensionless values.

Through the pseudo-inverse matrix calculation, we generated a PPI network encompassing distinct proteins according to gene symbol. By taking the threshold of ± 0.5, we obtained a matrix where 22,206 elements were non-zero. This reduces the number of genes showing significant interaction to 490. Specifically, five genes (Cdv3, Fbl, Got2, Hspb1 and Set) showed self inhibitory effect. The obtained PPI network is illustrated in figure 3.

Figure 4 shows the sub-graph build by the 50 most significant pair-wise protein interactions. Among them, 21 gene-pairs showed strong positive interaction, whereas 29 gene-pairs showed mutual inhibitory effect. However, the exact nature of inhibition is not to be deduced from our data.

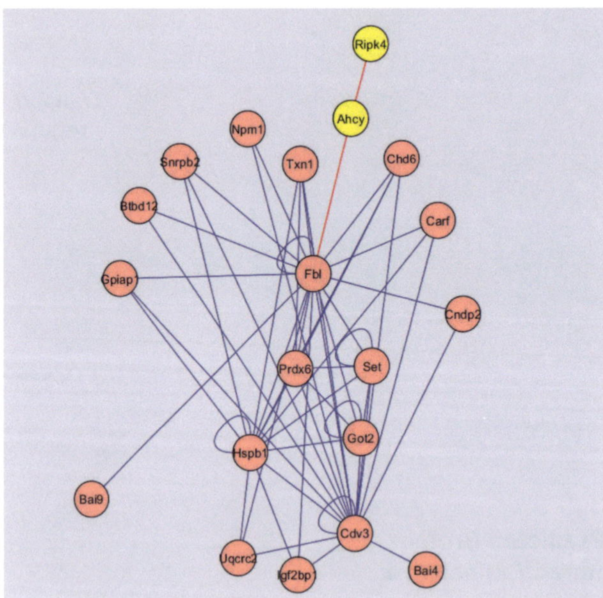

Fig. 4. Sub-graph showing the strongest pair-wise protein interactions in ES cells predicted by multiple genetic perturbations. Our result suggests that Ahcy represents one of the Ripk4 targets

3.2 Community structure analysis revealed key pathways involved in system rebalancing

In the next step, we investigated the community structures (sub-networks) inside our predict protein-protein interaction network. Community structures (CS) can be loosely defined as subsets of nodes that are more densely interconnected among each other than with the rest of the network [Newman and Girvan, 2004]. As shown in figure 3, three dominant community structures could be discovered in our predicted PPI network of ES cell. Respecting their molecular function, protein nodes of these community structures are representative for transcriptomic regulation, proteasome, stress-response pathways, respectively.

3.3 Predicted High degree nodes are reminiscent to "balancer" proteins

As can be seen in figure 3, our predicted PPI network demonstrates the small-worldness property and scale-free degree distribution. Specifically, the network nodes of our predicted PPI network vary substantially in their connectivity, with a small number of proteins exhibiting strong pair-wise interactions with many other genes. Table 2 lists the 20 protein nodes with more than three direct interaction partners.

Node	Degree	Overlap to Balancer proteins
Taldo1	3	Yes
Nudt16l1	3	Yes
Cndp2	3	
RGD1308600	3	
NSFL1C	4	
Cdv3	4	
Prdx6	4	
Psmb5	4	
Fbl	5	
Bai1	5	
Got2	5	Yes
Tpm1	7	Yes
Cfl1	8	
Mbd3	9	
Npm1	10	Yes
Alb	13	Yes
Mis12	13	
Hnrpa2b1	13	Yes
Eno1	15	Yes
Sod1	19	

Table 2. High degree nodes in the predicted PPI network show significant overlap to our previously documented "balancer" proteins

In our previous communication, we reported our hypothesis that upon genetic perturbations on mouse ES cells, there are so-turned "balancer" proteins, defined as proteins that buffer or cushion the system, that act against the system stimuli [Mao et al., 2007]. These central network elements could mediate network remodeling upon perturbation [Bode et al., 2007]. Consequently, we presume that part of these proteins with hub-like behavior could have system functionalities to prevent from severe proteomic shift. Indeed, among the 20 high degree nodes in our predicted PPI network (degree ≥ 3), eight of the proteins belong to our previously detected "balancer" proteins. A Gene Ontology functional enrichment analysis revealed their close involvement in RNA-binding, oxidative response and cellular transport processes (Figure 5).

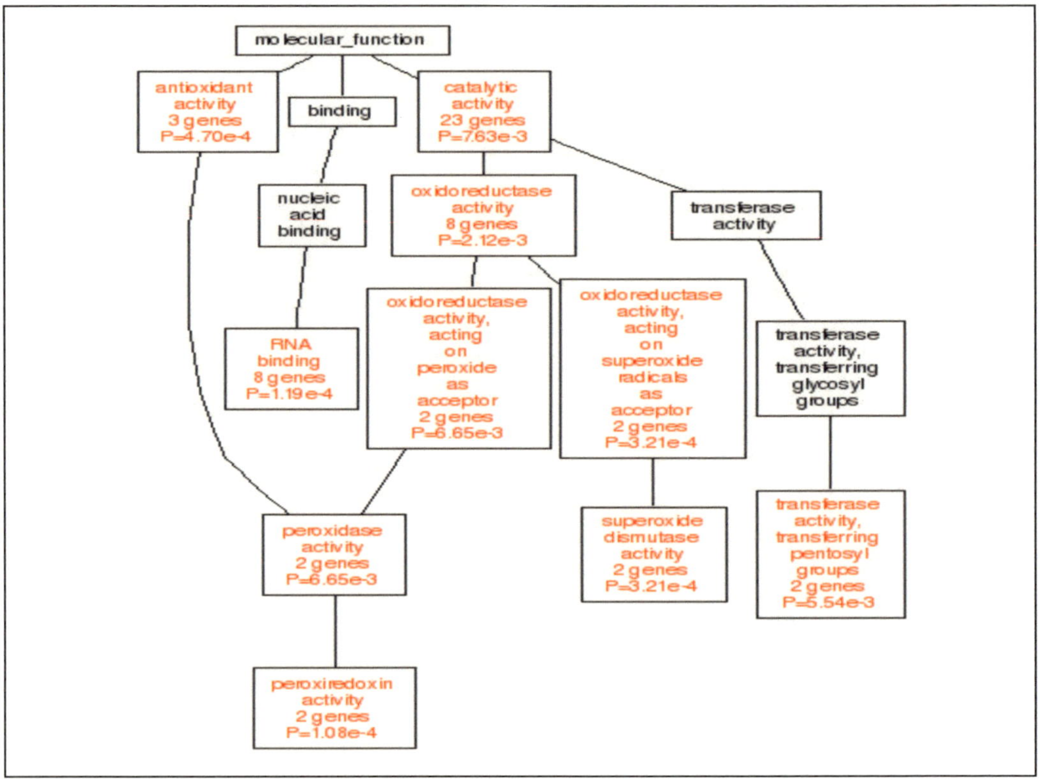

Fig. 5. Gene Ontology analysis revealed molecular function terms enriched in high degree protein nodes in our predicted PPI network

3.4 Comparison of our result to public PPI data reveals overlaps and novel predictions

In light of a comparison between our predicted PPI networks to publically available PPI databases, twelve direct pair-wise interactions predicted by our data-driven entropy maximization approach could be validated by public PPI databases (Table 3). For instance, one important protein interaction partner of the transcription factor single-minded homolog 2 (Sim2), the Ttc3 (tetratricopeptide repeat domain 3), was also predicted by our approach. However, another commonly known PPI partner of Sim2, aryl hydrocarbon receptor nuclear translocator (Arnt), did not appear in our list of predicted PPI partners of Sim2, although Arnt is present in the probset list of the microarray analysis.

In addition, 18 predicted links are similar to that documented in public databases (Table 4). For example, UniHI predicted the protein interaction of Fbl to many proteasome subunits including Psma4, Psma6, Psma7, Psma8, Psmb6 and Psmd8bp1. Here, we attempted to enrich this collection with an additional proteosome subunit, Psmb5, as a potential PPI partner for Fbl. Moreover, the pair-wise interactions Ahcy—Npm1 was predicted by our data. This is coherent to the UniHI protein interaction database, which has documented an indirect PPI relation of "Ahcy—Fbl —Nmp1".

Direct pair-wise protein interaction	Source database
Acad8 – Hadh	String
Cfl1 – Tmp1	String
Prdx1 – Sod2	String
Sod1 – Sod2	String
Ahcy – Sod2	ORTHO
Eno1– Eno1	ORTHO
HnrpA2B1– Snrpa1	Reactome
HnrpA2B1– Snrpb	Reactome
Hspb1 – Hspb1	IntAct
Mis12 – Mis12	HPRD-Binary
Ahcy – Mtap	String
Sim2 – Ttc3	String

Table 3. Consensus direct pair-wise interactions between our predicted PPI and public available PPI databases

Our prediction	Previous documentation	Source Database
Alb – Slc8A3	Alb – Slc1A5, Slc25A13, Slc9A8	IntAct
Cfl1 – Got2	Cfl1 – Got1	OPHID, ORTHO
Cfl1 – Hspb1	Cfl – Hsph1	HPRD-Binary, CCSB-LIT
Cfl1 – Psmb5	Cfl1 – Psme4	ORTHO
Eno1– Psmb5	Eno1– Psmd2	ORTHO
Fbl – Psmb5	Fbl – Psma4, Psma6, Psma7, Psma8, Psmb6 and Psmd8bp1	ORTHO, OPHID, BioGrid, HPRD-Binary
Fbl – Rpl23A	Fbl – Rpl30, Rpl4, Rpl6, Rpl8, Rplp0, Rplp2	HPRD-Complex, OPHID, ORTHO
Fbl – Snrpb2	Fbl – Snrpn	BioGrid, HPRD-Binary
Got2 – Psmb5	Got2 – Psmd10	ORTHO
HnrpA2B1–Rbm14	HnrpA2B1– Rbm5, Rbm8A	Reactome
HnrpA2B1–Txn1	HnrpA2B1–Txndc10, TxnL4A	ORTHO, Reactome
Ahcy – Got2	Ahcy – Got1	ORTHO, OPHID
Ahcy – Prdx6	Ahcy – Prdx1, Prdx2, Prdx4	ORTHO
Got2 – Prdx6	Got2 – Prdx5	ORTHO
Pdxk – Kcna7	Pdxk – Kcnma1	String
Pdxk – Prkg2	Pdxk – Prkab	String
Pdxk – Zfp469	Pdxk – Zfp295	String
Ahcy – Npm1	Ahcy – Fbl – Npm1	OPHID, ORTHO

Table 4. Similar protein pair-wise interactions between our predicted PPI and public available databases

Similarly, our approach predicted three PPI partners for Pdxk: Kcna7 (potassium voltage-gated channel, member 7), Prkg2 (Protein kinase cGMP dependent Type II), and Zfp469 (Zinger finger protein 469), whereas analogous protein PPI partners have been documented in public PPI databases: Kcnma1 (potassium large conductance calcium-activated channel, subfamily M, alpha member 1), Prkab (protein kinase, AMP-activated) and Zfp295 (Zinger finger protein 295), respectively. Such overlaps increase the confidence of our predicted interaction list, and thus support the entropy-maximization approach as a useful method for the *de novo* PPI prediction.

4. Discussion

We predicted a PPI network that contains a list of possible pair-wise protein interactions in mouse ES cells. Notice that this PPI network resulted solely from the gene co-regulation experiments under multiple genetic perturbations. This demonstrates the usefulness of ES cells as a uniform, standardized cell system for system perturbation experiments. Intrinsically, our predicted PPI information is by default weighted, which infers the strength and nature of protein-protein interaction. This could be more superior to some other experimental approaches.

In contrast to random network, the presence of community structures in our predicted PPI network is a signature of the hierarchical nature of intrinsic cellular PPI network. Being able to identify such community structures could help us explore the interplay inside the networks upon genetic perturbation. It should be noted that such perturbation approach reveals predominantly those part of the network structure that is affected by the system stimuli. Thus, the detected community structures in our predicted PPI network reflect the most significantly involved cellular pathways under genetic mutations.

In addition, our data predicted significant protein-protein trans-interaction between Ripk4, a protein kinase originally located on human chromosome 21, and Ahcy (S-adenosylhomocystein hydrolase). This could represent novel knowledge. Unfortunately, Ahcy was not revealed by the proteomic analysis, whereas the expression profile of Ahcy in microarray analysis was heterogeneous among different probesets. In order to in-depth analyse this issue, we performed an additional proteomic analysis. Indeed, using the label-free mass spectrometry protein quantification, we observed a 53 % concentration decrease in the Ripk4-overexpression ES cells. This supports our predicted PPI relation between Ahcy and Ripk4.

It has been reported previously that in rats fed by the Ahcy enzyme inhibitor, the total plasma cholesterol level decreases significantly [Yamada et al., 2007]. Moreover, drug-induced Ahcy inhibition can also lead to anemia due to low erythrocyte membrane fluidity [Altintas and Sezgin, 2004]. However, the direct link between Ripk4 and Ahcy has not been documented explicitly so far. Deduced from our *de novo* inferred PPI network topology, the overexpressing of Ripk4 could inevitably lead to the inhibition of Ahcy enzyme activity, which in turn leads to steroid metabolism disturbance.

In line with this, it has been previously shown that drug induced blocking of sterol conversion to cholesterol in *C. elegans* causes serious defect in germ cell development and motor function [Choi et al., 2003]. This suggests the significance of cholesterol synthesis in neuronal function.

Indeed, an additional *post hoc* functional analysis on the expression profile of Ripk4 transgenic ES cells showed significant down-regulation in proteins involved in lipid metabolism. This includes several key nuclear receptors such as retinoic acid receptor, retinoid X receptor, peroxisome proliferators-activated receptor and steroid hormone receptor ERR2.

Taken together, we may hypothesize that the overexpression of Ripk4 severely inhibits Ahcy's activity. Moreover, this interaction may be important for the cholesterol synthesis pathways. How the interaction of Rikp4 with Ahcy could be correlated to Down syndrome pathology and neuronal dysfunction need to be further investigated.

5. Conclusion

In conclusion, this study demonstrated the feasibility of *de novo* tracking the structure of protein interaction network using a combined experimental and entropy-maximization approach using mouse ES cells as a model system. Albeit useful, this approach also has some limitations: Firstly, biological organisms are rather complex systems with vast number of system components. This engenders high experimental and calculation workloads. Secondly, the entropy maximization approach applied in this study considers only pair-wise protein-protein interaction. More compound interactions, such as triple node interactions or loop effects, which could also be relevant for the cellular PPI network, are not considered.
Finally, it needs to be noted that like all other data-dependent modeling approaches, the performance of this method is highly dependent on data quality. In particular, too much of system reaction such as oscillation leads to high system noise, and can deleteriously influence the modeling outcome. In this sense, mouse ES cells could represent a warrant mammalian cell model of such system biological approaches due to their homogeneous and stable system behavior.

6. Acknowledgements

This Study was co-supported by the European Research Grant 37627 and the German Research Society (DFG) grant KL237/12-1.

7. References

Adamcsek, B., Palla, G., Farkas, I. J., Derenyi, I., and Vicsek, T. (2006). CFinder: locating cliques and overlapping modules in biological networks. *Bioinformatics* 22, 1021-1023.

Altintas, E., and Sezgin, O. (2004). S-adenosylhomocysteine hydrolase, S-adenosylmethionine, S-adenosylhomocysteine: correlations with ribavirin induced anemia. *Med Hypotheses* 63, 834-837.

Amit, M., Carpenter, M. K., Inokuma, M. S., Chiu, C. P., Harris, C. P., Waknitz, M. A., Itskovitz-Eldor, J., and Thomson, J. A. (2000). Clonally derived human embryonic stem cell lines maintain pluripotency and proliferative potential for prolonged periods of culture. *Dev Biol* 227, 271-278.

Antal, M. A., Bode, C., and Csermely, P. (2009). Perturbation waves in proteins and protein networks: applications of percolation and game theories in signaling and drug design. *Curr Protein Pept Sci* 10, 161-172.

Bansal, M., Belcastro, V., Ambesi-Impiombato, A., and di Bernardo, D. (2007). How to infer gene networks from expression profiles. *Mol Syst Biol* 3, 78.

Bode, C., Kovacs, I. A., Szalay, M. S., Palotai, R., Korcsmaros, T., Csermely, P., Szalay, M. S., Kovacs, I. A., Korcsmaros, T., Bode, C., and Csermely, P. (2007). Network analysis of protein dynamics. *FEBS Lett* 581, 2776-2782.

Choi, B. K., Chitwood, D. J., and Paik, Y. K. (2003). Proteomic changes during disturbance of cholesterol metabolism by azacoprostane treatment in Caenorhabditis elegans. *Mol Cell Proteomics* 2, 1086-1095.

De Cegli, R., Romito, A., Iacobacci, S., Mao, L., Lauria, M., Fedele, A. O., Klose, J., Borel, C., Descombes, P., Antonarakis, S. E., et al. (2010). A mouse embryonic stem cell bank for inducible overexpression of human chromosome 21 genes. *Genome Biol* 11, R64.

Dhadialla, P. S., Ohiorhenuan, I. E., Cohen, A., and Strickland, S. (2009). Maximum-entropy network analysis reveals a role for tumor necrosis factor in peripheral nerve development and function. *Proc Natl Acad Sci U S A* 106, 12494-12499.

di Bernardo, D., Thompson, M. J., Gardner, T. S., Chobot, S. E., Eastwood, E. L., Wojtovich, A. P., Elliott, S. J., Schaus, S. E., and Collins, J. J. (2005). Chemogenomic profiling on a genome-wide scale using reverse-engineered gene networks. *Nat Biotechnol* 23, 377-383.

Ishihama, Y., Oda, Y., Tabata, T., Sato, T., Nagasu, T., Rappsilber, J., and Mann, M. (2005). Exponentially modified protein abundance index (emPAI) for estimation of absolute protein amount in proteomics by the number of sequenced peptides per protein. *Mol Cell Proteomics* 4, 1265-1272.

Janes, K. A., and Yaffe, M. B. (2006). Data-driven modelling of signal-transduction networks. *Nat Rev Mol Cell Biol* 7, 820-828.

Klose, J. (1999). Large-gel 2-D electrophoresis. Methods Mol Biol 112, 147-172.

Lauria, M., Iorio, F., and di Bernardo, D. (2009). NIRest: a tool for gene network and mode of action inference. *Ann N Y Acad Sci* 1158, 257-264.

Lezon, T. R., Banavar, J. R., Cieplak, M., Maritan, A., and Fedoroff, N. V. (2006). Using the principle of entropy maximization to infer genetic interaction networks from gene expression patterns. *Proc Natl Acad Sci U S A* 103, 19033-19038.

Mao, L., Romer, I., Nebrich, G., Klein, O., Koppelstatter, A., Hin, S. C., Hartl, D., and Zabel, C. (2010). Aging in mouse brain is a cell/tissue-level phenomenon exacerbated by proteasome loss. *J Proteome Res* 9, 3551-3560.

Mao, L., Zabel, C., Herrmann, M., Nolden, T., Mertes, F., Magnol, L., Chabert, C., Hartl, D., Herault, Y., Delabar, J. M., et al. (2007). Proteomic shifts in embryonic stem cells with gene dose modifications suggest the presence of balancer proteins in protein regulatory networks. *PLoS One 2*, e1218.

Margolin, A. A., Nemenman, I., Basso, K., Wiggins, C., Stolovitzky, G., Dalla Favera, R., and Califano, A. (2006). ARACNE: an algorithm for the reconstruction of gene regulatory networks in a mammalian cellular context. *BMC Bioinformatics 7* Suppl 1, S7.

Newman, M. E., and Girvan, M. (2004). Finding and evaluating community structure in networks. *Phys Rev E Stat Nonlin Soft Matter Phys* 69, 026113.

Shannon, C. E. (1948). Prediction and entropy of printed English. *The Bell System Technical Journal* 30, 50-64.

Shmulevich, I., Dougherty, E. R., Kim, S., and Zhang, W. (2002). Probabilistic Boolean Networks: a rule-based uncertainty model for gene regulatory networks. *Bioinformatics* 18, 261-274.

Stelzl, U., and Wanker, E. E. (2006). The value of high quality protein-protein interaction networks for systems biology. *Curr Opin Chem Biol* 10, 551-558.

Yamada, T., Komoto, J., Lou, K., Ueki, A., Hua, D. H., Sugiyama, K., Takata, Y., Ogawa, H., and Takusagawa, F. (2007). Structure and function of eritadenine and its 3-deaza analogues: potent inhibitors of S-adenosylhomocysteine hydrolase and hypocholesterolemic agents. *Biochem Pharmacol* 73, 981-989.

Part 5

Methods for the Generation of Embryonic Stem Cell Lines for Production of Transgenic and Chimeric Rodents

21

Establishment of Embryonic Stem Cells and Generation of Genetically Modified Rats

Masaki Kawamata and Takahiro Ochiya
National Cancer Center Research Institute, Tokyo
Japan

1. Introduction

Authentic embryonic stem (ES) cells are derived from the inner cell mass (ICM) of preimplantation blastocysts in rodents. ES cells have been routinely derived since 1981 (Evans & Kaufman, 1981; Martin, 1981). They are capable of generating germline chimeras following injection into blastocysts. A very large number of knockin/knockout mice have been produced so far, leading to significant progress in both basic research and clinical investigation. However, recent reports indicate that the phenotypes of knockout mice sometimes do not correspond to human diseases (Rogers et al., 2008). Thus, the ES cells of other species, especially rats, have been desired for the generation of new animal models for human diseases.

In 2008, we successfully established rat ES cells with a chimeric contribution (Ueda et al., 2008). Soon after our report, authentic rat ES cells that could complete a germline transmission were established (Buehr et al., 2008; Li et al., 2008). These reports suggest that a removal of serum from a culture medium is necessary for maintaining the pluripotency of rat ES cells (Kawamata & Ochiya, 2010a). However, despite the assertion in these reports, we recently established high-quality rat ES cells by using a combination of 20% serum and signaling inhibitors. Furthermore, this culture condition enabled the ES cells to receive gene manipulation, leading to obtaining genetically modified rats via germline transmission. We also discovered an indispensable technique during a blastocyst injection process for the generation of germline chimeras (Kawamata & Ochiya, 2010b).

This new technology should provide valuable animal models for the study of human diseases by the induction of gene-targeting manipulations in the rat ES cells. In this chapter, we discuss the techniques for the establishment of rat ES cells compared to mouse ES cells and the creation of genetically modified rats.

2. Mouse ES cells

ES cells are derived from the inner cell mass (ICM) of blastocysts and are capable of unlimited, undifferentiated proliferation *in vitro*. Mouse ES cell lines were first established by culturing ICM (Evans & Kaufman, 1981; Martin, 1981) in the presence of serum and a feeder cell layer made of mouse embryonic fibroblast (MEF). Later, it was shown that the leukemia inhibitory factor (LIF) is the key cytokine secreted by feeders to support mouse ES cell self-renewal and that LIF was able to replace the requirement for feeders in propagation (Smith et al., 1988; Williams et al., 1988). These cells have a stable developmental potential to form derivatives of all three embryonic germ layers even after prolonged culture (Thomson & Marshall, 1998) and have been used to study the mechanism of cell differentiation. Moreover, they are capable of generating germ-line chimeras following injection into the blastocyst (Bradley et al., 1984). Thus, the creation of targeted mutation in the mouse has been a valuable source of animal models of human disease.

In 2003, Ying et al. demonstrated that bone morphogenetic proteins (BMPs) could replace serum and act together with LIF to maintain mouse ES cell self-renewal (Ying et al., 2003). Furthermore, in 2008, they found that a combination of mitogen-activated protein kinase kinase (MEK) inhibitor PD0325901 and glycogen synthase kinase-3 (GSK3) inhibitor CHIR99021, termed 2i, could replace serum, MEF, and LIF in mouse ES cells (Ying et al., 2008). Under this condition, PD0325901 shields inductive differentiation stimuli including autocrine fibroblast growth factor-4 (FGF4). CHIR99021 enhances ES cell propagation, which might be due to its exerting a global modulation of the ES cell metabolic and biosynthetic capacity rather than having a direct anti-apoptotic action (Ying et al., 2008) (Fig. 1).

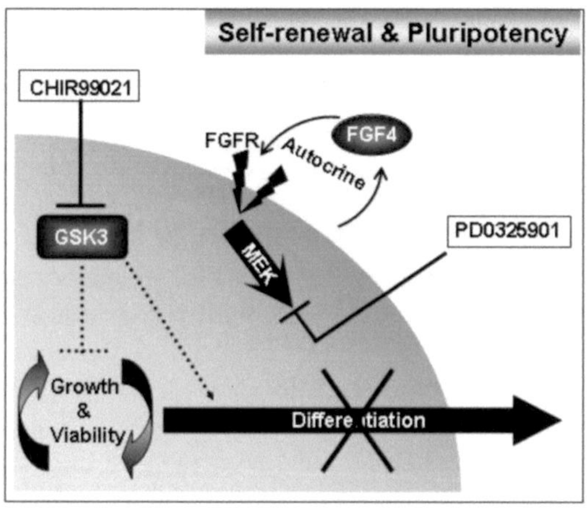

Fig. 1. Mechanism of maintenance of self-renewal and pluripotency

3. Rat

The laboratory rat was the first mammalian species domesticated for scientific research, and it has been used as an animal model in physiology, toxicology, nutrition, behavior, immunology, and neoplasia for over 150 years (Jacob, 1999). The physiology is easier to monitor in the rat, and, over time, a volume of data has developed that will take years to be replicated in the mouse. Moreover, in many cases, the physiology is more similar to the corresponding human condition. The size of the animal enhances its use as a disease model, not just because of the ability to perform surgical procedures but also because of the proportional size of important structures in organs that affects both the degree to which the organ is involved in an experimental lesion and the effects of the distance from the drug administration to specific anatomical areas (Iannaccone & Jacob, 2009).

3.1 Background of rat ES cells

After the first mouse ES cell lines were derived 29 years ago (Evans & Kaufman, 1981; Martin, 1981), many efforts were made to establish rat ES cells. Although a culture medium of the mouse ES cells, composed of either serum and LIF or BMP and LIF, has been thought to be available in all species, several groups have failed to establish rat ES cells under similar conditions (Brenin et al., 1997; Buehr et al., 2003; Demers et al., 2007; Fandrich et al., 2002; Vassilieva et al., 2000). We have succeeded in cloning complete rat LIF cDNA and demonstrated that rat LIF has an effect on rat ES cells for the maintenance of a stem-cell phenotype (Takahama et al., 1998). However, their self-renewal potential was temporal in early passages, which may be due to the fact that a similar culture medium to that of mouse ES cells was used.

In 2008, we established new lines of rat ES cells with chimeric contribution by using a devised culture medium and passaging method (Ueda et al., 2008). However, our ES cells could not achieve germline transmission. While a general culture medium for mouse ES cells contains 15 or 20% FBS, our culture medium contains only 3% FBS. Details of the method for establishing rat ES cells are described below.

Methods; The rat ES-cell culture medium consisted of DMEM/F12 supplemented with 3% (vol/vol) FBS, 0.1 mM 2-mercaptoethanol, 1% nonessential amino acid, 2 mM L-glutamine, 1 mM sodium pyruvate, antibiotic antimycotic and a nucleoside solution. Frozen embryos obtained at 4.5 days post-coitum of Wistar rats were used for the establishment of ES cells. After removal of the zona pellucida by treating with tyrode's solution, seven to ten embryos were placed on a plate pre-seeded with mitotically inactivated MEFs by a treatment with mytomycin C. After 2 or 3 days, ICM-derived cells were dissociated into clumps mechanically or by exposure to 0.05% collagenase type IV before transfer onto new MEFs. The propagated cells were routinely passaged every 3-4 days up to 5 passages in the culture medium in the presence of rat LIF at 1000 U/ml and were then cultured in a medium supplemented with 0, 250, 500, or 1000 U/ml rat LIF.

These rat ES cells showed marker gene expression of ES cells, such as Oct4, Nanog, and SSEA1, and formed embryoid bodies (EBs) after the ES-cell colonies were dissociated by treatment with collagenase IV. Teratomas were formed by subcutaneous, intratesticular or intraperitoneal injection of rat ES cells into SCID mice. Finally, chimeric rats were generated from embryos in which the rat ES cells, cultured in the presence of rat LIF, were injected (Ueda et al., 2008; Kawamata & Ochiya, 2010a). However, germline transmission was not achieved in the chimeras.

Soon after this report, other groups succeeded in establishing authentic ES cells completing germline transmission (Buehr et al., 2008; Li et al., 2008; Hirabayashi et al., 2010). The common technique for maintaining the pluripotency of rat ES cells was to remove the content of fetal bovine serum (FBS) in the culture medium. The two groups suggested that failure in the establishment of authentic rat ES cells over the two past decades was due to the presence of serum (Buehr et al., 2008; Li et al., 2008). Indeed, serum may contain various kinds of nutrient factors as well as differentiation ones for rat ES cells (Kawamata & Ochiya, 2010a).

The two groups used the 2i, MEK and GSK inhibitors and LIF to overcome the difficulty in the generation of germline-competent rat ES cells (Buehr et al., 2008; Li et al., 2008; Hirabayashi et al., 2010). These studies suggest that cell-signaling inhibitors play a critical role in the maintenance of rat ES cells as well as rat iPS cells.

3.2 Maintenance of pluripotency and self-renewal by signal inhibitors

Recent reports suggest that small molecules, which inhibit GSK3, FGF4 through the MAPK pathway, TGFβ, or ROCK signaling, have effects on ES cells for the maintenance of pluripotency and self-renewal. GSK3 is a central node for the negative modulation of a range of anabolic processes and generally acts to suppress the cellular biosynthetic capacity

(Frame & Cohen, 2001). GSK3 is inhibited by phosphorylation downstream of growth factors that activate phosphatidyl inositol 3 kinase and Akt. GSK3 is also a key component of the β-catenin destruction complex, and pharmacological inhibition of GSK3 increases cytoplasmic and nuclear β-catenin, mimicking canonical Wnt signaling (Ding et al., 2000). The Wnt pathway was assumed to maintain self-renewal of ES cells because the main components of the canonical Wnt pathway were detected in undifferentiated human ES cells (Sato et al., 2003). Indeed, Wnt pathway activation by 6-bromoindirubin-3′-oxime (BIO), a specific pharmacological inhibitor of GSK3 (Meijer et al., 2003), maintained an undifferentiated phenotype in mouse and human ES cells and sustained expression of the pluripotent state-specific transcription factors Oct4, Rex1 and Nanog even in the absence of LIF and MEF (Sato et al., 2004). However, BIO is not highly selective and cross-reacts with cyclin-dependent kinases and other kinases, while CHIR99021 was defined as a more selective inhibitor of GSK3 (Bain et al., 2007; Murray et al., 2004; Zhen et al., 2007). Ying et al. found that the activity of mouse ES cells was reduced by BIO but not by CHIR99021 (Ying et al., 2008). In a report relating to the Wnt pathway, a high-throughput cell-based assay showed that a small molecule IQ-1 allowed for long-term expansion of mouse ES cells and inhibited spontaneous differentiation to prevent β–catenin from switching coactivator usage from CBP to p300 (Miyabayashi et al., 2007). These reports suggest that the addition of the GSK3 inhibitor or Wnt recombinants in the ES culture medium might be a useful method to continuously propagate undifferentiated ES cells.

FGF signaling is a conserved initiator of vertebrate neural development (Bertrand et al., 2003; Delaune et al., 2005; Launay et al., 1996; Streit et al., 2000; Wilson et al., 2000). Activation of FGF receptors (FGFRs) can initiate transduction via three major intracellular pathways: classical MAPk, phosphatidylinositol 3′-OH kinase (PI3K), and phospholipase C gamma (PLCγ), the last two of which can activate protein kinase C proteins (PKCs), which can in turn stimulate ERK1/2 signaling (Schonwasser et al., 1998). A high-throughput chemical screen with a library of 50,000 compounds revealed that the compound SC-1 dually inhibited RasGAP and ERK1, which propagate mouse ES cells in an undifferentiated, pluripotent state even in the absence of MEF, serum and LIF (Chen et al., 2006). Treatment of ES cells with the specific inhibitor for MEK, PD098059 (Burdon et al., 1999), ERK, PD184352 or FGFR, PD173074 and SU5402 also suppressed differentiation of ES cells (Kunath et al., 2007; Stavridis et al., 2007). Furthermore, the majority of Fgf null (Fgf-/-) ES cells (Wilder et al., 1997) or Erk2-/- ES cells were able to retain expression of Oct4 under a differentiation condition without LIF (Kunath et al., 2007). Since Fgf4 mRNA is expressed specifically in ES cells of various animals, FGF4 has been considered as a marker gene of ES cells. On the other hand, these reports suggest that an autoinductive stimulation of the MAPk by FGF4 enhances differentiation of ES cells, especially into neural cells. Thus, the MAPk inhibition might be a key method for suppressing differentiation of ES cells.

An addition of type 1 TGFβ receptor Alk5 (A-83-01) to the 2i plus LIF medium enabled the generation of rat-induced pluripotent stem (iPS) cells with chimeric contribution, although germline transmission was not accomplished (Li et al., 2009). Furthermore, a combination of MEK and the ALK5 inhibitors dramatically improved the efficiency of iPS cell generation from human fibroblasts (Lin et al., 2009). These reports indicate that the inhibition of TGFβ signaling also plays a key role in pluripotency.

Recently, Watanabe et al. found that a ROCK inhibitor, Y-27632, caused human ES cells to block apoptosis after dissociation into single cells by enzymatic treatment. Characteristically, human ES cells need to be subcultured by the bulk-passage method since single ES cells form scant colonies. The propagated ES cells cultured by Y-27632 were positive for alkaline phosphatase (ALP), marker genes, such as E-cadherin, Oct4, and SSEA4, and the number of chromosomes was normally kept during a long-term culture (Watanabe et al., 2007). Although the mechanism that allows Y-27632 to form a human ES-cell colony with an undifferentiated state is unknown, the compound was recently used for a single-cell-passaging method.

3.3 Problems with current rat ES cells

Although the 2i plus LIF medium enables the establishment of authentic rat ES cells, the event of germline transmission is rarely achieved. A main reason for the failure is chromosomal instability in rat ES cells during long-term culture. It is known that a chromosomal abnormality is one of the major causes for the loss of germline competence of mouse ES cells (Liu et al., 1997). The germline competence seems to depend on a rat strain for donor ES cell-derivation and host blastocysts. Thus, trials to produce more potent cell lines and to find the optimal combination of rat strains for donor ES cells and host blastocysts remain to be addressed (Buehr et al., 2008; Li et al., 2008).

The rat ES cells cultured in the serum free-2i plus LIF medium are sensitive to the stimuli of genetic manipulation by electroporation and drug-selection, which lead to cell death. To overcome this problem, the temporal use of serum is necessary to protect rat ES cells from the death by such stimuli (Buehr et al., 2008; Li et al., 2008).

3.4 Importance to establish high-quality rat ES cells

To produce genetically modified rats, especially in knockout/knockin rats, it is necessary to establish high-quality rat ES cells that retain normal karyotype and pluripotency during long-term culture and should be strongly resistant to stimuli during the process of genetic manipulation. The use of serum is one way to overcome this problem because it generally enables culturing cells to be vigorous.

4. Establishment of rat ES cells

The use of serum is a way to overcome the problem of weakness in rat ES cells because it generally enables culturing cells to be vigorous. We addressed suitable combinations of signaling inhibitors based on a culture medium that included 20% (vol/vol) FBS, DMEM (including 110 mg/L sodium pyruvate and 200 mM GlutaMAX), 0.1 mM 2-mercaptoethanol, 1% nonessential amino acid stock, and 1 x antibiotic antimycotic. Mitomycin C-treated MEFs resistant to neomycin (Millipore) were used as feeders and maintained in DMEM / 10% FBS medium with 1 x antibiotic antimycotic.

4.1 An effect of Rho kinase inhibitor

We cultured Wistar rat blastocysts in a basic medium containing 20% FBS, which is generally used for mouse ES cell culture. Although the inner cell mass (ICM) outgrew and showed mouse ES cell-like morphology, an ES cell colony did not appear after dissociation and replating. An addition of the Rho kinase inhibitor Y27632 (10 μM) remarkably improved their outgrowth, leading to continuous expansion by performing a clump-passaging method. The clump included 5 to 20 cells. Once the colonies were dissociated into single cells, most of them immediately differentiated. The ES cell-like cells formed domed colonies and were positive for alkaline phosphatase activity and pluripotency markers such as *Oct4, Nanog* and *Sox2* mRNAs (Fig. 2).

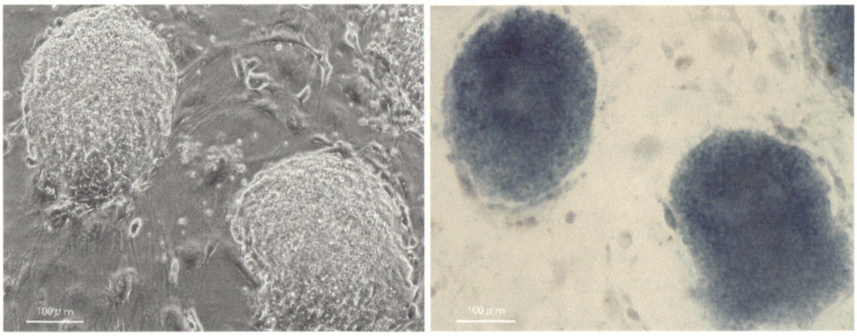

Fig. 2. Colonies cultured in a Y27632-containing medium. At passage 6, the colonies were stained blue, which means that they were positive for alkaline phosphatase activity (right). (scale bar, 100 μm)

Generating chimeric animals achieving germline transmission is the gold standard for documenting authentic ES cells. Thus, we tried to produce germline chimeras via microinjection with the rat ES cells into blastocysts. However, the cells did not contribute any tissues or organs. Thus, the cells established by an addition of Y-27632 alone were not authentic ES cells. Details of the blastocyst injection method are described below.

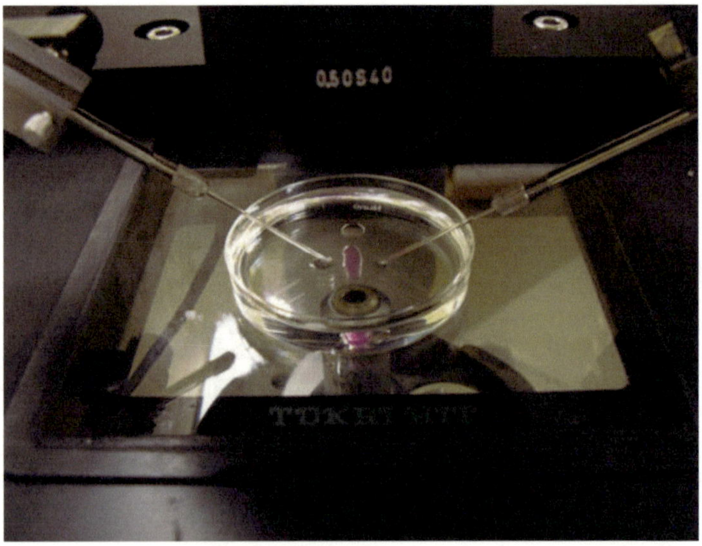

Fig. 3. Device for blastocyst injection with ES cells

Methods; The blastocysts from E4.5 timed-pregnant rats were placed into 500 l of an injection medium, a basal ES cell medium without antibiotic antimycotic, and then were incubated for 2-3 hr. The well-expanded blastocysts were used for microinjection. For ES cell preparation, 10 to 20 domed or floated colonies were picked up by hand-made capillary and treated with Accutase droplet for 5 min at 37 °C, followed by splitting into single cells in a droplet of the injection medium. The cells were transferred in 500 µl of the injection medium and incubated for 30-60 min at room temperature. After centrifugation, ES cells were transferred into a droplet of the injection medium in mineral oil (SIGMA). Ten to 15 ES cells were injected into each blastocyst (Fig. 3) and incubated at 37 °C for 3-5 hr in the injection medium to allow the embryos to recover. Ten to 20 embryos were then transferred into the uterine horn of each E3.5 pseudopregnant female rat.

4.2 Four signaling inhibitors are necessary to establish rat ES cells

We tried various combinations of signaling inhibitors to establish authentic ES cells. Finally, we found that the use of a combination of 4 inhibitors, 10 μM Y-27632, 1 μM PD0325901, 0.5 μM A-83-01, and 3 μM CHIR99021 (termed YPAC) allowed the establishment of authentic rat ES cells. PD0325901, A-83-01, and CHIR99021 are the signaling inhibitors for MEK, Tgfβ, and GSK, respectively.

Cell proliferation of ICM outgrowth was quite rapid under the YPAC condition. The ICMs were picked up using a hand-made needle, followed by extraction of mRNA for the analysis of gene expression. The expression levels of ES cell-marker genes, Oct4, Nanog, Sox2, and Rex1, in ICM cells with YPAC were over 100-fold higher than those without YPAC. Under the YPAC condition, blastocyst outgrowths were observed in 51 samples for all the tested embryos regardless of the strains. The blastocyst strains were derived from Wistar (albino), Long-Evans Agouti (LEA, agouti), or a hybrid of Wistar and LEA (agouti).

A total of Six ES cell lines were established from the blastocysts derived from the three strains. Details of the method are described below.

Methods: After approximately 7 days, the blastocyst outgrowths were cut into pieces and replated under the same YPAC conditions. Emerging ES cell colonies were then dissociated using Accutase and expanded. Established ES cell lines were routinely maintained under MEF-YPAC conditions and passaged every 3-4 days. Floated colonies were also passaged.

Domed colonies were formed from dissociated single cells and could be expanded infinitely. The morphology of their domed colony (Fig. 4) was similar to that of the mouse ES cell colony but slightly different from that of the rat ES cell under the condition of a Y medium (Fig. 2).

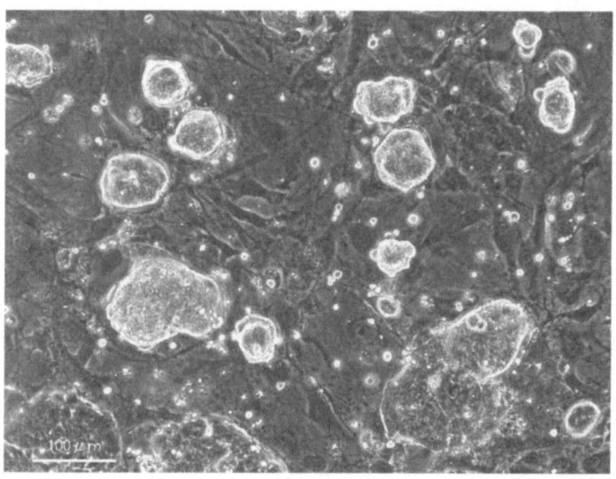

Fig. 4. Colonies cultured in a YPAC-containing medium (passage 3). (scale bar, 100 μm)

4.3 Characteristics of rat ES cells

The YPAC condition was indispensable for maintenance of the rat ES cells. Y was necessary for adherence on MEF, while PAC was necessary for maintaining their pluripotency. If Y was removed from the YPAC medium, few colonies appeared with proliferating. Thus, Y is the most important cell-signaling inhibitor in the YPAC medium.

The rat ES cell colonies tend to detach from MEF, differently from mouse ones. The domed colonies are easily detached by pipetting, followed by dissociation and passaging steps.

Recent reports have indicated that the domed colony possesses a normal karyotype, while the tightly adherent monolayer colony shows chromosomal instability (Kawamata & Ochiya, 2010; Tong et al., 2010). In fact, our rat ES cells possessed a normal karyotype during long-term culturing due to the passaging of floated colonies.

An alkaline phosphatase activity was positive in the rat ES cells. A microarray analysis showed the rat ES cells kept high levels of marker gene expression such as *Oct4, Nanog, Sox2, Rex1, Dppa3, Cdh1, and Tbx1*. Immunocytochemistry also showed that Oct4, Nanog, and Sox2 proteins were expressed in undifferentiated cells.

The classical method to induce ES cell differentiation is to allow the cells to grow in suspension and to form three-dimensional aggregates known as embryoid bodies (EBs) (Keller 1995). Dissociated ES cells were plated into Low-Cell-Binding-Dishes in the basal (without YPAC) medium. EBs were formed from the ES cells at a much lower efficiency compared with their formation from mouse ES cells. The expression of marker genes decreased during the process of EB differentiation. In the presence of PAC, cells aggregated with high efficiency and formed a clear three-dimensional structure. The EBs with PAC at day 7 still sustained high expression levels of the marker genes.

The rat ES cells formed teratomas 34 days after transplantation under the skin of an immunodeficient SCID mouse. A histological examination showed that the teratomas contained all three germ layers, namely, the intestinal epithelium (endoderm), cartilage (mesoderm), and neuronal rosette (ectoderm).

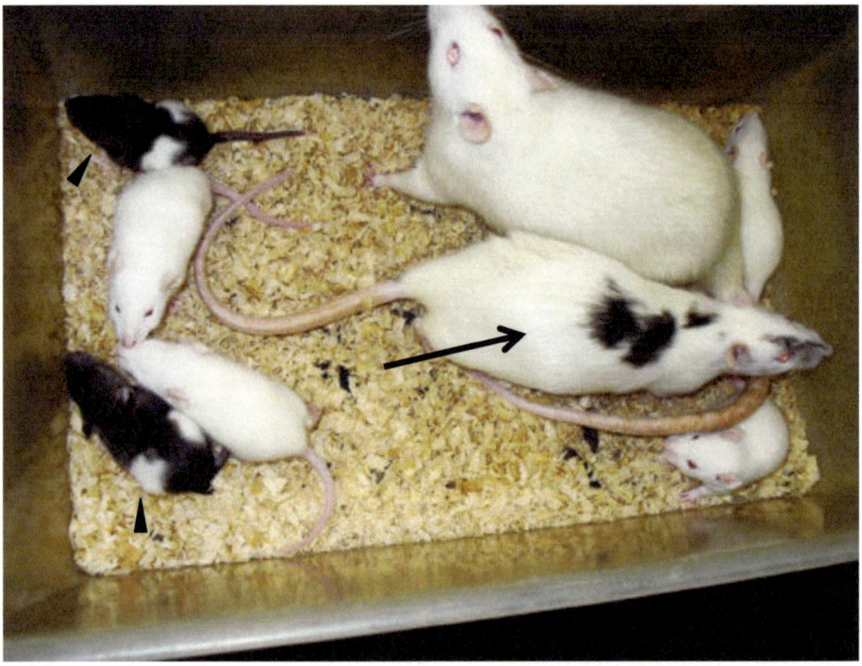

Fig. 5. Germline transmission from chimeric rats. The ES cell line derived from the LEA rat strain was injected into Wistar rat blastocysts, leading to the production of a female chimeric rat with an agouti coat color (arrow). The chimera was mated with a male Wistar rat, and ES cells were transmitted to the next generation with agouti coat color as a Wistar and LEA hybrid strain (arrowheads)

4.4 Production of germline chimeras

We first used a basic ES cell medium without YPAC. However, a coat-color chimera was hardly produced despite the fact that the ES cell line at early passages (6 to 8) was used. Only one male chimera out of 44 pups was obtained, but the chimerism was very sparse.
Next, we added YPAC to the medium during the process of microinjection and blastocyst incubation. This idea was came from the results showing that PAC maintains the pluripotency of cultured cells or EBs. Indeed, PAC addition significantly improved the efficacy to produce chimeric rats. Eight of 23 coat-color chimeras were obtained from the same ES cell line, as reported above, at passage 11 or 12. The generation of coat-color chimeras was successful in all 6 cell lines. Moreover, after mating with male rats, germline transmission was accomplished in adult female chimeras derived from all the 6 cell lines independently of the rat strain (Fig. 5). This efficiency is considerably higher than that seen in previous reports (Buehr et al., 2008; Li et al., 2008), which might be due to the maintenance of a normal karyotype in the ES cells during long-term culture.

4.5 Generation of genetically modified rats

To monitor the ability of stem cells by observing fluorescence, we introduced a transgene in which a Venus protein was expressed by an *Oct4* promoter/enhancer. Venus is a mutant protein of yellow fluorescent protein (YFP) (Nagai et al., 2002). The 3.9 kb *Oct4* promoter was obtained from Wistar rat genomic DNA. This region is known to include both the proximal and the distal enhancer, which gives *Oct4* expression in the morula, inner cell mass (ICM), epiblast, primordial germ cells (PGCs), and ES cells (Chew et al., 2005). Oct4 mRNA is slightly expressed in somatic stem cells in adult mice (Lengner et al., 2007). Details of the gene introduction are described below.
Methods; For gene introduction, a nucleofector was used. After dissociating ES cells with Accutase, they were washed with PBS (-). Ten µg pOct4-Venus transgene linearized by SalI was transfected into 3x 10^6 LEA rat ES cells with the Mouse ES Cell Nucleofector Kit (Amaxa Inc.). The cells were plated on MEFs in the YPAC medium with 2% matrigel (BD Biosciences). Here, the use of 2% matrigel is important to maintain the attachment of colonies on MEF. A single colony of a Venus-positive transfectant was picked up using a hand-made capillary and expanded without drug selection.
In this expansion process of each clone, we found that dominant clones showed a Venus expression pattern with heterogeneity, indicating a complex with a strong positive, a weak positive, and a negative. In addition, a small number of clones possessed a Venus expression pattern with homogeneity. A homogeneous expression pattern was also demonstrated by immunocytochemistry for the Oct4 protein. Thus, we injected this clone into Wistar rat blastocysts, leading to the production of chimeric rats. After mating with LEA male rats, Oct4-Venus transgenic rats were delivered from the female chimera via germline transmission. Genotyping for the Oct4-Venus transgene was performed by PCR analysis to amplify the Venus DNA fragment. The transgenic rats were healthy and could produce a new generation. Venus fluorescence was detected in PGCs in fetal gonad at 17.0 days post-coitum (Fig. 6).
We further investigated the Oct4-Venus expression during the outgrowth of ICM and expansion of ES cells. The Venus-positive blastocyst was plated on MEFs in the YPAC medium. ICM cells rapidly expanded, and Venus fluorescence was observed in some of the cells. After replating the ICM by dissociation with Accutase, domed colonies possessing homogeneous expression of Oct4-Venus appeared. The domed colonies could be infinitely

continued to passage (Fig. 7). This result suggests that we had generated Oct4-Venus transgenic rats, which enabled us to monitor authentic rat ES cells with Venus fluorescence.

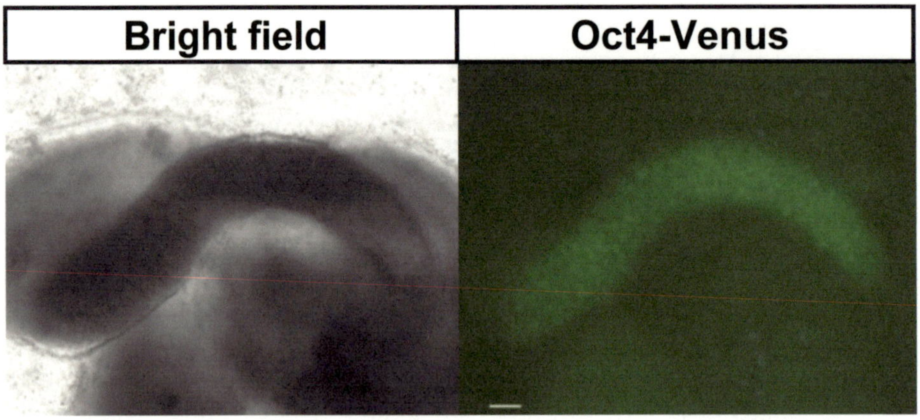

Fig. 6. Oct4-Venus expression in PGCs in fetal gonad. Fetal gonad of Oct4-Venus transgenic female embryo at 16.0 days of gestation was dissected, and Venus fluorescence was observed. (scale bar, 100 μ m)

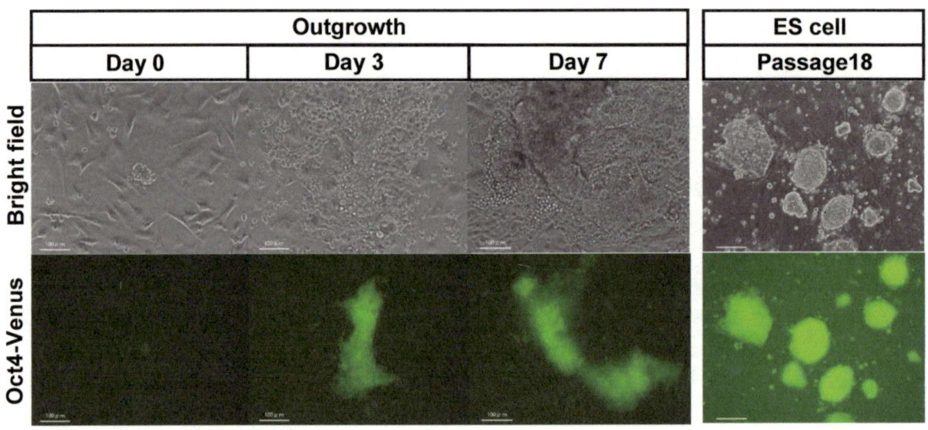

Fig. 7. Oct4-Venus expression during outgrowth of ICM and expansion of ES cells. Blastocyst derived from Oct4-Venus transgenic rats was outgrown on MEFs in the YPAC medium for 7 days. Venus fluorescence was detected in the ICM cells, while it was not detected in differentiated cells. Homogenous Venus fluorescence was observed in ES cells at passage 18. (scale bar, 100 μ m)

5. Conclusion

Our results demonstrated that the use of a combination of serum and cell-signaling inhibitors during outgrowth, cell culture, and blastocyst injection leads to the generation of germline chimeras with extremely high efficiency. Furthermore, we generated genetically modified rats from ES cells. The complete generation of Tg rats might be based on the use of a culture medium containing 20% serum and YPAC, which might strongly protect from cell damage during gene introduction with electric stimuli and maintain pluripotency with a stable karyotype during the cloning and expansion process. Previous works suggested that failure in the establishment of authentic rat ES cells over the two past decades was due to the presence of serum (Buehr et al, 2008; Li et al., 2008). Indeed, serum may contain various kinds of nutrient factors as well as differentiation ones for rat ES cells. Our reason for the present success in the establishment of such significant pluripotent cell lines might be not only the signaling inhibitors shielding ES cells from differentiation but also the utilization of the nutrients in the serum.

Although two groups have reported the establishment of authentic rat ES cells, only one out of several cell lines accomplished germline transmission in each group (Buehr et al., 2008; Li et al., 2008). Thus, trials to produce more potent cell lines and to find the optimal combination of rat strains for donor ES cells, host blastocysts, and recipient foster females remain to be addressed (Buehr et al., 2008; Li et al., 2008). In this study, our YPAC-culture and -injection method overcame the difficulty of completing germline transmission in all the six ES cell lines independently of the rat strain. The YPAC condition will enable the selection of preferable rat strains for the generation of genetically modified rats from ES cells, bringing great advantages to research for strain-specific disease models. We believe that the availability of our rat ES cells and the YPAC-injection technique will also open up a valuable platform for routinely generating knockout/knockin rats, holding out the promise for the generation of new disease models.

6. Acknowledgements

This work was supported by a Grant-in-Aid for the Third-Term Comprehensive 10-Year Strategy for Cancer Control.

7. References

Bain, J.; Plater, L.; Elliott, M.; Shpiro, N.; Hastie, C.J.; McLauchlan, H.; Klevernic, I.; Arthur, J.S.; Alessi, D.R. & Cohen, P. (2007). The selectivity of protein kinase inhibitors: a further update. *Biochem J*, Vol. 408: 297-315.

Bertrand, V.; Hudson, C.; Caillol, D.; Popovici, C. & Lemaire, P. (2003). Neural tissue in ascidian embryos is induced by FGF9/16/20, acting via a combination of maternal GATA and Ets transcription factors. *Cell*, Vol. 115: 615-627.

Bradley, A.; Evans, M.; Kaufman, M.H. & Robertson, E. (1984). Formation of germ-line chimaeras from embryo-derived teratocarcinoma cell lines. *Nature*, Vol. 309: 255–256.

Brenin, D.; Look, J.; Bader, M.; Hubner, N.; Levan, G. & Iannaccone, P. (1997). Rat embryonic stem cells: a progress report. *Transplant Proc*, Vol. 29: 1761–1765.

Buehr, M.; Nichols, J.; Stenhouse, F.; Mountford, P.; Greenhalgh, C.J.; Kantachuvesiri, S.; Brooker, G.; Mullins, J. & Smith, A.G. (2003). Rapid loss of Oct-4 and pluripotency in cultured rodent blastocysts and derivative cell lines. *Biol Reprod*, Vol. 68: 222–229.

Buehr, M.; Meek, S.; Blair, K.; Yang, J.; Ure, J.; Silva, J.; McLay, R.; Hall, J.; Ying, Q.L. & Smith, A. (2008). Capture of authentic embryonic stem cells from rat blastocysts. *Cell*, Vol. 135: 1287-1298.

Burdon, T.; Stracey, C.; Chambers, I.; Nichols, J. & Smith, A. (1999). Suppression of SHP-2 and ERK signalling promotes self-renewal of mouse embryonic stem cells. *Dev Biol*, Vol. 210: 30-43.

Chen, S.; Do, J.T.; Zhang, Q.; Yao, S.; Yan, F.; Peters, E.C.; Schöler, H.R.; Schultz, P.G. Ding, S. (2006). Self-renewal of embryonic stem cells by a small molecule. *Proc Natl Acad Sci U S A*, Vol. 103: 17266-17271.

Chew, J.L.; Loh, Y.H.; Zhang, W.; Chen, X.; Tam, W.L.; Yeap, L.S.; Li, P.; Ang, Y.S.; Lim, B.; Robson, P. & Ng, H.H. (2005). Reciprocal transcriptional regulation of Pou5f1 and Sox2 via the Oct4/Sox2 complex in embryonic stem cells. *Mol Cell Biol*, Vol. 25: 6031-6046.

Delaune, E.; Lemaire, P. & Kodjabachian, L. (2005). Neural induction in Xenopus requires early FGF signalling in addition to BMP inhibition. *Development*, Vol. 132: 299-310.

Demers, S.P.; Yoo, J.G.; Lian, L.; Therrien, J. & Smith, L.C. (2007). Rat embryonic stem-like (ES-like) cells can contribute to extraembryonic tissues in vivo. Cloning Stem Cells, Vol. 9: 512–522.

Ding, V.W.; Chen, R.H. & McCormick, F. (2000). Differential regulation of glycogen synthase kinase 3beta by insulin and Wnt signaling. *J Biol Chem*, Vol. 275: 32475-32481.

Evans, M.J. & Kaufman, M.H. (1981). Establishment in culture of pluripotential cells from mouse embryos. *Nature*, Vol. 292: 154-156.

Fandrich, F.; Lin, X.; Chai, G.X.; Schulze, M.; Ganten, D.; Bader, M.; Holle, J.; Huang, D.S.; Parwaresch, R.; Zavazava, N. & Binas, B. (2002). Preimplantation-stage stem cells induce long-term allogeneic graft acceptance without supplementary host conditioning. *Nat Med*, Vol. 8: 171-178.

Frame, S. & Cohen, P. (2001). GSK3 takes centre stage more than 20 years after its discovery. *Biochem J*, Vol. 359: 1-16.

Hirabayashi, M.; Kato, M.; Kobayashi, T.; Sanbo, M.; Yagi, T.; Hochi, S. & Nakauchi, H. (2010). Establishment of rat embryonic stem cell lines that can participate in germline chimerae at high efficiency. *Mol Reprod Dev*, Vol. 77: 94.

Iannaccone, P.M. & Jacob, H.J. Rats! (2009). *Dis Model Mech*, Vol. 2: 206-10.

Jacob, H.J. (1999). Functional genomics and rat models. *Genome Res*, Vol. 9: 1013-1016.

Kawamata, M. & Ochiya, T. (2010a). Establishment of embryonic stem cells from rat blastocysts. *Methods Mol Biol*, Vol. 597: 169-177.

Kawamata, M. & Ochiya, T. (2010b). Generation of genetically modified rats from embryonic stem cells. *Proc Natl Acad Sci U S A*. Vol. 107: 14223-14228.

Keller, G.M. (1995). In vitro differentiation of embryonic stem cells. *Curr Opin Cell Biol*, Vol. 7: 862-869.

Kunath, T.; Saba-El-Leil, M. K.; Almousailleakh, M.; Wray, J.; Meloche, S. & Smith, A. (2007). FGF stimulation of the Erk1/2 signalling cascade triggers transition of pluripotent embryonic stem cells from self-renewal to lineage commitment. *Development*, Vol. 134: 2895-2902.

Launay, C.; Fromentoux, V.; Shi, D.L. & Boucaut, J.C. (1996). A truncated FGF receptor blocks neural induction by endogenous Xenopus inducers. *Development*, Vol. 122: 869-880.

Lengner, C.J.; Camargo, F.D.; Hochedlinger, K.; Welstead, G.G.; Zaidi, S.; Gokhale, S.; Scholer, H.R.; Tomilin, A. & Jaenisch, R. (2007). Oct4 expression is not required for mouse somatic stem cell self-renewal. *Cell Stem Cell*, Oct Vol. 11: 403-15.

Li, P.; Tong, C.; Mehrian-Shai, R.; Jia, L.; Wu, N.; Yan, Y.; Maxson, R.E.; Schulze, E.N.; Song, H.; Hsieh, C.L.; Pera, M.F. & Ying, Q.L. (2008). Germline competent embryonic stem cells derived from rat blastocysts. *Cell*, Vol. 135: 1299-1310.

Li, W.; Wei, W.; Zhu, S.; Zhu, J.; Shi, Y.; Lin, T.; Hao, E.; Hayek, A.; Deng, H. & Ding, S. (2009). Generation of rat and human induced pluripotent stem cells by combining genetic reprogramming and chemical inhibitors. *Cell Stem Cell*, Vol. 4: 16-19.

Lin, T.; Ambasudhan, R.; Yuan, X.; Li, W.; Hilcove, S.; Abujarour, R.; Lin, X.; Hahm, H.S.; Hao, E.; Hayek, A. & Ding, S. (2009). Chemical platform for improved induction of human iPSCs. *Nat Methods*, Vol. 6: 805-808.

Liu, X.; Wu, H.; Loring, J.; Hormuzdi, S.; Disteche, C.M.; Bornstein, P. & Jaenisch, R. (1997). Trisomy eight in ES cells is a common potential problem in gene targeting and interferes with germ line transmission. *Dev Dyn*, Vol. 209, 85–91.

Martin, G.R. (1981). Isolation of a pluripotent cell line from early mouse embryos cultured in medium conditioned by teratocarcinoma stem cells. *Proc Natl Acad Sci U S A*, Vol. 78: 7634-8.

Meijer, L.; Skaltsounis, A.L.; Magiatis, P.; Polychronopoulos, P.; Knockaert, M.; Leost, M.; Ryan, X.P.; Vonica, C.A.; Brivanlou, A.; Dajani, R.; Crovace, C.; Tarricone, C.; Musacchio, A.; Roe, S.M.; Pearl, L. & Greengard, P. (2003). GSK-3 selective inhibitors derived from Tyrian purple indirubins. *Chem Biol*, Vol. 10: 1255-1266.

Miyabayashi, T.; Teo, J.L.; Yamamoto, M.; McMillan, M.; Nguyen, C. & Kahn, M. (2007). Wnt/betacatenin/CBP signaling maintains long-term murine embryonic stem cell pluripotency. *Proc Natl Acad Sci U S A*, Vol. 104: 5668-5673.

Murray, J.T.; Campbell, D.G.; Morrice, N.; Auld, G.C.; Shpiro, N.; Marquez, R.; Peggie, M.; Bain, J.; Bloomberg, G.B.; Grahammer, F.; Lang, F.; Wulff, P.; Kuhl, D. & Cohen, P. (2004). Exploita-tion of KESTREL to identify NDRG family members as physiological substrates for SGK1 and GSK3. *Biochem J*, Vol. 384: 477–488.

Nagai, T.; Ibata, K.; Park, E.S.; Kubota, M.; Mikoshiba, K. & Miyawaki, A. (2002). A variant of yellow fluorescent protein with fast and efficient maturation for cell-biological applications. *Nat Biotechnol*, Vol. 20 :87-90.

Rogers, C.S.; Stoltz, D.A.; Meyerholz, D.K.; Ostedgaard, L.S.; Rokhlina, T.; Taft, P.J.; Rogan, M..P; Pezzulo, A.A.; Karp, P.H.; Itani, O.A.; Kabel, A.C.; Wohlford-Lenane, C.L.; Davis, G.J.; Hanfland, R.A.; Smith, T.L.; Samuel, M.; Wax, D.; Murphy, C.N.; Rieke, A.; Whitworth, K.; Uc, A.; Starner, T.D.; Brogden, K.A.; Shilyansky, J.; McCray, P.B. Jr.; Zabner, J.; Prather, R.S. & Welsh, M.J. (2008). Disruption of the CFTR gene produces a model of cystic fibrosis in newborn pigs. *Science*, Vol. 321: 1837-1841.

Sato, N.; Sanjuan, I.M.; Heke, M.; Uchida, M.; Naef, F. & Brivanlou, A.H. (2003). Molecular signature of human embryonic stem cells and its comparison with the mouse. *Dev Biol*, Vol. 260: 404–413.

Sato, N.; Meijer, L.; Skaltsounis, L.; Greengard, P. & Brivanlou, A.H. (2004). Maintenance of pluripotency in human and mouse embryonic stem cells through activation of Wnt signaling by a pharmacological GSK-3-specific inhibitor. *Nature Med*, Vol. 10: 55–63.

Schonwasser, D.C.; Marais, R.M.; Marshall, C.J. & Parker, P.J. (1998). Activation of the mitogen-activated protein kinase/extracellular signal-regulated kinase pathway by conventional, novel, and atypical protein kinase C isotypes. *Mol Cell Biol*, Vol. 18: 790-798.

Smith, A.G.; Heath, J.K.; Donaldson, D.D.; Wong, G.G.; Moreau, J.; Stahl, M. & Rogers, D. (1988). Inhibition of pluripotential embryonic stem cell differentiation by purified polypeptides. *Nature*, Vol. 336: 688-690.

Stavridis, M.P.; Lunn, J.S.; Collins, B.J. & Storey, K.G. (2007). A discrete period of FGF-induced Erk1/2 signalling is required for vertebrate neural specification. *Development*, Vol. 134: 2889-2894.

Streit, A.; Berliner, A.J.; Papanayotou, C.; Sirulnik, A. & Stern, C.D. (2000). Initiation of neural induction by FGF signalling before gastrulation. *Nature*, Vol. 406: 74-78.

Takahama, Y.; Ochiya, T.; Sasaki, H.; Baba-Toriyama, H.; Konishi, H.; Nakano, H. & Terada, M. (1998). Molecular cloning and functional analysis of cDNA encoding a rat leukemia inhibitory factor: towards generation of pluripotent rat embryonic stem cells. *Oncogene*, Vol. 16: 3189–3196.

Thomson, J.A. & Marshall, V.S. (1998). Primate embryonic stem cells. *Curr Top Dev Biol*, Vol. 38: 133–165.

Ueda, S.; Kawamata, M.; Teratani, T.; Shimizu, T.; Tamai, Y.; Ogawa, H.; Hayashi, K.; Tsuda, H. & Ochiya, T. (2008). Establishment of rat embryonic stem cells and making of chimera rats. *PLoS ONE*, Vol. 3: e2800.

Vassilieva, S.; Guan, K.; Pich, U. & Wobus, A.M. (2000). Establishment of SSEA-1- and Oct-4-expressing rat embryonic stem-like cell lines and effects of cytokines of the IL-6 family on clonal growth. *Exp Cell Res*, Vol. 258: 361–373.

Watanabe, K.; Ueno, M.; Kamiya, D.; Nishiyama, A.; Matsumura, M.; Wataya, T.; Takahashi, J.B.; Nishikawa, S.; Nishikawa, S.; Muguruma, K. & Sasai, Y. (2007). A ROCK inhibitor permits survival of dissociated human embryonic stem cells. *Nat Biotechnol*, Vol. 25: 681-686.

Wilder, P.J.; Kelly, D.; Brigman, K.; Peterson, C.L.; Nowling, T.; Gao, Q.S.; McComb, R.D.; Capecchi, M.R. & Rizzino, A. (1997). Inactivation of the FGF-4 gene in embryonic stem cells alters the growth and/or the survival of their early differentiated progeny. *Dev Biol*, Vol. 192: 614-629.

Williams, R.L.; Hilton, D.J.; Pease, S.; Willson, T.A.; Stewart, C.L.; Gearing, D.P.; Wagner, E.F.; Metcalf, D.; Nicola, N.A. & Gough, N.M. (1988). Myeloid leukaemia inhibitory factor maintains the developmental potential of embryonic stem cells. *Nature*, Vol. 336: 684–687.

Wilson, S.I.; Graziano, E.; Harland, R.; Jessell, T.M. & Edlund, T. (2000). An early requirement for FGF signalling in the acquisition of neural cell fate in the chick embryo. *Curr Biol*, Vol. 10: 421-429.

Ying, Q.L.; Nichols, J.; Chambers, I. & Smith, A. (2003). BMP induction of Id proteins suppresses differentiation and sustains embryonic stem cell self-renewal in collaboration with STAT3. *Cell*, Vol. 115: 281–292.

Ying, Q.L.; Wray, J.; Nichols, J.; Batlle-Morera, L.; Doble, B.; Woodgett, J.; Cohen, P. & Smith, A. (2008). The ground state of embryonic stem cell self-renewal. *Nature*, Vol. 453: 281-292.

Zhen, Y.; Sørensen, V.; Jin, Y.; Suo, Z. & Wiedłocha, A. (2007). Indirubin-3'-monoxime inhibits autophosphorylation of FGFR1 and stimulates ERK1/2 activity via p38 MAPK. *Oncogene*, 26: 6372-6385.

22

Rat Embryonic Stem Cells: Establishment and Their Use for Transgenesis

Masumi Hirabayashi[1] and Shinichi Hochi[2]
[1]*Center for Genetic Analysis of Behavior, National Institute for Physiological Sciences*
[2]*Faculty of Textile Science and Technology, Shinshu University*
Japan

1. Introduction

Rats (*Rattus norvegicus*) have been used more extensively than mice in the research fields of neuroscience, pharmacology and toxicology. There are more than 100 rat strains with various genetic backgrounds, including some useful models for human diseases. For example, the SHR (spontaneously hypertensive rat) strain and the BB (rats spontaneously developing insulin-dependent diabetes mellitus) strain are well-established models for studying cardiovascular diseases and endocrinopathy (Okamoto, 1969; Like et al., 1982). Because of the well-understood mapping of brain functions, rats are often used for physiological studies on memory and emotion (Wood et al., 1999; van Erp et al., 2000). Furthermore, experimental studies on mammary tumors require the use of rats in which symptoms of the disease are distinct from those caused by mouse mammary tumor viruses; mouse mammary tumor viruses cause tumors in the mammary glands of mice but not those in humans or rats (Carr et al., 1981; Gould, 1986). In addition, transgenic rats have been used as model animals for human diseases (e.g., Alzheimer's disease, autoimmunity and high-density lipoprotein [HDL] metabolism) and organ transplantation, and as animal bioreactors for protein production (Heideman, 1991; Charreau et al., 1996; Ganten, 1998). One advantage of using rats rather than mice in transgenic studies is the ease of continuous or repeated sample collection (e.g., of blood or urine) and surgery, due to their larger size, while litter size, gestation length, maturation rate, estrous cycle length, and life span of rats are all very similar to those of mice. Thus, the rat has the advantage of being a reasonably well-characterized and intermediate-sized rodent that can be maintained much more cheaply than larger animals and can often be manipulated much more easily than smaller rodents.

On the other hand, the reverse genetic approach using the rats (precise and conditional gene replacements [knock-in] or loss of gene function [knock-out] at the specific locus) was considered impossible because any protocols to establish embryonic stem (ES) cell lines conventionally used in mice were not applicable to the rats. However in 2008, functional germline-competent ES cell lines have been reported (Buehr et al., 2008; Li et al., 2008). Very recently, successful production of p53 gene knock-out rats by homologous recombination in the ES cells has been achieved at last (Tong et al., 2010). The present chapter deals with an overview of attempts at producing gene-modified rats (with or without using ES cells), followed by detailed protocols for establishment of rat ES cell lines, and for successful use of the rat ES cells in transgenesis via electroporation (Hirabayashi et al., 2010a; 2010b).

2. Approaches to produce gene-modified rats

Transgenic rats integrated with foreign gene into their genomes can be routinely produced by conventional pronuclear microinjection (Hammer et al., 1990; Hochi et al., 1990; Mullins et al., 1990) or by an intracytoplasmic sperm injection (ICSI)-mediated gene transfer protocol (Kato et al., 2004; Hirabayashi et al., 2005). As for production of rats genetically modified (knock-in or knock-out of endogenous gene function), the most convenient tool, ES cell line, was not available for this rodent. Despite of numerous efforts, all attempts at establishing rat ES cell lines have resulted in total failures (Brenin et al., 1997; Vassilieva et al., 2000; Buehr et al., 2003); some ES-like cell lines have been derivated, but phenotype of the ES-like cell origin was not contributed to chimera or was not transmitted to G1 generation offspring, even though they could proliferate while maintaining undifferentiated status and show multipotency under certain conditions for inducing differentiation. Ueda et al. (2008) reported the production of chimeric rats derived from blastocyst injection of transgenic ES cells (established by classical method), without data for germline transmission of the exogenous gene.

Therefore alternatively, sophisticated approaches were used to induce genetic mutations in the rats. Gene-modified rats have been successfully produced by the following three approaches. (1) The yeast-based screening assay of N-ethyl-N-nitrosourea (ENU) mutated G1 offspring (Zan et al., 2003; Chen & Gould, 2004) allows to produce serial allelic variations of targeted genes, in addition to knock-out mutants expected. In another words, targeted gene modifications at the specific locus (Scn1a; a missense mutantion presumptively responsible to epilepsy) are difficult to achieve in the ENU-mutagenesis (Mashimo et al., 2008). (2) Sleeping beauty (SB) transposon system was found effective to induce insertional mutatation in rats (Kitada et al., 2007) as well as in mice (Keng et al., 2005). This system includes the production of two independent transgenic lines carrying the SB transposon vector (eqipped with promoter trap system and poly-[A] trap system) and SB transposase, respectively, and the subsequent interline coupling to produce double-transgenic rats. Transposition events occurs not only on the same chromosome but also between different chromosomes. This SB transposon-tagged mutagenesis is disadvantageous in controlling the frequency of mutation at the specific locus, too. (3) Zinc-finger nucleases (ZFNs) can create genome-specific double-stranded breaks and therefore more likely to induce targeted gene mutation. Microinjection of DNA or mRNA that encodes specific ZFNs into pronuclear zygotes derived from green fluorescent protein (GFP) transgenic rats resulted in a high frequency of rat offspring that do not express the transgenic marker, due to a consequence of homologous recombination at the GFP locus (Geurts et al., 2009). Mutation in rat genome has been induced in interleukin 2 receptor gamma (IL2rγ) locus by this approach (Mashimo et al., 2010). One of the obstacles in the ZFNs technology is the difficult customization of the DNA/mRNA motif for ZFNs specific to the desired locus.

Cloning by somatic cell nuclear transplantation (SCNT) is an alternative approach to produce gene-modified rats. Successful production of cloned rats with somatic cells was first reported by Zhou et al. (2003). However, the reproducibility of the data is still questionable (Hirabayashi & Hochi, 2006). As to reasons why rat SCNT is difficult, rat oocytes are likely to activate spontaneously in vitro after recovery from the oviductal ampullae (Zernicka-Goetz, 1991; Ito et al., 2005). The male germline stem (GS) cell line, which is capable of culturing and differentiating in vitro, is also a source of cells for gene targeting and production of knockout animals via ICSI or other microinsemination techniques (Feng et al., 2002; Toyooka et al., 2003; Geijsen et al., 2004). When rat spermatogonial stem cells were transplanted into busulfan-treated nude mouse testes, microinsemination of the retrieved rat spermatids and spermatozoa into rat oocytes resulted in viable rat offspring (Shinohara et al., 2006). The SCNT/GS cell technologies combined with homologous recombination-based gene targeting will become a powerful tool for production of knock-out rats, once they can be established and widespread.

Breakthrough in establishing rat ES cell lines was appeared at the end of 2008 year (26th December, 2008). Functional germline-competent ES cell lines have been reported by using a few inhibitors for fibroblast growth factor (FGF) receptor, mitogen activated protein kinase kinase (MEK) and glycogen synthase kinase 3 (GSK3) in differentiation-related signaling pathways (Buehr et al., 2008; Li et al., 2008). This protocol, the so-called 3i/2i culture system, was originated from ES cell research in mouse (Ying et al., 2008) and was proved reproducible even after slight modificatons were added to the culture system (Hirabayashi et al., 2010a; Kawamata & Ochiya, 2010). The modification made by Hirabayashi et al. (2010a) is to replace MEK activation inhibitor PD1843521 with MEK inhibitor PD325901 and to add rat leukemia inhibitory factor (LIF) instead of LIF-secreting feeder cells in the 3i system, while that made by Kawamata & Ochiya (2010) is to add fetal bovine serum (FBS), ß-mercaptoethanol, rat LIF, and inhibitors for Rho-associated coiled-coil kinase (ROCK) and transforming growth factor-beta (TGF-ß) type-I receptor ALK5 kinase to the 2i culture medium containing MEK and GSK3 inhibitors. Very recently, successful production of p53 gene knock-out rats by homologous recombination in the ES cells has been achieved (Tong et al., 2010). The p53 is tumour suppressor gene located on rat chromosome 10, and mutations in the p53 gene are highly associated with genetic lesions in human cancers. Soon or later, the homologous recombination-based gene targeting in rat ES cell lines will allow for practical production and analysis of individual rats with targeted mutation in every research fields, as routinely done in mice.

3. Establishment of rat ES cells

Development of a methodology for reverse genetic research has been long desired, because the rat is the most widely used animal model in biological research. The most convenient method for production of knockout rats is to use gene-targeted ES cells in germline chimera production. Although rat ES cell lines have not been established despite of numerous efforts, recent application of 2i/3i culture system enabled the derivation of germline-competent rat ES cell lines (Buehr et al., 2008; Li et al., 2008). In this section, rat ES cell lines that can participate in germline chimeras at high efficiency (Hirabayashi et al., 2010a; with a few additional data) are described.

3.1 Isolating undifferentiated colonies from blastocysts

Blastocysts at E4.5 (E0 was defined as the initiation of coupling) were recovered by uterine flushing of Wistar females couplated with a homogenous CAG/venus transgenic male rat. Green fluorescence of the venus gene was used as the transgenic marker. Their zonae pellucidae were removed in acid tyrodes solution (5 sec at an ambient temperature), and then each blastocyst was placed on mitomycin (MMC)-treated mouse embryonic fibroblasts (MEF; concentraton of the feeder cells, 3 to 4 x 10^5 cells/ml; see below note-1) in a well of 4-well dish. The culture medium consisted of 2 μM FGF receptor inhibitor, SU5402 (Calbiochem), 1 μM MEK inhibitor, PD0325901 (Stemgent), 3 μM GSK3 inhibitor, CHIR99021 (Axon), and 1,000 U/ml rat LIF (ESGRO®, Millipore) in N2B27 medium (see below note-2). After 7 days culture at 37°C in a humidified atmosphere of 5% CO_2 in air, the outgrowths of the blastocysts (Fig. 1a) were disaggregated by gentle pipetting and transferred to the same MEF/3i conditions (first passage). However, time required until the first passage (5 days to 2 weeks) may depend on rat strains for blastocysts. When ES cell-like colonies were emerged (Fig. 1b), they were trypsinized in 0.25% trypsin/0.6 mM EDTA solution (Sigma) and then expanded (Fig. 1c). The tentative ES cell lines were maintained in MEF/3i conditions, with medium exchange every other day and trypsinization/expansion (passage) every three days. Otherwise, the ES cell lines were cryopreserved to prevent senescence (see below note-3). It was checked whether the derivated cells are alkaline phosphatase (AP)-positive (Fig. 1d). Nine ES cell lines (69.2%) were established from 13 transgenic blastocysts. Among them, two lines with excellent growth rate (rESWIv3i-1 and rESWIv3i-5) were selected, and both lines were found female by PCR analysis to detect rat *Sry* gene. Attachment of ES cell colonies with the feeder cells was not so strong, and morphological appearance of the rat ES cell colonies was similar to that of mouse ES cells. As the passage number of the ES cells increased, signs of differentiation into extraembryonic cell-like cells were observed in cultures, especially when the culture medium was exchanged every other day rather than every day (Fig. 1e and 1f).

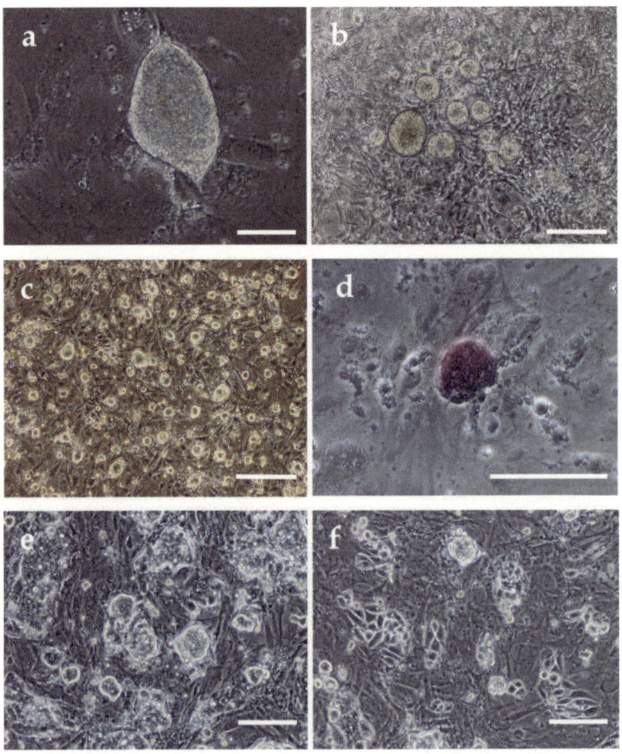

Fig. 1. Establishment of rat ES cell line. (a) Outgrowth of a blastocyst on MEF feeders 7 days after plating. (b) Formation of colonies 3 days after the first passage. (c) Expanded ES cell colonies 2 days after fourth passage. (d) Alkaline phosphatase-positive colony at passage 12. (e) The ES cells at passage 17. Under the daily exchange of culture medium, very little cells showed any signs of differentiation. (f) The ES cells at passage 17, maintained with medium exchange every other day. Differentiated extra-embryonic cells were observed. Scale bars: a, d, e and f; 100-μm, b and c; 500-μm

Notes: (1) MEF Preparation; One-ml of neomycin (G418)-resistant MEF cell cryostock (2 to 3 x 10^7 cells/vial, Kitayama Labes) were thawed in 37˚C waterbath, and resuspended in Dulbecco's modified Eagle's medium (DMEM) supplemented with 10% FBS (ES cell qualified). The cell suspension was centrifuged for 190 x *g* for 5 min at 4˚C, resuspended in DMEM/10% FBS, and then cultured for 2 to 3 days. Confluently proliferated cells were liberated by trypsin/EDTA treatment, centrihuged for 190 x *g* for 5 min at 4˚C, resuspended, and replated (passage). After several passages, the cells were treated with 10 µg/ml MMC for 2 h at 37˚C in 5% CO_2 in air. After washing twice by centrifugation, the MCC-treated MEFs were used as the feeder cells. The DMEM and FBS were purchased from Gibco. (2) N2B27 medium; A 1:1 mixture of DMEM/F12 (1:1) and Neurobasal medium was supplemented with B-27 serum-free supplement, N-2 serum-free supplement, 2 mM L-glutamine, 50 U/ml penicillin, and 50 mg/ml streptomycin. Media and reagents for N2B27 medium were all purchased from Gibco. (3) Cryopreservation of ES cells; ES cells were liberated by the trypsin/EDTA treatment for 5 min at 37˚C in 5% CO_2 in air, and were collected in a 15 ml conical tube after gentle pipetting. After centrifugation for 190 x *g* for 5 min at 4˚C, the cell pellet was resuspended with 1 ml of CellBanker®. The cryovials (1 x 10^6 cells/ml/cryovial) were transferred into a BICELL®, and kept overnight in a -80˚C deep-freezer. The cryovials were then stored in a liquid nitrogen tank until use.

3.2 Confirmation of characteristics essential for ES cells

In addition to AP activity of the tentative ES cells (Fig. 1d), expression of stem cell marker genes, such as Oct-4, Nanog, Fgf-4, or Rex-1, should be confirmed by reverse transcription (RT)-PCR analysis. Each primer sets were designed as shown in Table 1. Total RNA was prepared using the RNeasy mini Kit (Qiagen), and the RT-PCR was performed using the SuperScript™III One-Step RT-PCR System with Platinum®Taq High Fidelity kit (Invitrogen™), with the reaction conditions for cDNA synthesis (55°C for 30 min and 94°C for 2 min) and amplification (30 cycles at 94°C for 15 sec, 55°C for 30 sec, and 68°C for 60 sec). As shown in Fig. 2, both rESWIv3i-1 and rESWIv3i-5 lines expressed Oct-4, rNanog, Fgf-4 and Rex-1, indicating that these cell lines had undifferentiated characteristics. Weakly detectable bands of Gata6 may be due to the presence of differentiated cells in ES cell cultures (Fig. 1f).

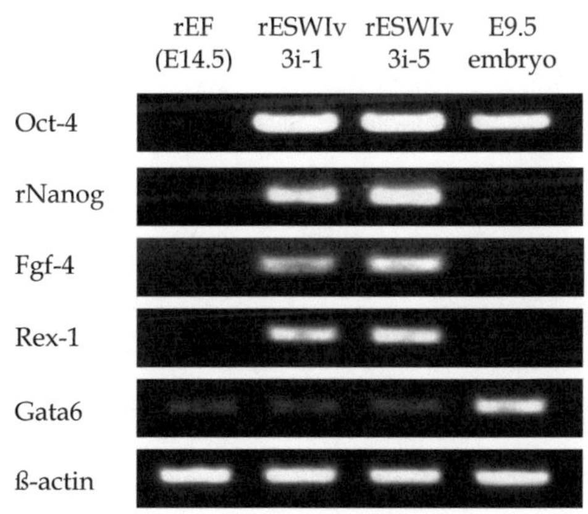

Fig. 2. Reverse transcription-PCR analysis for rat embryonic fibroblast cells at E14.5 (rEF; negative control for stem cell markers), rat embryos at E9.5 (positive control for Gata6), and two rat ES cell lines (rESWIv3i-1 and rESWIv3i-5). Both ES cell lines expressed stem cell markers, including Oct-4, rNanog, Fgf-4 and Rex-1

Furthermore, multipotency of the tentative ES cells (whether the cells can differentiate into endoderm, mesoderm, and ectoderm) should be confirmed. The tentative rESWIv3i-1 line (2.5 x 10^5 cells) has been subcutaneously transplanted into an adult male F344 nude rat. Five weeks after the transplntation, a tumor was observed (Fig. 3a), and its histological section was evaluated ater hematoxylin-eosin staining. As shown in Fig. 3b, the tumor was found to be a teratoma with various tissues including gut-like epithelium or hepatic cells (endoderm), bone, cartilage or muscle (mesoderm), and neural tissues (edctoderm).

Marker gene		Primer sequence	Product size
Oct-4	forward	5′- GGGATGGCATACTGTGGAC-3′	412 bp
	reverse	5′- CTTCCTCCACCCACTTCTC-3′	
rNanog	forward	5′- GCCCTGAGAAGAAAGAAGAG-3′	356 bp
	reverse	5′- CGTACTGCCCCATACTGGAA-3′	
Fgf-4	forward	5′- CGGGGTGTGGTGAGCATCTTC-3′	202 bp
	reverse	5′- CCTTCTTGGTCCGCCCGTTC-3′	
Rex-1	forward	5′- TTCTTGCCAGGTTCTGGAAGC-3′	297 bp
	reverse	5′- TTTCCCACACTCTGCACACAC-3′	
Gata6	forward	5′- TCATCACGACGGCTTGGACTG-3′	467 bp
	reverse	5′- GCCAGAGCACACCAAGAATCC-3′	
ß-actin	forward	5′- CATGGCATTGTGATGGACT-3′	427 bp
	reverse	5′- ACGGATGTCAACGTCACACT-3′	

Table 1. Primers used for PCR amplification

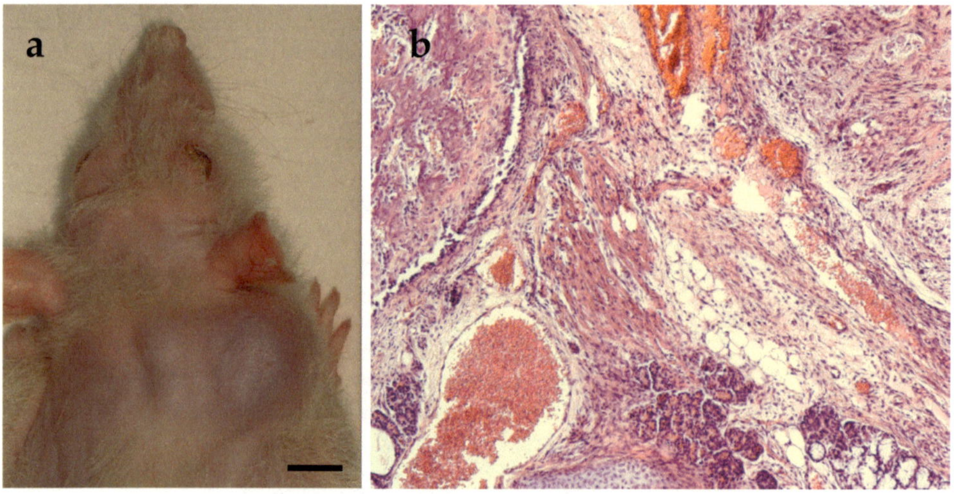

Fig. 3. Tumor observed 5 weeks after subcutaneous injection of rESWIv3i-1line ES cells into a nude rat (a), and histological section of the tumor stained with hematoxylin-eosin (b). The tumor was found to be a telatoma with various tissues including gut-like epithelium or hepatic cells (endoderm: bottom right), bone, cartilage or muscle (mesoderm: center), and neural tissues (ectoderm: upper right). Scale bar; 1-cm

3.3 Production of germline chimera rats by blastocyst injection

To generate ES cell-derived chimeras, host blastocysts at E4.5 derived from Wistar females or Wistar x Dark-Agouti F_1 females were microinjected with 10 ES cells each at passage 6 or 8 (Fig. 4a and 4b). Briefly, using blunt-ended injection pipette with an outer diamether of 15 µm, the zona pellucida of host blastocysts was penetrated by a piezo-pulse. The pipette tip was forwarded to inner cell mass (ICM) through thin junction area between the two trophoblastic cells (without any piezo force), and then the ES cells were deposited. Collapsed blastocysts (Fig. 4c) were re-blasturated 1 to 2 h after the microinjection, and allowed to develop to fetus (E15.5) or full-term pups in pseudopregnant Wistar recipients. All of the E15.5 fetuses (100%; Fig. 4d) and majority of the newborn pups (81.8 to 100%) were chimeric and expressed the venus gene (Table 2). The characteristics of the ES cells was successfully transmitted to their next generations in both lines (Fig. 5); Among 7 chimeras derived from passage-6 rESWIv3i-1 cells and developed to adult, three were female. Two of the 3 female chimeras transmitted the Venus gene through the germline (4/15 and 2/14, respectively). Among 5 chimeras derived from passage-8 rESWIv3i-5 cells and developed to adult, only one rat was female. This female chimera also transmitted the transgene to the next generations (1/26, two litters). The ability of the ES cells to participate in chemeras was still high (78.6 to 100%) at advanced passage numbers (17 or 18; Table 2). Overall efficiency of producing chimeric rats (50.3%, 94 chimeras/187 injected embryos) was higher than 8.2% (20/245) as reported in Buehr et al. (2008) and 11.0% (26/237) in Li et al. (2008). This higher efficiency of chimera rat production is probably due to the rat strain combination used for donor (ES cells) and host (blastocysts) and/or modification of culture medium.

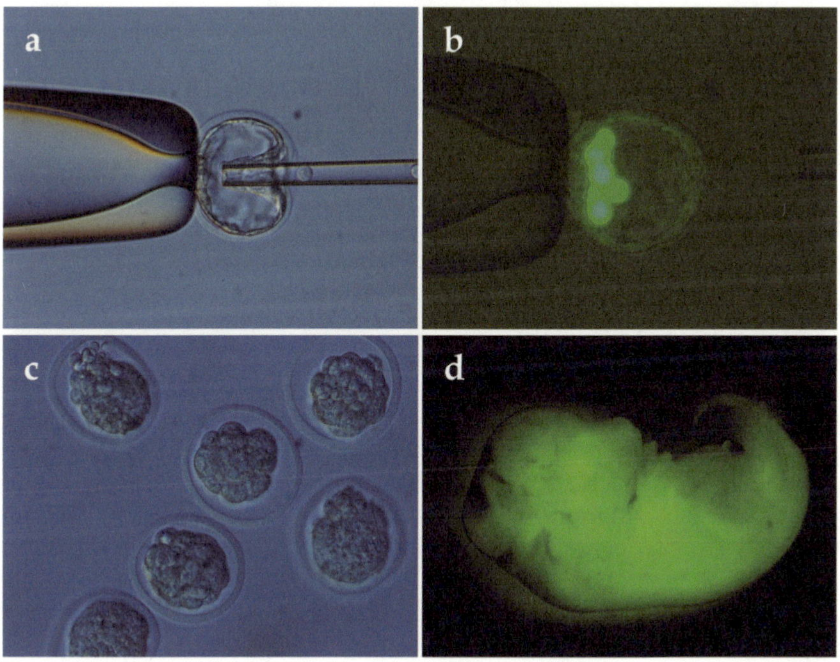

Fig. 4. Microinjection of 10 ES cells into a blastocoele of E4.5 blastocyst (a). Semi-bright, fluorescent image of venus-positive ES cells in the blastocyst (b). Collapsed blastocysts immediately after microinjection (c). Venus-positive fetal rat at E15.5 (d)

Cell line	Passage number	No. (%) of embryos			No. (%) of fetus or pups	
		Injected	Developed to fetus	Developed to pups	Analyzed	Identified as chimera
rESWIv 3i-1	6	17	13 (76.5)	–	13	13 (100)
	6	34	–	21 (61.8)	17	17 (100)
	17	28	17 (60.7)	–	17	14 (82.4)
	17	28	–	14 (50.0)	14	11 (78.6)
rESWIv 3i-5	8	12	9 (75.0)	–	9	9 (100)
	8	13	–	11 (84.6)	11	9 (81.8)
	18	22	7 (31.8)	–	7	7 (100)
	18	33	–	14 (42.4)	14	14 (100)

Table 2. Contribution of rat ES cells for chimera production

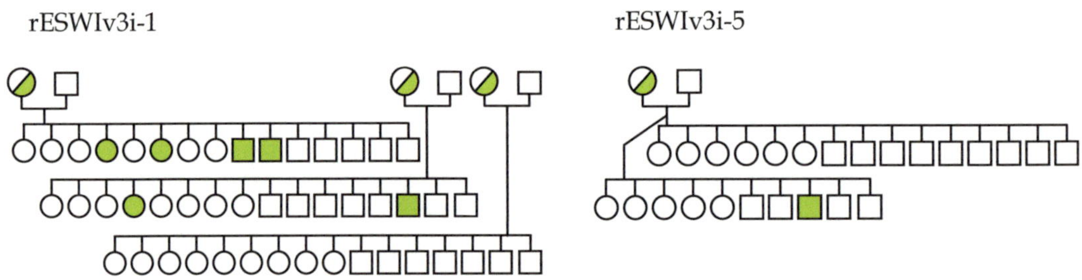

Fig. 5. Progeny test for germline transmission from rESWIv3i-1 or rESWIv3i-5-derived chimeric rats (half-green circle; female). Green circle and square indicate germline-transmitted (venus-positive) females and males, respectively

4. Transgenesis via rat ES cells

Functional germline-competent rat ES cell lines have been established by applying 2i/3i culture system, and the minimal essential materials for conducting transgenic studies including reverse genetic approaches, are now ready for this rodents. In this section, the production efficiency of chimeric rats by blastocyst injection of ES cells electroporated with a humanized Kusabira-Orange (huKO) gene and the germline transmission of the huKO gene from the chimeras to next generation (Hirabayashi et al., 2010b; with a few additional data) are described.

4.1 Electroporation of ES cells with foreign gene

Rat ES cell lines were newly established from E4.5 blastocysts derived from Brown-Norway (BN) females (couplated with BN males), as described above, by applying 2i culture system. The culture medium used for establishment of ES cell lines from BN rats was in N2B27 medium supplemented with 2i (MEK inhibitor PD0325901, 1 μM; GSK3 inhibitor CHIR99021, 3 μM) and rat LIF (1,000 U/ml ESGRO). One of the established lines, named as rESBN2i-4, was derived from a male embryo, based on a PCR analysis using a primer set to detect the rat *Sry* gene. At passage 8, the ES cells in N2B27 medium supplemented with 10% FBS (1 x 10^6 cells/0.5 ml) were electroporated with 25 μg huKO gene (CAG/huKO-neo plasmid; 4.5 kb) using the Bio-Rad Genepulser-II apparatus at 800 V, 10 μF. The electroporated ES cells were plated into a 60 mm petri dish containing 3 ml of 2i medium + 5% FBS (passage 9), and the next day the medium was changed to serum-free 2i medium. Two days after electroporation, 200 μg/ml G418 was added to the medium. The number of neomycin-resistant colonies (passage 10) was counted 8 days after electroporation (Fig. 6). The huKO-positive ES cells were passaged twice in G418-free medium.

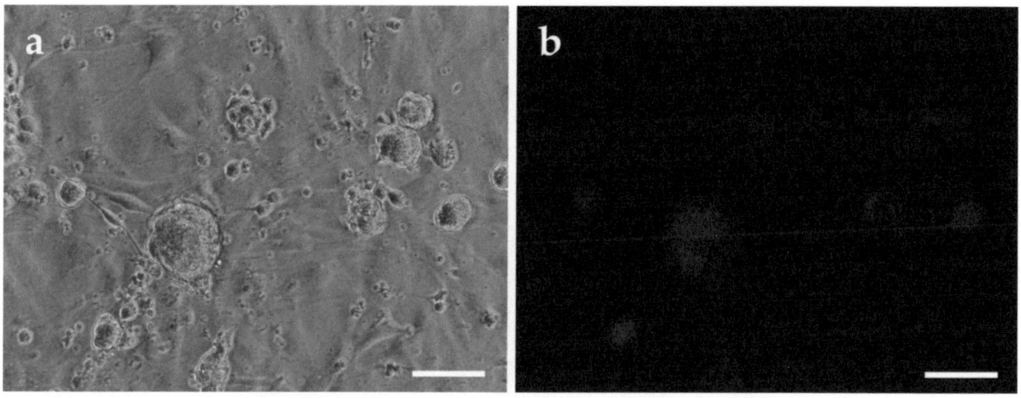

Fig. 6. Rat ES cell colonies stably expressing huKO gene, generated by electroporation, selection with neomycin, and the subsequent expansion. Bright field (a) and fluorescent image (b). Scale bars; 100-μm

4.2 Production of G1 generation transgenic rats

Host blastocysts derived from Wistar/ST or Wistar-Hannover females were microinjected with 10 each of the G418-resistant huKO-positive ES cells, and allowed to develop to full-term in pseudopregnant Wistar recipients. As shown in Table 3, transfer of 116 and 97 blastocysts resulted in 31 and 44 new-born offspring (26.7 and 45.4%), and 22 (70.9%; male 12, female 10) and 34 (77.3%; male 15, female 17, not-identified 2) out of the offspring were judged as chimeras by their coat color, respectively (Fig, 7a). Using non-electroporated control ES cells, similar offspring rate (37.5%, 9/24) and chimera production efficiency (88.9%, 8/9) were obtained. Rat strain for host blastocysts may be a factor influencing the the overall efficiency of chimera production, due to different preference for full-term development. Some male chimeras (n=9) were coupled with wild-type Wistar females, resulting in the birth of total 118 G1 offspring. Germline transmission of the CAG/huKO-neo gene was confirmed in 6 out of 25 G1 offspring (Fig. 7b and 7c) derived from 1 chimeric male with >95% brown-colored coat. Thus, integration of exogenous DNA into rat ES cells did not affect the production efficiency of chimera offspring. In addition, the result described in this section (slightly expanded from Hirabayashi et al., 2010b) achieved the first successful production of transgenic rats via electroporated ES cells, followed by Kawamata & Ochiya (2010) and Tong et al. (2010).

Transfected DNA	Rat strains for host blastocysts	No. (%) of embryos		No. (%) [%]* of pups
		Injected	Developed to pups	Identified as chimera
CAG/huKO	Wistar/ST	116	31 (26.7)	22 (70.9) [19.0]
	Wistar-Hannover	97	44 (45.4)	34 (77.3) [35.1]
Control (No DNA)	Wistar/ST	24	9 (37.5)	8 (88.9) [33.3]

Table 3. Effect of electroporation of rESBN2i-4 cells with huKO gene on the production of chimeric rats.
* Calculated from injected embryos

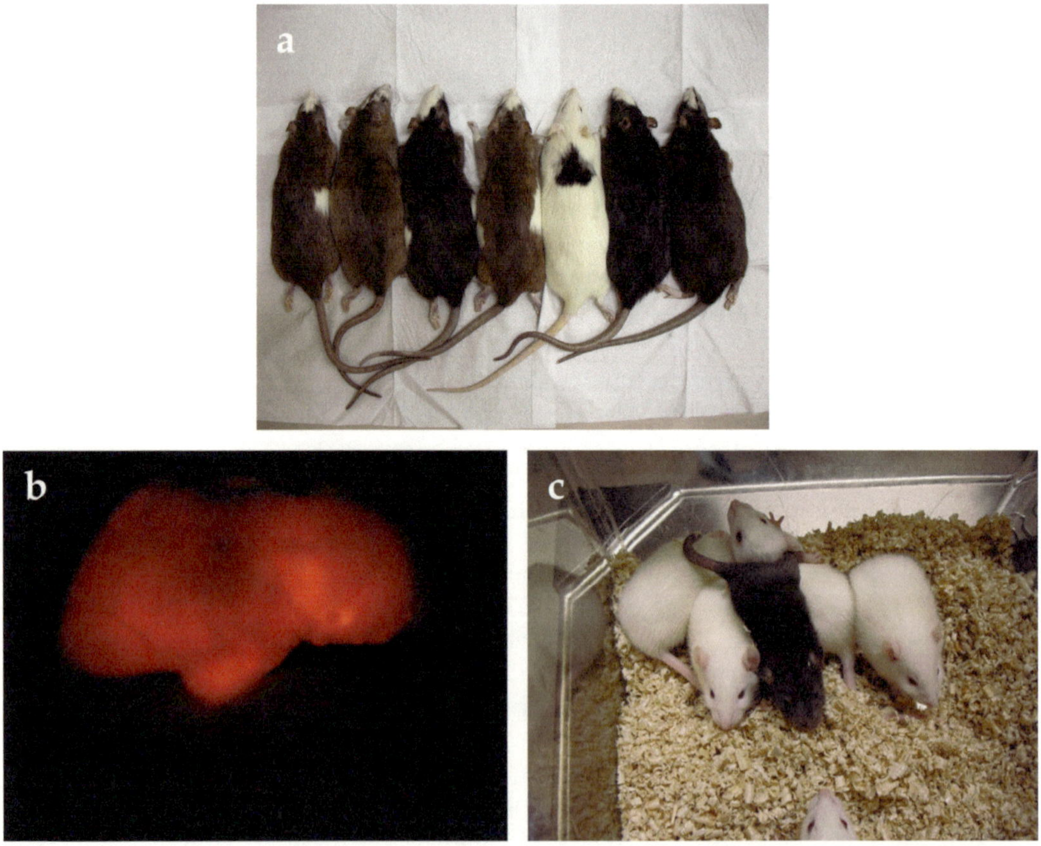

Fig. 7. Chimeric rats with different contribution of brown-colored coat at G0 generation, 17-days old (a), a huKO transgenic newborn young at G1 generation, 3-days old (b), and a huKO transgenic offspring (brown-colored) with huKO-negative littermates, 3-weeks old (c)

5. Conclusion

Functional germline-competent rat ES cell lines have been first reported in 2008 December (Buehr et al., 2008; Li et al., 2008). The difficulty in establishing rat ES cells until those time has been overcome by using a few inhibitors for FGF receptor, MEK and GSK3 in differentiation-related signaling pathways. Reproducibility of this 2i/3i culture system in establishing rat ES cells has been proved by two independent groups (Hirabayashi et al., 2010a; Kawamata & Ochiya, 2010) and both groups successfully applied their own cell lines for producing transgenic rats (Hirabayashi et al., 2010b; Kawamata & Ochiya, 2010). Very recently, the production of p53 gene knock-out rats via homologous recombination in ES cells has been reported by Tong et al. (2010). Although knockout rats were also successfully produced by ENU-induced, SB transposon-tagged, or ZFNs-based mutagenesis, such successes may not give impacts on creating rat models for human diseases due to limitation in genome modification by these technologies. Thus, the homologous recombination-based gene targeting technology in „ES cells" would provide a practical breakthrough for wide range of biomedical research in the laboratory rats.

6. References

Brenin, D.; Look, J.; Bader, M.; Hubner, N.; Levan, G. & Iannaccone, P. (1997). Rat embryonic stem cells: a progress report. *Transplantation Proceedings*, 29, 3, 1761-1765, ISSN 0041-1345 (print).

Buehr, M.; Nichols, J.; Stenhouse, F.; Mountford, P.; Greenhalgh, C.J.; Kantachuvesiri, S.; Brooker, G.; Mullins, J. & Smith, A.G. (2003). Rapid loss of Oct-4 and pluripotency in cultured rodent blastocysts and derivative cell lines. *Biology of Reproduction*, 68, 1, 222-229, ISSN 0006-3363 (print).

Buehr, M.; Meek, S.; Blair, K.; Yang, J.; Ure, J.; Silva, J.; McLay, R.; Hall, J.; Ying Q-L. & Smith, A. (2008). Capture of anthentic embryonic stem cells from rat blastocysts. *Cell*, 135, 7, 1287-1298, ISSN 0092-8674 (print).

Carr, I.; Carr, J. & Dreher, B. (1981). Lymphatic metastasis of mammary adenocarcinoma: an experimental study in the rat with a brief review of the literature. *Invasion Metastasis*, 1, 1, 34-53, ISSN 0251-1789 (print).

Charreau, B.; Tesson, L.; Soulillou, J.P.; Pourcel, C. & Anegon, I. (1996). Transgenesis in rats : technical aspects and models. *Transgenic Research*, 5, 4, 223-234, ISSN 0962-8819 (print).

Chen, K.S. & Gould, M.N. (2004). Development of a universal gap repair vector for yeast-based screening of knockout rodents. *Biotechniques*, 37, 3, 383-388, ISSN 0736-6205 (print).

Feng, L.X.; Chen, Y.; Dettin, L.; Pera, R.A.; Herr, J.C.; Goldberg, E. & Dym, M. (2002). Generation and in vitro differentiation of a spermatogonial cell line. *Science*, 297, 5580, 392-395, ISSN 0036-8075 (print).

Ganten, D. (1998). A transgenic rat model for Alzheimer's disease, In : *Human Genome Analysis*, Hallen, M. (Ed.), 450-457, IOS press, ISBN 978-90-5199-409-4, Amsterdam, The Netherlands.

Geijsen, N.; Horoschak, M.; Kim, K.; Gribnau, J.; Eggan, K. & Daley, G.Q. (2004). Derivation of embryonic germ cells and male gametes from embryonic stem cells. *Nature*, 427, 6970, 148-154, ISSN 0028-0836 (print).

Geurts, A.M.; Cost, G.J.; Freyvert, Y.; Zeitler, B.; Miller, J.C.; Choi, V.M.; Jenkins, S.S.; Wood, A.; Cui, X.; Meng, X.; Vincent, A.; Lam, S.; Michalkiewicz, M.; Schilling, R.; Foeckler, J.; Kalloway, S.; Weiler, H.; Ménoret, S.; Anegon, I.; Davis, G.D.; Zhang, L.; Rebar, E.J.; Gregory, P.D.; Urnov, F.D.; Jacob, H.J. & Buelow, R. (2009). Knockout rats via embryo microinjection of zinc-finger nucleases. *Science*, 325, 5939, 433, ISSN 0036-8075 (print).

Gould, M.N. (1986). Inheritance and site of expression of genes controlling susceptibility to mammary cancer in an inbred rat model. *Cancer Research*, 46, 3, 1199-1202, ISSN 0008-5472 (print).

Hammer, R.E.; Maika, S.D.; Richardson, J.A.; Tang, J.P. & Taurog, J.D. (1990). Spontaneous inflammatory disease in transgenic rats expressing HLA-B27 and human ß2m: an animal model of HLA-B27-associated human disorders. *Cell*, 63, 5, 1099-1112, ISSN 0092-8674 (print).

Heideman, J. (1991). Transgenic rats: a discussion. *Biotechnology*, 16, 325-332, ISSN 0740-7378 (print).

Hirabayashi, M.; Kato, M.; Ishikawa, A.; Kaneko, R.; Yagi, T. & Hochi, S. (2005). Factors affecting production of transgenic rats by ICSI-mediated DNA transfer : effects of sonication and freeze-thawing of spermatozoa, rat strains for sperm and oocyte donors, and different constructs of exogenous DNA. *Molecular Reproduction & Development*, 70, 6, 422-428, ISSN 1040-452X (print).

Hirabayashi, M. & Hochi, S. (2006). Cloning in the rat, In : *Epigenetic Risks of Cloning*, Inui, A. (Ed.), 165-175, Taylor & Francis, ISBN 0-8493-2527-7, Boca Raton, FL.

Hirabayashi, M.; Kato, M.; Kobayashi, T.; Sanbo, M.; Yagi, T.; Hochi, S. & Nakauchi, H. (2010a). Establishment of rat embryonic stem cell lines that can participate in germline chimerae at high efficiency. *Molecular Reproduction & Development*, 77, 2, 94, ISSN 1040-452X (print).

Hirabayashi, M.; Kato, M.; Sanbo, M.; Kobayashi, T.; Hochi, S. & Nakauchi, H. (2010b). Rat transgenesis via embryonic stem cells electroporated with the Kusabira-Orange gene. *Molecular Reproduction & Development*, 77, 6, 474, ISSN 1040-452X (print).

Hochi, S.; Ninomiya, T.; Honma, M. & Yuki, A. (1990). Successful production of transgenic rats. *Animal Biotechnology*, 1, 2, 175-184, ISSN 1049-5398 (print).

Ito, J.; Hirabayashi, M.; Kato, M.; Takeuchi, A.; Ito, M.; Shimada, M. & Hochi, S. (2005). Contribution of high $p34^{cdc2}$ kinase activity to premature chromosome condensation of injected somatic cell nuclei in rat oocytes. *Reproduction*, 129, 2, 171-180, ISSN 1470-1626 (print).

Kato, M.; Ishikawa, A.; Kaneko, R.; Yagi, T.; Hochi, S. & Hirabayashi, M. (2004). Production of transgenic rats by ooplasmic injection of spermatogenic cells exposed to exogenous DNA : a preliminary study. *Molecular Reproduction & Development*, 69, 2, 153-158, ISSN 1040-452X (print).

Kawamata, M. & Ochiya, T. (2010). Generation of genetically modified rats from embryonic stem cells. *Proceedings of the National Academy of Sciences of the United States of America*, 107, 32, 14223-14228, ISSN 0027-8424 (print).

Keng, V.W.; Yaee, K.; Hayakawa, T.; Mizuno, S.; Uno, Y.; Yusa, K.; Kokubu, C.; Kinoshuta, T.; Akagi, K.; Jenkins, N.A.; Copeland, N.G.; Horie, K. & Takeda, J. (2005). Region-specific saturation germline mutagenesis in mice using the Sleeping Beauty transposon system. *Nature Methods*, 2, 10, 763-769, ISSN 1548-7091 (print).

Kitada, K.; Ishishita, S.; Tosaka, K.; Takahashi, R.; Ueda, M.; Keng, V.W.; Horie, K. & Takeda, J. (2007). Transposon-tagged mutagenesis in the rat. *Nature Methods*, 4, 2, 131-133, ISSN 1548-7091 (print).

Li, P.; Tong, C.; Mehrian-Shai, R.; Jia, L.; Wu, N.; Yan, Y.; Maxson, R.E.; Schulze, E.N.; Song, H.; Hsieh, C.L.; Pera, M.F. & Ying, Q.L. (2008). Germline competent embryonic stem cells derived from rat blastocysts. *Cell*, 135, 7, 1299-1310, ISSN 0092-8674 (print).

Like, A.A.; Kislauskis, E.; Williams, R.R. & Rossini, A.A. (1982). Neonatal thymectomy prevents spontaneous diabetes mellitus in the BB/W rat. *Science*, 216, 4546, 644-646, ISSN 0036-8075 (print).

Mashimo, T.; Yanagihara, K.; Tokuda, S.; Voigt, B.; Takizawa, A.; Nakajima, R.; Kato, M.; Hirabayashi, M.; Kuramoto, T. & Serikawa, T. (2008). An ENU-induced mutant archive for gene targeting in rats. *Nature Genetics*, 40, 5, 514-515, ISSN 1061-4036.

Mashimo, T.; Takizawa, A.; Voigt, B.; Yoshimi, K.; Hirai, H.; Kuramoto, T. & Serikawa, T. (2010). Generation of knockout rats with X-linked severe combined immunodeficiency (X-SCID) using zinc-finger nucleases. *PLoS One*, 5, 1, e8870, ISSN 1923-6203 (electronic).

Mullins, J.J.; Peters, J. & Ganten, D. (1990). Fulminant hypertension in transgenic rats harboring the mose Ren-2 gene. *Nature*, 344, 6266, 541-544, ISSN 0028-0836 (print).

Okamoto, K. (1969). Spontaneous hypertension in rats. *International Review of Experimental Pathology*, 7, 227-270, ISSN 0074-7718 (print).

Shinohara, T.; Kato, M.; Takehashi, M.; Lee, J.; Chuma, S.; Nakatsuji, N.; Kanatsu-Shinohara, M. & Hirabayashi, M. (2006). Rats produced by interspecies spermatogonial transplantation in mice and in vitro microinsemination. *Proceedings of the National Academy of Sciences of the United States of America*, 103, 37, 13624-13628, ISSN 0027-8424 (print).

Tong, C.; Li, P.; Wu, N.L.; Yan, Y. & Ying, Q-L. (2010). Production of p53 gene knockout rats by homologous recombination in embryonic stem cells. *Nature*, 467, 7312, 211-213, ISSN 0028-0836 (print).

Toyooka, Y.; Tsunekawa, N.; Akasu, R. & Noce, T. (2003). Embryonic stem cells can form germ cells in vitro. *Proceedings of the National Academy of Sciences of the United States of America*, 100, 20, 11457–11462, ISSN 0027-8424 (print).

Ueda, S.; Kawamata, M.; Teratani, T.; Shimizu, T.; Tamai, Y.; Ogawa, H.; Hayashi, K.; Tsuda, H. & Ochiya, T. (2008). Establishment of rat embryonic stem cells and making of chimera rats. *PloS One*, 3, 7, e2800, ISSN 1923-6203 (electronic).

van Erp, A.M.M. & Miczek, K.A. (2000). Aggressive behavior, increased accumbal dopamine, and decreased cortical serotonin in rats. *The Journal of Neuroscience*, 20, 24, 9320-9325, ISSN 0270-6474 (print).

Vassilieva, S.; Gyan, K.; Pich, U. & Wobus, A.M. (2000). Establishment of SSEA-1- and Oct-4-expressing rat embryonic stem-like cell lines and effects of cytokines of the IL-6 family on clonal growth. *Experimental Cell Research*, 258, 2, 361-373, ISSN 0014-4827 (print).

Wood, E.R.; Dudchenko, P.A. & Eichenbaum, H. (1999). The global record of memory in hippocampal neuronal activity. *Nature*, 397, 6720, 613-616, ISSN 0028-0836 (print).

Ying, Q.L.; Wray, J.; Nichols, J.; Batlle-Morera, L.; Doble, B.; Woodgett, J.; Cohen, P. & Smith, A. (2008). The ground state of embryonic stem cell self-renewal. *Nature*, 453, 7194, 519-523, ISSN 0028-0836 (print).

Zan, Y.; Haag, J.D.; Chen, K.S.; Shepel, L.A.; Wigington, D.; Wang, Y.R.; Hu, R.; Lopez-Guajardo, C.C.; Brose, H.L.; Porter, K.I.; Leonard, R.A.; Hitt, A.A.; Schommer, S.L.; Elegbede, A.F. & Gould, M.N. (2003). Production of knockout rats using ENU-mutagenesis and a yeast-based screening assay. *Nature Biotechnology*, 21, 6, 645-651, ISSN 1087-0156 (print).

Zernicka-Goetz, M. (1991). Spontaneous and induced activation of rat oocytes. *Molecular Reproduction & Development*, 28, 2, 169-176, ISSN 1040-452X (print).

Zhou, Q.; Renard, J.P.; Le Friec, G.; Brochard, V.; Beaujean, N.; Cherifi, Y.; Fraichard, A. & Cozzi, J. (2003). Generation of fertile cloned rats by regulating oocyte activation. *Science*, 302, 5648, 1179, ISSN 0036-8075 (print).

Part 6

Methods of Using Embryonic Stem Cells for Toxicology

23

Assessment of Embryotoxicity and Teratogenicity by the Embryonic Stem Cell Test

Kazuaki Nakamura, Shinji Kusakawa and Akito Tanoue
National Research Institute for Child Health and Development
Japan

1. Introduction

During pregnancy, a fetus can be affected by exposure to a variety of chemicals and pharmaceuticals. These effects can occur through exposure of the mother and subsequent placental transport. Thus, it is important to assess embryotoxicity (developmental toxicity without the observation of maternal toxicity) and teratogenicity (irreversible structure developmental effects without the observation of maternal adversities) prior to the marketing of compounds. The complexity of the reproductive system and the vast number of tissue targets for the exogenic induction of malformations during embryonic development are the rationale underlying the toxicity testing of chemicals in highly standardized animal experiments such as screening tests or multigenerational studies, according to specific Organization for Economic Co-operation and Development (OECD) test guidelines. All of these tests provide information on the biological effects of industrial chemicals or the pharmacological side effects. These guidelines generally specify time-consuming and expensive *in vivo* experiments, most of which are performed with mammalian species such as rats or rabbits. However, for both economical and ethical reasons, there is a great demand for alternatives to living mammals in the testing of chemical-induced adverse effects on reproduction and development. Over the past 30 years, various *in vitro* models have been developed to detect the teratogenic effects of chemicals. These test systems utilize either dissociated cells from the limb buds and brains of rat embryos (micromass test; (Flint & Orton 1984)) or whole embryos of rats (whole embryo culture test; (Freeman & Steele 1986)). Thus, these test systems must sacrifice living animals in order to obtain cells or embryos for each experiment. In recent years, stem cells have become important new tools for the development of *in vitro* model systems to test drugs and chemicals; they have also shown potential to predict or estimate toxicity. Among various stem cells, embryonic stem (ES) cells are the most valuable in developing *in vitro* model systems because they are able to self-renew and differentiate into every cell type of the mammalian organism; they therefore have higher plasticity than adult stem cells. The fact that stem cells are able to self-renew means that they can be continuously cultured in an undifferentiated stage, giving rise to more specialized cells such as heart, liver, bone marrow, blood vessels, pancreatic islets, or neuronal cells upon addition or removal of certain growth factors (Hoffman & Merrill 2007). In 1997, Spielmann *et al.* developed an *in vitro* model for the screening of embryotoxicity based on mouse (m) ES cells. This is termed the "embryonic stem cell test" (EST; (Spielmann et al. 1997)). The EST is based on the assessment of three toxicological endpoints: (1) the morphological analysis of beating cardiomyocytes in embryoid body (EB) outgrowths compared to cytotoxic effects on (2) undifferentiated mES cells, and (3) differentiated NIH-3T3 fibroblasts. As an *in vitro* system, which mirrors both proliferation and differentiation, the EST was proved in an international European Centre for the Validation of Alternative Methods (ECVAM) validation study to be a reliable assay for the prediction of embryotoxicity *in vivo* (Genschow et al. 2004). Using a set of 20 reference compounds with different embryotoxic potencies (nonembryotoxic, weakly embryotoxic, and strongly embryotoxic), the EST was demonstrated to provide a correct judgment in 78% of all experiments. Remarkably, a predictive performance of 100% was obtained for strong embryotoxicants. As a consequence, the validated EST has been accepted and successfully introduced by many pharmaceutical companies as a tool for testing the developmental toxicity of lead compounds at an early stage in the research and development of new drug candidates (Whitlow et al. 2007). A major drawback of the classical EST is its reliance on a morphological endpoint (beating cardiomyocytes) and the need for experienced personnel to ensure reliable assessments of this endpoint (Buesen et al. 2009).

2. Technical details of EST

2.1 Classical EST

Murine ES cells are maintained in an undifferentiated state in culture under conditions that inhibit differentiation by supplementing the culture medium with murine leukemia inhibitory factor (mLIF; (Williams et al. 1988)). Differentiation of ES cells is then induced by the withdrawal of mLIF. Using the "hanging drop" culture technique described by Rudnicki and McBurney (Rudnicki & Mc Burney 1987), ES cells form multicellular aggregates called embryoid bodies (EBs; Fig. 1). Within the EBs, the three germ layers (endo-, meso-, and

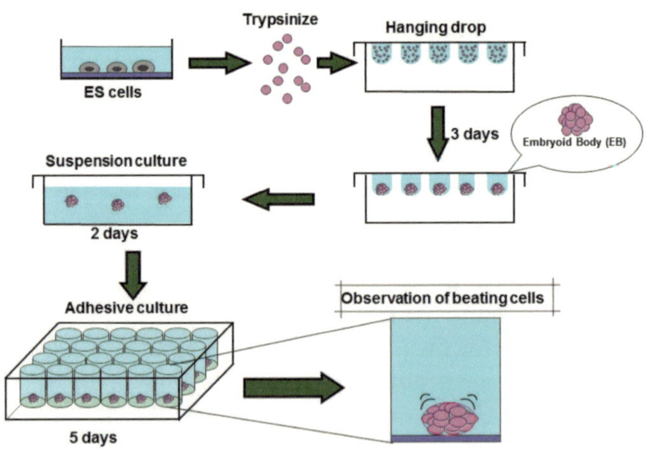

Fig. 1. Differentiation assay of the EST. ES cells are differentiated as EBs in hanging drops in differentiation medium. Differentiation medium containing 750 ES cells is placed on the lids of petri dishes filled with phosphate buffered saline (PBS) for 3 days. The EBs are transferred to suspension culture dishes and cultured for 2 days. They are plated onto 24-well tissue culture plates on day 5 and incubated for 5 additional days. To estimate the efficiency of differentiation from ES cells to cardiomyocytes, the distinctive beating movements of differentiated cardiomyocytes are analyzed under an inverted phase-contrast microscope.

ectoderm) can develop and the further differentiation into several cell types, including beating myocardial cells, can take place (Doetschman et al. 1985; Rudnicki & Mc Burney 1987; Maltsev et al. 1994; Hescheler et al. 1997). The EST benefits from the fact that differentiation into beating myocardial cells can be easily detected by microscopic inspection of EB outgrowths at day 10 of differentiation. In addition to the differentiation analysis, the cytotoxic effects of the test substance on ES cells and NIH-3T3 cells are analyzed. By using stem cells and differentiated fibroblasts, the assay takes embryonic as well as maternal toxicity into account. To assess the concentration of a substance which inhibits the development (inhibition of differentiation: ID_{50}) and proliferation ($IC_{50}ES$ and $IC_{50}3T3$) by 50% compared to the untreated control, dose-response profiles are collected. A biostatistical prediction model was developed to assign test compounds to three classes of embryotoxicity: *non-embryotoxic, weakly embryotoxic,* and *strongly embryotoxic* (Genschow et al. 2002; Genschow et al. 2004).

2.2 Modification of EST

Arguments have been made that the classical, validated EST might be sufficient to assess the embryotoxic potential of chemicals (Marx-Stoelting et al. 2009). The differentiation of ES cells into cardiomyocytes is easy to achieve experimentally and is also highly reproducible. It has been reported that the spontaneous beating of cardiomyocytes represents a crucial first step during embryonic development. However, there is the possibility that the embryotoxic effects of chemicals might be overlooked if their underlying mechanisms of action consist of a very specific interaction with molecules that is expressed during the differentiation of ES cells into a cell type not well represented in the classical EST, such as a neuronal cell. It has also been argued that the EBs in the EST do not simply consist of beating cardiomyocytes, but also contain cells of other germ layers, including neuronal cells. Therefore, several suggestions have been made in regard to how to add further new endpoints (Buesen et al. 2004; Marx-Stoelting et al. 2009). These include additional ES cell differentiation endpoints, and the development of new molecular markers for the detection of toxic effects on embryonic development. One suggestion for the extension of endpoints was to differentiate ES cells into several different tissues, including nervous tissue, bone, cartilage, and epithelia. This approach would ensure that the effects on multilineage differentiation, and on tissue of endodermal, mesodermal, and ectodermal origin, would become detectable. It has been reported that the EC_{50} of thalidomide is ~30-fold lower when the effects were studied on the differentiation of ES cells triggered in the direction of bone tissue than in the classical EST with cardiac differentiation as the endpoint. Similarly, valproic acid (VPA) was more active (by a factor of ~10) in a regimen involving the induction of neuronal differentiation of ES cells (Marx-Stoelting et al. 2009). These results suggest that a multilineage analysis could increase the predictive performance of the EST. It is important to note that the observed differences in sensitivity correspond well with the two different developmental target tissues of thalidomide and VPA, which are primarily bone and the nervous system, respectively, but not the heart. Hence, the additional differentiation endpoints enabled the successful classification of the two agents, which was not possible with the classical EST (zur Nieden et al. 2004). This example not only demonstrates the relevance of additional differentiation endpoints, but also suggests that an increase in predictive value may be achievable using a combination of new differentiation endpoints with the introduction of new molecular markers (Marx-Stoelting et al. 2009). In order to characterize the neural-tissue-specific toxicity of drugs, the stromal cell-derived inducing activity (SDIA) method, which promotes the differentiation of mES cells (Kawasaki et al. 2000; Kitajima et al. 2005), was introduced in the EST (Kusakawa et al. 2010) (Fig. 2). SDIA accumulates on the surface of PA6 stromal cells and induces efficient neuronal differentiation of cocultured ES cells in serum-free conditions without the use of either retinoic acid or EBs. A high proportion of tyrosine hydroxylase-positive neurons producing dopamine are obtained from SDIA-treated ES cells. (Kawasaki et al. 2000). This is one strategy for adding differentiation endpoints in the EST.

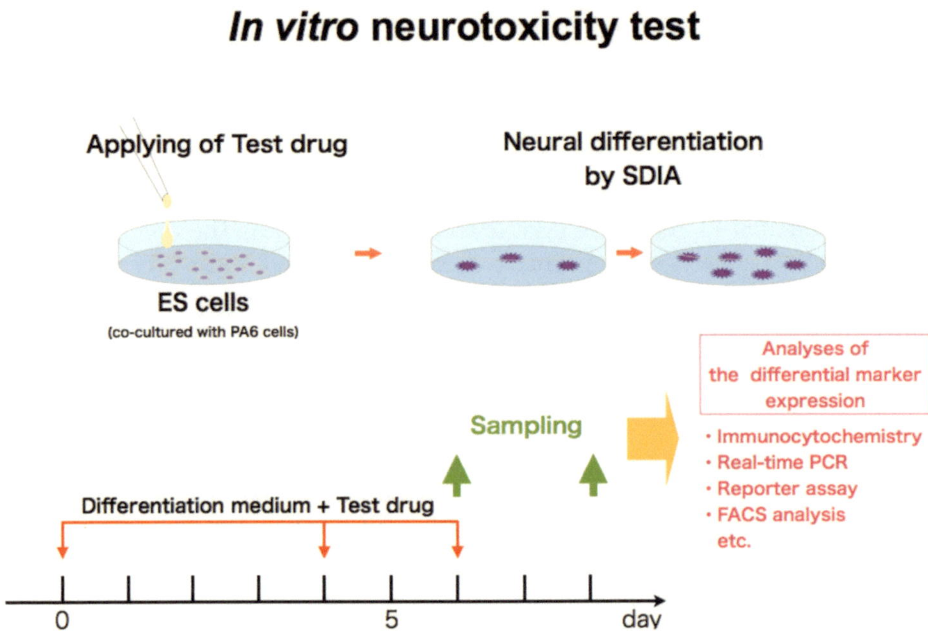

Fig. 2. An *in vitro* neurotoxicity test established using a neural differentiation system induced by the stromal cell-derived inducing activity (SDIA) method. By performing various analyses of neural marker expressions, the effects of test drugs on the developing neural system can be evaluated.

Several target genes and target proteins that could be used as molecular markers for the differentiation of specific embryonic tissues have been suggested and partially explored by methods employing quantitative real-time polymerase chain reaction (PCR) or fluorescence-activated cell sorting (FACS) analysis (Seiler et al. 2004; Seiler et al. 2006; Buesen et al. 2009). In addition to the analysis of endogenous molecular markers, reporter-based systems may represent an attractive alternative. For example, mES cells that have been stably transfected with a neural-tissue-specific promoter-driven green fluorescence protein (GFP) or luciferase reporter can be utilized as a detection system. With this system, the effects on neural cell differentiation can be detected in a high-throughput assay based on GFP expression or luciferase activity (Kusakawa et al. 2010) (Fig. 3). Thus, reporter-based systems are regarded as promising, and may offer potential new endpoints for the EST (Marx-Stoelting et al. 2009).

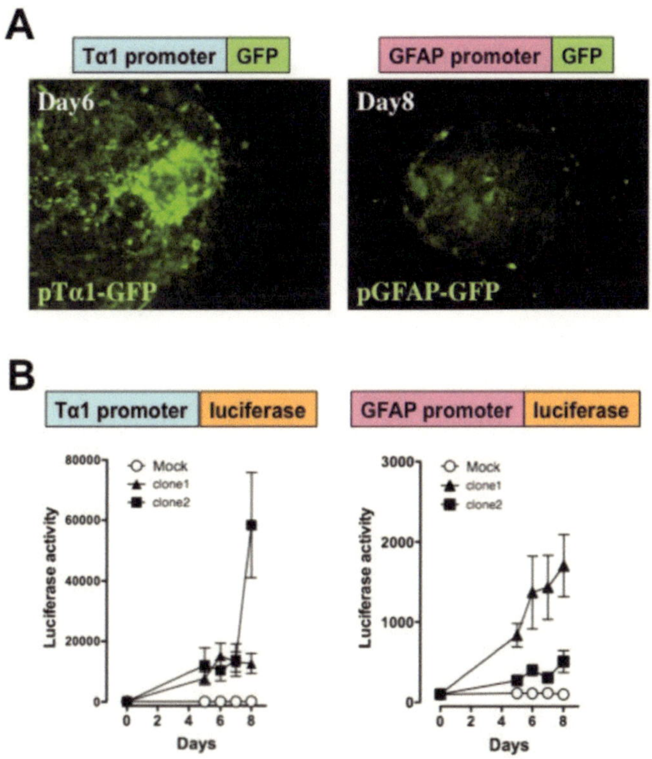

Fig. 3. Established ES cell lines that express reporter genes driven by each promoter of neuron marker gene, tubullin alpha 1 (Tα1), or glial cell marker gene, GFAP. In these stable cell lines, the expression of the reporter gene such as GFP (A) or luciferase (B) is observed, followed by neural differentiation.

3. Actual case using modified EST

3.1 Estimation of carbamazepine embryotoxicity using our modified EST

Carbamazepine (CBZ) is one of the most widely used antiepileptic drugs (AEDs), and is also used in the treatment of neuropathic pains and psychiatric disorders (Albani et al. 1995; Sindrup & Jensen 1999). CBZ is known to be a teratogen (Jones et al. 1989; Shepard et al. 2002). Pregnant women who undergo CBZ drug therapy have increased rates of congenital anomalies in the fetus, in particular, neural tube defects (NTDs), cardiovascular and urinary tract anomalies, and cleft palate. CBZ also induces a pattern of minor congenital anomalies and developmental retardation (Jones et al. 1989; Matalon et al. 2002). CBZ is structurally similar to tricyclic antidepressants, but shares remarkably similar clinical features to the structurally unrelated AED VPA, which has a short-chained fatty acid structure. CBZ is embryotoxic but is less teratogenic than VPA. It causes a spina bifida (an NTD) rate of approximately 0.5 to 1% and a cardiovascular anomaly rate of 1.5 to 2.0% (Ornoy 2006). VPA causes an NTD rate of approximately 2% and an increase of 4 to 8% in major congenital anomalies (Nau et al. 1991; Ornoy 2006). We attempted to characterize the tissue-specific embryotoxicity of CBZ using our modified EST (Murabe et al. 2007b).

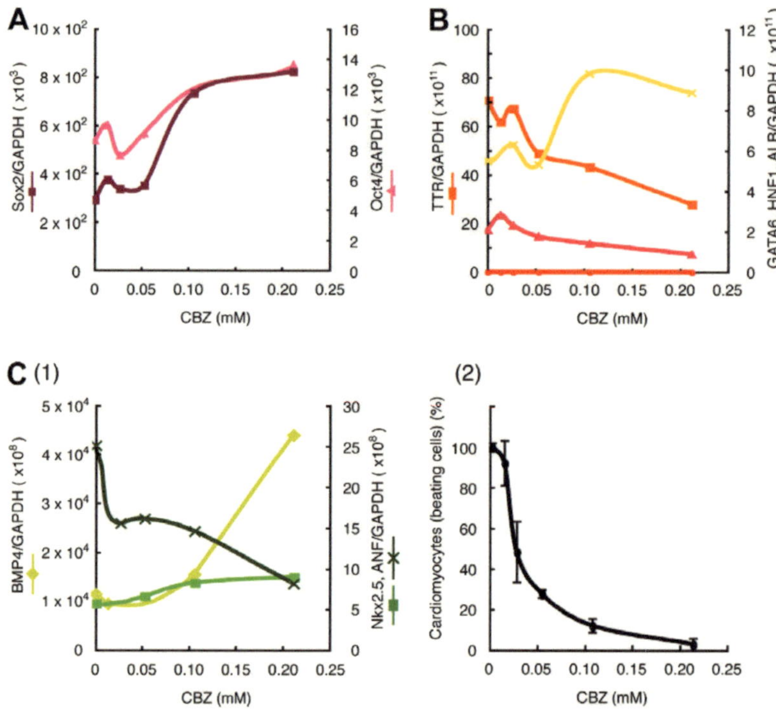

Fig. 4. Analysis of expression levels for mesodermal and endodermal differentiation with CBZ. The expression level of markers of the undifferentiated state, Sox2 and Oct4 (A), endodermal markers, GATA6, TTR, HNF1, and ALB (B), and mesodermal markers, BMP4, Nkx2.5, and ANF (C-1), were quantified at each concentration of CBZ with real-time RT-PCR. The frequencies of cardiomyocytes, identified by their distinctive beating movements, derived from ES cells were quantified at each concentration of CBZ (C-2) (Murabe et al., 2007b).

The tissue-specific effects of CBZ in the ES cell differentiation system were characterized at the molecular level. We used real-time reverse transcription (RT)-PCR on samples from day 5 of culture to determine the expression levels of tissue-specific genes in undifferentiated cells, and in cells differentiating into endodermal and mesodermal lineages (Fig. 4). The expression levels of Sox2 and Oct4 (undifferentiated markers) increased at high CBZ concentrations (Fig. 4A). In the differentiating endodermal lineage, only the expression level of the primitive marker, GATA6, increased (Fig. 4B); the expression levels of the markers of late differentiation stages, TTR and HNF1, decreased in a concentration-dependent manner. Albumin (ALB), a definitive endodermal hepatic marker, was not detected, suggesting that CBZ promoted initial endodermal differentiation but inhibited differentiation into mature endodermal lineages. In the mesodermal lineage, the expression level of the primitive marker, BMP4, increased in a dose-dependent manner (Fig. 4C-1). The expression level of an early cardiac marker, Nkx2.5, showed a slight increase. However, expression of a later stage cardiac marker, ANF, was reduced in a concentration-dependent manner. Under the same culture conditions but omitting CBZ, cardiomyocytes normally differentiate from EBs. We screened for cardiomyocyte differentiation at different concentrations of CBZ. We found that CBZ decreased the rate of undifferentiated ES cells differentiating to cardiomyocytes in a dose-dependent manner (Fig. 4C-2). Thus, based on the gene expression data and rates of cardiomyocyte differentiation, it is evident that CBZ promoted the initial differentiation into mesodermal lineages, including primitive cardiomyocytes, but inhibited later differentiation into the mature mesodermal lineages. We also examined the expression levels of various neural markers: Nestin, which is a marker of early differentiation, and Synaptophysin (Syn) and Neurofilament H (NFH), which are later stage neuron-specific markers. The expression of all three markers increased in a dose-dependent manner in the presence of CBZ (Fig. 5A-1/2). The glial markers GFAP, an astrocyte-specific marker, and Oligo2 and DM20, oligodendrocyte-specific markers, were also elevated in a concentration-dependent manner (Fig. 5A-3). These results suggest that CBZ induces ES cells to differentiate into neurons and glial cells. This contrasts with our observation that VPA induces ES cells to differentiate into neurons but not glial cells (Murabe et al. 2007a). In order to compare the expression profiles of neuronal and glial markers between CBZ and VPA, we performed RT-PCR with samples on days 5 and 7. RT-PCR analysis revealed that the Nestin expression levels induced by CBZ were higher than those induced by VPA (Fig. 5B). Syn and NFH showed higher levels of expression after CBZ than VPA at day 5, but no expression of either marker could be detected at day 7 in the CBZ cultures. In contrast, the expression of Syn and NFH induced by VPA was increased in the day 7 culture. The DM20 expression levels in CBZ cultures were higher on days 5 and 7 than in the VPA cultures. The expression levels of other typical glial markers, such as GFAP, and Olig2, were very low (data not shown). Our immunocytochemical study with an antibody against β-III tubulin (a neuronal marker)

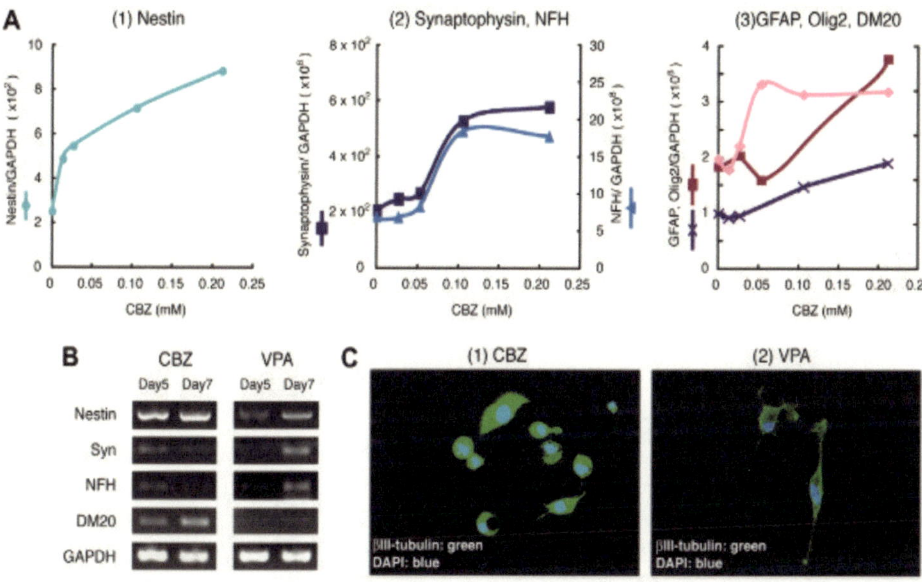

Fig. 5. Analysis of expression levels for ectodermal differentiation with CBZ. (A) gene expression levels of the ectodermal markers, Nestin (1), Synaptophysin, NFH (2), and GFAP, Olig2, DM20 (3), were quantified by real-timeRT-PCR. (B) The expression levels of ectodermal markers were quantified by RT-PCR in cultures treated with CBZ (0.11 mM) or VAP (1.50 mM). (C) Neural cells derived from ES cells cultured in the presence of CBZ (0.05 mM) or VPA (0.19 mM) were immunostained with an anti-βIII-tubulin antibody (Murabe et al., 2007b).

revealed that many positive cells were detected in samples on day 10 of cultures with either 0.05 mM CBZ (Fig. 5C, left) or 0.19 mM VPA (Fig. 5C, right). In the CBZ-administered group, the positive cells had an almost spherical shape and few had nerve processes. These cells had the appearance of immature neurons. In contrast, in the VPA-administered group, many of the positive cells had long nerve processes and had the appearance of mature neurons. These results suggest that: (1) CBZ induces neural lineage differentiation in ES cells but that the potential for neuronal differentiation is lower compared with VPA; and (2) CBZ induces differentiation of both neuronal and glial lineages, whereas VPA induces neuronal but not glial cells.

A cell viability assay was used to study the cytotoxic effect of CBZ on ES cells and NIH-3T3 fibroblasts. In both cell lines, CBZ inhibited the survival of cells in a dose-dependent manner (Fig. 6). There was no significant difference between the cytotoxic sensitivities of ES cells and NIH-3T3 fibroblasts to CBZ. The IC_{50} values were calculated as 0.24 and 0.31 mM for NIH-3T3 fibroblasts and ES cells, respectively. The therapeutic range of CBZ is 0.02 to 0.05 mM in serum. Thus, the IC_{50} values of NIH-3T3 fibroblasts and of ES cells were approximately 5- 15-fold larger than the therapeutic concentration. Neither cell line appears to show any significant response to CBZ within the therapeutic range. The cytotoxicity of CBZ for ES cells was much lower than that obtained for VPA in our system (Murabe et al. 2007a). To observe the cytotoxic and morphological effects of CBZ, ES cells and NIH-3T3 fibroblasts stained in MTT were observed on day 5 of the cytotoxicity assay (Fig. 6, bottom). In both cell types, cell densities were reduced in a concentration-dependent manner. In the high-dose CBZ group, NIH-3T3 fibroblasts showed strong indications of shrinkage or shape changes. In contrast, the ES cells contained many small, presumably undifferentiated cells, suggesting that CBZ strongly inhibited differentiation. This observation is almost the same as that found using VPA (Murabe et al. 2007a).

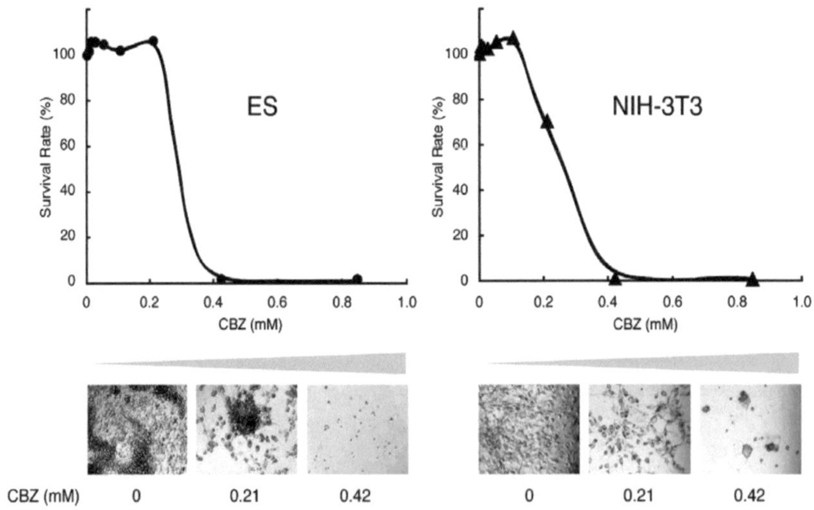

Fig. 6. Cytotoxicity assay on ES cells and NIH-3T3 fibroblasts with CBZ. Cells on day 10 of the assay were stained with MTT and solubilized. The activity of the mitochondrial enzyme of living cells was examined. The violet color of the MTT formazan, which is the enzyme product, was measured at an absorbance of 520 nm. On day 5, cells were stained with MTT (Murabe et al., 2007b).

3.2 Estimation of fluoxetine embryotoxicity using our modified EST

Twenty years have passed since fluoxetine, which is one of the selective serotonin-reuptake inhibitors (SSRIs), was introduced into clinical use; the class of antidepressants known as SSRIs is now used worldwide. Since their introduction, SSRIs have been recognized to be more effective and to have fewer side effects than older tricyclic antidepressants. For these reasons, treatment with SSRIs has become very popular, even among pregnant women, because approximately 10% of pregnant women exhibit symptoms of clinical depression and many are treated with antidepressants. Maternal use of SSRIs during pregnancy is of increasing public health concern, due to its wide prescriptive base for the treatment of depression and other disorders and its potential teratogenic effects on the developing fetus. Thus, we attempted to characterize the tissue-specific embryotoxicity of fluoxetine using our modified EST (Kusakawa et al. 2008).

A cell viability assay was used to study the cytotoxic effect of fluoxetine on ES cells and NIH-3T3 fibroblasts. In both cell lines, fluoxetine inhibited survival of cells in a dose-dependent manner (Fig. 7A and B), indicating that fluoxetine affected cell viability. The IC_{50} values were calculated at 1.79 µM for ES cells and 4.67 µM for NIH-3T3 fibroblasts. There was a significant difference between these two cell lines in their cytotoxic sensitivities to fluoxetine, indicating that ES cells were more sensitive to the toxicity of fluoxetine than NIH-3T3 fibroblasts were.

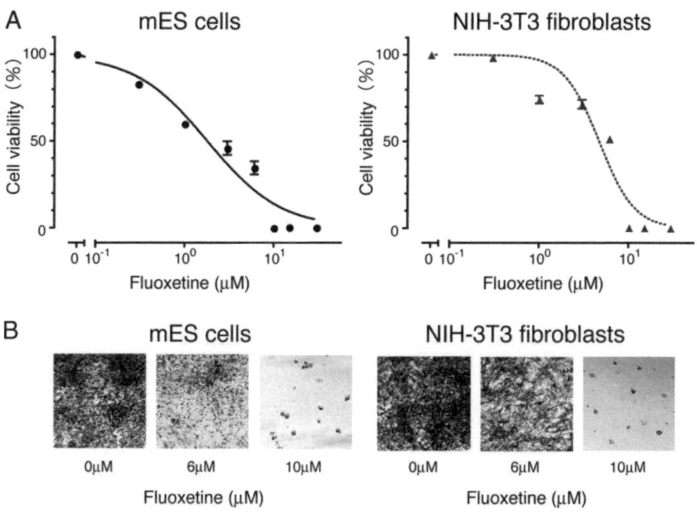

Fig. 7. Cytotoxicity assay on ES cells and NIH-3T3 fibroblasts with fluoxetine. (A) Cells on day 10 of the assay were stained with MTT and solubilized. The activity of the mitochondrial enzyme of living cells was examined. The violet color of the MTT formazan, which is the enzyme product, was measured at an absorbance of 520 nm. (B) On day 5, cells were stained with MTT (Kusakawa et al. 2008).

To characterize the tissue-specific effects of fluoxetine on the ES differentiation system at the molecular level, we examined the expression levels of typical tissue-specific genes by performing real-time RT-PCR analysis on our samples on days 5 and 10 of the differentiation assay (Figs. 8 - 10). Under a control culture condition without the drug (0 µM fluoxetine), the undifferentiated markers Oct3/4 and Sox2 were highly expressed on day 5, and their expression levels decreased between day 5 and day 10. In fluoxetine-treated ES

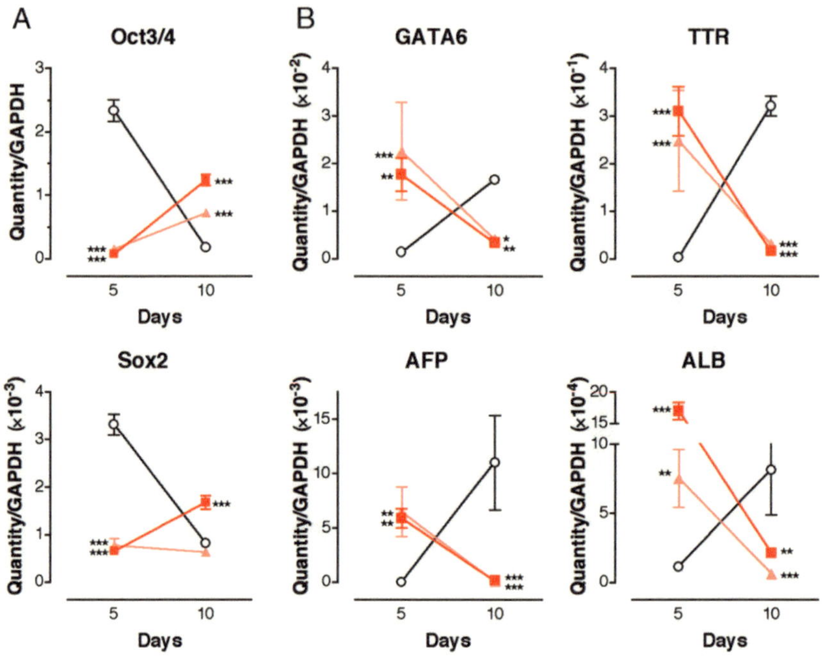

Fig. 8. Analysis of expression levels for the differentiated state and endodermal differentiation. The expression levels of the undifferentiated markers, sox2 and Oct3/4 (A), and endoderm markers, GATA6, TTR, AFP, and ALB (B), were quantified at each concentration of fluoxetine (0 μM: open circle; 1μM: closed triangle; 3μM: closed square) with real-time RT-PCR (Kusakawa et al. 2008).

cells, on the other hand, the expression levels of Oct3/4 and Sox2 were lower on day 5, but increased between day 5 and day 10 (Fig. 8A). There were significant interactions between fluoxetine treatment and the expression pattern of each undifferentiated marker. These results suggested that fluoxetine treatment affected the expression of the undifferentiated markers Oct3/4 and Sox2 in differentiating ES cells. Under the control culture condition, the expression levels of endodermal markers such as GATA6, TTR, AFP, and albumin (ALB) were low on day 5 and up-regulated from day 5 to day 10 (Fig. 8B). In fluoxetine-treated ES cells, in contrast, these endodermal markers were more strongly expressed on day 5, and down-regulated from day 5 to day 10. There were significant interactions between fluoxetine treatment and the expression pattern of each endodermal marker, suggesting that fluoxetine could also affect the differentiating endodermal lineage. The expression levels of BMP4 (a primitive marker), Nkx2.5 (an early cardiac marker), MLC-2v, and ANF (both later cardiac markers) were low on day 5 and increased between day 5 and day 10 under the control culture condition (Fig. 9A). In the fluoxetine-treated cells, on the other hand, these mesodermal markers were more highly expressed on day 5, and decreased between day 5 and day 10 (Fig. 9A). There were significant interactions between fluoxetine treatment and the expression pattern of each endodermal marker. Thus, fluoxetine treatment down-regulated the expressions of mesodermal markers as well as endodermal markers. In addition to studying gene expression levels in the mesodermal lineages, we examined how fluoxetine affected mesodermal cell differentiation. Ordinarily, in the absence of fluoxetine, cardiomyocyte can differentiate from EBs at different concentrations

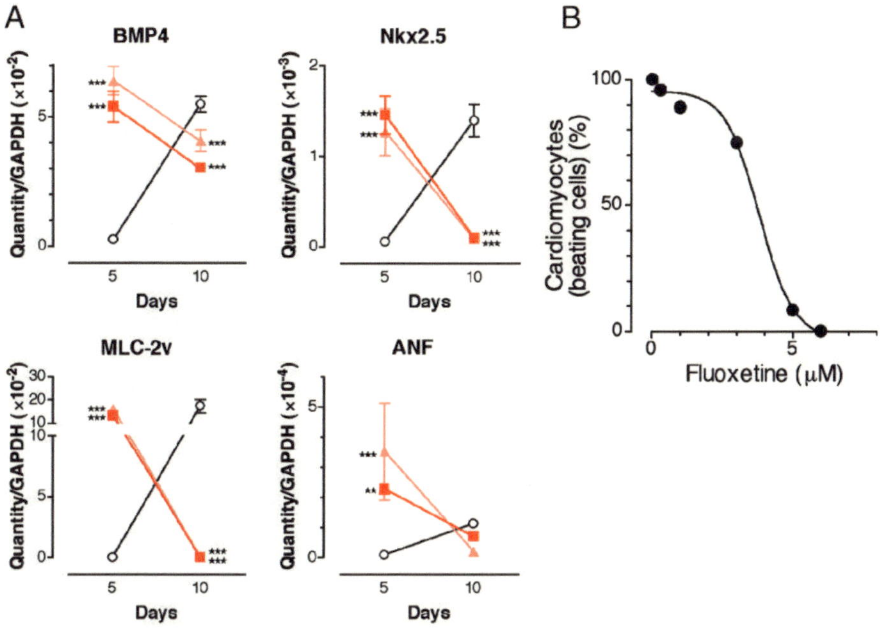

Fig. 9. Analysis of expression levels for mesodermal differentiation. The expression levels of the mesodermal markers, BMP4, Nkx2.5, MLC-2v, and ANF were quantified at each concentration of fluoxetine (0 μM: open circle; 1 μM: closed triangle; 3 μM: closed square) with real-time RT-PCR (A). The frequencies of cardiomyocytes derived from ES cells were identified by their distinctive beating movement and quantified at each concentration of fluoxetine (B) (Kusakawa et al. 2008).

of fluoxetine. We found that fluoxetine decreased the rate at which undifferentiated ES cells differentiated to cardiomyocytes in a concentration-dependent manner (Fig. 9B). The ID_{50} value was calculated at 3.79 μM. Thus, based on the gene expression data and the cardiomyocyte differentiation data, it was confirmed that fluoxetine could inhibit differentiation into mesodermal lineages.

In the absence of fluoxetine, the expression level of Nestin (a primitive neural stem cell marker) was low on day 5, and increased slightly on day 10. In fluoxetine-treated cells, in contrast, the expression level of Nestin was higher on day 5 and decreased between day 5 and day 10. Similarly, the expression levels of synaptophysin (a later neuron-specific marker), GFAP (an astrocyte-specific marker), and Olig2 (an oligodendrocyte-specific marker) in fluoxetine-treated cells were lower on day 5 and increased from day 5 to day 10 (Fig. 10A). There were significant interactions between fluoxetine treatment and the expression pattern of each ectodermal marker. The expression levels of GFAP and Olig2 increased in a concentration-dependent manner (GFAP: 3-fold increase at 1 μM and 18-fold increase at 3 μM; Olig2: 2-fold increase at 1 μM and 20-fold increase at 3 μM). We also performed an immunocytochemical study with antibodies against Neurofilament H (NFH) (a later neuron-specific marker) and GFAP (an astrocyte-specific marker). Among ES cells that had been treated with 3 μM fluoxetine, we detected many positive cells, both for NFH and for GFAP, on day 10 (Fig. 10B). In this fluoxetine-treated group, the positive cells had an almost spherical shape, and few had any nerve processes (Fig. 10B). These cells had the appearance of immature neurons. These results suggest that fluoxetine induces ectodermal lineage differentiation in ES cells, but that the potential for neuronal differentiation is lower compared with that for glial lineages.

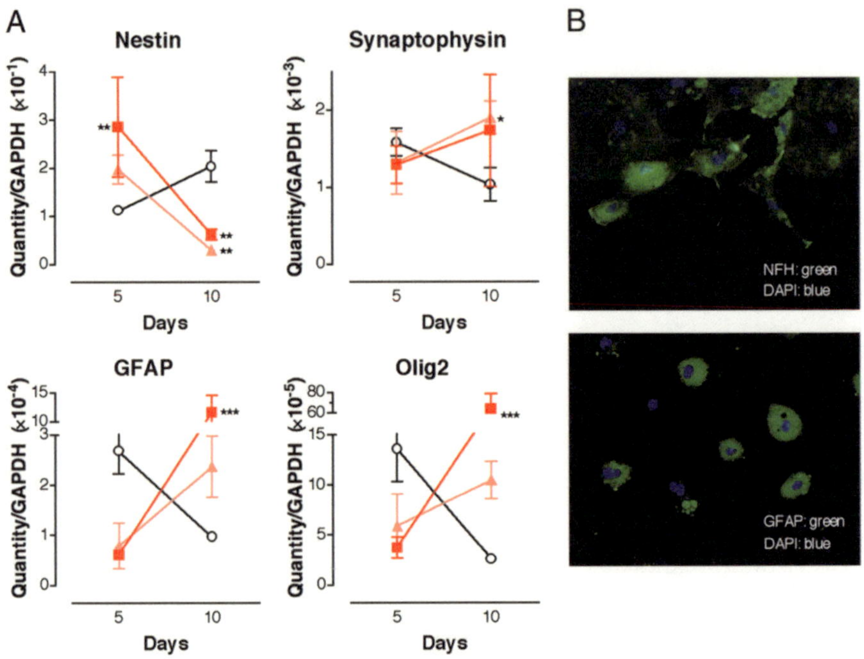

Fig. 10. Analysis of expression levels for ectodermal differentiation. The expression levels of the ectodermal markers, Nestin, Synaptophysin, GFAP, and Olig2 were quantified at each concentration of fluoxetine (0 µM: open circle; 1 µM: closed triangle; 3 µM: closed square) with real-time RT-PCR (A). Neural and glial cells derived from ES cells were cultured in the presence of fluoxetine (3 µM) and immunostained with NFH antibody and GFAP antibody (B) (Kusakawa et al. 2008).

4. Further improvement for EST

The issue of the necessity of adding a metabolic system to the EST to ensure the metabolic activation of potential embryotoxicants has been actively discussed (Marx-Stoelting et al. 2009). Potential experimental procedures are the use of cultured hepatocytes, especially human primary hepatocytes. Hepatocytes may be useful in systems when a preincubation step is applied to a chemical or a pharmaceutical and subsequently added to the ES cell culture with the active metabolite. Whether or not the development of a metabolic activation system for the EST is truly required has been the subject of much discussion, given the difficulties associated with it (Marx-Stoelting et al. 2009). Many hepatotoxicities arise from metabolites derived from drugs (Park et al. 2005). For this reason, it is important to use an assay system with cells that harbor drug-metabolizing enzymes to assess the drug toxicity accurately. In addition, maternal metabolism and fetoplacental interaction have to be taken into account, because some proteratogens require bioactivation to provide the active molecule (Brown et al. 1986). Therefore, the inclusion of a reliable metabolizing system would extend the usefulness of *in vitro* test procedures.

5. Conclusion

Among the various toxic tests, an EST can determine embryotoxicity and teratogenicity, and it does not require experimental animals. The EST has thus far been modified by being combined with real-time PCR, FACS analysis, GFP reporter analysis, or luciferase reporter analysis to detect molecular markers as new endpoints for the EST. In addition, some methods for ES cell differentiation into several cell types such as neural cells have been attempted to identify additional differentiation endpoints. However, further improvement of the EST is needed in order to investigate the harmful effects of chemicals and pharmaceuticals more exhaustively. For instance, a metabolic system designed to detect proteratogenic compounds has to be integrated in order to extend the applicability. Furthermore, a main advantage of embryotoxicity testing by the EST is the availability of human ES cells. Not all mammalian species are equally susceptible or sensitive to the toxic influences of a chemical. A compound that brings out defects in one species can have other or no effects on another species. Using two or more species for regulatory developmental toxicity testing currently covers the detection of genetic differences that influence the response to a chemical. A test system based on mES cells can now be adapted to human ES cells. The use of a humanized test system will have much greater predictive ability because some developmental pathways that could act as targets for chemicals are specific to human development. Thus, the EST can be significantly improved by combining tissue differentiation systems and/or metabolic systems, or by using human ES cells in the quick and accurate estimation of the *in vivo* embryotoxic effects of various medicines.

6. Acknowledgments

Reprinted from Biochemical and Biophysical Research Communications, Vol. 356, Mayu Murabe, Junji Yamauchi, Yoko Fujiwara, Yuki Miyamoto, Masami Hiroyama, Atsushi Sanbe, and Akito Tanoue, Estimation of the embryotoxic effect of CBZ using an ES cell differentiation system, Pages No. 739–744, Copyright (2007), with permission from Elsevier. Reprinted from Life Sciences, Vol. 83, Shinji Kusakawa, Junji Yamauchi, Yuki Miyamoto, Atsushi Sanbe, and Akito Tanoue, Estimation of embryotoxic effect of fluoxetine using embryonic stem cell differentiation system, Pages No. 871–877, Copyright (2008), with permission from Elsevier.

7. References

Albani, F.; Riva, R. & Baruzzi, A. (1995). Carbamazepine clinical pharmacology: a review. *Pharmacopsychiatry*, Vol. 28, pp. 235-244

Brown, L. P.; Flint, O. P.; Orton, T. C. & Gibson, G. G. (1986). Chemical teratogenesis: testing methods and the role of metabolism. *Drug metabolism reviews*, Vol. 17, pp. 221-260

Buesen, R.; Visan, A.; Genschow, E.; Slawik, B.; Spielmann, H. & Seiler, A. (2004). Trends in improving the embryonic stem cell test (EST): an overview, *ALTEX*, Vol. 21, pp. 15-22

Buesen, R.; Genschow, E.; Slawik, B.; Visan, A.; Spielmann, H.; Luch, A. & Seiler, A. (2009). Embryonic stem cell test remastered: comparison between the validated EST and the new molecular FACS-EST for assessing developmental toxicity in vitro. *Toxicological sciences*, Vol. 108, pp. 389-400

Doetschman, T. C.; Eistetter, H.; Katz, M.; Schmidt, W. & Kemler, R. (1985). The in vitro development of blastocyst-derived embryonic stem cell lines: formation of visceral yolk sac, blood islands and myocardium. *Journal of embryology and experimental morphology*, Vol. 87, pp. 27-45

Flint, O. P. & Orton, T. C. (1984). An in vitro assay for teratogens with cultures of rat embryo midbrain and limb bud cells. *Toxicology and applied pharmacology*, Vol. 76, pp. 383-395

Freeman, S. J. & Steele, C. E. (1986). Post-implantation whole embryo culture and the study of teratogenesis. *Food and chemical toxicology*, Vol. 24, pp. 619-622

Genschow, E.; Spielmann, H.; Scholz, G.; Pohl, I.; Seiler, A.; Clemann, N.; Bremer, S. & Becker, K. (2004). Validation of the embryonic stem cell test in the international ECVAM validation study on three in vitro embryotoxicity tests. *Alternatives to laboratory animals*, Vol. 32, pp. 209-244

Genschow, E.; Spielmann, H.; Scholz, G.; Seiler, A.; Brown, N.; Piersma, A.; Brady, M.; Clemann, N.; Huuskonen, H.; Paillard, F.; Bremer, S. & Becker, K. (2002). The ECVAM international validation study on in vitro embryotoxicity tests: results of the definitive phase and evaluation of prediction models. European Centre for the Validation of Alternative Methods. *Alternatives to laboratory animals*, Vol. 30, pp. 151-176

Hescheler, J.; Fleischmann, B. K.; Lentini, S.; Maltsev, V. A.; Rohwedel, J.; Wobus, A. M. & Addicks, K. (1997). Embryonic stem cells: a model to study structural and functional properties in cardiomyogenesis, *Cardiovascular research*. Vol. 36, pp. 149-162

Hoffman, J. A. & Merrill, B. J. (2007). New and renewed perspectives on embryonic stem cell pluripotency. *Frontiers in bioscience*, Vol. 12, pp. 3321-32

Jones, K. L.; Lacro, R. V.; Johnson, K. A. & Adams, J. (1989). Pattern of malformations in the children of women treated with carbamazepine during pregnancy. *The New England journal of medicine*, Vol. 320, pp. 1661-1666

Kawasaki, H.; Mizuseki, K.; Nishikawa, S.; Kaneko, S.; Kuwana, Y.; Nakanishi, S.; Nishikawa, S. I. & Sasai, Y. (2000). Induction of midbrain dopaminergic neurons from ES cells by stromal cell-derived inducing activity. *Neuron*, Vol. 28, pp. 31-40

Kitajima, H.; Yoshimura, S.; Kokuzawa, J.; Kato, M.; Iwama, T.; Motohashi, T.; Kunisada, T. & Sakai, N. (2005). Culture method for the induction of neurospheres from mouse embryonic stem cells by coculture with PA6 stromal cells. *Journal of neuroscience research*, Vol. 80, pp. 467-474

Kusakawa, S.; Yamauchi, J.; Miyamoto, Y.; Sanbe, A. & Tanoue, A. (2008). Estimation of embryotoxic effect of fluoxetine using embryonic stem cell differentiation system. *Life sciences*, Vol. 83, pp. 871-877

Kusakawa, S.; Nakamura, K.; Miyamoto, Y.; Sanbe, A.; Torii, T.; Yamauchi, J. & Tanoue, A. (2010). Fluoxetine promotes gliogenesis during neural differentiation in mouse ES cells. *Journal of neuroscience research*, Vol. 88, pp. 3479-3487

Maltsev, V. A.; Wobus, A. M.; Rohwedel, J.; Bader, M. & Hescheler, J. (1994). Cardiomyocytes differentiated in vitro from embryonic stem cells developmentally express cardiac-specific genes and ionic currents. *Circulation research*, Vol. 75, pp. 233-244

Marx-Stoelting, P.; Adriaens, E.; Ahr, H. J.; Bremer, S.; Garthoff, B.; Gelbke, H. P.; Piersma, A.; Pellizzer, C.; Reuter, U.; Rogiers, V.; Schenk, B.; Schwengberg, S.; Seiler, A.; Spielmann, H.; Steemans, M.; Stedman, D. B.; Vanparys, P.; Vericat, J. A.; Verwei, M.; van der Water, F.; Weimer, M. & Schwarz, M. (2009). A review of the implementation of the embryonic stem cell test (EST). The report and recommendations of an ECVAM/ReProTect Workshop. *Alternatives to laboratory animals*, Vol. 37, pp. 313-28

Matalon, S.; Schechtman, S.; Goldzweig, G. & Ornoy, A. (2002). The teratogenic effect of carbamazepine: a meta-analysis of 1255 exposures, *Reproductive toxicology*. Vol. 16, pp. 9-17

Murabe, M.; Yamauchi, J.; Fujiwara, Y.; Hiroyama, M.; Sanbe, A. & Tanoue, A. (2007a). A novel embryotoxic estimation method of VPA using ES cells differentiation system. *Biochemical and biophysical research communications*, Vol. 352, pp. 164-169

Murabe, M.; Yamauchi, J.; Fujiwara, Y.; Miyamoto, Y.; Hiroyama, M.; Sanbe, A. & Tanoue, A. (2007b). Estimation of the embryotoxic effect of CBZ using an ES cell differentiation system. *Biochemical and biophysical research communications*, Vol. 356, pp. 739-744

Nau, H.; Hauck, R. S. & Ehlers, K. (1991). Valproic acid-induced neural tube defects in mouse and human: aspects of chirality, alternative drug development, pharmacokinetics and possible mechanisms. *Pharmacology & toxicology*, Vol. 69, pp. 310-321

Ornoy, A. (2006). Neuroteratogens in man: an overview with special emphasis on the teratogenicity of antiepileptic drugs in pregnancy. *Reproductive toxicology*, Vol. 22, pp. 214-226

Park, B. K.; Kitteringham, N. R.; Maggs, J. L.; Pirmohamed, M. & Williams, D. P. (2005). The role of metabolic activation in drug-induced hepatotoxicity. *Annual review of pharmacology and toxicology*, Vol. 45, pp. 177-202

Rudnicki, M. A. & Mc Burney, M. W. (1987). Cell culture methods and induction of differentiation of embryonal carcinoma cell lines. Teratocarcinoma and Embryonic Stem Cells: A Practical Approach. E. J. Robertson. Washington D.C., IRL Press: pp. 19-49

Seiler, A.; Visan, A.; Buesen, R.; Genschow, E. & Spielmann, H. (2004). Improvement of an in vitro stem cell assay for developmental toxicity: the use of molecular endpoints in the embryonic stem cell test. *Reproductive toxicology*, Vol. 18, pp. 231-240

Seiler, A. E.; Buesen, R.; Visan, A. & Spielmann, H. (2006). Use of murine embryonic stem cells in embryotoxicity assays: the embryonic stem cell test. *Methods in molecular biology*, Vol. 329, pp. 371-395

Shepard, T. H.; Brent, R. L.; Friedman, J. M.; Jones, K. L.; Miller, R. K.; Moore, C. A. & Polifka, J. E. (2002). Update on new developments in the study of human teratogens. *Teratology*, Vol. 65, pp. 153-161

Sindrup, S. H. & Jensen, T. S. (1999). Efficacy of pharmacological treatments of neuropathic pain: an update and effect related to mechanism of drug action. *Pain*, Vol. 83, pp. 389-400

Spielmann, H.; Pohl, I.; Droing, B.; Liebsch, M. & Moldenhauser, F. (1997). The embryonic stem cell test, in vitro embryo toxicity test using two permanent mouse cell line: 3T3 fibroblast and embryonic stem cells. *In Vitro Toxicology*, Vol. 10, pp. 119-127

Whitlow, S.; Burgin, H. & Clemann, N. (2007). The embryonic stem cell test for the early selection of pharmaceutical compounds. *ALTEX*, Vol. 24, pp. 3-7

Williams, R. L.; Hilton, D. J.; Pease, S.; Willson, T. A.; Stewart, C. L.; Gearing, D. P.; Wagner, E. F.; Metcalf, D.; Nicola, N. A. & Gough, N. M. (1988). Myeloid leukaemia inhibitory factor maintains the developmental potential of embryonic stem cells. *Nature*, Vol. 336, pp. 684-687

zur Nieden, N. I.; Kempka, G. & Ahr, H. J. (2004). Molecular multiple endpoint embryonic stem cell test-a possible approach to test for the teratogenic potential of compounds. *Toxicology and applied pharmacology*, Vol. 194, pp. 257-269

24

Application of Embryonic Stem Cells as a Novel Tool in Drug Screening

Gi Jin Kim, Ph.D
Department of Biomedical Science, CHA University
Republic of Korea

1. Introduction

Toxicological screening using animals are necessary for drug development registration. This approach is time-consuming, costly, labour intensive, stressful for the animals and susceptible to inaccuracies due to individual differences between animals. So, the screening of candidate chemicals in early development is often replaced with *in vitro* cell culture systems (Pearson, 1986; Liebsch & Spielmann, 2002). *In vitro* studies using cell lines were capable of providing more rapid, precise, relevant information than some animal studies, and economical approach for the evaluation of the pharmaco-toxicological profiling of target drugs, characterised by a low compound requirement and short duration (Pearson, 1986; Kari et al., 2007). Also, it is possible to include mechanistic studies, and to test for toxicity that is specific to humans: sensitivity differences between humans and rodents can affect animals (Kari et al., 2007). Among the *in vitro* screening systems, primary cell cultures and/or target organ-specific cell lines can be used to measure the general toxicity of a test compound (Zhou et al., 2006).

However, the sensitivity of hepatotoxicity using primary human hepatocytes or the HepG2 cell line cannot predict effects in early development and toxicological differences, which depend on the state of differentiation in hepatocytes (Knasmuller et al., 2004; Xu et al., 2004). In addition to, primary cells such as hepatocytes in particular and many transformed human hepatocyte-derived cell lines (immortalized cultures, i.e Fa2-N4 cells, HepaRG cells) have limitations in their life span and can have donor-dependent variations (Mills et al., 2004). Also, they have disadvantages such as discontinuous phenotypic characteristics, functional properties and genetic instability. Therefore, more promised future is waiting for hepatocyte-like cells as the source of hepatocytes regarding the approach of stem cells use in the high throughput testing (Duret et al., 2007).

Stem cells are defined functionally as cells that have the capacity to self-renew as well as the ability to generate differentiated cells that specialized functions in specific tissues and make up the organ (Thomson et al., 1998; Zhang & Wang, 2008; Schnerch et al., 2010). The classification of stem cells divided into embryonic stem cells (ESCs) and adult stem cells (ASCs) according to derivative origins. ESCs, which are derived from the inner cell mass of blastocysts after fertilization, can unlimited self-renewal and have pluripotent could be rise to cells derived from all three germ lineages. Otherwise, ASCs derived from the specialized cell types of the tissue from which are originated have limited self-replicate and mutilpotent could be giving rise to specialized cells into multiple-lineages not all three germ lineages.

Since Thomson and colleagues were firstly reported to potentials for therapeutic and drug discovery using embryonic stem cells derived from inner cell mass of human blastocysts in 1998, stem cell research have been spotlighted more public and professional interest than other subject in biology field (Thomson et al., 1998; Fuh & Brinton, 2009; Astori et al., 2010; Rashid et al., 2010).

Although the field of stem cell research has grown rapidly, there are focused on the cell therapy using stem cells in regenerative medicine because stem cells have a great fascinating potential could be differentiation into the specialized cells to replace damaged cells or tissues. ESCs is an attractive source for cell therapy, however, there are several obstacles including ethical problems to clinical application using ESCs in degenerative medicine. Especially, the optimal guideline for differentiation of ESCs have been shown to be controversy in the condition of differentiation into target cells due to various factors used to induce the differentiation in ESCs (Denham & Dottori, 2009; Zagami et al., 2009; Bourzac et al., 2010).

Due to the reason, many scientists have been found a research field to make the best use of the potential benefits of stem cells. Stem cell-based systems have striking advantages to select of lead candidates and development of new therapeutic drugs because their proliferation ability and plasticity to generate several cell types. In addition, it is possible to analyze effects or toxicities of target drugs according to differentiation steps from immature to mature cells could be function. *In vitro* assaying of embryotoxicity using embryonic stem cells for the early determination of the teratogenic potential of a compound have been attempted (Kim et al., 2006; Kulkarni & Khanna, 2006). Also, in vitro screening system using mouse embryonic stem cells have been suggested by the European Center for Validation of Alternative Methods (ECVAM) committee (Spielmann et al., 2001; Genschow et al., 2002). Therefore, drug screening system using stem cells is a large range of opinion on this issue. The application of ESCs in drug screening is a promising, innovative alternative that appears to have early efficacy, while reducing the adverse outcomes at later stages of development. Furthermore, it is very sensitive in toxicity screening, as compared to other cell lines. For this reason, the application of ESCs could be an important new tool for developing a unique *in vitro* model with the potential to predict genotoxicity in humans. Drug screening using ESCs focuses on reproductive biology and embryo development. The embryonic stem cell test (ESCT) can detect developmental abnormalities such as abnormal embryogenesis and malformations caused by mutagenic or embryotoxic substances in early embryogenesis.

With this intention, this chapter review outlines the limitation of traditional toxicity screening including in vitro and in vivo screening system. This brief review outlines the construction of more efficient mouse ESC-derived embryoid bodies (EBs) for use as a vasculogenesis model and as a tool for screening the cytotoxicity of new compounds, and the cytotoxicity of 5-fluorouracil (5-FU), and to examine its effects on cell viability, proliferation, and differentiation in mouse ESC-derived endothelial differentiation. Also, we also consider the obstacles that need to be overcome to make the embryonic stem cells cost-effective, to enable guided differentiation to target cells, and to establish a reproducible *in vitro* assay to increase its efficacy as a screening system. Finally, we introduce the latest research on the ESCT and the potential of using ESCs for drug screening.

2. Characterization of embryonic stem cells

Stem cells have the capacity to self-renew as well as the ability to generate differentiated cells that specialized functions in specific tissues and make up the organ. The classification of stem cells divided into embryonic stem cells (ESCs) and adult stem cells (ASCs) according to derivative origins. ESCs, which are derived from the inner cell mass of blastocysts after fertilization, can unlimited self-renewal and have pluripotent could be rise to cells derived from all three germ lineages. Otherwise, ASCs derived from the specialized cell types of the tissue from which are originated have limited self-replicate and mutilpotent could be giving rise to specialized cells into multiple-lineages not all three germ lineages.

Since mouse embryonic stem cells (mESCs) and human embryonic stem cells (ESCs) were established in 1981 and in 1998, ESCs are a hot issue in stem cell research (Martin, 1981; Thomson et al., 1998). Especially, unique characterization of human ESCs shows unlimited proliferation activity through the expression of stemness markers such as Oct-4, alkaline phospatase, Nanog, SSEA-4, TRA-1-60, and TRA − 81 in condition of co-cultured with feeder cells (Figure 1).

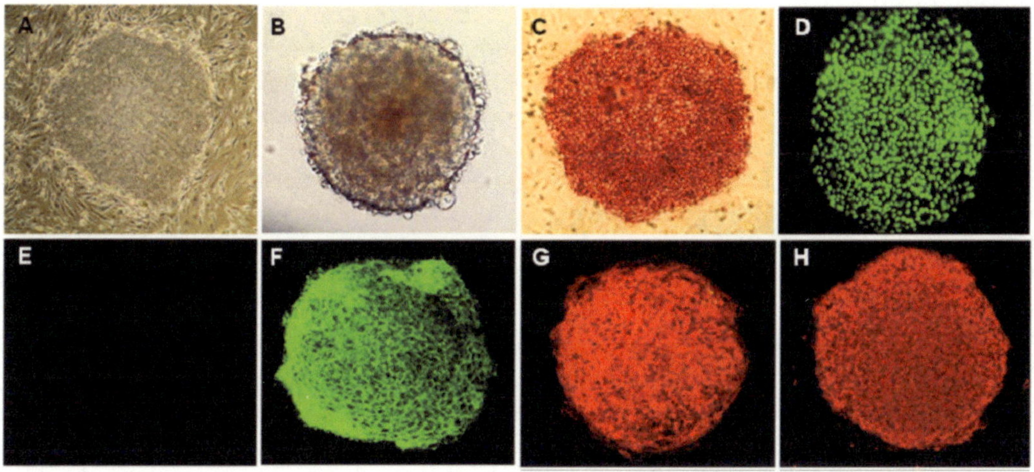

Fig. 1. Characterization of human embryonic stem cells. Morphology of human ES cells cultured with feeder cells (A) and embryonic body (B), Expression of stemness markers, Alkaline phosphatase (C), Oct-4 (D), SSEA-1 (E), SSEA-4 (F), TRA-1-60 (G), and TRA-1-80

Also, human ESCs in vitro can be expanded indefinitely in the undifferentiated state and still retain the capacity for differentiation into endodermal, mesodermal, and ectodemal lineages cells under specific condition in vitro. The potential for differentiation of human ESCs could be confirmed by teratoma formation in SCID mice transplanted human ESCs. Teratoma is an encapsulated benign tumor with tissues or organ components resembling normal derivatives of all three germ layers (Figure 2). The potentials offer a therapeutic intervention in the treatment of degenerative diseases that affect various tissues and their therapeutic effects may be influenced by numerous factors such as cell count, differentiation potential, transplant method, and disease model (Petersen et al., 1999; Huttmann et al., 2003; Kuo et al., 2008).

The characterization of human ESCs derived from early embryos that seem to share many of the properties of mouse ESCs has refocused attention on the in vitro properties of ESCs (Thomson et al., 1998). Therefore, mouse and human ESCs will provide new insights into embryonic development as well as its translation into therapeutic outcomes.

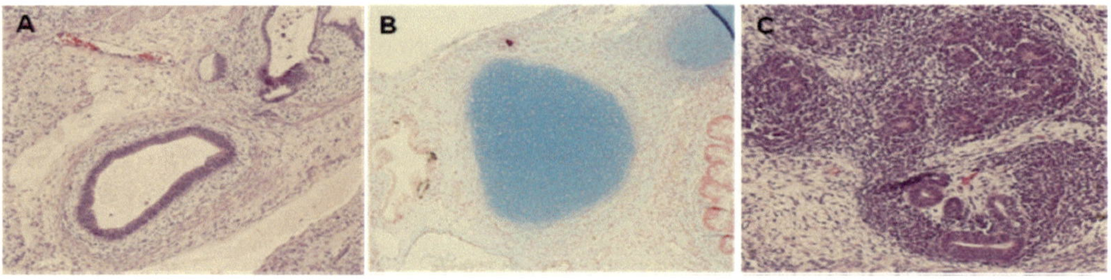

Fig. 2. Differentiation potential of human ESCs by teratoma formation in SCID mice engrafted human ESCs. Teratoma showing tissues from all three germ layers; endoderm (gut epithelium, A), mesoderm (cartilage, B), and ectoderm (neural rosettes, C)

3. Limitation of traditional toxicity screening

Toxicogenomics, which combines toxicology and genomics, is a scientific field that studies how the gene and protein activity within particular cell or tissues of an organism is involved in responses to toxicants. Also, they studies for genetics, mRNA expression, cell and tissue-wide protein expression and metabolomics to understand the role of gene-toxicants interactions in disease. From these processing, new biomarkers for toxicity can discover as well as predictive in toxicology. However, the relationship between dose and its effects on the exposed organism still controversy because there are several factors including age, sex, individual variability, and species differences affected susceptibility and variability in toxicology.

Toxicological screening using animals are necessary for drug development registration. Although there are several animal models from zebra fish to monkey for in vivo toxicity screening as well as drug development, however, there are many kinds of considerations in vivo toxicity screening using animal model. Especially, embryotoxicity tests using animals are a traditional strategy to identify potentially hazardous chemicals. They can also be used to confirm the absence of toxic properties in the development of potentially useful new substances (Spielmann, 2009). Because this approach is time-consuming, costly, labour intensive, stressful for the animals and susceptible to inaccuracies due to individual differences between animals, the screening of candidate chemicals in early development is often replaced with *in vitro* cell culture systems (Pearson, 1986; Liebsch & Spielmann, 2002; Bremer & Hartung, 2004; Knight, 2007). Animal ethics in vivo screening system have been becoming emphasized by Institutional Animal Care and Use Committees (IACUC). In addition, it is difficult to check the correct time point for toxicity testing in screening using animal model. Therefore, there is a need for alternative methods to evaluate the potential reproductive toxicity of chemical substances, by *in vitro* systems. To develop a new alternative screening test, many scientists have tried to use cell lines, primary cell cultures of dissociated cells from mice or rat embryo limb buds, midbrains for micromass tests, or whole embryos from rat (Steele et al., 1983).

In vitro studies using cell lines were capable, or potentially capable, of providing more rapid, precise, relevant information than some animal studies, and economical approach for the evaluation of the pharmaco-toxicological profiling of target drugs, characterised by a low compound requirement and short duration (Pearson, 1986; Kari et al., 2007). Also, it is possible to include mechanistic studies, and to test for toxicity that is specific to humans: sensitivity differences between humans and rodents can affect animals (Kari et al., 2007). Among the *in vitro* screening systems, primary cell cultures and/or target organ-specific cell lines can be used to measure the general toxicity of a test compound (Zhou et al., 2006). However, the sensitivity of hepatotoxicity using primary human hepatocytes or the HepG2 cell line cannot predict effects in early development and toxicological differences, which depend on the state of differentiation in hepatocytes (Knasmuller et al., 2004; Xu et al., 2004). In addition to, primary cells such as hepatocytes in particular and many transformed human hepatocyte-derived cell lines (immortalized cultures, i.e Fa2-N4 cells, HepaRG cells) have limitations in their life span and can have donor-dependent variations (Mills et al., 2004). However, they have disadvantages such as discontinuous phenotypic characteristics, functional properties and genetic instability.

In toxicological research, the development of alternative *in vitro* toxicity screening systems to replace *in vivo* screening methods using animal experiments and conventional screening systems is important. Although applications using primary cells derived from rodent embryo or tissue-specific cell lines originating from humans have been tried, the validation of cytotoxicity during cellular biological processing must be improved in order to establish new and alternative *in vitro* screening methods, and to increase the efficiency of toxicological analysis through these methods (Tiffany-Castiglioni et al., 1999). Therefore, more promised future is waiting for hepatocyte-like cells as the source of hepatocytes regarding the approach of stem cells use in the high throughput testing (Duret et al., 2007).

4. In vitro toxicity screening using embryonic stem cells

Embryonic stem cells are generally obtained from inner cells mass in blastocysts of either mice or humans. Especially, drug or toxicity screening using human ESCs have many advantages over primary cells and immortalized cell lines for *in vitro* toxicity screening, including unique properties such as unlimited self-renewal, plasticity to generate various cell types and availability of cells of human origin (Davila et al., 2004; Kulkarni & Khanna, 2006; Zhang & Wang, 2008). However, there are some disadvantages as they do not grow as well and are more difficult to cultivation than mouse ESCs. Also, the stem cell technologies for directing them to differentiate are less defined than those for mouse ESCs and it is necessary to provide human ESCs that can be employed in a practical manner for compound screening.

Due to the reason, an Embryonic Stem Cell Test (EST), which mirrors growth and differentiation, is an *in vitro* test system well-suited for the evaluation of the embryotoxic potential of substances (Evans & Kaufman, 1981; Martin, 1981; Smith, 1992; Spielmann et al., 1997; Ramalho-Santos et al., 2002; Wobus & Boheler, 2005). The regulation of the differentiation of mouse ES cells is controlled by critical transcription factors, such as STAT3 (Niwa et al., 1998), Oct4 (Nichols et al., 1998), and Nanog (Chambers et al., 2003): Previous studies have shown that embryonic stem (ES) cells require the expressions of different transcription factors to specify the stem cell state, and that these contribute to the generation of entirely different lineages on changing culture conditions and altering the expression levels of Oct4, a key determinant of the pluripotency for *in vitro* systems (Wobus & Boheler, 2005). In addition, these mES cell systems were reported to be very useful for measuring embryotoxicity (Seiler et al., 2004). Moreover, mouse embryonic stem cells can be routinely employed to screen chemical compounds for teratogenic effects (Scholz et al., 1999; Rohwedel et al., 2001; Vanparys, 2002).

In this chapter, we introduce the possibility of endothelial-like cells derived from mouse embryonic bodies as an in vitro vasculogenesis model and the usefulness of mouse ESCs for in vitro toxicity screening for 5-fluorouracil (5-FU), anti-angiogenesis agents.

5. Endothelial-like cells derived from mouse embryonic bodies as an in vitro vasculogenesis model

The formation of new blood vessels (vasculogenesis) during embryonic development is an important basic step. This process involves the differentiation of angioblasts from mesoderm during the early developmental stages and their organization from a primitive vascular network (Vittet et al., 1996), and is characterized by the expression of endothelial cell specific molecules during the formation of vascular structures in ES-derived embryoid bodies (EBs). A number of markers including vascular endothelial growth factor receptors-2 (Flk-1) (Yamaguchi et al., 1993), platelet endothelial cell adhesion molecule (PECAM) (Redick & Bautch, 1999), and vascular endothelial (VE) cadherin as markers of vasculogenesis have been reported (Dejana et al., 1999). Endothelial, endothelial-like cells, and endothelial precursor cells (EPCs) derived from stem cells have been explored to establish a toxicity screening system for endothelial-specific toxicants (Kim & von Recum, 2008). The feasibility of these screening systems depends on the differentiation processes of the ESCs used; guided differentiation into target cell types and accurate investigation of the mechanisms of endothelial toxicity are necessary. Recently, we reported that endothelial-like and endothelial cells derived from mouse ESCs using EGM medium and optimal protocols are more sensitive to 5-FU toxicity than undifferentiated endothelial cells as well as a mouse endothelial cell line (Kim & von Recum, 2008).

Doetschman et al (1985) were the first to show an *in vitro* mouse model based on the differentiated ES cells. When mES cells are grown in the absence of feeder cells and myeloid leukaemia inhibitory factor (LIF), they are able to differentiate spontaneously (Doetschman et al., 1985). EBs were formed by spheroid aggregation during post-implantation in embryonic tissues and maintained in conditioned medium containing a cocktail of vascular endothelial growth factor (VEGF), basic fibroblast growth factor (bFGF), interleukin-6 (IL-6) and erythropoietin (EPO) in 1% methylcellulose to induce the formation of EBs (Wang et al., 1992; Vittet et al., 1996). This model system, which forms a primitive vascular plexus, provides an attractive tool for investigating the mechanisms involved in vasculogenesis, i.e., angioblast differentiation, proliferation, migration, endothelial cell-cell adhesion, and vascular morphogenesis (Doetschman et al., 1985; Risau et al., 1988; Wang et al., 1992). In addition, co-cultures with stromal cells, even adherent monolayer cultures in the absence of LIF (Ying et al., 2003), have been used to differentiate mES cells *in vitro*. Moreover, the expression levels of markers of endothelial differentiation at the mRNA and protein levels differ according to endothelial cell differentiation and culture conditions (Doetschman et al., 1985; Risau et al., 1988; Vittet et al., 1996). For these reasons, the mechanism of endothelial cell differentiation involved in the regulation of vasculogenesis remains unclear. However, the exact molecular biological role of 5-FU on cell cycle regulation in the endothelial differentiation of mouse ESCs has not been fully explained yet.

Therefore, we demonstrated that the population of endothelial cells derived from mES cells and to apply this as a tool for the screening of agents with vascular developmental toxicity. Also, we investigated that the specific action of 5-FU on the endothelial differentiation of cells derived from mouse ESCs, finally, we demonstrated that the correlation between cell cycle regulation and endothelial differentiation in mouse ESCs exposed by 5-FU.

5.1 Materials and methods

5.1.1 Cell line and culture conditions

Mouse D_3 ES cells (ATCC Cat. No. CRL-1934, Rockville, MD, USA) were co-cultured with mitomycin C-treated mouse embryonic fibroblast (MEF) cells in high glucose DMEM (Gibco-BRL, Invitrogen, Carlsbad, CA) containing 15% fetal bovine serum (FBS; Hyclone, Ogden, UT), 1000 U/ ml of LIF/ESGRO (Chemicon, Temecula, CA), and basic ES medium components [50 U/ml of penicillin and 50 µg/ml streptomycin (Gibco-BRL, Invitrogen, Carlsbad, CA), 1% non-essential amino acids (Gibco-BRL, Invitrogen, Carlsbad, CA) and 0.1 mM β-mercaptoethanol (Gibco-BRL, Invitrogen, Carlsbad, CA)]. Mouse endothelial cells (C166) (ATCC Cat. No. CRL-2581, Rockville, MD), used as a control for toxicity testing, were cultured in endothelial cell basal medium-2 (EBM-2) containing 5% FBS and cytokine cocktail. The hanging drops method (20 µl per drop; $1x10^5$ cells ml^{-1}) was used to induced differentiation as described by Heuer and colleagues (Heuer et al., 1993) with minor modifications. After incubation for 3 more days, EBs were transferred to gelatin-coated wells of Chamber slides (Nunc, Denmark) or 60mm dishes to allow attachment. To promote endothelial cell differentiation, 3-day-old EBs were placed in DMEM containing 10% FBS and medium consisting of EBM-2, 5% FBS, growth factor cocktail, and ascorbic acid (EGM2-MV Bullet Kit; Clonetics/BioWhittaker, Walkersville, MD).

5.1.2 Cell viability and proliferation analysis

The viability and proliferative activity of ES cells were analyzed using the 3-(4,5-dimethylthiazol-2yl)-2,5,-diphenyl tetrazolium bromide (MTT) assay and by 5-bromo-2′-deoxyuridine (BrdU) incorporation for 10 days, respectively. For the MTT assay, mEBs were hanging drop cultured for 2 days and then grown in standard culture medium (DMEM containing 10% FBS) for 24 hours to allow attachment. 20 µl of MTT (5mg/ml) was then added to 200 µl of the culture medium on days 3, 4, 6, 8, and 10, followed by incubation at 37°C for 4 hr. After incubation, the MTT solution was carefully removed and 150 µl of DMSO (Sigma, St Louis, MO) was added to each well. The plates were then shaken on a plate mixer until the crystals dissolved. The absorbance of the resulting colored solution was measured at 570 nm in a Genios luminometer (TECAN, Austria) at a reference wavelength of 630nm. BrdU detection kits (Roche Molecular Biochemicals, Indianapolis, IN) were used to measure BrdU incorporation. For BrdU assays, mEBs were hanging drop cultured for 2 days and then grown in standard culture medium (DMEM containing 10% FBS) for 24 hours to allow attachment. On days 3, 4, 6, 8, and 10, 10 µl of 100 µM BrdU was added to each well, and the cells were incubated for 4 h. After removing the medium, the cells were fixed and DNA was denatured using FixDenat reagent (Roche Applied Sciences) for 30 min. The reagent was then removed, anti-BrdU-POD solution was added and the plates were incubated for 90 min at room temperature. The cells were then washed three times with washing solution, after which 100 µl of substrate solution was added and the absorbance was measured at 370 and 490 nm using a Genios luminometer (TECAN, Austria).

5.1.3 RNA isolation and reverse transcription-polymerase chain reaction (RT-PCR)

Cells were directly sorted into tubes containing Trizol (Gibco-BRL, Invitrogen, Carlsbad, CA) and mRNA was extracted according to the manufacturer's protocol. The isolated RNAs were quantified using a spectrophotometer (SmartSpec 3000, Bio-Rad). First-strand cDNA was synthesized from 2 µg of total RNA using an oligo (dT) primer and a SuperScript First-

Strand Synthesis System for RT-PCR (Invitrogen, Carlsbad, CA), according to the manufacturer's instructions. First-strand cDNAs were amplified in a final volume of 25 µl containing 0.5 U Taq DNA polymerase (TaKaRa Biotechnology, Korea) and 10 pmol of each target primer. PCR conditions were as follows: 5 minutes at 94°C, 30 amplification cycles (denaturation at 94°C for 1 minute, annealing at 55°C or 60°C for 1 minute, and extension at 72°C for 1 minute), followed by a final extension at 72°C for 5 minutes. The PCR primers and the size of the amplified products are shown in Table 1. The amplified products were separated on 1.5% agarose gels and visualized by ethidium bromide staining. cDNA samples were adjusted to yield equal GAPDH amplifications.

Gene	Sequences	Size (bp)
FLk-1	F: 5′-CAGCTTCCAAGTGGCTAAGG-3′ R: 5′-CAGAGCAACACACCGAAAGA-3′	264
PECAM	F: 5′- GCCTGGAGAGGTTGTCAGAG-3′ R: 5′- GGTGCTGAGACCTGCTTTTC-3′	357
VE-Cadherin	F: 5′- ACCGGATGACCAAGTACAGC-3′ R: 5′- TTCTGGTTTTCTGGCAGCTT-3′	292
GAPDH	F: 5′- TGTTCCTACCCCCAATGTGT-3′ R: 5′- TGTGAGGGAGATGCTCAGTG-3′	396

Table 1. Sequences of oligonucleotide primers used for RT-PCR analysis

5.1.4 Immunocytochemistry

After inducing differentiation, the cells were exposed to 5-FU with/without probucol for 24 hours and fixed with freshly prepared MeOH/DMSO (4:1) overnight at 4°C. Cells were blocked with blocking solution containing 1% BSA and 0.1% Tween 20 for 30 min, and then incubated with rabbit anti-mouse FLK-1 (1:100) (Santa Cruz Biotechnology, Inc., Santa Cruz, CA), rat anti-mouse PECAM (1:100) (MEC 13.3, Santa Cruz), or goat anti-mouse VE-cadherin (1:100) (Santa Cruz), or rabbit anti-mouse PCNA (1:100) (Santa Cruze, Biotechnology, Inc) at 4°C overnight. After washing, cells were incubated with goat anti-rabbit IgG-TRITC (1:100) (Chemicon, Temecula, CA), goat anti-rat IgG-FITC (1:100) (Santa Cruz), or donkey anti-goat IgG-FITC (1:200) (Santa Cruz) as secondary antibodies, respectively. Stained slides were embedded in 30% Mowiol (Calbiochem-Novabiochem, Schwalbach, Germany). Images were obtained and analyzed using a Bio-Rad confocal microscope (Radiance 2000 FCMP, Bio-Rad, USA)

5.1.5 Flow cytometry

Cultured cells were harvested using cell dissociation buffer (Sigma, St Louis, MO). Cells were resuspended at 10^6 cells/100 µl in suspension buffer, and then were incubated with 1 µg/100 µl of PE anti-mouse FLK-1 (Avas 12a1, PharMingen) and FITC anti-mouse PECAM (MEC 13.3, PharMingen), or (FITC) rat anti-mouse VE-cadherin (11D4.1, PharMingen) for 30 minutes at 4°C. Negative controls were incubated for 30 min at 4°C with fluorochrome-labeled irrelevant isotype control antibodies: 1 µg/100 µl PE rat IgG2a′ (PharMingen, San Diego, CA), 1 µg/100µl FITC rat IgG2a′ (PharMingen), or 1 µg/ 100 µl anti-rat FITC-conjugated IgG (Santa Cruz). In order to analyse how Oct-4 expression varies in mouse

ESCs through different stages of endothelial cell differentiation, endothelial differentiation induced cells for 0, 4, 7 and 10 days were harvested using cell dissociation buffer (Sigma, St Louis, MO). To analyse PCNA expression, the cells were exposed to 5-FU with/without probucol for 24 hours at day 9. After 24 hours incubation, control cells and treated cells were harvested using cell dissociation buffer (Sigma, St Louis, MO). Cells were re-suspended at 10^6 cells/100 µl in suspension buffer and then incubated with 1µg/100 µl of rabbit anti-mouse Oct-4 (Santa Cruz, Biotechnology, Inc), or rabbit anti-mouse PCNA (1:100) (Santa Cruze, Biotechnology, Inc) for 1 hour at 4°C. Negative controls were incubated for 1 hour at 4°C with fluorochrome labelled irrelevant isotype control antibodies: 1 µg/ 100 µl goat anti-rabbit FITC-conjugated IgG (Chemicon) or goat anti-rabbit IgG-TRITC (1:100) (Chemicon, Temecula, CA) for primary antibodies, respectively. After staining, cells were analyzed without fixation using a FACS Calibur flow cytometer (Becton Dickinson, MA) using 5 µg/ml of propidium iodide (Sigma) to exclude dead cells. Data was analyzed using CellQuest software (Becton Dickinson, MA).

For cell cycle analysis in differentiation into endothelial cell of mouse ESCs, the treated cells were trypsinised and fixed in cooled 70% ethanol at 4°C. The cells were then incubated in 0.3 ml of DNA staining solution (100 µg/ml PI, 1 mg/ml RNase A (DNase-free) in PBS). The cells were then transferred into D-Hank's solution. The cell suspension was stored on ice in a dark room for a minimum of 30 min and analysed within 2 hours. Data analysis was carried out using CellQuest software (Becton Dickinson, MA).

5.1.6 Capillary tube formation by endothelial cells on matrigel

To induce tube formation on matrigel, mES cells were grown in EGM-2 medium for 10 days and then prepared as a single-cell suspension. 5×10^4 ES cells were then plated on Matrigel (Becton Dickinson, MA), incubated at 37°C, and analyzed using a phase contrast microscope (Nikon, Eclipse TE 2000-U, Japan).

5.1.7 Cytotoxicity analysis

The cytotoxic effects of 5-Fluorouracil on mES cells and mouse endothelial cell (C166) were analyzed using a validated embryonic stem cell test protocol and MTT assay (Spielmann et al., 1997; Scholz et al., 1999). Briefly, 1000 cells were seeded into each well of a 96-well microtitre plate and grown in the presence of a concentration range of 5-FU and probucol. A negative control containing solvent diluted in medium was also included. At day 9, the cells were exposed to 5-FU (10 µM) with/without probucol (50 µM) in a total volume of 200 µl for 24 hours. The 5-FU and probucol were dissolved in cell culture medium and ethanol, respectively. The final ethanol concentration in the wells was 0.1%. The controls were incubated with equal volumes of drug solvents to avoid changes that could be due to solvent. 20 µl of MTT (5mg/ml) was added to 200 µl culture medium on day 10, followed by incubation at 37°C for 4 hrs. After incubation, the MTT solution was carefully removed and 150 µl of DMSO (Sigma, St Louis, MO) was added to each well. The plates were shaken on a plate mixer until all crystals had dissolved. The absorbance of the resulting coloured solution was measured at 570 nm with a Genios luminometer (TECAN, Austria) at a reference wavelength of 630nm. Cytotoxicity was expressed as a percentage of cells surviving, relative to untreated cultures, and the concentration required to inhibit cell growth by 50% (IC_{50}) was calculated. Each experiment was performed using six replicates for each drug concentration and repeated in triplicate.

5.1.8 Microarray

For cDNA microarray analysis, differentiated mouse ESCs treated with/without 5-FU (10 μM) for 24 hours were collected. Their total RNA was extracted using Trizol (Gibco-BRL, Invitrogen, USA) and mRNA was extracted according to the manufacturer's protocol. The quantity and quality of total RNA and amplified RNA were assessed by using a Bioanalyzer 2100 (Agilent Technologies). The Applied Biosystems Mouse Genome Survey Microarray contains 32,996 60-mer oligonucleotide probes, representing 32,181 individual mouse genes. Digoxigenin-UTP labelled cRNA was generated and linearly amplified from 2 μg of total RNA using Applied Biosystems Chemiluminescent RT-IVT Labelling Kit v.2.0 and manufacturer's protocol (Applied Biosystems). Array hybridization (five arrays per sample), chemiluminescence detection, image acquisition and analysis were performed using Applied Biosystems Chemiluminescence Detection Kit (Applied Biosystems) and Applied Biosystems 1700 Chemiluminescent Microarray Analyzer (Applied Biosystems), following the manufacturer's protocol. Images were auto-gridded, then spot and spatially normalised. Chemiluminescent signals were quantified, corrected for background, and the final images and feature data were processed using the Applied Biosystems 1700 Chemiluminescent Microarray Analyzer software v1.1. Data and images were collected through an automated process for each microarray using the 1700 analyzer. A total of 10 arrays were run for the two groups (five technical replicates for each group). A global median normalisation, which normalises signal intensities across all microarrays to achieve the same median signal intensities for each array, was performed on the Applied Biosystems data sets. For Applied Biosystems arrays, the detection threshold was set as $S/N > 3$ with a quality flag < 100. Correlation and coefficient of variation (CV) analyses were performed using Matlab® software (Mathworks, Natick, MA). Differential expression analysis was done using two different statistical methods: (1) ANOVA analysis was performed using Avadis software. Differentially expressed genes between control and 5-FU (10 μM) treated groups were determined based on the following criteria: (a) $p < 0.001$ in ANOVA analysis; (b) average change between control and 5-FU (10 μM) treated groups > 2 fold; (c) detectable in more than 50% samples. (2) Significance Analysis of Microarray (SAM: http://www-stat.stanford.edu/~tibs/SAM), a supervised learning statistical software that performs a modified *t*-test to identify genes with significant changes in expression, and uses permutations to estimate the false discovery rate (FDR). Hierarchical clustering of log ratios was performed using the software (http://rana.lbl.gov/EissenSoftware.htm) Cluster and Treeview; Euclidean correlation, median centring and complete linkage were applied in all clustering applications.

5.1.9 Western blot analysis

Control cells and treated Cells were lysed in RIPA buffer (1% NP-40, 150 mM NaCl, 0.05% DOC, 1% SDS, 50 mM Tris) containing protease inhibitor for 1 h at 4°C. The supernatant was separated by centrifugation, and protein concentration was determined with a Bradford protein assay kit II (Bio-rad). Proteins (25 μg/well) denatured with Laemmli sample buffer (Sigma) were separated by 10% SDS-polyacrylamide gel (Bio-Rad) under a constant current of 50 mV. Proteins were transferred onto nitrocellulose membranes (0.45 mm, Amersham Life Sciences). The membranes were blocked with a 5% BSA solution for 3 hrs, washed with PBS containing 0.2% Tween 20, then incubated with the primary antibody overnight at 4°C. Human specific antibodies against CDK-2, CDK-4, Cyclin D1, Cyclin E, $p21^{WAF1/CIP1}$, p27 ^{Kip1}, p53, and β-actin from Santa Cruz Biotechnology (Santa Cruz Biotechnology) were used to probe the separate membranes. The immunoreaction was continued with the secondary goat anti-rabbit horseradish-peroxidase conjugated antibody after washing for 2 hours at room temperature. The specific protein bands were detected by enhanced chemiluminescence (Pierce), with X-Omat AR films (Kodak). As a protein loading control, parallel gels were subjected to Western blot analysis using a β-actin antibody (Santa Cruz Biotechnology).

5.1.10 Statistical analysis

All results are expressed as percentages of untreated control values or as the means ± SD of three independent experiments, each with six replicates. Statistical significance was determined using the Student's *t*-test for paired data. A *P* value of < 0.05 was regarded as significant, and IC_{50} values were calculated using Sigmaplot version 9.0.

5.2 Results

5.2.1 Expressions of endothelial markers at the mRNA and protein levels during the early stages of differentiation

We analyzed the mRNA expression levels of endothelial markers, i.e., FLK-1, PECAM, and VE-Cadherin, during different culture conditions and differentiation stages within 7 days of plating embryoid bodies by RT-PCR (Figure 3A). In case of the expression of the FLK-1 gene, no difference was found between DMEM containing FBS and EGM-2 media.

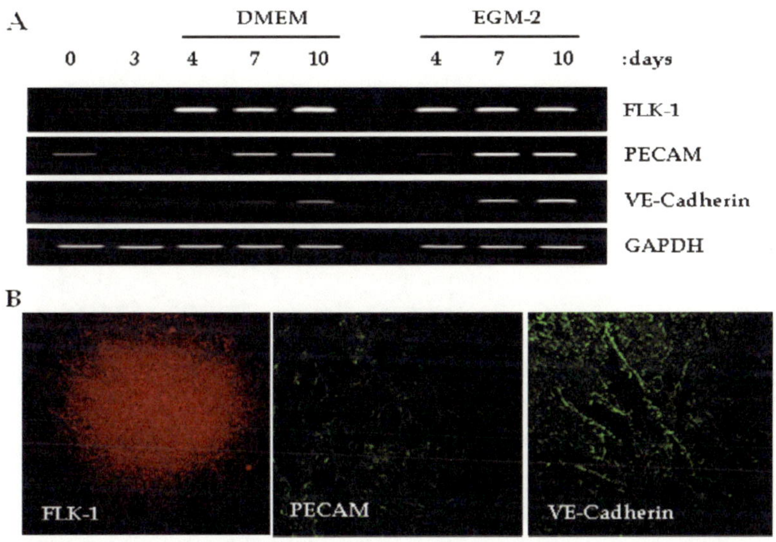

Fig. 3. Expression patterns of endothelial cell-specific markers at the mRNA and protein levels in mES-derived embryoid bodies. Gene expression in ES-derived EBs by RT-PCR (A). Expressions of FLK-1, PECAM, and VE-cadherin in differentiated mES cells at day 10 by immunofluorescence (B) (× 40)

However, the expressions of PECAM and VE-Cadherin mRNA were higher in EGM-2 than in DMEM containing FBS, and their expressions in EGM-2 medium were higher during the early stages. Expression of endothelial cell differentiation markers at the mRNA and protein levels were similar with respect to time course (data not shown). We carried out immunocytochemistry for endothelial cell markers in mEB cells from EGM-2 medium (Figure 3B), and observed capillary-like structures on day 10. These were stained with specific antibodies for FLK-1, PECAM and VE-Cadherin. The immunoreactivity of each marker was strong and was observed throughout EBs cells.

In addition, we performed flow cytometry analysis for the expressions of FLK-1, PECAM, and VE-Cadherin with time (Figure 4). Cells which are positive for these markers were observed to increase with time. For cells grown in DMEM containing 10% FBS medium and EGM-2 medium for 10 days the percentages expressing these markers were; FLK-1 4.0% and 5.3%, PECAM 10.9% and 44.5%, and VE-Cadherin 5.2% and 6.4%, respectively. These results demonstrate that the expressions of endothelial cell markers at the mRNA and protein levels were greater in cells grown in EGM-2 medium than in DMEM containing 10% FBS medium.

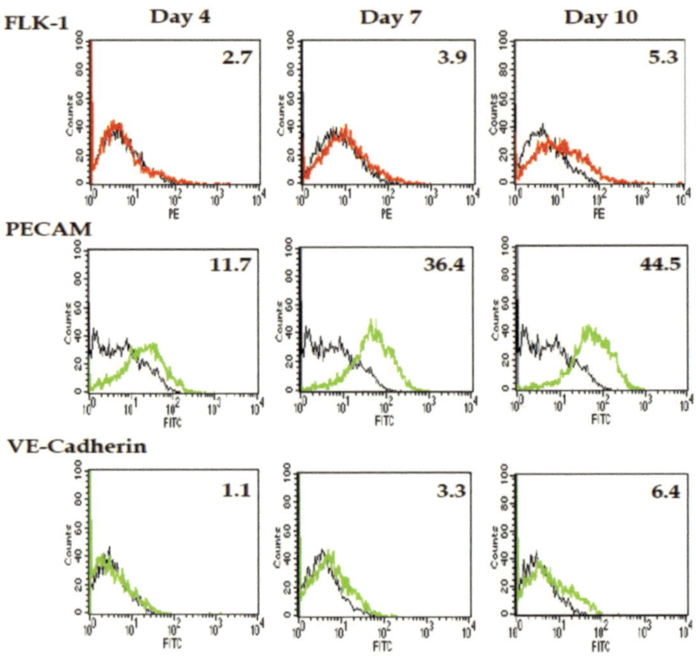

Fig. 4. Kinetics of the expressions of FLK-1, PECAM and VE-cadherin during the differentiation of mES cells according to flow cytometric analysis. Numbers indicate the percentages of target antibody-positive cells

5.2.2 Formation of capillary structures by differentiated mES cells on matrigel

Mouse ES cells cultured in EGM-2 medium for 10 days differentiated into the endothelial cell lineage. Thus, we used gelatin coated dishes and a Matrigel system to confirm whether the mES cells cultured EGM-2 medium effectively construct capillary structures. We cultured mES derived endothelial cells in EGM-2 medium on gelatin coated dishes and Matrigel. After 5 days, we found that mES cells on the gelatin coated dishes could not construct capillary-like structures (Figure 5A), on the other hand, mES cells on Matrigel rapidly formed capillary-like structures, which resembled those formed by HUVEC and other endothelial cell populations (Figure 5B).

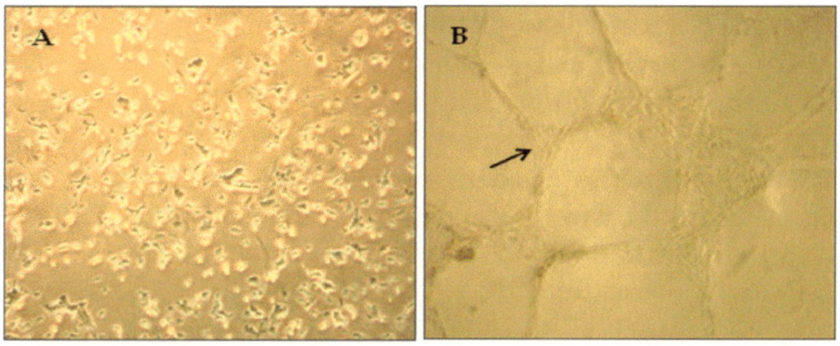

Fig. 5. Formation patterns of vascular-like structures of mES cells. Morphologies of mouse ES cell-derived endothelial cells on a gelatin-coated plate (A) and a matrigel-coated plate (B) (×100). Arrow indicates vascular-like structure derived from mouse ESCs

5.2.3 Differentiated mouse ES cells were more sensitive than mouse endothelial cells (C166) to 5-Fluorouracil

In the sensitivity to expose 5-FU in mouse ES cells, the respective concentrations of 5-FU for 50% reduction (IC_{50}) in early stage (day 3) and late stage (day 9) assays for 24hr were 126 uM and 7.9 uM, respectively (data not shown). This data suggests that the sensitivity of differentiated endothelial cells to toxicity of 5-FU are more sensitive than undifferentiated endothelial cells. In order to confirm the usefulness of the vasculogenesis model produced from differentiated mouse ESCs using EGM-2 medium, we performed vasculogenesis cytotoxicity assays, by treating cells with 5-fluorouracil, a strong inhibitor of vessel

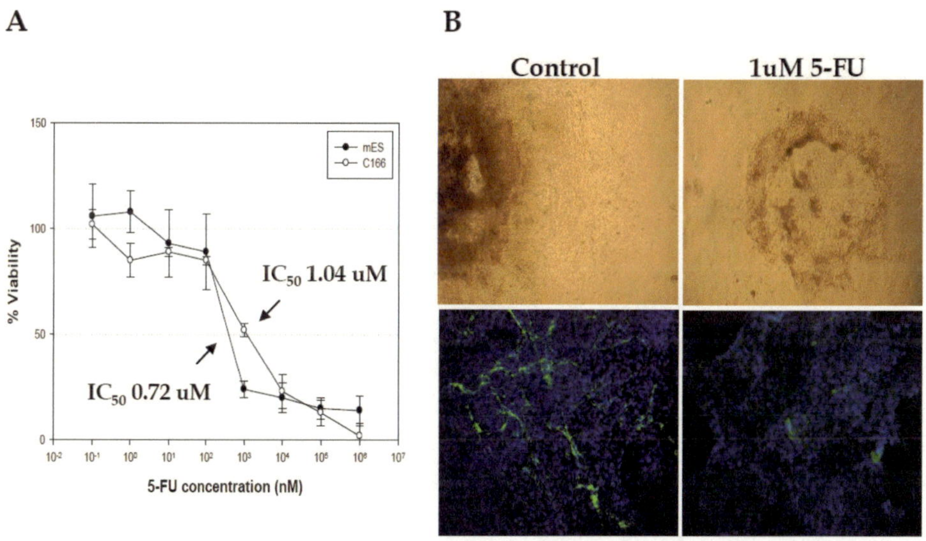

Fig. 6. Illustration of the effects of 5-Fluorouracil on the 72-hour survival of mouse embryonic stem cells and C166 mouse endothelial cells as determined by MTT assays (A). The inhibitory concentrations (IC_{50}) of 5-Fluorouracil in mES cells and C166 cells were 0.72 μM and 1.04 μM, respectively. The morphology and PECAM expression of mouse ES cells exposed to 5- Fluorouracil (1 μM) (B). The expression of PECAM was decreased by 5-fluorouracil treatment in mouse ES cells

formation, to both mouse ES cells and C166 cells. Generally, the growth rates of mouse ES and C166 cells were inhibited in a dose-dependent manner by 5-fluorouracil (Figure 6A). However, the concentrations 5-fluorouracil that caused a 50% reduction in mouse ES and C166 cells were 0.72 and 1.04 uM, respectively. In addition, the expression of PECAM significantly decreased in mouse ES derived endothelial cells which are exposed 5-FU (Figure 6B). From these results, we confirm that differentiated mouse ES cells are sensitive to 5-fluorouracil comparing traditional screening systems (C166) for cell toxicity, so that endothelial cells derived mES can use as an in vitro model for vasculogenesis and toxicity screening.

5.2.4 Effect of 5-FU on the endothelial differentiation of cells derived from mouse ESCs

To study the role of 5-FU in endothelial differentiation, mouse ESCs were differentiated into endothelial precursor cells and treated with 5-FU (10 μM). Oct-4 expression and cell viability were analysed. The expression of Oct-4 gradually decreased, with 100%, 90.6%, 81.3%, 50.5%

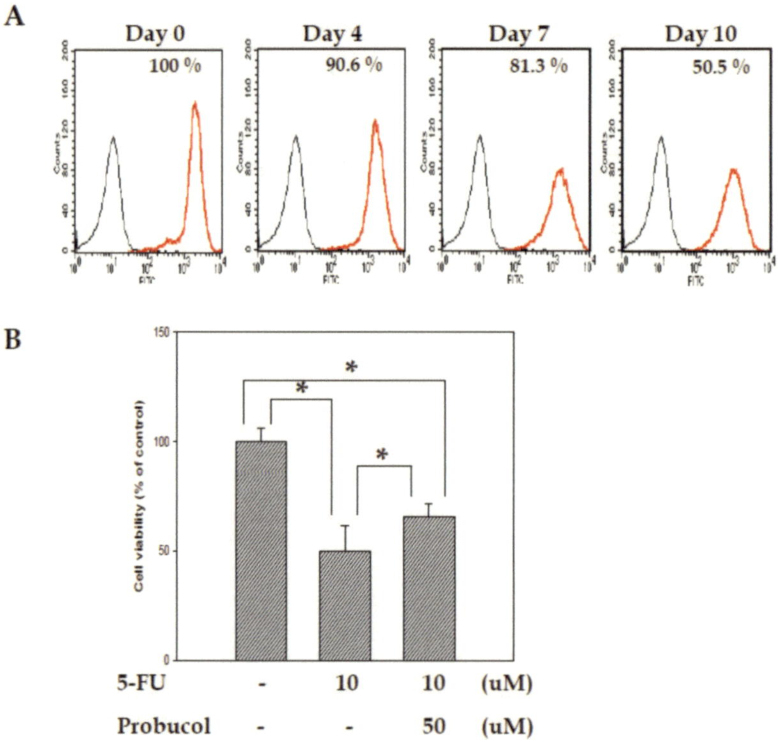

Fig. 7. The expression of Oct4 and the cytotoxicity of 5-Fluorouracil in endothelial precursor cells, derived from mouse embryonic stem cells. The expression of Oct4 gradually decreased during endothelial differentiation (A). After inducing endothelial differentiation for 9 days with EGM-2 medium, the cells were exposed to 5-fluorouracil (10 μM) with/without probucol (50 μM) for 24 hrs. Analysis of the cell viability of the endothelial differentiated cells exposed to 5-fluorouracil were assessed by MTT assays (B). The experiments were performed in 3 independent runs (n = 6). Standard error bars are shown. Significance was tested by the student *t-test* (* $p < 0.01$). Numbers indicate the percentages of target antibody-positive cells

expression on differentiation days 0, 4, 7, and 10, respectively (Figure 7A). In a previous study, we found that the percentage of PECAM expression increased by up to 44.5% in mouse ES-derived endothelial precursor cells, grown in EGM-2 medium for 10 days. There is a strong correlation between the decreased Oct-4 expression and increased PECAM expression during the endothelial differentiation of cells derived from mouse ESCs. Next, we examined the viability of endothelial precursor cells when they were exposed to 5-FU (10 μM), using MTT assays. As shown in the supplemental data, cells exposed to various concentrations of 5-FU and probucol for 24 hours displayed a decrease in cell viability in a concentration dependent manner. The IC_{50} of 5-FU and the concentration of probucol needed to avoid affecting cell viability were 10 μM, 50 μM, respectively. The viability was significantly decreased to 49.8% of the control in the endothelial differentiated cells exposed to 5-FU (10 μM) for 24 hours. The viability was significantly recovered to 65.7% in cells exposed to 5-FU (10 μM) combined with probucol (50 μM) ($P < 0.01$) (Figure 7B).

5.2.5 5-FU inhibits cell proliferation in endothelial differentiated cells derived from mouse ESCs.

The morphology of mouse ES-derived endothelial precursor cells after treatment with the IC_{50} values of 5-FU (10 μM) for 24 hours was changed and detached, compared to the

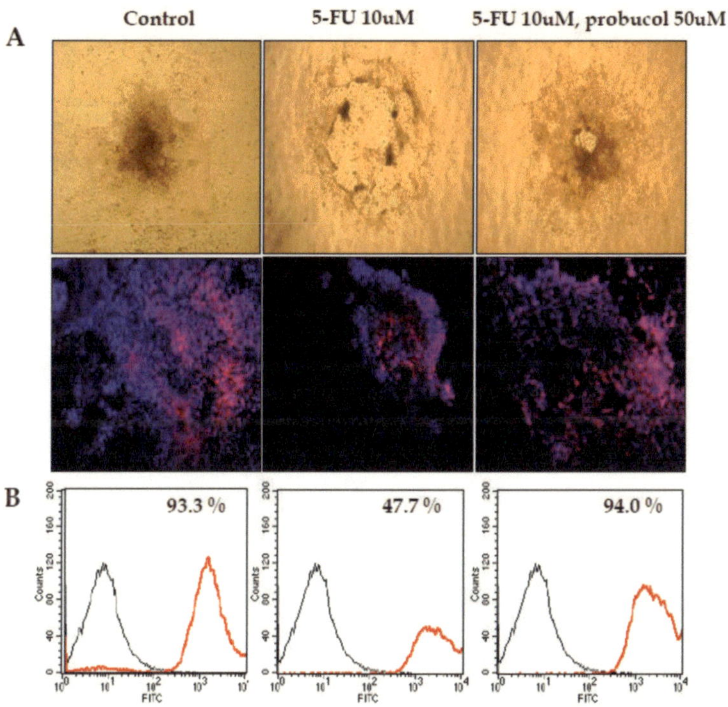

Fig. 8. Anti-proliferative affect of 5-fluorouracil in endothelial precursor cells. Morphology (A, top panel) and PCNA expression (A, bottom panel) of control and cells exposed to 5-fluorouracil (10 μM) with/without probucol (50 μM) for 24hrs. Cell nuclei were stained with DAPI. The percentages of cells with PCNA expression were analysed by FACS (B). Numbers indicate the percentages of target antibody-positive cells. Original Magnification: × 40 for (A)

control cells which grew as confluent aggregates with rounded and polygonal cell morphology. However, probucol (50 μM) treatment to the 5-FU treatment groups for 24 hours induced a morphology similar to that of the control cells (Figure 8A, top panel). In order to investigate the effects of 5-FU on the proliferation of endothelial precursor cells derived from mouse ESCs, PCNA expression was assessed by immunocytochemistry and FACS analysis. After inducing endothelial differentiation for 9 days, we observed that the endothelial precursor cells exposed to 5-FU (10 μM) for 24 hours had decreased PCNA expression, whilst the probucol (50 μM) treated group had recovered its PCNA expression (Figure 8A, bottom panel). These results correlated with the PCNA expression of the endothelial precursor cells exposed to 5-FU (47.7%), as shown by FACS analysis (Figure 8B).

5.2.6 Expression of PECAM in endothelial differentiated cells is down-regulated by 5-FU

To test whether the treatment of mouse ES-derived endothelial precursor cells with 5-FU can influence endothelial differentiation through expression of endothelial specific genes, the expression levels of PECAM were analysed by immunocytochemistry and RT-PCR. The expression of PECAM was dramatically decreased in the 5-FU treatment group, compared to the control group. However, the expression of PECAM was maintained in the probucol treatment group (Figure 9). These findings demonstrate that, in accordance with the morphological differentiation analysis, 5-FU decreases endothelial-specific mRNA levels and so inhibits the expression of genes involved in endothelial differentiation.

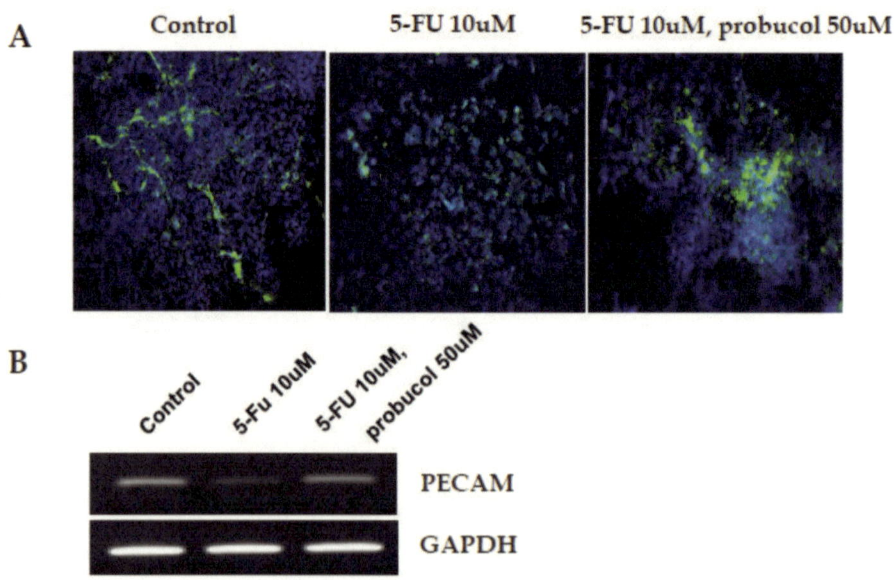

Fig. 9. PECAM expression of endothelial precursor cells, derived from mouse embryonic stem cells. Expression of PECAM in control and cells exposed to 5-fluorouracil (10 μM) with/without probucol (50 μM) for 24hrs were analysed by immunocytochemistry. Cell nuclei were stained with DAPI (A). mRNA levels of PECAM and GAPDH were evaluated by RT-PCR (B). GAPDH used as internal standard. Original Magnification: × 40 for (A)

5.2.7 Gene profiling of mouse embryonic stem cells exposed to 5-FU using Microarray analysis

In order to determine gene expression changes in cells exposed to 5-FU (10 μM), the RNA contained in both the control cells, and the cells exposed to 5-FU treatment for 24 hours at differentiation day 9 was collected. As shown a Fig. 4, a total of 11,668 genes out of 32,996

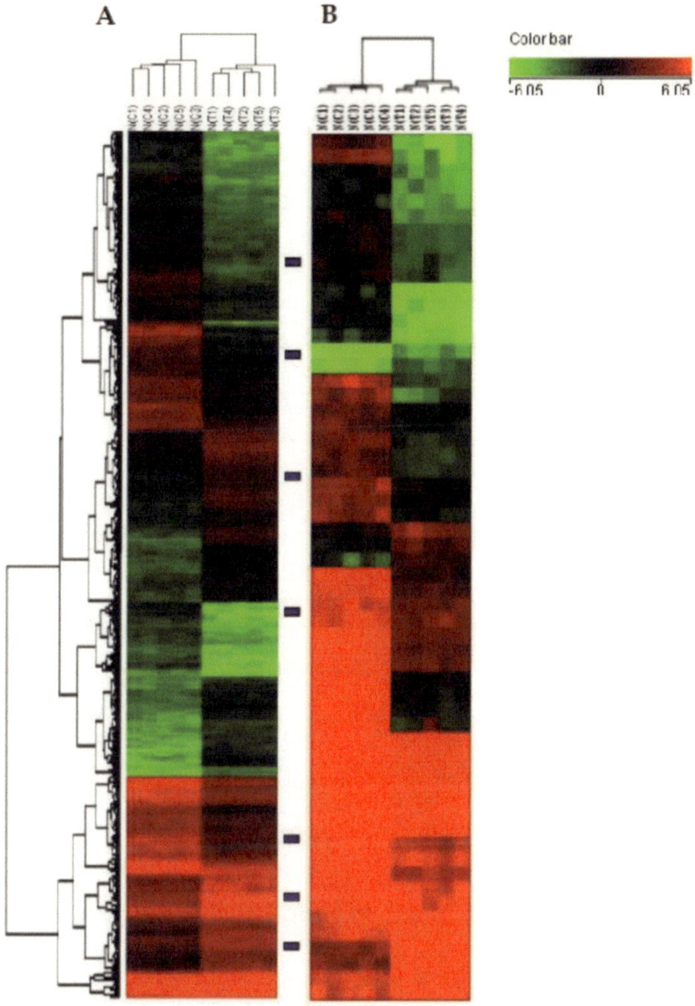

Fig. 10. Gene expression profiling of endothelial precursor control and cells exposed to 5-fluorouracil using microarray analysis. Hierarchical clustering analysis was conducted using control (n=5) and 5-FU treated samples (n=5). The black bars on right side of *A* illustrate the location of clusters shown. The dendrogram in *B* shows the samples identified as being in the cell cycle category, between control and 5-FU treated cells. The intensity of red and green colour is proportional to the relative up-regulation (red) or down-regulation (green) of gene expression in the differentiated samples, compared to that in the undifferentiated reference. Gene names and accession numbers are from Unigene (http://www.ncbi.nlm.nih.gov/entrez/query.fcgi?db=unigene)

cDNAs were selected as differentially expressed genes by the ANOVA test ($p < 0.001$). A hierarchical cluster analysis yielded major clusters in the 11,668 expressed genes (Figure 10). These gene expression patterns were classified into functional groups, based on their biological process as defined by the Gene Ontology (GO) annotation system. Most of the observed genes were related to physiological processes, including apoptosis, cell cycle, developmental processes, and signal transduction in all clusters (Table 2). These results suggest that 5-FU treatment affects the expression of numerous genes via the alteration of several processes. Notably, expression of 58 genes out of the 1,439 genes in the cell cycle category was modulated by more than 2-fold, between the control cells and endothelial differentiated cells exposed to 5-fluorouracil (data not shown).

Function (Panther classcification system)	Number
Angiogenesis	4
Apoptosis	30
Cell adhesion	22
Cell cycle	58
Cell proliferation and differentiation	9
Cell structure and motility	36
Developmental processes	80
Homeostasis	7
Immunity and defence	45
Signal transduction	126
Cell adhesion molecule	20
Cell junction protein	8
The others	994/1439

Table 2. Gene content list of the AB 1700 mouse chip

5.2.8 5-FU induces arrest of G1/S phase in endothelial-like cells derived from mouse ESCs

To confirm the effects of 5-fluorouracil (10 μM) with and without probucol (50 μM) in the cell cycle, the cell cycle distribution was analysed by flow cytometry (Figure 11). Generally, the frequencies of G0/G1 phase and S-phase were 62.7% and 33.5% in mouse ESCs, respectively. In contrast, mouse ES-derived endothelial precursor cells exposed to 5-FU (10 μM) for 24 hours showed a decrease to 31.1% in G0/G1 phase and an increase to 60.6% in S phase. Treatment with 50 μM probucol for 24 hours showed 35.0% of cells in G0/G1 phase and 62.8% in S-phase (Figure 11A). These data suggest that 5-FU arrests cells at the G1/S phase boundary in endothelial precursor cells, derived from mouse ESCs. This was followed by decreased proliferation of mouse ES-derived endothelial precursor cells, similar to that reported previously with other cell types (Vittet et al., 1996).To determine whether the expression levels of cell cycle related proteins were changed, we analysed Cyclins, CKDs,

CKDIs, and p53 expression after treatment with 5-FU (10 μM), with and without probucol (50 μM), by Western blot analysis. The expression of Cyclin E, CDK2, p21$^{WAF1/CIP1}$, and p53 was up-regulated in ES-derived endothelial precursor cells exposed to 5-FU, compared to those in the control group and cells treated with 5-FU (10 μM) and probucol (50 μM) (Figure 11B). Although the expression of cyclin D1 remained constant in control cells and mouse ES-derived endothelial precursor cells exposed to 5-FU, they were remarkably decreased by probucol treatment. Otherwise, there were no differences in the expression of CKD4 and *p27*^{Kip1}, the tumour suppressor and inhibitors of Cyclin E / CDK2 in mouse ES-derived endothelial precursor cells, according to 5-FU and probucol treatments (Figure 11B). From these results, we suggest that 5-FU might inhibit G1-related Cyclin/CDK activities through the augmentation of p21$^{WAF1/CIP1}$ expression and by binding to cyclin D/CDK complexes.

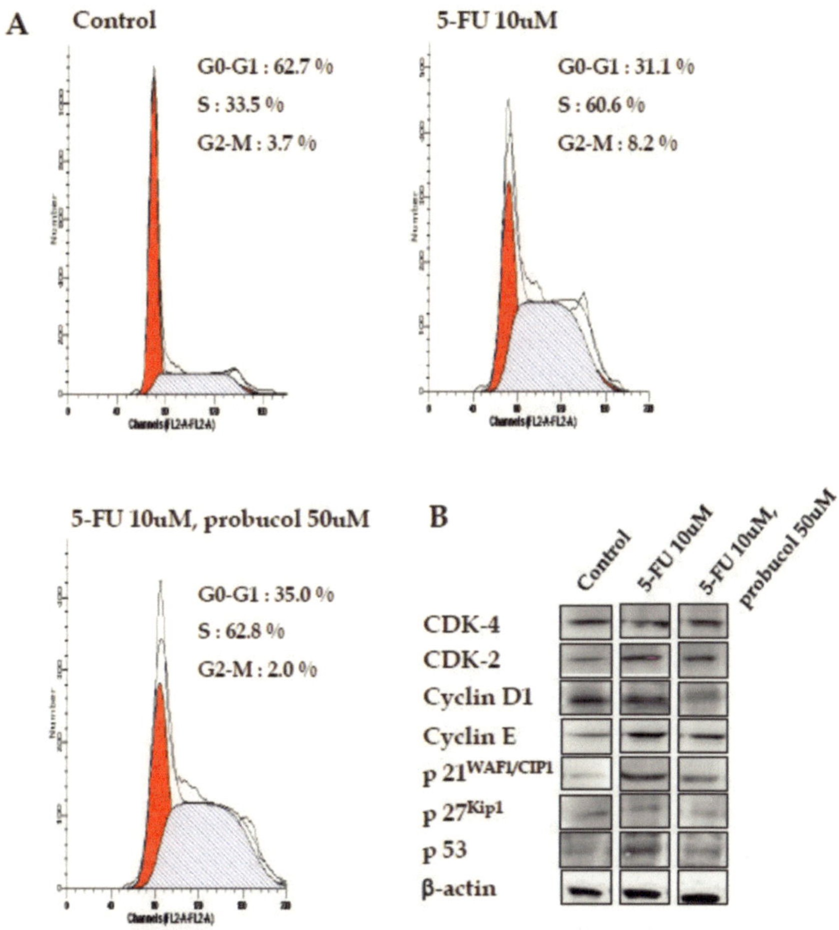

Fig. 11. Cell cycle arrest in endothelial precursor cells after 5-Fluorouracil exposure. DNA distribution histogram, using PI labelling (x-axis) and total number of cells in each channel (y-axis) in control and cells treated with 5-fluorouracil (10μM) with/without probucol (50 μM) (A). Western blot analysis of the expression of G1/S phase-related Cyclins, CDKs, and CDKIs in control and cells exposed to 5-fluorouracil (10μM) with/without probucol (50 μM) (B)

6. Conclusion

In toxicological research, the development of alternative *in vitro* toxicity screening systems to replace *in vivo* screening methods using animal experiments and conventional screening systems is important. Although applications using primary cells derived from rodent embryo or tissue-specific cell lines originating from humans have been tried, the validation of cytotoxicity during cellular biological processing must be improved in order to establish new and alternative *in vitro* screening methods, and to increase the efficiency of toxicological analysis through these methods (Tiffany-Castiglioni et al., 1999). So, toxicity screening using stem cells is highlighted as an alternative cell source for *in vitro* toxicity screening because of several limitations of traditional *in vitro* assays using primary cells or cell lines. For examples, they couldn't demonstrating the biological process involved in a toxic response to xenobiotics comparing to in vivo toxicity testing using animal model. In comparison to *in vivo* studies, the screening system using embryonic stem cell is highly accurate at predicting cellular toxicity, and outperforms classical assays, such as, the fetal limb micromass and post-implantation whole rat embryos culture assays (Scholz et al., 1999).

Embryonic stem cell testing (EST) using the EB system provides a useful tool for analyzing the embryotoxic effects of chemical compounds (Spielmann et al., 1997; Scholz et al., 1999; Huuskonen, 2005). Whole embryo culture showed the best concordance between in vivo classification and in vitro test results with 80% correct classifications versus 78% for EST and 71% for limb bud micromass. Moreover, strong embryotoxicants showed a predictivity of 100% in each of the test systems (Genschow et al., 2002). The powerful advantages of ESCs when they are applied to toxicology come as a result of their unique properties compared to primary cells: self-renewal, plasticity to generate various cells types, and that they are a readily available alternative source to replace primary cells. Therefore, stem cell based screening systems for toxicants offer a very promising technology; it is possible to obtain large numbers of cells for consistent analysis and the study at the different stages of differentiation. In addition, toxicity screenings using ESCs have been validated as a reliable source for *in vitro* developmental toxicology studies (Rohwedel et al., 2001).

In the present study, we developed a vasculogenesis model for the effective differentiation of mES cells into the endothelial lineage using EGM-2 medium. In addition, the availability of differentiated mouse embryonic bodies was confirmed by examining vascular toxicity using 5-Fluorouracil. Since Hirashima and colleagues reported that mES cells are important for studies for murine development, several systems for mES cell differentiation have been reported (Hirashima et al., 1999; Feraud & Vittet, 2003; Davila et al., 2004; Wobus & Boheler, 2005). The majority of reports issued during the past decade have focused on murine ES cells derived embryoid body (EB) formation assays. Differentiation system using embryoid bodies that have the potential to generate various embryonic cell lineages spontaneously induce differentiation (Doetschman et al., 1985; Risau et al., 1988; Wang et al., 1992). These studies show that the differentiation of ES derived EBs is more effective than ES cells in studies of vasculogenesis alone (Risau et al., 1988; Wang et al., 1992; Vittet et al., 1996; Bloch et al., 1997; Wartenberg et al., 1998; Hirashima et al., 1999). Many strategies have been used for endothelial differentiation, i.e., the suspension method (Risau et al., 1988; Wang et al., 1992), a method using semisolid medium (Vittet et al., 1996), and ES cell aggregation by the hanging drop method (Goumans et al., 1999). Vittet and colleagues confirmed that ES-derived endothelial cells acquire cell-specific markers in a time-dependent manner after LIF removal, which suggests that endothelial markers are expressed in sequential steps, which closely resemble endothelial cell differentiation *in vitro* during embryonic development (Vittet et al., 1996). In the present study, when mEBs formed by the hanging drop method for 3 days were cultured in EGM-2 medium containing 5% FBS and cocktailed cytokines for 10 days, the mRNA expressions of endothelial markers were found to show patterns similar to Vittet's groups. However, the expression levels of endothelial markers of PECAM by FACS analysis were higher in the early stage than reported by Vittet. In addition, the expression levels of these markers were maintained to day 10. Although the proliferation activity of mES cells cultured in EGM-2 medium was lower than that of cells cultured in DMEM containing 10% FBS, the differentiation activity of those grown in EGM-2 medium was higher. These results suggest that serum contained potent stimulators of cell growth, such as, amino acids, growth factors, vitamins, proteins, hormones, lipids, and minerals can affect both proliferation and differentiation. Of these components, albumin protein has been previously reported to stimulate the proliferation of cells and to suppress their differentiation (Ishida & Yamaguchi, 2004). Therefore, we suggested that factors that stimulate differentiation were more abundant in EGM-2 medium than in DMEM medium containing 10% FBS.

A tyrosine kinase receptor for VEGF (FLK-1) is the marker of the lateral plate mesoderm and the earliest differentiation marker of endothelial cells and blood cells (Eichmann et al., 1997). Thus, an early stage defined by FLK-1 and PECAM expression might reflect commitment towards the endothelial lineage. PECAM is a member of the immunoglobulin superfamily expressed by endothelial cells and a subset of hematopoietic cells in the adult organism (Watt et al., 1995; DeLisser et al., 1997). It also was recently reported that undifferentiated ES cells express PECAM (Vittet et al., 1996; Ling & Neben, 1997). Because ES cells resemble the inner cell mass of the blastocyst, this suggested that PECAM may be expressed embryonically earlier than the described post-implantation stages. From these results, we confirmed that undifferentiated ES cells express PECAM, and we showed for the first time that this early expression mirrors PECAM expression in the mouse blastocyst. PECAM is expressed continuously during ES cell differentiation and characterizes a population of cells that are $PECAM^+$ but not part of patent blood vessels. These cells have a possible counterpart *in vivo* in the yolk sac mesodermal cells that express PECAM before blood island formation. These findings are consistent with the model that ES cell differentiation recapitulates yolk sac development. (Redick & Bautch, 1999). These results define PECAM as the first cell adhesion molecule to be ICM-specific and suggest a potential role for this molecule prior to vascular differentiation. (Robson et al., 2001).

Otherwise, expression of VE-cadherin was observed in later maturation (Vittet et al., 1996). VE-cadherin is expressed at the earliest stages (E7.5) of vascular development in the mesodermal cells of the yolk mesenchyme. At later embryonic stages, VE-cadherin expression is restricted to the peripheral layer of blood islands that gives rise to endothelial cells and to the endothelium of most vessel types (Breier et al., 1996). According to our data, the mRNA expressions of Flk-1, PECAM, and VE-cadherin in mES cells cultured in EGM-2 medium were detected on day 4, day 4, and day 7, respectively. These results are similar to those of previous reports (Vittet et al., 1996; Hirashima et al., 1999; Magnusson et al., 2004).

Festag and their colleagues reported that the exposed six test compounds such as all-trans-retinoic acid (RA), 5-FU, diphenylhydantoin (DPH), valproic acid (Val), saccharin (Sacch), and penicillin G (Pen G), with known embryotoxic potential could disturb differentiation potential of murine embryonic stem cell as well as inhibit the differentiation of ES cells into endothelial cells (Festag et al., 2007). Because the sensitivity of differentiated murine embryonic stem cells to 5-FU comparing to endothelial cell line did not mentioned, the guideline as a correct mathematic model would not applied to a different evaluation of the experiments. However, we demonstrate that the differentiation of mES cells into endothelial cells as a measure of 5-FU, which is an anti-angiogenesis chemical toxicity is more sensitive than mouse endothelial cells (C166). In addition, we confirmed that the sensitivity to 5-FU for in the processing of endothelial differentiation from mouse embryonic bodies. These results suggest that EGM-2 medium is suitable to differentiate the endothelial cells from mES cells. In addition, the differentiated endothelial cells derived mES cells were more sensitive to vascular toxicity testing than traditional method. This devised system could be used as a tool for understanding of mechanism of vasculogenesis and for the screening of mesodermal-derived target organs toxicant.

Furthermore, we demonstrate that 5-fluorouracil, an anti-cancer drug, can induce cytotoxicity in endothelial differentiation of mouse ESCs via inhibition of processes involved in cell viability, proliferation, and the cell cycle. During the induction of endothelial differentiation in cells derived from mouse ESCs, the expression of mesodermal lineage-related genes was up-regulated at 7 days after differentiation using EBs, followed by the up-regulation of endoderm lineage related genes at 14 days (Heo et al., 2005). Based on these reports, we induced endothelial differentiation from mouse ESCs for 10 days using EGM-2 to maximise endothelial differentiation, and confirmed the expression patterns of Oct-4 and PECAM. The frequencies of cells undergoing endothelial differentiation gradually increased through subsequent differentiation days; the expression of Oct-4 was 50.5% at endothelial differentiation day 10. These findings show that the characteristics of cells at endothelial differentiation day 10 are similar to endothelial precursor cells and endothelial-like cells. Because of this, these models can be used to evaluate the cytotoxicity of 5-FU at different stages of endothelial differentiation.

A number of anti-cancer agents have been implicated in vascular toxicity and their effects have been attributed to direct toxicity to the endothelium, such as the HUVEC and C166 cell lines. 5-FU gives an increase in the permeability of endothelial monolayers, as well as inducing vascular collapse and tumour necrosis (Watts et al., 1997). 5-FU, as a cytostatic agent with a strong embryotoxic potential, is known to have an effect in rapidly proliferating cells (Parker & Cheng, 1990; Shimizu et al., 2001). These reports are well matched with our results. The cytotoxicity of 5-FU in mouse ESCs was relieved by the addition of probucol (50 μM), an antagonist of 5-FU. This prevented the endothelial injury usually caused by 5-FU, but the rate of recovery for damaged endothelial cells did not completely return to normal (Kita et al., 1987; Kaneko et al., 1996).

In order to understand the biological mechanism of target cell specific toxicants, DNA microarrays have been used to analyse gene profiling, to show the alteration of gene expression (Gunji et al., 2004; Mori et al., 2007; Fumoto et al., 2008). However, studies of gene profiling and the mechanisms within ESCs exposed to special toxicants, including 5-FU, are still rare. Thus, we analysed expression patterns of genes involved in inducing endothelial differentiation, using a DNA microarray with untreated and 5-FU-treated cells. As shown in the DNA microarray data, a huge number of changes in gene expression were identified by 5-FU treatment. Among these changes, we focused on genes involved in the cell cycle, as our data indicated that 5-FU induces cell proliferation inhibition. In contrast to the somatic cell cycle, ESCs have a unique feature, an abbreviated cell cycle (Becker et al., 2006). This is thought to be controlled through an unusual mechanism of CDK regulation, followed by a very short period of cell cycle. This allows ESCs to preserve their unlimited differentiation potential (Stead et al., 2002). In ESCs, the expressions of cyclin D1 and Cyclin D3 are low, and Cyclin D2 is not expressed, moreover, the expression of CDK-4 and the CDK 4-associated kinase activity is also weak, compared to somatic cells (Burdon et al., 2002). These previous reports lacked a consensus with our data, which showed increased expression of CDK4 and Cyclin D1 during endothelial differentiation. This discrepancy might originate from the differences between cell sources.

Until now, the mechanism that has been reported to explain the cell cycle arrest by 5-FU treatment is a G1/S phase arrest in cancer cell lines, induced by blocking DNA synthesis through the inhibition of thymidylate synthase (Pinedo & Peters, 1988). When damaged by 5-FU treatment, in addition, the expression of the tumour suppressor *p53* is increased *p53* has been identified as a participant in the DNA damage response, resulting in either cell cycle arrest or death. It activates and regulates the transcription of cell cycle arrest or apoptosis related genes (Lane, 1992). The expression of *p21,* which a member of the WAF1/CIP1 family, and activates CKI which works as a linker with the *p53*-dependent pathway is up-regulated (Gartel et al., 1998). Once activated, the *p53* gene product works together with the $p21^{WAF1/CIP1}$ protein, and binds to the cyclin D-CDK4 and cyclin E-CDK2 complexes to inhibit their kinase activities (el-Deiry et al., 1993; Sherr & Roberts, 1999). Finally, increased expression of $p21^{WAF1/CIP1}$ and *p53* is involved in the G1/S-phase arrest of the cell cycle. These reports are in accordance with our data. In addition, the probucol treatment in ESCs, which blocks the cell cycle arrest by 5-FU, appears to aid the repair of damaged cells. Taken together, 5-FU affects endothelial differentiation by decreasing cell viability, proliferation and differentiation, as well as inducing the G1/S phase arrest. These toxicity screening are capable to use the mouse ESC system, therefore, mouse ESCs might be a useful model to use as a tool for screening the cytotoxicity of new compounds.

However, there are still obstacles to overcome in toxicity screening using ESCs. For example, they do not grow as well and are more difficult to maintain and expand than other cell types and the procedures for directing them to differentiate are limited until now. Although stem cells allow for optimal preclinical evaluation of compounds directly on "relevant" human cells prior to clinical testing, it is important because relatively little information is usually obtained during preclinical development procedures on the manner in which novel drugs act on human tissues. Therefore, we should develop new sources of stem cells could be alternative to the use of ESCs in drug screening.

7. References

Astori, G.; Soncin, S. Lo Cicero, V. Siclari, F. Surder, D. Turchetto, L. Soldati, G. & Moccetti, T. (2010). Bone marrow derived stem cells in regenerative medicine as advanced therapy medicinal products. *Am J Transl Res* 2. 3. 285-295, 1943-8141 (Electronic)

Becker, K.A.; Ghule, P.N. Therrien, J.A. Lian, J.B. Stein, J.L. van Wijnen, A.J. & Stein, G.S. (2006). Self-renewal of human embryonic stem cells is supported by a shortened G1 cell cycle phase. *J Cell Physiol* 209. 3. 883-893, 0021-9541 (Print) 0021-9541 (Linking)

Bloch, W.; Forsberg, E. Lentini, S. Brakebusch, C. Martin, K. Krell, H.W. Weidle, U.H. Addicks, K. & Fassler, R. (1997). Beta 1 integrin is essential for teratoma growth and angiogenesis. *J Cell Biol* 139. 1. 265-278, 0021-9525 (Print) 0021-9525 (Linking)

Bourzac, C.; Smith, L.C. Vincent, P. Beauchamp, G. Lavoie, J.P. & Laverty, S. (2010). Isolation of equine bone marrow-derived mesenchymal stem cells: a comparison between three protocols. *Equine Vet J* 42. 6. 519-527, 0425-1644 (Print) 0425-1644 (Linking)

Breier, G.; Breviario, F. Caveda, L. Berthier, R. Schnurch, H. Gotsch, U. Vestweber, D. Risau, W. & Dejana, E. (1996). Molecular cloning and expression of murine vascular endothelial-cadherin in early stage development of cardiovascular system. *Blood* 87. 2. 630-641, 0006-4971 (Print) 0006-4971 (Linking)

Bremer, S.; & Hartung, T. (2004). The use of embryonic stem cells for regulatory developmental toxicity testing in vitro--the current status of test development. *Curr Pharm Des* 10. 22. 2733-2747, 1381-6128 (Print) 1381-6128 (Linking)

Burdon, T.; Smith, A. & Savatier, P. (2002). Signalling, cell cycle and pluripotency in embryonic stem cells. *Trends Cell Biol* 12. 9. 432-438, 0962-8924 (Print) 0962-8924 (Linking)

Chambers, I.; Colby, D. Robertson, M. Nichols, J. Lee, S. Tweedie, S. & Smith, A. (2003). Functional expression cloning of Nanog, a pluripotency sustaining factor in embryonic stem cells. *Cell* 113. 5. 643-655, 0092-8674 (Print) 0092-8674 (Linking)

Davila, J.C.; Cezar, G.G. Thiede, M. Strom, S. Miki, T. & Trosko, J. (2004). Use and application of stem cells in toxicology. *Toxicol Sci* 79. 2. 214-223, 1096-6080 (Print) 1094-2025 (Linking)

Dejana, E.; Bazzoni, G. & Lampugnani, M.G. (1999). Vascular endothelial (VE)-cadherin: only an intercellular glue? *Exp Cell Res* 252. 1. 13-19, 0014-4827 (Print) 0014-4827 (Linking)

DeLisser, H.M.; Christofidou-Solomidou, M. Strieter, R.M. Burdick, M.D. Robinson, C.S. Wexler, R.S. Kerr, J.S. Garlanda, C. Merwin, J.R. Madri, J.A. & Albelda, S.M. (1997). Involvement of endothelial PECAM-1/CD31 in angiogenesis. *Am J Pathol* 151. 3. 671-677, 0002-9440 (Print) 0002-9440 (Linking)

Denham, M. & Dottori, M. (2009). Signals involved in neural differentiation of human embryonic stem cells. *Neurosignals* 17. 4. 234-241, 1424-8638 (Electronic) 1424-862X (Linking)

Doetschman, T.C.; Eistetter, H. Katz, M. Schmidt, W. & Kemler, R. (1985). The in vitro development of blastocyst-derived embryonic stem cell lines: formation of visceral yolk sac, blood islands and myocardium. *J Embryol Exp Morphol* 87. 27-45, 0022-0752 (Print) 0022-0752 (Linking)

Duret, C.; Gerbal-Chaloin, S. Ramos, J. Fabre, J.M. Jacquet, E. Navarro, F. Blanc, P. Sa-Cunha, A. Maurel, P. & Daujat-Chavanieu, M. (2007). Isolation, characterization, and differentiation to hepatocyte-like cells of nonparenchymal epithelial cells from adult human liver. *Stem Cells* 25. 7. 1779-1790, 1066-5099 (Print) 1066-5099 (Linking)

Eichmann, A.; Corbel, C. Nataf, V. Vaigot, P. Breant, C. & Le Douarin, N.M. (1997). Ligand-dependent development of the endothelial and hemopoietic lineages from embryonic mesodermal cells expressing vascular endothelial growth factor receptor 2. *Proc Natl Acad Sci U S A* 94. 10. 5141-5146, 0027-8424 (Print) 0027-8424 (Linking)

El-Deiry, W.S.; Tokino, T. Velculescu, V.E. Levy, D.B. Parsons, R. Trent, J.M. Lin, D. Mercer, W.E. Kinzler, K.W. & Vogelstein, B. (1993). WAF1, a potential mediator of p53 tumor suppression. *Cell* 75. 4. 817-825, 0092-8674 (Print) 0092-8674 (Linking)

Evans, M.J. & Kaufman, M.H. (1981). Establishment in culture of pluripotential cells from mouse embryos. *Nature* 292. 5819. 154-156, 0028-0836 (Print) 0028-0836 (Linking)

Feraud, O. & Vittet, D. (2003). Murine embryonic stem cell in vitro differentiation: applications to the study of vascular development. *Histol Histopathol* 18. 1. 191-199, 0213-3911 (Print) 0213-3911 (Linking)

Festag, M.; Viertel, B. Steinberg, P. & Sehner, C. (2007). An in vitro embryotoxicity assay based on the disturbance of the differentiation of murine embryonic stem cells into endothelial cells. II. Testing of compounds. *Toxicol In Vitro* 21. 8. 1631-1640, 0887-2333 (Print) 0887-2333 (Linking)

Fuh, E. & Brinton, T.J. (2009). Bone marrow stem cells for the treatment of ischemic heart disease: a clinical trial review. *J Cardiovasc Transl Res* 2. 2. 202-218, 1937-5395 (Electronic)

Fumoto, S.; Shimokuni, T. Tanimoto, K. Hiyama, K. Otani, K. Ohtaki, M. Hihara, J. Yoshida, K. Hiyama, E. Noguchi, T. & Nishiyama, M. (2008). Selection of a novel drug-response predictor in esophageal cancer: a novel screening method using microarray and identification of IFITM1 as a potent marker gene of CDDP response. *Int J Oncol* 32. 2. 413-423, 1019-6439 (Print) 1019-6439 (Linking)

Gartel, A.L.; Goufman, E. Tevosian, S.G. Shih, H. Yee, A.S. & Tyner, A.L. (1998). Activation and repression of p21(WAF1/CIP1) transcription by RB binding proteins. *Oncogene* 17. 26. 3463-3469, 0950-9232 (Print) 0950-9232 (Linking)

Genschow, E.; Spielmann, H. Scholz, G. Seiler, A. Brown, N. Piersma, A. Brady, M. Clemann, N. Huuskonen, H. Paillard, F. Bremer, S. & Becker, K. (2002). The ECVAM international validation study on in vitro embryotoxicity tests: results of the definitive phase and evaluation of prediction models. European Centre for the Validation of Alternative Methods. *Altern Lab Anim* 30. 2. 151-176, 0261-1929 (Print) 0261-1929 (Linking)

Goumans, M.J.; Zwijsen, A. van Rooijen, M.A. Huylebroeck, D. Roelen, B.A. & Mummery, C.L. (1999). Transforming growth factor-beta signalling in extraembryonic mesoderm is required for yolk sac vasculogenesis in mice. *Development* 126. 16. 3473-3483, 0950-1991 (Print) 0950-1991 (Linking)

Gunji, W.; Kai, T. Takahashi, Y. Maki, Y. Kurihara, W. Utsugi, T. Fujimori, F. & Murakami, Y. (2004). Global analysis of the regulatory network structure of gene expression in Saccharomyces cerevisiae. *DNA Res* 11. 3. 163-177, 1340-2838 (Print) 1340-2838 (Linking)

Heo, J.; Lee, J.S. Chu, I.S. Takahama, Y. & Thorgeirsson, S.S. (2005). Spontaneous differentiation of mouse embryonic stem cells in vitro: characterization by global gene expression profiles. *Biochem Biophys Res Commun* 332. 4. 1061-1069, 0006-291X (Print) 0006-291X (Linking)

Heuer, J.; Bremer, S. Pohl, I. & Spielmann, H. (1993). Development of an in vitro embryotoxicity test using murine embryonic stem cell cultures. *Toxicol In Vitro* 7. 4. 551-556, 0887-2333 (Print) 0887-2333 (Linking)

Hirashima, M.; Kataoka, H. Nishikawa, S. & Matsuyoshi, N. (1999). Maturation of embryonic stem cells into endothelial cells in an in vitro model of vasculogenesis. *Blood* 93. 4. 1253-1263, 0006-4971 (Print) 0006-4971 (Linking)

Huttmann, A.; Li, C.L. & Duhrsen, U. (2003). Bone marrow-derived stem cells and "plasticity". *Ann Hematol* 82. 10. 599-604, 0939-5555 (Print) 0939-5555 (Linking)

Huuskonen, H. (2005). New models and molecular markers in evaluation of developmental toxicity. *Toxicol Appl Pharmacol* 207. 2 Suppl. 495-500, 1096-0333 (Electronic) 0041-008X (Linking)

Ishida, K. & Yamaguchi, M. (2004). Role of albumin in osteoblastic cells: enhancement of cell proliferation and suppression of alkaline phosphatase activity. *Int J Mol Med* 14. 6. 1077-1081, 1107-3756 (Print) 1107-3756 (Linking)

Kaneko, M.; Hayashi, J. Saito, I. & Miyasaka, N. (1996). Probucol downregulates E-selectin expression on cultured human vascular endothelial cells. *Arterioscler Thromb Vasc Biol* 16. 8. 1047-1051, 1079-5642 (Print) 1079-5642 (Linking)

Kari, G.; Rodeck, U. & Dicker, A.P. (2007). Zebrafish: an emerging model system for human disease and drug discovery. *Clin Pharmacol Ther* 82. 1. 70-80, 0009-9236 (Print) 0009-9236 (Linking)

Kim, S. & von Recum, H. (2008). Endothelial stem cells and precursors for tissue engineering: cell source, differentiation, selection, and application. *Tissue Eng Part B Rev* 14. 1. 133-147, 1937-3368 (Print)

Kim, S.K.; Kim, B.K. Shim, J.H. Gil, J.E. Yoon, Y.D. & Kim, J.H. (2006). Nonylphenol and octylphenol-induced apoptosis in human embryonic stem cells is related to Fas-Fas ligand pathway. *Toxicol Sci* 94. 2. 310-321, 1096-6080 (Print) 1094-2025 (Linking)

Kita, T.; Nagano, Y. Yokode, M. Ishii, K. Kume, N. Ooshima, A. Yoshida, H. & Kawai, C. (1987). Probucol prevents the progression of atherosclerosis in Watanabe heritable hyperlipidemic rabbit, an animal model for familial hypercholesterolemia. *Proc Natl Acad Sci U S A* 84. 16. 5928-5931, 0027-8424 (Print) 0027-8424 (Linking)

Knasmuller, S.; Mersch-Sundermann, V. Kevekordes, S. Darroudi, F. Huber, W.W. Hoelzl, C. Bichler, J. & Majer, B.J. (2004). Use of human-derived liver cell lines for the detection of environmental and dietary genotoxicants; current state of knowledge. *Toxicology* 198. 1-3. 315-328, 0300-483X (Print) 0300-483X (Linking)

Knight, A.; (2007). Systematic reviews of animal experiments demonstrate poor human clinical and toxicological utility. *Altern Lab Anim* 35. 6. 641-659, 0261-1929 (Print) 0261-1929 (Linking)

Kulkarni, J.S. & Khanna, A. (2006). Functional hepatocyte-like cells derived from mouse embryonic stem cells: a novel in vitro hepatotoxicity model for drug screening. *Toxicol In Vitro* 20. 6. 1014-1022, 0887-2333 (Print) 0887-2333 (Linking)

Kuo, T.K.; Hung, S.P. Chuang, C.H. Chen, C.T. Shih, Y.R. Fang, S.C. Yang, V.W. & Lee, O.K. (2008). Stem cell therapy for liver disease: parameters governing the success of using bone marrow mesenchymal stem cells. *Gastroenterology* 134. 7. 2111-2121, 2121 e2111-2113, 1528-0012 (Electronic) 0016-5085 (Linking)

Lane, D.P.; (1992). Cancer. p53, guardian of the genome. *Nature* 358. 6381. 15-16, 0028-0836 (Print) 0028-0836 (Linking)

Liebsch, M. & Spielmann, H. (2002). Currently available in vitro methods used in the regulatory toxicology. *Toxicol Lett* 127. 1-3. 127-134, 0378-4274 (Print) 0378-4274 (Linking)

Ling, V. & Neben, S. (1997). In vitro differentiation of embryonic stem cells: immunophenotypic analysis of cultured embryoid bodies. *J Cell Physiol* 171. 1. 104-115, 0021-9541 (Print) 0021-9541 (Linking)

Magnusson, P.; Rolny, C. Jakobsson, L. Wikner, C. Wu, Y. Hicklin, D.J. & Claesson-Welsh, L. (2004). Deregulation of Flk-1/vascular endothelial growth factor receptor-2 in fibroblast growth factor receptor-1-deficient vascular stem cell development. *J Cell Sci* 117. Pt 8. 1513-1523, 0021-9533 (Print) 0021-9533 (Linking)

Martin, G.R. (1981). Isolation of a pluripotent cell line from early mouse embryos cultured in medium conditioned by teratocarcinoma stem cells. *Proc Natl Acad Sci U S A* 78. 12. 7634-7638, 0027-8424 (Print) 0027-8424 (Linking)

Mills, J.B.; Rose, K.A. Sadagopan, N. Sahi, J. & de Morais, S.M. (2004). Induction of drug metabolism enzymes and MDR1 using a novel human hepatocyte cell line. *J Pharmacol Exp Ther* 309. 1. 303-309, 0022-3565 (Print) 0022-3565 (Linking)

Mori, N.; Glunde, K. Takagi, T. Raman, V. & Bhujwalla, Z.M. (2007). Choline kinase down-regulation increases the effect of 5-fluorouracil in breast cancer cells. *Cancer Res* 67. 23. 11284-11290, 1538-7445 (Electronic) 0008-5472 (Linking)

Nichols, J.; Zevnik, B. Anastassiadis, K. Niwa, H. Klewe-Nebenius, D. Chambers, I. Scholer, H. & Smith, A. (1998). Formation of pluripotent stem cells in the mammalian embryo depends on the POU transcription factor Oct4. *Cell* 95. 3. 379-391, 0092-8674 (Print) 0092-8674 (Linking)

Niwa, H.; Burdon, T. Chambers, I. & Smith, A. (1998). Self-renewal of pluripotent embryonic stem cells is mediated via activation of STAT3. *Genes Dev* 12. 13. 2048-2060, 0890-9369 (Print) 0890-9369 (Linking)

Parker, W.B. & Cheng, Y.C. (1990). Metabolism and mechanism of action of 5-fluorouracil. *Pharmacol Ther* 48. 3. 381-395, 0163-7258 (Print) 0163-7258 (Linking)

Pearson, R.M. (1986). In-vitro techniques: can they replace animal testing? *Hum Reprod* 1. 8. 559-560, 0268-1161 (Print) 0268-1161 (Linking)

Petersen, B.E.; Bowen, W.C. Patrene, K.D. Mars, W.M. Sullivan, A.K. Murase, N. Boggs, S.S. Greenberger, J.S. & Goff, J.P. (1999). Bone marrow as a potential source of hepatic oval cells. *Science* 284. 5417. 1168-1170, 0036-8075 (Print) 0036-8075 (Linking)

Pinedo, H.M. & Peters, G.F. (1988). Fluorouracil: biochemistry and pharmacology. *J Clin Oncol* 6. 10. 1653-1664, 0732-183X (Print) 0732-183X (Linking)

Ramalho-Santos, M.; Yoon, S. Matsuzaki, Y. Mulligan, R.C. & Melton, D.A. (2002). "Stemness": transcriptional profiling of embryonic and adult stem cells. *Science* 298. 5593. 597-600, 1095-9203 (Electronic) 0036-8075 (Linking)

Rashid, S.T.; Corbineau, S. Hannan, N. Marciniak, S.J. Miranda, E. Alexander, G. Huang-Doran, I. Griffin, J. Ahrlund-Richter, L. Skepper, J. Semple, R. Weber, A. Lomas,

D.A. & Vallier, L. (2010). Modeling inherited metabolic disorders of the liver using human induced pluripotent stem cells. *J Clin Invest* 120. 9. 3127-3136, 1558-8238 (Electronic) 0021-9738 (Linking)

Redick, S.D.; & Bautch, V.L. (1999). Developmental platelet endothelial cell adhesion molecule expression suggests multiple roles for a vascular adhesion molecule. *Am J Pathol* 154. 4. 1137-1147, 0002-9440 (Print) 0002-9440 (Linking)

Risau, W.; Sariola, H. Zerwes, H.G. Sasse, J. Ekblom, P. Kemler, R. & Doetschman, T. (1988). Vasculogenesis and angiogenesis in embryonic-stem-cell-derived embryoid bodies. *Development* 102. 3. 471-478, 0950-1991 (Print) 0950-1991 (Linking)

Robson, P.; Stein, P. Zhou, B. Schultz, R.M. & Baldwin, H.S. (2001). Inner cell mass-specific expression of a cell adhesion molecule (PECAM-1/CD31) in the mouse blastocyst. *Dev Biol* 234. 2. 317-329, 0012-1606 (Print) 0012-1606 (Linking)

Rohwedel, J.; Guan, K. Hegert, C. & Wobus, A.M. (2001). Embryonic stem cells as an in vitro model for mutagenicity, cytotoxicity and embryotoxicity studies: present state and future prospects. *Toxicol In Vitro* 15. 6. 741-753, 0887-2333 (Print) 0887-2333 (Linking)

Schnerch, A.; Cerdan, C. & Bhatia, M. (2010). Distinguishing between mouse and human pluripotent stem cell regulation: the best laid plans of mice and men. *Stem Cells* 28. 3. 419-430, 1549-4918 (Electronic) 1066-5099 (Linking)

Scholz, G.; Pohl, I. Genschow, E. Klemm, M. & Spielmann, H. (1999). Embryotoxicity screening using embryonic stem cells in vitro: correlation to in vivo teratogenicity. *Cells Tissues Organs* 165. 3-4. 203-211, 1422-6405 (Print) 1422-6405 (Linking)

Seiler, A.; Visan, A. Buesen, R. Genschow, E. & Spielmann, H. (2004). Improvement of an in vitro stem cell assay for developmental toxicity: the use of molecular endpoints in the embryonic stem cell test. *Reprod Toxicol* 18. 2. 231-240, 0890-6238 (Print) 0890-6238 (Linking)

Sherr, C.J. & Roberts, J.M. (1999). CDK inhibitors: positive and negative regulators of G1-phase progression. *Genes Dev* 13. 12. 1501-1512, 0890-9369 (Print) 0890-9369 (Linking)

Shimizu, N.; Aoyama, H. Hatakenaka, N. Kaneda, M. & Teramoto, S. (2001). An in vitro screening system for characterizing the cleft palate-inducing potential of chemicals and underlying mechanisms. *Reprod Toxicol* 15. 6. 665-672, 0890-6238 (Print) 0890-6238 (Linking)

Smith, A.G. (1992). Mouse embryo stem cells: their identification, propagation and manipulation. *Semin Cell Biol* 3. 6. 385-399, 1043-4682 (Print) 1043-4682 (Linking)

Spielmann, H. (2009). The way forward in reproductive/developmental toxicity testing. *Altern Lab Anim* 37. 6. 641-656, 0261-1929 (Print) 0261-1929 (Linking)

Spielmann, H.; Genschow, E. Scholz, G. Brown, N.A. Piersma, A.H. Brady, M. Clemann, N. Huuskonen, H. Paillard, F. Bremer, S. & Becker, K. (2001). Preliminary results of the ECVAM validation study on three in vitro embryotoxicity tests. *Altern Lab Anim* 29. 3. 301-303, 0261-1929 (Print) 0261-1929 (Linking)

Spielmann, H.; Pohl, I. Doering, B. Liebsch, M. & Moldenhauer, F. (1997). The embryonic stem cell test, an in vitro embryotoxicity test using two permanent mouse cell lines:

3T3 fibroblasts and embryonic stem cells. *In. Vitro. Toxicol.* 10. 119-127, 0887-2333 (Print) 1879-3177 (Linking)

Stead, E.; White, J. Faast, R. Conn, S. Goldstone, S. Rathjen, J. Dhingra, U. Rathjen, P. Walker, D. & Dalton, S. (2002). Pluripotent cell division cycles are driven by ectopic Cdk2, cyclin A/E and E2F activities. *Oncogene* 21. 54. 8320-8333, 0950-9232 (Print) 0950-9232 (Linking)

Steele, C.E.; New, D.A. Ashford, A. & Copping, G.P. (1983). Teratogenic action of hypolipidemic agents: an in vitro study with postimplantation rat embryos. *Teratology* 28. 2. 229-236, 0040-3709 (Print) 0040-3709 (Linking)

Thomson, J.A.; Itskovitz-Eldor, J. Shapiro, S.S. Waknitz, M.A. Swiergiel, J.J. Marshall, V.S. & Jones, J.M. (1998). Embryonic stem cell lines derived from human blastocysts. *Science* 282. 5391. 1145-1147, 0036-8075 (Print) 0036-8075 (Linking)

Tiffany-Castiglioni, E.; Ehrich, M. Dees, L. Costa, L.G. Kodavanti, P.R. Lasley, S.M. Oortgiesen, M. & Durham, H.D. (1999). Bridging the gap between in vitro and in vivo models for neurotoxicology. *Toxicol Sci* 51. 2. 178-183, 1096-6080 (Print) 1094-2025 (Linking)

Vanparys, P. (2002). ECVAM and pharmaceuticals. *Altern Lab Anim* 30 Suppl 2. 221-223, 0261-1929 (Print) 0261-1929 (Linking)

Vittet, D.; Prandini, M.H. Berthier, R. Schweitzer, A. Martin-Sisteron, H. Uzan, G. & Dejana, E. (1996). Embryonic stem cells differentiate in vitro to endothelial cells through successive maturation steps. *Blood* 88. 9. 3424-3431, 0006-4971 (Print) 0006-4971 (Linking)

Wang, R.; Clark, R. & Bautch, V.L. (1992). Embryonic stem cell-derived cystic embryoid bodies form vascular channels: an in vitro model of blood vessel development. *Development* 114. 2. 303-316, 0950-1991 (Print) 0950-1991 (Linking)

Wartenberg, M.; Gunther, J. Hescheler, J. & Sauer, H. (1998). The embryoid body as a novel in vitro assay system for antiangiogenic agents. *Lab Invest* 78. 10. 1301-1314, 0023-6837 (Print) 0023-6837 (Linking)

Watt, S.M.; Gschmeissner, S.E. & Bates, P.A. (1995). PECAM-1: its expression and function as a cell adhesion molecule on hemopoietic and endothelial cells. *Leuk Lymphoma* 17. 3-4. 229-244, 1042-8194 (Print) 1026-8022 (Linking)

Watts, M.E.; Woodcock, M. Arnold, S. & Chaplin, D.J. (1997). Effects of novel and conventional anti-cancer agents on human endothelial permeability: influence of tumour secreted factors. *Anticancer Res* 17. 1A. 71-75, 0250-7005 (Print) 0250-7005 (Linking)

Wobus, A.M. & Boheler, K.R. (2005). Embryonic stem cells: prospects for developmental biology and cell therapy. *Physiol Rev* 85. 2. 635-678, 0031-9333 (Print) 0031-9333 (Linking)

Xu, J.J.; Diaz, D. & O'Brien, P.J. (2004). Applications of cytotoxicity assays and pre-lethal mechanistic assays for assessment of human hepatotoxicity potential. *Chem Biol Interact* 150. 1. 115-128, 0009-2797 (Print) 0009-2797 (Linking)

Yamaguchi, T.P.; Dumont, D.J. Conlon, R.A. Breitman, M.L. & Rossant, J. (1993). flk-1, an flt-related receptor tyrosine kinase is an early marker for endothelial cell precursors. *Development* 118. 2. 489-498, 0950-1991 (Print) 0950-1991 (Linking)

Ying, Q.L.; Stavridis, M. Griffiths, D. Li, M. & Smith, A. (2003). Conversion of embryonic stem cells into neuroectodermal precursors in adherent monoculture. *Nat Biotechnol* 21. 2. 183-186, 1087-0156 (Print) 1087-0156 (Linking)

Zagami, C.J.; Zusso, M. & Stifani, S. (2009). Runx transcription factors: lineage-specific regulators of neuronal precursor cell proliferation and post-mitotic neuron subtype development. *J Cell Biochem* 107. 6. 1063-1072, 1097-4644 (Electronic) 0730-2312 (Linking)

Zhang, H. & Wang, Z.Z. (2008). Mechanisms that mediate stem cell self-renewal and differentiation. *J Cell Biochem* 103. 3. 709-718, 1097-4644 (Electronic) 0730-2312 (Linking)

Zhou, B.; Liu, C. Wang, J. Lam, P.K. & Wu, R.S. (2006). Primary cultured cells as sensitive in vitro model for assessment of toxicants--comparison to hepatocytes and gill epithelia. *Aquat Toxicol* 80. 2. 109-118, 0166-445X (Print) 0166-445X (Linking)

25

Human Pluripotent Stem Cell-Derived Neuronal Networks: Their Electrical Functionality and Usability for Modelling and Toxicology

Riikka Äänismaa[1], Laura Ylä-Outinen[1], Jarno E. Mikkonen[2] and Susanna Narkilahti[1]
[1]*Regea – Institute for Regenerative Medicine, University of Tampere, Tampere,*
[2]*Department of Psychology, University of Jyväskylä, Jyväskylä,*
Finland

1. Introduction

Micro electrode array (MEA)-based platforms have been used to study neuronal networks for decades. The used cells have, for the most part, been rodent primary neurons. The gained knowledge has indeed increased the understanding of neuronal network development and maturation both *in vitro* and *in vivo*. If aiming to understand the development of human brain, however, the used cell type should preferably be of human origin due to difficult interpolation from the rodent cell data. In addition, the development of functional human neuronal networks would open up a new era for, e.g., toxicology testing, drug screening and disease modelling.

The use of MEA with bioelectrically active cells was first reported by Thomas et al. 1972. Gross et al. (1977; 1979) extended the applicability of the MEA platform to long-term recordings of neuronal cells and Pine (1980) reported the first recordings from dissociated neurons. Interestingly, although the solutions related to the MEA fabrication and design were similar in the above mentioned papers, all three groups came to their conclusions independently demonstrating the drive and need to record neurons *in vitro*. MEAs have been successfully tested in several species and cell types using organotypic and primary dissociated cell cultures. Due to their ease of use, the long-term survival of the cultures and vast possibilities of manipulation of the culture, MEAs provide an attractive tool to explore basic neuroscience as well as for pharmacology and toxicology.

Functional neuronal networks have been derived from mouse embryonic stem cells (Evans & Kaufman 1981; Martin 1981), but the invention of culturing human embryonic stem cells (Thomson et al., 1998) and human induced pluripotent stem cells (Takahashi et al., 2007) have enabled the researchers to build up human cell-derived neuronal cells and networks (Ban et al., 2006; Carpenter et al., 2001; Heikkilä et al., 2009; Karumbayaram et al., 2009; Nat et al., 2007). These studies have proved that human pluripotent stem cell-derived neuronal cells are indeed capable of forming functional neuronal networks which most likely resemble the networks in human brain. These human-derived neuronal networks have increased and will further increase the understanding of human brain development and functions. More importantly, with these networks it is possible to model the "normal" brain

by using human embryonic stem cell-derived neuronal cells and various diseases can be modelled when using patient specific induced pluripotent stem cell-derived neuronal cells. In addition, these cell types are suitable for drug screening and neurotoxicological studies.
This chapter will shortly introduce the neuronal differentiation of both human embryonic stem cells and human induced pluripotent stem cells. Also, MEA platforms will be discussed concentrating especially on the use of MEAs with human pluripotent stem cell-derived neuronal networks. Finally, some future aspects will be discussed.

2. MEA technology

The structure of the MEAs is simple. They consist of a glass slide that has wire electrodes photo-etched onto it. Each electrode is insulated from each other (e.g. by silicon nitrate) and only the tip of the electrode is coated with conductive material, such as titanium, platinum, or indium-tin oxide. Typically, the electrode size is 10 - 30 μm and they are aligned in a grid with 50 - 500 μm separation from the centres of the electrodes (Johnstone et al., 2010; Pine 2006), MEAs measure the potential difference between a common ground electrode and the measurement electrodes. Because the measured voltage differences between the ground and the measurement electrodes are small and the locations of the electrodes are static, the reduction of noise and the amplification of the recorded signals are crucial. Often, the whole MEA chip is placed inside of the amplifier to ensure minimal distance between the electrodes and the amplifiers.
To make hydrophobic MEA surface more appealing to neurons, the surface is often coated with different protein solutions like extracellular proteins (e.g. laminin, fibronectin, polylysines). Surfaces can be also pre-treated with Poly(ethyleneimine) (PEI) that enhances the coating procedure (Heikkilä et al., 2009; Illes et al., 2007; Wagenaar et al., 2006).
Cells grown on MEAs can be continuously perfused with fresh medium, but because this is a cumbersome and laborious task, the medium is usually replaced on a weekly basis. Different laboratories have different routines for medium replacement depending on the available staff and type and amount of the cells used in the experiments. Sensitive cells, such

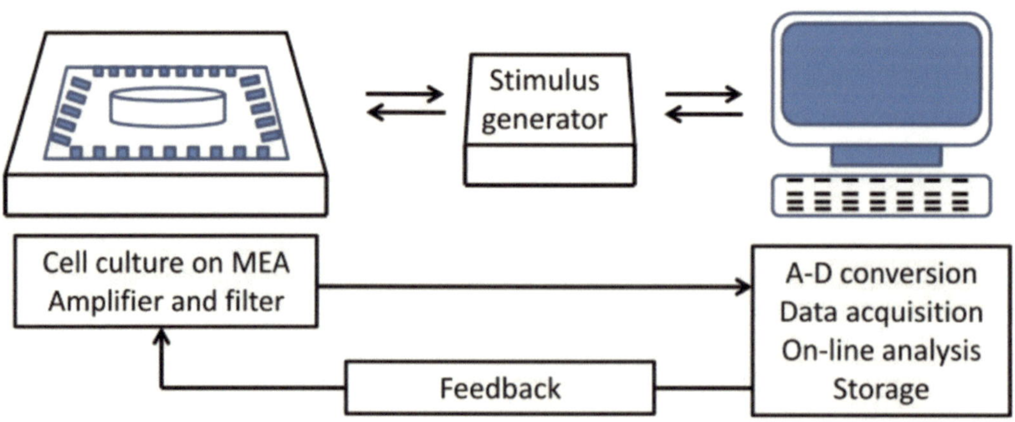

Fig. 1. A typical MEA setup. The Array is located inside an amplifier that is connected to an analog to digital converter on a computer. Bidirectional information transfer allows timing of the stimulations and more sophisticated study designs.

as hESC-derived neuronal cells, require fresh medium at least three times a week (Heikkilä et al., 2009). Typically, only one third or half of the medium is replaced in order to maintain more stable culturing conditions. Change of full medium would generate a massive change in the cellular microenvironment due to differences in the ionic and osmotic concentrations in the fresh and used growth media. Frequent change of the growth medium requires continuous handling of the MEAs and exposes the cultured neurons to an additional stress and contaminants in the laboratory. Thus, there is always a trade-off between the optimal medium composition and risk of contamination. After the introduction of the semi-permeable transparent hydrophobic membrane (Potter & DeMarse 2001) that seals the MEA from the surroundings allowing only gas exchange it has been possible to record and maintain neuronal cultures on MEAs for months or even years without evaporation or contamination problems (Heikkilä et al., 2009; Illes et al., 2009; Potter & DeMarse 2001). These long and stable experiments are a prerequisite e.g. for chronic toxicology experiments.

2.1 Functional MEA studies

One of the first functional MEA studies was performed by Corner & Ramkers (1991) who evaluated the activity changes of maturing neurons grown on MEAs using tetrodotoxin and picrotoxin in long-term culture. After inclusion of the stimulation electrode into the MEA design (Gross et al., 1993), the applicability of MEAs to conduct functional studies improved significantly. Modern imaging methods, such as gated neurotransmitters, calcium imaging and voltage sensitive dyes (Mennerick et al., 2010) have further enhanced the usability of the MEA platform.

Although neurons grown on a MEA chip form 2-dimensional networks that lack physiological input from the environment, they form functional spontaneously active networks that typically develop in 3 phases. First, activity is detected as single spiking that can reflect both axonal and dendritic signalling in the developing network. Second, as network maturates, it starts to express a train-like spiking activity that can further mature into burst -like activity that is considered as mature signalling activity of the network (Heikkilä et al., 2009; Wagenaar et al., 2006). The described types of signalling are represented in Figure 2. These functional networks can interact with the environment by selectively responding to the artificial cues presented by the stimulation apparatus. Jimbo and co-workers demonstrated that repetitive stimulation reinforces the synchronicity on MEA grown networks (Jimbo et al., 1998). However, firing of the neurons is pathway-specific and stimulation may either promote or suppress the network firing activity depending on the stimulation location (Jimbo et al., 1999). Stimulations can also be used to

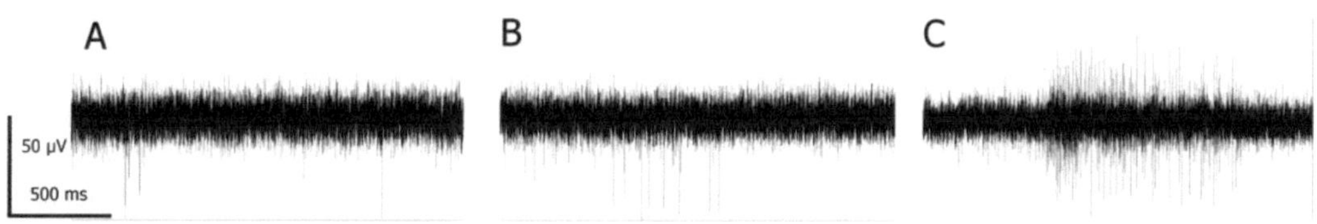

Fig. 2. Development of signalling in MEA. (A) First, after the neuronal cells have grown few days on MEA, first signals are detected. They are individual single spikes. (B) Then, about after 1-2 weeks of culturing spiking activity matures into training phase. There are several spikes organised together. (C) Finally, bursting activity occurs. Burst is a complex signal package that happens in several channels at the same time.

tune network firing activity. Shahaf & Marom (2001) introduced a method of learning by stimulus removal on MEA grown dissociated cortical neuronal networks. It seems that MEA grown neurons can communicate in an activity-dependent manner that resembles the behaviour of intact neurons *in vivo*. On MEAs, this communication typically occurs in bursts of synchronous spikes, whose pathways (Jimbo et al., 1999), timing (Shahaf & Marom 2001), type (Wagenaar et al., 2006) and initiation (Eytan & Marom 2006) can be considered as variables for systemic activity changes.

2.2 MEA analyses

Most commonly, routines of MEA research follow the paths of decades of research in *in vivo* and *in vitro* electrophysiology. Due to the ease of use of MEAs and the benefits of dissociated culturing, the speed of the experiments is, however, much faster. Furthermore, due to elevated control over experimental conditions and variables in culture, the reproducibility of the experiments is more reliable and large sample sizes easily attainable.
Raw data is usually high-pass filtered to extract spikes from the raw data. Additional sorting may be applied to obtain unit data. Although many laboratories have in-house MATLAB scripts to do the task, commercial softwares also exist. Traditional scatter plots, time histograms and time-frequency analyses are commonly used as analysis tools for MEA data. As in psychology and medicine, stimulus averaged methods are common for triggered data. On a network scale, separation of activity regimes (Tanskanen et al., 2005), analysis of bursts (Mazzoni et al., 2007) and pattern clustering (Madhavan et al., 2007) are challenging the traditional estimations of network cross-correlations. Synfire chains (Abeles 1982; Tetzlaff et al., 2002), avalanches (Beggs & Plenz 2003; 2004) and unitary events (Gruen et al., 2002; Gruen 2009) are emerging as computational tools for more complex MEA network analysis. MEA data presentation can be found in Figure 3.

3. Human pluripotent stem cells and their neural differentiation

Short-term *in vitro* culturing of the inner cell mass of human blastocysts was first reported on 1994 (Bongso et al., 1994), but the successful isolation, culturing and characterization of human embryonic stem (hES) cells was reported 4 years later (Thomson et al., 1998). These cells are referred as human pluripotent stem cells due to their capability to form every cell type of the human body. After the discovery, hES cell research was intensively conducted for almost 10 years before remarkable progress took place with the invention of human induced pluripotent stem (hiPS) cells. These cells are formed from somatic cells by viral reprogramming (Takahashi et al., 2007; Yu et al., 2007) and they closely resemble hES cells by their characteristics. Thus, currently term human pluripotent stem cells includes both hES and hiPS cells (Figure 4). Most of the work cited here have been conducted using hES cells and their neural derivatives.
Originally hES cells were cultured in their undifferentiated stage on mouse embryonic fibroblasts as feeder cells (Thomson et al., 1998) but shortly after also on human foreskin fibroblasts (Hovatta et al., 2003). Several feeder-free systems have also been studied with success on maintaining hES cells in undifferentiated form (Benzing et al., 2006; Gerrard et al., 2005; Hakala et al., 2009). There are clear indications, however, that hES cells cultured without feeder cells exhibit more abnormalities caused by suboptimal culture conditions and enzymatic passaging in long-term cultures (Imreh et al., 2006; Mitalipova et al., 2005).

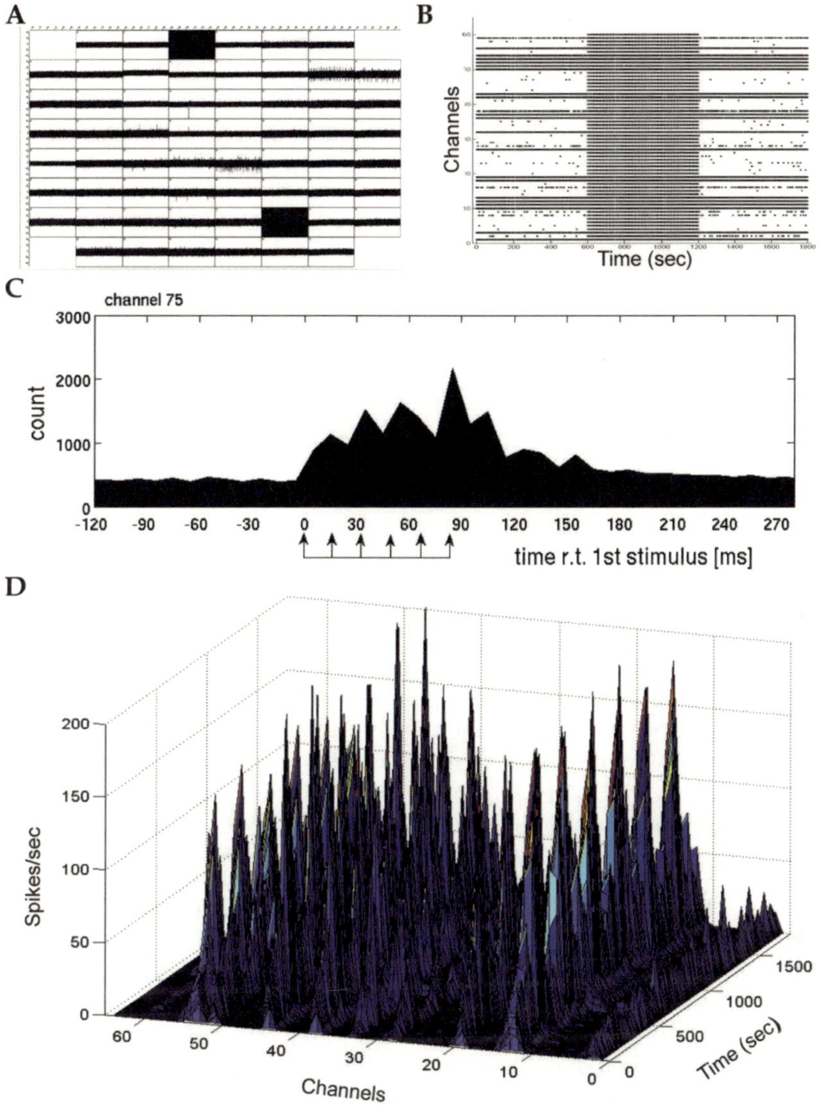

Fig. 3. MEA data and its analysis. A) Typical recorded MEA signal streams on 60 electrodes. Each small window represents a signal from one MEA electrode. Note the noisy electrodes 41 and 67 which appear all black. B) MEA activity plotted as a channel wise scatter plot. Each vertical line represents an electrode and each dot marks a spike recorded on that electrode. The easiest way to separate spikes is to draw a threshold based on standard deviation of the signal (e.g. 5*SD) and accept all crossings of the threshold as spikes. Note the stimulus responses and artefacts present on all the channels from 600 s to 1200 s. C) Peri-stimulus time histogram of the selected channels in B. This histogram sums stimulus aligned responses and thus presents an average response to a given stimulus. Arrows below the figures show the timing of individual stimulus in a train of stimuli. Note the elevation of the spike rate in response to repetitive stimulus. D) Rate plot of the whole MEA recording. This 3D plot clarifies the parameters of time, electrode channels and spikes/sec in a single figure.

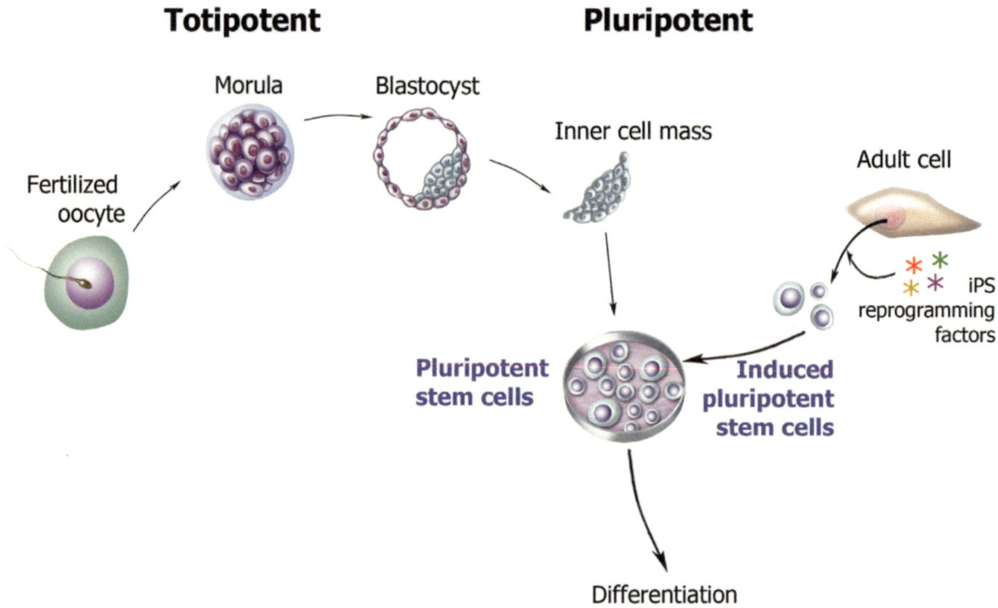

Fig. 4. Stem cells. Stem cells can be divided in groups in accordance to their differentiation capacity. Embryos in zygote and morula stages are defined as totipotent. In blastocyst stage the inner cell mass is capable of producing the three germ layers and primordial germ cells, thus defined as pluripotent embryonic stem cells (ESCs). Adult cells can be re-programmed to produce embryonic stem cell-like pluripotent cells (iPS cells). Original images prepared by Cathrine Twomey from the National Academies *Understanding stem cells: An Overview of the Science and Issues*, http://www.nationalacademies.org/stemcells.

Thus, the culture conditions of hES cells have been systematically improved towards containing only human or synthetic components. Similarly, hiPS cells need to be cultured on top of a feeder cell layer to avoid the spontaneous differentiation of these cells. This aspect has been widely studied due to ultimate aim to elude the use of feeder cells altogether.

Undifferentiated human pluripotent stem cells can not be used for disease modelling or regenerative medicine as such due to their massive teratoma formation capacity (Adewumi et al., 2007; Skottman et al., 2007; Thomson et al., 1998). Indeed, neuronal differentiation of human pluripotent stem cells has been widely studied. The first articles on neural differentiation of hES cells were published 2001 (Carpenter et al., 2001; Reubinoff et al., 2001; Zhang et al., 2001) and of hiPS cells 2009 (Chambers et al., 2009; Karumbayaram et al., 2009). All of these protocols relied on embryoid body (EB) formation and/or further replating of the cells on appropriately coated surfaces in neural differentiation medium. Regardless of the differentiation methods used, these studies all showed that neural progenitors, specific neuronal cells, astrocytes and, to a lesser extent, oligodendrocytes could be produced. Since then, several methods and protocols for neural differentiation of hES and also hiPS cells have been published. Some groups utilize co-culture method with e.g. PA-6 stromal cells (Aberdam et al., 2008; Pomp et al., 2008) whereas other groups differentiate the cells with medium including minimum amounts of animal material (Lappalainen et al., 2010; Nat et al., 2007) or in totally defined animal-component free medium (Erceg et al., 2008; Yao et al., 2006).

Methodologically hES cells can be differentiated towards neural lineages using adherent and suspension culture systems or their various combinations with the help of specific factors facilitating differentiation. These factors include basic fibroblast growth factor (bFGF) (Benzing et al., 2006; Lappalainen et al., 2010), bone morphogenic protein signalling blocker noggin (Gerrard et al., 2005; Itsykson et al., 2005; Li et al., 2008) or retinoic acid (Baharvand et al., 2007; Erceg et al., 2008). Thus, the production of neural progenitors and specific neuronal phenotypes from hES cells appears to be possible with many methods, growth factors and inducing agents. The functionality of the produced neuronal cells has not, however, been taken into consideration on a large scale in relation to the differentiation protocol. HiPS cells have been neurally differentiated with the same methods as hES cells but the electrical functionality of hiPS cell-derived neuronal cells has so far been studied only with patch clamp technology (Karumbayaram et al., 2009).

4. MEA and pluripotent stem cell-derived neuronal networks

In neural applications MEA setup has traditionally been used with rodent primary cells. With pluripotent stem cell-derived neural cells the research has been conducted just over a decade but for now only a few groups have reported the usage of MEA with neuronal networks derived from pluripotent stem cells (Ban et al., 2006; Heikkilä et al., 2009; Illes et al., 2007). To date, little is known about the electrical properties of the pluripotent stem cell-derived neuronal cells, their maturation, and network development and function.

The first articles describing the measurement of murine ES cell-derived neuronal networks were published 5 years ago (Ban et al., 2006; Illes et al., 2007). Both of these studies showed that the cells were capable of forming spontaneously active networks. Further, the activity developed from single spikes into more complex trains and bursts over time (Ban et al., 2006; Illes et al., 2007). The first study reporting successful measurement of human ES cell-derived neuronal networks with MEA was reported two years later (Heikkilä et al., 2009). This is a vital step in human research (Table 1) due to possible interspecies differences of developed neuronal networks. Especially with toxicological and drug studies, the specific pathways play important role in evaluating results and the translation between different species have to be take into account. For example, mouse embryonic stem cell-derived neuronal cells need JAK/STAT pathway activation whereas for the human ones activin/nodal and FGF signalling are important. These differences may cause large variation to results in drug responses. Thus, it might be meaningful to test toxicological effects directly on human cells (Hardingham et al., 2010). Indeed, we recently performed a neurotoxicological evaluation of methyl mercury in human neuronal cell-based MEA platform (Ylä-Outinen et al., 2010). In this study we reported that at subcytotoxic levels methyl mercury caused changes in functionality of neuronal networks whereas no alterations were detected with standard molecule biological analyses (Ylä-Outinen et al., 2010).

MEAs with ES cell-derived neuronal networks can also be used for studying the fundamental properties of these networks by pharmaceutical modulation. It has been shown that these networks, from human or mouse origin, contain functional glutamate and GABA receptors as they respond to AMPA/kainate, NMDA and $GABA_A$ receptor blockers (Heikkilä et al., 2009; Illes et al., 2007). These results are similar to results from rodent dissociated primary cultures (Kamioka et al., 1996; Corner & Ramakers 1991). Thus, human cell-based MEA platforms offer intriguing possibilities to conduct large scale *in vitro* drug screening and testing thereby reducing the need to use experimental animals. In addition, usage of human cells makes the translation into clinical setting more direct.

Research field	Benefits of using MEA networks from human origin
Basic research	Cell-cell signalling Cellular and network connectivity Receptor expression and functionality
Developmental studies	Time course of development of cell types and networks Developmental disorders Intervention in development
Toxicological studies	Study of toxicological mechanisms On-site-sampling Continuous screening Acute and chronic responses
Drug screening	High throughput Standardized environment Animal-free testing
Regenerative medicine	Functionality of cell grafts Verification of uniform graft quality Graft-host tissue interactions

Table 1. Research fields that benefit from the information gained with human cell-based MEA measurements.

Dissociated neuronal networks have capacity and plasticity to learn *in vitro*. This has been shown by electrical stimulation of neuronal populations on MEA (*e.g.* Shahaf & Marom 2001). Also, mES cell-derived neuronal networks have the ability to process information as shown by Ban and co-workers (Ban et al., 2006). They varied the intensity of the stimulus and showed that the neuronal networks responded to the intensity change. We have gained similar results with hES cell-derived neuronal cells (Heikkilä et al., 2009).

The electrophysiological properties of hiPS cell-derived neuronal cells have not been extensively studied. So far, a few groups have described hiPS cell-derived neuronal cell functions with patch clamp on a single cell level (Karumbayaram et al., 2009; Swistowski et al., 2010). The ability of these cells to form spontaneously active networks has been somewhat questionable. Here, for the first time we show that neuronal cells derived from normal hiPS cell line, indeed, form functional neuronal networks. This functionality was evidenced on MEAs already two days after cell plating. The activity evolved from single spikes to complex burst patterns within four weeks similarly to hES cell-derived counterparts as shown in Figure 5. In future, neurological diseases with genetic background can be modelled with patient-specific hiPS cell lines as already published with e.g. inherited metabolic disorders in liver (Rashid et al., 2010) or long QT-syndrome in heart with patch clamp (Moretti et al., 2010). Because neuronal networks rely on electrical communication, especially the modelling of neurological diseases would crucially benefit from MEA platform. MEA-based disease models have already been introduced. For example, Otto et al. show that mES cell-derived neuronal networks respond by lowered signalling when cultured in cerebrospinal fluid of brain trauma patients (Otto et al., 2009).

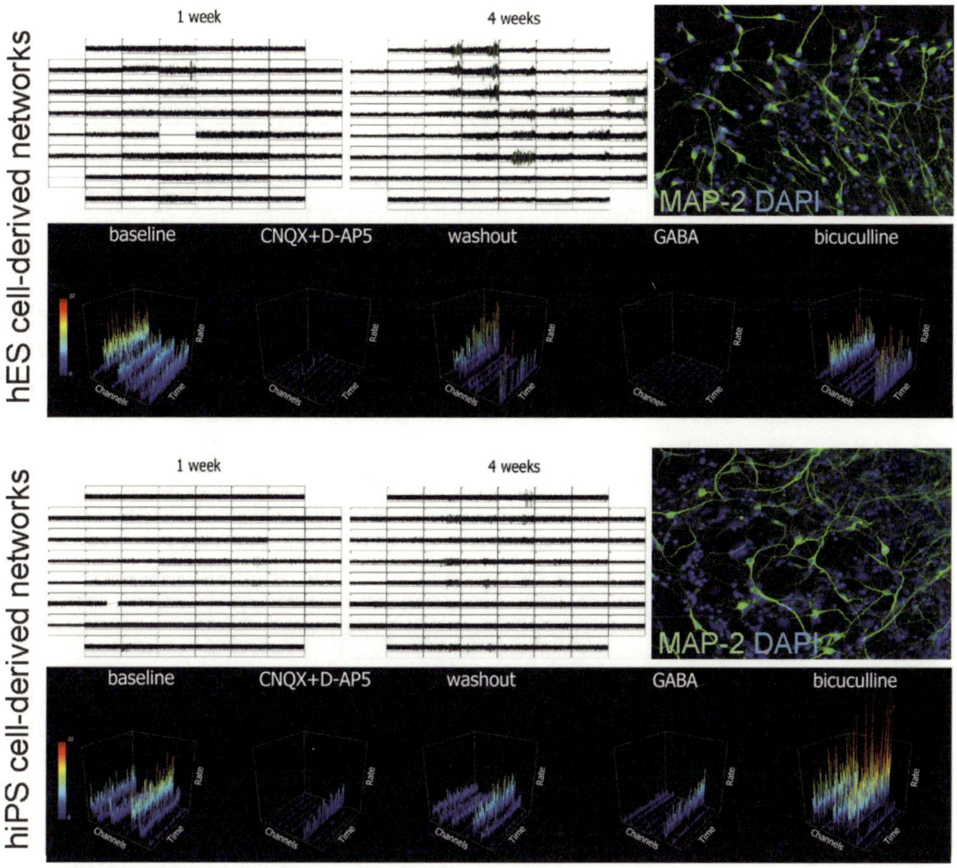

Fig. 5. Comparison of electrical and molecular properties of hES and hiPS cell-derived neuronal networks. Array wide plots of spontaneous activity of hES and hiPS cell-derived neuronal networks on MEA. After one week single spiking and training activity is occurring in both hES and hiPS cell-derived cultures. After four weeks both cultures show developed bursting activity. The cells are verified as neuronal cells with immunostaining. 3D-histograms present the responses of the hES and hiPS cell-derived neuronal networks to pharmaceuticals. First, the baseline activity was measured. Thereafter, the addition of 30 μM CNQX and 20 μM D-AP5 suppressed the activity. The activity of the networks reappeared when CNQX and D-AP5 were washed out. GABA-addition (100 μM) blocks the signalling whereas 30 μM bicuculline restored the signalling again. In 3D-histograms recording time [s] is represented by the y-axis and recorded channels in x-axis. The spike rate, spikes/second, is displayed on the z-axis.

5. Future perspectives

MEA-based platforms have been fairly satisfactorily established, yet several challenges still remain. For example, the large variation between individual MEA plates after cell seeding as well as the poor repeatability of connectivity in network development are areas that require improvement. One of the fruitful new approaches to reduce trial variability is a cell-cage (Erickson et al., 2008) where single cells on a network can be individually monitored. Future of

MEA technology will be 3-dimensional when moving towards stem cell technologies and cell therapies. It is essential to construct more *in vivo*-like environment where the whole culture can be screened in 3D. Similar approach is also important for toxicology and pharmacology to mimic tissues as accurately as possible. The current research has used mainly young, heterogeneous neuronal networks and in the future more efforts should be put into producing and measuring specific, well-characterized neuronal subtypes such as glutaminergic, GABAergic or dopaminergic neurons. This approach would bring MEA platforms closer to clinical relevance. For example, controlled co-cultures of various neuronal subtypes and glial cells would offer a tailored tool for clinical pharmacology and toxicology.

Also more relevant, standardized and more robust data analysis protocols should be developed. In addition to existing analysis protocols, companies, such as NeuroProof GmbH (Rostock, Germany), have developed protocols where hundreds of different parameters are evaluated from the functional networks (Schröder, et al., 2010). In practice, it seems that analysis focusing on different parameters on overall spiking activity and burst amount, duration and mean frequency are quite good indicators of networks' responses to different treatments (Hardingham, 2010), but these protocols are not standardised between laboratories. Clearly there is need for new kinds of analysis platforms as thousands of substances in EU area only should be screened for toxicological effects. Thus, MEA-based human cell platforms can offer new tools to take end point testing from animals (LD50, lethal dose for 50%) to cells (EC50, effective concentration in 50%) for testing more subtle and sensitive end points, such as alterations in neuronal activity patterns. Importantly, as there is a great need to reduce, replace and refine animal experiments, these functional level platforms will help to achieve this important goal. This goal requires that standardization and validation are seen as critical development points for the reproducibility and reliability of the MEA technology.

6. Conclusions

As MEA platforms have been shown to enable more accurate and refined analyses of for example developmental biology and neurotoxicity, the use of MEAs is expected to expand in the future. Some issues still, however, remain and efforts should be concentrated on improving e.g. the repeatability of the measurement platforms and setups.

7. Acknowledgements

Professor Timo Otonkoski, Professor Anu Wartiovaara, Riikka Hämäläinen and Ras Trokovic are acknowledged for collaboration with hiPS cells. The results presented here have been funded by Academy of Finland (projects 122959 and 123233), TEKES Stem in Clin, BioneXt Tampere, Competitive funding of Pirkanmaa hospital district and InterBrain University Alliance of Finland.

8. References

Abeles, M. (1982) Local Cortical Circuits: An Electrophysiological study. Springer, Berlin.

Aberdam, E.; Barak, E.; Rouleau, M.; de LaForest, S.; Berrih-Aknin, S.; Suter, D.M.; Krause, K.H.; Amit, M.; Itskovitz-Eldor, J. & Aberdam, D. (2008). A pure population of ectodermal cells derived from human embryonic stem cells. *Stem Cells*, 26. 2.; 440-4.

Adewumi, O.; Aflatoonian, B.; Ahrlund-Richter, L.; Amit, M.; Andrews, P.W.; Beighton, G.; Bello, P.A.; Benvenisty, N.; Berry, L.S.; Bevan, S.; Blum, B.; Brooking, J.; Chen, K.G.; Choo, A.B.; Churchill, G.A.; Corbel, M.; Damjanov, I.; Draper, J.S.; Dvorak, P.; Emanuelsson, K.; Fleck, R.A.; Ford, A.; Gertow, K.; Gertsenstein, M.; Gokhale, P.J.; Hamilton, R.S.; Hampl, A.; Healy, L.E.; Hovatta, O.; Hyllner, J.; Imreh, M.P.; Itskovitz-Eldor, J.; Jackson, J.; Johnson, J.L.; Jones, M.; Kee, K.; King, B.L.; Knowles, B.B.; Lako, M.; Lebrin, F.; Mallon, B.S.; Manning, D.; Mayshar, Y.; McKay, R.D.; Michalska, A.E.; Mikkola, M.; Mileikovsky, M.; Minger, S.L.; Moore, H.D.; Mummery, C.L.; Nagy, A.; Nakatsuji, N.; O'Brien, C.M.; Oh, S.K.; Olsson, C.; Otonkoski, T.; Park, K.Y.; Passier, R.; Patel, H.; Patel, M.; Pedersen, R.; Pera, M.F.; Piekarczyk, M.S.; Pera, R.A.; Reubinoff, B.E.; Robins, A.J.; Rossant, J.; Rugg-Gunn, P.; Schulz, T.C.; Semb, H.; Sherrer, E.S.; Siemen, H.; Stacey, G.N.; Stojkovic, M.; Suemori, H.; Szatkiewicz, J.; Turetsky, T.; Tuuri, T.; van den Brink, S.; Vintersten, K.; Vuoristo, S.; Ward, D.; Weaver, T.A.; Young, L.A. & Zhang, W. (2007). Characterization of human embryonic stem cell lines by the International Stem Cell Initiative. *Nat Biotechnol*, 25. 7.; 803-16.

Baharvand, H.; Mehrjardi, N.Z.; Hatami, M.; Kiani, S.; Rao, M. & Haghighi, M.M. (2007). Neural differentiation from human embryonic stem cells in a defined adherent culture condition. *Int J Dev Biol*, 51. 5.; 371-8.

Ban, J.; Bonifazi, P.; Pinato, G.; Broccard, F.; Studer, L.; Torre, V. & Ruaro, M.E. (2006). ES-derived neurons form functional networks in vitro. *Stem Cells*, 25. 3.; 738-49.

Beggs, J. M. & D. Plenz (2003). Neuronal avalanches in neocortical circuits. *J Neurosci*, 23. 35.; 11167-77.

Beggs, J. M. & D. Plenz (2004). Neuronal avalanches are diverse and precise activity patterns that are stable for many hours in cortical slice cultures. *J Neurosci*, 24. 22.; 5216-29.

Benzing, C.; Segschneider, M.; Leinhaas, A.; Itskovitz-Eldor, J. & Brustle, O. (2006). Neural conversion of human embryonic stem cell colonies in the presence of fibroblast growth factor-2. *Neuroreport*, 17. 16.; 1675-81.

Bongso, A.; Fong, C.Y.; Ng, S.C. & Ratnam, S. (1994). Isolation and culture of inner cell mass cells from human blastocysts. *Hum Reprod*, 9. 11.; 2110-7.

Carpenter, M.K.; Inokuma, M.S.; Denham, J.; Mujtaba, T.; Chiu, C.P. & Rao, M.S. (2001). Enrichment of neurons and neural precursors from human embryonic stem cells. *Exp Neurol*, 172. 2.; 383-97.

Chambers, S.M.; Fasano, C.A.; Papapetrou, E.P.; Tomishima, M.; Sadelain, M. & Studer, L. (2009). Highly efficient neural conversion of human ES and iPS cells by dual inhibition of SMAD signaling. *Nat Biotechnol*, 27. 3.; 275-80.

Corner, M.A. & Ramakers, G.J. (1991) Spontaneous bio- electric activity as both dependent and independent variable in cortical maturation. Chronic tetrodotoxin versus picrotoxin effects on spike-train patterns in developing rat neocortex neurons during long-term culture. *Ann. NY Acad. Sci*, 627.; 349–353.

Erceg, S.; Lainez, S.; Ronaghi, M.; Stojkovic, P.; Perez-Arago, M.A.; Moreno-Manzano, V.; Moreno-Palanques, R.; Planells-Cases, R. & Stojkovic, M. (2008). Differentiation of human embryonic stem cells to regional specific neural precursors in chemically defined medium conditions. *PLoS ONE*, 3. 5. e2122.

Erickson, J.C.; Tooker, A.; Tai, Y.-C. & Pine, J. (2008). Caged neuron MEA: A system for long-term investigation of cultured neural network connectivity. *J Neurosci Meth*, 175., 1-16.

Evans, M.J. & Kaufman, M.H. (1981). Establishment in culture of pluripotential cells from mouse embryos. *Nature*, 292. 5819.; 154-6.

Eytan, D. & Marom, S. (2006). Dynamics and effective topology underlying synchronization in networks of cortical neurons. *J Neurosci*, 26.; 8465–76.

Gerrard, L.; Rodgers, L. & Cui, W. (2005). Differentiation of human embryonic stem cells to neural lineages in adherent culture by blocking bone morphogenetic protein signaling. *Stem Cells*, 23. 9.; 1234-41.

Gross, G.W.; Rhoades, B.K.; Reust, D.L. & Schwalm FU. (1993) Stimulation of monolayer networks in culture through thin film indium-tin oxide recording electrodes. *J Neurosci Methods*, 50.; 131–43.

Gross, G.W. (1979). Simultaneous single unit recording in vitro with a photoetched laser deinsulated gold multi-microelectrode surface. *IEEE Trans. Biomed. Eng*, 26.; 273–279.

Gross, G.W.; Reiske, E.; Kreutzberg, G.W.; & Mayer, A. (1977). A new fixed-array multimicroelectrode system designed for long-term recording of extracellular single unit activity in vitro. *Neurosci. Lett*, 6.; 101–105.

Hakala, H.; Rajala, K.; Ojala, M.; Panula, S.; Areva, S.; Kellomäki, M.; Suuronen, R. & H, S. (2009). Comparison of biomaterials and extracellular matrices as a culture platform for multiple, indeoendently derived human embryonic stem cell lines. *Tissue Eng Part A*, 15.; 1-12.

Hardingham, G.E.; Patani, R.; Baxter, P.; Wyllie, D.J. & Chandran, S. (2010). Human embryonic stem cell-derived neurons as a tool for studying neuroprotection and neurodegeneration. *Mol Neurobiol*, 42. 1.; 97-102.

Heikkilä, T.J.; Ylä-Outinen, L.; Tanskanen, J.M.; Lappalainen, R.S.; Skottman, H.; Suuronen, R.; Mikkonen, J.E.; Hyttinen, J.A. & Narkilahti, S. (2009). Human embryonic stem cell-derived neuronal cells form spontaneously active neuronal networks in vitro. *Exp Neurol*, 218. 1.; 109-16.

Hovatta, O.; Mikkola, M.; Gertow, K.; Stromberg, A.M.; Inzunza, J.; Hreinsson, J.; Rozell, B.; Blennow, E.; Andang, M. & Ahrlund-Richter, L. (2003). A culture system using human foreskin fibroblasts as feeder cells allows production of human embryonic stem cells. *Hum Reprod*, 18. 7.; 1404-9.

Illes, S.; Fleischer, W.; Siebler, M.; Hartung, H.P. & Dihne, M. (2007). Development and pharmacological modulation of embryonic stem cell-derived neuronal network activity. *Exp Neurol*, 207. 1.; 171-6.

Illes, S.; Theiss, S.; Hartung, H.P.; Siebler, M. & Dihne, M. (2009). Niche-dependent development of functional neuronal networks from embryonic stem cell-derived neural populations. *BMC Neurosci*, 10. 93.

Imreh, M.P.; Gertow, K.; Cedervall, J.; Unger, C.; Holmberg, K.; Szoke, K.; Csoregh, L.; Fried, G.; Dilber, S.; Blennow, E. & Ahrlund-Richter, L. (2006). In vitro culture conditions favoring selection of chromosomal abnormalities in human ES cells. *J Cell Biochem*, 99. 2.; 508-16.

Itsykson, P.; Ilouz, N.; Turetsky, T.; Goldstein, R.S.; Pera, M.F.; Fishbein, I.; Segal, M. & Reubinoff, B.E. (2005). Derivation of neural precursors from human embryonic stem cells in the presence of noggin. *Mol Cell Neurosci*, 30. 1.; 24-36.

Jimbo, Y.; Robinson, H.P. & Kawana, A. (1998). Strengthening of synchronized activity by tetanic stimulation in cortical cultures: application of planar electrode arrays. *IEEE Trans Biomed Eng*, 45. 11.; 1297-304.

Jimbo, Y.; Tateno, T. & Robinson, H.P. (1999). Simultaneous induction of pathway-specific potentiation and depression in networks of cortical neurons. *Biophys J*, 76. 2.; 670-8.

Johnstone, A.F.; Gross, G.W.; Weiss, D.G.; Schroeder, O.H.; Gramowski, A. & Shafer, T.J. (2010). Microelectrode arrays: a physiologically based neurotoxicity testing platform for the 21st century. *Neurotoxicology*, 31. 4.; 331-50.

Kamioka, H; Maeda, E; Jimbo, Y; Robinson, H.P. & Kawana A. (1996) Spontaneous periodic synchronized bursting during formation of mature patterns of connections in cortical cultures. *Neurosci Lett.* 206(2-3):109–112

Karumbayaram, S.; Novitch, B.G.; Patterson, M.; Umbach, J.A.; Richter, L.; Lindgren, A.; Conway, A.E.; Clark, A.T.; Goldman, S.A.; Plath, K.; Wiedau-Pazos, M.; Kornblum, H.I. & Lowry, W.E. (2009). Directed differentiation of human-induced pluripotent stem cells generates active motor neurons. *Stem Cells*, 27. 4.; 806-11.

Lappalainen, R.S.; Salomäki, M.; Ylä-Outinen, L.; Heikkilä, J.T.; Hyttinen, J.A.K.; Suuronen, R.; Skottman, H. & Narkilahti, S. (2010). Similarly derived and cultured hESC lines show variation in their developmental potential towards neuronal cells in long-term culture. *Regen Med*, 5(5):749-762

Li, X.J.; Hu, B.Y.; Jones, S.A.; Zhang, Y.S.; Lavaute, T.; Du, Z.W. & Zhang, S.C. (2008). Directed differentiation of ventral spinal progenitors and motor neurons from human embryonic stem cells by small molecules. *Stem Cells*, 26. 4.; 886-93.

Madhavan, R.; Chao, Z.C. & Potter, S.M. (2007). Plasticity of recurring spatiotemporal activity patterns in cortical networks. *Phys. Biol*, 4.; 181–193.

Martin, G.R. (1981). Isolation of a pluripotent cell line from early mouse embryos cultured in medium conditioned by teratocarcinoma stem cells. *Proc Natl Acad Sci U S A*, 78. 12.; 7634-8.

Mazzoni, A.; Broccard, F.D.; Garcia-Perez, E.; Bonifazi, P.; Ruaro, M.E. & Torre, V. (2007). On the Dynamics of the Spontaneous Activity in Neuronal Networks. *PLoS ONE*, 2. 5.; e439.

Mennerick, S.; Chisari, M.; Shu, H.-J.; Taylor, A.; Vasek, M.; Eisenman, L.N. & Zorumski, C.F. (2010). Diverse Voltage-Sensitive Dyes Modulate GABAAReceptor Function. *J Neurosci*, 30.8.; 2871-2879.

Mitalipova, M.M.; Rao, R.R.; Hoyer, D.M.; Johnson, J.A.; Meisner, L.F.; Jones, K.L.; Dalton, S. & Stice, S.L. (2005). Preserving the genetic integrity of human embryonic stem cells. *Nat Biotechnol*, 23. 1.; 19-20.

Moretti, A.; Bellin, M.; Welling, A.; Jung, C.B.; Lam, J.T.; Bott-Flugel, L.; Dorn, T.; Goedel, A.; Hohnke, C.; Hofmann, F.; Seyfarth, M.; Sinnecker, D.; Schomig, A. & Laugwitz, K.L. (2010). Patient-Specific Induced Pluripotent Stem-Cell Models for Long-QT Syndrome. *N Engl J Med*.

Nat, R.; Nilbratt, M.; Narkilahti, S.; Winblad, B.; Hovatta, O. & Nordberg, A. (2007). Neurogenic neuroepithelial and radial glial cells generated from six human embryonic stem cell lines in serum-free suspension and adherent cultures. *Glia*, 55. 4.; 385-99, 0894-1491.

Otto, F.; Illes, S.; Opatz, J.; Laryea, M.; Theiss, S.; Hartung, H.P.; Schnitzler, A.; Siebler, M. & Dihne, M. (2009). Cerebrospinal fluid of brain trauma patients inhibits in vitro neuronal network function via NMDA receptors. *Ann Neurol*, 66. 4.; 546-55.

Pine, J. (2006). A history of MEA development. *Advances in network electrophysiology*.

Pine, J. (1980). Recording action potentials from cultured neurons with extracellular microcircuit electrodes. *J Neurosci Methods*, 2. 1.; 19-31.

Pomp, O.; Brokhman, I.; Ziegler, L.; Almog, M.; Korngreen, A.; Tavian, M. & Goldstein, R.S. (2008). PA6-induced human embryonic stem cell-derived neurospheres: a new source of human peripheral sensory neurons and neural crest cells. *Brain Res*, 1230.; 50-60.

Potter, S.M. & DeMarse, T.B. (2001). A new approach to neural cell culture for long-term studies. *J Neurosci Methods*, 110. 1-2.; 17-24.

Rashid, S.T.; Corbineau, S.; Hannan, N.; Marciniak, S.J.; Miranda, E.; Alexander, G.; Huang-Doran, I.; Griffin, J.; Ahrlund-Richter, L.; Skepper, J.; Semple, R.; Weber, A.; Lomas, D.A. & Vallier, L. (2010). Modeling inherited metabolic disorders of the liver using human induced pluripotent stem cells. *J Clin Invest*, 120. 9.; 3127-36.

Reubinoff, B.E.; Itsykson, P.; Turetsky, T.; Pera, M.F.; Reinhartz, E.; Itzik, A. & Ben-Hur, T. (2001). Neural progenitors from human embryonic stem cells. *Nat Biotechnol*, 19. 12.; 1134-40.

Shahaf, G. & Marom, S. (2001). Learning in networks of cortical neurons. *J Neurosci*, 21.; 8782–8788.

Schröder, O.H-U.; Weiss D.G.; Jügelt, K. & Gramowski, A. (2010). Profiling of acute neurotoxic effects of insect repellents with MEA-neurochip technology. *Conf. Proc. 7th International Meeting on Subtrate-Integrated Micro Electrode Arrays*, Reutlingen, Germany, June-July 2010.

Skottman, H.; Narkilahti, S. & Hovatta, O. (2007). Challenges and approaches to the culture of pluripotent human embryonic stem cells. *Regen Med*, 2. 3.; 265-73.

Swistowski, A.; Peng, J.; Liu, Q.; Mali, P.; Rao, M.S.; Cheng, L. & Zeng, X. (2010). Efficient Generation of Functional Dopaminergic Neurons from Human Induced pluripotent Stem Cells under Defined Conditions. *Stem Cells*.

Takahashi, K.; Tanabe, K.; Ohnuki, M.; Narita, M.; Ichisaka, T.; Tomoda, K. & Yamanaka, S. (2007). Induction of pluripotent stem cells from adult human fibroblasts by defined factors. *Cell*, 131. 5.; 861-72.

Tanskanen, J.M.A.; Mikkonen, J.E. & Penttonen, M. (2005). Independent component analysis of neural populations from multielectrode field potential measurements. *J Neurosci Methods*, 145. 1-2.; 213-232.

Tetzlaff, T.; Geisel, T. & Diesmann M. (2002). The ground state of cortical feed-forward networks. *Neurocomputing*, 44-46.; 673-678.

Thomas, C.A.; Springer, P.A.; Loeb, G.E.; Berwald-Netter, Y. & Okun, L.M. (1972). A miniature microelectrode array to monitor the bioelectric activity of cultured cells. *Exp Cell Res*, 74.; 61–66.

Thomson, J.A.; Itskovitz-Eldor, J.; Shapiro, S.S.; Waknitz, M.A.; Swiergiel, J.J.; Marshall, V.S. & Jones, J.M. (1998). Embryonic stem cell lines derived from human blastocysts. *Science*, 282. 5391.; 1145-7.

Wagenaar, D.A.; Pine, J. & Potter, S.M. (2006). An extremely rich repertoire of bursting patterns during the development of cortical cultures. *BMC Neurosci*, 7. 11.

Yao, S.; Chen, S.; Clark, J.; Hao, E.; Beattie, G.M.; Hayek, A. & Ding, S. (2006). Long-term self-renewal and directed differentiation of human embryonic stem cells in chemically defined conditions. *Proc Natl Acad Sci U S A*, 103. 18.; 6907-12.

Ylä-Outinen, L.; Heikkilä, J.; Skottman, H.; Suuronen, R.; Äänismaa, R. & Narkilahti, S. (2010). Human cell-based micro electrode array platform for studying neurotoxicity. *Front Neuroeng*, 3:111

Yu, J.; Vodyanik, M.A.; Smuga-Otto, K.; Antosiewicz-Bourget, J.; Frane, J.L.; Tian, S.; Nie, J.; Jonsdottir, G.A.; Ruotti, V.; Stewart, R.; Slukvin, II & Thomson, J.A. (2007). Induced pluripotent stem cell lines derived from human somatic cells. *Science*, 318. 5858.; 1917-20.

Zhang, S.C.; Wernig, M.; Duncan, I.D.; Brustle, O. & Thomson, J.A. (2001). In vitro differentiation of transplantable neural precursors from human embryonic stem cells. *Nat Biotechnol*, 19. 12.; 1129-33.

26

A Novel Strategy for Drug Discovery and Development by Analyzing the Behavior of ES Cells Cultured on TOSHI (Tissue/Organ Sections for Histopathology)–Substrata

Toshiaki Takezawa
National Institute of Agrobiological Sciences, Japan

1. Introduction

To create new *in vitro* culture models for extrapolating the cell response *in vivo*, Takezawa *et al.* have attempted to devise culture substrata of anchorage-dependent cells (Takezawa et al., 1990; Takezawa et al., 1992; Takezawa & Yoshizato, 1997; Takezawa et al., 2000; Takezawa et al., 2002; Takezawa, 2003; Takezawa et al., 2004; Takezawa et al., 2007a; Takezawa et al., 2007b; Takezawa et al., 2008a; Takezawa et al., 2008b; Takezawa et al., 2010). One model is a culture system utilizing substrata made of tissue/organ sections for histopathology (TOSHI), which was found to conserve both tissue components and microarchitecture in an *in vivo* environment (Takezawa et al., 2002; Takezawa, 2003).

Meanwhile, it is reported that mouse embryonic stem (ES) cells injected into the tail veins of carbon tetrachloride (CCl_4) liver-injured mouse were differentiated into hepatocyte-like cells in the host liver (Yamamoto et al., 2003). Therefore, we investigated whether the ES cells could also be differentiated into hepatocyte-like cells when they were cultured on the TOSHI-substrata prepared from livers in various stages after CCl_4 administration into mice. Consequently, it was found that the substrata derived from regenerating livers enhanced cell attachment, supported growth as clusters, and induced differentiation into cells expressing albumin, although the substrata from injured livers did not. In particular, the cells cultured on the most proliferative regenerating liver-derived substratum reconstructed the hepatic cord-like structures with bile canaliculus-like aspects in which some binucleated cells were involved, secreted albumin, and expressed cytochrome P450IA1 activity within a few days (Takeuchi et al., 2008; Takezawa et al., 2008; Takezawa et al., 2008b).

These data suggest two advantages based on the behavior of ES cells in a culture system utilizing TOSHI-substrata; one is that TOSHI-substratum derived from regenerating livers with high proliferative potential efficiently induced the differentiation of ES cells toward hepatic lineage and another is that ES cells functioned as a sensor recognizing liver toxicity in TOSHI-substratum derived from injured livers after CCl_4 administration into mice. Therefore, it is considered that the former would be available for the novel approach of drug discovery to find bioactive factors and the latter provide a new alternative method of animal experiments in toxicology to support drug development.

In this chapter, I propose a novel strategy for drug discovery and development by analyzing the behavior of ES cells cultured on TOSHI-substrata. Section 2 describes the basics and advantages of a cell culture system utilizing TOSHI-substrata. Section 3 presents behavior of mouse ES cells cultured on TOSHI-substrata derived from livers in various stages after CCl_4 administration into mice. Section 4 proposes a novel concept for drug discovery and development utilizing the behavior of ES cells cultured on TOSHI-substrata. Section 5 presents a discussion and summarizes the paper.

2. The basics and advantages of a cell culture system utilizing TOSHI-substrata

In this section I describe the basic outline of a cell culture system utilizing TOSHI-substrata and the advantages of investigating cell behavior on TOSHI-substrata. A novel idea for diagnosing characteristics of undefined cells and tissues will be introduced as applied concepts for "cellomics" and "histomics", respectively.

2.1 The basic outline of a cell culture system utilizing TOSHI-substrata

A lot of cell culture substrata that control multicellular behaviors have been developed from natural or synthetic or their hybrid materials and are currently utilized for reconstructing organoids as cellular scaffold(s) in tissue/organ engineering (Takezawa et al., 2000). However, no one has succeeded in producing a culture substratum reflecting the complex architecture of various cellular microenvironments in tissues *in vivo*. On the other hand, thin tissue/organ sections commonly prepared on a glass slide for histopathology retain many *in vivo* biochemical attributes related not only to structure but also function. We hypothesized that such tissue/organ sections might serve as novel cell culture substrata that would reflect conditions *in vivo*.

In the beginning study, such novel culture substrata made of tissue/organ sections for histopathology (TOSHI-substrata) were prepared from a frozen bovine placenta embedded in an OCT compound, and subsequently four different types of cells were cultured on the substrata. As the results, the labyrinth region of the substratum induced unique cell behaviors to form multicellular spheroids of BeWo cells (human choriocarcinoma cell line), a capillary network-like structure of CPAE cells (bovine pulmonary artery endothelial cell line), and a neuronal network-like structure of PC-12 cells (rat pheochromocytoma cell line). The substratum provided a microenvironment that maintained the viability of PC-12 cells under serum-free conditions. Also, we succeeded in preparing a multicellular mass of NHDFs (normal human dermal fibroblasts) with acellularized section-derived components (Takezawa et al., 2002).

The typical experimental procedure for a cell culture system utilizing TOSHI-substrata was shown in Figure 1.

2.2 The advantages of investigating cell behavior on TOSHI-substrata

TOSHI-substrata can be prepared various tissues and every cell line can be cultured on the TOSHI-substrata, and consequently we can analyze the behavior of each cell line induced by each TOSHI-substratum. Therefore, we consider that it will be available to construct the database cumulated diverse interactions between culture cells and tissues/organs of the substrata. As the first step, we recently prepared TOSHI-substrata from normal rat mature organs (cerebrum, thymus, heart, liver, kidney and testis) in a frozen state and cultured two

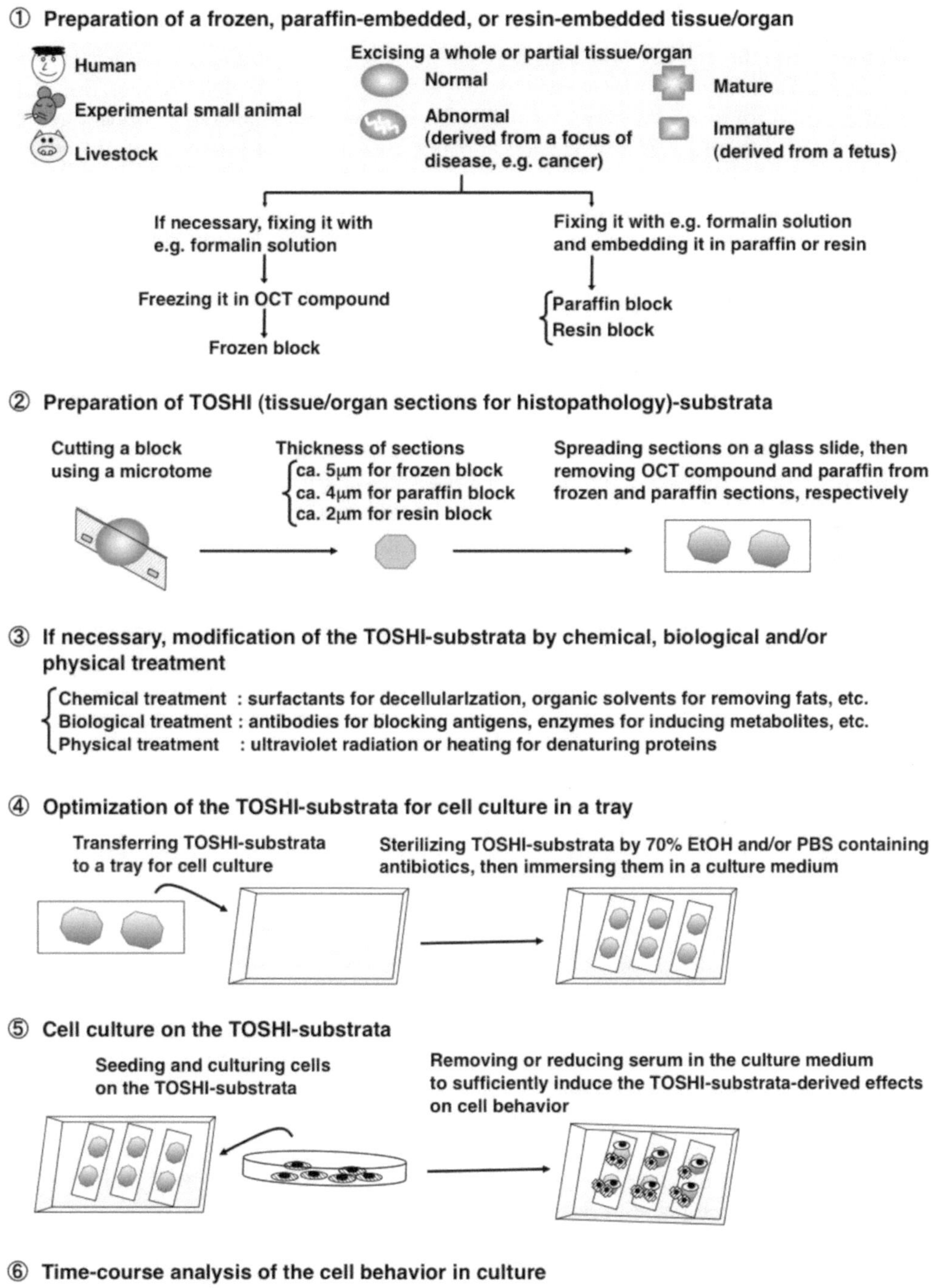

Fig. 1. Schematic procedure for preparing TOSHI (tissue/organ sections for histopathology) -substrata and culturing cells on it. Reproduced and partially-modified with permission of the publisher (Takezawa et al., 2008a).

different cell lines (RIN5F cells of rat insulinoma and HepG2 cells of human hepatoma) to compare their behaviors on the substrata. As the results, RIN5F cells showed high attachability onto the TOSHI-substrata derived from heart and kidney, standard growth on the substrata from heart, kidney, and liver, and high level of insulin secretion on the substratum from thymus. Whereas HepG2 cells showed low attachability onto the TOSHI-substrata derived from liver, and kidney and not only standard growth but also standard level of albumin secretion on the substrata from cerebrum, thymus, heart, liver, kidney and testis. Also, the time-course profiles for cell growth and secretion level of insulin or albumin on each TOSHI-substratum were successfully converted into a three-dimensional graph chart, i.e. a mathematical model. In conclusion, it was demonstrated that the mathematical profile describing cell behavior was specific for the combination of a cell line and a tissue/organ of the substrata (Yanagihara et al., 2007). These findings suggest that we can provide novel research concept for cellomics and histomics (Fig. 2). Here, cellomics and histomics facilitate comprehensive analyses for various cell lines using one tissue/organ-derived TOSHI-substratum and for various tissue/organ-derived TOSHI-substrata using one cell line, respectively. In future, the database construction for cellomics and histomics would lead a new system for diagnosing characteristic-unknown cell line or a characteristic-unknown TOSHI-substratum by feedback of the cell behavior profile obtained from the combinations of "a characteristic-unknown cell line and characteristic-known TOSHI-substrata" or "characteristic-known cell lines and a characteristic-unknown TOSHI-substratum" to the database.

Further, applied researches such as induction of cell differentiation, serum-free culture, exploration and/or production of bioactive molecules, estimation of gene function, and reconstruction of tissues will be achieved by designing an appropriate combination of a cell line and a TOSHI-substratum.

3. Behavior of mouse ES cells cultured on TOSHI-substrata derived from livers in various stages after CCl_4 administration into mice

In this section I mainly describe our previous study to differentiate mouse ES cells towards hepatic lineage utilizing TOSHI-substrata derived from regenerating livers after CCl_4 administration into mice (Takeuchi et al., 2008), then briefly introduce a new strategy for predicting the toxicity of chemicals.

3.1 Background for the study to differentiate mouse ES cells towards hepatic lineage

Embryonic stem (ES) cells differentiate into tissues of all the three germ layers, and the effort is to restrict or direct their development to specific cell type or lineage. A number of studies have attempted to produce hepatocytes from ES cells and have successfully induced hepatic development *in vitro* (Ishi et al., 2005; Lavon & Benvenisty, 2005; Novik et al., 2006; Teratani et al., 2005; Zhou et al., 2007). However, all these studies have used culture systems supplemented by a temporal series of growth factors similar to those known to control embryonic liver development and therefore limited success, e.g. low efficacy of hepatic differentiation or it takes long period. Meanwhile, mouse ES cells injected into the tail vein of carbon tetrachloride (CCl_4) liver-injured mouse were differentiated into hepatocyte-like cells in the host liver (Yamamoto et al., 2003). Here, we aimed to investigate whether the ES cells could also be differentiated into hepatocyte-like cells when they were cultured on the TOSHI-substrata prepared from livers in various stages after CCl_4-administration into mice.

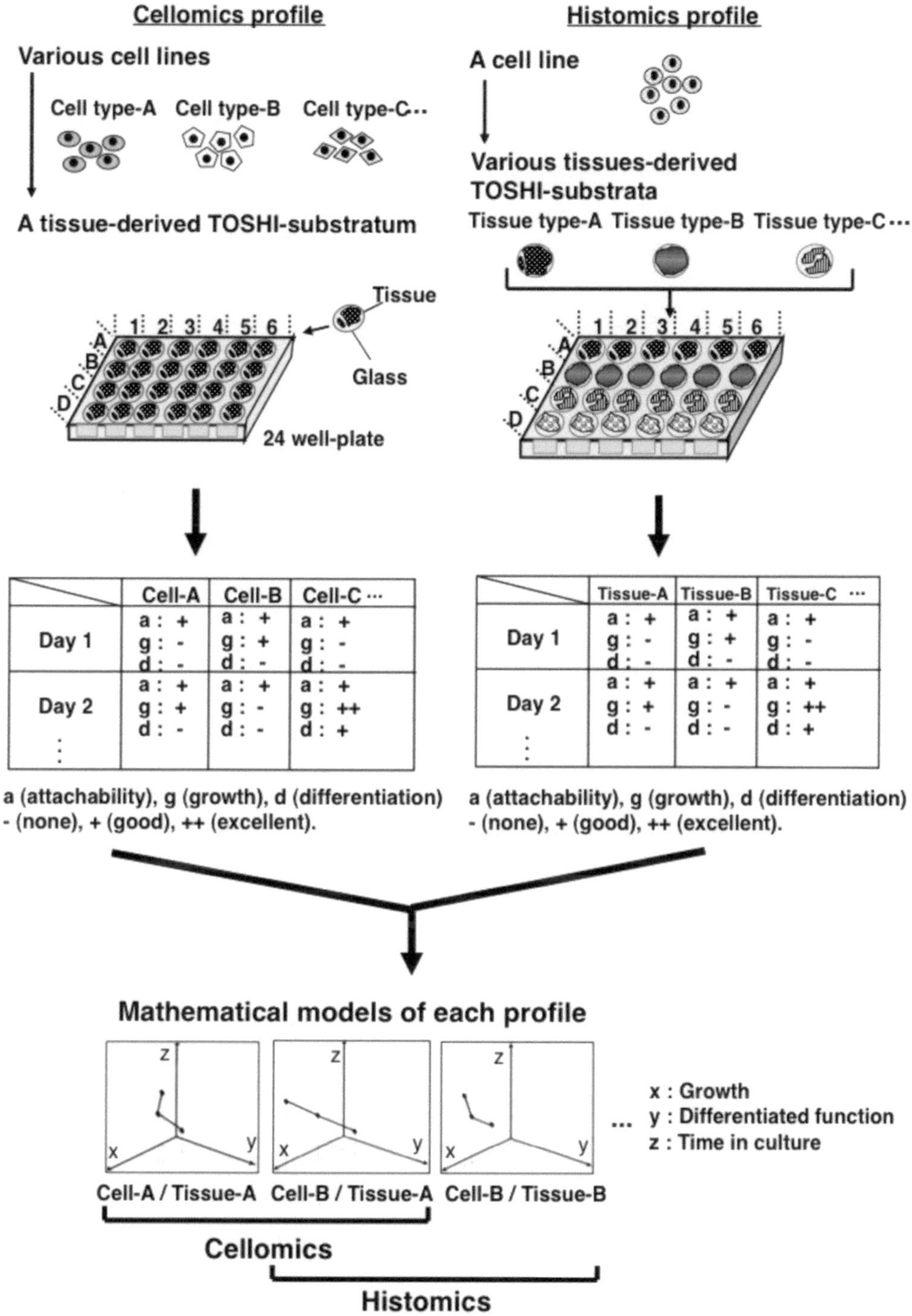

Fig. 2. Novel research concept for cellomics and histomics. Reproduced and partially-modified with permission of the publisher (Takezawa et al., 2008a).

3.2 Preparation of TOSHI-substrata reflecting liver failure and regeneration after CCl_4-administration into mice

Male mice (129SV) were purchased from CLEA. Light liver injury was caused by intraperitoneally injecting 100 μl of olive oil containing 10 μl of CCl_4 into 8 week-old mice weighing between 18 and 23 g to induce liver regeneration. As a control, 100 μl of olive oil alone was intraperitoneally injected into other mice. The blood sera and livers of the mice at 12, 24 and 48 h, 4, 7 and 15 days after CCl_4-administration and at 24 h after olive oil-administration were subjected to the measurement of aspartate aminotransferase (AST) leakage level and the preparation of TOSHI-substrata, respectively. Livers were excised immediately after sacrificing the mice, embedded in an OCT compound and rapidly frozen with liquid nitrogen. The frozen livers were cut into sections with a thickness of 5 μm. The liver-derived sections were spread onto a glass slide and air-dried to yield TOSHI-substrata. All animal experiments were performed in accordance with the guidelines of the Experimental Animal Committee at the National Institute of Agrobiological Sciences.

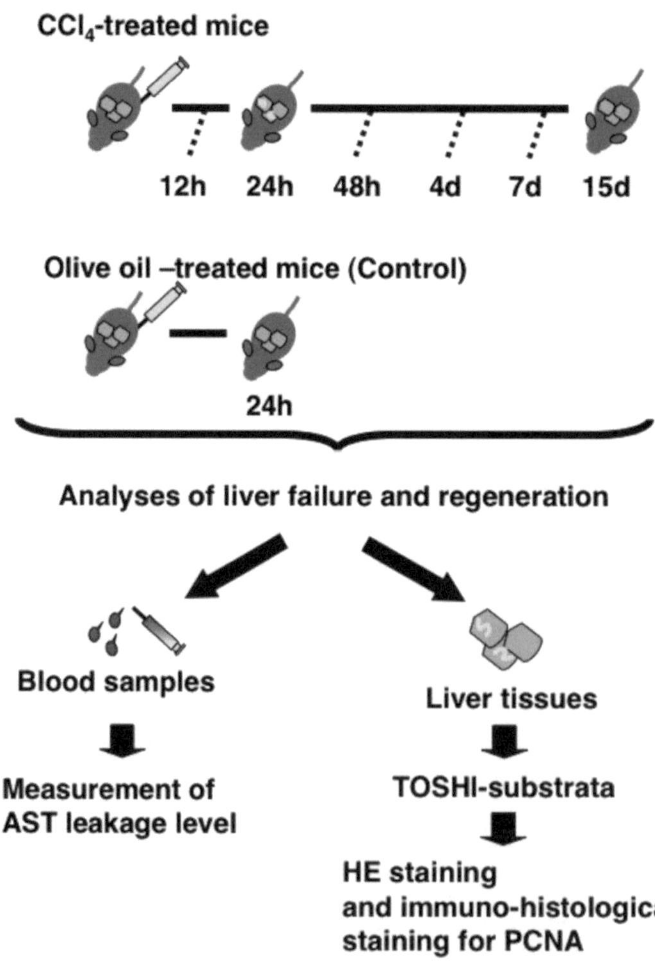

Fig. 3. Analytical protocol of liver failure and regeneration after CCl_4-administration into mice.

To investigate the time-course changes of liver failure and regeneration, the AST leakage level in blood sera was measured by utilizing an assay kit for AST (Wako). Also, TOSHI-substrata without culturing cells were subjected to standard hematoxylin and eosin (HE) staining and immuno-histological staining for proliferating cell nuclear antigen (PCNA), and observed by a light microscope. Subsequently, the PCNA-positive cell number in a unit area of 1.31 mm^2 on each substratum was counted from three independent microphotographs (Fig. 3).

As a result, the serum AST level of liver-injured mice increased more than three times from 12 to 48 h after CCl_4-administration, then gradually decreased and finally recovered on day 15 (Table 1). Meanwhile, immuno-histological observations of liver-derived TOSHI-substrata revealed that the PCNA-positive cell number increased in mice from 24 h to 7 days after CCl_4-administration and that the peak representing about 27-fold of the control mice was seen in mice on day 4 (Table 1). Also, this progression process of liver failure and regeneration was clearly confirmed by observation of the TOSHI-substrata stained with HE (Takeuchi et al., 2008).

3.3 Culture of mouse ES cells on TOSHI-substrata reflecting liver failure and regeneration after CCl_4-administration into mice

Mouse ES cells, J1 cell clones of 129SV male origin, and pALB-EGFP/ES cells [a stable mouse ES J1 cell line carrying a plasmid vector of pALB-EGFP which drives expression of albumin promoter-induced green fluorescent protein (GFP)] were sub-cultured on feeder fibroblasts according to the previous method (Teratani et al., 2005; Yamamoto et al., 2003). TOSHI-substrata reflecting liver failure and regeneration after CCl_4-administration, a control TOSHI-substratum prepared from a liver of the mice at 24 h after olive oil-administration, and a TOSHI-free glass slide substratum were subjected to the optimization for cell culture. Each substratum was inserted into a well of 4-well culture plates (Greiner), and immersed twice in phosphate buffered saline (PBS) containing 200 units/ml penicillin and 200 µg/ml streptomycin for 10 min and once in the culture medium for 5 min before use.

To investigate cell attachability, growth, morphology and differentiation into the cells expressing albumin, the suspension of pALB-EGFP/ES cells prepared in a culture medium [Dulbecco's Modified Eagle Medium with 4.5 g/l glucose (DMEM) containing 20% heat-inactivated fetal bovine serum (FBS), 1,000 units/ml leukemia inhibitory factor (LIF) (ESGRO; Chemicon), 100 µM 2-mercaptoethanol (2-ME), 0.1 mM non-essential amino acids, 30 µM Adenosine, 30 µM Guanosine, 30 µM Cytidine, 30 µM Uridine, 10 µM Thymidine, 100 units/ml penicillin, 100 µg/ml streptomycin and 250 ng/ml Amphotericin B] was seeded on each substratum at an initial cell density of 1.0 x 10^5 cells/cm^2 (Fig. 4). The cells were cultured at 37°C in a humidified atmosphere of 5% CO_2 in air and the culture media were once changed to basal media of a DMEM alone at 2 h in culture. The morphology and GFP expression of pALB-EGFP/ES cells cultured on each substratum were observed over time by a phase-contrast microscope and a fluorescent one under the same visual field, respectively. For analyzing the attachability, growth and differentiation of pALB-EGFP/ES cells, the area of the regions occupied by cells in a unit area of 1.18 mm^2 on each substratum at 2, 8, 16, 24 and 48 h in culture and that of their GFP-positive parts at 24 h were measured by applying NIH image software to computer images converted from three independent microphotographs, and consequently the ratio of the area occupied by cells and the appearance ratio of GFP-positive cells among the cells on each substratum were calculated, respectively. Similarly, each area of ten cell colonies on the CCl_4-4d substratum and the appearance ratio of GFP-positive cells in each colony were calculated at 2, 8, 16, 24, 48, 96 and 168 h in culture.

As a result, the initial cell attachment onto the control TOSHI-substratum prepared from livers at 24 h after olive oil-administration into mice (cont.-substratum) was 1.7-hold higher than that onto the glass slide-substratum (glass-substratum). As compared to the cont.-substratum, the TOSHI-substratum prepared from livers at 12 h after CCl_4-administration into mice (CCl_4-12h substratum) suppressed the attachment of ES cells whereas the CCl_4-4d and CCl_4-7d substrata promoted it (Table 1). The population of pALB-EGFP/ES cells cultured in the basal media gradually decreased on any substratum except for the CCl_4-4d substratum as the culture period increased. Most colonies on the CCl_4-4d substratum expanded their size until 16 or 24 h in culture, and subsequently reduced the size due to spontaneous detachment of the colony cells from the substratum. About 10 % of the colonies on the CCl_4-4d substratum enlarged its size more than 4-fold at 24 h, suggesting the colony cells divided about 2 times within 24 h. Furthermore, GFP-positive cells indicating the albumin expression appeared in some colonies formed on the CCl_4-4d substratum at 16 h in culture and were still observed at 168 h. To the contrary, pALB-EGFP/ES cells cultured on the glass-, cont.-, CCl_4-12h, CCl_4-24h substrata did not express GFP. The extent of GFP positive cells on the CCl_4-48h, CCl_4-4d, CCl_4-7d, CCl_4-15d substrata was about 46, 70, 24, 26% of the cells at 24 h in culture, respectively (Table 2). Morphological observation revealed

Time after CCl_4-administration	12h	24h	48h	4d	7d	15d
AST leakage level	3.12	3.38	3.12	2.37	1.37	1.07
PCNA-positive cells	0.66	3.31	2.66	26.6	6.06	0.38
Cell attachability	0.69	0.89	1.07	1.20	1.20	0.94

Table 1. Ratio changes of AST leakage level in sera, PCNA-positive cells and cell attachability in TOSHI-substrata at 12h, 24h, 48h, 4d, 7d and 15d after CCl_4-administration in comparison to the control serum and TOSHI-substratum at 24 h after olive oil-administration. The initial cell attachability onto each substratum were examined after culturing the cells in a culture medium mainly supplemented with fetal bovine serum, leukemia inhibitory factor, and 2-mercaptoethanol for 2 h.

Toshi-free substratum	Toshi-substrata						
Glass	Control	CCl_4-12h	CCl_4-24h	CCl_4-48h	CCl_4-4d	CCl_4-7d	CCl_4-15d
0 %	0 %	0 %	0 %	45.9 %	69.8 %	23.8 %	25.8 %

Table 2. The ratio of GFP positive cells in the cells cultured on the indicated substrata for 24h.

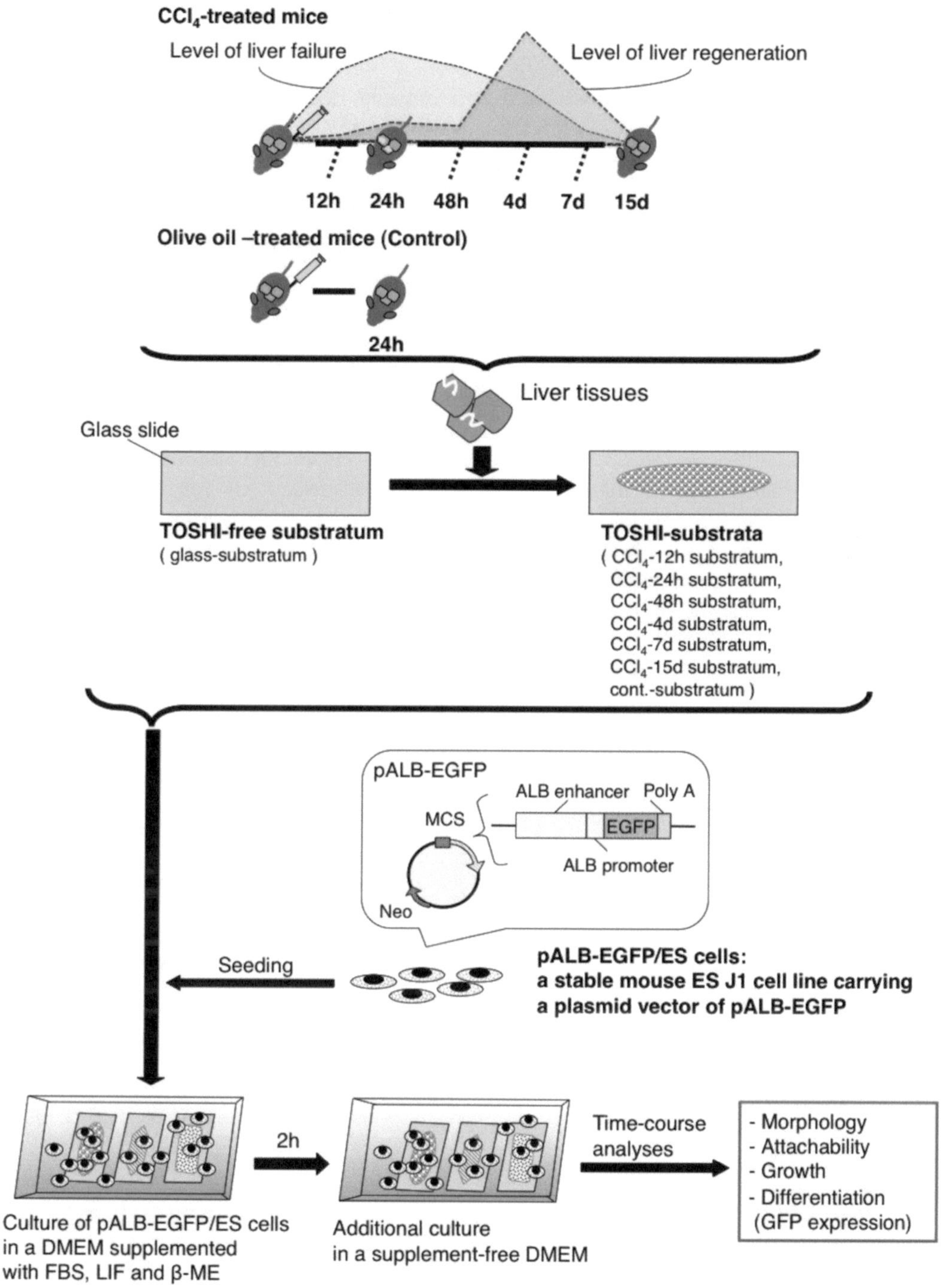

Fig. 4. Experimental protocol of investigating the behavior of pALB-EGFP/ES cells cultured on TOSHI-substrata reflecting liver failure and regeneration after CCl_4-administration.

that pALB-EGFP/ES cells cultured for 24 h on the glass-, cont.-, CCl_4-12h, CCl_4-24h substrata were remarkably reduced in number due to spontaneous detachment from each substratum, and that the retained cells relatively small and round in shape were individually scattered on each substratum and entirely negative for GFP expression. In contrast, those on the CCl_4-48h, CCl_4-4d, CCl_4-7d, CCl_4-15d substrata mostly participated in forming clusters in which the cells tended to express GFP and relatively large colonies were observed on the sinusoidal capillary region of the CCl_4-4d and CCl_4-7d substrata. In particular, the colonies on the CCl_4-4d substratum showed reconstruction of hepatic cord-like structures with bile canaliculus-like aspects (Harada et al., 2003) between the cells in polygonal shape and most of the colony cells expressed GFP.

Further, to investigate any changes in hepatic morphology and function ES cells were seeded and cultured on the CCl_4-4d substratum (Fig. 5). The culture medium mainly containing FBS, LIF and 2-ME was changed to a basal medium without the supplements at 2 h to estimate the effect of the substratum on cell growth and differentiation after inducing cell adherence under the presence of serum. ES cells cultured on the CCl_4-4d substratum for 24 h were stained with HE and observed by a light microscope. The albumin secretion level into a culture medium was measured by utilizing an assay kit for mouse albumin (Shibayagi). The cytochrome P450IA1 activity was analyzed by adding ethoxyresorufin (ER) into the culture media at a final concentration of 10 mM, incubating for 60 min and measuring the resorufin production level (Sakai et al., 2002).

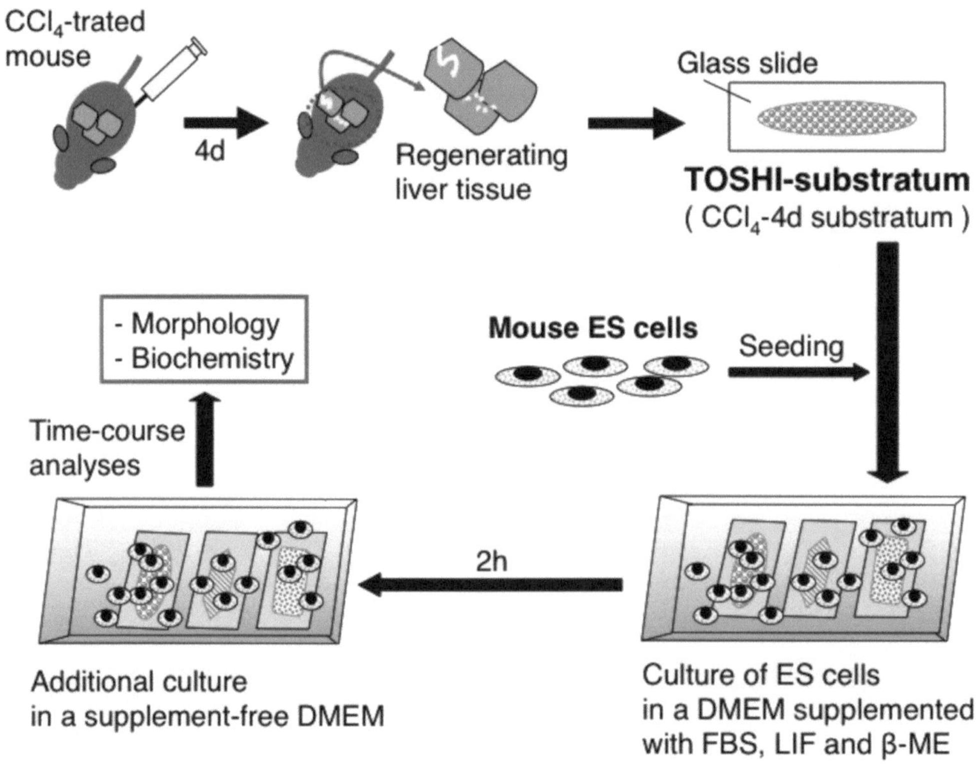

Fig. 5. Experimental protocol of investigating the hepatic differentiation of ES cells cultured on TOSHI-substratum derived from regenerating livers at 4 days after CCl_4 administration.

Consequently, binucleated cells in a polygonal shape appeared in the colonies cultured for 24 h. Albumin secretion into a culture medium and the cytochrome P450IA1 activity metabolizing ER to the dealkylated resorufin were confirmed at 48 and 62 h in culture, respectively (Fig. 6). Also, the CCl_4-4d substratum alone released albumin and reached to the plateau level of 0.1 ng/ml until 24 h, however it did not possess the cytochrome P450IA1 activity.

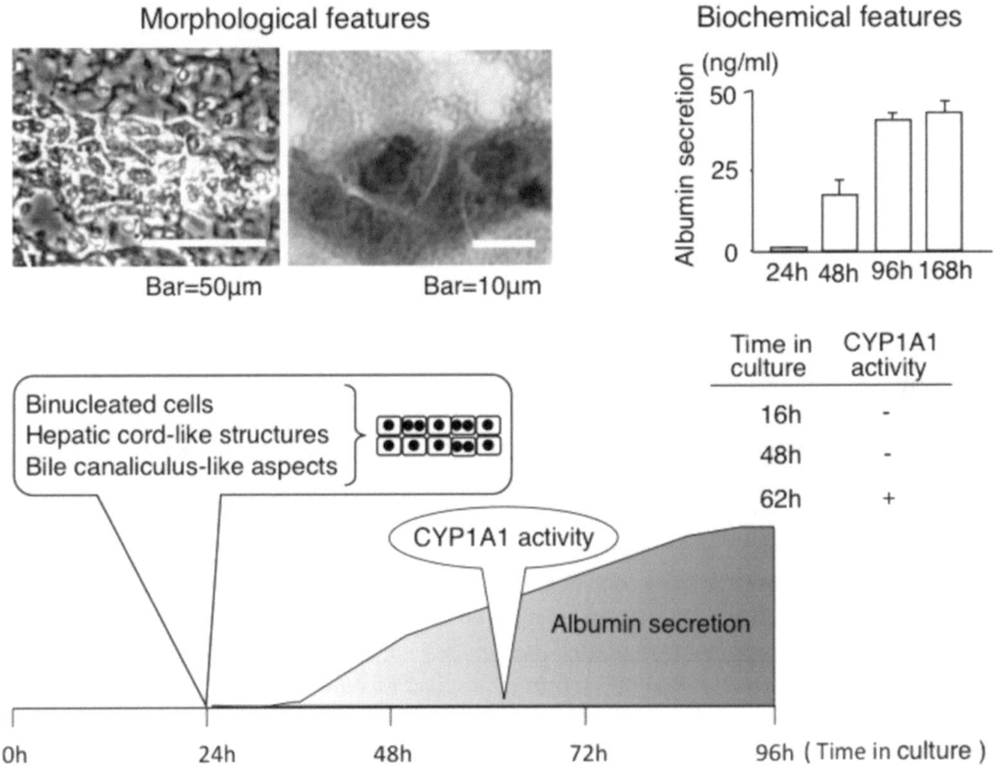

Fig. 6. Hepatic differentiation of ES cells cultured on TOSHI-substratum derived from regenerating livers at 4 days after CCl_4 administration.

In this study, we demonstrated that the CCl_4-4d substratum had a high potential to efficiently differentiate ES cells into hepatocyte-like clonies with some functional and morphological characteristics of mature parenchymal hepatocytes within a few days. Thus, we succeeded in directly differentiating ES cells into hepatic cells without the use of any soluble growth factor and also the process of embryoid body formation although either was essential for conventional methods (Ishi et al., 2005; Lavon & Benvenisty, 2005; Novik et al., 2006; Teratani et al., 2005; Zhou et al., 2007). Also, each TOSHI-substratum prepared from livers in various stages after CCl_4-administration into mice had different activities for inducing behavior of ES cells. It is quite interesting that the CCl_4-15d substratum was different from the cont.-substratum and retained the induction activity of differentiation although the mice on 15 days after CCl_4-administration showed the normal level of both serum AST and liver PCNA-positive cell index, suggesting that liver regeneration-related factor(s) was deposited in the CCl_4-15d substratum. Such ES cell behavior-regulating factor(s) that was sustained on each TOSHI-substratum prepared from CCl_4-administrated mouse livers is unclear at present. To investigate the molecular mechanism inducing the different profile of ES cell behavior on each TOSHI-substratum, we are planning to treat the TOSHI-substratum with antibody or enzyme in biological approach, fixation agent or detergent in chemical approach, and/or heat or ultra violet irradiation in physical approach.

3.4 A new strategy for predicting the toxicity of chemicals

In the above experiment, the serum AST level indicating liver injury and the number of PCNA-positive cells indicating liver regeneration increased more than twice of the control from 12 h to 4 d with the peak (about 3.4-fold of the control) at 24 h and from 24 h to 7 d with the peak (about 27-fold of the control) at 4 d after CCl_4-administration, respectively (Table 1). On the other hand, the initial cell attachability of pALB-EGFP/ES cells to TOSHI-substrata prepared from livers after CCl_4 administration into mice represented -31% on the CCl_4-12h and -11% on CCl_4-24h substrata whereas +20% on the CCl_4-4d and CCl_4-7d substrata in comparison to the control TOSHI-substratum (Table 1). Also, the CCl_4-48h, CCl_4-4d, CCl_4-7d and CCl_4-15d substrata showed the induction activity to differentiate the pALB-EGFP/ES cells to the cells expressing GFP (Table 2). These findings suggest that time-dependent toxic changes of CCl_4 in liver *in vivo* can be estimated by the cell behaviors of pALB-EGFP/ES cells cultured on the TOSHI-substrata prepared from livers after CCl_4 administration into mice.

I think that an organ-specific toxicity level after drug administration or chemical exposure can be defined as not only injury intensity of differentiated cells composing the organ but also regeneration activity of stem cells in the organ. Here, the above study demonstrated that the injury intensity and regeneration activity was predicted by the initial cell attachability and differentiation efficiency of ES cells cultured on the TOSHI-substrata derived from livers after administrating chemical into mice, respectively. Therefore, I propose as a new strategy for predicting the toxicity of drugs or chemicals in human by utilizing the time-course cell behavior of human stem cells cultured on the TOSHI-substrata derived from various organs after exposing some chemical into experimental animals. The human stem cells include ES cells, iPS (induced pluripotent stem) cells and somatic stem cells (e.g. mesenchymal stem cells), (Fig. 7).

4. A novel concept for drug discovery and development utilizing the behavior of ES cells cultured on TOSHI-substrata

In this section I describe the recent progress of ES cell-based assay systems for drug discovery and development, then introduce advantages of the ES cell-based assay system utilizing TOSHI-substrata in drug discovery and development.

4.1 Recent progress of ES cell-based assay systems for drug discovery and development

ES cells are pluripotent stem cells possessing the capacity for self-renewal of proliferating indefinitely and multilineage differentiation. Therefore, ES cells generally function as a ready source of large quantities of partially differentiated progenitors or terminally differentiated specialized cells, which serve as cell-based assay systems (Sartipy et al., 2007). Somatic stem cell marker prominin-1/CD133, a plasma membrane marker is expressed in

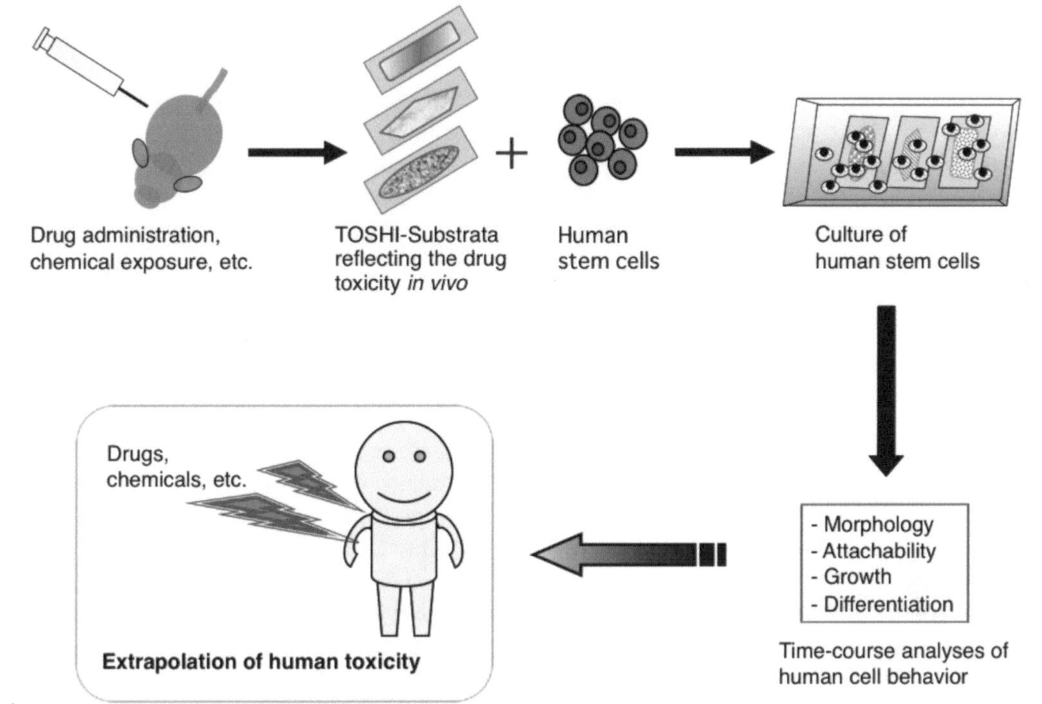

Fig. 7. A novel strategy for extrapolating the toxicity *in vivo* utilizing TOSHI-substrata.

ES-derived progenitors but not in differentiated cells (Kania et al., 2005). ES-derived somatic stem cells such as hematopoietic and neural stem cells and also ES-derived differentiated mature cells such as cardiomyocytes and hepatocytes have been utilized for drug screening and toxicity assessment as ES cell-based assay systems *in vitro* (Kettenhofen & Bohlen, 2008; Kuegler et al., 2010; Ma et al., 2008; Sartipy et al., 2007). Recently, various high-throughput cellular microarray platforms have been developed in a viewpoint for applying ES cells to drug discovery and development (Derda et al., 2007; Flaim et al., 2008). Consequently, it has been realized that the ability to track stem cell fate decisions and to quantify specific stem cell markers on microarray platforms has the potential to increase our understanding of the cellular mechanisms involved. Here, signals emanating from the stem cell microenvironment, or niche, are crucial in regulating stem cell fate. These signals include physical cues (e.g. matrix elasticity, cell–cell and cell–ECM interactions) and soluble factors (e.g. growth factors and small molecules) (Fernandes et al., 2009).

4.2 Advantages of the ES cell-based assay system utilizing TOSHI-substrata in drug discovery and development

The greatest merit of utilizing TOSHI-substrata is to provide niches reflecting various conditions in tissue(s)/organ(s) *in vivo* towards ES cells, resulting in the generation of assay systems for estimating not only biological activity but also the drug efficacy and chemical toxicity remained in the TOSHI-substrata by analyzing the behavior of the ES cells. Therefore, it is available to propose new research strategies for exploring bioactive molecule(s) involved in the TOSHI-substrata and/or for extrapolating pharmacological and toxicological effect(s) of chemicals in human. As the exploration research, bioactive molecules such as ligand, growth factor, growth inhibitor and differentiation-inducing factor would be isolated from specific regions in the TOSHI-substrata where ES cells revealed behaviors such as adhesion, proliferation, apoptosis and differentiation, respectively. Also, as the pharmacological and toxicological research, the TOSHI-substrata would provide two applications. One is means for efficiently differentiating human ES cells to sufficient amount of mature differentiated cells (e.g. hepatocytes) by utilizing the TOSHI-substrata derived from a regenerating tissue/organ. The other is a novel extrapolation system of the drug efficacy and/or toxicity by investigating the behavior of human ES cells cultured on the TOSHI-substrata derived from every tissue/organ of experimental animals after administrating drugs and/or exposing chemicals, respectively (Fig. 7).

5. Conclusions

I presented the basics and advantages of a cell culture system utilizing TOSHI (tissue/organ sections for histopathology)-substrata, and introduced a new platform for orienting differentiation of mouse ES cells towards hepatic lineage utilizing TOSHI-substrata derived from regenerating liver after CCl_4-administration into mice. Also, I proposed a new strategy for applying the ES cell-based assay system utilizing TOSHI-substrata to the researches in drug discovery and development. I hope that such a new strategy could contribute not only to expand the utility of ES cells but also to generate a high-throughput assay system utilizing TOSHI-substrata as a tissue/organ microarray useful for extrapolating drug efficacy and chemical toxicity in human body.

6. Acknowledgments

This work is being partly supported by grants for the Pioneer Research from the Ministry of Agriculture, Forestry and Fisheries of Japan and for general research from the National Institute of Agrobiological Sciences. I would like to thank Dr. T. Ochiya, Dr. T. Takeuchi, Dr. K. Yanagihara, Dr. S. Terada, Dr. M. Watanabe, Mr. N. Suzuki, Mr. Y. Hamanaka and Ms. Y. Ohsumi.

7. References

Derda, R.; Li, L.; Orner, B. P.; Lewis, R. L.; Thomson, J. A. & Kiessling, L. L. (2007). Defined substrates for human embryonic stem cell growth identified from surface arrays. *ACS Chemical Biology*, Vol.2, No.5, 347–355.

Fernandes, T. G.; Diogo, M. M.; Clark, D. S.; Dordick, J. S. & Cabral, J. M. (2009). High-throughput cellular microarray platforms: applications in drug discovery, toxicology and stem cell research. *Trends in Biotechnology*, Vol.27, No.6, 324-329.

Flaim, C. J.; Teng, D.; Chien, S. & Bhatia, S. N. (2008). Combinatorial signaling microenvironments for studying stem cell fate. *Stem Cells and Development*, Vol.17, No.1, 29–39.

Harada, K.; Mitaka, T.; Miyamoto, S.; Sugimoto, S.; Ikeda, S.; Takeda, H.; Mochizuki, Y. & Hirata, K. (2003). Rapid formation of hepatic organoid in collagen sponge by rat small hepatocytes and hepatic nonparenchymal cells. *Journal of Hepatology*, Vol.39, No.5, 716-723.

Ishi, T.; Yasuchika, K.; Fujii, H.; Hoppo, T.; Baba, S.; Naito, M.; Machimoto, T.; Kamo, N.; Suemori, H.; Nakatsuji, N. & Ikai, I. (2005). In vitro differentiation and maturation of mouse embryonic stem cells into hepatocytes. *Experimental Cell Research*, Vol.309, No.1, 68-77.

Kania, G.; Corbeil, D.; Fuchs, J.; Tarasov, K. V.; Blyszczuk, P.; Huttner, W. B.; Boheler, K. R. & Wobus, A. M. (2005). Somatic stem cell marker prominin-1/CD133 is expressed in embryonic stem cell-derived progenitors. *Stem Cells*, Vol.23, No.6, 791-804.

Kettenhofen, R. & Bohlen, H. (2008). Preclinical assessment of cardiac toxicity. *Drug Discovery Today*, Vol.13, No.15-16, 702-707.

Kuegler, P. B.; Zimmer, B.; Waldmann, T.; Baudis, B.; Ilmjärv, S.; Hescheler, J.; Gaughwin, P.; Brundin, P.; Mundy, W.; Bal-Price, A. K.; Schrattenholz, A.; Krause, K. H.; van Thriel, C.; Rao, M. S.; Kadereit, S. & Leist, M. (2010). Markers of murine embryonic and neural stem cells, neurons and astrocytes: reference points for developmental neurotoxicity testing. *ALTEX*, Vol.27, No.1, 17-42.

Lavon, N. & Benvenisty, N. (2005). Study of hepatocyte differentiation using embryonic stem cells. *Journal of Cellular Biochemistry*, Vol.96, No.6, 1193-1202.

Ma, F.; Ebihara, Y.; Umeda, K.; Sakai, H.; Hanada, S.; Zhang, H.; Zaike, Y.; Tsuchida, E.; Nakahata, T.; Nakauchi, H. & Tsuji, K. (2008). Generation of functional erythrocytes from human embryonic stem cell-derived definitive hematopoiesis. *Proceedings of the National Academy of Sciences of the United States of America*, Vol.105, No.35, 13087-13092.

Novik, E. I.; Maguire, T. J.; Orlova, K.; Schloss, R. S. & Yarmush, M. L. (2006). Embryoid body-mediated differentiation of mouse embryonic stem cells along a hepatocyte lineage: Insights from gene expression profiles. *Tissue Engineering*, Vol.12, No.6, 1515-1525.

Sakai, Y.; Jiang, J.; Kojima, N.; Kinoshita, T. & Miyajima, A. (2002). Enhanced in vitro maturation of fetal mouse liver cells with oncostatin M, nicotinamide, and dimethyl sulfoxide. *Cell Transplantation*, Vol.11, No.5, 435-441.

Sartipy, P.; Björquist, P.; Strehl, R. & Hyllner, J. (2007). The application of human embryonic stem cell technologies to drug discovery. *Drug Discovery Today*, Vol.12, No.17-18, 688-699.

Takeuchi, T.; Ochiya, T. & Takezawa, T. (2008). Tissue array substratum composed of histological sections: a new platform for orienting differentiation of embryonic stem cells towards hepatic lineage. *Tissue Engineering Part A*, Vol.14, No.2, 267-274.

Takezawa, T.; Mori, Y. & Yoshizato, K. (1990). Cell culture on a thermo-responsive polymer surface. *Bio/Techonology (N.Y.) (currently, Nature Biotechnology)*, Vol.8, No.9, 854-856.

Takezawa, T.; Yamazaki, M.; Mori, Y.; Yonaha, T. & Yoshizato, K. (1992). Morphological and immuno-cytochemical characterization of a hetero-spheroid composed of fibroblasts and hepatocytes. *Journal of Cell Science*, Vol.101, No.3, 495-501.

Takezawa, T. & Yoshizato, K. (1997). Mass transport via naturally branched scaffolds maintains viability of a reconstituted model of connective tissue. *Tissue Engineering*, Vol.3, No.4, 329-343.

Takezawa, T.; Inoue, M.; Aoki, S.; Sekiguchi, M.; Wada, K.; Anazawa, H. & Hanai, N. (2000). Concept for organ engineering: A reconstruction method of rat liver for *in vitro* culture. *Tissue Engineering*, Vol.6, No.6, 641-500.

Takezawa, T.; Takenouchi, T.; Imai, K.; Takahashi, T. & Hashizume, K. (2002). Cell culture on thin tissue sections commonly prepared for histopathology. *The FASEB Journal,* Vol.16, No.13, 1847-1849. (http://www.fasebj.org/cgi/doi/10.1096/fj.02-0405fje)

Takezawa, T. (2003). A strategy for the development of tissue engineering scaffolds that regulate cell behavior. *Biomaterials,* Vol.24, No.13, 2267-2275.

Takezawa, T.; Ozaki, K.; Nitani, A.; Takabayashi, C. & Shimo-Oka, T. (2004). Collagen vitrigel: A novel scaffold that can facilitate a three-dimensional culture for reconstructing organoids. *Cell Transplantation,* Vol.13, No.4, 463-473.

Takezawa, T.; Ozaki, K. & Takabayashi, C. (2007a). Reconstruction of a hard connective tissue utilizing a pressed silk sheet and type-I collagen as the scaffold for fibroblasts. *Tissue Engineering,* Vol.13, No.6, 1357–1366.

Takezawa, T.; Takeuchi, T.; Nitani, A.; Takayama, Y.; Kino-oka, M.; Taya, M. & Enosawa, S. (2007b). Collagen vitrigel membrane useful for paracrine assays *in vitro* and drug delivery systems *in vivo*. *Journal of Biotechnology,* Vol.131, No.1, 76–83.

Takezawa, T.; Takeuchi, T.; Yanagihara, K.; Nakazawa, Y.; Nitani, A.; Terada, S.; Ochiya, T. & Ueno, K. (2008a). Advantages of culture models utilizing substrata made of TOSHI (tissue/organ sections for histopathology) or collagen vitrigel membrane and their application concept for drug development researches. *Yakugaku Zasshi,* Vol.128, No.1, 51–60 [in Japanese].

Takezawa, T.; Takeuchi, T.; Ochiya, T.; Yanagihara, K.; Terada, S.; Suzuki, N. & Watanabe, M. (2008b). Advanced research utilizing histopathological sections as a culture substratum of animal cells. *Organ Biology,* Vol.15, No.2, 107–114 [in Japanese].

Takezawa, T.; Fukuda, M.; McIntosh-Ambrose, W.; Ko, J-A.; Elisseeff, J.; Haga, S.; Ozaki, M.; Kato, K.; Wang, P.C.; Uchino, T. & Nishida, T. (2010). Development of novel cell culture systems utilizing the advantages of collagen vitrigel membrane. *Yakugaku Zasshi,* Vol.130, No.4, 565–574 [in Japanese].

Teratani, T.; Yamamoto, H.: Aoyagi, K.; Sasaki, H.; Asari, A.; Quinn, G.; Sasaki, H.; Terada, M. & Ochiya, T. (2005). Direct hepatic fate specification from mouse embryonic stem cells. *Hepatology*, Vol.41, No.4, 836-846.

Yamamoto, H.; Quinn, G.; Asari, A.; Yamanokuchi, H.; Teratani, T.; Terada, M. & Ochiya T. (2003). Differentiation of embryonic stem cells into hepatocytes: biological functions and therapeutic application. *Hepatology,* Vol.37, No.5, 983-993.

Yanagihara, K.; Takeuchi, T.; Terada, S.; Miki, M. & Takezawa, T. (2007) Cell behaviors of two different cell lines on the section substrata prepared from rat organs and their mathematical models. *Alternatives to Animal Testing and EXperimentation,* Vol.12, Supplement, 126.

Zhou, Q. J.; Xiang, L. X.; Shao, J. Z.; Hu, R. Z.; Lu, Y. L.; Yao, H. & Dai, L. C. (2007). In vitro differentiation of hepatic progenitor cells from mouse embryonic stem cells induced by sodium butyrate. *Journal of Cellular Biochemistry*, Vol.100, No.1, 29-42.